Understanding the National Electrical Code®

Volume 1 Articles 90 - 450

2008 Edition

Mike Holt Enterprises, Inc.

1.888.NEC.CODE • www.MikeHolt.com • Info@MikeHolt.com

NOTICE TO THE READER

Mike Holt's Illustrated Guide to
Understanding the National Electrical Code. Volume 1

First Printing: March 2008

Technical Illustrator: Mike Culbreath
Cover Design: Tracy Jette
Layout Design and Typesetting: Cathleen Kwas and Tara Moffitt

ISBN 978-1-932685-33-6

For more information, call 1.888.NEC.CODE (888-632-2633), or E-mail Info@MikeHolt.com.

I dedicate this book to the
Lord Jesus Christ,
my mentor and teacher.
Proverbs 16:3

One Team

To Our Instructors and Students:

We are committed to providing you the finest product with the fewest errors, but we are realistic and know that there will be errors found and reported after the printing of this book. The last thing we want is for you to have problems finding, communicating, or accessing this information. It is unacceptable to us for there to be even one error in our textbooks or answer keys. For this reason, we are asking you to work together with us as One Team.

Students: Please report any errors that you may find to your instructor.
Instructors: Please communicate these errors to us.

Our Commitment:

We will continue to list all of the corrections that come through for all of our textbooks and answer keys on our Website. We will always have the most up-to-date answer keys available to instructors to download from our instructor Website. We do not want you to have problems finding this updated information, so we're outlining where to go for all of this below:

To view textbook and answer key corrections: Students and instructors go to our Website, www.MikeHolt.com, click on "Books" in the sidebar of links, and then click on "Corrections."

To download the most up-to-date answer keys: Instructors go to our Website, www.MikeHolt.com, click on "Instructors" in the sidebar of links and then click on "Answer Keys." On this page you will find instructions for accessing and downloading these answer keys.

If you are not registered as an instructor you will need to register. Your registration will be sent to our educational director who in turn reviews and approves your registration. In your approval E-mail will be the login and password so you can have access to all of the answer keys. If you have a situation that needs immediate attention, please contact the office directly at 1.888.NEC.CODE.

Call 1.888.NEC.CODE or visit us online at www.MikeHolt.com

Table of Contents

Introduction

Mike Holt's Illustrated Guide to Understanding the National Electrical Code, Volume 1

This edition of *Mike Holt's Illustrated Guide to Understanding the National Electrical Code, Volume 1* textbook is intended to provide you with the tools necessary to understand the technical requirements of the *National Electrical Code (NEC)®*. The writing style of this textbook, and in all of Mike Holt's products, is meant to be informative, practical, useful, informal, easy to read, and applicable for today's electrical professional. Also, just like all of Mike Holt's textbooks, it contains hundreds of full-color illustrations to help you see the safety requirements of the *NEC* in practical use, as they apply to today's electrical installations.

This illustrated textbook contains advice, cautions about possible conflicts or confusing *Code* requirements, tips on proper electrical installations, and warnings of dangers related to improper electrical installations. In spite of this effort, some rules are unclear or need additional editorial improvement.

This textbook can't eliminate confusing, conflicting, or controversial *Code* requirements, but it does try to put these requirements into sharper focus to help you understand their intended purpose. Sometimes a requirement is so confusing nobody really understands its actual application. When this occurs, this textbook will point the situation out in an upfront and straightforward manner.

The *NEC* is updated every three years to accommodate new electrical products and materials, changing technologies, and improved installation techniques, along with editorial improvements. While the uniform adoption of each new edition of the *Code* is the best approach for all involved in the electrical industry, many inspection jurisdictions modify the *NEC* when it's adopted. In addition, the *Code* allows the authority having jurisdiction, also known as the "AHJ," typically the electrical inspector, the authority to waive *NEC* requirements or permit alternative wiring methods contrary to the *Code* requirements when assured the completed electrical installation is equivalent in establishing and maintaining effective safety [90.4].

Keeping up with the *NEC* should be the goal of all those who are involved in the safety of electrical installations. This includes electrical installers, contractors, owners, inspectors, engineers, instructors, and others concerned with electrical installations.

About the 2008 *NEC*

The actual process of changing the *Code* took about two years, and it involved thousands of individuals making an effort to have the *NEC* as current and accurate as possible. Let's review how this process worked:

Step 1. *Proposals—November, 2005.* Anybody can submit a proposal to change the *Code* before the proposal closing date. Over 3,600 proposals were submitted to modify the 2008 *NEC*; of these proposals, 300 rules were revised that significantly affect the electrical industry. Some changes were editorial revisions, while others were more significant, such as new articles, sections, exceptions, and fine print notes.

Step 2. *Code-Making Panels Review Proposals—January, 2006.* All *Code* proposals were reviewed by Code-Making Panels (there were 20 panels in the 2008 *Code* process) who voted to accept, reject, or modify them.

Step 3. *Report on Proposals (ROP)—July, 2006.* The voting of the 20 Code-Making Panels on the proposals was published for public review in a document called the "Report on Proposals," frequently referred to as the "ROP."

Step 4. *Public Comments—October, 2006.* Once the ROP was available, public comments were submitted asking the Code-Making Panels members to revise their earlier actions on change proposals, based on new information. The closing date for "Comments" was October, 2006.

Step 5. *Comments Reviewed by Code-Making Panels—December, 2006.* The Code-Making Panels met again to review, discuss, and vote on public comments.

Step 6. *Report on Comments (ROC)—April, 2007.* The voting on the "Comments" was published for public review in a document called the "Report on Comments," frequently referred to as the "ROC."

Step 7. *Electrical Section—June, 2007.* The NFPA Electrical Section discussed and reviewed the work of the Code-Making Panels. The Electrical Section developed recommendations on last-minute motions to revise the proposed *NEC* draft that would be presented at the NFPA annual meeting.

Step 8. *NFPA Annual Meeting—June, 2007.* The 2008 *NEC* was officially adopted at the annual meeting, after a number of motions (often called "floor actions") were voted on.

Step 9. *Standards Council Review Appeals and Approves the 2008 NEC—July, 2007.* The NFPA Standards Council reviewed the record of the *Code*-making process and approved publication of the 2008 *NEC.*

Step 10. *2008 NEC Published—September, 2007.* The 2008 *National Electrical Code* was published, following the NFPA Board of Directors review of appeals.

> **Author's Comment:** Submitting proposals and comments online can be accomplished by going to the NFPA Website (www.nfpa.org), click on "Codes and Standards" at the top of the page, and once on the Codes and Standards page click on "Proposals and Comments" in the box on the right-hand side of the page. The deadline for proposals to create the 2011 *National Electrical Code* is November 5, 2008.

The Scope of this Textbook

This textbook, *Understanding the National Electrical Code, Volume 1,* covers the general installation requirements contained in Articles 90 through 460 (*NEC* Chapters 1 through 4) that Mike considers to be of critical importance. This textbook contains the following stipulations:

- **Power Systems and Voltage.** All power-supply systems are assumed to be solidly grounded and of any of the following voltages: 120V single-phase, 120/240V single-phase, 120/208V three-phase, 120/240V three-phase, or 277/480V three-phase, unless identified otherwise.
- **Electrical Calculations.** Unless the question or example specifies three-phase, the questions and examples are based on a single-phase power supply.
- **Rounding.** All calculations are rounded to the nearest ampere in accordance with 220.5(B).
- **Conductor Material.** All conductors are considered copper, unless aluminum is identified or specified.
- **Conductor Sizing.** All conductors are sized based on a THHN copper conductor terminating on a 75°C terminal in accordance with 110.14(C), unless the question or example identifies otherwise.
- **Overcurrent Device.** The term "overcurrent device" in this textbook refers to a molded case circuit breaker, unless identified otherwise. Where a fuse is identified, it's to be of the single-element type, also known as a "one-time fuse," unless identified otherwise.

How to Use This Textbook

This textbook is to be used with the *NEC*, not as a replacement for the *Code* book, so be sure to have a copy of the 2008 *National Electrical Code* handy. Compare what Mike is explaining to the text in your *Code* book, and discuss those topics that you find difficult to understand with others.

You'll notice that all *NEC* text has been paraphrased, as well as some of the article and section titles being different than they appear in the actual *Code*. Mike believes by doing so makes it easier to understand the content of the rule, so keep this in mind when comparing this textbook against the actual *NEC.*

As you read through this textbook, be sure to take the time to review the text along with the outstanding graphics and examples provided.

Textbook Format

This textbook follows the *NEC* format, but it doesn't cover every *Code* requirement. For example, it doesn't include every article, section, subsection, exception, or fine print note. So don't be concerned if you see the textbook contains Exception No. 1 and Exception No. 3, but not Exception No. 2.

Graphics with red borders are graphics that contain a 2008 change; graphics without a red border are graphics that support the concept being discussed, but nothing in the graphic was affected by a 2008 *Code* change.

Special Sections and Examples. Additional information to better help you understand a concept is identified with light green shading. In addition, examples are highlighted with a yellow background.

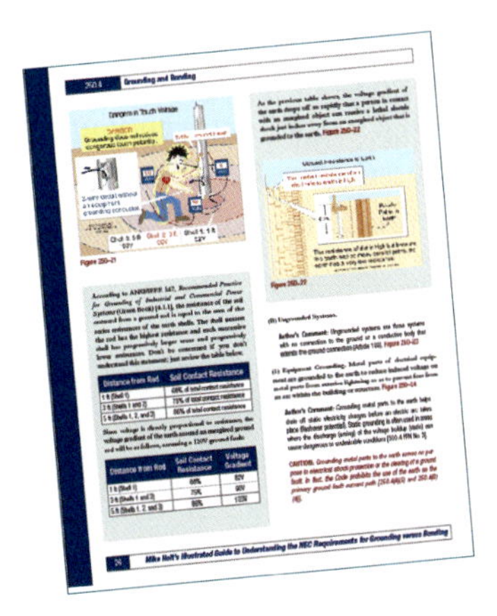

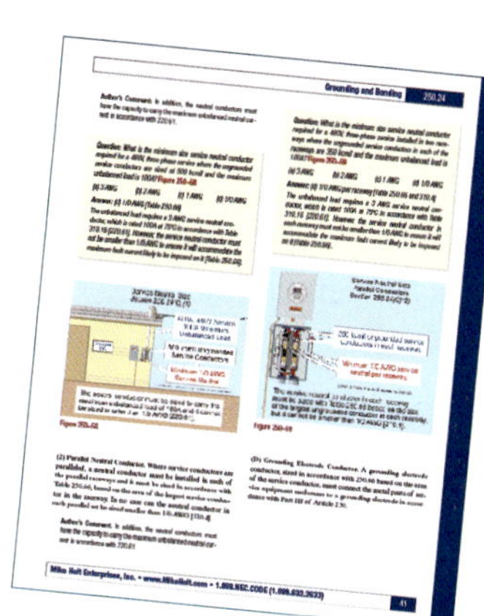

How to Use This Textbook

Cross-References and Author's Comments

This textbook contains several *NEC* cross-references to other related *Code* requirements to help you develop a better understanding of how the *NEC* rules relate to one another. These cross-references are identified by *Code* section numbers in brackets, an example of which is "[90.4]."

Author's Comments were written by Mike to help you (the reader) better understand the *NEC* by bringing to your attention items you should be aware of.

Difficult Concepts

As you progress through this textbook, you might find you don't understand every explanation, example, calculation, or comment. Don't get frustrated, and don't get down on yourself. Remember, this is the *National Electrical Code,* and sometimes the best attempt to explain a concept isn't enough to make it perfectly clear. If you're still confused, visit www.MikeHolt.com, and post your question on the *Code* Forum for help.

Different Interpretations

Some electricians, contractors, instructors, inspectors, engineers, and others enjoy the challenge of discussing the *Code* requirements, hopefully in a positive and a productive manner. This give-and-take is important to the process of better understanding *NEC* requirements and applications. However, if you're going to get into an *NEC* discussion, please don't spout out what you think without having the actual *Code* in your hand. The professional way of discussing an *NEC* requirement is by referring to a specific section, rather than talking in vague generalities.

Textbook Errors and Corrections

If you believe there's an error of any kind in this textbook (typographical, grammatical, or technical), no matter how insignificant, please let us know.

Any errors found after printing are listed on our Website, so if you find an error, first check to see if it has already been corrected. Go to www.MikeHolt.com, click on the "Books" link, and then the "Corrections" link (www.MikeHolt.com/bookcorrections.htm).

If you don't find the error listed on the Website, contact us by E-mailing us at Corrections@MikeHolt.com. Be sure to include the book title, page number, and any other pertinent information.

Understanding the *National Electrical Code* Library

The *NEC* Library includes the *Understanding the NEC Volumes 1 & 2* textbooks, the *NEC Practice Questions* book and ten videos or DVDs (a total of 41.5 hours). This option allows you to learn at the most cost effective price.

For more information, visit www.mikeholt.com or call 1-888-NEC-CODE (1-888-632-2633).

How to Use the *National Electrical Code*

This textbook is to be used with the *NEC*, not as a replacement for the *NEC*, so be sure to have a copy of the 2008 *National Electrical Code* handy. Compare what Mike explains in the text to your *Code* book, and discuss those topics that you find difficult to understand with others. As you read through this textbook, be sure to take the time to review the text with the outstanding graphics and examples.

The *National Electrical Code* is written for persons who understand electrical terms, theory, safety procedures, and electrical trade practices. These individuals include electricians, electrical contractors, electrical inspectors, electrical engineers, designers, and other qualified persons. The *Code* is not written to serve as an instructive or teaching manual for untrained individuals [90.1(C)].

Learning to use the *NEC* is somewhat like learning to play the game of chess; it's a great game if you enjoy mental warfare. When learning to play chess, you must first learn the names of the game pieces, how the pieces are placed on the board, and how each piece moves.

Once you understand the fundamentals of the game of chess, you're ready to start playing the game. Unfortunately, at this point all you can do is make crude moves, because you really don't understand how all the information works together. To play chess well, you'll need to learn how to use your knowledge by working on subtle strategies before you can work your way up to the more intriguing and complicated moves.

Not a Game

Electrical work isn't a game, and it must be taken very seriously. Learning the basics of electricity, important terms and concepts, as well as the basic layout of the *NEC* gives you just enough knowledge to be dangerous. There are thousands of specific and unique applications of electrical installations, and the *Code* doesn't cover every one of them. To safely apply the *NEC*, you must understand the purpose of a rule and how it affects the safety aspects of the installation.

NEC Terms and Concepts

The *NEC* contains many technical terms, so it's crucial for *Code* users to understand their meanings and their applications. If you don't understand a term used in a *Code* rule, it will be impossible to properly apply the *NEC* requirement. Be sure you understand that Article 100 defines the terms that apply to two or more articles. For example, the term "Dwelling Unit" applies to many articles. If you don't know what a dwelling unit is, how can you apply the *Code* requirements for it?

In addition, many articles have terms unique for that specific article. This means that the definitions of those terms are only applicable for that given article. For example, Section 250.2 contains the definitions of terms that only apply to Article 250, Grounding and Bonding.

Small Words, Grammar, and Punctuation

It's not only the technical words that require close attention, because even the simplest of words can make a big difference to the intent of a rule. The word "or" can imply alternate choices for equipment wiring methods, while "and" can mean an additional requirement. Let's not forget about grammar and punctuation. The location of a comma "," can dramatically change the requirement of a rule.

Slang Terms or Technical Jargon

Electricians, engineers, and other trade-related professionals use slang terms or technical jargon that isn't shared by all. This makes it very difficult to communicate because not everybody understands the intent or application of those slang terms. So where possible, be sure you use the proper word, and don't use a word if you don't understand its definition and application. For example, lots of electricians use the term "pigtail" when describing the short conductor for the connection of a receptacle, switch, luminaire, or equipment. Although they may understand this, not everyone does.

NEC Style and Layout

Before we get into the details of the *NEC*, we need to take a few moments to understand its style and layout. Understanding the structure and writing style of the *Code* is very important before it can be used effectively. If you think about it, how can you use something if you don't know how it works? The *National Electrical Code* is organized into ten components.

1. Table of Contents
2. Article 90 (Introduction to the *Code*)
3. Chapters 1 through 9 (major categories)
4. Articles 90 through 830 (individual subjects)
5. Parts (divisions of an article)
6. Sections and Tables (*Code* requirements)
7. Exceptions (*Code* permissions)
8. Fine Print Notes (explanatory material)
9. Annexes (information)
10. Index

1. Table of Contents. The Table of Contents displays the layout of the Chapters, Articles, and Parts as well as the page numbers. It's an excellent resource and should be referred to periodically to observe the interrelationship of the various *NEC* components. When attempting to locate the rules for a particular situation, knowledgeable *Code* users often go first to the Table of Contents to quickly find the specific *NEC* part that applies.

2. Introduction. The *NEC* begins with Article 90, the introduction to the *Code*. It contains the purpose of the *NEC*, what is covered and what is not covered along with how the *Code* is arranged. It also gives information on enforcement and how mandatory and permissive rules are written as well as how explanatory material is included. Article 90 also includes information on formal interpretations, examination of equipment for safety, wiring planning, and information about formatting units of measurement.

3. Chapters. There are nine chapters, each of which is divided into articles. The articles fall into one of four groupings: General Requirements (Chapters 1 through 4), Specific Requirements (Chapters 5 through 7), Communications Systems (Chapter 8), and Tables (Chapter 9).

- Chapter 1 General
- Chapter 2 Wiring and Protection
- Chapter 3 Wiring Methods and Materials
- Chapter 4 Equipment for General Use
- Chapter 5 Special Occupancies
- Chapter 6 Special Equipment
- Chapter 7 Special Conditions
- Chapter 8 Communications Systems (Telephone, Data, Satellite, and Cable TV)
- Chapter 9 Tables–Conductor and Raceway Specifications

4. Articles. The *NEC* contains approximately 140 articles, each of which covers a specific subject. For example:

- Article 110 General Requirements
- Article 250 Grounding and Bonding
- Article 300 Wiring Methods
- Article 430 Motors and Motor Controllers
- Article 500 Hazardous (Classified) Locations
- Article 680 Swimming Pools, Fountains, and Similar Installations
- Article 725 Remote-Control, Signaling, and Power-Limited Circuits
- Article 800 Communications Systems

5. Parts. Larger articles are subdivided into parts.

Author's Comment: Because the parts of a *Code* article aren't included in the section numbers, we have a tendency to forget what "Part" the *NEC* rule is relating to. For example, Table 110.34(A) contains the working space clearances for electrical equipment. If we aren't careful, we might think this table applies to all electrical installations, but Table 110.34(A) is located in Part III, which contains the requirements for Over 600 Volts, Nominal installations. The rules for working clearances for electrical equipment for systems 600V, nominal, or less are contained in Table 110.26(A)(1), which is located in Part II—600 Volts, Nominal, or Less.

6. Sections and Tables.

Sections. Each *NEC* rule is called a *Code* section. A *Code* section may be broken down into subsections by letters in parentheses "(A), (B)," etc. Numbers in parentheses (1), (2), etc., may further break down a subsection, and lowercase letters (a), (b), etc., further break the rule down to the third level. For example, the rule requiring all receptacles in a dwelling unit bathroom to be GFCI protected is contained in Section 210.8(A)(1). Section 210.8(A)(1) is located in Chapter 2, Article 210, Section 8, subsection (A), sub-subsection (1).

Many in the industry incorrectly use the term "Article" when referring to a *Code* section. For example, they say "Article 210.8," when they should say "Section 210.8."

Tables. Many *Code* requirements are contained within tables, which are lists of *NEC* requirements placed in a systematic arrangement. The titles of the tables are extremely important; you must read them carefully in order to understand the contents, applications, limitations, etc., of each table in the *Code*. Many times notes are provided in or below a table; be sure to read them as well since they are also part of the requirement. For example, Note 1 for Table 300.5 explains how to measure the cover when burying cables and raceways, and Note 5 explains what to do if solid rock is encountered.

7. Exceptions. Exceptions are *Code* requirements or allowances that provide an alternative method to a specific requirement. There are two types of exceptions—mandatory and permissive. When a rule has several exceptions, those exceptions with mandatory requirements are listed before the permissive exceptions.

Mandatory Exception. A mandatory exception uses the words "shall" or "shall not." The word "shall" in an exception means that if you're using the exception, you're required to do it in a particular way. The phrase "shall not" means it isn't permitted.

Permissive Exception. A permissive exception uses words such as "shall be permitted," which means it's acceptable (but not mandatory) to do it in this way.

8. Fine Print Note (FPN). A fine print note contains explanatory material intended to clarify a rule or give assistance, but it isn't a *Code* requirement [90.5(C)].

9. Annexes. Annexes aren't a part of the *NEC* requirements, and are included in the *Code* for informational purposes only.

10. Index. The Index at the back of the *NEC* is helpful in locating a specific rule.

Author's Comment: Changes to the *NEC* since the previous edition(s), are identified by shading, but rules that have been relocated aren't identified as a change. A bullet symbol "•" is located on the margin to indicate the location of a rule that was deleted from a previous edition.

How to Locate a Specific Requirement

How to go about finding what you're looking for in the *Code* depends, to some degree, on your experience with the *NEC*. *Code* experts typically know the requirements so well they just go to the correct rule without any outside assistance. The Table of Contents might be the only thing very experienced *NEC* users need to locate the requirement they're looking for. On the other hand, average *Code* users should use all of the tools at their disposal, and that includes the Table of Contents and the Index.

Table of Contents. Let's work out a simple example: What *NEC* rule specifies the maximum number of disconnects permitted for a service? If you're an experienced *Code* user, you'll know Article 230 applies to "Services," and because this article is so large, it's divided up into multiple parts (actually eight parts). With this knowledge, you can quickly go to the Table of Contents and see that it lists Service Equipment Disconnecting Means requirements in Part VI.

Author's Comment: The number 70 precedes all page numbers because the *NEC* is NFPA standard number 70.

Index. If you use the Index, which lists subjects in alphabetical order, to look up the term "service disconnect," you'll see there's no listing. If you try "disconnecting means," then "services," you'll find the Index specifies the rule is located in Article 230, Part VI. Because the *NEC* doesn't give a page number in the Index, you'll need to use the Table of Contents to find the page number, or flip through the *Code* to Article 230, then continue to flip through pages until you find Part VI.

Many people complain that the *NEC* only confuses them by taking them in circles. As you gain experience in using the *Code* and deepen your understanding of words, terms, principles, and practices, you will find the *NEC* much easier to understand and use than you originally thought.

Customizing Your *Code* Book

One way to increase your comfort level with the *Code* is to customize it to meet your needs. You can do this by highlighting and underlining important *NEC* requirements, and by attaching tabs to important pages.

Highlighting. As you read through this textbook, be sure you highlight those requirements in the *Code* that are the most important or relevant to you. Use yellow for general interest and orange for important requirements you want to find quickly. Be sure to highlight terms in the Index and Table of Contents as you use them.

Underlining. Underline or circle key words and phrases in the *NEC* with a red pen (not a lead pencil) and use a six-inch ruler to keep lines straight and neat. This is a very handy way to make important requirements stand out. A small six-inch ruler also comes in handy for locating specific information in the many *Code* tables.

Tabbing the *NEC*. By placing tabs on *Code* articles, sections, and tables, it will make it easier for you to use the *NEC*. However, too many tabs will defeat the purpose. You can order a custom set of *Code* tabs online at www.MikeHolt.com, or by calling 1.888.NEC.CODE.

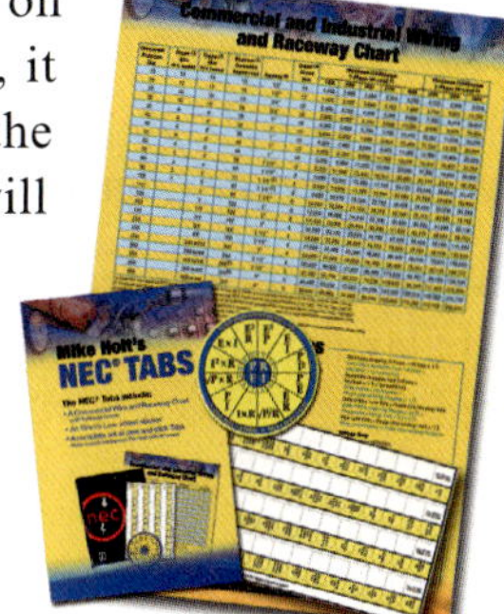

Acknowledgments

About the Author

Mike Holt worked his way up through the electrical trade from an apprentice electrician to become one of the most recognized experts in the world as it relates to electrical power installations. He was a Journeyman Electrician, Master Electrician, and Electrical Contractor. Mike came from the real world, and he has a unique understanding of how the *NEC* relates to electrical installations from a practical standpoint. You'll find his writing style to be simple, nontechnical, and practical.

Did you know that he didn't finish high school? So if you struggled in high school or if you didn't finish it at all, don't let this get you down, you're in good company. As a matter of fact, Mike Culbreath, Master Electrician, who produces the finest electrical graphics in the history of the electrical industry, didn't finish high school either. So two high school dropouts produced the text and graphics in this textbook! However, realizing success depends on one's continuing pursuit of education, Mike immediately attained his GED (as did Mike Culbreath) and ultimately attended the University of Miami's Graduate School for a Master's degree in Business Administration (MBA).

Mike Holt resides in Central Florida, is the father of seven children, and has many outside interests and activities. He is a five-time National Barefoot Water-Ski Champion (1988, 1999, 2005, 2006, and 2007); he has set many national records and continues to train year-round at a World competition level [www.barefootwaterskier.com].

What sets him apart from some is his commitment to living a balanced lifestyle; he places God first, then family, career, and self.

Special Acknowledgments

First, I want to thank God for my godly wife who is always by my side and my children, Belynda, Melissa, Autumn, Steven, Michael, Meghan, and Brittney.

A special thank you must be sent to the staff at the National Fire Protection Association (NFPA), publishers of the *NEC*—in particular Jeff Sargent for his assistance in answering my many *Code* questions over the years. Jeff, you're a "first class" guy, and I admire your dedication and commitment to helping others understand the *NEC*. Other former NFPA staff members I would like to thank include John Caloggero, Joe Ross, and Dick Murray for their help in the past.

A personal thank you goes to Sarina, my long-time friend and office manager. It has been wonderful working side-by-side with you for over 25 years nurturing this company's growth from its small beginnings.

Mike Holt Enterprises Team

Graphic Illustrator

Mike Culbreath devoted his career to the electrical industry and worked his way up from an apprentice electrician to master electrician. While working as a journeyman electrician, he suffered a serious on-the-job knee injury. With a keen interest in continuing education for electricians, he completed courses at Mike Holt Enterprises, Inc. and then passed the exam to receive his Master Electrician's license. In 1986, after attending classes at Mike Holt Enterprises, Inc. he joined the staff to update material and later studied computer graphics and began illustrating Mike Holt's textbooks and magazine articles. He's worked with the company for over 20 years and, as Mike Holt has proudly acknowledged, has helped to transform his words and visions into lifelike graphics.

Technical Editorial Director

Steve Arne has been involved in the electrical industry since 1974 working in various positions from electrician to full-time instructor and department chair in technical postsecondary education. Steve has developed curriculum for many electrical training courses and has developed university business and leadership courses. Currently, Steve offers occasional exam prep and continuing education *Code* classes. Steve believes that as a teacher he understands the joy of helping others as they learn and experience new insights. His goal is to help others understand more of the technological marvels that surround us. Steve thanks God for the wonders of His creation and for the opportunity to share it with others.

Steve and his lovely wife Deb live in Rapid City, South Dakota where they are both active in their church and community. They have two grown children and five grandchildren.

Technical *Code* Consultant

Ryan Jackson is a combination inspector for Draper City, Utah. He is certified as a building, electrical, mechanical, and plumbing inspector. He's also certified as a building plans examiner and electrical plans examiner. Ryan is the senior electrical inspector for Draper City, and also teaches seminars on the *NEC*. Ryan is very active in the Utah Chapter of IAEI, where he is currently president. He also enjoys staying active in the *NEC* change process, and loves to help people with their *Code* problems. On Mike Holt's *Code* Forum, he has been involved in nearly 5,000 topics.

Ryan enjoys reading, going to college football games, and spending time with his wife Sharie and their two children, Kaitlynn and Aaron.

Editorial Team

I would like to thank Toni Culbreath and Barbara Parks who worked tirelessly to proofread and edit the final stages of this publication. Their attention to detail and dedication to this project is greatly appreciated.

Production Team

I would like to thank Tara Moffitt and Cathleen Kwas who worked as a team to do the layout and production of this book. Their desire to create the best possible product for our customers is appreciated.

Video Team Members

Steve Arne, Mike Culbreath and **Ryan Jackson** (members of the Mike Holt Enterprises Team) were video team members, along with the following highly qualified professionals.

Tarry Baker

Chief Electrical Code Compliance Officer
Broward County Board of Rules & Appeals
Fort Lauderdale, Florida

Tarry Baker has been the Chief Electrical Code Compliance Officer for the Broward County Board of Rules and Appeals (Florida) for the last 16 years, standardizing enforcement in 32 municipalities and the unincorporated area of Broward County. He has served on the Electrical Technical Committee to the Board of Rules and Appeals for over 10 years, and as an advisor to the committee for 10 years. He was a principal member of the *NEC* Code-Making Panel-13 (2002, 2005 and 2008 *NEC*) and 20 (2008 *NEC*) for the IAEI, Electrical/Alarms Technical Advisory Committee to the Florida Building Commission, and is a former member of the Education, Disciple, and Licensing Work Group for the Governor's Building Code Study Commission for the State of Florida. He serves as an electrical coordinator and instructor for continuing education for the Broward County Board of Rules and Appeals and IAEI Maynard Hamilton/Fort Lauderdale Division.

He currently serves as Chaplain and past Chairman of the *Code* Question Committee for the IAEI Southern Section, and is a past President of the IAEI Florida Chapter. Tarry is also currently serving as Chairman of the Central Examining Board of Electricians of Broward County. He has been an Electrical Inspector since 1977 and an Electrical Plans Examiner since 1978. He is a State of Florida Certified Electrical Contractor, Building Code Administrator, Electrical Plans Examiner, Electrical Inspector, and One and Two Family Combination Inspector.

Doug Douty
D & D Resources, Owner
Fresno City College
Fresno, California

Doug Douty has been a consultant in the electrical and building fields for many years. Doug's vision is to help you maximize your talents and abilities and to provide solutions to difficult problems. He is a member of the National Fire Protection Association, the International Association of Electrical Inspectors, and a 2006 winner of Mike Holt's Top Gun presentation award.

Doug holds a Master of Science degree in Industrial Technology from California State University, Fresno, and California State Contractor's license in electrical general contracting. He is an instructor in the electronics department at Fresno City College, an Adjunct Professor at California State University, Fresno, and a nationally recognized speaker in the electrical industry.

Doug resides in Fresno in the Central Valley of California with his wife. He has four grown children and five grandchildren. Doug has a passion for helping people reach their full potential in life. He also enjoys cycling, running, reading, and continual learning as well as family vacations.

Doug's motivational style of teaching captures his audience's attention. At Doug's seminars, students not only increase their knowledge, but their confidence as well. Seminar topics include Electrical Safety in the Workplace (NFPA 70E), the *National Electrical Code* (NFPA 70), AC Fundamentals, AC Power Systems, Control Systems, Electrical Certification Preparation, and Continuing Education.

Eric Stromberg
Electrical Engineer/Instructor
Dow Chemical
Lake Jackson, TX

Eric Stromberg enrolled in the University of Houston in 1976, with Electrical Engineering as his major. During the first part of his college years, Eric worked for a company that specialized in installing professional sound systems. Later, he worked for a small electrical company and eventually became a journeyman electrician. After graduation from college in 1982, Eric went to work for an electronics company that specialized in fire alarm systems for high-rise buildings. He became a state licensed Fire Alarm Installation Superintendent and was also a member of IBEW local union 716. In 1989, Eric began a career with the Dow Chemical Company as an Electrical Engineer designing power distribution systems for large industrial facilities. In 1997, Eric began teaching *National Electrical Code* classes.

Eric currently resides in Lake Jackson, Texas, with his wife Jane and three children: Ainsley, Austin, and Brieanna.

Kevin Vogel
Professional Engineer/Instructor
Crescent Electric Supply
Dalton Gardens, ID

Kevin Vogel graduated from Santa Clara University in 1964 with a Bachelor of Science degree in Mechanical Engineering. A licensed Professional Engineer since 1969, he formerly worked as chief engineer for a manufacturer of electric heating products and of (the first) thermoplastic electrical outlet boxes. He has also worked as an electrician and holds a Master Electrician's license.

In 1978, Kevin was co-founder of an electrical wholesale distribution company in Coeur d'Alene, ID that was sold in 1991 to Crescent Electric Supply Company. Kevin continues to work in that part of the industry. He also provides expert witness testimony in civil and criminal cases, and serves as a certified instructor for *National Electrical Code* classes at North Idaho College. He also provides occasional assistance to Trindera Engineering, an electrical engineering consulting firm located in Coeur d'Alene.

Kevin married his beloved wife, Linda, in 1966, and they have been blessed with 13 wonderful children and 17 (so far) grandchildren. Kevin is extremely grateful to God for all the gifts He has bestowed on him and on his loved ones.

Advisory Committee

Rahe Loftin P.E.
Fire Protection Engineer
U.S. General Services Administration
Fort Worth, TX

Introduction to the *National Electrical Code*

INTRODUCTION TO ARTICLE 90—INTRODUCTION TO THE *NATIONAL ELECTRICAL CODE*

Many *NEC* violations and misunderstandings wouldn't occur if people doing the work simply understood Article 90. For example, many people see *Code* requirements as performance standards. In fact, the *NEC* requirements are bare minimums for safety. This is exactly the stance electrical inspectors, insurance companies, and courts take when making a decision regarding electrical design or installation.

Article 90 opens by saying the *NEC* isn't intended as a design specification or instruction manual. The *National Electrical Code* has one purpose only, and that is the "practical safeguarding of persons and property from hazards arising from the use of electricity." It goes on to indicate that the *Code* isn't intended as a design specification or instruction manual. Yet, the necessity to study, study, and study the *NEC* rules some more can't be overemphasized. Understanding where to find the rules in the *Code* that apply to the installation is invaluable. Rules in several different articles often apply to even a simple installation.

Article 90 then describes the scope and arrangement of the *NEC*. A person who says, "I can't find anything in the *Code*," is really saying, "I never took the time to review Article 90." The balance of Article 90 provides the reader with information essential to understanding those items you do find in the *NEC*.

Typically, electrical work requires you to understand the first four chapters of the *Code* which apply generally, plus have a working knowledge of the Chapter 9 tables. That knowledge begins with Article 90. Chapters 5, 6, and 7 make up a large portion of the *NEC*, but they apply to special occupancies, special equipment, or other special conditions. They build on, modify, or amend the rules in the first four chapters. Chapter 8 contains the requirements for communications systems, such as telephone, antenna wiring, CATV, and network-powered broadband systems. Communications systems aren't subject to the general requirements of Chapters 1 through 4, or the special requirements of Chapters 5 through 7, unless there's a specific reference in Chapter 8 to a rule in Chapters 1 through 7.

90.1 Purpose of the *NEC*.

(A) Practical Safeguarding. The purpose of the *NEC* is to ensure that electrical systems are installed in a manner that protects people and property by minimizing the risks associated with the use of electricity.

(B) Adequacy. The *Code* contains requirements considered necessary for a safe electrical installation. When an electrical installation is installed in compliance with the *NEC*, it will be essentially free from electrical hazards. The *Code* is a safety standard, not a design guide.

NEC requirements aren't intended to ensure the electrical installation will be efficient, convenient, adequate for good service, or suitable for future expansion. Specific items of concern, such as electrical energy management, maintenance, and power quality issues aren't within the scope of the *Code*. Figure 90–1

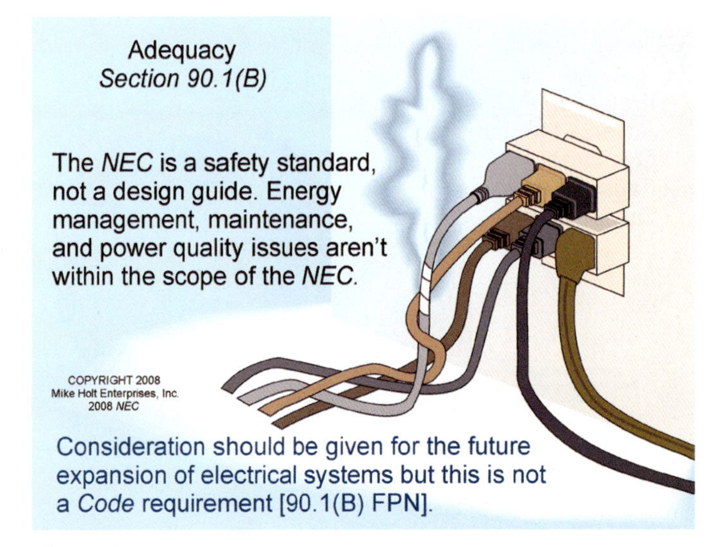

Figure 90–1

FPN: Hazards in electrical systems often occur because circuits are overloaded or not properly installed in accordance with the *NEC*. These often occur when the initial wiring did not provide reasonable provisions for system changes or for the increase in the use of electricity.

Author's Comments:

- See the definition of "Overload" in Article 100.
- The *NEC* does not require electrical systems to be designed or installed to accommodate future loads. However, the electrical designer, typically an electrical engineer, is concerned with not only ensuring electrical safety (*Code* compliance), but also with ensuring the system meets the customers' needs, both of today and in the near future. To satisfy customers' needs, electrical systems are often designed and installed above the minimum requirements contained in the *NEC*.

(C) Intention. The *Code* is intended to be used by those skilled and knowledgeable in electrical theory, electrical systems, construction, and the installation and operation of electrical equipment. It is not a design specification standard or instruction manual for the untrained and unqualified.

(D) Relation to International Standards. The requirements of the *NEC* address the fundamental safety principles contained in the International Electrotechnical Commission (IEC) standards, including protection against electric shock, adverse thermal effects, overcurrent, fault currents, and overvoltage. Figure 90–2

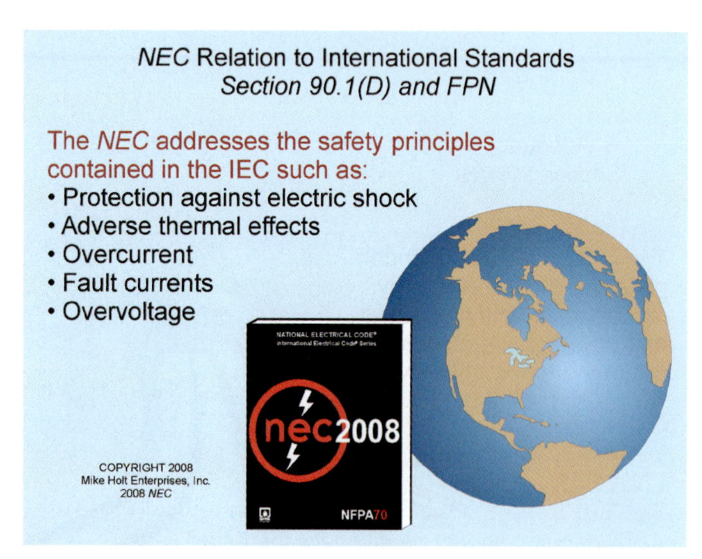

Figure 90–2

Author's Comments:

- See the definition of "Overcurrent" in Article 100.
- The *NEC* is used in Chile, Ecuador, Peru, and the Philippines. It's also the electrical code for Colombia, Costa Rica, Mexico, Panama, Puerto Rico, and Venezuela. Because of these adoptions, the *NEC* is available in Spanish from the National Fire Protection Association, 1.617.770.3000, www.NFPA.Org.

90.2 Scope of the *NEC*.

(A) What is Covered. The *NEC* contains requirements necessary for the proper installation of electrical conductors, equipment, and raceways; signaling and communications conductors, equipment, and raceways; as well as optical fiber cables and raceways for the following locations: Figure 90–3

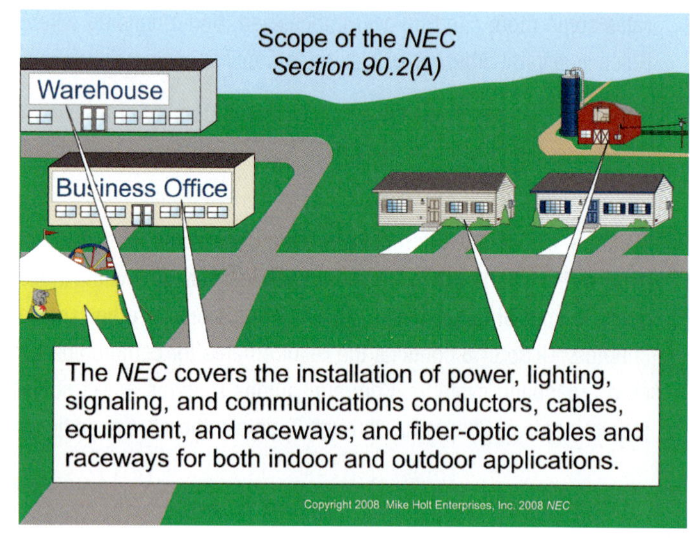

Figure 90–3

(1) Public and private premises, including buildings or structures, mobile homes, recreational vehicles, and floating buildings.

(2) Yards, lots, parking lots, carnivals, and industrial substations.

(3) Conductors and equipment connected to the utility supply.

(4) Installations used by an electric utility, such as office buildings, warehouses, garages, machine shops, recreational buildings, and other electric utility buildings that are not an integral part of a utility's generating plant, substation, or control center. Figure 90–4

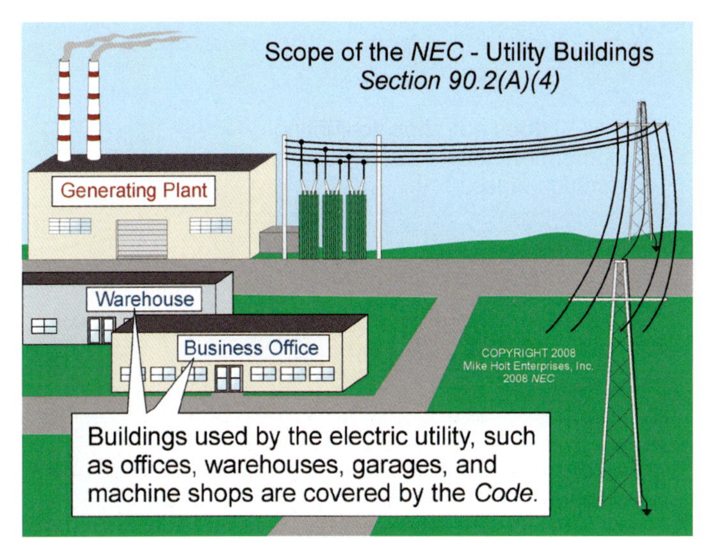

Figure 90–4

(B) What isn't Covered. The *NEC* doesn't apply to:

(1) Transportation Vehicles. Installations in cars, trucks, boats, ships and watercraft, planes, electric trains, or underground mines.

(2) Mining Equipment. Installations underground in mines and self-propelled mobile surface mining machinery and its attendant electrical trailing cables.

(3) Railways. Railway power, signaling, and communications wiring.

(4) Communications Utilities. The installation requirements of the *NEC* don't apply to communications (telephone), Community Antenna Television (CATV), or network-powered broadband utility equipment located in building spaces used exclusively for these purposes, or outdoors if the installation is under the exclusive control of the communications utility. Figure 90–5

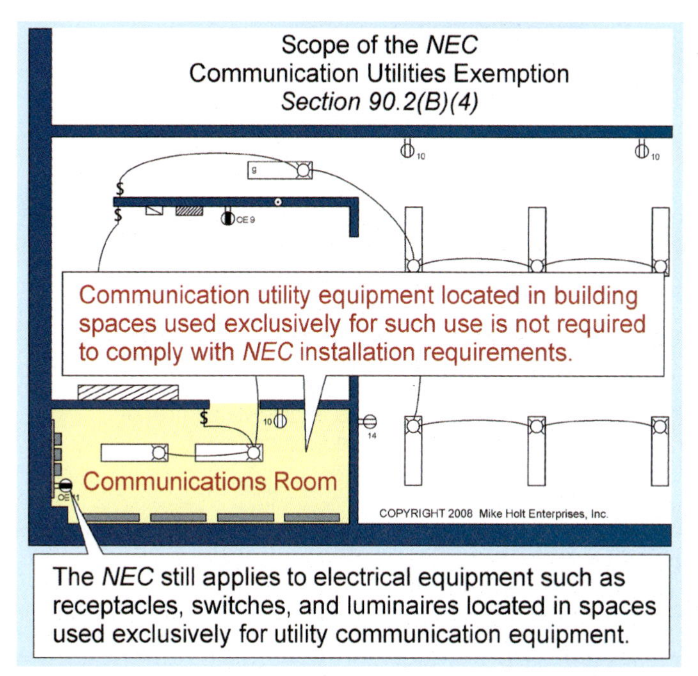

Figure 90–5

Author's Comment: Interior wiring for communications systems, not in building spaces used exclusively for these purposes, must be installed in accordance with the following Chapter 8 requirements: Figure 90–6

- Telephone and Data, Article 800
- CATV, Article 820
- Network-Powered Broadband, Article 830

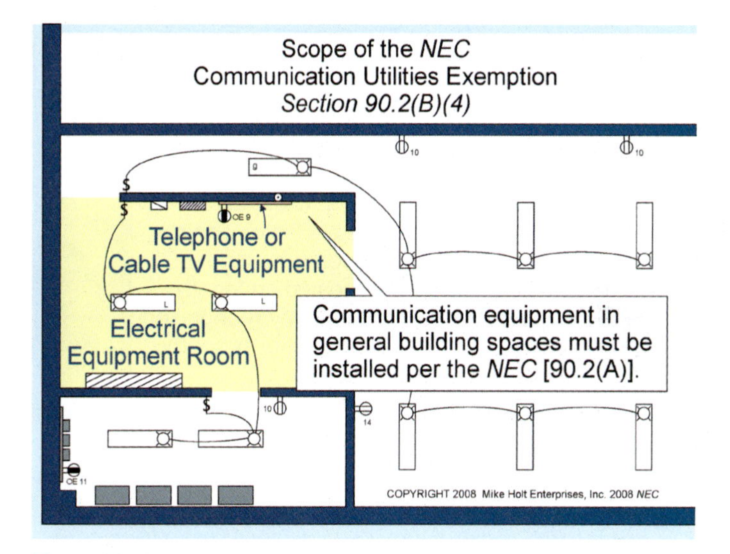

Figure 90–6

(5) Electric Utilities. The *NEC* doesn't apply to installations under the exclusive control of an electric utility where such installations:

a. Consist of service drops or service laterals and associated metering. Figure 90–7

b. Are located on legally established easements, or rights-of-way recognized by public/utility regulatory agencies, or property owned or leased by the electric utility. Figure 90–8

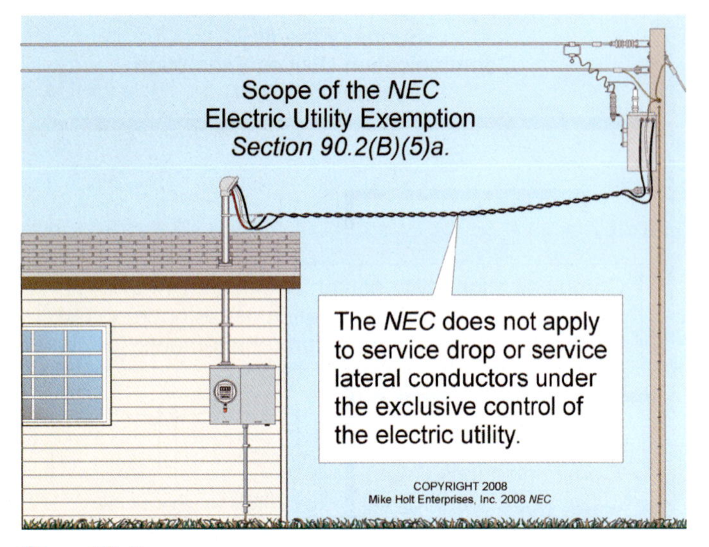

Figure 90–7

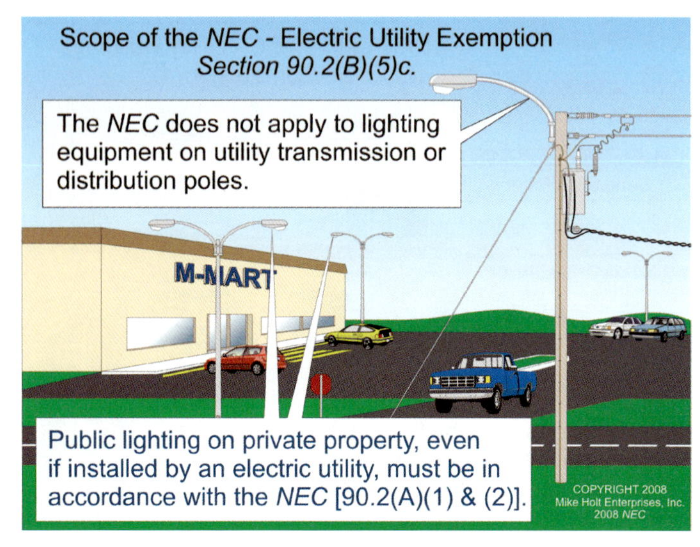

Figure 90–9

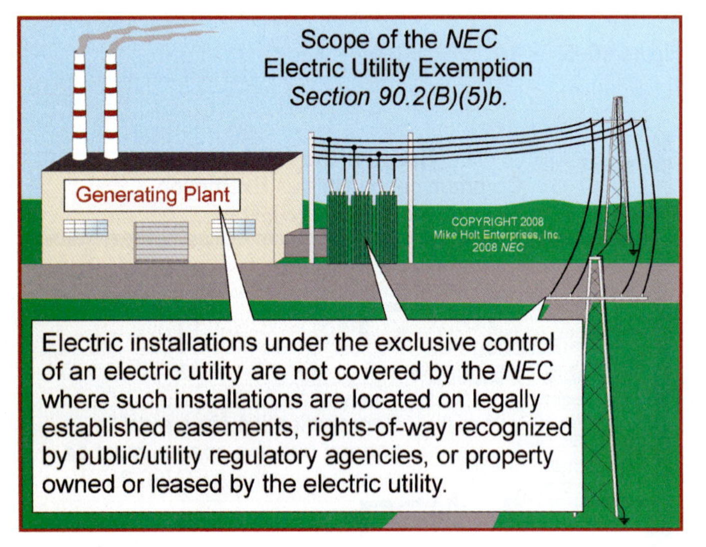

Figure 90–8

c. Are on property owned or leased by the electric utility for the purpose of generation, transformation, transmission, distribution, or metering of electric energy. Figure 90–9

Author's Comment: Luminaires located in legally established easements, or rights-of-way, such as at poles supporting transmission or distribution lines, are exempt from the *NEC*. However, if the electric utility provides site and public lighting on private property, then the installation must comply with the *NEC* [90.2(A)(4)].

FPN to 90.2(B)(4) and (5): Utilities include entities that install, operate, and maintain communications systems (telephone, CATV, Internet, satellite, or data services) or electric supply systems (generation, transmission, or distribution systems) and are designated or recognized by governmental law or regulation by public service/utility commissions. Utilities may be subject to compliance with codes and standards covering their regulated activities as adopted under governmental law or regulation.

90.3 *Code* Arrangement. The *Code* is divided into an introduction and nine chapters. Figure 90–10

General Requirements. The requirements contained in Chapters 1, 2, 3, and 4 apply to all installations.

Special Requirements. The requirements contained in Chapters 5, 6, and 7 apply to special occupancies, special equipment, or other special conditions. They can supplement or modify the requirements in Chapters 1 through 4.

Communications Systems. Chapter 8 contains the requirements for communications systems, such as telephone, antenna wiring, CATV, and network-powered broadband systems. Communications systems aren't subject to the general requirements of Chapters 1 through 4, or the special requirements of Chapters 5 through 7, unless there's a specific reference in Chapter 8 to a rule in Chapters 1 through 7.

Author's Comment: An example of how Chapter 8 works is the rules for working space about equipment. The typical 3 ft working space isn't required in front of communications equipment, because Table 110.26(A)(1) isn't referenced in Chapter 8.

Code Arrangement
Section 90.3

General Requirements

- Chapter 1 - General
- Chapter 2 - Wiring and Protection
- Chapter 3 - Wiring Methods and Materials
- Chapter 4 - Equipment for General Use

Chapters 1 through 4 apply to all applications.

Special Requirements

- Chapter 5 - Special Occupancies
- Chapter 6 - Special Equipment
- Chapter 7 - Special Conditions

Chapters 5 through 7 can supplement or modify the general requirements of Chapters 1 through 4.

• Chapter 8 - Communications Systems

Chapter 8 requirements are not subject to requirements in Chapters 1 through 7, unless there is a specific reference in Chapter 8 to a rule in Chapters 1 through 7.

• Chapter 9 - Tables

Chapter 9 tables are applicable as referenced in the *NEC* and are used for calculating raceway sizes, conductor fill, and voltage drop.

• Annexes A through H

Annexes are for information only and not enforceable.

Figure 90–10

Tables. Chapter 9 consists of tables applicable as referenced in the *NEC.* The tables are used to calculate raceway sizing, conductor fill, the radius of conduit and tubing bends, and conductor voltage drop.

Annexes. Annexes aren't part of the *Code*, but are included for informational purposes. There are eight Annexes:

- Annex A. Product Safety Standards
- Annex B. Application Information for Ampacity Calculation
- Annex C. Raceway Fill Tables for Conductors and Fixture Wires of the Same Size
- Annex D. Examples
- Annex E. Types of Construction
- Annex F. Critical Operations Power Systems (COPS)
- Annex G. Supervisory Control and Data Acquisition (SCADA)
- Annex H. Administration and Enforcement

90.4 Enforcement. The *Code* is intended to be suitable for enforcement by governmental bodies that exercise legal jurisdiction over electrical installations for power, lighting, signaling circuits, and communications systems, such as: Figure 90–11

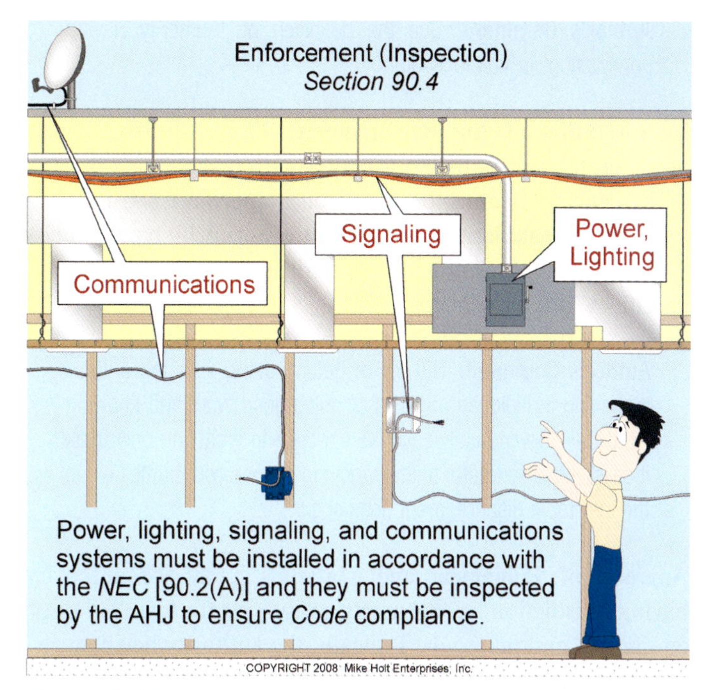

Figure 90–11

Signaling circuits which include:

- Article 725 Class 1, Class 2, and Class 3 Remote-Control, Signaling, and Power-Limited Circuits
- Article 760 Fire Alarm Systems
- Article 770 Optical Fiber Cables and Raceways

Communications systems which include:

- Article 800 Communications Circuits (twisted-pair conductors)
- Article 810 Radio and Television Equipment (satellite dish and antenna)
- Article 820 Community Antenna Television and Radio Distribution Systems (coaxial cable)
- Article 830 Network-Powered Broadband Communications Systems

Author's Comment: The installation requirements for signaling circuits and communications circuits are covered in Mike Holt's *Understanding the National Electrical Code, Volume 2* textbook.

The enforcement of the *NEC* is the responsibility of the authority having jurisdiction (AHJ), who is responsible for interpreting requirements, approving equipment and materials, waiving *Code* requirements, and ensuring equipment is installed in accordance with listing instructions.

Author's Comment: See the definition of "Authority Having Jurisdiction" in Article 100.

Interpretation of the Requirements. The authority having jurisdiction is responsible for interpreting the *NEC*, but his or her decisions must be based on a specific *Code* requirement. If an installation is rejected, the authority having jurisdiction is legally responsible for informing the installer which specific *NEC* rule was violated.

Author's Comment: The art of getting along with the authority having jurisdiction consists of doing good work and knowing what the *Code* actually says (as opposed to what you only think it says). It's also useful to know how to choose your battles when the inevitable disagreement does occur.

Approval of Equipment and Materials. Only the authority having jurisdiction has authority to approve the installation of equipment and materials. Typically, the authority having jurisdiction will approve equipment listed by a product testing organization, such as Underwriters Laboratories, Inc. (UL), but the *NEC* doesn't require all equipment to be listed. See 90.7, 110.2, 110.3, and the definitions for "Approved," "Identified," "Labeled," and "Listed" in Article 100. Figure 90–12

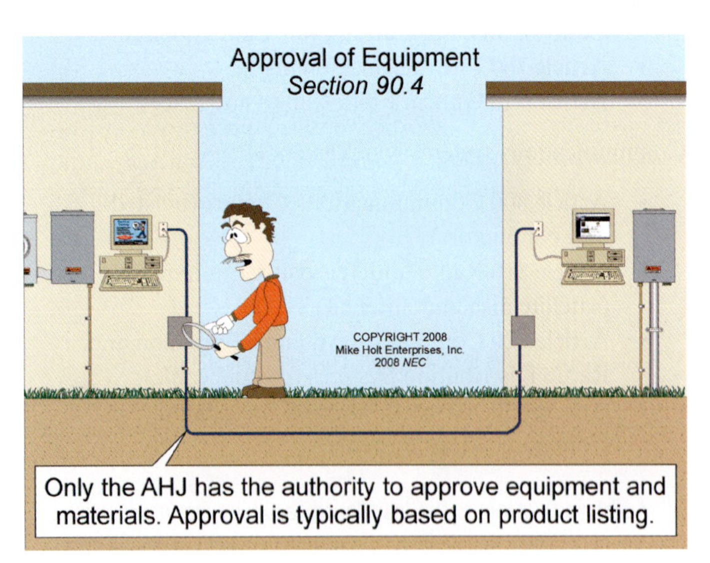

Figure 90–12

Author's Comment: According to the *NEC*, the authority having jurisdiction determines the approval of equipment. This means he/she can reject an installation of listed equipment and he/she can approve the use of unlisted equipment. Given our highly litigious society, approval of unlisted equipment is becoming increasingly difficult to obtain.

Waiver of Requirements. By special permission, the authority having jurisdiction can waive specific requirements in the *Code* or permit alternative methods where it's assured equivalent safety can be achieved and maintained.

Author's Comment: Special permission is defined in Article 100 as the written consent of the authority having jurisdiction.

Waiver of New Product Requirements. If the 2008 *NEC* requires products that aren't yet available at the time the *Code* is adopted, the authority having jurisdiction can allow products that were acceptable in the previous *Code* to continue to be used.

Author's Comment: Sometimes it takes years before testing laboratories establish product standards for new *NEC* requirements, and then it takes time before manufacturers can design, manufacture, and distribute these products to the marketplace.

Compliance with Listing Instructions. It's the authority having jurisdiction's responsibility to ensure electrical equipment is installed in accordance with equipment listing and/or labeling instructions [110.3(B)]. In addition, the authority having jurisdiction can reject the installation of equipment modified in the field [90.7].

Author's Comment: The *NEC* doesn't address the maintenance of electrical equipment because the *Code* is an installation standard, not a maintenance standard. See NFPA 70B—*Recommended Practice for Electrical Equipment Maintenance.*

90.5 Mandatory Requirements and Explanatory Material.

(A) Mandatory Requirements. In the *NEC* the words "shall" or "shall not," indicate a mandatory requirement.

Author's Comment: For the ease of reading this textbook, the word "shall" has been replaced with the word "must," and the words "shall not" have been replaced with "must not."

(B) Permissive Requirements. When the *Code* uses "shall be permitted" it means the identified actions are allowed but not required, and the authority having jurisdiction is not allowed to restrict an installation from being done in that manner. A permissive rule is often an exception to the general requirement.

Author's Comment: For ease of reading, the phrase "shall be permitted," as used in the *Code*, has been replaced in this textbook with the phrase "is permitted" or "are permitted."

(C) Explanatory Material. References to other standards or sections of the *NEC*, or information related to a *Code* rule, are included in the form of Fine Print Notes (FPNs). Fine Print Notes are for information only and are not enforceable.

For example, Fine Print Note No. 4 in 210.19(A)(1) recommends that the circuit voltage drop should not exceed 3 percent. This isn't a requirement; it's just a recommendation.

90.6 Formal Interpretations. To promote uniformity of interpretation and application of the provisions of the *NEC*, formal interpretation procedures have been established and are found in the NFPA Regulations Governing Committee Projects.

Author's Comment: This is rarely done because it's a very time-consuming process, and formal interpretations from the NFPA are not binding on the authority having jurisdiction.

90.7 Examination of Equipment for Product Safety. Product evaluation for safety is typically performed by a testing laboratory, which publishes a list of equipment that meets a nationally recognized test standard. Products and materials listed, labeled, or identified by a testing laboratory are generally approved by the authority having jurisdiction.

Author's Comment: See Article 100 for the definition of "Approved."

Listed, factory-installed, internal wiring and construction of equipment need not be inspected at the time of installation, except to detect alterations or damage [300.1(B)]. Figure 90–13

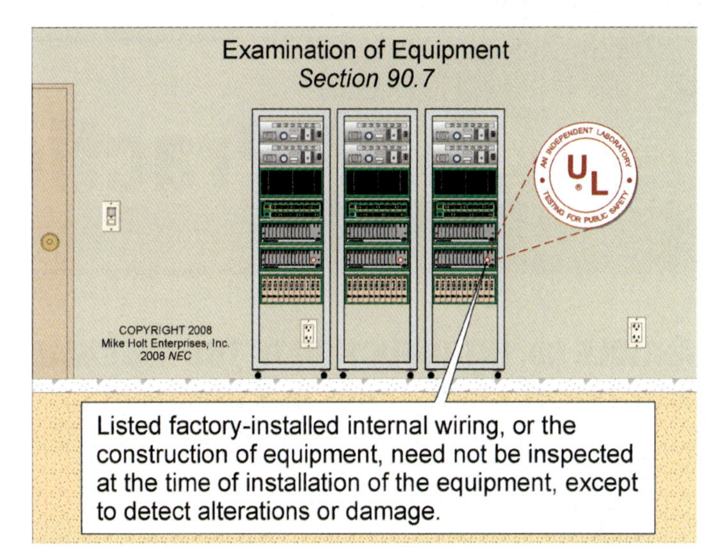

Figure 90–13

90.9 Units of Measurement.

(B) Dual Systems of Units. Both the metric and inch-pound measurement systems are shown in the *NEC*, with the metric units appearing first and the inch-pound system immediately following in parentheses.

Author's Comment: This is the standard practice in all NFPA standards, even though the U.S. construction industry uses inch-pound units of measurement.

(D) Compliance. Installing electrical systems in accordance with the metric system or the inch-pound system is considered to comply with the *Code*.

Author's Comment: Since compliance with either the metric or the inch-pound system of measurement constitutes compliance with the *NEC*, this textbook uses only inch-pound units.

ARTICLE 90 Practice Questions

ARTICLE 90. INTRODUCTION TO THE *NATIONAL ELECTRICAL CODE*—PRACTICE QUESTIONS

1. The *NEC* is _____.

   (a) intended to be a design manual
   (b) meant to be used as an instruction guide for untrained persons
   (c) for the practical safeguarding of persons and property
   (d) published by the Bureau of Standards

2. Hazards often occur because of _____.

   (a) overloading of wiring systems by methods or usage not in conformity with the *NEC*
   (b) initial wiring not providing for increases in the use of electricity
   (c) a and b
   (d) none of these

3. The *NEC* applies to the installation of _____.

   (a) electrical conductors and equipment within or on public and private buildings
   (b) outside conductors and equipment on the premises
   (c) optical fiber cables
   (d) all of these

4. The *Code* covers underground mine installations and self-propelled mobile surface mining machinery and its attendant electrical trailing cable.

   (a) True
   (b) False

5. Utilities may be subject to compliance with codes and standards covering their regulated activities as adopted under governmental law or regulation.

   (a) True
   (b) False

6. Chapters 1 through 4 of the *NEC* apply _____.

   (a) generally to all electrical installations
   (b) only to special occupancies and conditions
   (c) only to special equipment and material
   (d) all of these

7. The authority having jurisdiction shall not be required to enforce any requirements of Chapter 7 (Special Conditions) or Chapter 8 (Communications Systems).

   (a) True
   (b) False

8. The authority having jurisdiction has the responsibility _____.

   (a) for making interpretations of rules
   (b) for deciding upon the approval of equipment and materials
   (c) for waiving specific requirements in the *Code* and permitting alternate methods and material if safety is maintained
   (d) all of these

9. When the *Code* uses ____, it means the identified actions are allowed but not required, and the authority having jurisdiction is not to restrict an installation from being done in that manner.

   (a) shall
   (b) shall not
   (c) shall be permitted
   (d) a or b

10. Compliance with either the metric or the inch-pound unit of measurement system shall be permitted.

    (a) True
    (b) False

CHAPTER 1

GENERAL

INTRODUCTION TO CHAPTER 1—GENERAL

Many people skip Chapter 1 of the *NEC* because they want something prescriptive—they want something that tells them what to do, cookbook style. But electricity isn't a simple topic you can jump right into. You can't just follow a few simple steps to get a safe installation. You need a foundation from which you can apply the *Code*.

Consider Ohm's law. Will Ohm's Law make sense to you if you don't know what an ohm is? Similarly, you must be familiar with a few basic rules, concepts, definitions, and requirements that apply to the rest of the *NEC*, and you must maintain that familiarity as you continue to apply the *Code*.

Chapter 1 consists of two main topics. Article 100 provides definitions so people can understand one another when trying to communicate on *Code*-related matters. Article 110 provides general requirements needed to correctly apply the rest of the *NEC*.

Time spent learning this general material is a great investment. After understanding Chapter 1, some of the *Code* requirements that seem confusing to other people—those who don't understand Chapter 1—will become increasingly clear to you. That is, they will strike you as being "common sense," because you'll have the foundation from which to understand and apply them. Because you'll understand the principles upon which many *NEC* requirements in later chapters are based, you'll read those requirements and not be surprised at all. You'll read them and feel like you already know them.

- **Article 100—Definitions.** Part I of Article 100 contains the definitions of terms used throughout the *Code* for systems that operate at 600V nominal or less. The definitions of terms in Part II apply to systems that operate at over 600V nominal.

  Author's Comment: The requirements covered in this textbook apply to systems that operate at 600V nominal or less.

  Definitions of standard terms, such as volt, voltage drop, ampere, impedance, and resistance, aren't listed in Article 100. If the *NEC* doesn't define a term, then a dictionary suitable to the authority having jurisdiction should be consulted. A building code glossary might provide a better definition than a dictionary found at your home or school.

  Definitions at the beginning of an article apply only to that specific article. For example, the definition of a "Swimming Pool" is contained in 680.2, because this term applies only to the requirements contained in Article 680 Swimming Pools, Fountains, and Similar Installations.

- **Article 110—Requirements for Electrical Installations.** This article contains general requirements for electrical installations for the following:
  - PART I. GENERAL
  - PART II. 600V, NOMINAL, OR LESS
  - PART III. OVER 600V, NOMINAL
  - PART IV. TUNNEL INSTALLATIONS OVER 600V, NOMINAL
  - PART V. MANHOLES AND OTHER ELECTRIC ENCLOSURES INTENDED FOR PERSONNEL ENTRY

ARTICLE

100 Definitions

INTRODUCTION TO ARTICLE 100—DEFINITIONS

Have you ever had a conversation with someone, only to discover what you said and what he/she heard were completely different? This often happens when people in a conversation don't understand the definitions of the words being used, and that's why the definitions of key terms are located right at the beginning of the *NEC* (Article 100), or at the beginning of each article.

If we can all agree on important definitions, then we speak the same language and avoid misunderstandings. Because the *Code* exists to protect people and property, we can agree it's very important to know the definitions presented in Article 100.

Now, here are a couple of things you may not know about Article 100:

Article 100 contains the definitions of many, but not all, of the terms used throughout the *NEC*. In general, only those terms used in two or more articles are defined in Article 100.

- Part I of Article 100 contains the definitions of terms used throughout the *Code* for systems that operate at 600V, nominal, or less.
- Part II of Article 100 contains only terms that apply to systems that operate at over 600V nominal.

How can you possibly learn all of these definitions? There seem to be so many. Here are a few tips:

- Break the task down. Study a few words at a time, rather than trying to learn them all at one sitting.
- Review the graphics in the textbook. These will help you see how a term is applied.
- Relate them to your work. As you read a word, think about how it applies to the work you're doing. This will provide a natural reinforcement to the learning process.

DEFINITIONS

Accessible (as it applies to equipment). Admitting close approach and not guarded by locked doors, elevation, or other effective means.

Accessible (as it applies to wiring methods). Not permanently closed in by the building structure or finish and capable of being removed or exposed without damaging the building structure or finish. Figure 100–1

Author's Comments:

- Conductors in a concealed raceway are considered concealed, even though they may become accessible by withdrawing them. See the definition of "Concealed" in this article.

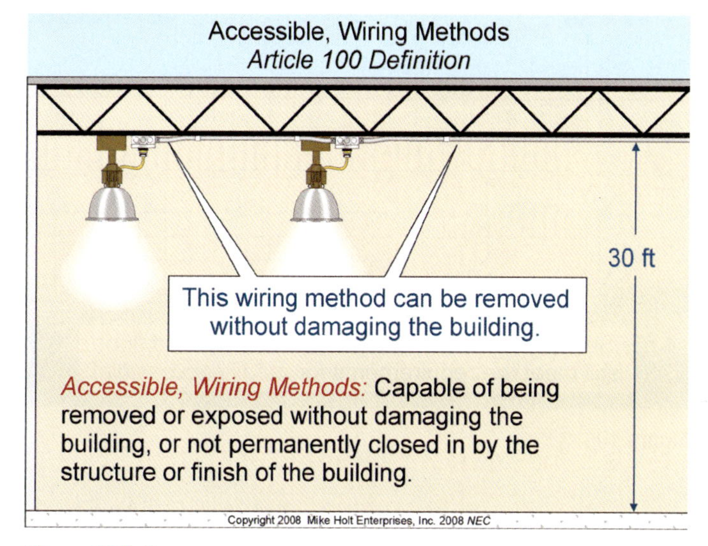

Figure 100–1

- Raceways, cables, and enclosures installed above a suspended ceiling or within a raised floor are considered accessible, because the wiring methods can be accessed without damaging the building structure. See the definitions of "Concealed" and "Exposed" in this article.

Accessible, Readily (Readily Accessible). Capable of being reached quickly without having to climb over or remove obstacles or resort to portable ladders. Figures 100–2 and 100–3

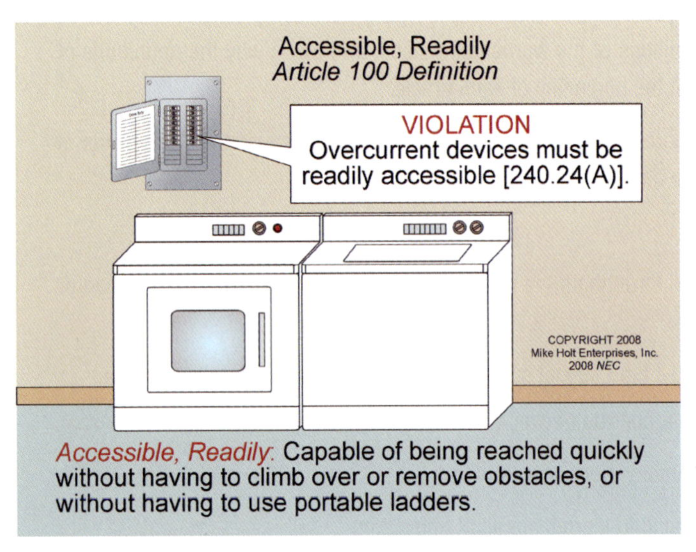

Figure 100–2

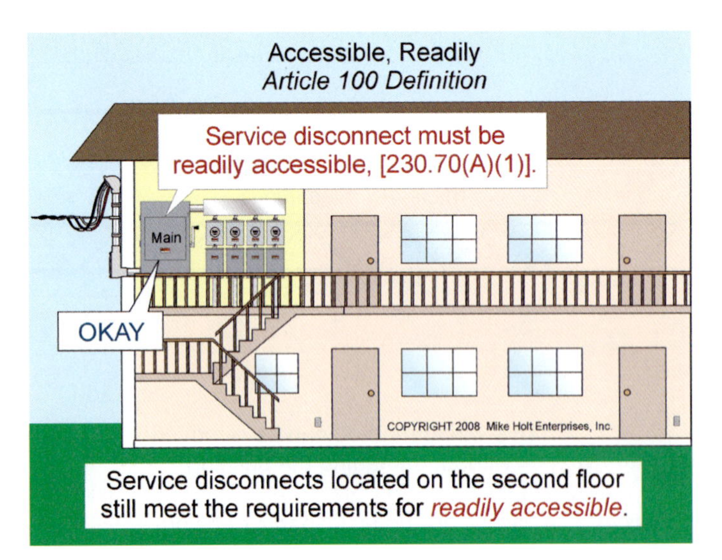

Figure 100–3

Ampacity. The current, in amperes, a conductor can carry continuously, where the temperature of the conductor will not be raised in excess of its insulation temperature rating. See 310.10 and 310.15 for details and examples. Figure 100–4

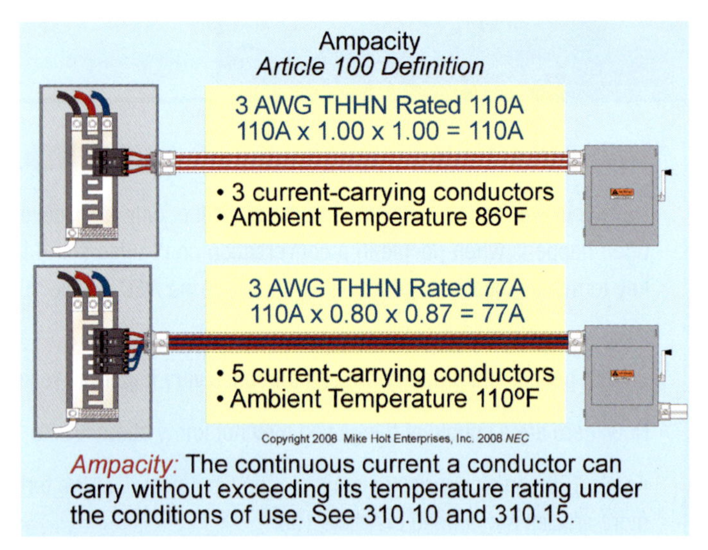

Figure 100–4

Appliance [Article 422]. Electrical equipment, other than industrial equipment, built in standardized sizes, such as ranges, ovens, cooktops, refrigerators, drinking water coolers, or beverage dispensers.

Approved. Acceptable to the authority having jurisdiction, usually the electrical inspector.

> **Author's Comment:** Product listing doesn't mean the product is approved, but it's a basis for approval. See 90.4, 90.7, 110.2, and the definitions in this article for "Authority Having Jurisdiction," "Identified," "Labeled," and "Listed."

Attachment Plug (Plug Cap)(Plug) [Article 406]. A wiring device at the end of a flexible cord intended to be inserted into a receptacle. Figure 100–5

> **Author's Comment:** The use of cords with attachment plugs is limited by 210.50(A), 400.7, 410.24, 410.62, 422.16, 422.33, 590.4, 645.5 and other sections.

Authority Having Jurisdiction (AHJ). The organization, office, or individual responsible for approving equipment, materials, an installation, or a procedure. See 90.4 and 90.7 for more information.

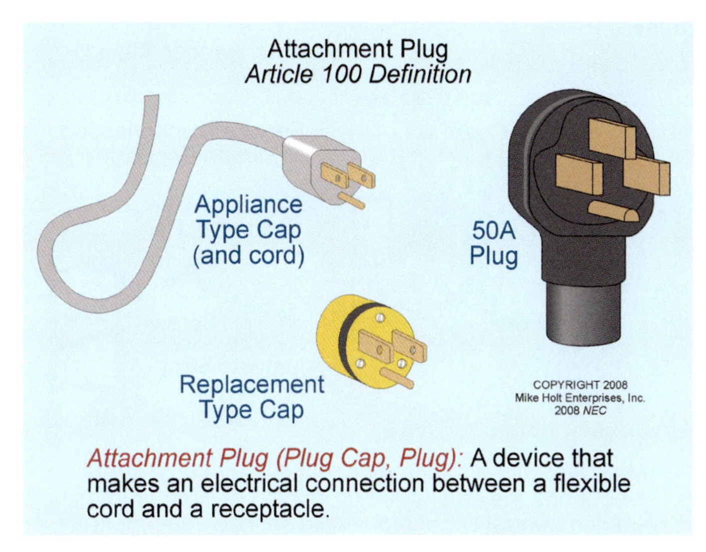

Figure 100–5

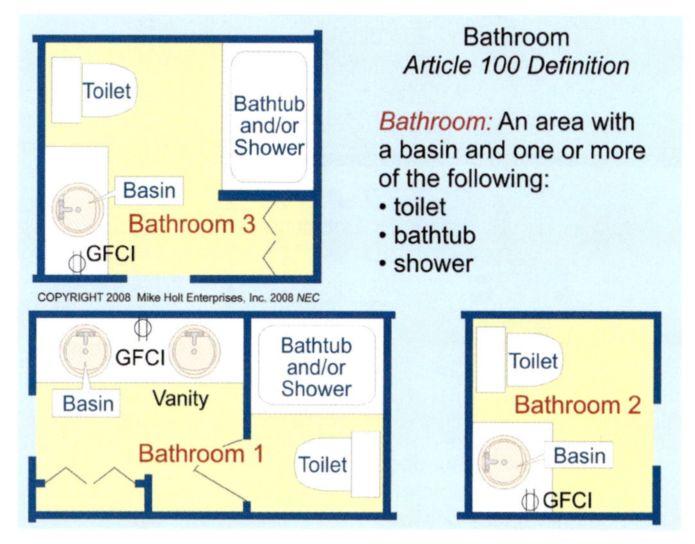

Figure 100–6

FPN: The authority having jurisdiction may be a federal, state, or local government, or an individual such as a fire chief, fire marshal, chief of a fire prevention bureau or labor department or health department, a building official or electrical inspector, or others having statutory authority. In some circumstances, the property owner or his/her agent assumes the role, and at government installations, the commanding officer, or departmental official may be the authority having jurisdiction.

Author's Comments:

- Typically, the authority having jurisdiction is the electrical inspector who has legal statutory authority. In the absence of federal, state, or local regulations, the operator of the facility or his/her agent, such as an architect or engineer of the facility, can assume the role.
- Some believe the authority having jurisdiction should have a strong background in the electrical field, such as having studied electrical engineering or having obtained an electrical contractor's license, and in a few states this is a legal requirement. Memberships, certifications, and active participation in electrical organizations, such as the International Association of Electrical Inspectors (IAEI) (www.IAEI.org), speak to an individual's qualifications.

Bathroom. A bathroom is an area that includes a basin with a toilet, tub, or shower. Figure 100–6

Author's Comment: All 15A and 20A, 125V receptacles located in bathrooms must be GFCI protected [210.8].

Bonded (Bonding). Connected to establish electrical continuity and conductivity. Figure 100–7

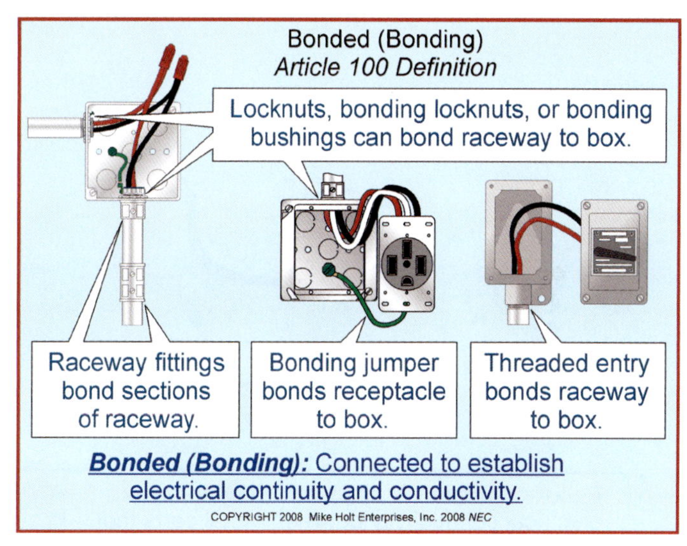

Figure 100–7

Author's Comment: The purpose of bonding is to connect two or more conductive objects together to ensure the electrical continuity of the fault current path, provide the capacity and ability to conduct safely any fault current likely to be imposed, and to minimize potential differences (voltage) between conductive components. Figure 100–8

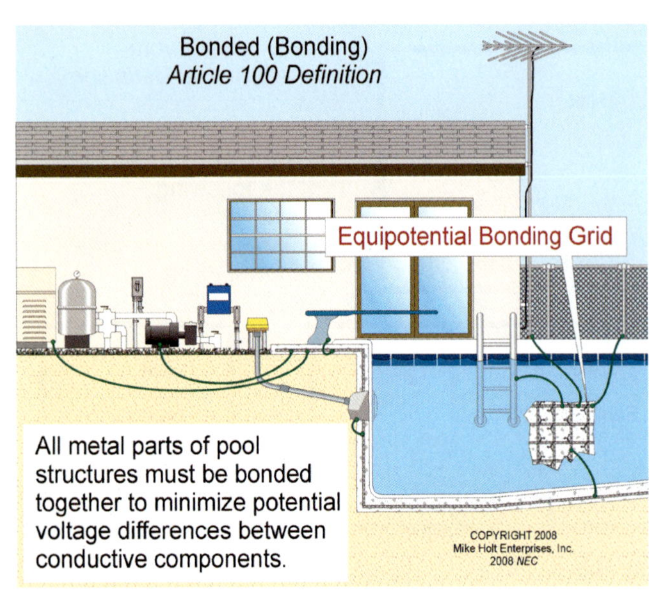

Figure 100–8

Bonding Jumper. A conductor that ensures electrical conductivity between metal parts of the electrical installation. Figure 100–9

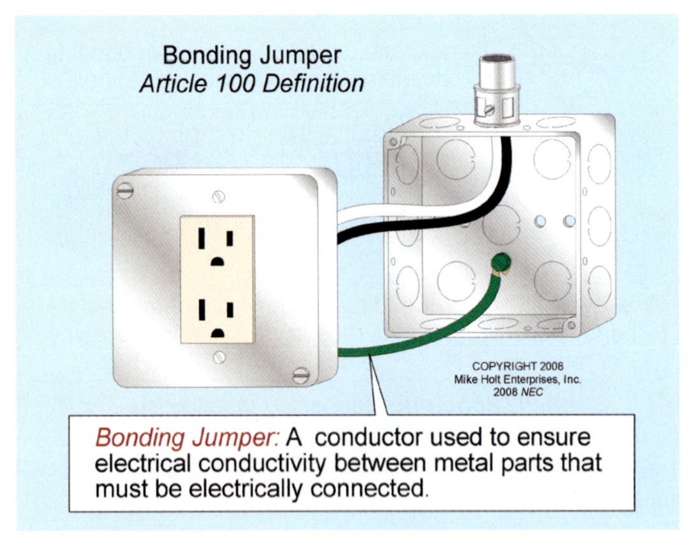

Figure 100–9

Bonding Jumper, Main. A conductor, screw, or strap that connects the circuit equipment grounding conductor to the neutral conductor at service equipment in accordance with 250.24(B) [250.24(A)(4), 250.28, and 408.3(C)]. Figure 100–10

Branch Circuit [Article 210]. The conductors between the final overcurrent device and the receptacle outlets, lighting outlets, or other outlets as defined in this article. Figure 100–11

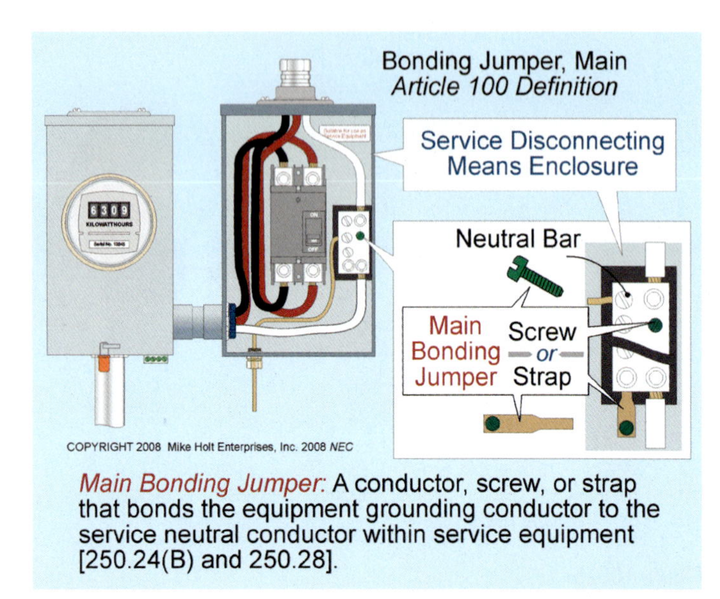

Figure 100–10

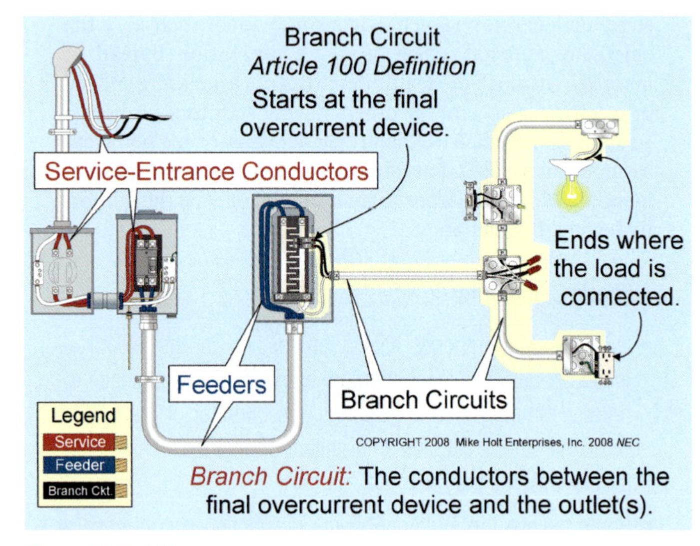

Figure 100–11

Branch Circuit, Individual. A branch circuit that only supplies one load.

Branch Circuit, Multiwire. A branch circuit that consists of two or more ungrounded circuit conductors with a common neutral conductor. There must be a difference of potential (voltage) between the ungrounded conductors and an equal difference of potential (voltage) from each ungrounded conductor to the neutral conductor. Figure 100–12

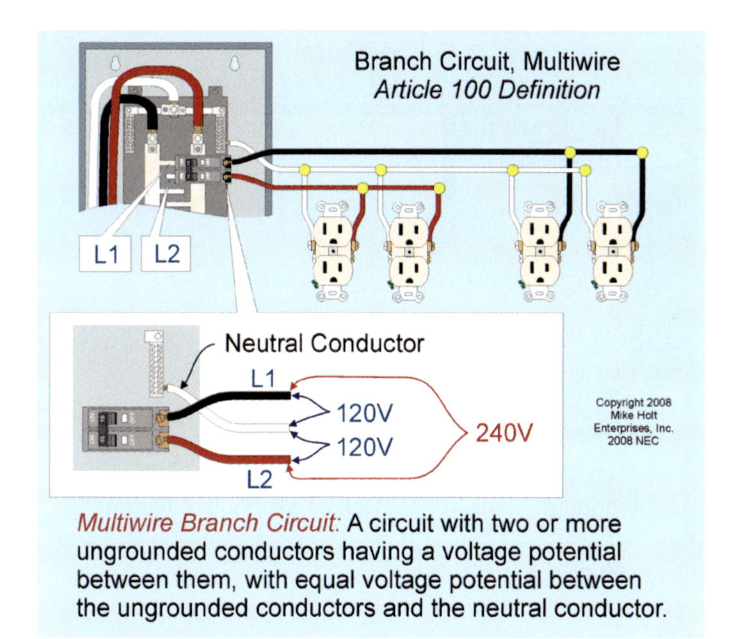

Figure 100–12

Author's Comment: Multiwire branch circuits offer the advantage of fewer conductors in a raceway, smaller raceway sizing, and a reduction of material and labor costs. In addition, multiwire branch circuits can reduce circuit voltage drop by as much as 50 percent. However, because of the dangers associated with multiwire branch circuits, the *NEC* contains additional requirements to ensure a safe installation. See 210.4, 300.13(B), and 408.41 in this textbook for details.

Building. A structure that stands alone or is cut off from other structures by firewalls with all openings protected by fire doors approved by the authority having jurisdiction. Figure 100–13

Author's Comment: Not all fire-rated walls are fire walls. Building codes describe fire barriers, fire partitions and other fire-rated walls, in addition to fire walls. Check with your local building inspector to determine if a rated wall creates a separate building (fire wall).

Cabinet [Article 312]. An enclosure for either surface mounting or flush mounting provided with a frame in which a door can be hung. Figure 100–14

Author's Comment: Cabinets are used to enclose panelboards. See the definition of "Panelboard" in this article.

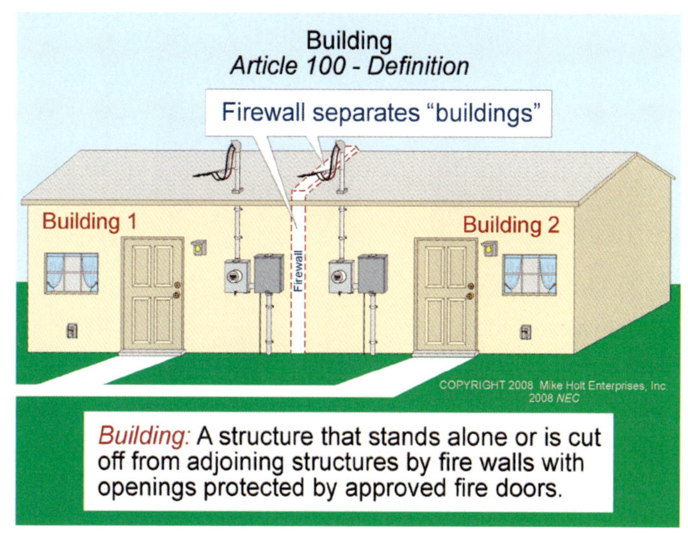

Figure 100–13

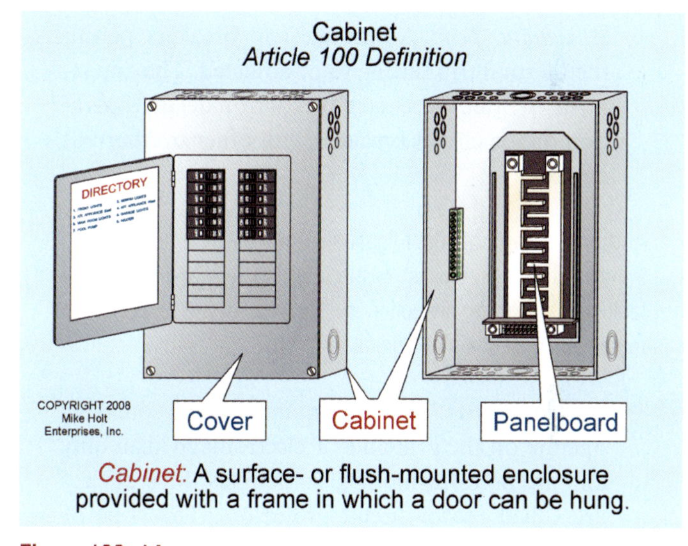

Figure 100–14

Circuit Breaker. A device designed to be opened and closed manually, and which opens automatically on a predetermined overcurrent without damage to itself. Circuit breakers are available in different configurations, such as inverse time molded case, adjustable (electronically controlled), and instantaneous trip/motor-circuit protectors. Figure 100–15

- *Inverse Time:* Inverse time breakers operate on the principle that as the current increases, the time it takes for the devices to open decreases. This type of breaker provides overcurrent protection (overload, short circuit, and ground fault).

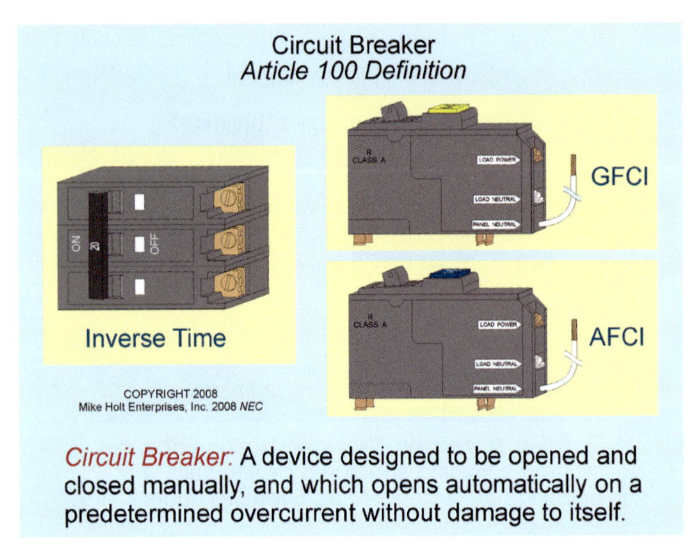

Figure 100–15

- *Adjustable Trip:* Adjustable trip breakers permit the thermal trip setting to be adjusted. The adjustment is often necessary to coordinate the operation of the circuit breakers with other overcurrent devices.

Author's Comment: Coordination means that the devices with the lowest ratings, closest to the fault, operate and isolate the fault and minimize disruption so the rest of the system can remain energized and functional.

- *Instantaneous Trip:* Instantaneous trip breakers operate on the principle of electromagnetism only and are used for motors. Sometimes these devices are called motor-circuit protectors. This type of overcurrent device doesn't provide overload protection. It only provides short-circuit and ground-fault protection; overload protection must be provided separately.

Author's Comment: Instantaneous trip circuit breakers have no intentional time delay and are sensitive to current inrush, and to vibration and shock. Consequently, they should not be used where these factors are known to exist.

Clothes Closet. A nonhabitable room or space intended primarily for storage of garments and apparel. Figure 100–16

Author's Comment: The definition of "Clothes Closet" provides clarification in the application of overcurrent devices [240.24(D)] and luminaires [410.16] in clothes closets.

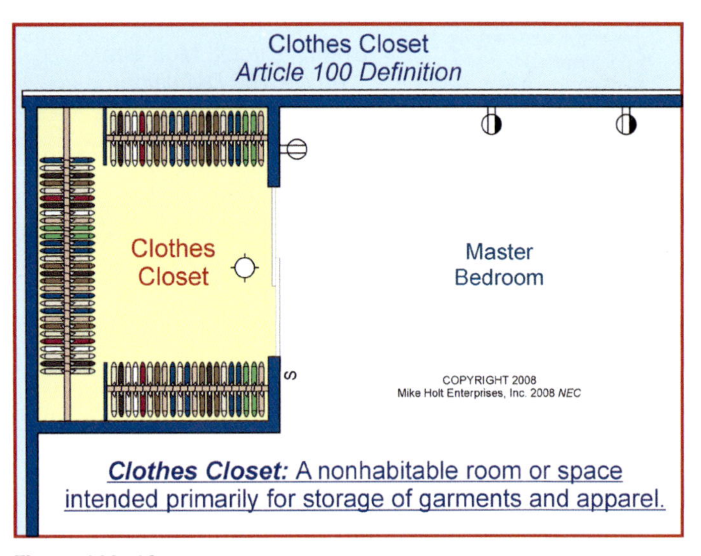

Figure 100–16

Communications Equipment. Electronic telecommunications equipment used for the transmission of audio, video, and data, including support equipment such as computers. Figure 100–17

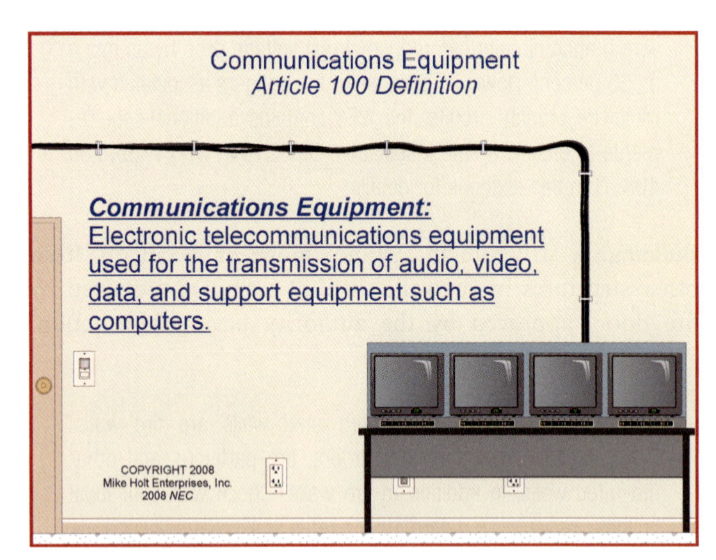

Figure 100–17

Concealed. Rendered inaccessible by the structure or finish of the building. Conductors in a concealed raceway are considered concealed, even though they may become accessible by withdrawing them. Figure 100–18

Author's Comment: Wiring behind panels designed to allow access is considered exposed.

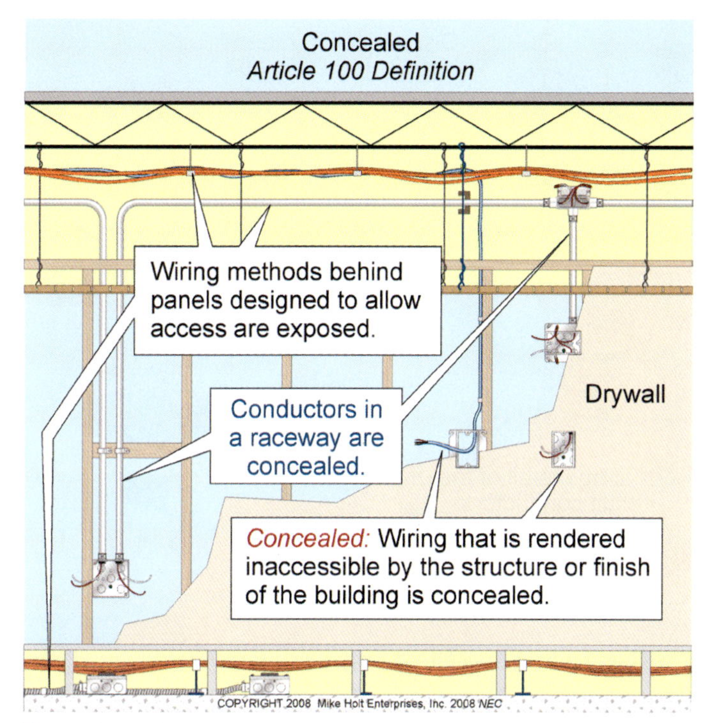

Figure 100–18

Conduit Body. A fitting that provides access to conductors through a removable cover. Figure 100–19

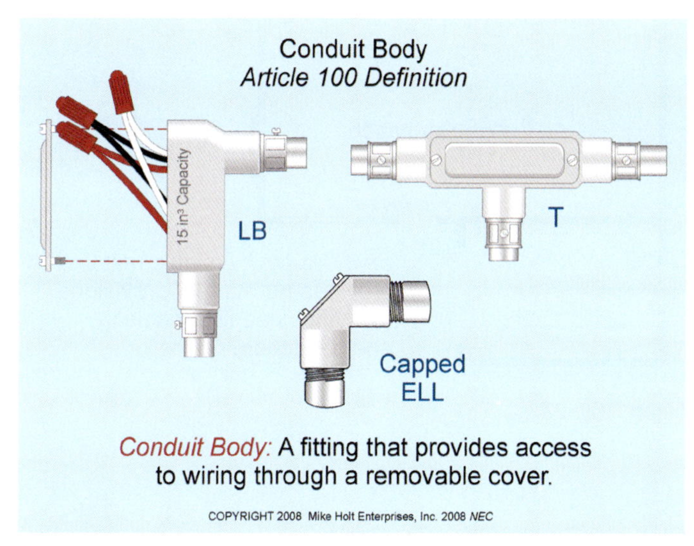

Figure 100–19

Connector, Pressure (Solderless). A device that establishes a conductive connection between conductors and a terminal by the means of mechanical pressure.

Continuous Load. A load where the current is expected to exist for three hours or more, such as store or parking lot lighting.

Controller. A device that controls, in some predetermined manner, the electric power delivered to electrical equipment. This includes time clocks, lighting contactors, photocells, etc. Figure 100–20

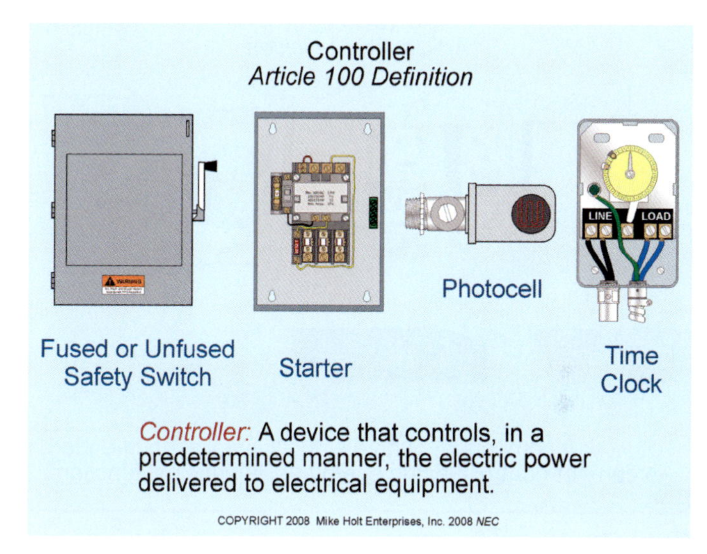

Figure 100–20

Author's Comment: For the definition of motor controller, see 430.2

Coordination (Selective). Localization of an overcurrent condition to restrict outages to the circuit or equipment affected, accomplished by the choice of overcurrent devices.

Author's Comment: Selective coordination means the overcurrent protection scheme confines the interruption to a particular area rather than to the whole system. For example, if someone plugs in a space heater and raises the total demand on a 20A circuit to 25A, or if a short circuit or ground fault occurs with selective coordination, the only breaker/fuse that will open is the one protecting just that branch circuit. Without selective coordination, an entire building can go dark!

Cutout Box [Article 312]. Cutout boxes are designed for surface mounting with a swinging door.

Demand Factor. The ratio of the maximum demand to the total connected load.

Author's Comment: This definition is primarily used in the application of the requirements of Article 220—Branch-Circuit, Feeder, and Service Calculations.

Device. A component of an electrical installation intended to carry or control electric energy as its principal function. Figure 100–21

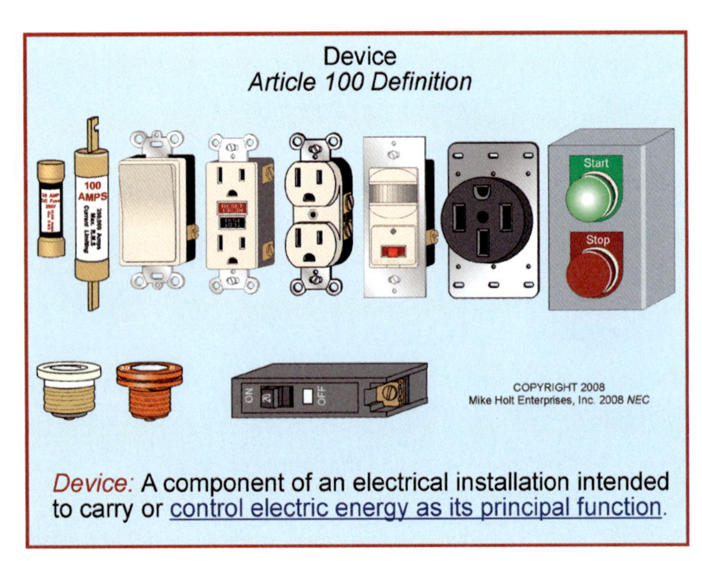

Figure 100–21

Author's Comment: Devices include wires and busbars, receptacles, switches, illuminated switches, circuit breakers, fuses, time clocks, controllers, etc., but not locknuts or other mechanical fittings.

Disconnecting Means. A device that opens all of the ungrounded circuit conductors from their power source. These devices include switches, attachment plugs and receptacles, and circuit breakers. Figure 100–22

Author's Comment: Review the following for the specific requirements for equipment disconnecting means:

- Air-conditioning and refrigeration equipment, 440.14
- Appliances, Article 422, Part III
- Building supplied by a feeder, Article 225, Part II
- Electric space-heating equipment, 424.19
- Electric duct heaters, 424.65
- Luminaires, 410.130(G)
- Motor control conductors, 430.75
- Motor controllers, 430.102(A)
- Motors, 430.102(B)(1)
- Refrigeration equipment, 440.14
- Services, Article 230, Part VI
- Swimming pool, spa, hot tub, and fountain equipment, 680.12

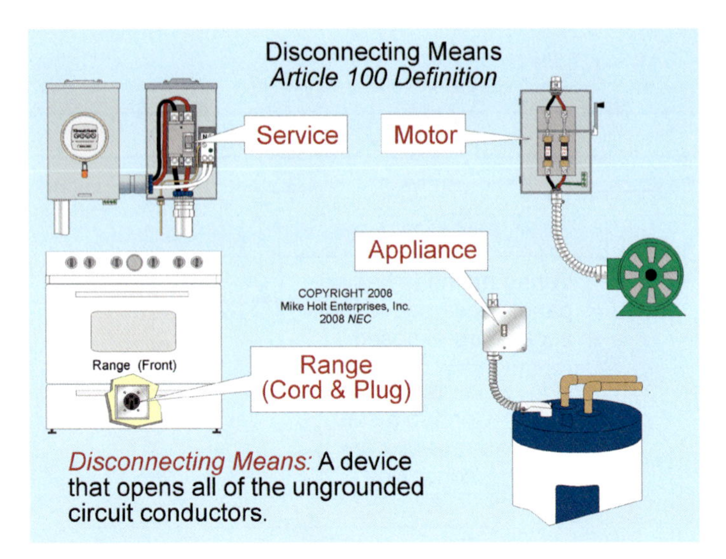

Figure 100–22

Dwelling Unit. A space that provides independent living facilities, with space for eating, living, and sleeping; as well as permanent facilities for cooking and sanitation. Figure 100–23

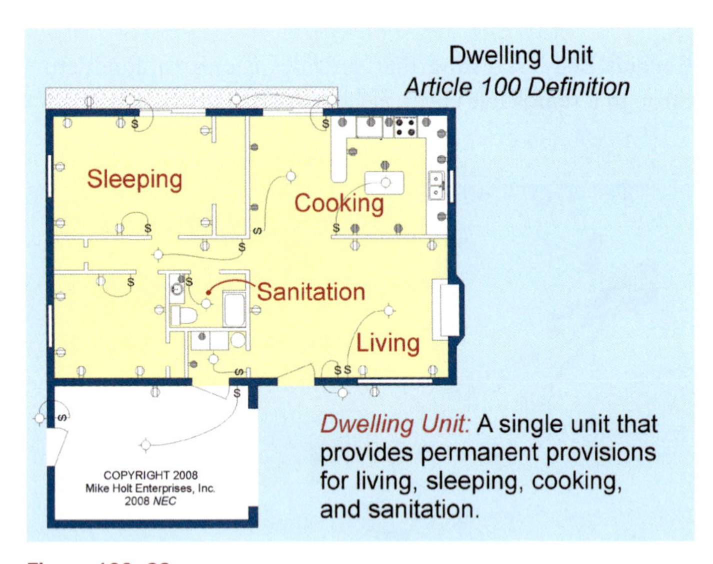

Figure 100–23

Dwelling, Multifamily. A building that contains three or more dwelling units.

Electric Sign [Article 600]. A fixed, stationary, or portable self-contained, electrically illuminated piece of equipment with words or symbols designed to convey information or attract attention.

Energized. Electrically connected to a source of voltage.

Equipment. A general term including material, fittings, devices, appliances, luminaires, apparatus, machinery, and the like. Figure 100–24

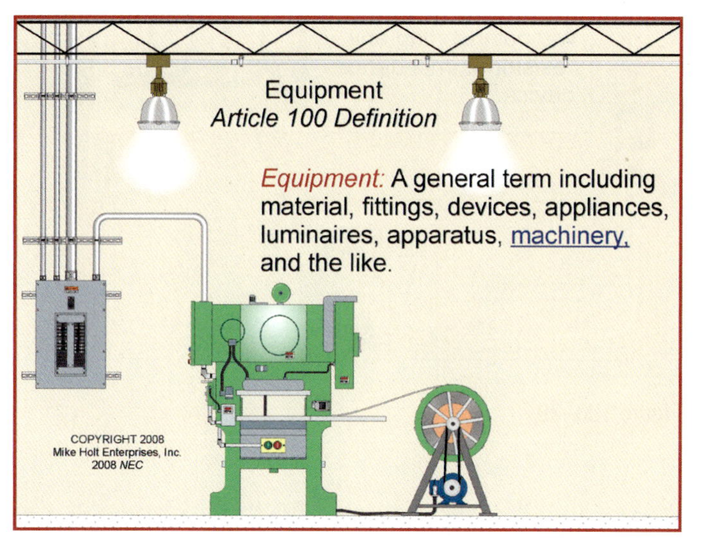

Figure 100–24

Exposed (as it applies to wiring methods). On or attached to the surface of a building, or behind panels designed to allow access.

> **Author's Comment:** An example is wiring located in the space above a suspended ceiling or below a raised floor. Figure 100–25

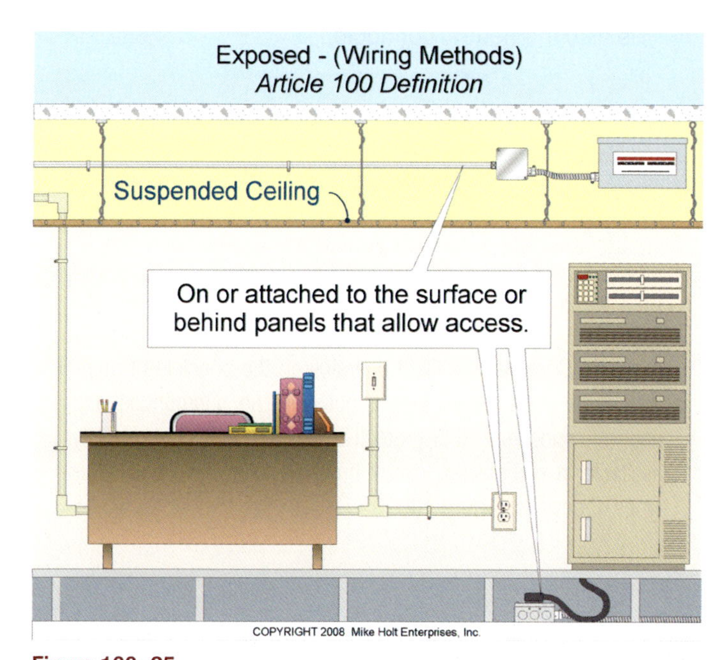

Figure 100–25

Feeder [Article 215]. The conductors between the service equipment, a separately derived system, or other power supply and the final branch-circuit overcurrent device. Figure 100–26

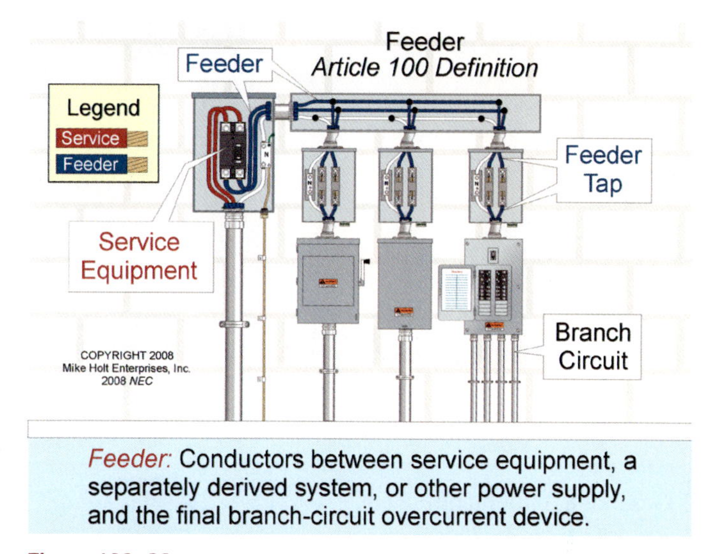

Figure 100–26

> **Author's Comments:**
>
> - An "other power source" includes a solar energy system (photovoltaic).
> - To have a better understanding of what a feeder is, be sure to review the definitions of "Service Equipment" and "Separately Derived System."

Fitting. An accessory, such as a locknut, intended to perform a mechanical function.

Garage. A building or portion of a building where self-propelled vehicles can be kept.

Ground. The earth. Figure 100–27

Grounded (Grounding). Connected to ground or to a conductive body that extends the ground connection.

> **Author's Comment:** An example of a "body that extends the ground (earth) connection" is the termination to structural steel that is connected to the earth either directly or by the termination to another grounding electrode in accordance with 250.52. Figure 100–28

Grounded, Solidly. Connected to ground without inserting any resistor or impedance device. Figure 100–29

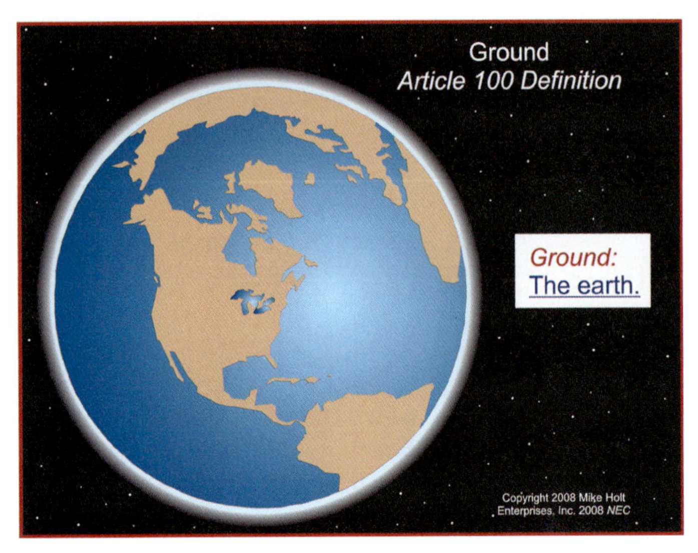

Figure 100–27

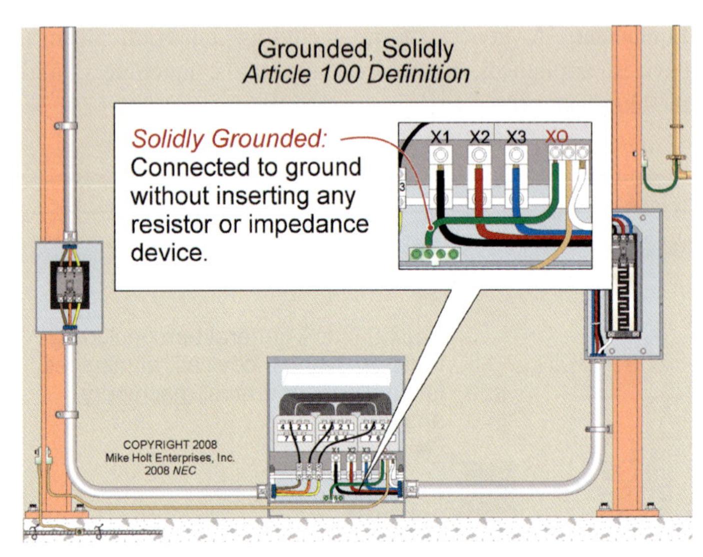

Figure 100–29

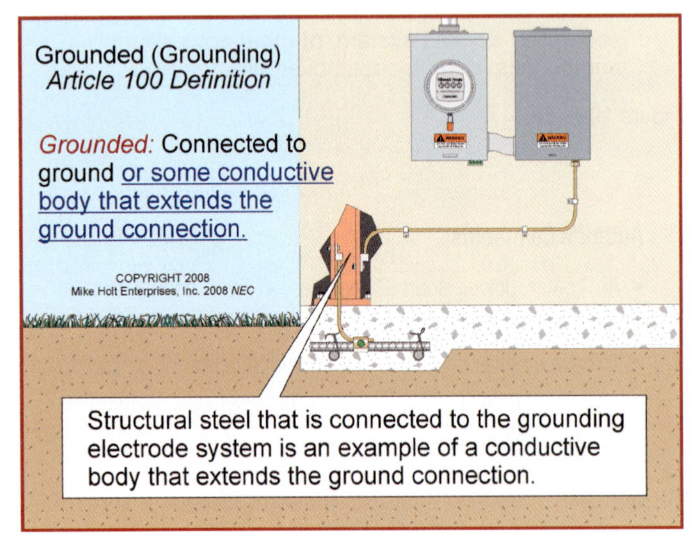

Figure 100–28

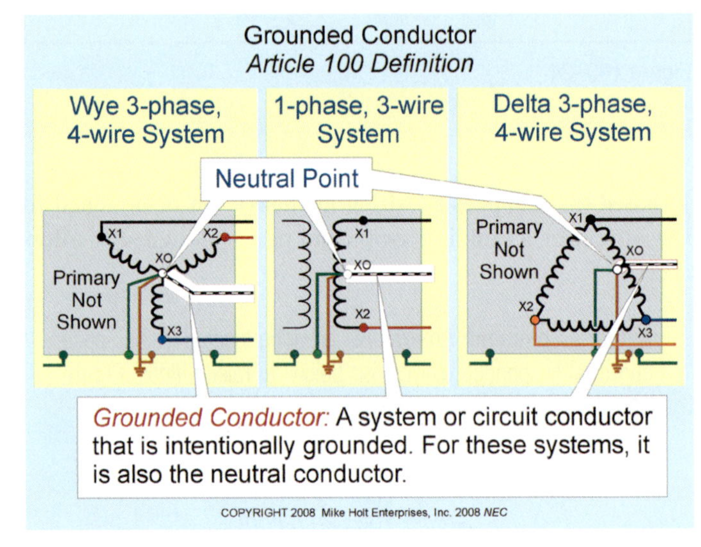

Figure 100–30

Grounded Conductor [Article 200]. The conductor that is intentionally grounded (connected to the earth). Figure 100–30

Author's Comment: Because the neutral conductor of a solidly grounded system is always grounded (connected to the earth), it's both a "grounded conductor" and a "neutral" conductor. To make it easier for the reader of this textbook, we will refer to the "grounded" conductor of a solidly grounded system as the "neutral" conductor.

Ground-Fault Circuit Interrupter (GFCI). A device intended to protect people by de-energizing a circuit when the current-to-ground exceeds the value established for a "Class A" device. Figure 100–31

FPN: A "Class A" ground-fault circuit interrupter opens the circuit when the current-to-ground has a value of 6 mA or higher and doesn't trip when the current-to-ground is less than 4 mA.

Author's Comment: A GFCI operates on the principle of monitoring the unbalanced current between the ungrounded and neutral conductor. GFCI-protective devices are commercially available in receptacles, circuit breakers, cord sets, and other types of devices. Figure 100–32

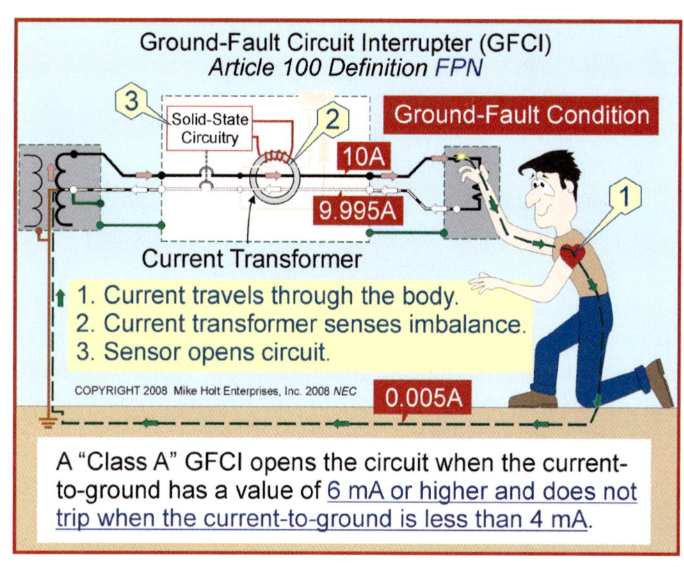

Figure 100–31

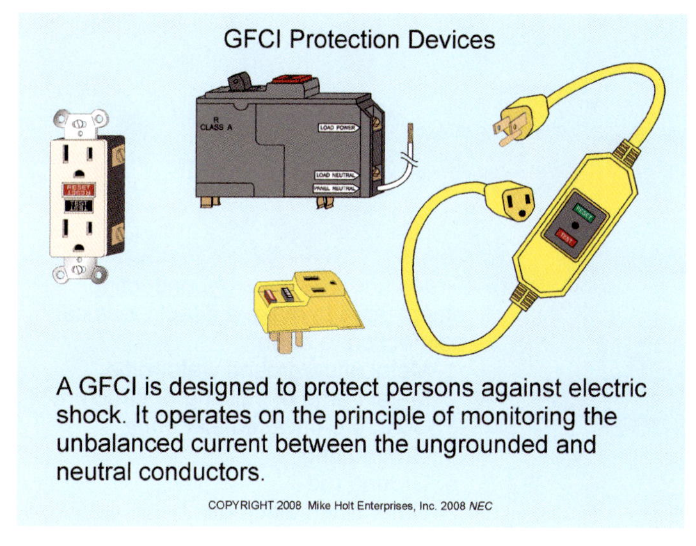

Figure 100–32

Ground-Fault Protection of Equipment. A system intended to provide protection of equipment from damaging ground-fault currents by opening all ungrounded conductors of the faulted circuit. This protection is provided at current levels less than those required to protect conductors from damage through the operation of a supply circuit overcurrent device [215.10, 230.95, and 240.13].

Author's Comment: This type of protective device is not a ground-fault circuit interrupter, because it's not intended to protect people.

Grounding Conductor. A conductor used to connect equipment or the neutral conductor of a wiring system to a grounding electrode. Figure 100–33

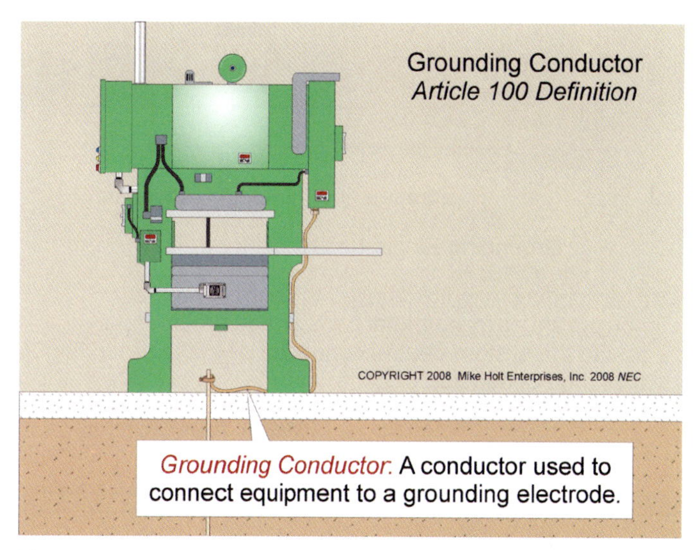

Figure 100–33

Author's Comment: Grounding conductors are used for auxiliary electrodes [250.54] and communications systems electrodes [770.100(A), 800.100(A), 810.21(A), 820.100(A), and 830.100(A)].

Grounding Conductor, Equipment (EGC). The conductive path that connects metal parts of equipment to the system neutral conductor or to the grounding electrode conductor, or both [250.110 through 250.126]. Figure 100–34

FPN No. 1: The circuit equipment grounding conductor also performs bonding.

Author's Comment: To quickly remove dangerous touch voltage on metal parts from a ground fault, the equipment grounding conductor must be connected to the source, and have sufficiently low impedance, so that fault current will quickly rise to a level that will open the branch-circuit overcurrent device [250.2 and 250.4(A)(3)].

FPN No. 2: An equipment grounding conductor can be any one or a combination of the types listed in 250.118. Figure 100–35

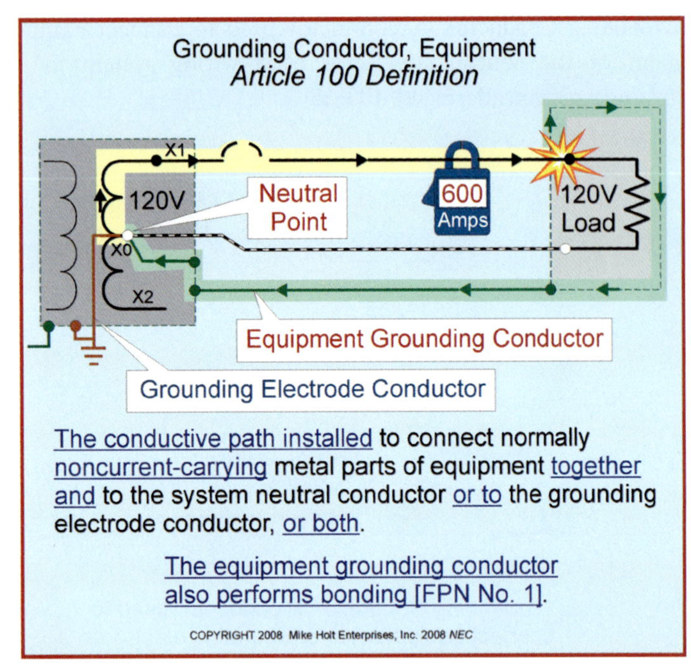

Figure 100–34

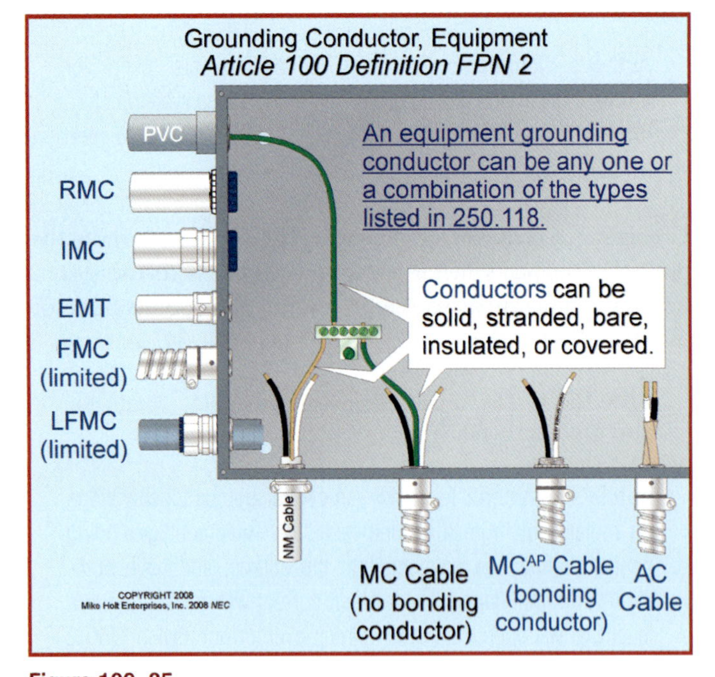

Figure 100–35

Author's Comment: Equipment grounding conductors include:

- Bare or insulated conductor
- Rigid Metal Conduit
- Intermediate Metal Conduit
- Electrical Metallic Tubing
- Listed Flexible Metal Conduit as limited by 250.118(5)
- Listed Liquidtight Flexible Metal Conduit as limited by 250.118(6)
- Armored Cable
- Copper metal sheath of Mineral Insulated Cable
- Metal Clad Cable as limited by 250.118(10)
- Metallic cable trays as limited by 250.118(11) and 392.7
- Electrically continuous metal raceways listed for grounding
- Surface Metal Raceways listed for grounding

Grounding Electrode. A conducting object used to make a direct electrical connection to the earth [250.50 through 250.70]. Figure 100–36

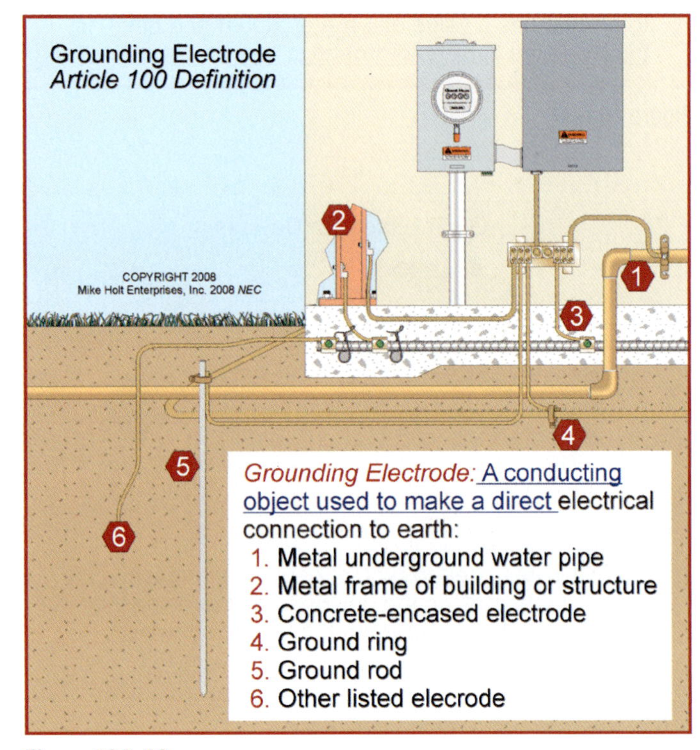

Figure 100–36

Grounding Electrode Conductor. The conductor used to connect the system neutral conductor or the metal parts of electrical equipment to a grounding electrode or to a point on the grounding electrode system. Figure 100–37

Author's Comment: For services see 250.24(A), for separately derived systems see 250.30(A), and for buildings or structures supplied by a feeder see 250.32(A).

Guest Room. An accommodation that combines living, sleeping, sanitary, and storage facilities. Figure 100–38

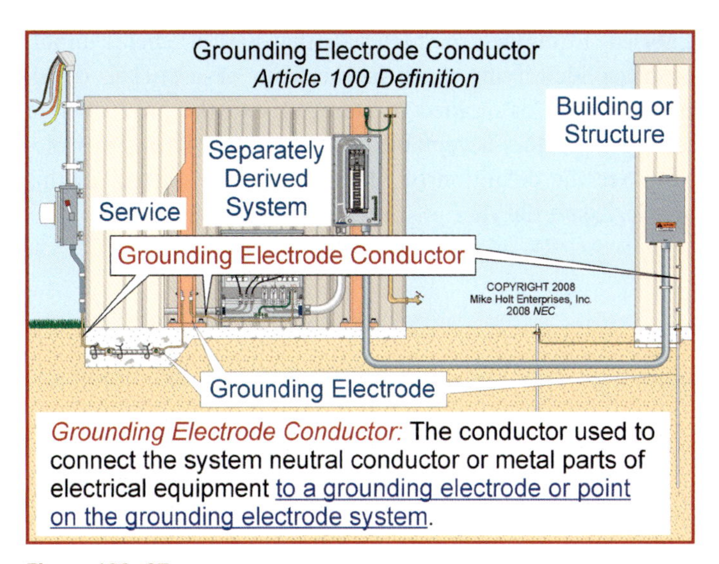

Figure 100–37

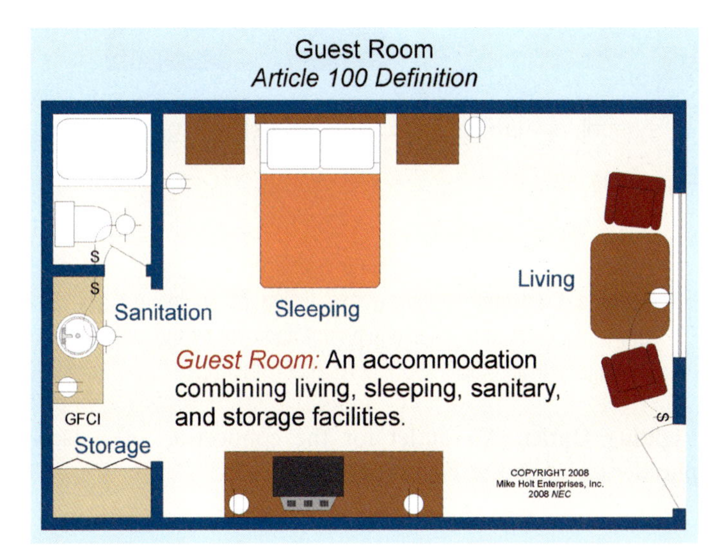

Figure 100–38

Guest Suite. An accommodation with two or more contiguous rooms comprising a compartment, with or without doors between such rooms, that provides living, sleeping, sanitary, and storage facilities.

Handhole Enclosure. An enclosure for underground system use, with an open or closed bottom, and sized to allow personnel to reach into, for the purpose of installing or maintaining equipment or wiring. Figure 100–39

> **Author's Comment:** See 314.30 for the installation requirements for handhole enclosures.

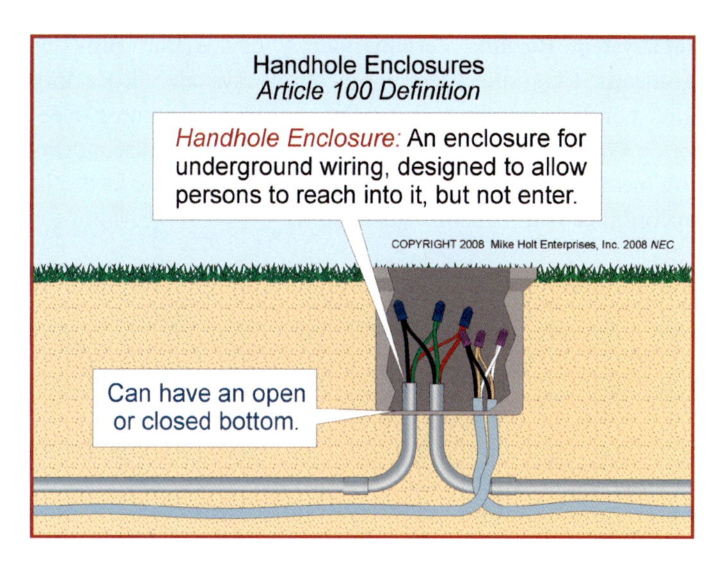

Figure 100–39

Identified Equipment. Recognized as suitable for a specific purpose, function, or environment by listing, labeling, or other means approved by the authority having jurisdiction. See 90.4, 90.7, 110.3(A)(1), and the definitions for "Approved," "Labeled," and "Listed" in this article.

In Sight From (Within Sight). Visible and not more than 50 ft away from the equipment. Figure 100–40

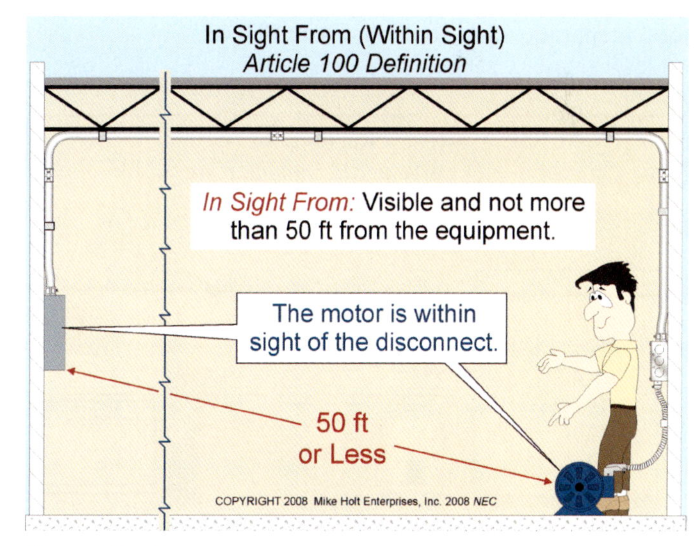

Figure 100–40

Interrupting Rating. The highest short-circuit current at rated voltage the device can safely interrupt. For more information, see 110.9 in this textbook.

Intersystem Bonding Termination. A device that provides a means to connect communications systems grounding and bonding conductors to the building grounding electrode system at the service equipment or at the disconnecting means for buildings or structures supplied by a feeder in accordance with 250.94. Figure 100–41

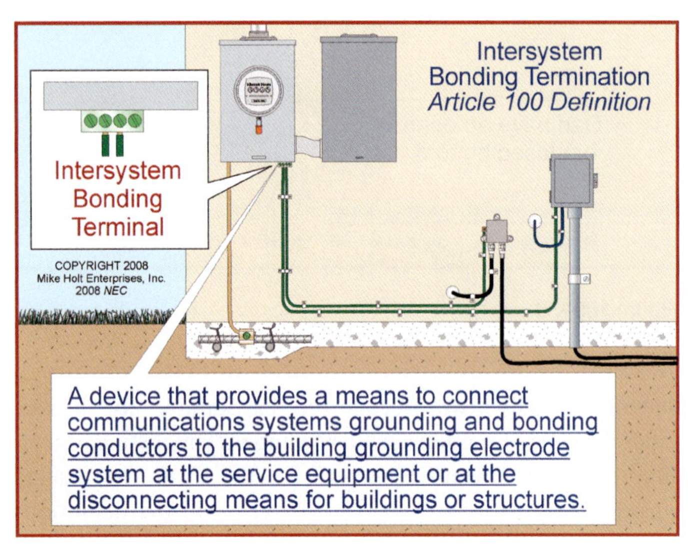

Figure 100–41

Kitchen. An area with a sink and permanent facilities for food preparation and cooking. Figure 100–42

Figure 100–42

Labeled. Equipment or materials that have a label, symbol, or other identifying mark in the form of a sticker, decal, printed label, or molded or stamped into the product by a testing laboratory acceptable to the authority having jurisdiction. See the definition of "Identified" and "Listed" in this article. Figure 100–43

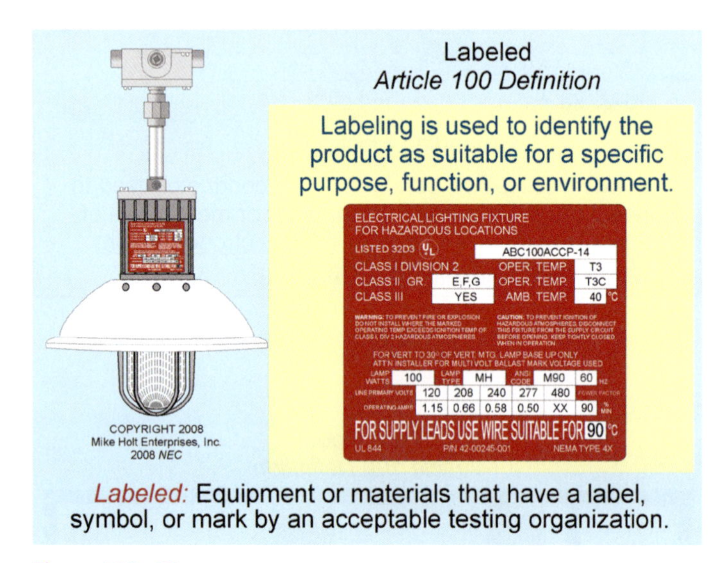

Figure 100–43

Author's Comment: Labeling and listing of equipment typically provides the basis for equipment approval by the authority having jurisdiction [90.4, 90.7, 110.2, and 110.3].

Lighting Outlet. An outlet for the connection of a lampholder or luminaire. Figure 100–44

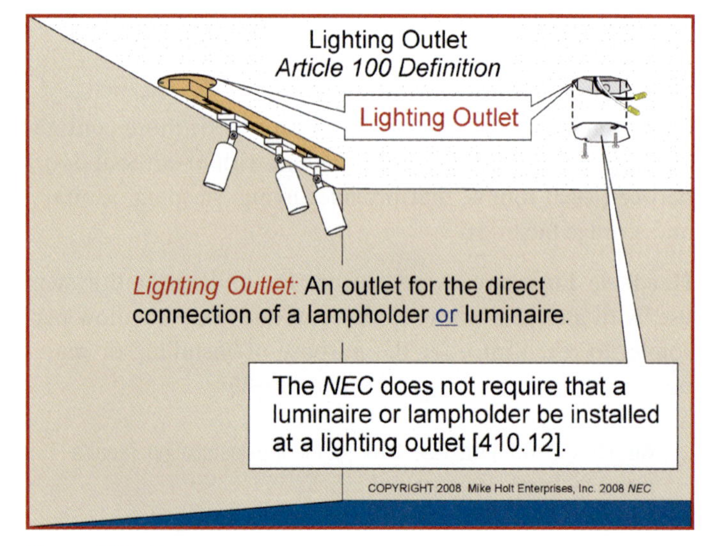

Figure 100–44

Listed. Equipment or materials included in a list published by a testing laboratory acceptable to the authority having jurisdiction. The listing organization must periodically inspect the production of listed equipment or material to ensure the equipment or material meets appropriate designated standards and is suitable for a specified purpose. See the definition of "Identified" and "Labeled" in this article.

> **Author's Comment:** The *NEC* doesn't require all electrical equipment to be listed, but some *Code* requirements do specifically require product listing. Organizations such as OSHA increasingly require that listed equipment be used when such equipment is available [90.7, 110.2, and 110.3].

Location, Damp. Locations protected from weather and not subject to saturation with water or other liquids. This includes locations partially protected under canopies, marquees, roofed open porches, and interior locations subject to moderate degrees of moisture, such as some basements, barns, and cold-storage warehouses.

Location, Dry. An area not normally subjected to dampness or wetness, but which may temporarily be subjected to dampness or wetness, such as a building under construction.

Location, Wet. An installation underground, in concrete slabs in direct contact with the earth, locations subject to saturation with water, and unprotected locations exposed to weather. Figure 100–45

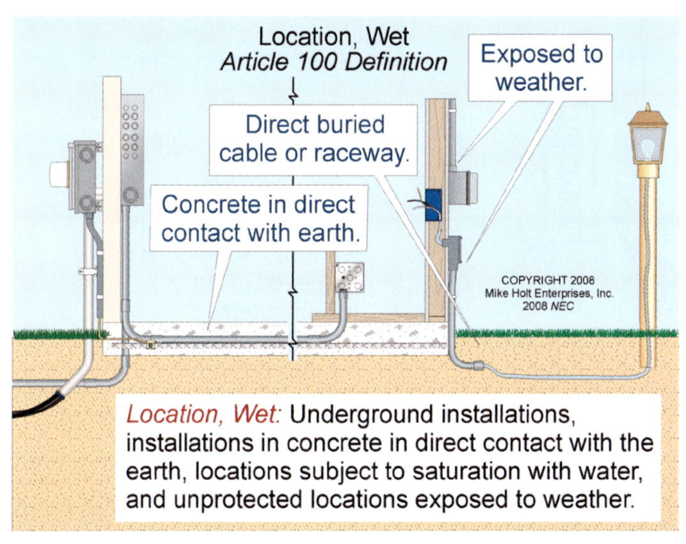

Figure 100–45

> **Author's Comment:** Where raceways are installed in wet locations, the interior of these raceways is considered to be a wet location [300.5(B) and 300.9].

Luminaire [Article 410]. A complete lighting unit consisting of a light source with the parts designed to position the light source and connect it to the power supply and distribute the light. A lampholder by itself is not a luminaire. Figure 100–46

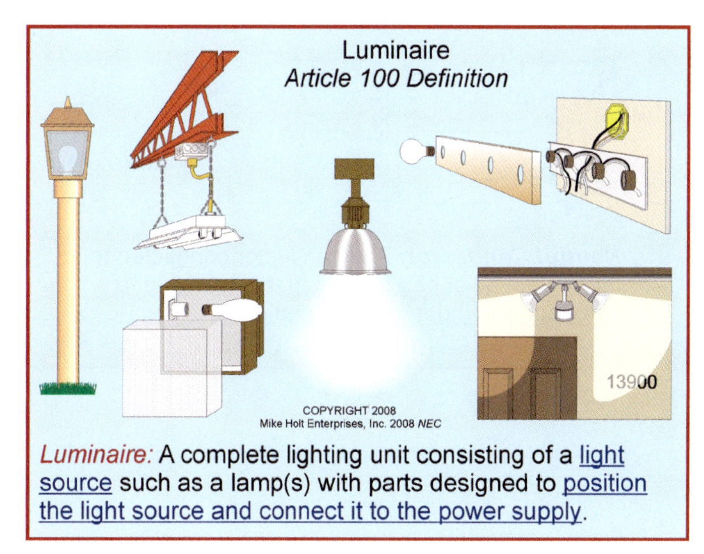

Figure 100–46

Multioutlet Assembly [Article 380]. A surface, flush, or freestanding raceway designed to hold conductors and receptacles. Figure 100–47

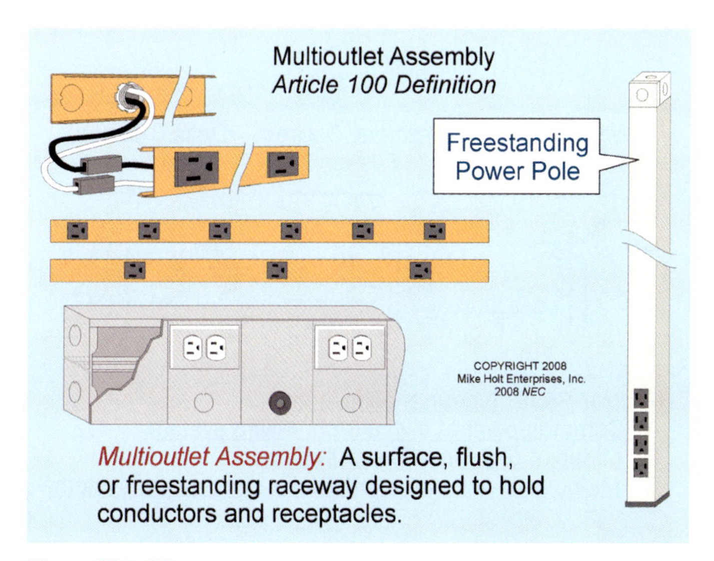

Figure 100–47

Neutral Conductor. The conductor connected to the neutral point of a system that is intended to carry current under normal conditions. Figure 100–48

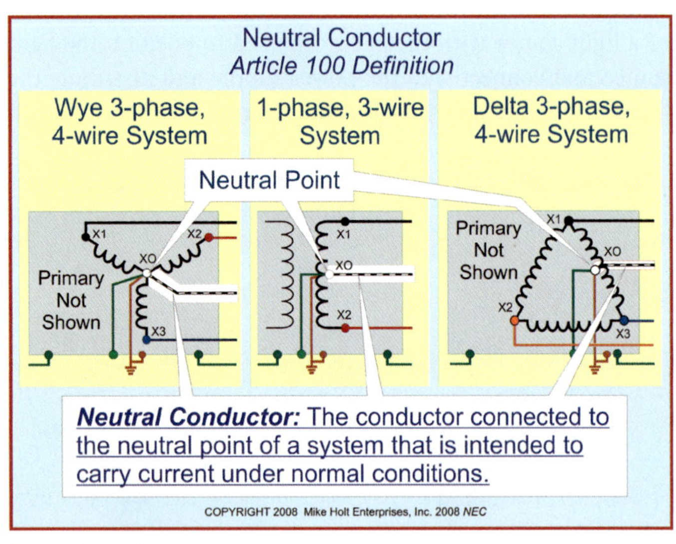

Figure 100–48

Author's Comment: The neutral conductor of a solidly grounded system is required to be grounded (connected to the earth), therefore this conductor is also called a "grounded conductor."

Neutral Point. The common point of a 4-wire, three-phase, wye-connected system; the midpoint of a 3-wire, single-phase system; or the midpoint of the single-phase portion of a three-phase, delta-connected system. Figure 100–49

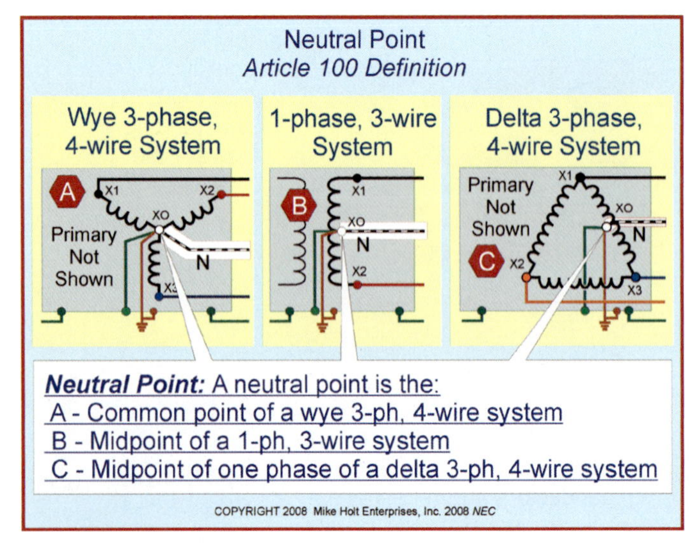

Figure 100–49

Nonlinear Load. A load where the current waveform doesn't follow the applied sinusoidal voltage waveform. Figure 100–50

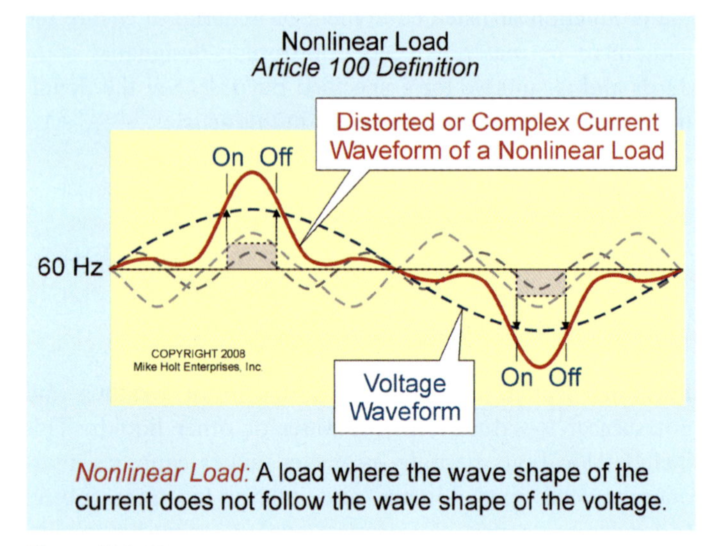

Figure 100–50

FPN: Single-phase nonlinear loads include electronic equipment, such as copy machines, laser printers, and electric-discharge lighting. Three-phase nonlinear loads include uninterruptible power supplies, induction motors, and electronic switching devices, such as adjustable speed drives (variable frequency drives). Figure 100–51

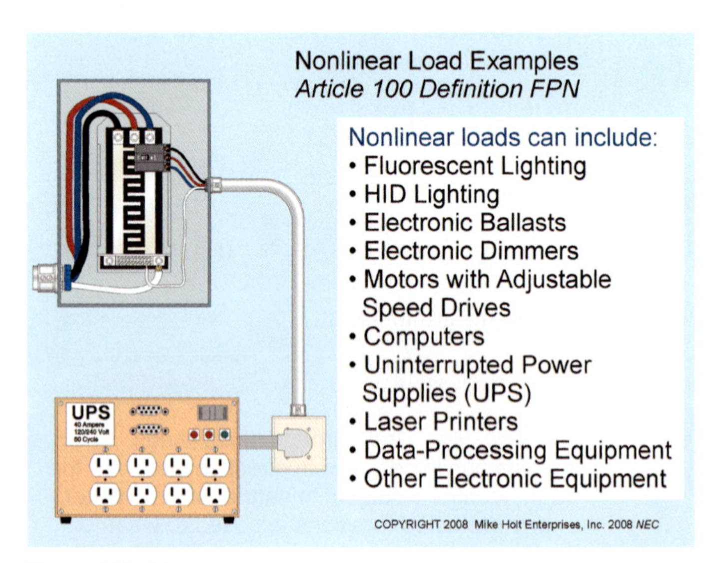

Figure 100–51

Author's Comment: The subject of nonlinear loads is outside the scope of this textbook. For more information on this topic, visit www.MikeHolt.com, click on the "Technical" link, then on the "Power Quality" link.

Outlet. A point in the wiring system where electric current is taken to supply a load. This includes receptacle outlets and lighting outlets, as well as outlets for ceiling paddle fans and smoke alarms. Figure 100–52

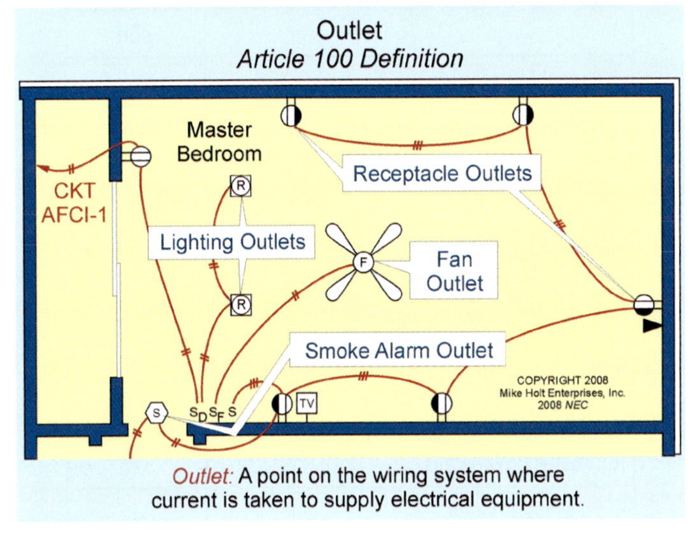

Outlet: A point on the wiring system where current is taken to supply electrical equipment.

Figure 100–52

Outline Lighting [Article 600]. An arrangement of incandescent lamps, electric-discharge lighting, or other electrically powered light sources to outline or call attention to certain features such as the shape of a building or the decoration of a window.

Overcurrent. Current, in amperes, greater than the rated current of the equipment or conductors resulting from an overload, short circuit, or ground fault. Figure 100–53

Author's Comment: See the definitions of "Ground Fault" in 250.2 and "Overload" in this article.

Overload. The operation of equipment above its current rating, or current in excess of conductor ampacity. When an overload condition persists for a sufficient length of time, it can result in equipment failure or in a fire from damaging or dangerous overheating. A fault, such as a short circuit or ground fault, isn't an overload.

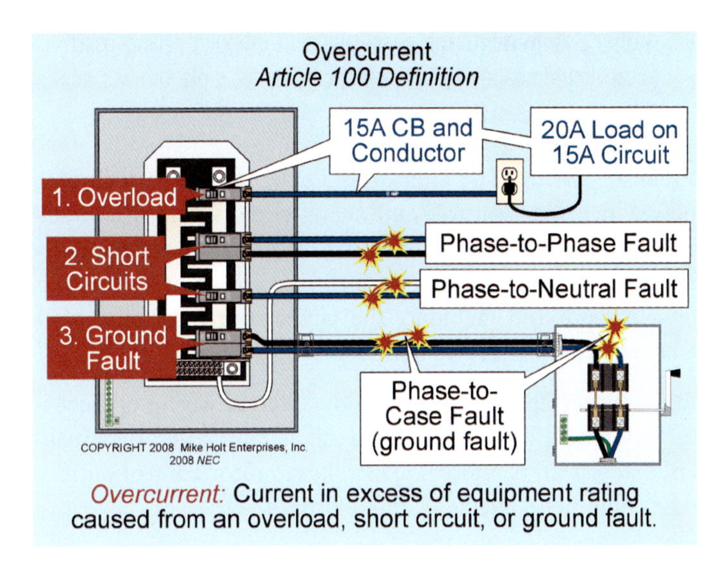

Overcurrent: Current in excess of equipment rating caused from an overload, short circuit, or ground fault.

Figure 100–53

Panelboard [Article 408]. A distribution point containing overcurrent devices and designed to be installed in a cabinet. Figure 100–54

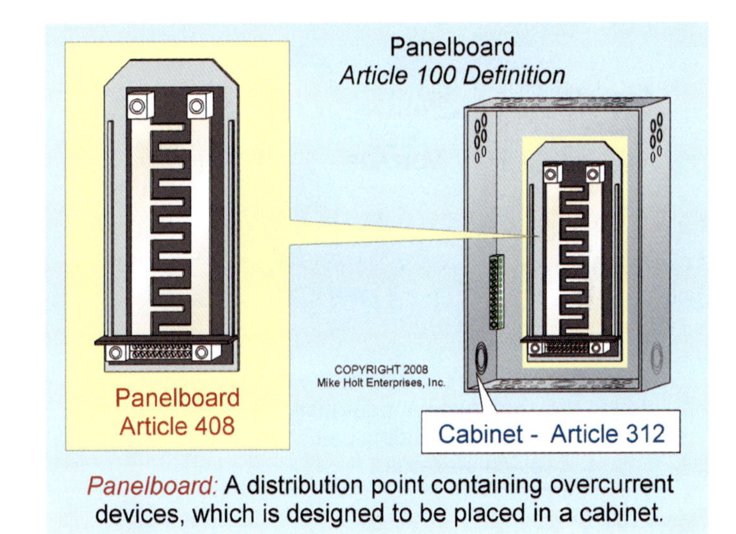

Panelboard: A distribution point containing overcurrent devices, which is designed to be placed in a cabinet.

Figure 100–54

Author's Comments:

- See the definition of "Cabinet" in this article.
- The slang term in the electrical field for a panelboard is "the guts."

Plenum. A chamber to which ducts are connected and form part of the air distribution system.

Author's Comment: The space above a dropped ceiling used for environmental air-handling purposes is not a plenum, it's an "other space used for environmental air" [300.22(C)].

Premises Wiring. The interior and exterior wiring, including power, lighting, control, and signal circuits, and all associated hardware, fittings, and wiring devices, both permanently and temporarily installed from the service point to the outlets, or from and including the power source to the outlets where there is no service point.

Premises wiring doesn't include the internal wiring of electrical equipment and appliances, such as luminaires, dishwashers, water heaters, motors, controllers, motor control centers, A/C equipment, etc. [90.7 and 300.1(B)].

Qualified Person. A person who has the skill and knowledge related to the construction and operation of electrical equipment and its installation. This person must have received safety training to recognize and avoid the hazards involved with electrical systems. Figure 100–55

Figure 100–55

Author's Comments:

- Examples of this safety training include, but aren't limited to, training in the use of special precautionary techniques, of personal protective equipment (PPE), of insulating and shielding materials, and of using insulated tools and test equipment when working on or near exposed conductors or circuit parts that can become energized.
- In many parts of the United States, electricians, electrical contractors, electrical inspectors, and electrical engineers must complete from 6 to 24 hours of *NEC* review each year as a requirement to maintain licensing. This in itself doesn't make one qualified to deal with the specific hazards involved.

Raceway. An enclosure designed for the installation of conductors, cables, or busbars. Raceways in the *NEC* include:

Raceway Type	Article
Busways	368
Electrical Metallic Tubing	358
Electrical Nonmetallic Tubing	362
Flexible Metal Conduit	348
Intermediate Metal Conduit	342
Liquidtight Flexible Metal Conduit	350
Liquidtight Flexible Nonmetallic Conduit	356
Metal Wireways	376
Multioutlet Assembly	380
Nonmetallic Wireways	378
Rigid Metal Conduit	344
PVC Conduit	352
Strut-Type Channel Raceways	384
Surface Metal Raceways	386
Surface Nonmetallic Raceways	388

Author's Comments: A cable tray system isn't a raceway; it's a support system for cables and raceways [392.2].

Receptacle [Article 406]. A contact device installed at an outlet for the connection of an attachment plug. A single receptacle contains one device on a strap (mounting yoke), and a multiple receptacle contains more than one device on a common yoke. Figure 100–56

Receptacle Outlet. An opening in an outlet box where receptacles have been installed.

Remote-Control Circuit [Article 725]. An electric circuit that controls another circuit by a relay or equivalent device installed in accordance with Article 725. Figure 100–57

Separately Derived System. A wiring system whose power is derived from a source of electric energy or equipment other than the electric utility service. This includes a generator, a

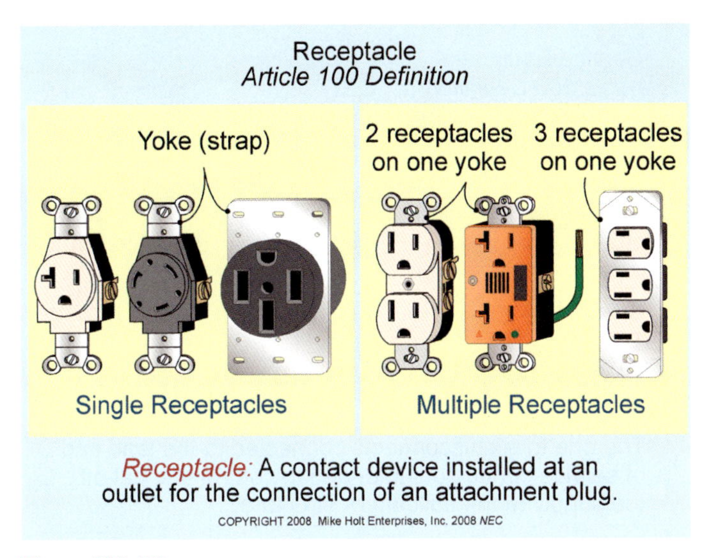

Figure 100–56

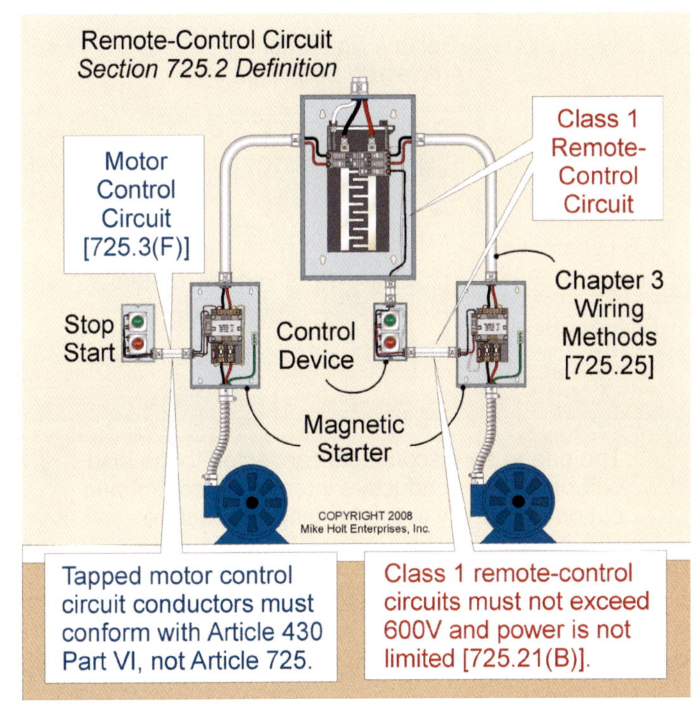

Figure 100–57

battery, a solar photovoltaic system, a transformer, or a converter winding, where there's no direct electrical connection to the supply conductors of another system. **Figures 100–58** and **100–59**

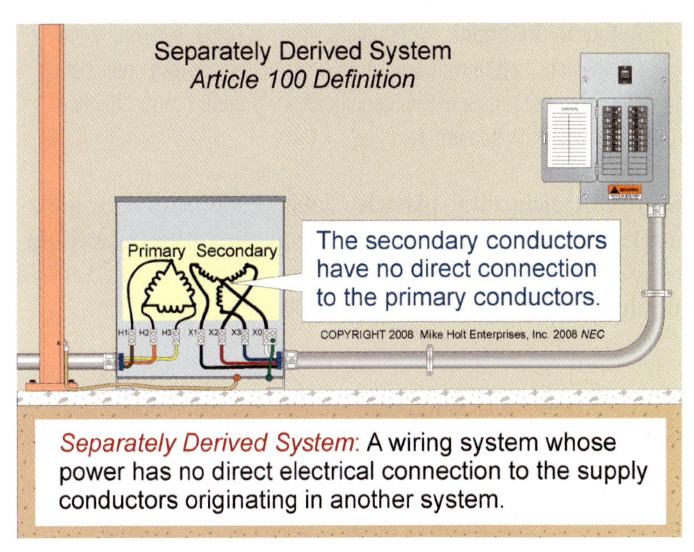

Figure 100–58

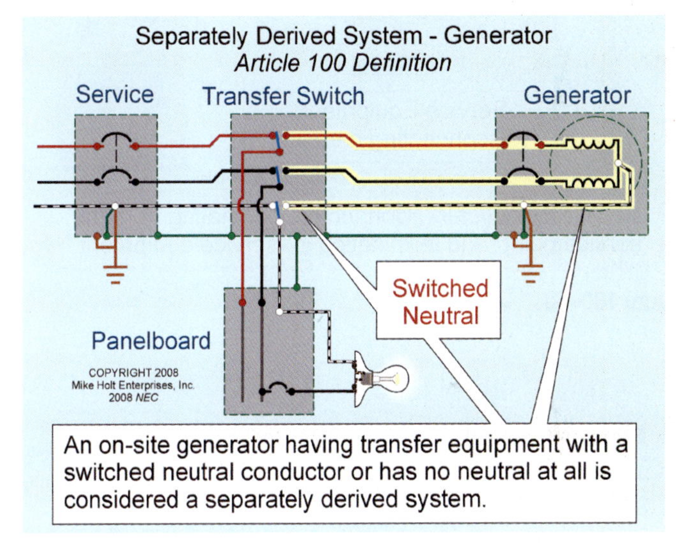

Figure 100–59

Author's Comments:

- The definition clarifies that separately derived systems also include equipment such as transformers, converters, and inverters, which might not be considered sources of energy.
- Separately derived systems are actually a lot more complicated than the above definition suggests, and understanding them requires additional study. For more information, see 250.20(D) and 250.30.

Service [Article 230]. The conductors from the electric utility that deliver electric energy to the premises.

Author's Comment: Conductors from a UPS system, solar photovoltaic system, generator, or transformer are not service conductors. See the definitions of "Feeder" and "Service Conductors" in this article.

Service Conductors [Article 230]. Conductors originating from the service point and terminating in service equipment. See the definitions of "Service Point" and "Service Equipment" in this article. Figure 100–60

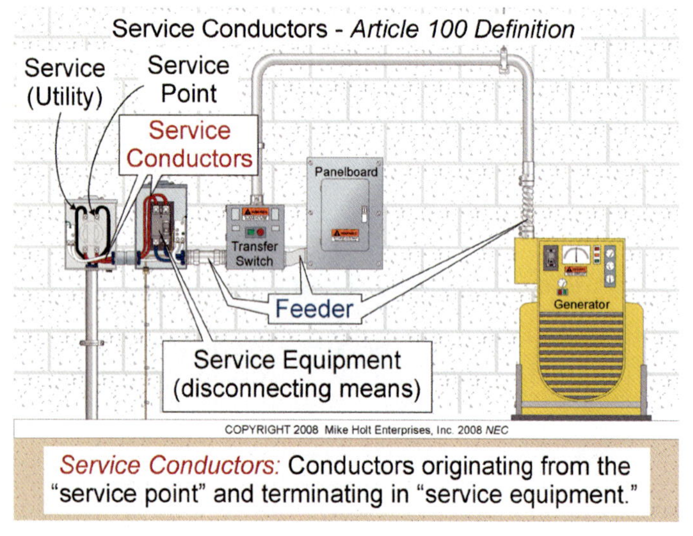

Figure 100–60

Author's Comment: Conductors from a generator, UPS system, or transformer are feeder conductors, not service conductors. See the definition of "Feeder" in this article.

Service Drop [Article 230]. Overhead service conductors from the last pole to and including the splices connecting to the premises service-entrance conductors.

Service Equipment [Article 230]. The one to six disconnects connected to the load end of service conductors intended to control and cut off the supply to the building or structure. Figure 100–61

Author's Comments:

- It's important to know where a service begins and where it ends in order to properly apply the *NEC* requirements. Sometimes the service ends before the metering equipment. Figure 100–62
- Service equipment is often referred to as the "service disconnect" or "service disconnecting means."

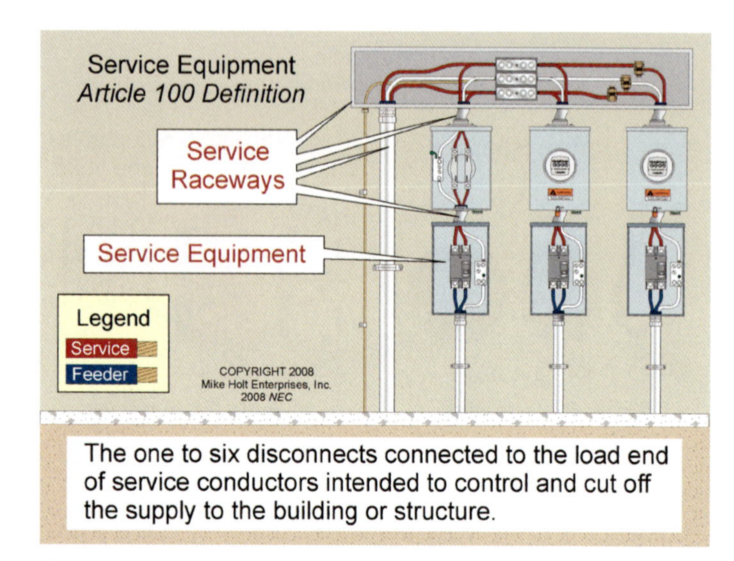

Figure 100–61

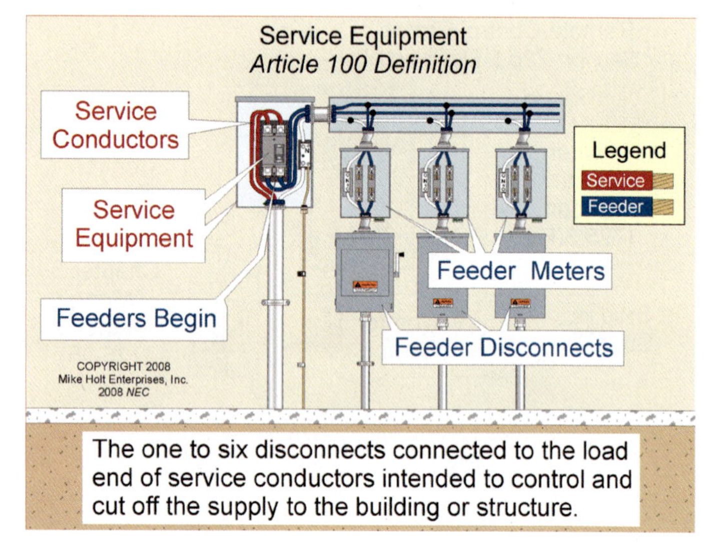

Figure 100–62

Service Lateral [Article 230]. Underground service conductors connecting to the premises service-entrance conductors.

Service Point [Article 230]. The point where the electrical utility conductors make contact with premises wiring. Figure 100–63

Author's Comments:

- See the definition of "Premises Wiring" in this article.
- The service point can be at the utility transformer, at the service weatherhead, or at the meter socket enclosure, depending on where the utility conductors terminate. Figure 100–64

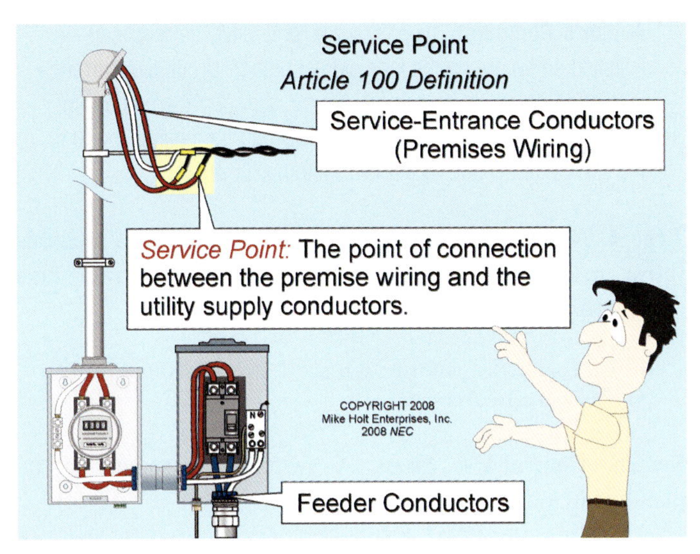

Figure 100–63

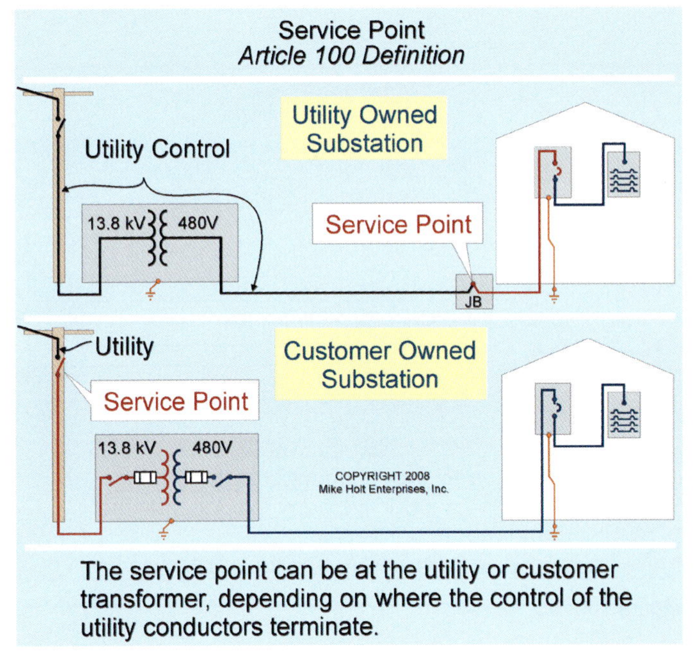

Figure 100–64

Short-Circuit Current Rating. The prospective symmetrical fault current at a nominal voltage that electrical equipment is able to be connected to without sustaining damage exceeding defined acceptance criteria. See 110.10.

Signaling Circuit [Article 725]. A circuit that energizes signaling equipment.

Special Permission. Written consent from the authority having jurisdiction.

> **Author's Comment:** See the definition of "Authority Having Jurisdiction" in this article.

Structure. That which is built or constructed.

Supplementary Overcurrent Device. A device intended to provide limited overcurrent protection for specific applications and utilization equipment, such as luminaires and appliances. This limited protection is in addition to the required protection provided in the branch circuit by the branch-circuit overcurrent device. Figure 100–65

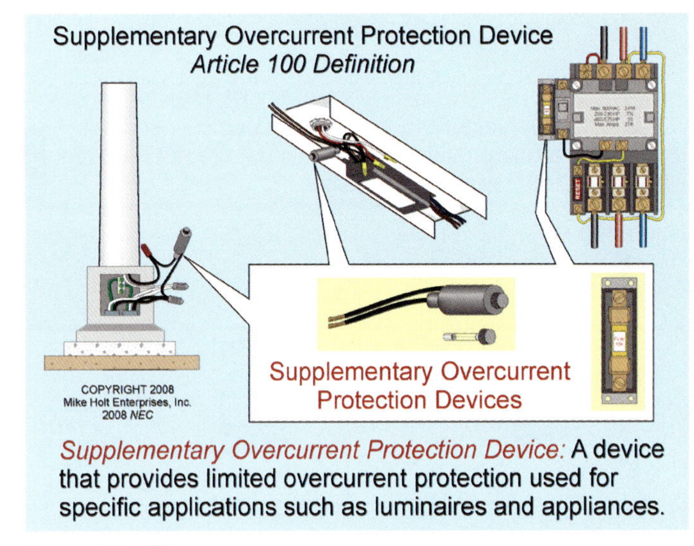

Figure 100–65

> **Author's Comment:** Supplementary overcurrent devices aren't required to be readily accessible [240.10 and 240.24(A)(2)]. Figure 100–66

Surge Protective Device (SPD) [Article 285]. A protective device intended to limit transient voltages by diverting or limiting surge current and preventing the continued flow of current while remaining capable of repeating these functions. Figure 100–67

Type 1. A permanently connected surge protective device listed for installation between the utility transformer and the service equipment.

> **Author's Comment:** The 2005 *NEC* referred to a Type 1 device as a surge arrester.

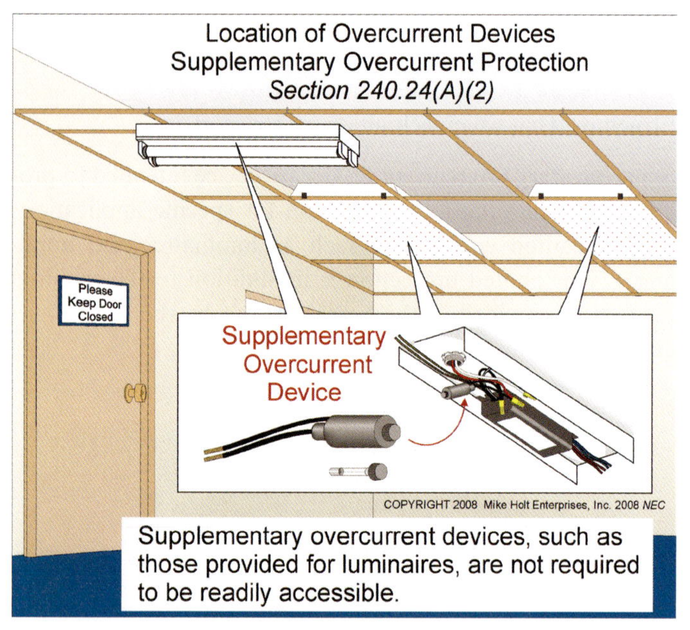

Figure 100–66

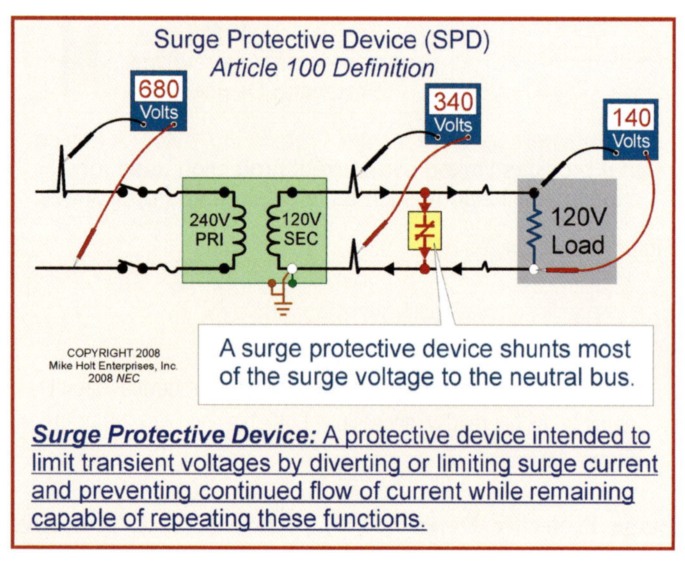

Figure 100–67

Type 2. A permanently connected surge protective device listed for installation on the load side of the service disconnecting means.

> **Author's Comment:** The 2005 *NEC* referred to a Type 2 device as a transient voltage surge suppressor (TVSS).

Type 3. A surge protective device listed for installation on branch circuits.

> **Author's Comment:** Type 3 surge protective devices can be installed anywhere on the load side of branch-circuit overcurrent protection up to the equipment served, provided there is a minimum of 30 ft of conductor length between the connection and the service or separately derived system [285.25].

Type 4. A component surge protective device; this includes those installed in receptacles and relocatable power taps (plug strips).

> **FPN:** For further information see UL 1449, *Standard for Surge Protective Devices.*

Switch, General-Use Snap. A switch constructed to be installed in a device box or a box cover.

Ungrounded. Not connected to the ground (earth) or a conductive body that extends the ground (earth) connection. Figure 100–68

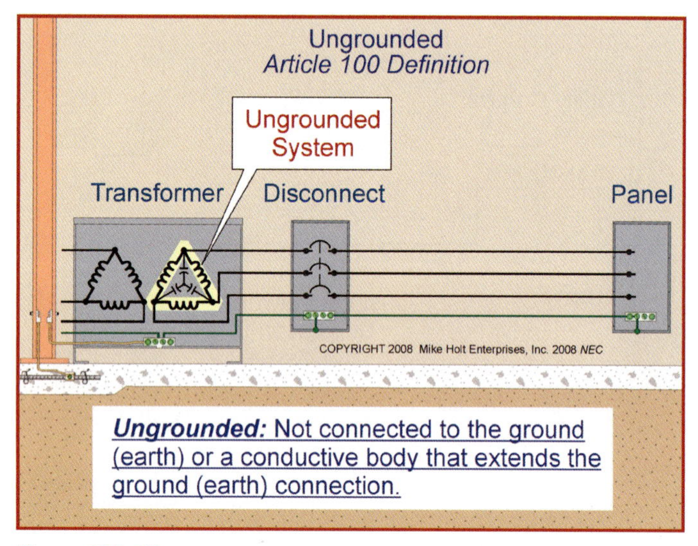

Figure 100–68

> **Author's Comment:** The use of this term relates to an ungrounded system, where the system windings are not grounded (connected to the earth) [250.4(B) and 250.30(B)].

Voltage of a Circuit. The greatest effective root-mean-square difference of potential between any two conductors of the circuit.

Voltage, Nominal. A value assigned for the purpose of conveniently designating voltage class, such as 120/240V, 120/208V, or 277/480V [220.5(A)]. Figure 100–69

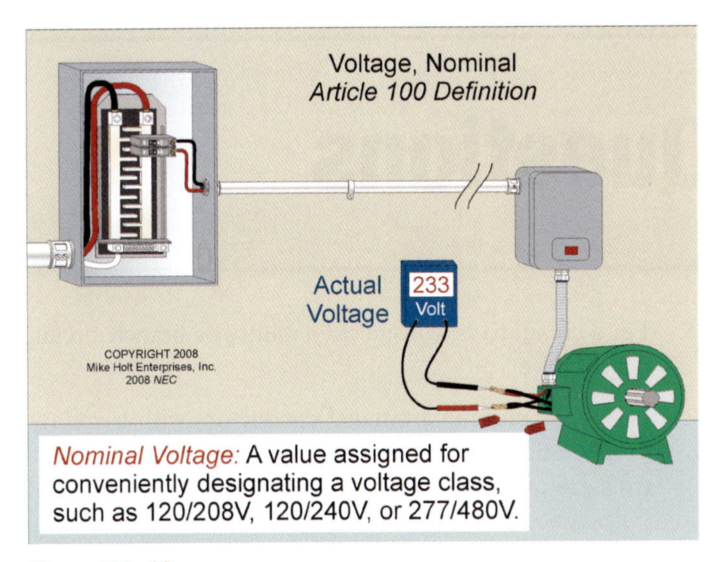

Figure 100–69

Author's Comment: The actual voltage at which a circuit operates can vary from the nominal within a range that permits satisfactory operation of equipment. In addition, common voltage ratings of electrical equipment are 115, 200, 208, 230, and 460. The electrical power supplied might be at the 240V, nominal voltage, but the voltage at the equipment will be less. Therefore, electrical equipment is rated at a value less than the nominal system voltage.

Voltage-to-Ground. The voltage between the ungrounded conductor and the neutral point; for ungrounded circuits, the voltage between any two conductors of the circuit. Figure 100–70

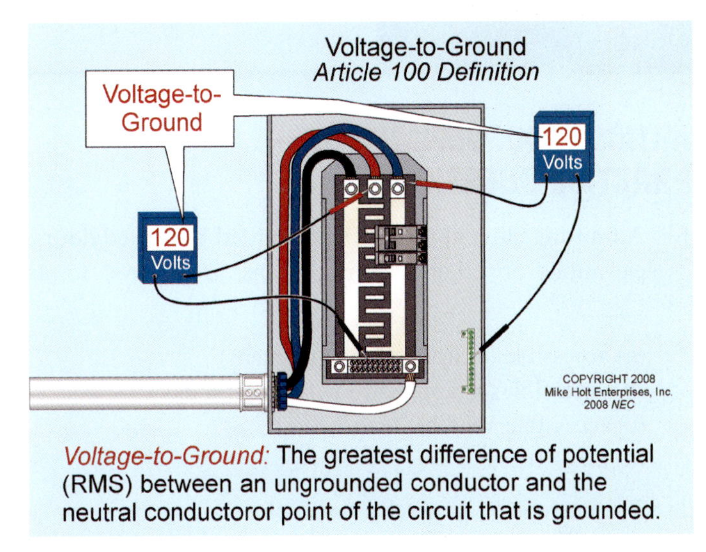

Figure 100–70

ARTICLE 100 Practice Questions

ARTICLE 100. DEFINITIONS— PRACTICE QUESTIONS

1. Admitting close approach, not guarded by locked doors, elevation, or other effective means, is referred to as _____.

   (a) accessible (as applied to equipment)
   (b) accessible (as applied to wiring methods)
   (c) accessible, readily
   (d) all of these

2. The current in amperes a conductor can carry continuously, where the temperature will not be raised in excess of the conductor's insulation temperature rating is called its _____.

   (a) short-circuit rating
   (b) ground-fault rating
   (c) ampacity
   (d) all of these

3. Where no statutory requirement exists, the authority having jurisdiction can be a property owner or his/her agent, such as an architect or engineer.

   (a) True
   (b) False

4. The connection between the grounded conductor and the equipment grounding conductor at the service is accomplished by installing a(n) _____ bonding jumper.

   (a) main
   (b) system
   (c) equipment
   (d) circuit

5. For a circuit to be considered a multiwire branch circuit, it shall have _____.

   (a) two or more ungrounded conductors with a voltage potential between them
   (b) a grounded conductor having equal voltage potential between it and each ungrounded conductor of the circuit
   (c) a grounded conductor connected to the neutral or grounded terminal of the system
   (d) all of these

6. A circuit breaker is a device designed to _____ the circuit automatically on a predetermined overcurrent without damage to itself when properly applied within its rating.

   (a) energize
   (b) reset
   (c) connect
   (d) open

7. Communications equipment includes equipment used for the transmission of _____.

   (a) audio
   (b) video
   (c) data
   (d) all of these

8. A solderless pressure connector is a device that _____ between conductors or between conductors and a terminal by means of mechanical pressure.

   (a) provides access
   (b) protects the wiring
   (c) is never needed
   (d) establishes a connection

9. The choice of overcurrent devices so that an overcurrent condition will be localized and restrict outages to the circuit or equipment affected, is called _____.

   (a) overcurrent protection
   (b) interrupting capacity
   (c) selective coordination
   (d) overload protection

10. A unit of an electrical system that carries or controls electric energy as its principal function is a(n) _____.

(a) raceway
(b) fitting
(c) device
(d) enclosure

11. A building that contains three or more dwelling units is called a _____.

(a) one-family dwelling
(b) two-family dwelling
(c) dwelling unit
(d) multifamily dwelling

12. On or attached to the surface or behind access panels designed to allow access is known as _____.

(a) open
(b) uncovered
(c) exposed
(d) bare

13. A _____ is a building or portion of a building in which one or more self-propelled vehicles can be kept for use, sale, storage, rental, repair, exhibition, or demonstration purposes.

(a) garage
(b) residential garage
(c) service garage
(d) commercial garage

14. A circuit conductor that is intentionally grounded is called a(n) _____.

(a) grounding conductor
(b) unidentified conductor
(c) grounded conductor
(d) grounding electrode conductor

15. A "Class A" GFCI protection device is designed to trip when the ground-fault current to ground is _____ or higher.

(a) 4 mA
(b) 5 mA
(c) 6 mA
(d) none of these

16. A conducting object through which a direct connection to earth is established is a _____.

(a) bonding conductor
(b) grounding conductor
(c) grounding electrode
(d) grounded conductor

17. A _____ is an accommodation with two or more contiguous rooms comprising a compartment that provides living, sleeping, sanitary, and storage facilities.

(a) guest room
(b) guest suite
(c) dwelling unit
(d) single-family dwelling

18. Within sight means visible and not more than _____ ft distant from the equipment.

(a) 10
(b) 20
(c) 25
(d) 50

19. A kitchen is defined as an area with a sink and _____ facilities for food preparation and cooking.

(a) listed
(b) labeled
(c) temporary
(d) permanent

20. Listed equipment or materials are included in a _____ published by a testing laboratory acceptable to the authority having jurisdiction.

(a) book
(b) digest
(c) manifest
(d) list

21. Conduit installed underground or encased in concrete slabs that are in direct contact with the earth are considered a _____ location.

(a) dry
(b) damp
(c) wet
(d) moist

22. The common point on a wye-connection in a polyphase system describes a neutral point.

(a) True
(b) False

23. Any current in excess of the rated current of equipment or the ampacity of a conductor is called _____.

    (a) trip current
    (b) faulted current
    (c) overcurrent
    (d) shorted current

24. A panel, including buses and automatic overcurrent devices, designed to be placed in a cabinet or cutout box and accessible only from the front, is known as a _____.

    (a) switchboard
    (b) disconnect
    (c) panelboard
    (d) switch

25. The *NEC* defines a(n) _____ as one who has skills and knowledge related to the construction and operation of electrical equipment and installations and has received safety training to recognize and avoid the hazards involved.

    (a) inspector
    (b) master electrician
    (c) journeyman electrician
    (d) qualified person

26. A contact device installed at an outlet for the connection of an attachment plug is known as a(n) _____.

    (a) attachment point
    (b) tap
    (c) receptacle
    (d) wall plug

27. When one electrical circuit controls another circuit through a relay, the first circuit is called a _____.

    (a) primary circuit
    (b) remote-control circuit
    (c) signal circuit
    (d) controller

28. Service conductors originate at the service point and terminate at service equipment.

    (a) True
    (b) False

29. Underground service conductors between the street main and the first point of connection to the service-entrance conductors are known as the _____.

    (a) utility service
    (b) service lateral
    (c) service drop
    (d) main service conductors

30. A signaling circuit is any electrical circuit that energizes signaling equipment.

    (a) True
    (b) False

31. A structure is that which is built or constructed.

    (a) True
    (b) False

32. A surge protective device (SPD) intended for installation on the load side of the service disconnect overcurrent device, including SPDs located at the branch panel, is a _____ SPD.

    (a) Type I
    (b) Type II
    (c) Type III
    (d) Type IV

33. The voltage of a circuit is defined by the *Code* as the _____ root-mean-square (effective) difference of potential between any two conductors of the circuit.

    (a) lowest
    (b) greatest
    (c) average
    (d) nominal

ARTICLE

110 Requirements for Electrical Installations

INTRODUCTION TO ARTICLE 110—REQUIREMENTS FOR ELECTRICAL INSTALLATIONS

Article 110 sets the stage for how you'll implement the rest of the *NEC*. This article contains a few of the most important and yet neglected parts of the *Code*. For example:

- What do you do with unused openings in enclosures?
- How should you terminate conductors?
- What kinds of warnings, markings, and identification does a given installation require?
- What's the right working clearance for a given installation?

It's critical you master Article 110, and that's exactly what this *Illustrated Guide to Understanding the National Electrical Code* is designed to help you do. As you read this article, remember that doing so helps you build your foundation for correctly applying much of the *NEC*. In fact, this article itself is a foundation for much of the *Code*. You may need to read something several times to understand it, but the time you take to do so will be well spent. The illustrations will also help. But if you find your mind starting to wander, take a break. What matters is how well you master the material and how safe your work is—not how fast you blazed through a book.

PART I. GENERAL REQUIREMENTS

110.1 Scope. Article 110 covers the general requirements for the examination and approval, installation and use, access to and spaces about electrical equipment; as well as general requirements for enclosures intended for personnel entry (manholes, vaults, and tunnels).

110.2 Approval of Conductors and Equipment. The authority having jurisdiction must approve all electrical conductors and equipment. Figure 110–1

Author's Comment: For a better understanding of product approval, review 90.4, 90.7, 110.3 and the definitions for "Approved," "Identified," "Labeled," and "Listed" in Article 100.

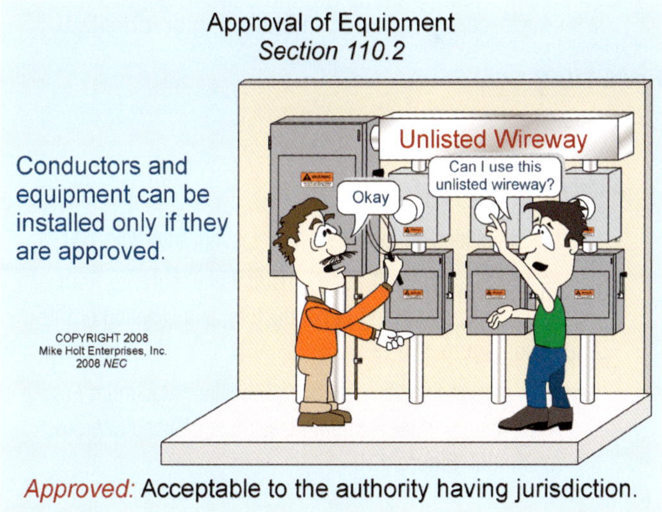

Approval of Equipment
Section 110.2

Conductors and equipment can be installed only if they are approved.

Unlisted Wireway

Okay

Can I use this unlisted wireway?

COPYRIGHT 2008 Mike Holt Enterprises, Inc. 2008 *NEC*

Approved: Acceptable to the authority having jurisdiction.

Figure 110–1

110.3 Examination, Identification, Installation, and Use of Equipment.

(A) Guidelines for Approval. The authority having jurisdiction must approve equipment, and consideration must be given to the following:

(1) Listing or labeling

(2) Mechanical strength and durability

(3) Wire-bending and connection space

(4) Electrical insulation

(5) Heating effects under all conditions of use

(6) Arcing effects

(7) Classification by type, size, voltage, current capacity, and specific use

(8) Other factors contributing to the practical safeguarding of persons using or in contact with the equipment

(B) Installation and Use. Equipment must be installed and used in accordance with any instructions included in the listing or labeling requirements. **Figure 110–2**

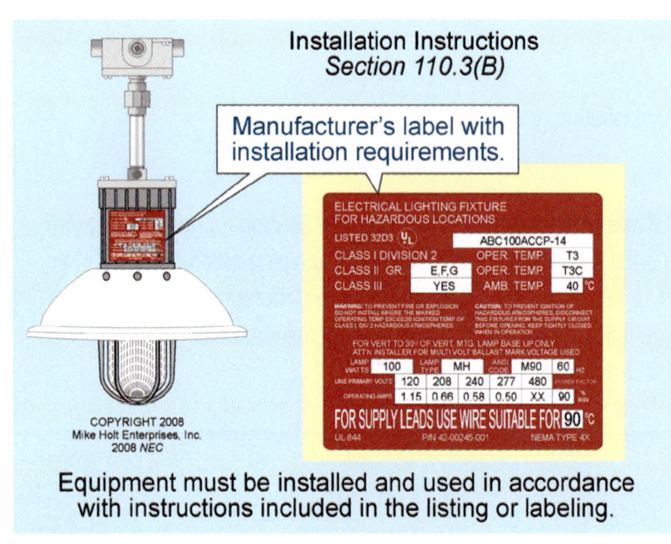

Figure 110–2

Author's Comments:

- See the definitions of "Labeling" and "Listing" in Article 100.
- Failure to follow product listing instructions, such as torquing of terminals and sizing of conductors, is a violation of this *Code* rule. **Figure 110–3**
- When an air conditioner nameplate specifies "Maximum Fuse Size," one-time or dual-element fuses must be used to protect the equipment. **Figure 110–4**

110.4 Voltages.

The voltage rating of electrical equipment must not be less than the nominal voltage of a circuit. Put another way, electrical equipment must be installed on a circuit where the nominal system voltage doesn't exceed the voltage rating of the equipment. This rule is intended to prohibit the installation of 208V rated motors on a 240V nominal voltage circuit. **Figure 110–5**

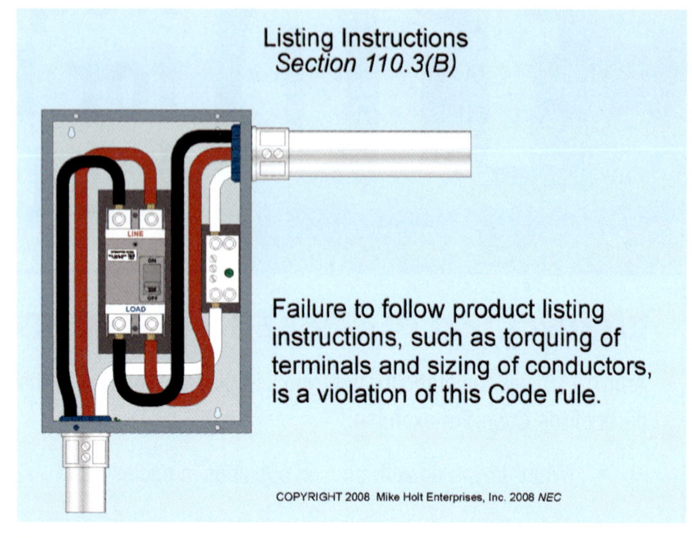

Figure 110–3

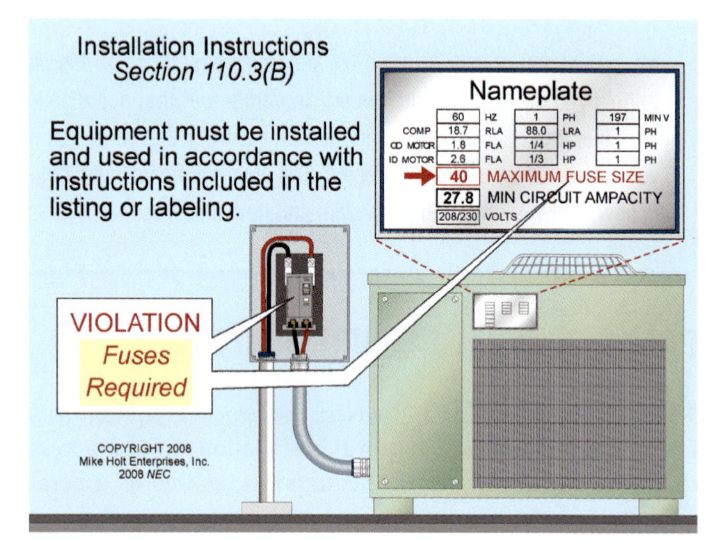

Figure 110–4

Author's Comments:

- See the definition of "Voltage, Nominal" in Article 100.
- According to 110.3(B), equipment must be installed in accordance with any instructions included in the listing or labeling. Therefore, equipment must not be connected to a circuit where the nominal voltage is less than the rated voltage of the electrical equipment. For example, you can't place a 230V rated motor on a 208V system. **Figure 110–6**

110.5 Copper Conductors.

Where the conductor material isn't specified in a rule, the material and the sizes given in the *Code* (and this textbook) are based on copper.

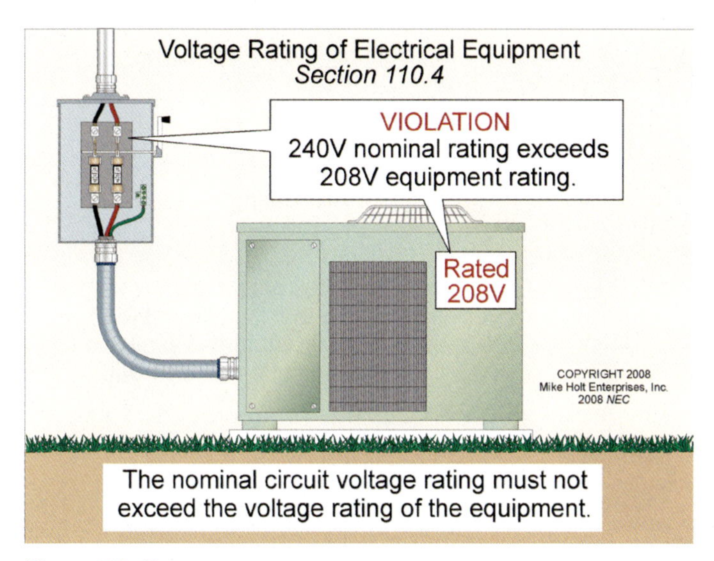

Figure 110–5

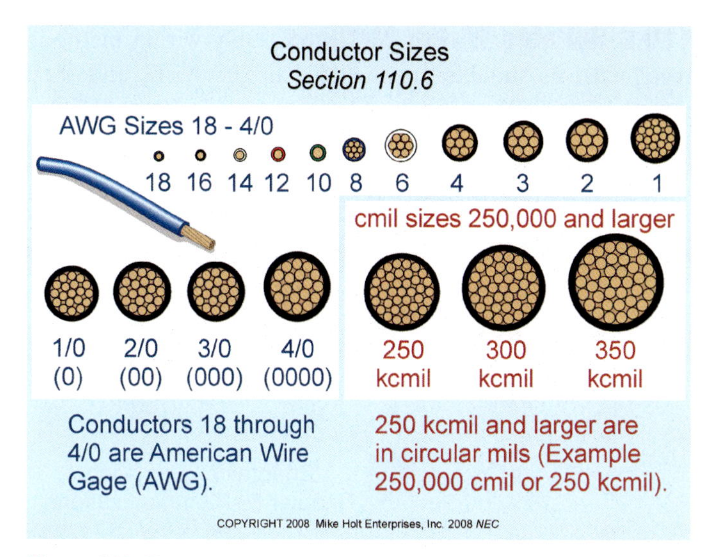

Figure 110–7

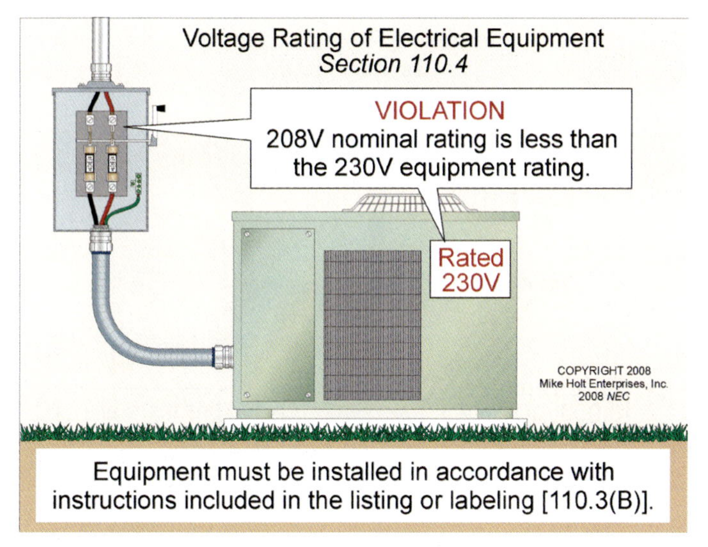

Figure 110–6

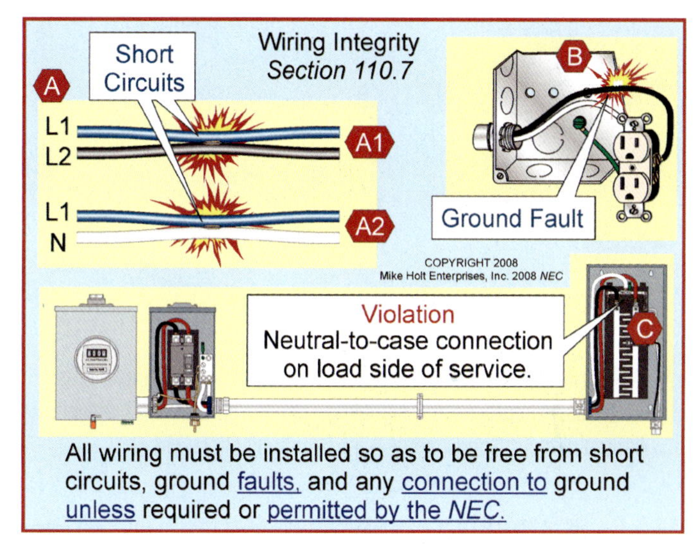

Figure 110–8

110.6 Conductor Sizes. Conductor sizes are expressed in American Wire Gage (AWG), typically from 18 AWG up to 4/0 AWG. Conductor sizes larger than 4/0 AWG are expressed in kcmil (thousand circular mils). Figure 110–7

110.7 Wiring Integrity. All wiring must be installed so as to be free from short circuits, ground faults, and any connection to ground (such as a neutral conductor connection to the circuit equipment grounding conductor), unless required or permitted by this *Code*. Figure 110–8

Author's Comments:

- Short circuits and ground faults often arise from insulation failure due to mishandling or improper installation. This happens when, for example, wire is dragged over a sharp edge, when insulation is scraped on boxes and enclosures, when wire is pulled too hard, when insulation is nicked while being stripped, or when cable clamps and/or staples are installed too tightly.
- To protect against accidental contact with energized conductors, the ends of unused conductors must be covered with an insulating device identified for the purpose, such as a twist-on or push-on wire connector [110.14(B)].

110.8 Suitable Wiring Methods. Only wiring methods recognized as suitable are included in the *NEC*, and they must be installed in accordance with the *Code*. Figure 110–9

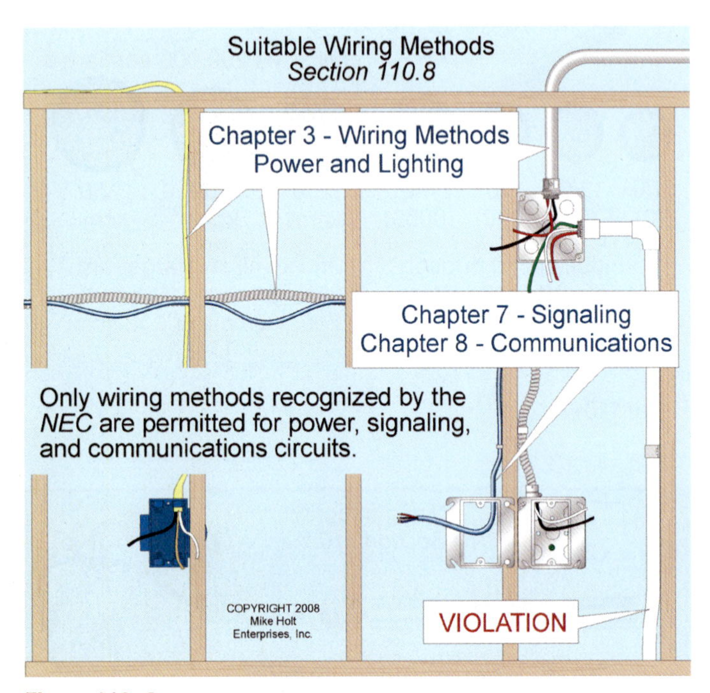

Figure 110–9

Author's Comment: See Chapter 3 for power and lighting wiring methods, Chapter 7 for signaling, remote-control, and power limited circuits, and Chapter 8 for communications circuits.

110.9 Interrupting Protection Rating. Overcurrent devices such as circuit breakers and fuses are intended to interrupt the circuit, and they must have an interrupting rating sufficient for the short-circuit current available at the line terminals of the equipment. Figure 110–10

Author's Comments:

- See the definition of "Interrupting Rating" in Article 100.
- Unless marked otherwise, the ampere interrupting rating for circuit breakers is 5,000A [240.83(C)], and for fuses it's 10,000A [240.60(C)(3)]. Figure 110–11

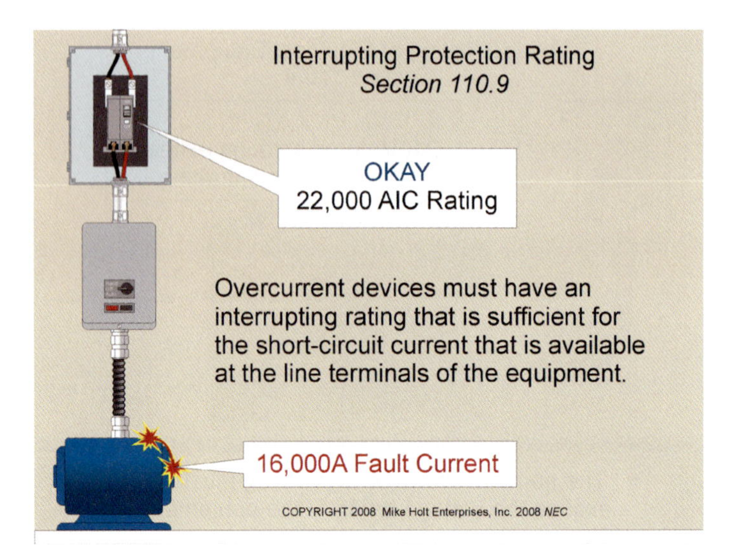

Figure 110–10

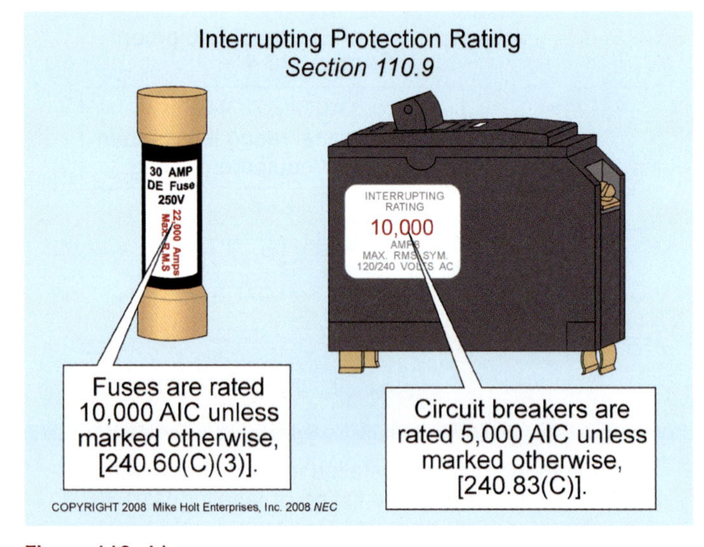

Figure 110–11

AVAILABLE SHORT-CIRCUIT CURRENT

Available Short-Circuit Current. Available short-circuit current is the current, in amperes, available at a given point in the electrical system. This available short-circuit current is first determined at the secondary terminals of the utility transformer. Thereafter, the available short-circuit current is calculated at the terminals of service equipment, then at branch-circuit panelboards and other equipment. The available short-circuit current is different at each point of the electrical system.

It's highest at the utility transformer and lowest at the branch-circuit load.

The available short-circuit current depends on the impedance of the circuit, which increases moving downstream from the utility transformer. The greater the circuit impedance (utility transformer and the additive impedances of the circuit conductors), the lower the available short-circuit current. Figure 110–12

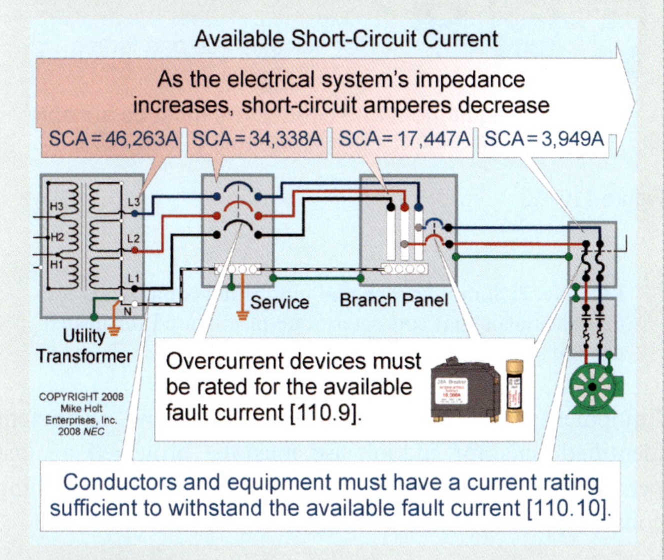

Figure 110–12

The factors that effect the available short-circuit current at the utility transformer include the system voltage, the transformer kVA rating, and the circuit impedance (expressed in a percentage on the equipment nameplate). Properties that have an impact on the impedance of the circuit include the conductor material (copper versus aluminum), conductor size, conductor length, and motor-operated equipment supplied by the circuit.

Author's Comment: Many people in the industry describe Amperes Interrupting Rating (AIR) as "Amperes Interrupting Capacity" (AIC).

DANGER: *Extremely high values of current flow (caused by short circuits or ground faults) produce tremendously destructive thermal and magnetic forces. If the circuit overcurrent device isn't rated to interrupt the current at the available fault values at its listed voltage rating, it can explode while attempting to open the circuit overcurrent device resulting from a short circuit or ground fault, which can cause serious injury or death, as well as property damage.* Figure 110–13

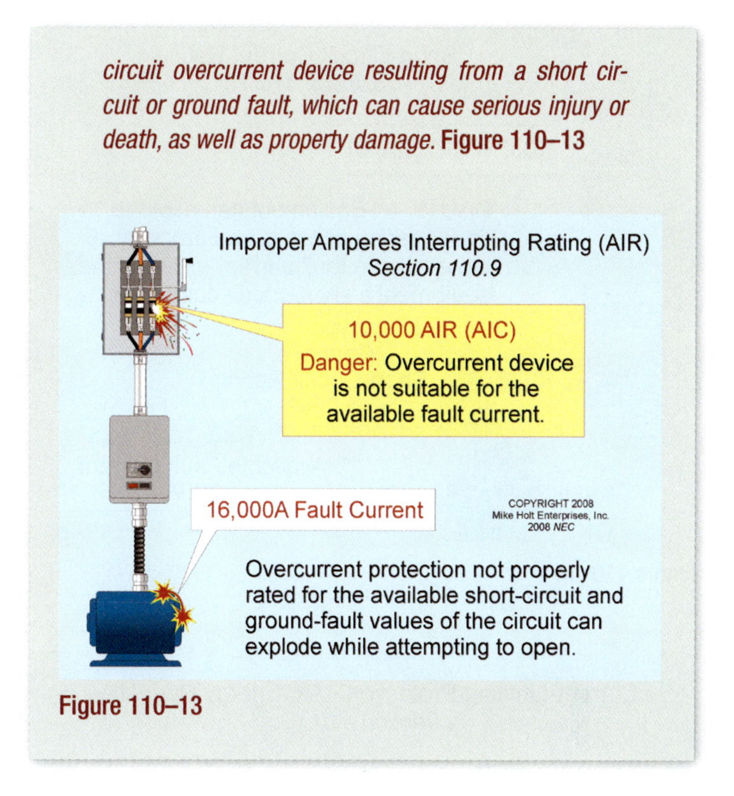

Figure 110–13

110.10 Short-Circuit Current Rating. Electrical equipment must have a short-circuit current rating that permits the circuit overcurrent device to open from a short circuit or ground fault without extensive damage to the electrical components of the circuit. For example, a motor controller must have a sufficient short-circuit rating for the available fault current.

Author's Comment: If the fault exceeds the controller's 5,000A short-circuit current rating, the controller can explode, endangering persons and property. Figure 110–14

To solve this problem, a current-limiting overcurrent device (fast-clearing fuse) can be used to reduce the let-through current to less than 5,000A. Figure 110–15

Author's Comment: For more information on the application of current-limiting devices, see 240.2 and 240.60(B).

110.11 Deteriorating Agents. Electrical equipment and conductors must be suitable for the environment and conditions of use. Consideration must also be given to the presence of corrosive gases, fumes, vapors, liquids, or other substances that can have a deteriorating effect on the conductors or equipment. Figure 110–16

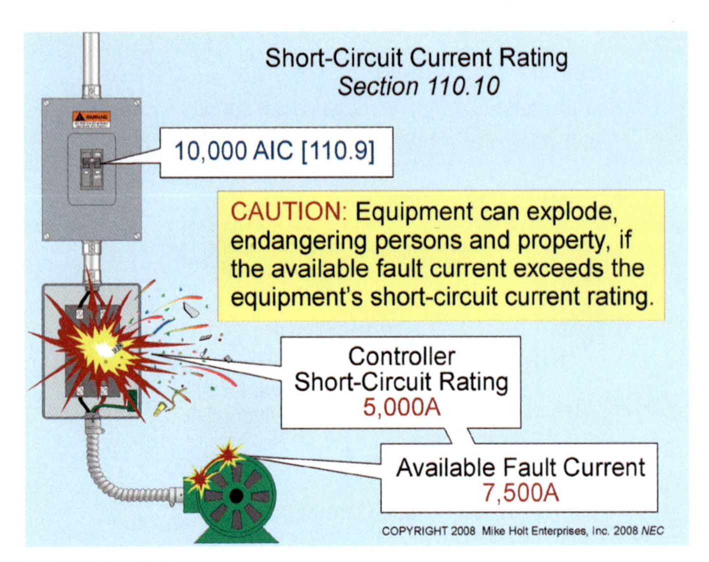

Figure 110–14

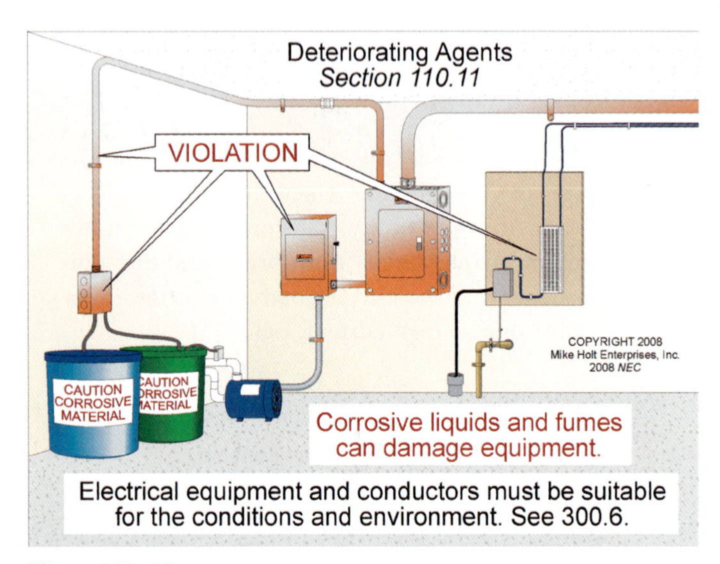

Figure 110–16

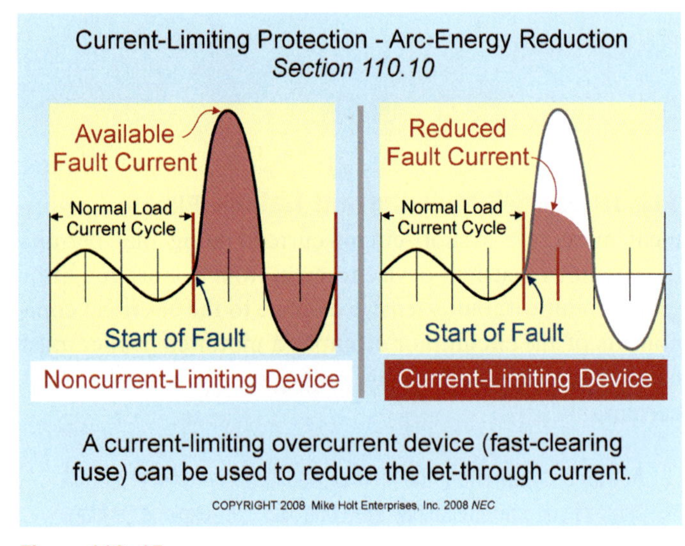

Figure 110–15

Author's Comment: Conductors must not be exposed to ultraviolet rays from the sun unless identified for the purpose [310.8(D)].

FPN No. 1: Raceways, cable trays, cablebus, cable armor, boxes, cable sheathing, cabinets, elbows, couplings, fittings, supports, and support hardware must be of materials suitable for the environment in which they are to be installed, in accordance with 300.6.

Author's Comment: See the definition of "Raceway" in Article 100.

FPN No. 2: Some cleaning and lubricating compounds contain chemicals that can deteriorate plastic used for insulating and structural applications in equipment.

Equipment not identified for outdoor use and equipment identified only for indoor use must be protected against permanent damage from the weather during building construction.

Author's Comment: This rule requires indoor-use equipment to be protected against "permanent damage," not incidental damage such as scratched paint.

FPN No. 3: See *NEC* Table 110.20 for appropriate enclosure-type designations.

110.12 Mechanical Execution of Work. Electrical equipment must be installed in a neat and workmanlike manner.

FPN: Accepted industry practices are described in ANSI/NECA 1, *Standard Practices for Good Workmanship in Electrical Contracting*.

Author's Comment: The National Electrical Contractors Association (NECA) has created a series of National Electrical Installation Standards (NEIS)® that established the industry's first quality guidelines for electrical installations. These standards define a benchmark or baseline of quality and workmanship for installing electrical products and systems. They explain what installing electrical products and systems in a "neat and workmanlike manner" means. For more information about these standards, visit www.neca-neis.org/.

(A) Unused Openings. Unused openings, other than those intended for the operation of equipment or mounting purposes, or part of the product listing, must be closed by fittings that provide protection substantially equivalent to the wall of the equipment. Figure 110–17

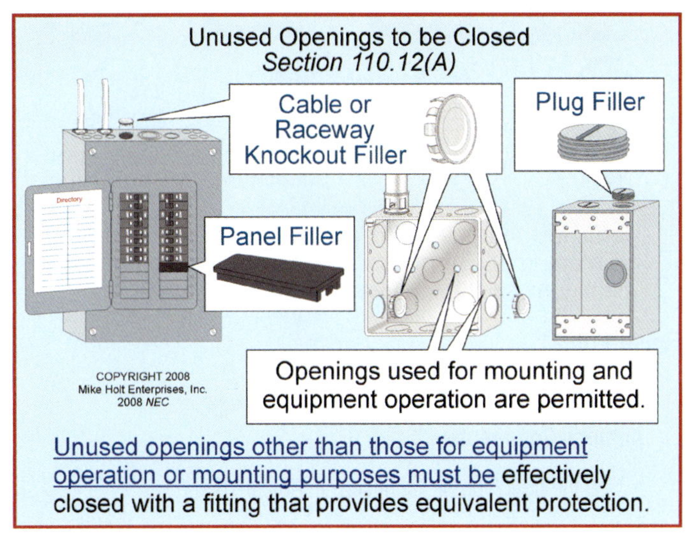

Figure 110–17

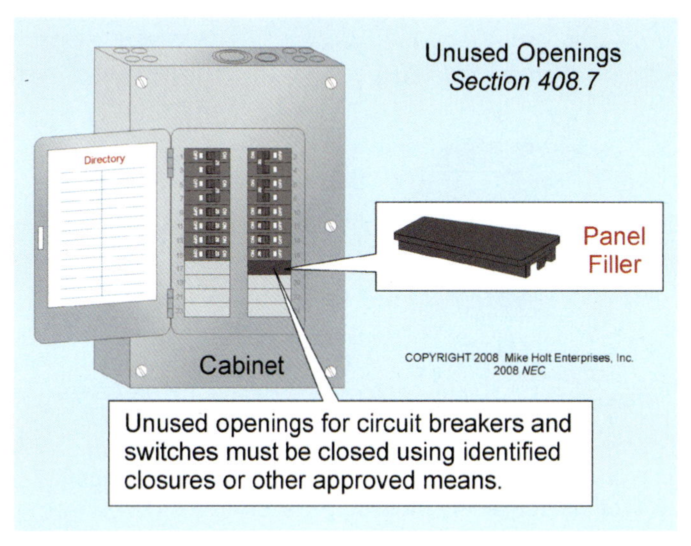

Figure 110–18

Author's Comments:

- See the definition of "Fitting" in Article 100.
- Unused openings for circuit breakers must be closed using identified closures, or other means approved by the authority having jurisdiction, that provide protection substantially equivalent to the wall of the enclosure [408.7]. Figure 110–18
- Openings intended to provide entry for conductors in cabinets, cutout boxes, and meter socket enclosures must be adequately closed [312.5(A)].

(B) Integrity of Electrical Equipment. Internal parts of electrical equipment must not be damaged or contaminated by foreign material, such as paint, plaster, cleaners, etc.

Author's Comment: Precautions must be taken to provide protection from the contamination of the internal parts of panelboards and receptacles during the building construction. Figure 110–19

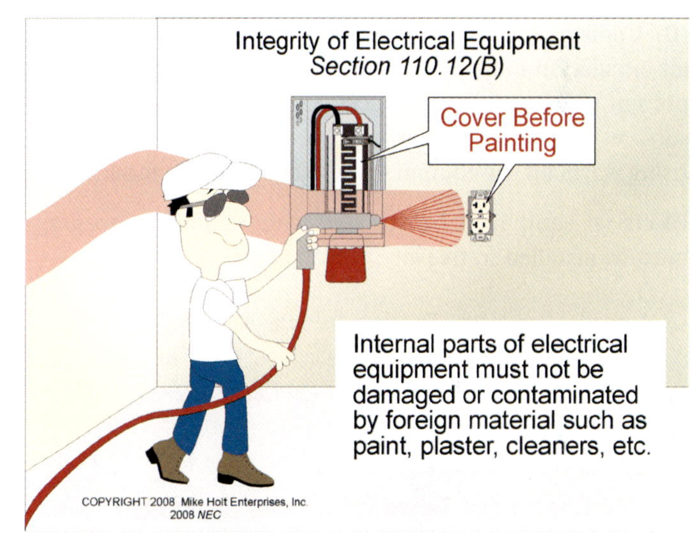

Figure 110–19

Electrical equipment that contains damaged parts may adversely affect safe operation or mechanical strength of the equipment and must not be installed. This includes parts that are broken, bent, cut, or deteriorated by corrosion, chemical action, or overheating.

Author's Comment: Damaged parts include cracked insulators, arc shields not in place, overheated fuse clips, and damaged or missing switch or circuit-breaker handles. Figure 110–20

110.13 Mounting and Cooling of Equipment.

(A) Mounting. Electrical equipment must be firmly secured to the surface on which it's mounted.

Author's Comment: See 314.23 for similar requirements for boxes.

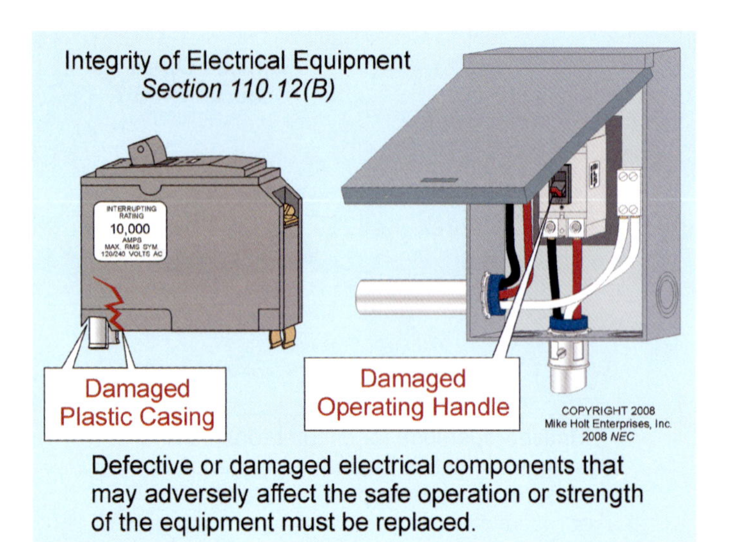

Figure 110–20

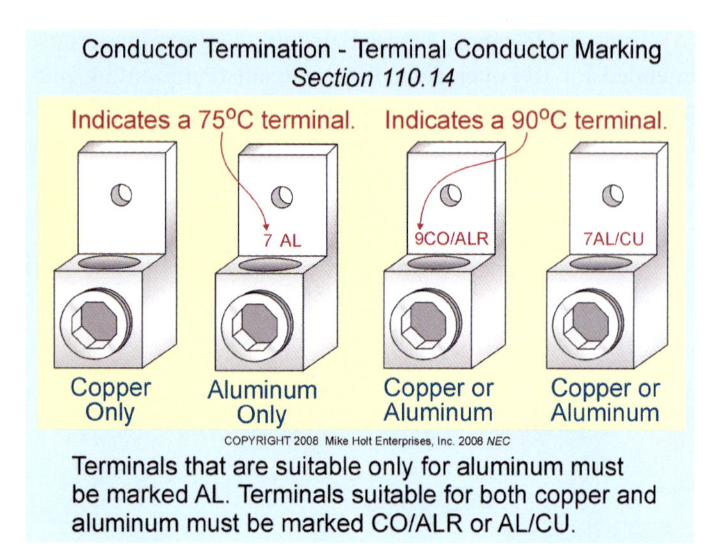

Figure 110–21

(B) Cooling. Electrical equipment that depends on natural air circulation must be installed so walls or equipment don't prevent airflow over the surfaces. The clearances between top surfaces and side surfaces must be maintained to dissipate rising warm air for equipment designed for floor mounting.

Electrical equipment constructed with ventilating openings must be installed so free air circulation isn't inhibited.

Author's Comment: Transformers with ventilating openings must be installed so that the ventilating openings aren't blocked, and the required wall clearances are clearly marked on the transformer case [450.9].

110.14 Conductor Termination and Splicing.

Conductor terminal and splicing devices must be identified for the conductor material and they must be properly installed and used. **Figure 110–21**

CAUTION: *When the insulation is stripped from an aluminum conductor and the conductor is exposed to air, an insulating film (aluminum oxide) immediately forms on the conductor. This film can create a poor connection and overheating at terminations. Unless the terminal or the device is manufactured with the right contacts designed to break through the film and ensure a good connection, overheating may occur.*

Switches and receptacles marked CO/ALR are designed to ensure a good connection through the use of the larger contact area and compatible materials. The terminal screws are plated with the element Indium. Indium is an extremely soft metal that forms a gas-sealed connection with the aluminum conductor.

Author's Comments:

- See the definition of "Identified" in Article 100.
- Conductor terminations must comply with the manufacturer's instructions as required by 110.3(B). For example, if the instructions for the device state "Suitable for 18-12 AWG Stranded," then only stranded conductors can be used with the terminating device. If the instructions state "Suitable for 18-12 AWG Solid," then only solid conductors are permitted, and if the instructions state "Suitable for 18-12 AWG," then either solid or stranded conductors can be used with the terminating device.

Copper and Aluminum Mixed. Copper and aluminum conductors must not make contact with each other in a device unless the device is listed and identified for this purpose.

Author's Comment: Few terminations are listed for the mixing of aluminum and copper conductors, but if they are, they will be marked on the product package or terminal device. The reason copper and aluminum should not be in contact with each other is because corrosion develops between the two different metals due to galvanic action, resulting in increased contact resistance at the splicing device. This increased resistance can cause overheating of the splice and cause a fire.

FPN: Many terminations and equipment are marked with a tightening torque.

Author's Comment: Conductors must terminate in devices that have been properly tightened in accordance with the manufacturer's torque specifications included with equipment instructions.

Failure to torque terminals can result in excessive heating of terminals or splicing devices (due to a loose connection), which can result in a fire because of a short circuit or ground fault. In addition, this is a violation of 110.3(B), which requires all equipment to be installed in accordance with listing or labeling instructions. Figure 110–22

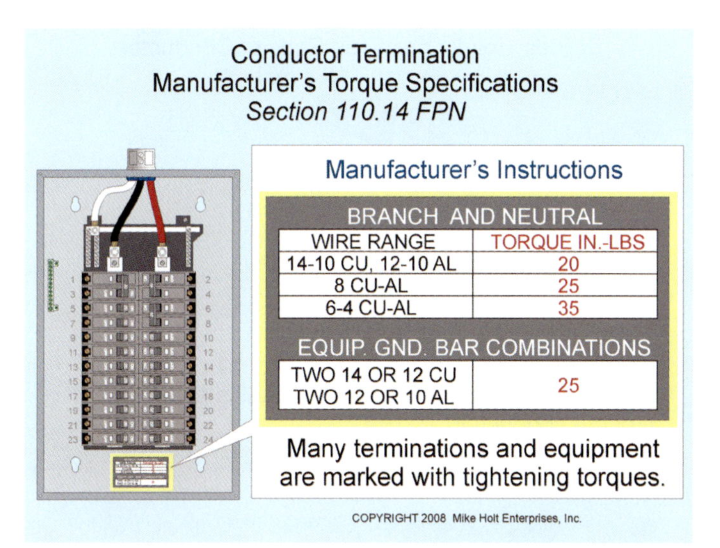

Figure 110–22

Question: *What do you do if the torque value isn't provided with the device?*

Answer: *Call the manufacturer, visit the manufacturer's Website, or have the supplier make a copy of the installation instructions.*

Author's Comment: Terminating conductors without a torque tool can result in an improper and unsafe installation. If a torque screwdriver is not used, there's a good chance the conductors are not properly terminated.

(A) Terminations. Conductor terminals must ensure a good connection without damaging the conductors and must be made by pressure connectors (including set screw type) or splices to flexible leads.

Author's Comments:

- See the definition of "Connector, Pressure" in Article 100.
- Grounding conductors and bonding jumpers must be connected by listed pressure connectors, terminal bars, exothermic welding, or other listed means [250.8(A)].

Question: *What if the conductor is larger than the terminal device?*

Answer: *This condition needs to be anticipated in advance, and the equipment should be ordered with terminals that will accommodate the larger conductor. However, if you're in the field, you should:*

- *Contact the manufacturer and have them express deliver you the proper terminals, bolts, washers, and nuts, or*
- *Order a terminal device that crimps on the end of the larger conductor and reduces the termination size.*

One Wire Per Terminal. Terminals for more than one conductor must be identified for this purpose, either within the equipment instructions or on the terminal itself. Figure 110–23

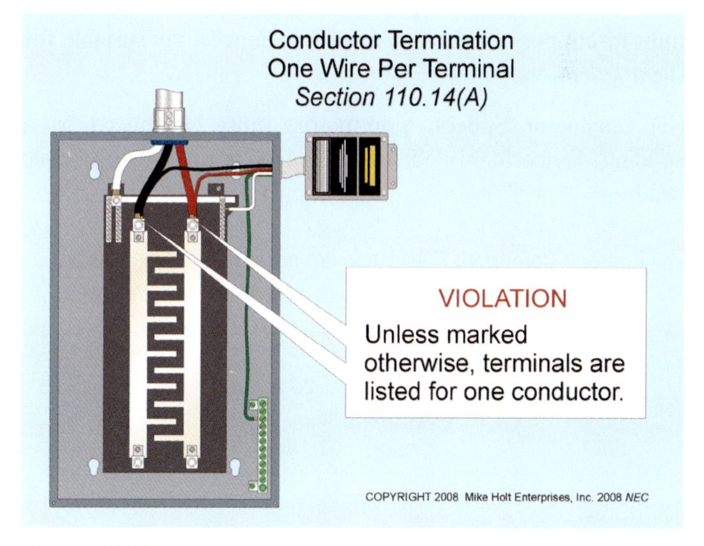

Figure 110–23

Author's Comments:

- Split-bolt connectors are commonly listed for only two conductors, although some are listed for three conductors. However, it's a common industry practice to terminate as many conductors as possible within a split-bolt connector, even though this violates the *NEC*. Figure 110–24
- Many devices are listed for more than one conductor per terminal. For example, some circuit breakers rated 30A or less can have two conductors under each lug. Grounding and bonding terminals are also listed for more than one conductor under the terminal.

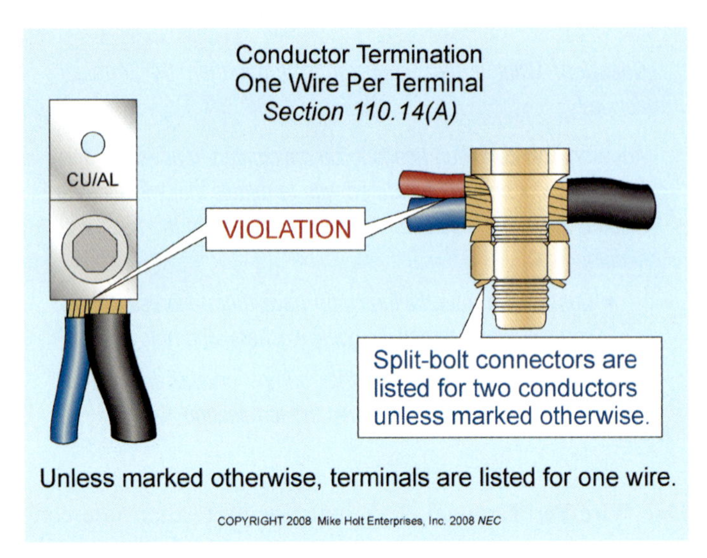

Figure 110–24

Split-bolt connectors for aluminum-to-aluminum or aluminum-to-copper conductors must be identified as suitable for the application.

(B) Conductor Splices. Conductors must be spliced by a splicing device identified for the purpose or by exothermic welding.

> **Author's Comment:** Conductors are not required to be twisted together prior to the installation of a twist-on wire connector. Figure 110–25

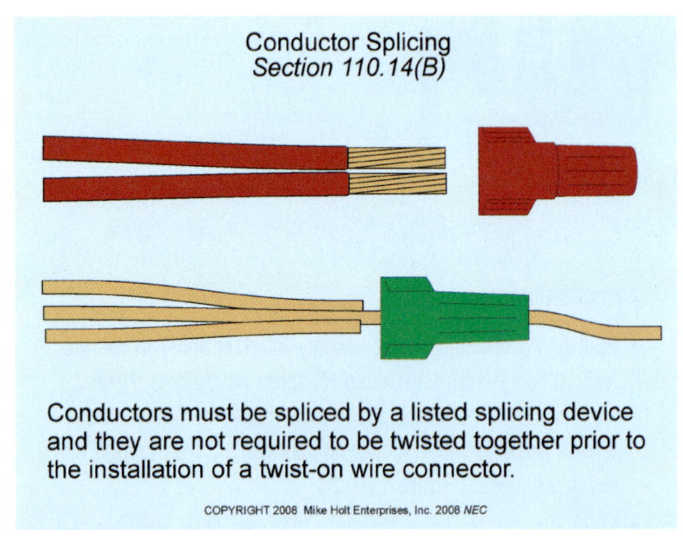

Figure 110–25

Unused circuit conductors are not required to be removed. However, to prevent an electrical hazard, the free ends of the conductors must be insulated to prevent the exposed end of the conductor from touching energized parts. This requirement can be met by the use of an insulated twist-on or push-on wire connector. Figure 110–26

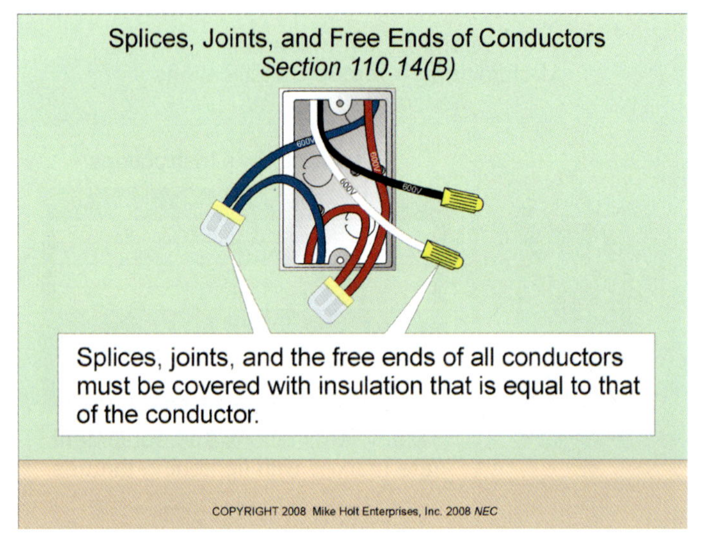

Figure 110–26

> **Author's Comment:** See the definition of "Energized" in Article 100.

Underground Splices:

Single Conductors. Single direct burial conductors of types UF or USE can be spliced underground without a junction box, but the conductors must be spliced with a device listed for direct burial [300.5(E) and 300.15(G)]. Figure 110–27

Multiconductor Cable. Multiconductor UF or USE cable can have the individual conductors spliced underground without a junction box, if a listed splice kit that encapsulates the conductors and the cable jacket is used.

(C) Temperature Limitations (Conductor Size). Conductors are to be sized to the lowest temperature rating of any terminal, device, or conductor of the circuit, in accordance with the equipment terminal temperature rating .

Conductor Ampacity. Conductors with insulation temperature ratings higher than the termination's temperature rating can be used for conductor ampacity adjustment, correction, or both.

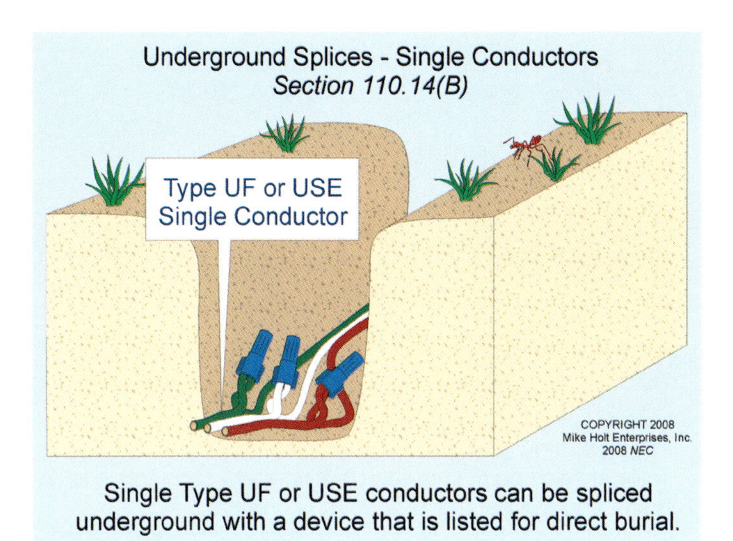

Figure 110–27

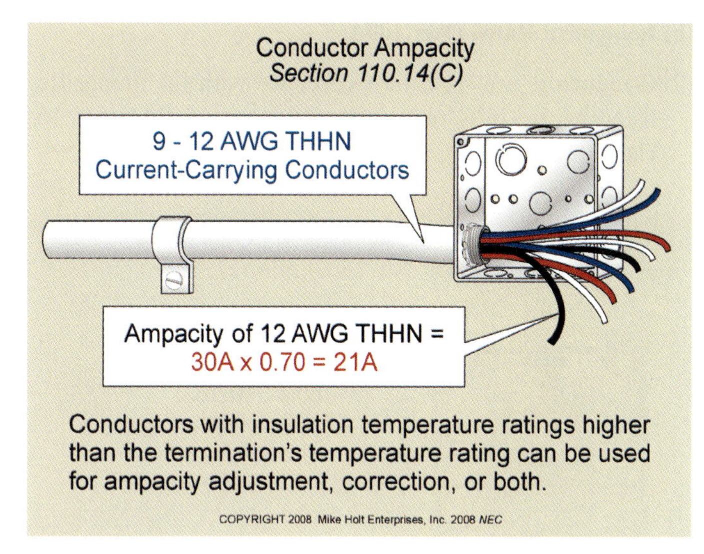

Figure 110–28

Author's Comments:

- See the definition of "Ampacity" in Article 100.
- This means conductor ampacity must be based on the conductor's insulation temperature rating listed in Table 310.16, as adjusted for ambient temperature correction factors, conductor bundling adjustment factors, or both.
- The location of conductors can influence their ampacity as well. For example, THHW is a 90°C conductor in a dry location, but it's a 75°C conductor in a wet location. THHN/THWN-2 is a 90°C conductor, in wet, dry, or damp locations [Table 310.13(A)].

Example: *The ampacity of each 12 THHW conductor in a dry location is 30A, based on the values listed in the 90°C column of Table 310.16. If we bundle nine current-carrying 12 THHN conductors, the ampacity for each conductor (30A at 90°C, Table 310.16) needs to be adjusted by a 70 percent adjustment factor [Table 310.15(B)(2)(a)].* Figure 110–28

Adjusted Conductor Ampacity = 30A x 0.70
Adjusted Conductor Ampacity = 21A

If the conductors are installed in a wet location, the ampacity of 12 THHW conductors is 25A according to the 75°C column of Table 310.16 [Table 310.13(A)].

Adjusted Conductor Ampacity = 25A x 0.70
Adjusted Conductor Ampacity = 17.50A, 18A [220.5(B)]

(1) Equipment Temperature Rating Provisions. Unless the equipment is listed and marked otherwise, conductor sizing for equipment terminations must be based on Table 310.16 in accordance with (a) or (b):

(a) Equipment Rated 100A or Less.

(3) Conductors terminating on terminals rated 75°C are sized in accordance with the ampacities listed in the 75°C temperature column of Table 310.16, provided the conductors have an insulation rating of at least 75°C. Figure 110–29

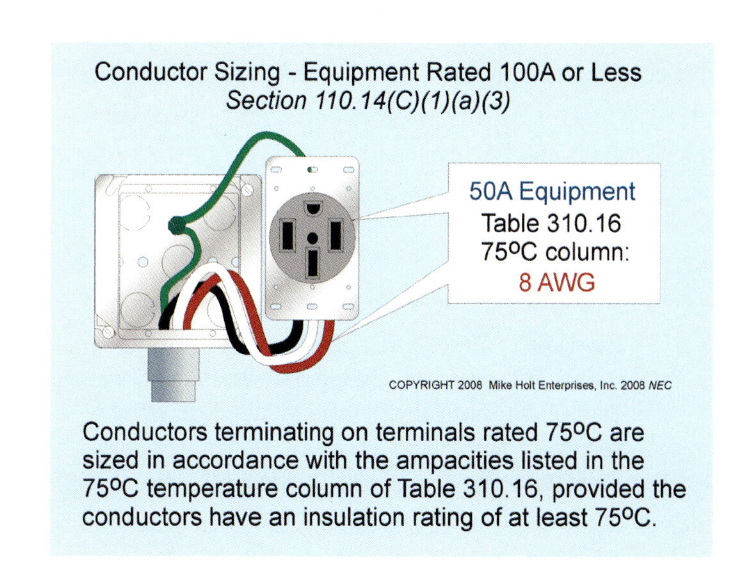

Figure 110–29

(b) Equipment Rated Over 100A

(2) Conductors are sized in accordance with the ampacities listed in the 75°C temperature column of Table 310.16. **Figure 110–30**

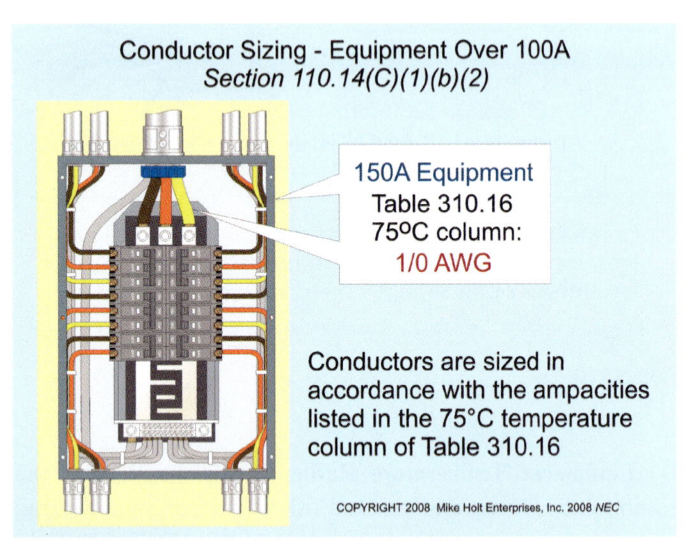

Figure 110–30

110.15 High-Leg Conductor Identification.

On a 4-wire, delta-connected, three-phase system, where the midpoint of one phase winding is grounded (high-leg system), the conductor with 208V to ground must be durably and permanently marked by an outer finish orange in color, or other effective means. Such identification must be placed at each point on the system where a connection is made if the neutral conductor is present [110.15, 215.8, and 230.56]. **Figure 110–31**

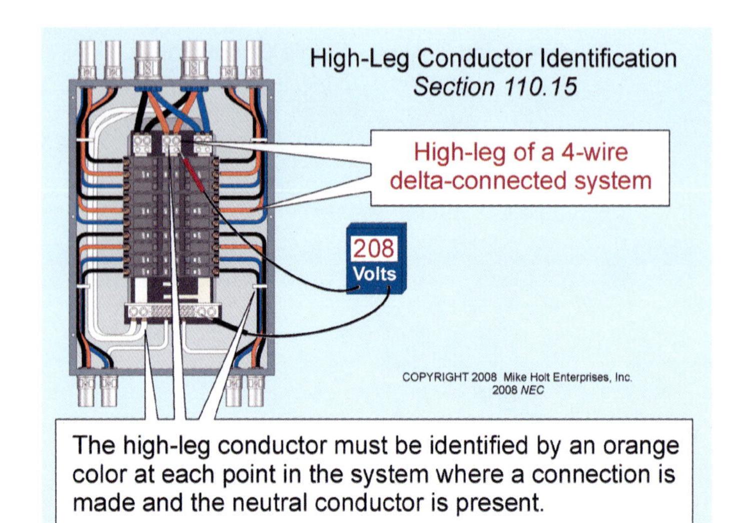

Figure 110–31

Author's Comments:

- The high-leg conductor is also called the "wild leg," "stinger leg," or "bastard leg."
- Other important *NEC* rules relating to the high leg are as follows:
  - Panelboards. Since 1975, panelboards supplied by a 4-wire, delta-connected, three-phase system must have the high-leg conductor terminate to the "B" phase of a panelboard [408.3(E)]. Section 408.3(F) requires panelboards to be field-marked with "Caution 208V to ground."
  - Disconnects. The *NEC* does not specify the termination location for the high-leg conductor in switch equipment (Switches—Article 404), but the generally accepted practice is to terminate this conductor to the "B" phase.
  - Utility Equipment. The ANSI standard for meter equipment requires the high-leg conductor (208V to neutral) to terminate on the "C" (right) phase of the meter socket enclosure. This is because the demand meter needs 120V, and it gets that voltage from the "B" phase.
- Also hope the utility lineman is not color blind and doesn't inadvertently cross the "orange" high-leg conductor (208V) with the red (120V) service conductor at the weatherhead. It's happened before...

WARNING: *When replacing equipment in existing facilities that contain a high-leg conductor, care must be taken to ensure the high-leg conductor is replaced in its original location. Prior to 1975, the high-leg conductor was required to terminate on the "C" phase of panelboards and switchboards. Failure to re-terminate the high leg in accordance with the existing installation can result in 120V circuits inadvertently connected to the 208V high leg, with disastrous results.*

110.16 Flash Protection Warning.

Electrical equipment such as switchboards, panelboards, industrial control panels, meter socket enclosures, and motor control centers in other than dwelling units that are likely to require examination, adjustment, servicing, or maintenance while energized must be field-marked to warn qualified persons of the danger associated with an arc flash from line-to-line or ground faults. The field-marking must be clearly visible to qualified persons before they examine, adjust, service, or perform maintenance on the equipment. **Figure 110–32**

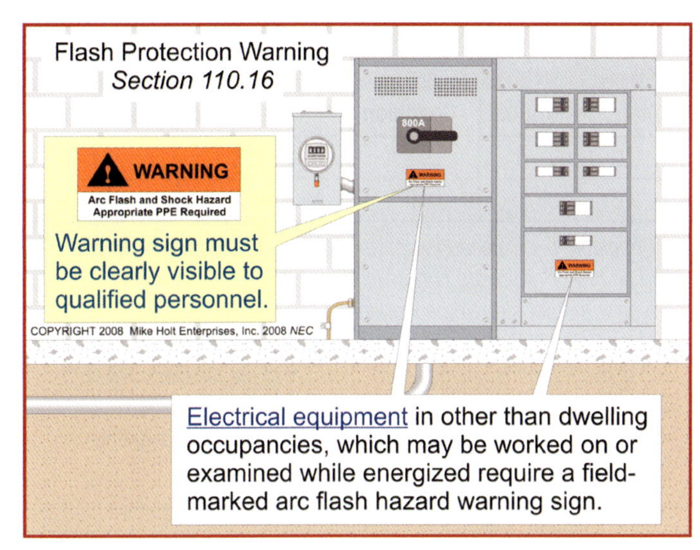

Figure 110–32

Author's Comments:

- See the definition of "Qualified Person" in Article 100.
- This rule is meant to warn qualified persons who work on energized electrical systems that an arc flash hazard exists so they will select proper personal protective equipment (PPE) in accordance with industry accepted safe work practice standards.

FPN No. 1: NFPA 70E, *Standard for Electrical Safety in the Workplace*, provides assistance in determining the severity of potential exposure, planning safe work practices, and selecting personal protective equipment.

110.20 Enclosure Types. Enclosures must be marked with an Enclosure Type number and be suitable for the location in accordance with the *NEC* Table 110.20.

The enclosures are not intended to protect against condensation, icing, corrosion, or contamination within the enclosure or that enters via the conduit or unsealed openings.

FPN: Raintight enclosures include Types 3, 3S, 3SX, 3X, 4, 4X, 6, 6P; rainproof enclosures are Types 3R, 3RX; watertight enclosures are Types 4, 4X, 6, 6P; driptight enclosures are Types 2, 5, 12, 12K, 13; and dusttight enclosures are Types 3, 3S, 3SX, 3X, 5, 12, 12K, 13.

110.21 Manufacturer's Markings. The manufacturer's name, trademark, or other descriptive marking must be placed on all electrical equipment and, where required by the *Code*, markings such as voltage, current, wattage, or other ratings must be provided. All marking must have sufficient durability to withstand the environment involved.

110.22 Identification of Disconnecting Means.

(A) General. Each disconnecting means must be legibly marked to indicate its purpose unless located and arranged so the purpose is evident. The marking must be of sufficient durability to withstand the environment involved. Figure 110-33

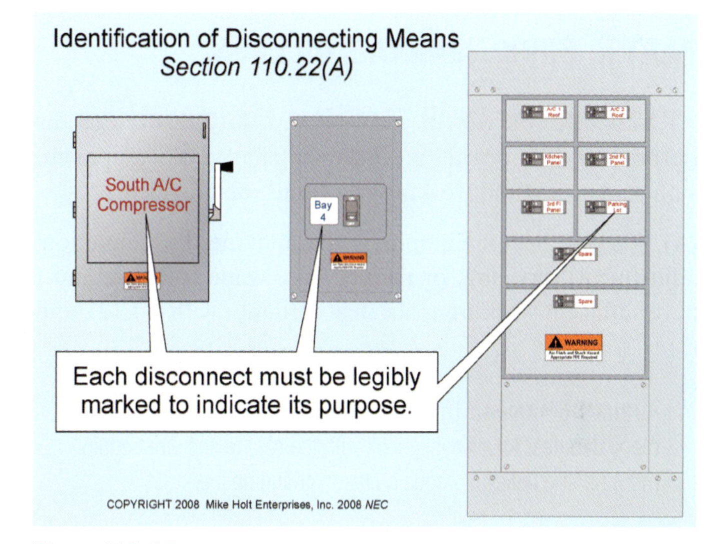

Figure 110–33

(B) Engineered Series Combination Systems. Where circuit breakers or fuses are applied in compliance with the series combination ratings selected under engineering supervision and marked on the equipment as directed by the engineer, the equipment enclosure(s) must be legibly marked in the field to indicate the equipment was applied with a series combination rating. The marking must be readily visible and state the following:

CAUTION—ENGINEERED SERIES COMBINATION SYSTEM RATED ____ AMPERES. IDENTIFIED REPLACEMENT COMPONENTS REQUIRED.

FPN: See 240.86(A) for Engineered Series Combination Systems.

(C) Tested Series Combination Systems. Where circuit breakers or fuses are applied in compliance with the series combination ratings marked on the equipment by the manufacturer, the equipment enclosure(s) must be legibly marked in the field to indicate the equipment was applied with a series combination rating. The marking must be readily visible and state the following:

CAUTION—SERIES COMBINATION
SYSTEM RATED ____ AMPERES.
IDENTIFIED REPLACEMENT COMPONENTS REQUIRED.

FPN: See 240.86(B) for Tested Series Combination Systems.

PART II. 600V, NOMINAL, OR LESS

110.26 Spaces About Electrical Equipment. For the purpose of safe operation and maintenance of equipment, sufficient access and working space must be provided.

(A) Working Space. Equipment that may need examination, adjustment, servicing, or maintenance while energized must have sufficient working space in accordance with (1), (2), and (3):

Author's Comment: The phrase "while energized" is the root of many debates. As always, check with the AHJ to see what equipment he/she believes needs a clear working space.

(1) Depth of Working Space. The working space, which is measured from the enclosure front, must not be less than the distances contained in Table 110.26(A)(1). Figure 110–34

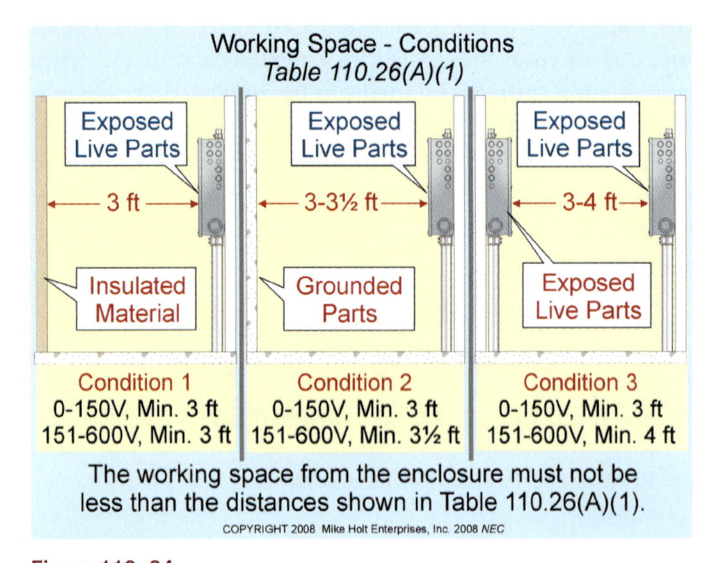

Figure 110–34

Table 110.26(A)(1) Working Space

Voltage-to-Ground	Condition 1	Condition 2	Condition 3
0–150V	3 ft	3 ft	3 ft
151–600V	3 ft	3½ft	4 ft

- *Condition 1—Exposed live parts on one side of the working space and no live or grounded parts, including concrete, brick, or tile walls are on the other side of the working space.*
- *Condition 2—Exposed live parts on one side of the working space and grounded parts, including concrete, brick, or tile walls are on the other side of the working space.*
- *Condition 3—Exposed live parts on both sides of the working space.*

(a) Rear and Sides. Working space isn't required for the back or sides of assemblies where all connections and all renewable or adjustable parts are accessible from the front. Figure 110–35

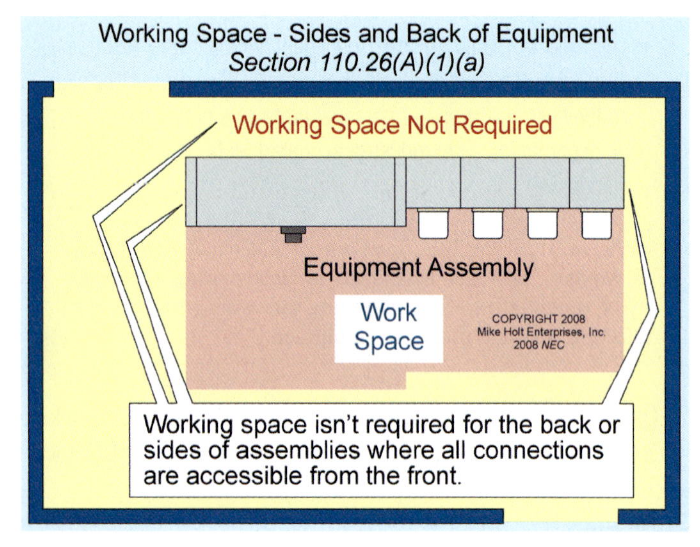

Figure 110–35

(b) Low Voltage. Where special permission is granted in accordance with 90.4, working space for equipment that operates at not more than 30V ac or 60V dc can be less than the distance in Table 110.26(A)(1). Figure 110–36

Author's Comment: See the definition of "Special Permission" in Article 100.

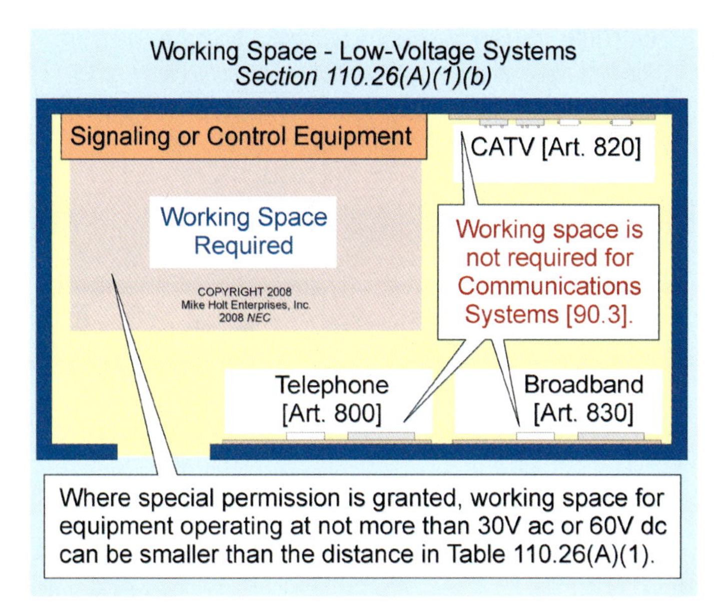

Figure 110–36

(c) **Existing Buildings.** Where electrical equipment is being replaced, Condition 2 working space is permitted between dead-front switchboards, panelboards, or motor control centers located across the aisle from each other where conditions of maintenance and supervision ensure that written procedures have been adopted to prohibit equipment on both sides of the aisle from being open at the same time, and only authorized, qualified persons will service the installation.

Author's Comment: The working space requirements of 110.26 don't apply to equipment included in Chapter 8—Communications Circuits [90.3].

(2) Width of Working Space. The width of the working space must be a minimum of 30 in., but in no case less than the width of the equipment. Figure 110–37

Author's Comment: The width of the working space can be measured from left-to-right, from right-to-left, or simply centered on the equipment, and the working space can overlap the working space for other electrical equipment. Figure 110–38

In all cases, the working space must be of sufficient width, depth, and height to permit all equipment doors to open 90 degrees. Figure 110–39

(3) Height of Working Space (Headroom). For service equipment, switchboards, panelboards, and motor control equipment, the height of the working space in front of equipment must not be less than 6½ ft, measured from the grade, floor, or platform [110.26(E)].

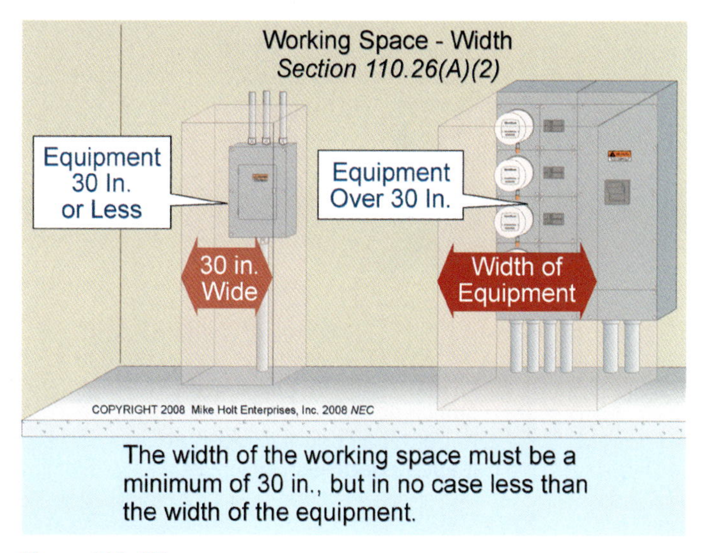

Figure 110–37

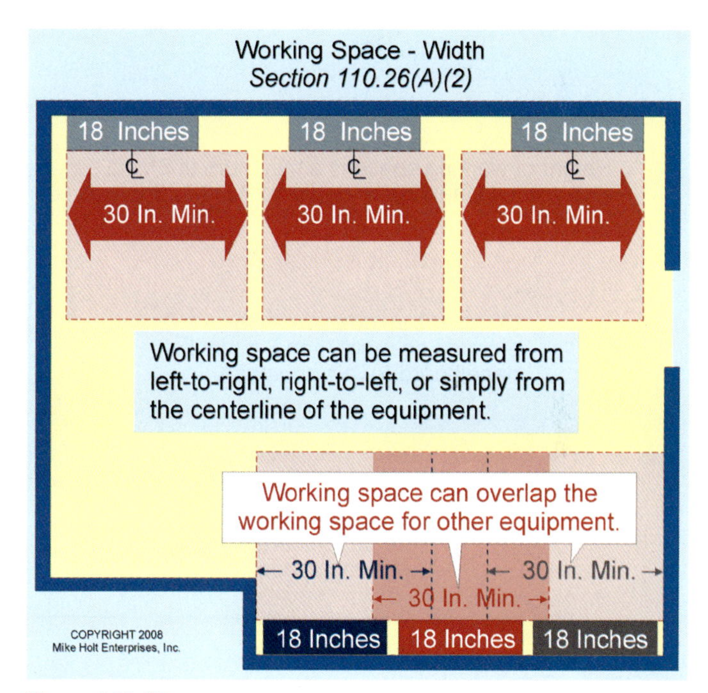

Figure 110–38

Equipment such as raceways, cables, wireways, cabinets, panels, etc., can be located above or below electrical equipment, but must not extend more than 6 in. into the equipment's working space. Figure 110–40

(B) Clear Working Space. The working space required by this section must be clear at all times. Therefore, this space is not permitted for storage. When normally enclosed live parts are exposed for inspection or servicing, the working space, if in a passageway or general open space, must be suitably guarded.

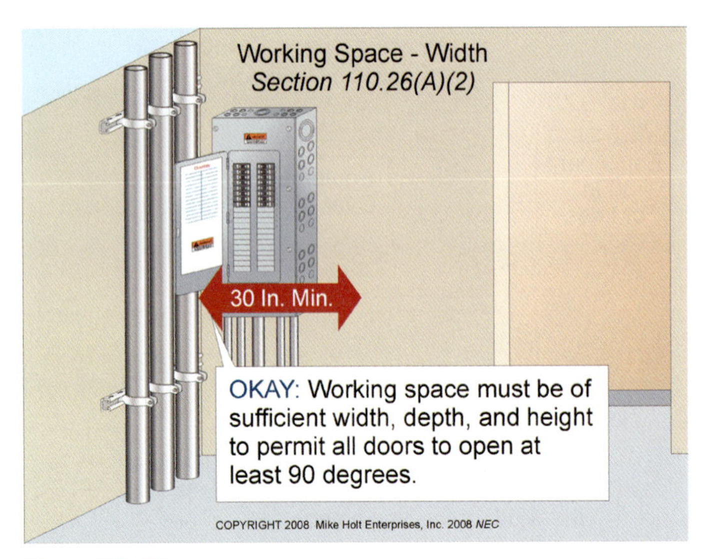

Figure 110–39

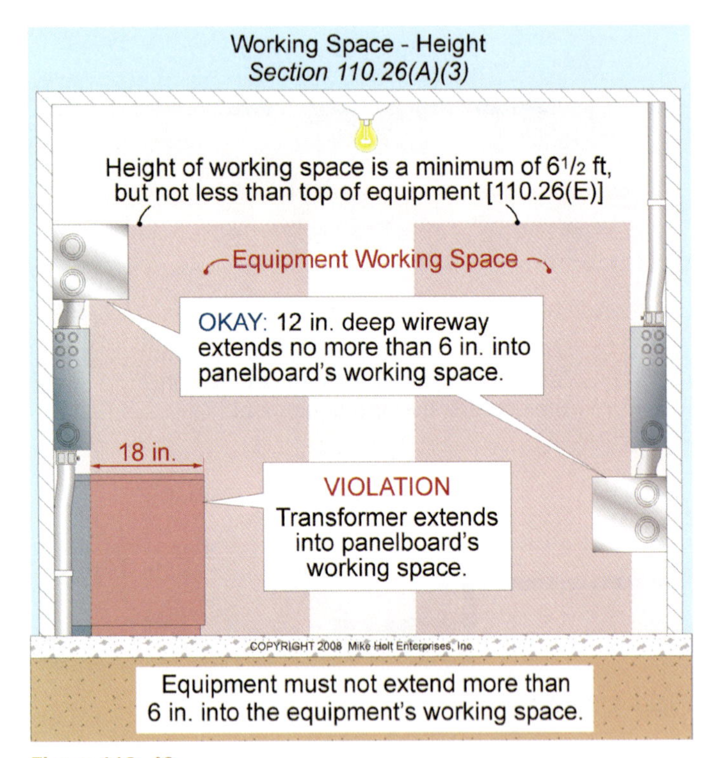

Figure 110–40

Author's Comment: When working in a passageway, the working space should be guarded from occupants using the passageway. When working on electrical equipment in a passageway one must be mindful of a fire alarm evacuation with numerous occupants congregated and moving through the passageway.

CAUTION: *It's very dangerous to service energized parts in the first place, and it's unacceptable to be subjected to additional dangers by working around bicycles, boxes, crates, appliances, and other impediments.* **Figure 110–41**

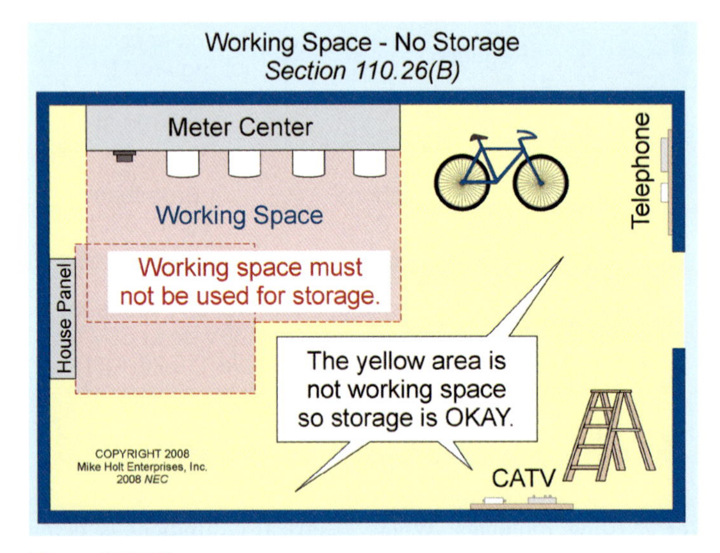

Figure 110–41

Author's Comment: Signaling and communications equipment must not be installed in a manner that encroaches on the working space of the electrical equipment. **Figure 110–42**

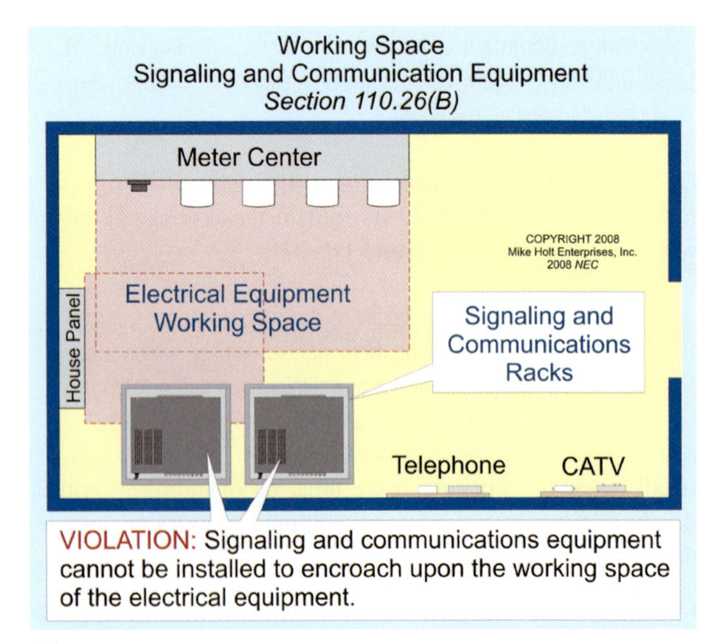

Figure 110–42

(C) Entrance to and Egress from Working Space.

(1) Minimum Required. At least one entrance of sufficient area must provide access to and egress from the working space.

> **Author's Comment:** Check to see what the authority having jurisdiction considers "Sufficient Area." Building codes contain minimum dimensions for doors and openings for personnel travel.

(2) Large Equipment. An entrance to and egress from each end of the working space of electrical equipment rated 1,200A or more that is over 6 ft wide is required. The opening must be not less than 24 in. wide and 6½ ft high. A single entrance to and egress from the required working space is permitted where either of the following conditions is met. Figure 110–43

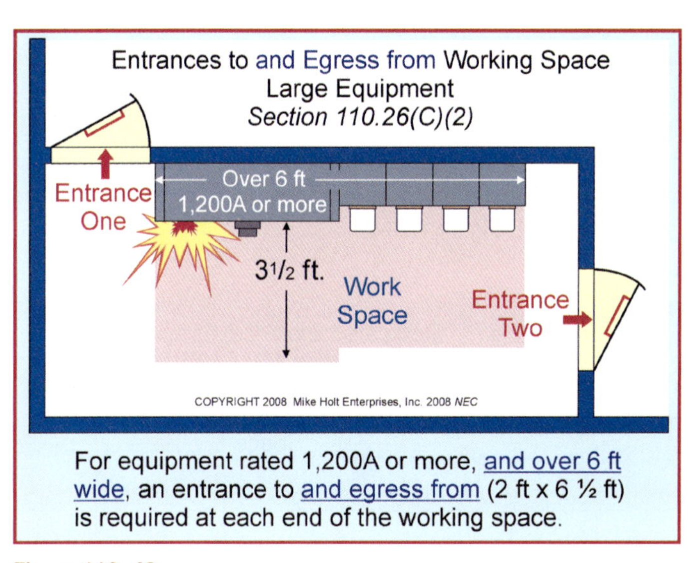

For equipment rated 1,200A or more, and over 6 ft wide, an entrance to and egress from (2 ft x 6 ½ ft) is required at each end of the working space.

Figure 110–43

(a) Unobstructed Egress. Only one entrance is required where the location permits a continuous and unobstructed way of egress travel.

(b) Double Workspace. Only one entrance is required where the required working space depth is doubled, and the equipment is located so the edge of the entrance is no closer than the required working space distance. Figure 110–44

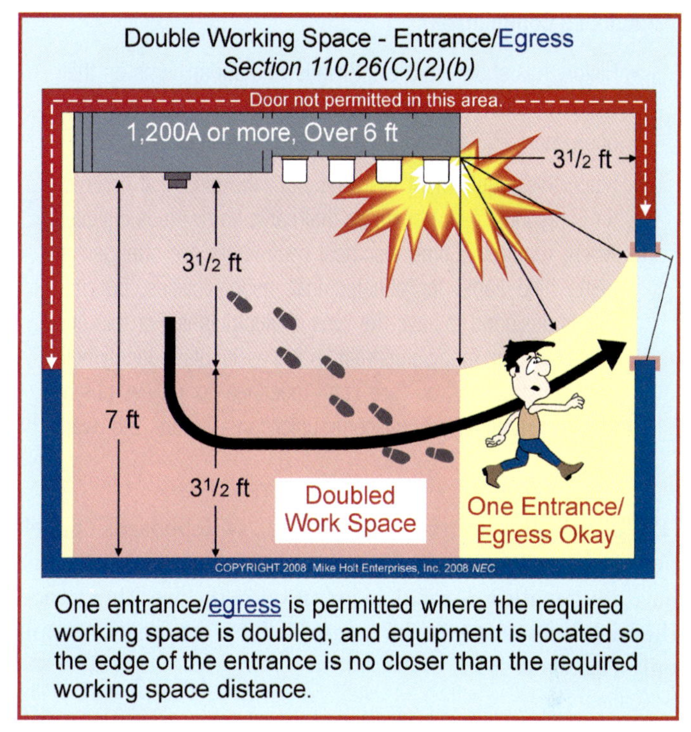

One entrance/egress is permitted where the required working space is doubled, and equipment is located so the edge of the entrance is no closer than the required working space distance.

Figure 110–44

(3) Personnel Doors. Where equipment with overcurrent or switching devices rated 1,200A or more is installed, personnel door(s) for entrance to and egress from the working space located less than 25 ft from the nearest edge of the working space must have the door(s) open in the direction of egress and be equipped with panic hardware or other devices that open under simple pressure. Figure 110–45

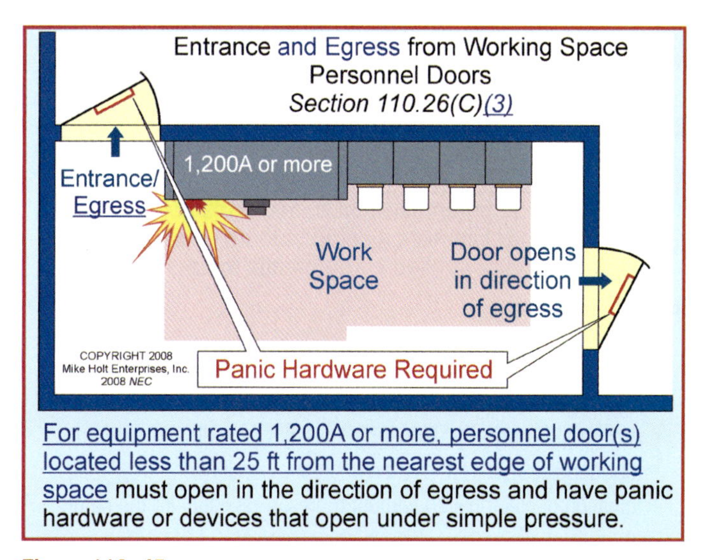

For equipment rated 1,200A or more, personnel door(s) located less than 25 ft from the nearest edge of working space must open in the direction of egress and have panic hardware or devices that open under simple pressure.

Figure 110–45

Author's Comments:

- History has shown that electricians who suffer burns on their hands in electrical arc flash or arc blast events often can't open doors equipped with knobs that must be turned.
- Since this requirement is in the *NEC*, the electrical contractor is responsible for ensuring that panic hardware is installed where required. Some electrical contractors are offended at being held liable for nonelectrical responsibilities, but this rule is designed to save the lives of electricians. For this and other reasons, many construction professionals routinely hold "pre-construction" or "pre-con" meetings to review potential opportunities for miscommunication—before the work begins.

(D) Illumination. Service equipment, switchboards, panelboards, as well as motor control centers located indoors must have illumination located in or next to the working space. Illumination must not be controlled by automatic means only. Figure 110–46

Working Space - Illumination Control
Section 110.26(D)

Service equipment, switchboards, panelboards, and motor control centers located indoors must not have illumination controlled by automatic means only.

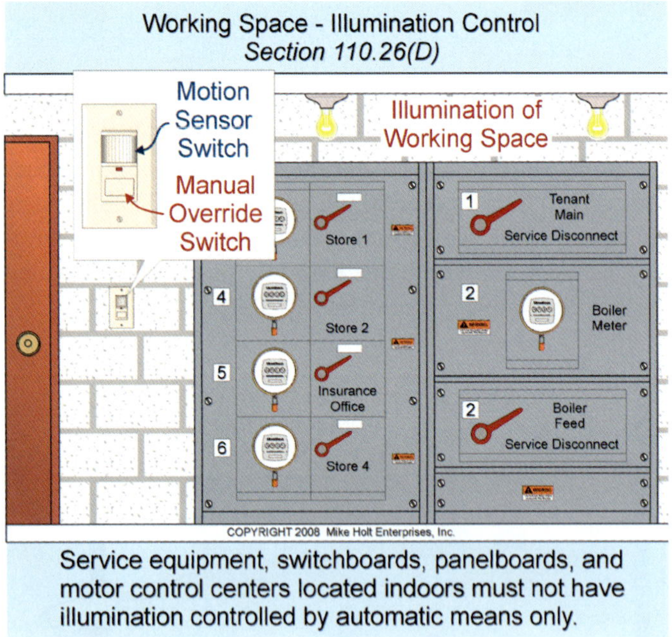

Figure 110–46

Author's Comment: The *NEC* does not provide the minimum foot-candles required to provide proper illumination. Proper illumination of electrical equipment rooms is essential for the safety of those qualified to work on such equipment.

(E) Headroom. For service equipment, panelboards, switchboards, or motor control centers, the minimum working space headroom must not be less than 6½ ft. When the height of the equipment exceeds 6½ ft, the minimum headroom must not be less than the height of the equipment.

Exception: The minimum headroom requirement doesn't apply to service equipment or panelboards rated 200A or less located in an existing dwelling unit.

Author's Comment: See the definition of "Dwelling Unit" in Article 100.

(F) Dedicated Equipment Space. Switchboards, panelboards, and motor control centers must have dedicated equipment space as follows:

(1) Indoors.

(a) Dedicated Electrical Space. The footprint space (width and depth of the equipment) extending from the floor to a height of 6 ft above the equipment or to the structural ceiling, whichever is lower, must be dedicated for the electrical installation. No piping, ducts, or other equipment foreign to the electrical installation can be installed in this dedicated footprint space. Figure 110–47

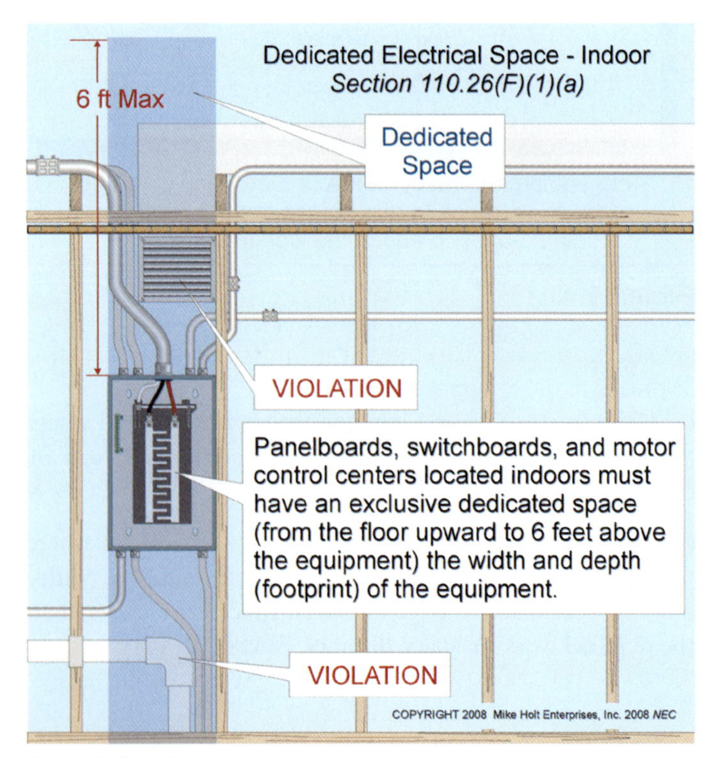

Figure 110–47

Exception: Suspended ceilings with removable panels can be within the dedicated footprint space [110.26(G)].

Author's Comment: Electrical raceways and cables not associated with the dedicated space can be within the dedicated space. These aren't considered "equipment foreign to the electrical installation." Figure 110–48

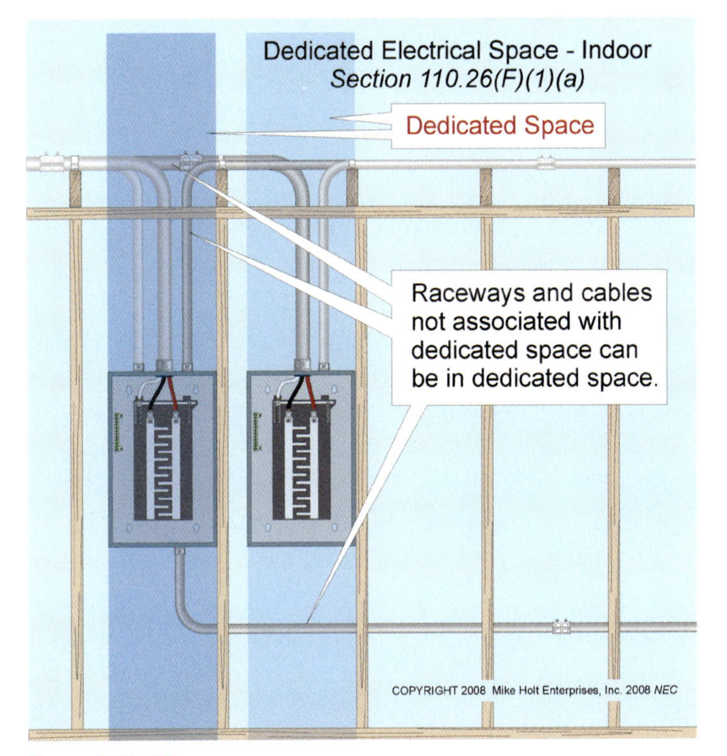

Figure 110–48

(b) Foreign Systems. Foreign systems can be located above the dedicated space if protection is installed to prevent damage to the electrical equipment from condensation, leaks, or breaks in the foreign systems. Figure 110–49

(c) Sprinkler Protection. Sprinkler protection piping isn't permitted in the dedicated space, but the *NEC* doesn't prohibit sprinklers from spraying water on electrical equipment.

(d) Suspended Ceilings. A dropped, suspended, or similar ceiling isn't considered a structural ceiling.

(G) Locked Electrical Equipment Rooms or Enclosures. Electrical equipment rooms and enclosures housing electrical equipment or apparatus controlled by locks are considered accessible to qualified persons who require access. Figure 110–50

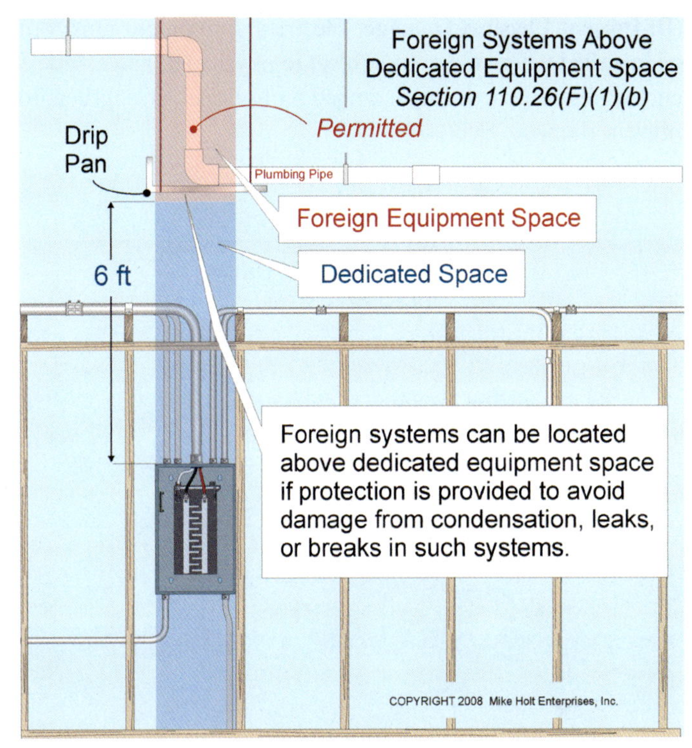

Figure 110–49

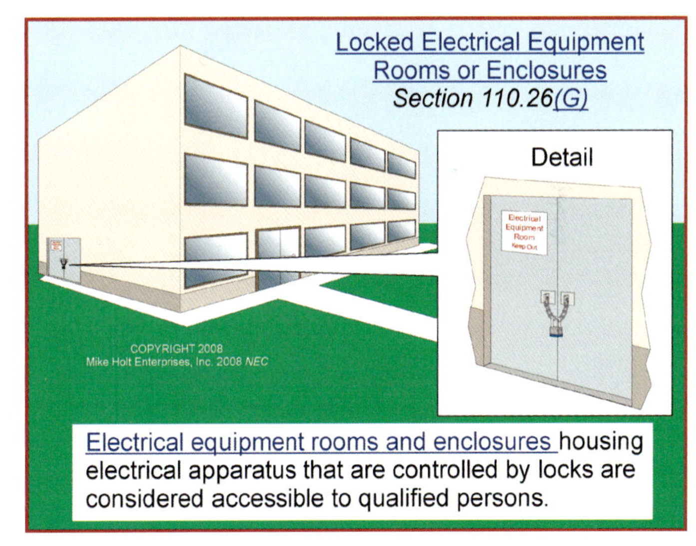

Figure 110–50

Author's Comment: See the definition of "Accessible as it applies to equipment" in Article 100.

110.27 Guarding.

(A) Guarding Live Parts. Live parts of electrical equipment operating at 50V or more must be guarded against accidental contact.

(B) Prevent Physical Damage. Electrical equipment must not be installed where it can be subject to physical damage, unless enclosures or guards are arranged and of such strength as to prevent damage. Figure 110–51

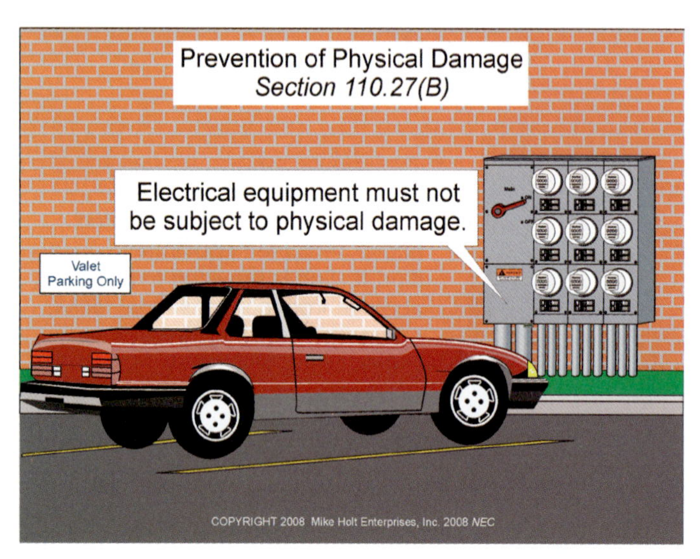

Figure 110–51

ARTICLE 110 Practice Questions

ARTICLE 110. REQUIREMENTS FOR ELECTRICAL INSTALLATIONS—PRACTICE QUESTIONS

1. In judging equipment for approval, considerations such as the following shall be evaluated:

   (a) Mechanical strength
   (b) Wire-bending space
   (c) Arcing effects
   (d) all of these

2. Conductor sizes are expressed in American Wire Gage (AWG) or _____.

   (a) inches
   (b) circular mils
   (c) square inches
   (d) none of these

3. Equipment intended to interrupt current at fault levels shall have an interrupting rating sufficient for the nominal circuit voltage and the current that is available at the line terminals of the equipment.

   (a) True
   (b) False

4. Some cleaning and lubricating compounds can cause severe deterioration of materials used for insulating and structural applications in equipment.

   (a) True
   (b) False

5. Unused openings other than those intended for the operation of equipment, intended for mounting purposes, or permitted as part of the design for listed equipment shall be _____.

   (a) filled with cable clamps or connectors only
   (b) taped over with electrical tape
   (c) repaired only by welding or brazing in a metal slug
   (d) effectively closed to afford protection substantially equivalent to the wall of the equipment

6. Conductor terminal and splicing devices must be _____ for the conductor material and they must be properly installed and used.

   (a) listed
   (b) approved
   (c) identified
   (d) all of these

7. The temperature rating associated with the ampacity of a _____ shall be so selected and coordinated so as not to exceed the lowest temperature rating of any connected termination, conductor, or device.

   (a) terminal
   (b) conductor
   (c) device
   (d) all of these

8. For circuits rated 100A or less, when the equipment terminals are listed for use with 75°C conductors, the _____ column of Table 310.16 shall be used to determine the ampacity of all THHN conductors installed.

   (a) 30°C
   (b) 60°C
   (c) 75°C
   (d) 90°C

9. On a 4-wire, delta-connected system where the midpoint of one phase winding is grounded, the conductor having the higher phase voltage-to-ground shall be durably and permanently marked by an outer finish that is _____ in color.

   (a) black
   (b) red
   (c) blue
   (d) orange

10. Where required by the *Code*, markings on all electrical equipment shall contain voltage, current, wattage, or other ratings with durability to withstand _____.

    (a) the voltages encountered
    (b) painting and other finishes applied
    (c) the environment involved
    (d) any lack of planning by the installer

11. The *NEC* requires tested series-rated installations of circuit breakers or fuses to be field-marked to indicate the current rating for which the system has been installed.

    (a) True
    (b) False

12. The minimum working space on a circuit that is 120 volts to ground, with exposed live parts on one side and no live or grounded parts on the other side of the working space, is _____ ft.

    (a) 1
    (b) 3
    (c) 4
    (d) 6

13. The dimension of working space for access to live parts operating at 300V, nominal-to-ground, where there are exposed live parts on both sides of the workspace is _____ ft according to Table 110.26(A)(1).

    (a) 3
    (b) 3½
    (c) 4
    (d) 4½

14. Equipment associated with the electrical installation can be located above or below other electrical equipment within their working space when the associated equipment does not extend more than _____ in. from the front of the electrical equipment.

    (a) 3
    (b) 6
    (c) 12
    (d) 30

15. For equipment rated 1,200A or more and over 6 ft wide that contains overcurrent devices, switching devices, or control devices, there shall be one entrance to and egress from the required working space not less than 24 in. wide and _____ high at each end of the working space.

    (a) 5 ft 6 in.
    (b) 6 ft
    (c) 6 ft 6 in.
    (d) any of these

16. The minimum headroom of working spaces about motor control centers is _____ ft.

    (a) 3
    (b) 5
    (c) 6
    (d) 6½

17. The dedicated equipment space for electrical equipment that is required for panelboards is measured from the floor to a height of _____ ft above the equipment, or to the structural ceiling, whichever is lower.

    (a) 3
    (b) 6
    (c) 12
    (d) 30

18. In locations where electrical equipment is likely to be exposed to _____, enclosures or guards shall be so arranged and of such strength as to prevent such damage.

    (a) corrosion
    (b) physical damage
    (c) magnetic fields
    (d) weather

WIRING AND PROTECTION

INTRODUCTION TO CHAPTER 2—WIRING AND PROTECTION

Chapter 2 provides general rules for wiring and protection of conductors. The rules in this chapter apply to all electrical installations covered by the *NEC*—except as modified in Chapters 5, 6, and 7 [90.3].

Communications Systems (Chapter 8 systems) aren't subject to the general requirements of Chapters 1 through 4, or the special requirements of Chapters 5 through 7, unless there's a specific reference in Chapter 8 to a rule in Chapters 1 through 7 [90.3].

As you go through Chapter 2, remember its purpose. Chapter 2 is primarily concerned with correctly sizing and protecting circuits. Every article in Chapter 2 deals with a different aspect of this purpose. This differs from the purpose of Chapter 3, which is to correctly install the conductors that make up those circuits.

Chapter 1 introduced you to the *NEC* and provided a solid foundation for understanding the *Code*. Chapters 2 (Wiring and Protection) and 3 (Wiring Methods and Materials) together form the heart of the *NEC*—a solid foundation for applying the *NEC*. Chapter 4 applies the preceding chapters to general equipment. So, once again, you find yourself needing to learn the *NEC* in a sequential manner because each chapter builds on the one before it. Once you've mastered the first four chapters, you can learn the next four in any order you wish.

- **Article 200.** Use and Identification of Grounded Conductors. This article contains the requirements for the use and identification of the grounded conductor and its terminals.

  Author's Comment: Because the neutral conductor of a solidly grounded system is always grounded to the earth, it's both a "grounded conductor" and a "neutral conductor." To make it easier for the reader of this textbook, we will refer to the "grounded conductor" as the "neutral conductor."

- **Article 210.** Branch Circuits. Article 210 contains the requirements for branch circuits, such as conductor sizing, identification, GFCI protection of receptacles, and receptacle and lighting outlet requirements.
- **Article 215.** Feeders. This article covers the requirements for the installation, minimum size, and ampacity of feeders.
- **Article 220.** Branch-Circuit, Feeder, and Service Calculations. Article 220 provides the requirements for calculating branch circuits, feeders, and services, for determining the number of receptacles on a circuit in nondwelling units, and the number of branch circuits required.

- **Article 225.** Outside Branch Circuits and Feeders. This article covers the installation requirements for equipment, including conductors located outside, on, or between buildings, poles, and other structures on the premises.

- **Article 230.** Services. Article 230 covers the installation requirements for service conductors and equipment. It's very important to know where the service begins and ends when applying Article 230.

  Author's Comment: Conductors from a battery, uninterruptible power supply, solar photovoltaic system, generator, or transformer are not considered service conductors; they are feeder conductors.

- **Article 240.** Overcurrent Protection. This article provides the requirements for overcurrent protection and overcurrent devices. Overcurrent protection for conductors and equipment is provided to open the circuit if the current reaches a value that will cause an excessive or dangerous temperature on the conductors or conductor insulation.

- **Article 250.** Grounding and Bonding. Article 250 covers the grounding requirements for providing a low-impedance path to the earth to reduce overvoltage from lightning, and the requirements for a low-impedance fault current path necessary to facilitate the operation of overcurrent devices in the event of a ground fault.

- **Article 285.** Surge Protective Devices (SPDs). This article covers general requirements, installation requirements, and connection requirements for surge protective devices (SPDs) permanently installed on both the line and load side of service equipment.

Use and Identification of Grounded Conductors

INTRODUCTION TO ARTICLE 200—USE AND IDENTIFICATION OF GROUNDED CONDUCTORS

This article contains the requirements for the identification of the grounded conductor and its terminals.

> **Author's Comment:** Because the neutral conductor of a solidly grounded system is always grounded to the earth, it's both a "grounded conductor" and a "neutral conductor." To make it easier for the reader of this textbook, we will refer to the "grounded conductor" as the "neutral conductor." **Figure 200–1**

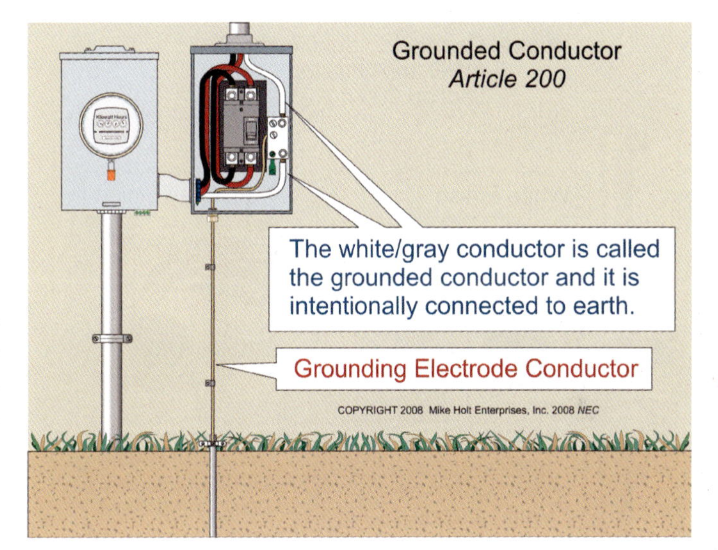

Figure 200–1

200.1 Scope. Article 200 contains requirements for the use and identification of neutral conductors and terminals.

200.2 General.

(B) Continuity. The continuity of the neutral conductor is not permitted to be dependent on a connection to metal enclosures, raceways, or cable armor. **Figure 200–2**

> **Author's Comment:** This requirement prohibits the practice of terminating the neutral conductor on the enclosure of a panel or other equipment, rather than on the neutral terminal bar. This ensures the metallic panelboard, raceway, or cable armor does not carry neutral current. Some panelboards have two terminal bars, one on either side of the panelboard with a strap connecting the terminal bars together. Caution must be taken to terminate the neutral conductor to the neutral terminal, not to the equipment grounding conductor terminal.

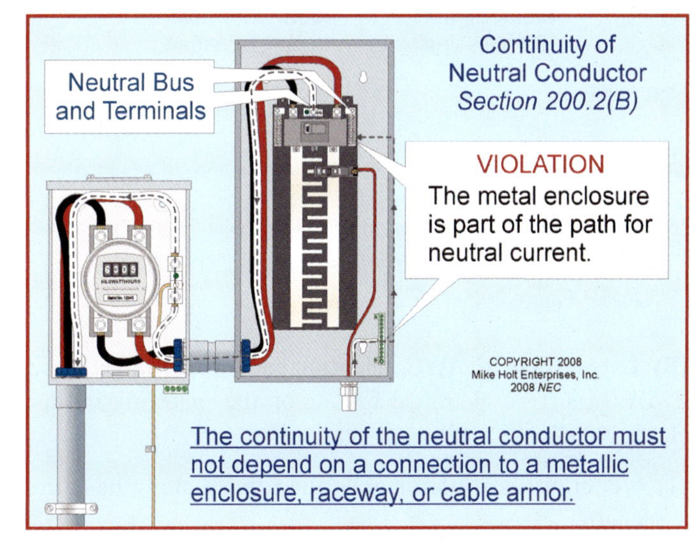

Figure 200–2

200.6 Neutral Conductor Identification.

(A) 6 AWG or Smaller. Neutral conductors 6 AWG and smaller must be identified by a continuous white or gray outer finish along their entire length, or by any color insulation (except green) with three white stripes, or by white or gray insulation with any color stripes (except green). Figure 200–3

Figure 200–3

Author's Comment: The use of white tape, paint, or other methods of identification isn't permitted for conductors 6 AWG or smaller. Figure 200–4

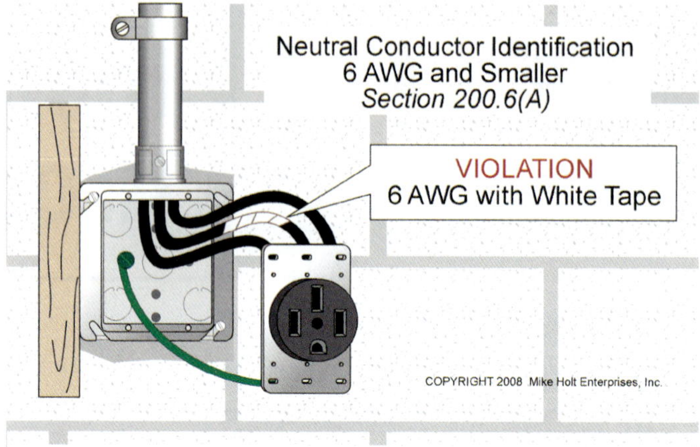

Figure 200–4

(B) Larger than 6 AWG. Neutral conductors larger than 6 AWG must be identified by one of the following means: Figure 200–5

(1) A continuous white or gray outer finish along its entire length.

(2) Three continuous white stripes along its length.

(3) White or gray tape or paint at terminations.

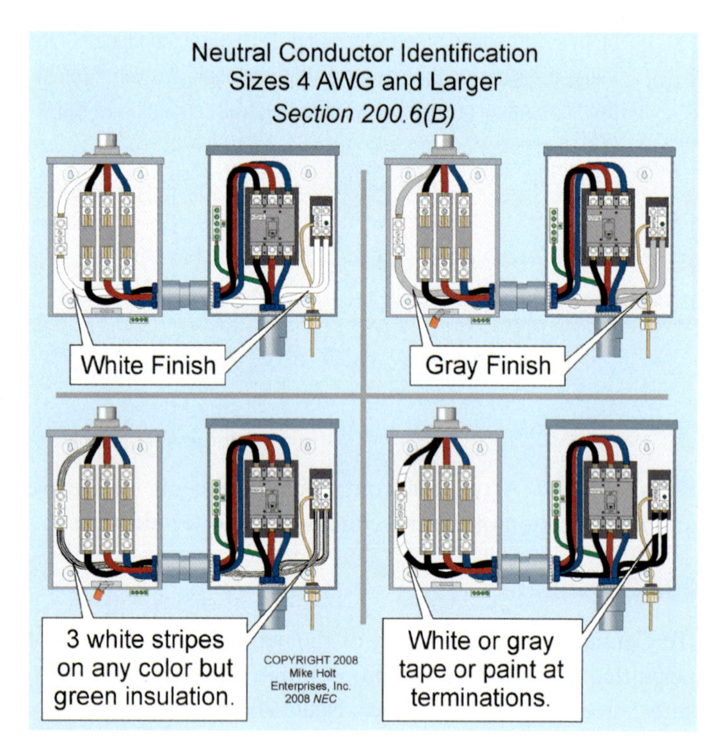

Figure 200–5

(D) Neutral Conductors of Different Systems. Where neutral conductors of different voltage systems are installed in the same raceway, cable, or enclosure, each neutral conductor must be identified to distinguish the systems by:

(1) A continuous white or gray outer finish along its entire length.

(2) The neutral conductor of the other system must have a different outer covering of continuous white or gray outer finish along its entire length. Figure 200–6

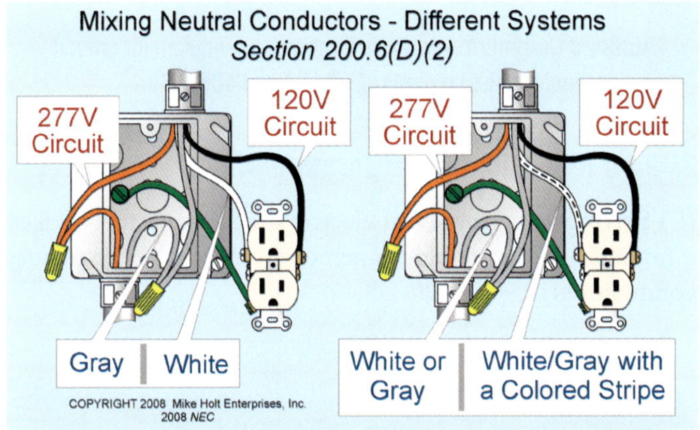

The neutral conductor of the different systems must have an outer covering of continuous white or gray finish along its entire length or by an outer covering of white or gray with a readily distinguishable colored stripe (other than green) along its entire length.

Figure 200–6

(3) The means of identification must be permanently posted at each branch-circuit panelboard. Figure 200–7

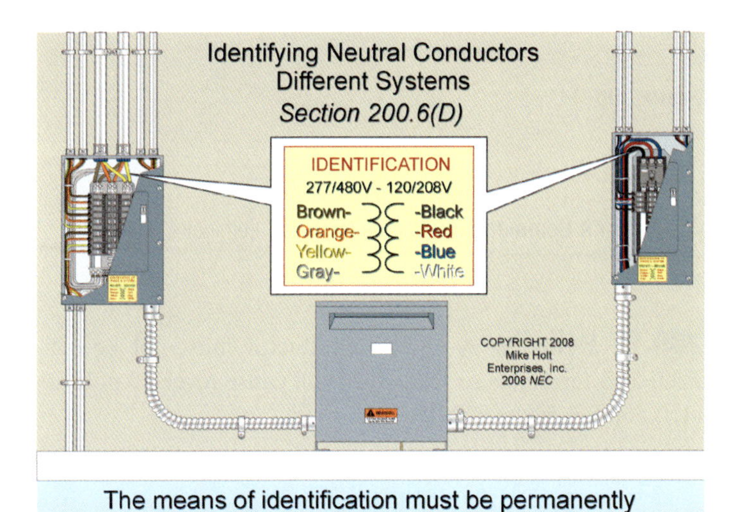

The means of identification must be permanently posted at each branch-circuit panelboard.

Figure 200–7

Author's Comment: Where a premises wiring system has circuits supplied from more than one voltage system, each ungrounded conductor must be identified by system [210.5(C) and 215.12(C)].

200.7 Use of White or Gray Color.

(C) Circuits of 50V or More. A conductor with white insulation can only be used for the ungrounded conductor as permitted in (1), (2), and (3) below.

(1) Cable Assembly. The white conductor within a cable can be used for the ungrounded conductor if permanently reidentified at each location where the conductor is visible to indicate its use as an ungrounded conductor. Identification must encircle the insulation and must be a color other than white, gray, or green. Figure 200–8

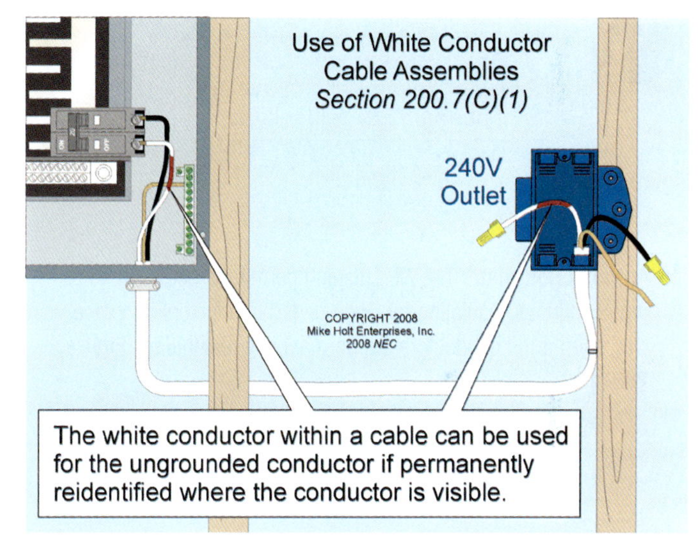

The white conductor within a cable can be used for the ungrounded conductor if permanently reidentified where the conductor is visible.

Figure 200–8

(2) Switches. The white conductor within a cable can be used for single-pole, 3-way or 4-way switch loops if permanently reidentified at each location where the conductor is visible to indicate its use as an ungrounded conductor. Figure 200–9

(3) Flexible Cord. The white conductor within a flexible cord can be used for the ungrounded conductor for connecting an appliance or equipment as permitted by 400.7.

FPN: Care should be taken when working on existing systems because a gray insulated conductor may have been used in the past as an ungrounded conductor.

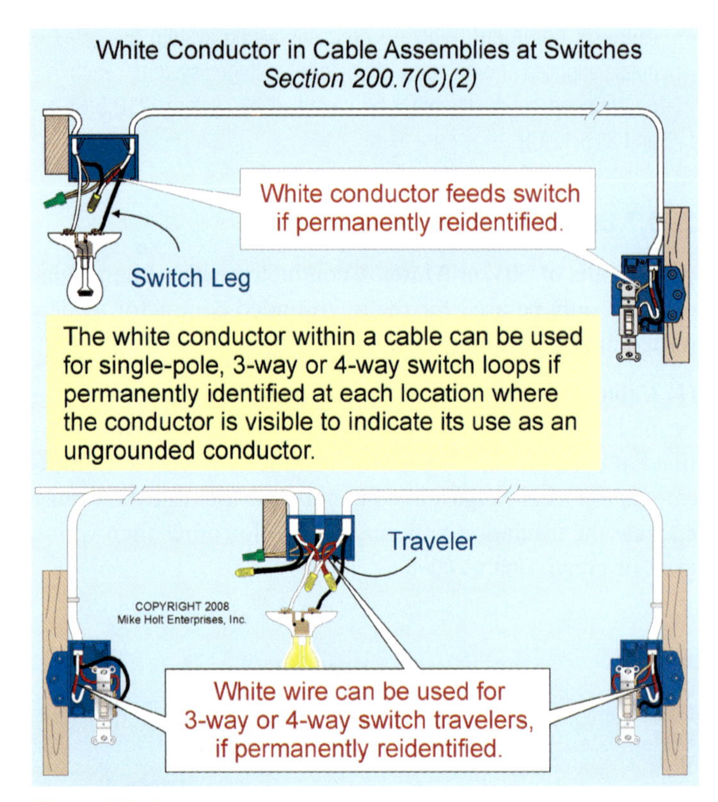

Figure 200–9

Author's Comment: The *NEC* doesn't permit the use of white or gray conductor insulation for ungrounded conductors in a raceway, even if the conductors are permanently reidentified. Figure 200–10

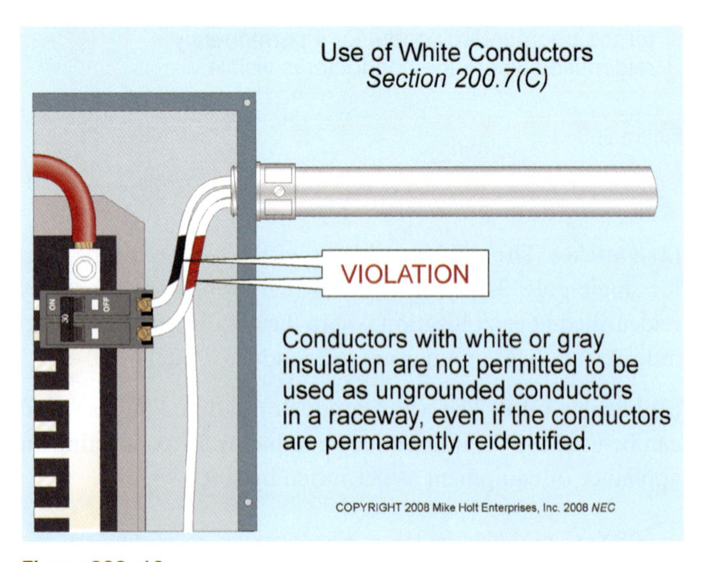

Figure 200–10

200.9 Terminal Identification.

The terminal for the neutral conductor must be colored white (actually silver). The terminal for the ungrounded conductor must be a color readily distinguishable from white (brass or copper).

Author's Comment: Terminals for the circuit equipment grounding conductor must be green [250.126 and 406.9(B)].

200.10 Identification of Terminals.

(C) Screw Shell. To prevent electric shock, the screw shell of a luminaire or lampholder must be connected to the neutral conductor [410.90]. Figure 200–11

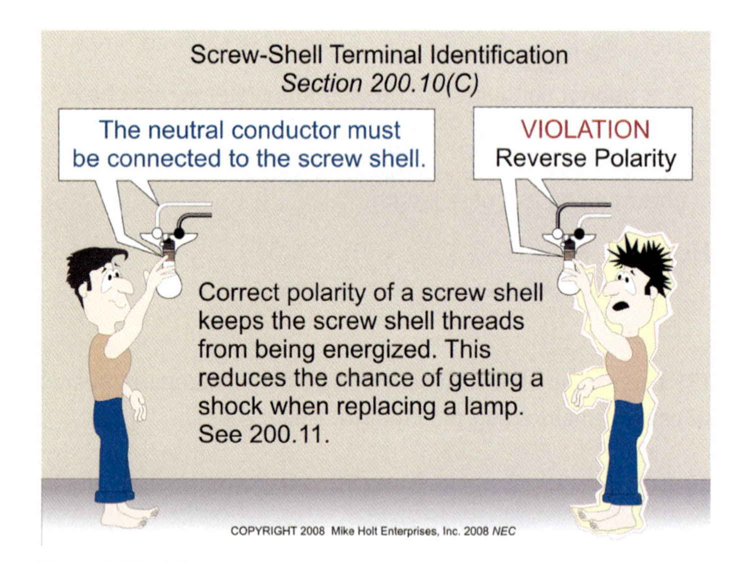

Figure 200–11

Author's Comment: See the definition of "Luminaire" in Article 100.

200.11 Polarity.

A neutral conductor must not be connected to terminals or leads that will cause reversed polarity [410.50]. See Figure 200–11.

ARTICLE 200 Practice Questions

ARTICLE 200. USE AND IDENTIFICATION OF GROUNDED CONDUCTORS—PRACTICE QUESTIONS

1. Article 200 contains the requirements for _____.

   (a) identification of terminals
   (b) grounded conductors in premises wiring systems
   (c) identification of grounded conductors
   (d) all of these

2. Grounded conductors larger than _____ can be identified by distinctive white or gray markings at their terminations.

   (a) 10 AWG
   (b) 8 AWG
   (c) 6 AWG
   (d) 4 AWG

3. The identification of _____ to which a grounded conductor is to be connected shall be substantially white in color.

   (a) wire connectors
   (b) circuit breakers
   (c) terminals
   (d) ground rods

4. No _____ shall be attached to any terminal or lead so as to reverse designated polarity.

   (a) grounded conductor
   (b) grounding conductor
   (c) ungrounded conductor
   (d) grounding connector

ARTICLE

210 Branch Circuits

INTRODUCTION TO ARTICLE 210—BRANCH CIRCUITS

This article contains the requirements for branch circuits, such as conductor sizing and identification, GFCI protection, and receptacle and lighting outlet requirements. It consists of three parts:

- PART I. GENERAL PROVISIONS
- PART II. BRANCH-CIRCUIT RATINGS
- PART III. REQUIRED OUTLETS

Table 210.2 of this article identifies specific-purpose branch circuits. When people complain that the *Code* "buries stuff in the last few chapters and doesn't provide you with any way of knowing where to find things," it's often because they didn't pay attention to this table.

The following sections and tables contain a few key items on which to spend extra time as you study Article 210:

- 210.4. Multiwire Branch Circuits. The conductors of these circuits must originate from the same panel. These circuits can supply only line-to-neutral loads.
- 210.8. GFCI Protection. Crawl spaces, unfinished basements, and boathouses are just some of the many locations that require GFCI protection.
- 210.11. Branch Circuits Required. With three subheadings, 210.11 gives summarized requirements for the number of branch circuits in a given system, states that a load calculated on a VA per area basis must be evenly proportioned, and covers rules for dwelling units.
- 210.12. Arc-Fault Circuit-Interrupter Protection. An arc-fault circuit interrupter is a device intended to de-energize the circuit when it detects the current waveform characteristics unique to an arcing fault. The purpose of an AFCI is to protect against a fire hazard, whereas the purpose of a GFCI is to protect people against electrocution.
- 210.19. Conductors. Minimum Ampacity and Size. This becomes complicated in a hurry, but we'll guide you through it.
- Table 210.21(B)(2) shows that the maximum cord-and plug-connected load on a given circuit is 80 percent of the receptacle rating and circuit rating. We'll explain more about the implications of this later.
- 210.23. Permissible Loads. This is intended to prevent a circuit overload from occurring just because someone plugs in a lamp or vacuum cleaner. We'll show you how to conform.
- 210.52. Dwelling Unit Receptacle Outlets. Receptacle spacing is a subject area that is very confusing. We cut through the confusion, and you'll understand the meaning of 210.52 and how to apply it correctly.

The rest of the material is also important, but mastering these key items will give you a decided edge in your ability to complete installations that are free of *Code* violations.

PART I. GENERAL PROVISIONS

210.1 Scope. Article 210 contains the requirements for conductor sizing, overcurrent protection, identification, and GFCI protection of branch circuits, as well as receptacle outlets and lighting outlet requirements.

Author's Comment: Article 100 defines a "branch circuit" as the conductors between the final overcurrent device and the receptacle outlets, lighting outlets, or other outlets. Figure 210–1

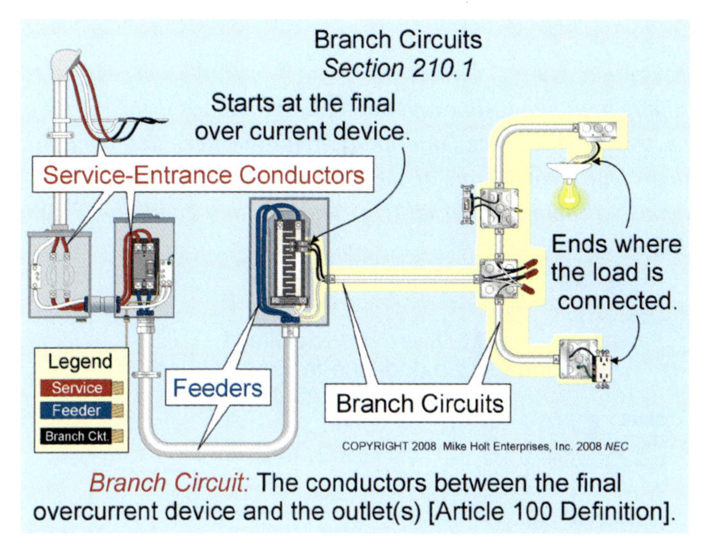

Figure 210–1

210.2 Other Articles. Other *NEC* sections that have specific requirements for branch circuits include:

- Air-Conditioning and Refrigeration, 440.6, 440.31, and 440.32
- Appliances, 422.10
- Data Processing (Information Technology) Equipment, 645.5
- Electric Space-Heating Equipment, 424.3(B)
- Motors, 430.22
- Signs, 600.5

210.3 Branch-Circuit Rating. The rating of a branch circuit is determined by the rating of the branch-circuit overcurrent device, not the conductor size.

Author's Comment: For example, the branch-circuit ampere rating of 10 THHN conductors on a 20A circuit breaker is 20A. Figure 210–2

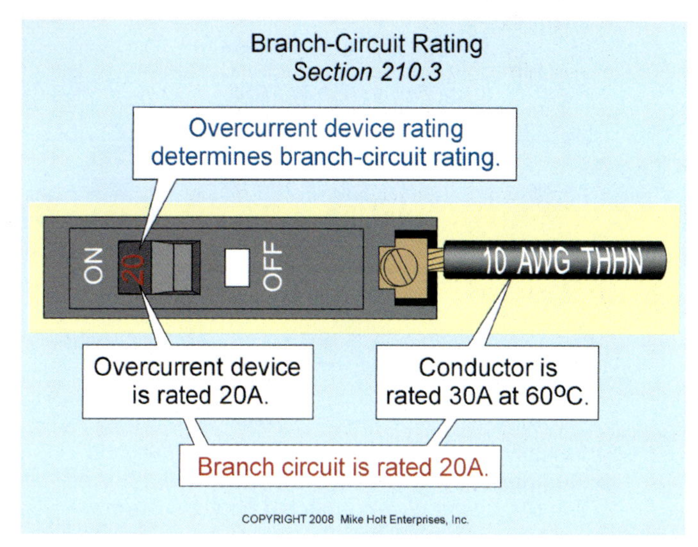

Figure 210–2

210.4 Multiwire Branch Circuits.

(A) General. A multiwire branch circuit can be considered a single circuit or a multiple circuit.

Author's Comments:

- See the definition of "Multiwire Branch Circuit" in Article 100.
- Two small-appliance circuits are required for receptacles that serve countertops in dwelling unit kitchens [210.11(C)(1) and 210.52(B)]. One 3-wire, single-phase, 120/240V multiwire branch circuit can be used for this purpose.

To prevent inductive heating and to reduce conductor impedance for fault currents, all conductors of a multiwire branch circuit must originate from the same panelboard.

Author's Comment: For more information on the inductive heating of metal parts, see 300.3(B), 300.5(I), and 300.20.

FPN: Unwanted and potentially hazardous harmonic neutral currents can cause additional heating of the neutral conductor of a 4-wire, three-phase, 120/208V or 277/480V wye-connected system, which supplies nonlinear loads.

Author's Comment: To prevent fire or equipment damage from excessive harmonic neutral currents, the designer should consider: (1) increasing the size of the neutral conductor, or (2) installing a separate neutral for each phase. Also see 220.61(C)(2) FPN No. 2, and 310.15(B)(4)(c) in this textbook. Figure 210–3

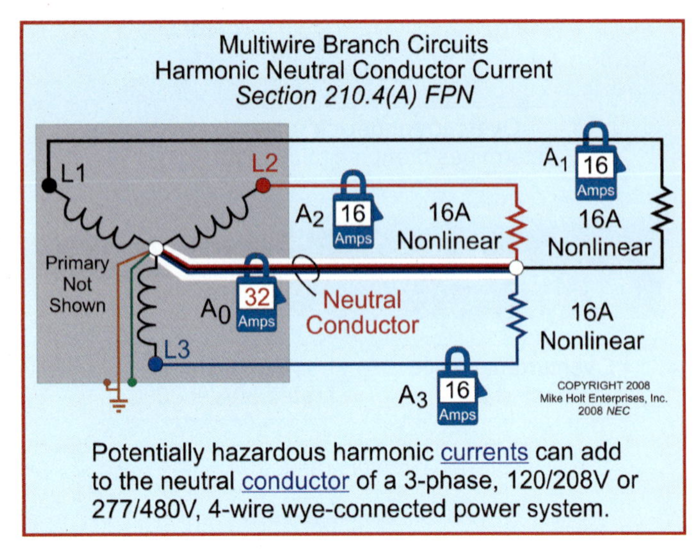

Figure 210–3

Author's Comments:

- See the definition of "Nonlinear Load" in Article 100.
- For more information, please visit www.MikeHolt.com. Click on "Technical Information" on the left side of the page, then select "Power Quality."

(B) Disconnecting Means. Each multiwire branch circuit must have a means to simultaneously disconnect all ungrounded conductors at the point where the branch circuit originates. Figure 210–4

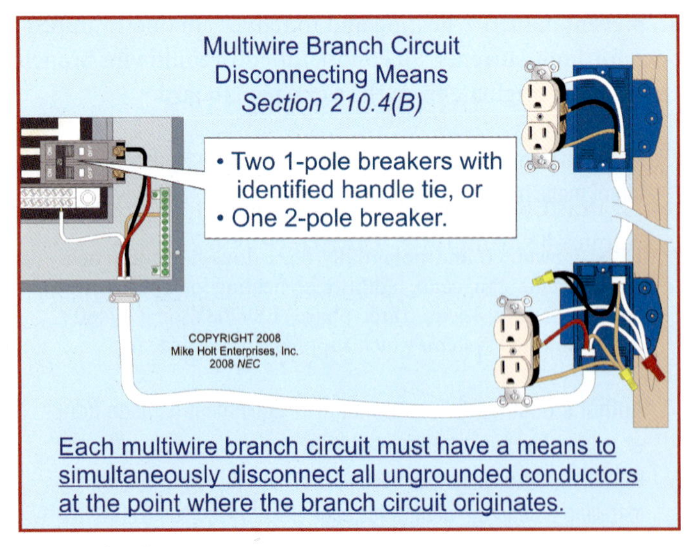

Figure 210–4

Author's Comment: Individual single-pole circuit breakers with handle ties identified for the purpose, or a breaker with common internal trip, can be used for this application [240.15(B)(1)].

CAUTION: *This rule is intended to prevent people from working on energized circuits they thought were disconnected.*

(C) Line-to-Neutral Loads. Multiwire branch circuits must supply only line-to-neutral loads.

Exception No. 1: A multiwire branch circuit is permitted to supply an individual piece of line-to-line utilization equipment, such as a range or dryer.

Exception No. 2: A multiwire branch circuit is permitted to supply both line-to-line and line-to-neutral loads if the circuit is protected by a device (multipole circuit breaker) that opens all ungrounded conductors of the multiwire branch circuit simultaneously (common internal trip) under a fault condition. Figure 210–5

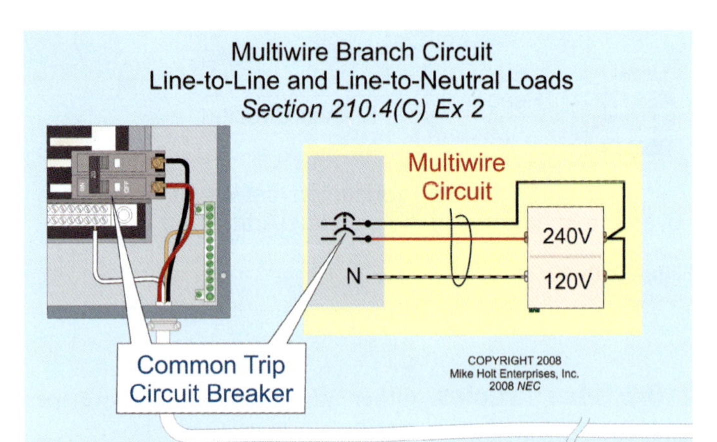

Figure 210–5

FPN: See 300.13(B) for the requirements relating to the continuity of the neutral conductor on multiwire branch circuits.

CAUTION: *If the continuity of the neutral conductor of a multiwire circuit is interrupted (open), the resultant over- or undervoltage can cause a fire and/or destruction of electrical equipment. For details on how this occurs, see 300.13(B) in this textbook.* Figure 210–6

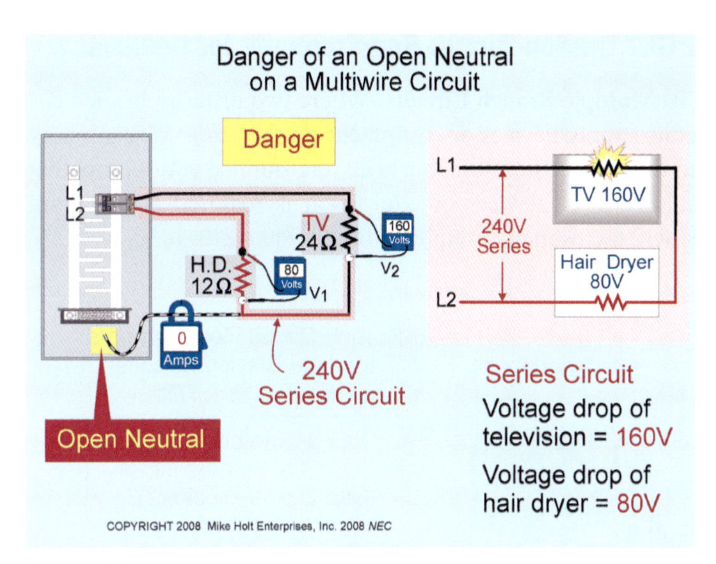

Figure 210–6

(D) Grouping. The ungrounded and neutral conductors of a multiwire branch circuit must be grouped together in at least one location by wire ties or similar means at the point of origination. Figure 210–7

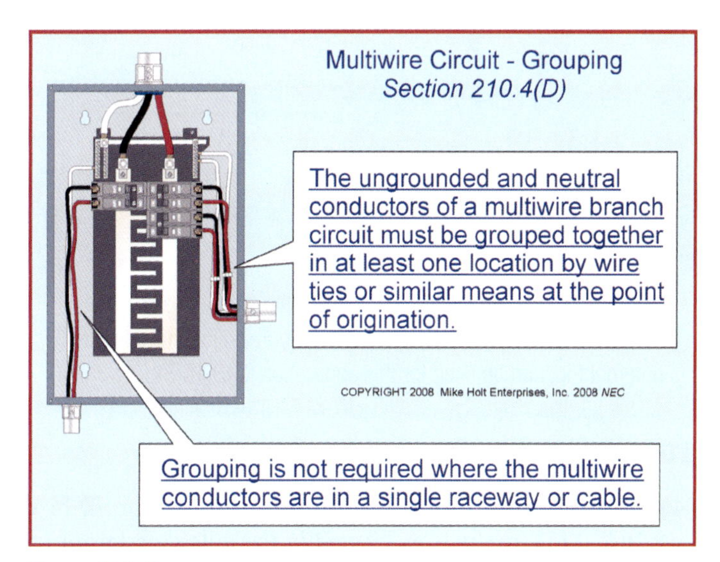

Figure 210–7

Exception: Grouping is not required where the circuit conductors are contained in a single raceway or cable that makes the grouping obvious.

Author's Comment: Grouping all associated conductors of a multiwire branch circuit together by wire ties or other means within the panel or origination point of the circuit makes it easier to visually identify the conductors of the multiwire branch circuit. The grouping will assist in connecting multiwire branch circuit conductors to circuit breakers correctly, particularly where twin breakers are used. If proper diligence is not exercised when making these connections, two circuit conductors could be accidentally connected to the same phase conductor.

CAUTION: *If the ungrounded conductors of a multiwire circuit are not terminated to different phases or lines, the currents on the neutral conductor will not cancel, but will add, which can cause an overload on the neutral conductor.* Figure 210–8

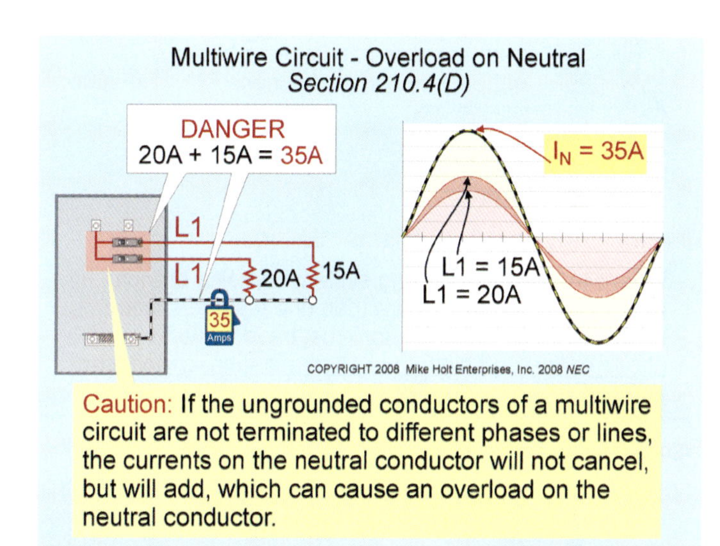

Figure 210–8

210.5 Identification for Branch Circuits.

(A) Neutral Conductor. The neutral conductor of a branch circuit must be identified in accordance with 200.6.

(B) Equipment Grounding Conductor. Equipment grounding conductors can be bare, covered, or insulated. Insulated equipment grounding conductors size 6 AWG and smaller must have a continuous outer finish either green or green with one or more yellow stripes [250.119].

On equipment grounding conductors larger than 6 AWG, insulation can be permanently reidentified with green marking at the time of installation at every point where the conductor is accessible [250.119(A)].

(C) Ungrounded Conductors—More Than One Voltage System. Where the premises wiring system contains branch circuits supplied from more than one voltage system, each ungrounded conductor, at all termination, connection, and splice points, must be identified by phase and system. Identification can be by color coding, marking tape, tagging, or other means approved by the authority having jurisdiction. The method of identification must be documented in a manner readily available or permanently posted at each branch-circuit panelboard. Figure 210–9

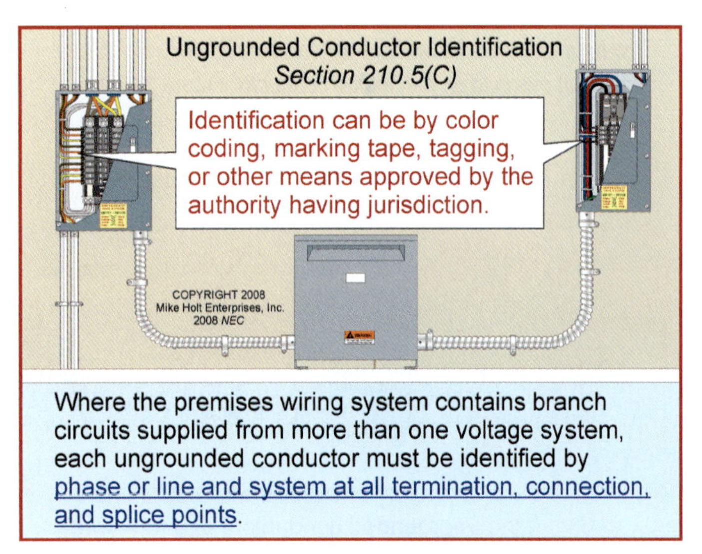

Figure 210–9

Author's Comment: Although the *NEC* doesn't require a specific color code for ungrounded conductors, electricians often use the following color system for power and lighting conductor identification:

- 120/240V, single-phase—black, red, and white
- 120/208V, three-phase—black, red, blue, and white
- 120/240V, three-phase—black, orange, blue, and white
- 277/480V, three-phase—brown, orange, yellow, and gray; or, brown, purple, yellow, and gray
- Conductors with insulation that is green or green with one or more yellow stripes can't be used for an ungrounded or neutral conductor [250.119].

210.7 Branch-Circuit Requirements for Devices.

(B) Multiple Branch Circuits. Where two or more branch circuits supply devices or equipment on the same yoke, a means to disconnect simultaneously all ungrounded conductors that supply those devices or equipment is required at the point where the branch circuit originates. Figure 210–10

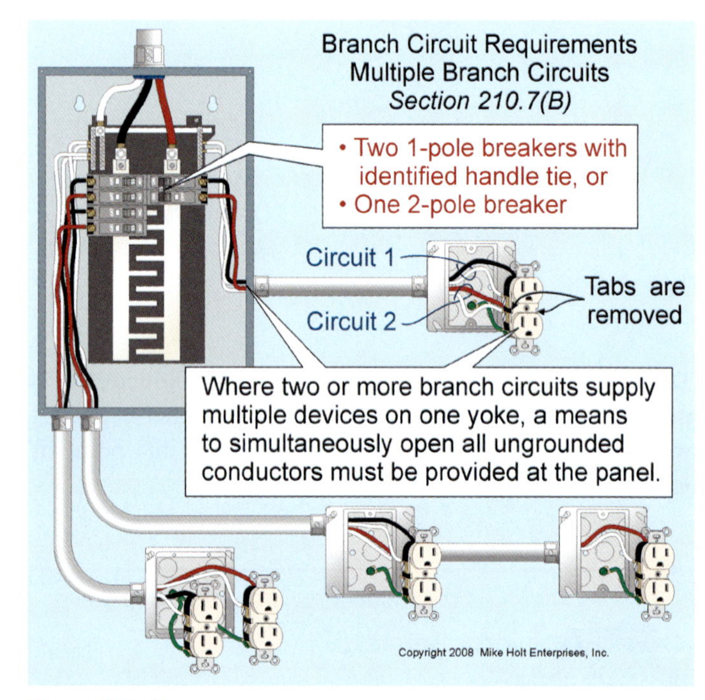

Figure 210–10

Author's Comment: Individual single-pole circuit breakers with handle ties identified for the purpose, or a breaker with a common internal trip, can be used for this application [240.15(B)(1)].

210.8 GFCI Protection.

(A) Dwelling Units. GFCI protection is required for all 15A and 20A, 125V receptacles located in the following locations of a dwelling unit:

Author's Comment: See the definitions of GFCI and Dwelling Unit in Article 100.

(1) Bathroom Area. GFCI protection is required for all 15A and 20A, 125V receptacles in the bathroom area of a dwelling unit. Figure 210–11

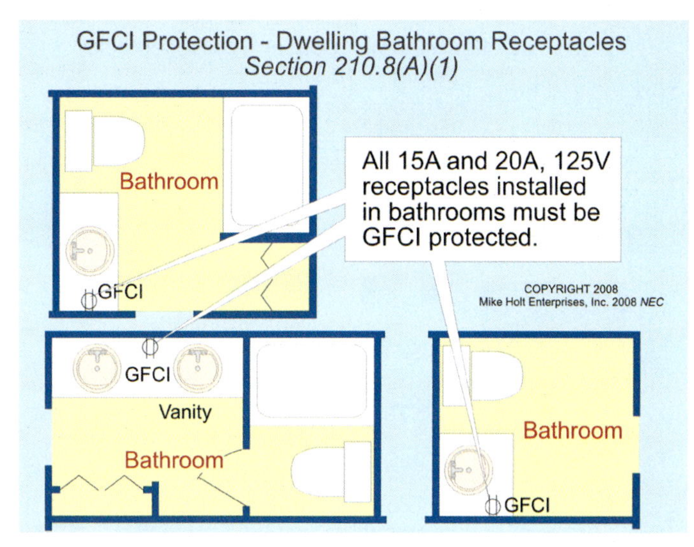

Figure 210–11

Author's Comments:

- See the definition of "Bathroom" in Article 100.
- In the continued interests of safety, proposals to allow receptacles for dedicated equipment in the bathroom area to be exempted from the GFCI protection requirements have been rejected.

(2) Garages and Accessory Buildings. GFCI protection is required for all 15A and 20A, 125V receptacles in garages, and in grade-level portions of accessory buildings used for storage or work areas of a dwelling unit. Figure 210–12

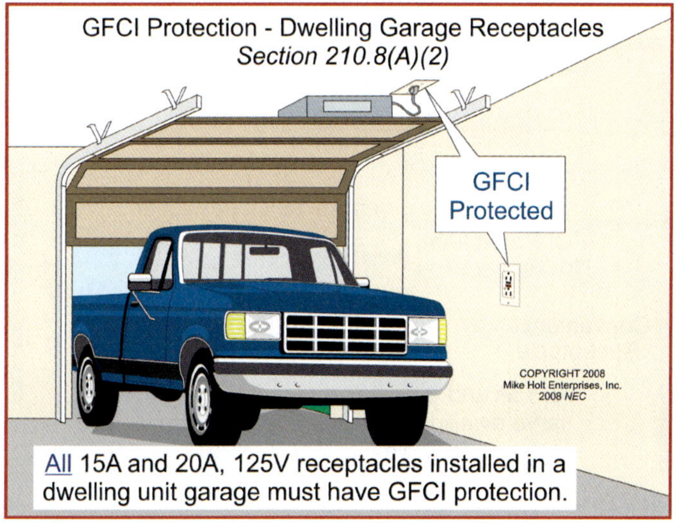

Figure 210–12

Author's Comments:

- See the definition of "Garage" in Article 100.
- A receptacle outlet is required in a dwelling unit attached garage [210.52(G)], but a receptacle outlet isn't required in an accessory building or a detached garage without power. If a 15A or 20A, 125V receptacle is installed in an accessory building, it must be GFCI protected. Figure 210–13

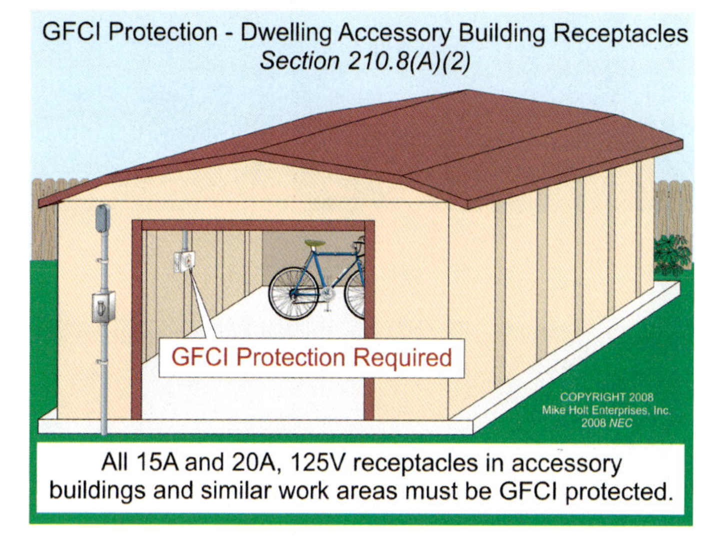

Figure 210–13

(3) Outdoors. All 15A and 20A, 125V receptacles located outdoors of dwelling units, including receptacles installed under the eaves of roofs, must be GFCI protected. Figure 210–14

Author's Comments:

- Each dwelling unit of a multifamily dwelling that has an individual entrance at grade level must have at least one GFCI-protected receptacle outlet accessible from grade level located not more than 6½ ft above grade [210.52(E)(2)].
- Balconies, decks, and porches over 20 sq ft that are attached to the dwelling unit and are accessible from inside the dwelling must have at least one GFCI-protected receptacle outlet accessible from the balcony, deck, or porch [210.52(E)(3)].

Exception: GFCI protection isn't required for a fixed electric snow-melting or deicing equipment receptacle supplied by a dedicated branch circuit, if the receptacle isn't readily accessible and the equipment or receptacle has ground-fault protection of equipment (GFPE) [426.28]. Figure 210–15

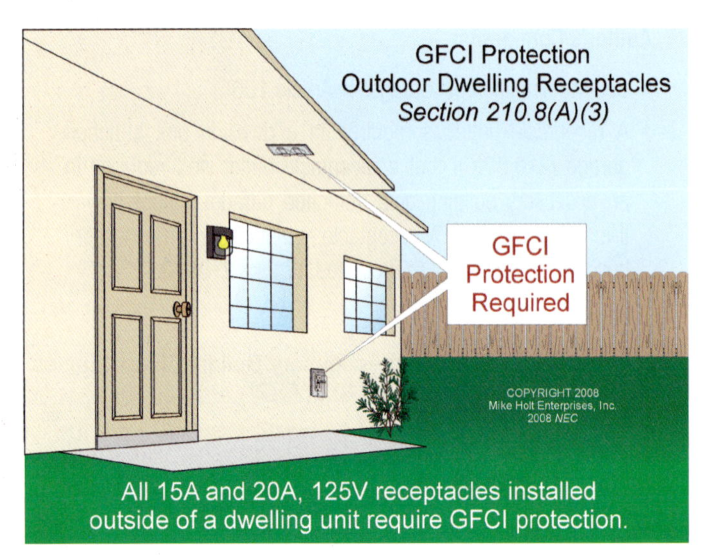

Figure 210–14

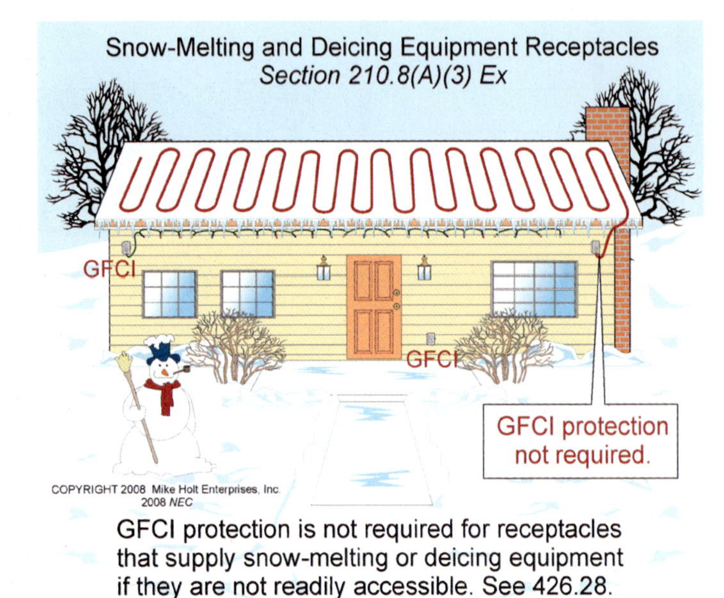

Figure 210–15

(4) Crawl Spaces. All 15A and 20A, 125V receptacles installed in crawl spaces at or below grade of a dwelling unit must be GFCI protected.

> **Author's Comment:** The *Code* doesn't require a receptacle to be installed in a crawl space, except when heating, air-conditioning, and refrigeration equipment is installed there [210.63].

(5) Unfinished Basements. GFCI protection is required for all 15A and 20A, 125V receptacles located in the unfinished portion of a basement not intended as a habitable room and limited to storage and work areas. Figure 210–16

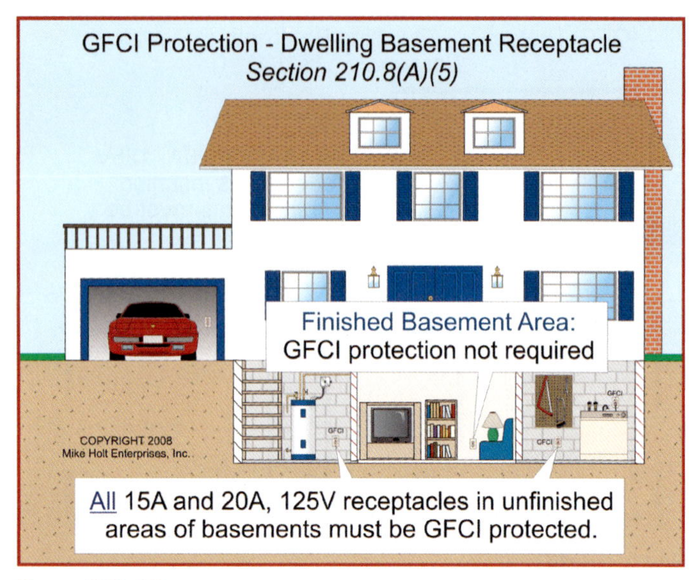

Figure 210–16

Exception: A receptacle supplying only a permanently installed fire alarm or burglar alarm system is not required to be GFCI protected [760.41(B) and 760.121(B)].

> **Author's Comment:** A receptacle outlet is required in each unfinished portion of a dwelling unit basement [210.52(G)].

(6) Kitchen Countertop Surfaces. GFCI protection is required for all 15A and 20A, 125V receptacles that serve countertop surfaces in a dwelling unit. Figure 210–17

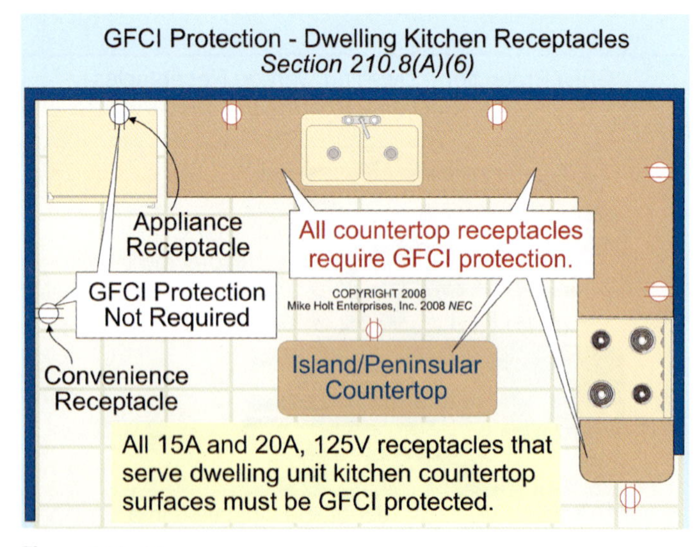

Figure 210–17

Author's Comments:

- GFCI protection is required for all receptacles that serve countertop surfaces. But GFCI protection isn't required for receptacles that serve built-in appliances, such as dishwashers or kitchen waste disposals.
- See 210.52(C) for the location requirements of countertop receptacles.

(7) Laundry, Utility, and Wet Bar Sinks. GFCI protection is required for all 15A and 20A, 125V receptacles located within an arc measurement of 6 ft from the sink. Figures 210–18 and 210–19

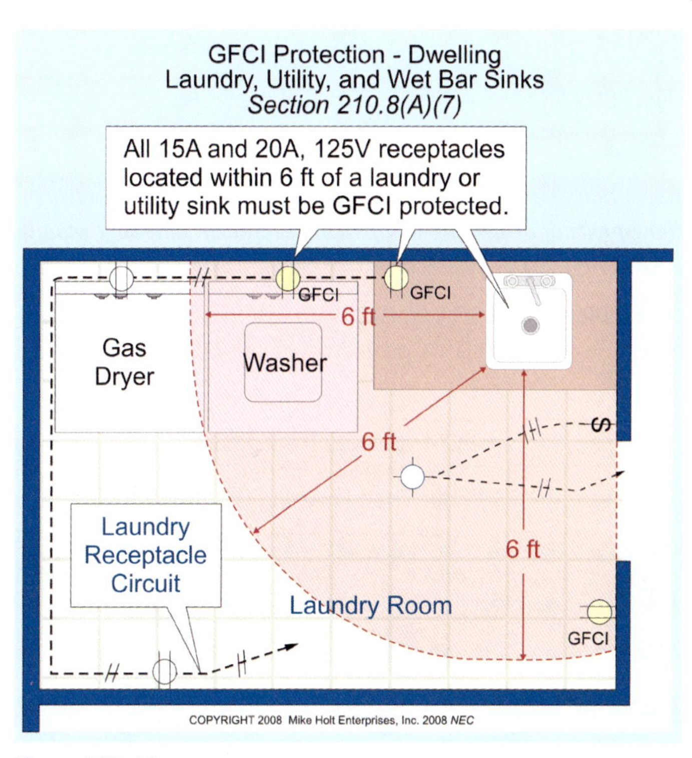

Figure 210–18

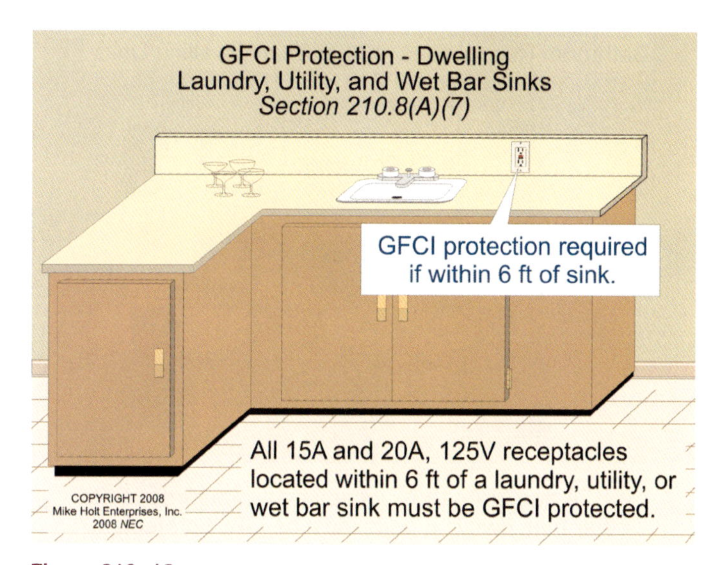

Figure 210–19

Figure 210–20

(8) Boathouses. GFCI protection is required for all 15A and 20A, 125V receptacles located in a dwelling unit boathouse. Figure 210–20

Author's Comment: The *Code* doesn't require a 15A or 20A, 125V receptacle to be installed in a boathouse, but if one is installed, it must be GFCI protected.

(B) Other than Dwelling Units. GFCI protection is required for all 15A and 20A, 125V receptacles installed in the following commercial/industrial locations:

(1) Bathrooms. All 15A and 20A, 125V receptacles installed in commercial or industrial bathrooms must be GFCI protected. Figure 210–21

Author's Comments:

- See the definition of a "Bathroom" in Article 100.
- A 15A or 20A, 125V receptacle isn't required in a commercial or industrial bathroom, but if one is installed, it must be GFCI protected.

(2) Kitchens. All 15A and 20A, 125V receptacles installed in an area with a sink and permanent facilities for food preparation and cooking [Article 100], even those that don't supply the countertop surface, must be GFCI protected. Figure 210–22

Figure 210–21

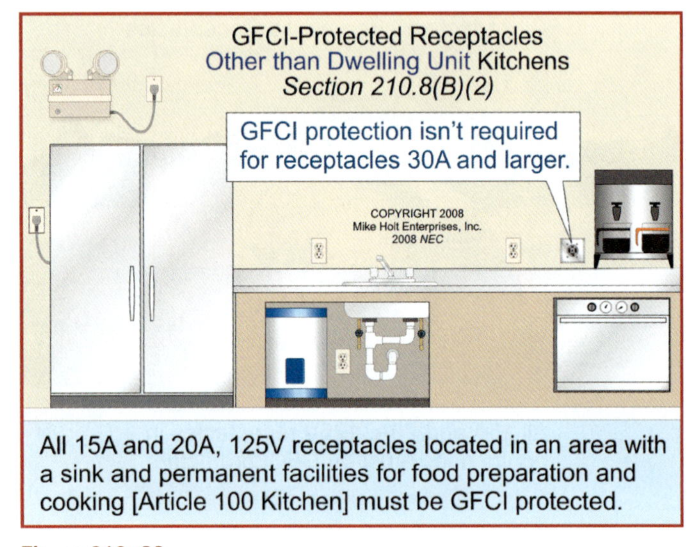

Figure 210–22

Author's Comments:

- GFCI protection is not required for receptacles rated other than 15A and 20A, 125V in these locations.
- GFCI protection is not required for hard-wired equipment in these locations.

(3) Rooftops. All 15A and 20A, 125V receptacles installed on rooftops must be GFCI protected. Figure 210–23

Author's Comment: A 15A or 20A, 125V receptacle outlet must be installed within 25 ft of heating, air-conditioning, and refrigeration equipment [210.63].

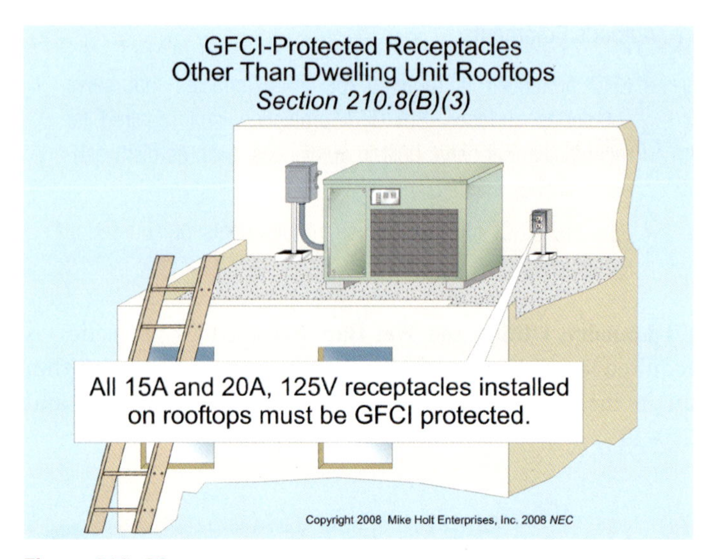

Figure 210–23

Exception: GFCI protection isn't required for a fixed electric snow-melting or deicing equipment receptacle that isn't readily accessible [426.28].

(4) Outdoors. All 15A and 20A, 125V receptacles installed outdoors must be GFCI protected. Figure 210–24

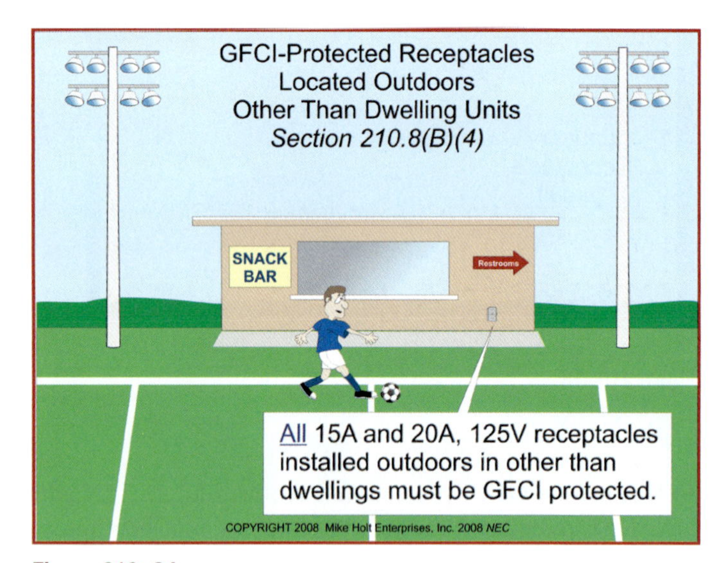

Figure 210–24

Exception No. 1 to (3) and (4): GFCI protection isn't required for a fixed electric snow-melting or deicing equipment receptacle supplied by a dedicated branch circuit, if the receptacle isn't readily accessible and the equipment or receptacle has ground-fault protection of equipment (GFPE) [426.28].

(5) Sinks. All 15A and 20A, 125V receptacles installed within 6 ft of the outside edge of a sink must be GFCI protected. Figure 210–25

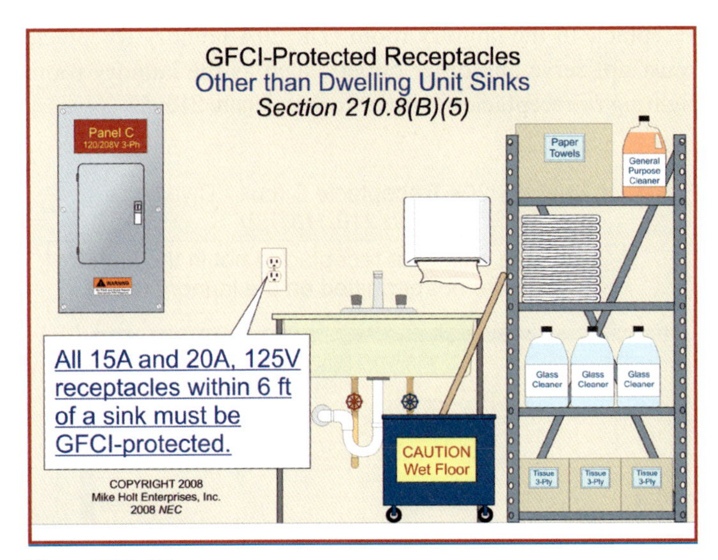

Figure 210–25

(C) Boat Hoists. GFCI protection is required for outlets not exceeding 240 volts that supply boat hoists in dwelling unit locations. Figure 210–26

Figure 210–26

Author's Comments:

- See the definition of "Outlet" in Article 100.
- This ensures GFCI protection regardless of whether the boat hoist is cord-and-plug-connected or hard-wired.

210.11 Branch Circuits Required.

(A) Number of Branch Circuits. The minimum number of general lighting and general-use receptacle branch circuits must be determined by dividing the total calculated load in amperes by the ampere rating of the circuits used. See Example D1(a) in Annex D.

Question: *How many 15A, 120V circuits are required for the general lighting and general-use receptacles of a 2,100 sq ft dwelling unit?* Figure 210–27

(a) 1 *(b) 2* *(c) 3* *(d) 4*

Answer: *(d) 4*

Step 1: Determine the total VA load:

VA = 2,100 sq ft x 3 VA per sq ft [Table 220.12]
VA = 6,300 VA

Step 2: Determine the amperes:

I = VA/E
I = 6,300VA/120V
I = 52.50A

Step 3: Determine the number of circuits:

Number of Circuits = 52.50A/15A
Number of Circuits = 4

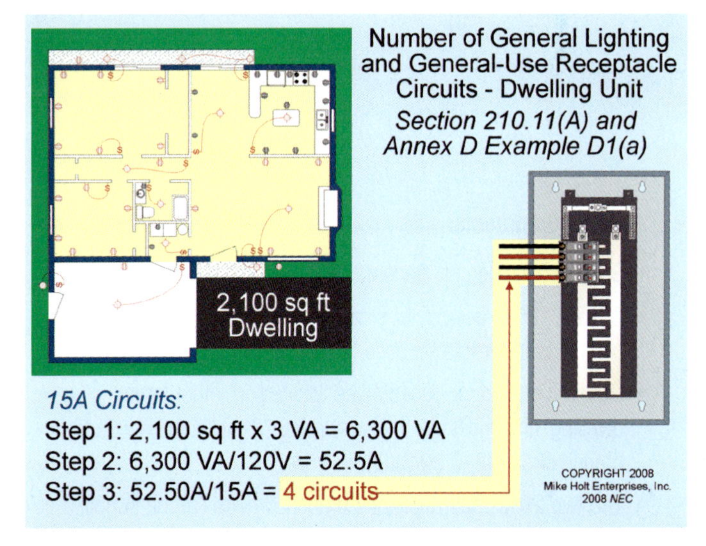

Figure 210–27

(B) Load Evenly Proportioned Among Branch Circuits. Where the load is calculated on the volt-amperes per square foot, the wiring system must be provided to serve the calculated load, with the loads evenly proportioned among multioutlet branch circuits within the panelboard.

(C) Dwelling Unit.

(1) Small-Appliance Branch Circuits. Two or more 20A, 120V small-appliance receptacle branch circuits are required for the 15A or 20A receptacle outlets in a dwelling unit kitchen, dining room, breakfast room, pantry, or in similar dining areas as required by 210.52(B). The 20A small-appliance receptacle circuits must supply no other outlets [210.52(B)(2)]. **Figure 210–28**

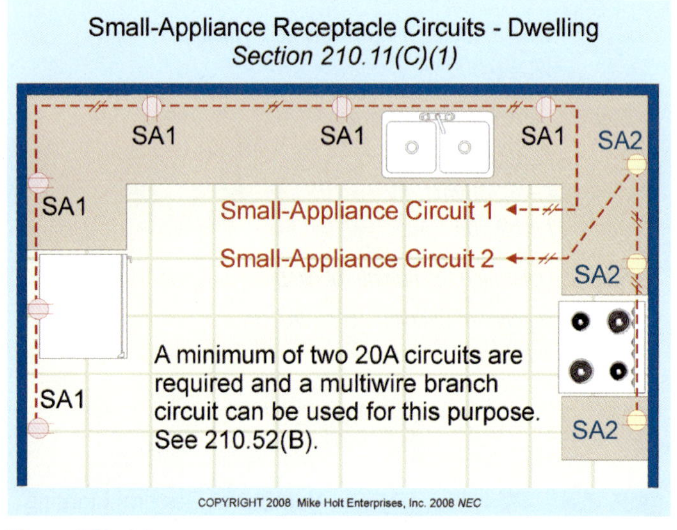

Figure 210–28

Author's Comments:

- See the definition of "Receptacle Outlet" in Article 100.
- A 15A, 125V receptacle is rated for 20A feed-through, so it can be used for this purpose [210.21(B)(3)].
- Lighting outlets or receptacles located in other areas of a dwelling unit must not be connected to the small-appliance branch circuit [210.52(B)(2)].
- The two 20A small-appliance branch circuits can be supplied by one 3-wire multiwire circuit or by two separate 120V circuits [210.4(A)].
- Each separate countertop must be supplied with two small-appliance circuits [210.52(B)(3)].

(2) Laundry Branch Circuit. One 20A, 120V branch circuit must be provided for the receptacle outlets required by 210.52(F) for a dwelling unit laundry room. The 20A laundry room receptacle circuit is permitted to supply more than one receptacle in the laundry room. The 20A laundry receptacle must not serve any other outlets, such as the laundry room lighting or receptacles in other rooms. **Figure 210–29**

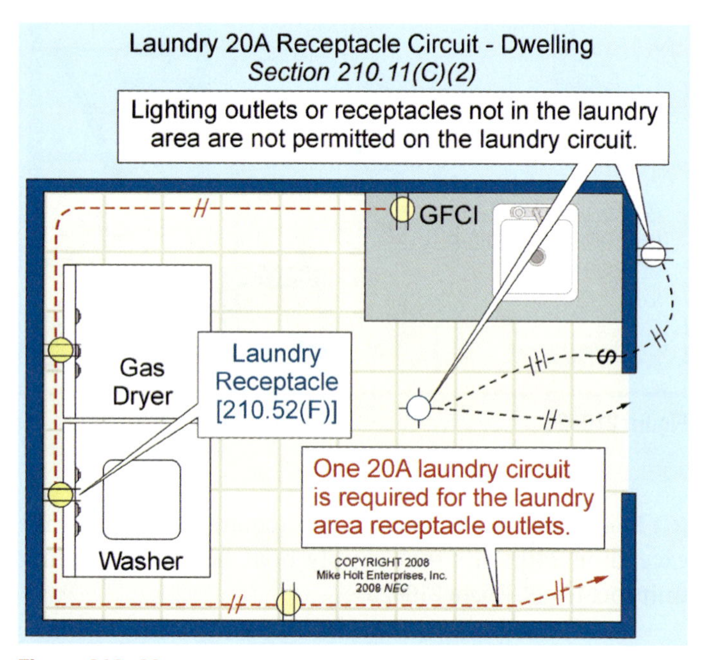

Figure 210–29

Author's Comments:

- The 20A, 120V laundry branch circuit is required, even if the laundry appliance installed is a 30A, 230V combination washer/dryer. **Figure 210–30**
- A 15A receptacle is rated for 20A feed-through, so it can be used for this purpose [210.21(B)(3)].
- GFCI protection isn't required for 15A and 20A, 125V receptacles located in a laundry room, unless they are within 6 ft of a sink [210.8(A)(7)]. **Figure 210–31**

(3) Bathroom Branch Circuit. One 20A, 120V branch circuit must be provided for the receptacle outlets required by 210.52(D) for a dwelling unit bathroom. This 20A bathroom receptacle circuit must not serve any other outlet, such as bathroom lighting outlets or receptacles in other rooms. **Figure 210–32**

Author's Comment: A 15A, 125V receptacle is rated for 20A feed-through, so it can be used for this purpose [210.21(B)(3)].

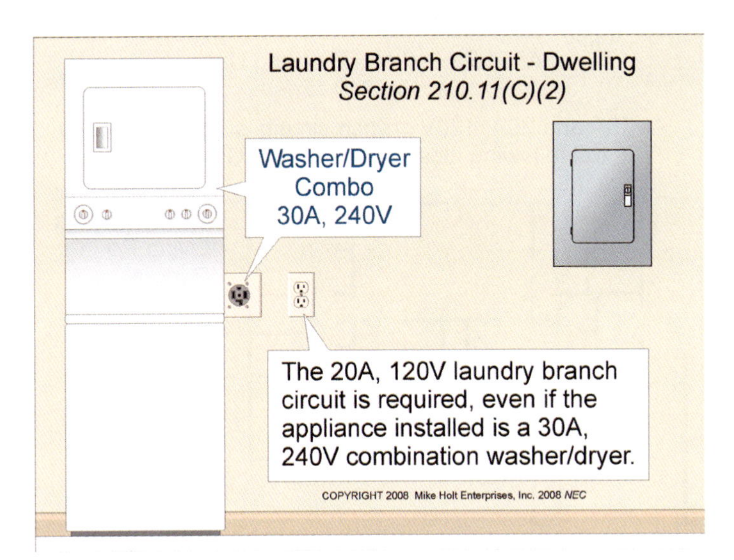

Figure 210–30

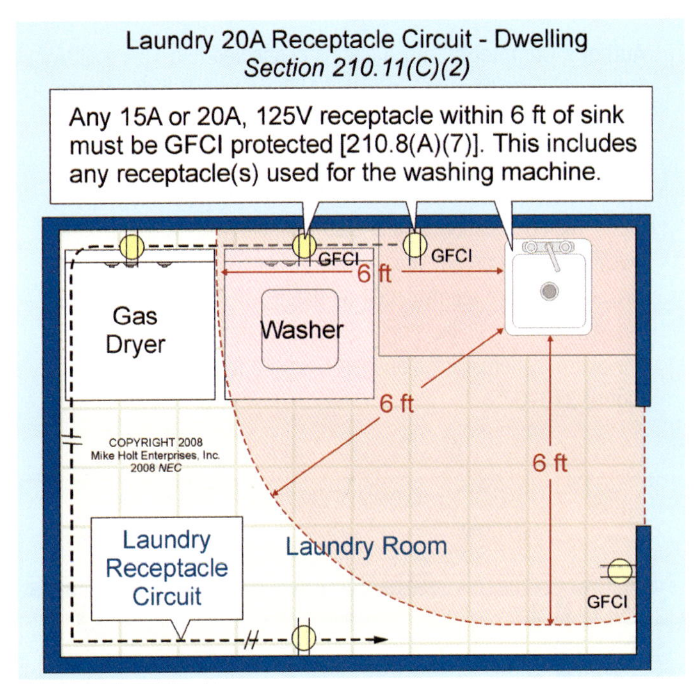

Figure 210–31

Exception: A single 20A, 120V branch circuit is permitted to supply all of the outlets in a single bathroom, as long as no single load fastened in place is rated more than 10A [210.23(A)]. Figure 210–33

Question: *Can a luminaire, ceiling fan, or bath fan be connected to the 20A, 120V branch circuit that supplies one bathroom?*

Answer: *Yes.*

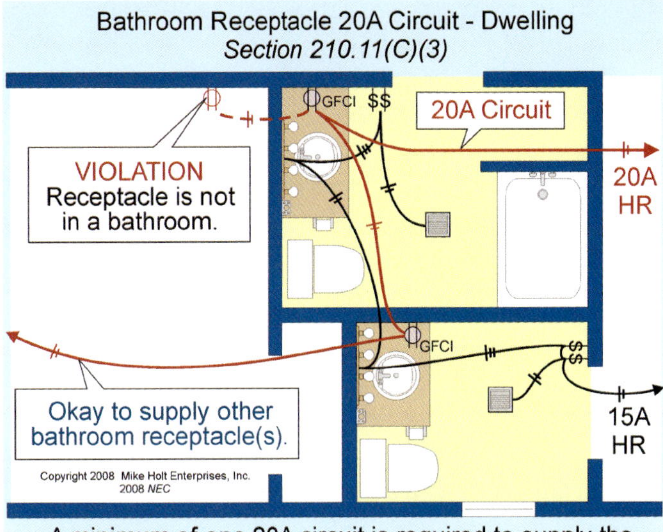

Figure 210–32

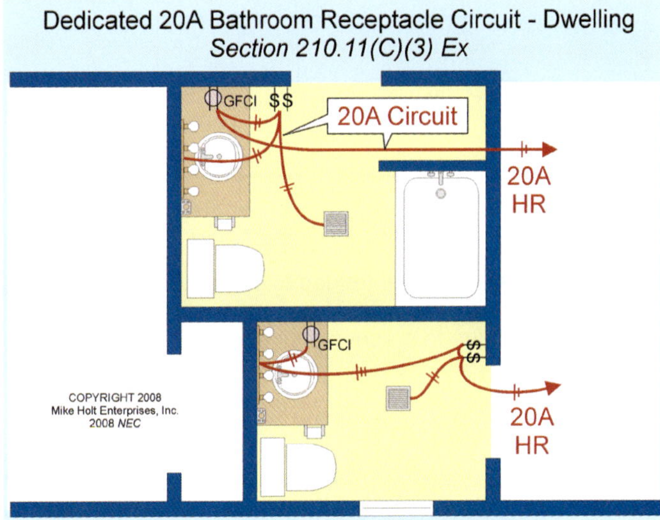

Figure 210–33

210.12 Arc-Fault Circuit-Interrupter—Protected Circuits.

(A) AFCI Definition. An arc-fault circuit interrupter is a device intended to de-energize the circuit when it detects the current waveform characteristics unique to an arcing fault. Figures 210–34 and 210–35

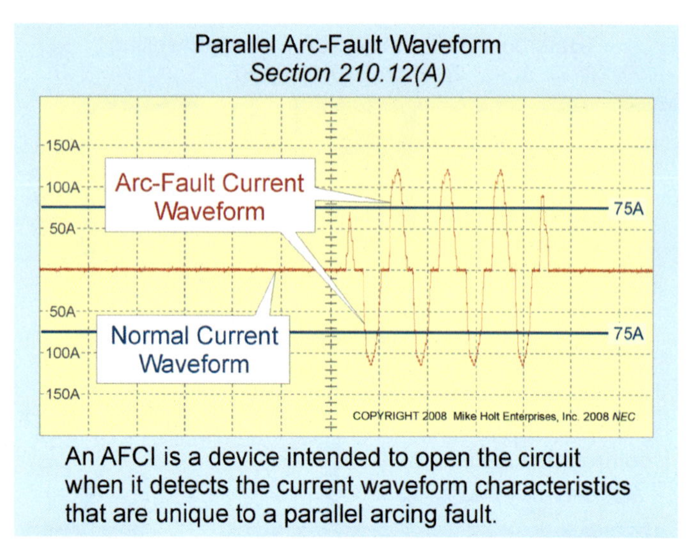

Figure 210–34

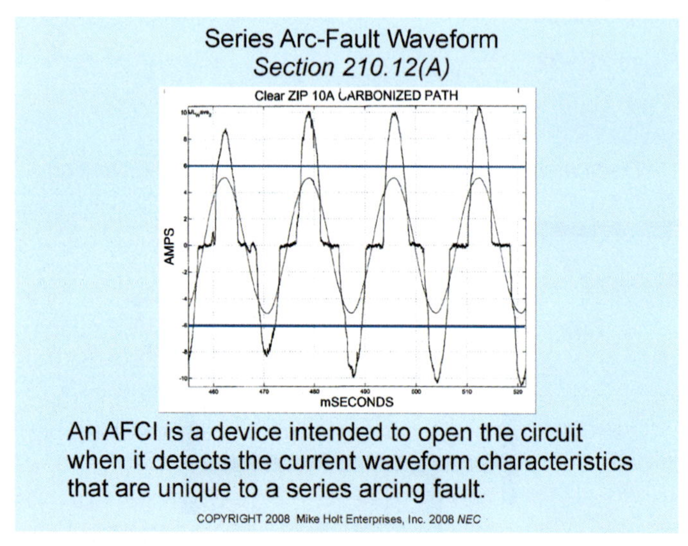

Figure 210–35

(B) Dwelling Unit Circuits. All 15A or 20A, 120V branch circuits in dwelling units supplying outlets in family rooms, dining rooms, living rooms, parlors, libraries, dens, bedrooms, sunrooms, recreation rooms, closets, hallways, or similar rooms or areas must be protected by a listed AFCI device of the combination type. Figure 210–36

> **Author's Comment:** The 120V circuit limitation means AFCI protection isn't required for equipment rated 230V, such as a baseboard heater or room air conditioner. For more information, visit www.MikeHolt.com, click on the "Search" link, and search for "AFCI."

> **FPN No. 3:** See 760.41(B) and 760.121(B) for power-supply requirements for fire alarm systems.

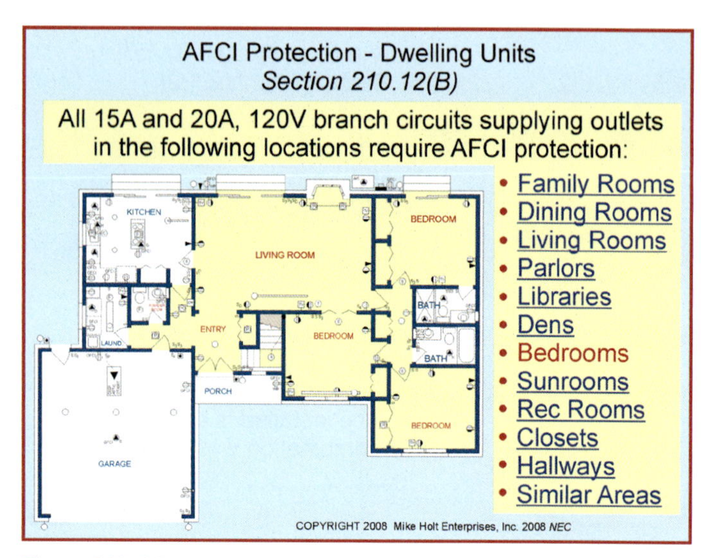

Figure 210–36

> **Author's Comment:** Smoke alarms connected to a 15A or 20A circuit of a dwelling unit must be AFCI protected if the smoke alarm is located in one of the areas specified In 210.12(B). The exemption from AFCI protection for the "fire alarm circuit" contained in 760.41(B) and 760.121(B) doesn't apply to the single-or multiple-station smoke alarm circuit typically installed in dwelling unit bedroom areas. This is becausc a smoke alarm circuit isn't a fire alarm circuit as defined in NFPA 72, *National Fire Alarm Code.* Unlike single-or multiple-station smoke alarms, fire alarm systems are managed by a fire alarm control panel. Figure 210–37

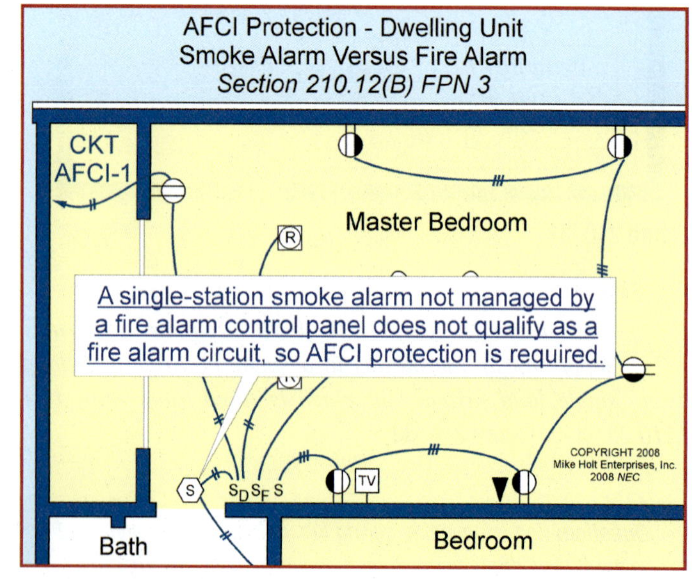

Figure 210–37

Exception No. 1: The AFCI protection can be located at the first outlet if the circuit conductors are installed in RMC, IMC, EMT or steel armored Type AC cable, and the AFCI device is contained in a metal outlet or junction box.

Exception No. 2: AFCI protection can be omitted for branch-circuit wiring to a fire alarm system in accordance with 760.41(B) and 760.121(B), if the circuit conductors are installed in RMC, IMC, EMT, or steel armored Type AC cable.

210.18 Guest Rooms and Guest Suites. Guest rooms and guest suites provided with permanent provisions for cooking must have branch circuits installed in accordance with the dwelling unit requirements of 210.11.

Author's Comment: See the definitions of "Guest Room" and "Guest Suite" in Article 100.

PART II. BRANCH-CIRCUIT RATINGS

210.19 Conductor Sizing.

(A) Branch Circuits.

(1) Continuous and Noncontinuous Loads. Conductors must be sized no less than 125 percent of the continuous loads, plus 100 percent of the noncontinuous loads, based on the terminal temperature rating ampacities as listed in Table 310.16, before any ampacity adjustment [110.14(C)].

Exception No. 1: Where the assembly and the overcurrent device are both listed for operation at 100 percent of its rating, the conductors can be sized at 100 percent of the continuous load.

Author's Comment: Equipment suitable for 100 percent continuous loading is rarely available in ratings under 400A.

Exception No. 2: Neutral conductors can be sized at 100 percent of the continuous and noncontinuous load.

Author's Comments:

- See the definition of "Continuous Load" in Article 100.
- See 210.20 for the sizing requirements for the branch-circuit overcurrent device for continuous and noncontinuous loads.

Question: *What size branch-circuit conductors are required for the ungrounded conductors of a 44A continuous load, if the equipment terminals are rated 75°C?* **Figure 210–38**

(a) 10 AWG (b) 8 AWG (c) 6 AWG (d) 4 AWG

Answer: *(c) 6 AWG*

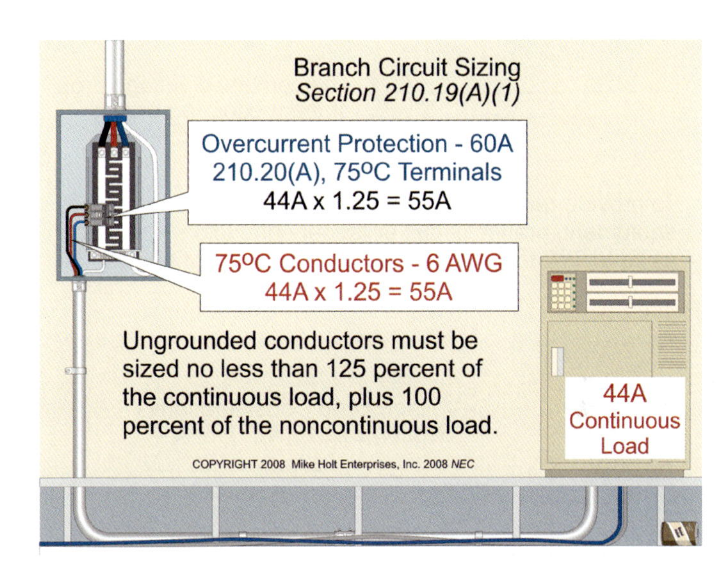

Figure 210–38

Since the load is 44A continuous, the ungrounded conductors must be sized to have an ampacity of not less than 55A (44A x 1.25). According to the 75°C column of Table 310.16, a 6 AWG conductor is suitable, because it has an ampere rating of 65A at 75°C before any conductor ampacity adjustment and/or correction.

FPN No. 4: To provide reasonable efficiency of operation of electrical equipment, branch-circuit conductors should be sized to prevent a voltage drop not to exceed 3 percent. In addition, the maximum total voltage drop on both feeders and branch circuits should not exceed 5 percent. **Figure 210–39**

Author's Comments:

- Many believe the *NEC* requires conductor voltage drop, as per Fine Print Note No. 4 to be applied when sizing conductors. Although this is often a good practice, it's not a *Code* requirement because FPNs are only advisory statements [90.5(C)]. **Figure 210–40**
- The *NEC* doesn't consider voltage drop to be a safety issue, except for sensitive electronic equipment [647.4(D)] and fire pumps [695.7].

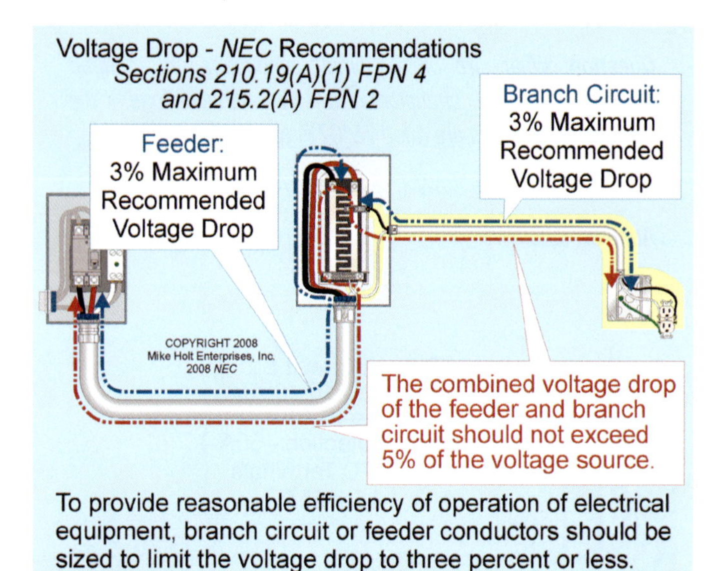

Figure 210–39

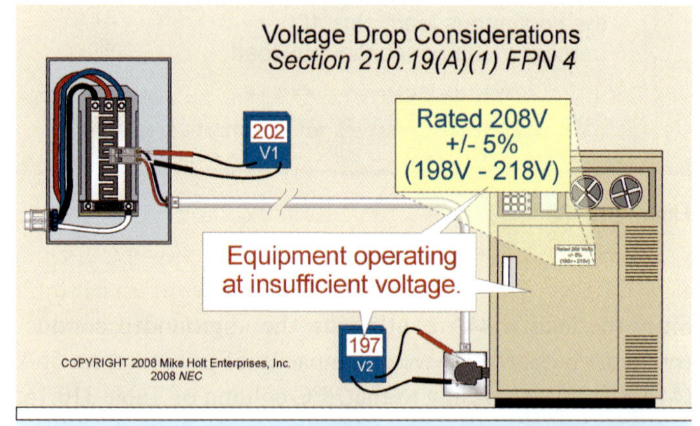

Figure 210–40

(2) Multioutlet Branch Circuits. Branch-circuit conductors that supply more than one receptacle for cord-and-plug-connected portable loads must have an ampacity not less than the rating of the circuit overcurrent device [210.3].

(3) Household Ranges and Cooking Appliances. Branch-circuit conductors that supply household ranges, wall-mounted ovens or counter-mounted cooking units must have an ampacity not less than the rating of the branch circuit, and not less than the maximum load to be served. For ranges of 8¾ kW or more rating, the minimum branch-circuit ampere rating is 40A.

Exception No. 1: Conductors tapped from a 50A branch circuit for electric ranges, wall-mounted electric ovens and counter-mounted electric cooking units must have an ampacity not less than 20A, and must have sufficient ampacity for the load to be served. The taps must not be longer than necessary for servicing the appliances.

(4) Other Loads. Branch-circuit conductors must have an ampacity sufficient for the loads served and must not be smaller than 14 AWG.

Exception No. 1: Tap conductors must have sufficient ampacity for the load to be served and have an ampacity not less than 15A for circuits rated less than 40A and not less than 20A for circuits rated at 40A or 50A for the following:

(b) Luminaires having tap conductors sized according to 410.117.

(c) Individual outlets, other than receptacle outlets, with taps not over 18 in. long.

Author's Comment: Branch-circuit tap conductors aren't permitted for receptacle outlets. Figure 210–41

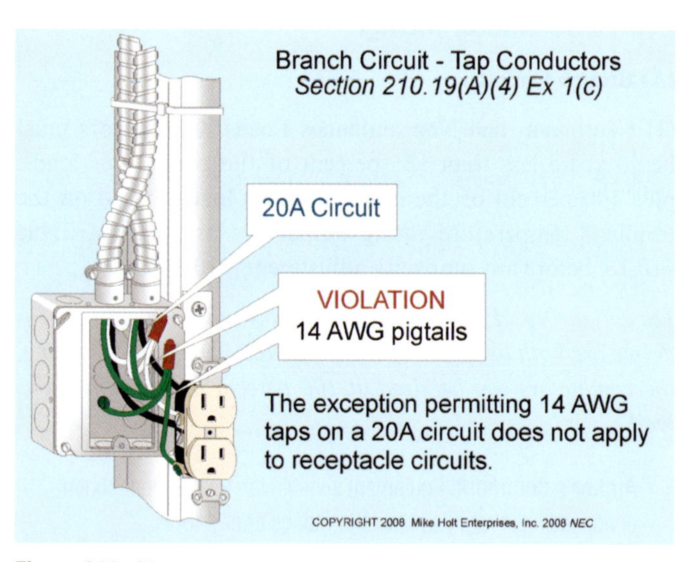

Figure 210–41

210.20 Overcurrent Protection.

(A) Continuous and Noncontinuous Loads. Branch-circuit overcurrent devices must have an ampacity of not less than 125 percent of the continuous loads, plus 100 percent of the noncontinuous loads. Figure 210–42

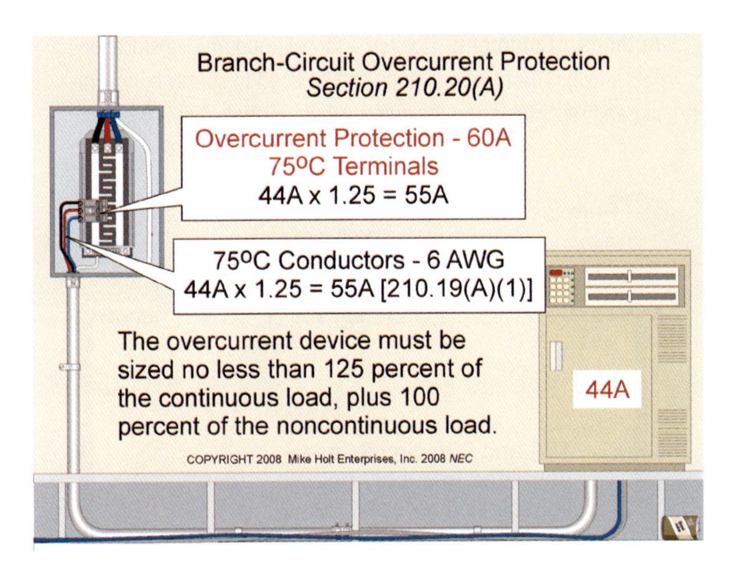

Figure 210–42

Author's Comment: See 210.19(A)(1) for branch-circuit conductor sizing requirements.

Exception: Where the assembly and the overcurrent devices are both listed for operation at 100 percent of their rating, the branch-circuit overcurrent device can be sized at 100 percent of the continuous load.

Author's Comment: Equipment suitable for 100 percent continuous loading is rarely available in ratings under 400A.

(B) Conductor Protection. Branch-circuit conductors must be protected against overcurrent in accordance with 240.4.

(C) Equipment Protection. Branch-circuit equipment must be protected in accordance with 240.3.

210.21 Outlet Device Rating.

(A) Lampholder Ratings. Lampholders connected to a branch circuit rated over 20A must be of the heavy-duty type.

Author's Comment: Fluorescent lampholders aren't rated heavy duty, so fluorescent luminaires must not be installed on circuits rated over 20A.

(B) Receptacle Ratings and Loadings.

(1) Single Receptacles. A single receptacle on an individual branch circuit must have an ampacity not less than the rating of the overcurrent device. Figure 210–43

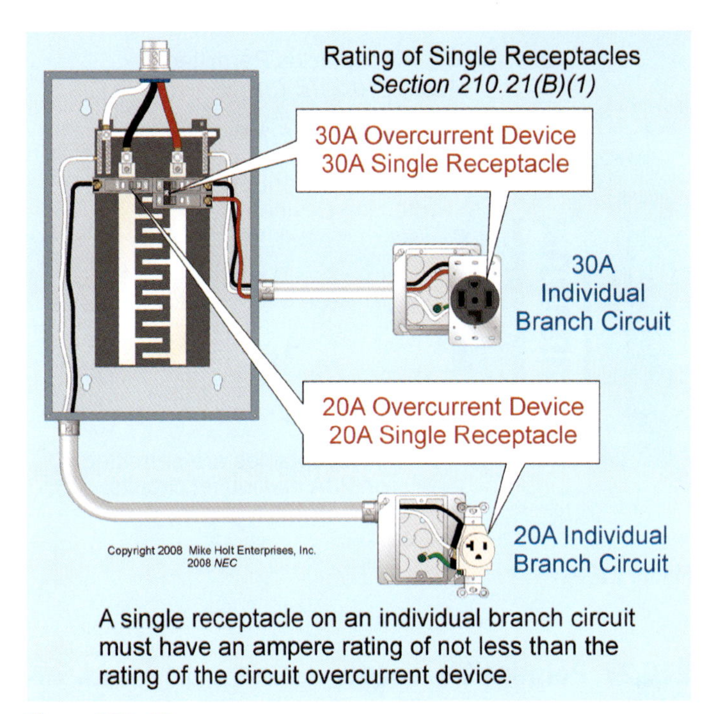

Figure 210–43

FPN: A single receptacle has only one contact device on its yoke [Article 100]. This means a duplex receptacle is considered two receptacles.

(2) Multiple Receptacle Loading. Where connected to a branch circuit that supplies two or more receptacles, the total cord-and-plug-connected load must not exceed 80 percent of the receptacle rating.

Author's Comment: A duplex receptacle has two contact devices on the same yoke [Article 100]. This means even one duplex receptacle on a circuit makes that circuit a multioutlet branch circuit.

(3) Multiple Receptacle Rating. Where connected to a branch circuit that supplies two or more receptacles, receptacles must have an ampere rating in accordance with the values listed in Table 210.21(B)(3). Figure 210–44

Table 210.21(B)(3) Receptacle Ratings

Circuit Rating	Receptacle Rating
15A	15A
20A	15A or 20A
30A	30A
40A	40A or 50A
50A	50A

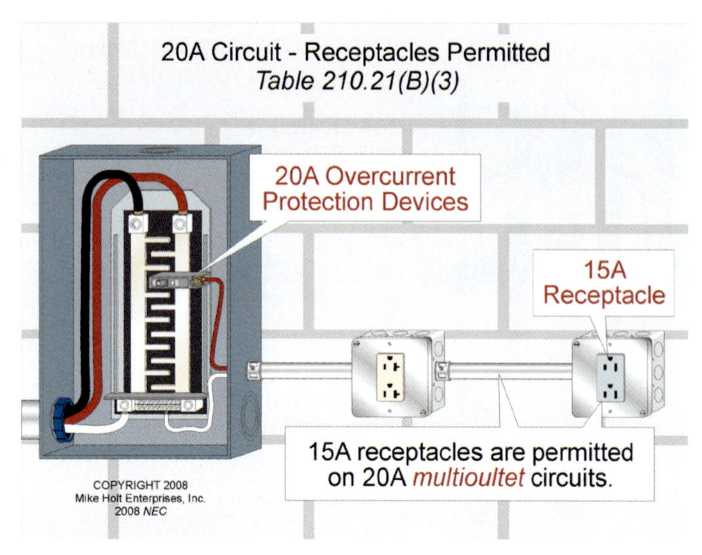

Figure 210–44

210.23 Permissible Loads. An individual branch circuit is permitted to supply any load for which it's rated, but in no case is the load permitted to exceed the branch-circuit ampere rating. Figure 210–45

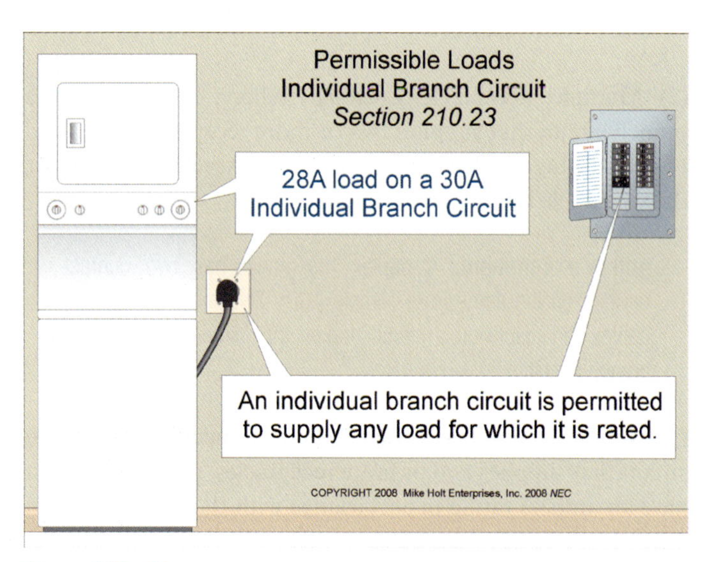

Figure 210–45

Branch circuits rated 15A or 20A supplying two or more outlets must only supply loads in accordance with 210.23(A).

(A) 15A and 20A Circuit. A 15A or 20A branch circuit is permitted to supply lighting, equipment, or any combination of both.

Author's Comment: Except for temporary installations [590.4(D)], 15A or 20A circuits can be used to supply both lighting and receptacles on the same circuit. Figure 210–46

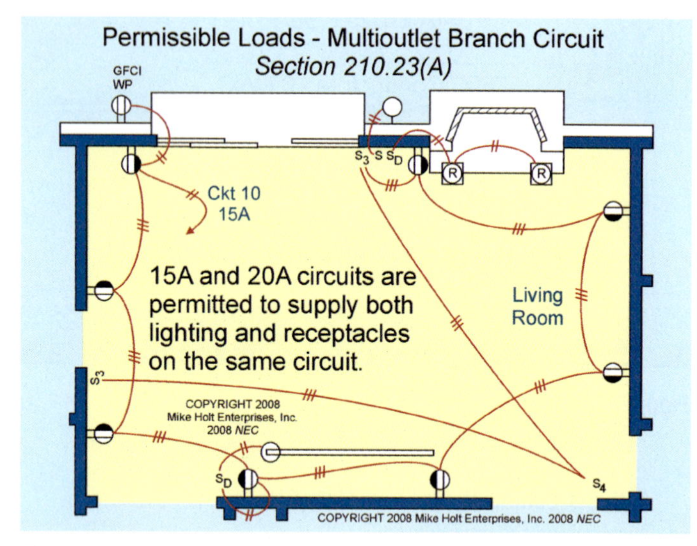

Figure 210–46

(1) Cord-and-Plug-Connected Equipment Not Fastened in Place. Cord-and-plug-connected equipment not fastened in place, such as a drill press or table saw, must not have an ampere rating more than 80 percent of the branch-circuit rating. Figure 210–47

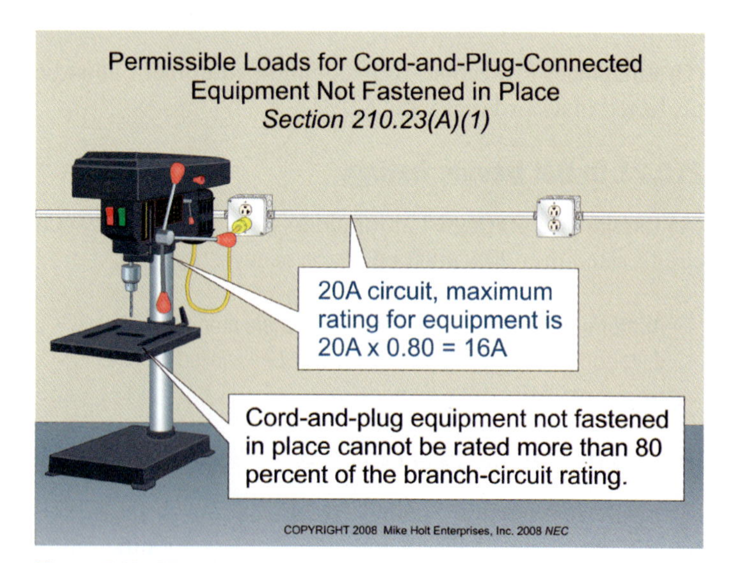

Figure 210–47

Author's Comment: UL and other testing laboratories list portable equipment (such as hair dryers) up to 100 percent of the circuit rating. The *NEC* is an installation standard, not a product standard, so it can't prohibit this practice. There really is no way to limit the load to 80 percent of the branch-circuit rating if testing laboratories permit equipment to be listed for 100 percent of the circuit rating.

(2) Fixed Equipment. Equipment fastened in place (other than luminaires) must not be rated more than 50 percent of the branch-circuit ampere rating if this circuit supplies luminaires, receptacles, or both. Figure 210–48

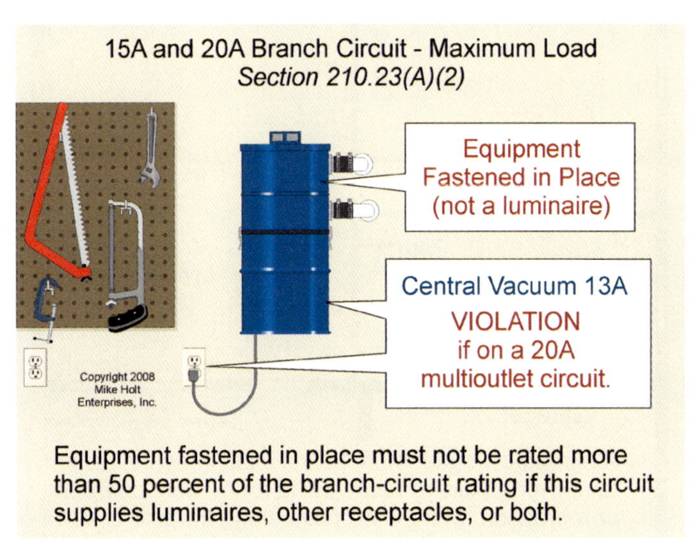

Figure 210–48

Question: *Can a whole house (central) vacuum motor rated 13A be installed on an existing 20A circuit that supplies more than one receptacle outlet?*

Answer: *No, an individual 15A or 20A branch circuit would be required.*

210.25 Branch Circuits in Buildings with Multiple Occupancies.

(A) Dwelling Unit Branch Circuits. Dwelling unit branch circuits are only permitted to supply loads within or associated with the dwelling unit.

(B) Common Area Branch Circuits. Branch circuits for public or common areas of a multi-occupancy building must not originate from equipment that supplies an individual dwelling unit or tenant space.

Author's Comment: This rule prohibits common area branch circuits from being supplied from an individual dwelling unit or tenant space to prevent common area circuits from being turned off by tenants or by the utility due to nonpayment of electric bills.

PART III. REQUIRED OUTLETS

210.50 General. Receptacle outlets must be installed in accordance with 210.52 through 210.63.

(A) Cord Pendant Receptacle Outlet. A permanently installed flexible cord pendant receptacle is considered a receptacle outlet. Figure 210–49

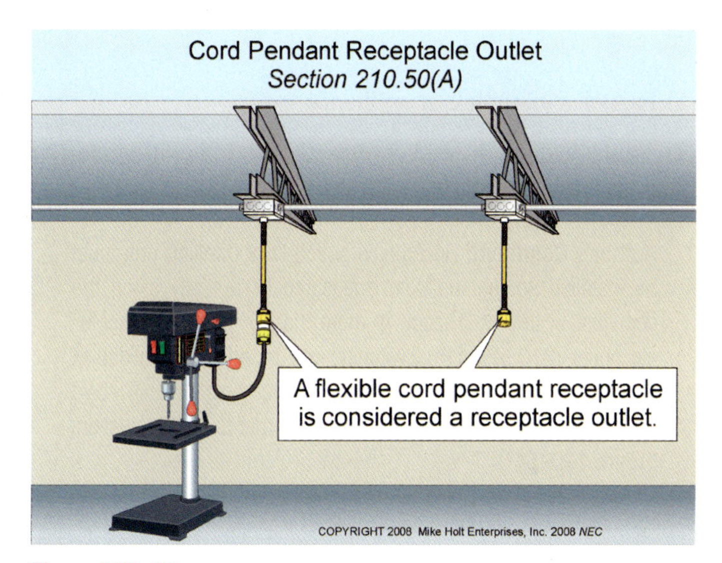

Figure 210–49

Author's Comment: See Article 400 for the installation requirements for flexible cords, and 314.23(H) for the requirements for boxes at pendant outlets.

(C) Appliance Receptacle Outlets. Receptacle outlets installed for a specific appliance, such as a clothes washer, dryer, range, garage door opener or refrigerator, must be within 6 ft of the intended location of the appliance. Figure 210–50

210.52 Dwelling Unit Receptacle Outlet Requirements. Receptacles rated 15A and 20A, 125V must be installed as required in (A) through (H), and the receptacles required by this section are in addition to any receptacle that is:

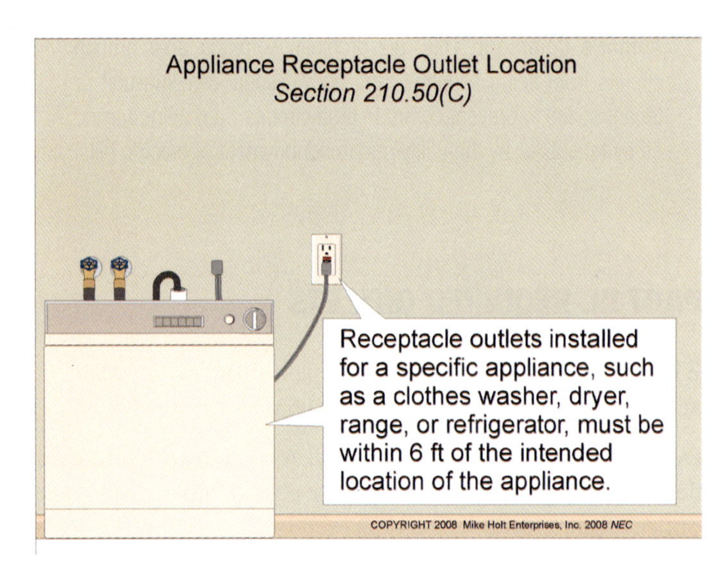

Figure 210–50

(1) Part of a luminaire or appliance,

(2) Controlled by a wall switch to meet the illumination requirements of 210.70(A)(1) Ex 1,

Author's Comment: Receptacle outlets in a dwelling unit must be installed so that no point measured horizontally along the floor line in any wall space is more than 6 ft from a receptacle outlet [210.52(A)(1)]. Switching one receptacle of a duplex receptacle can meet the lighting requirements of 210.70(A)(1) and the receptacle placement requirements of this section. Figure 210–51

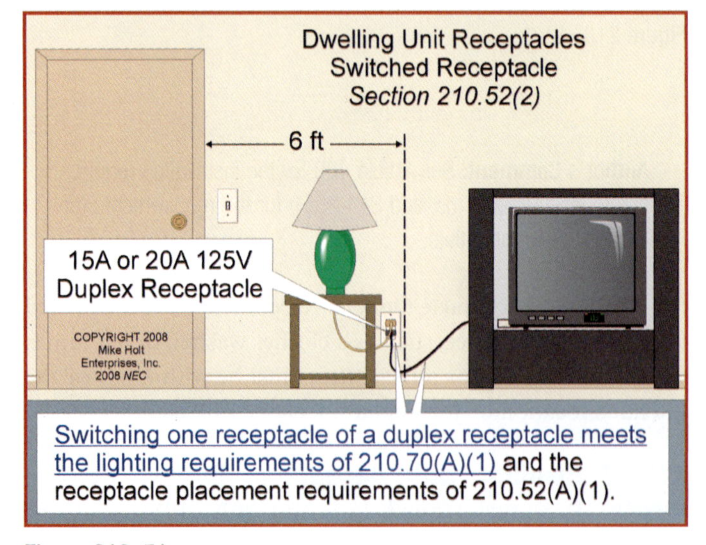

Figure 210–51

(3) Located within cabinets or cupboards, or

(4) Located more than 5½ ft above the floor.

(A) General Requirements—Dwelling Unit. A receptacle outlet must be installed in every kitchen, family room, dining room, living room, sunroom, parlor, library, den, bedroom, recreation room, and similar room or area in accordance with (1), (2), and (3): Figure 210–52

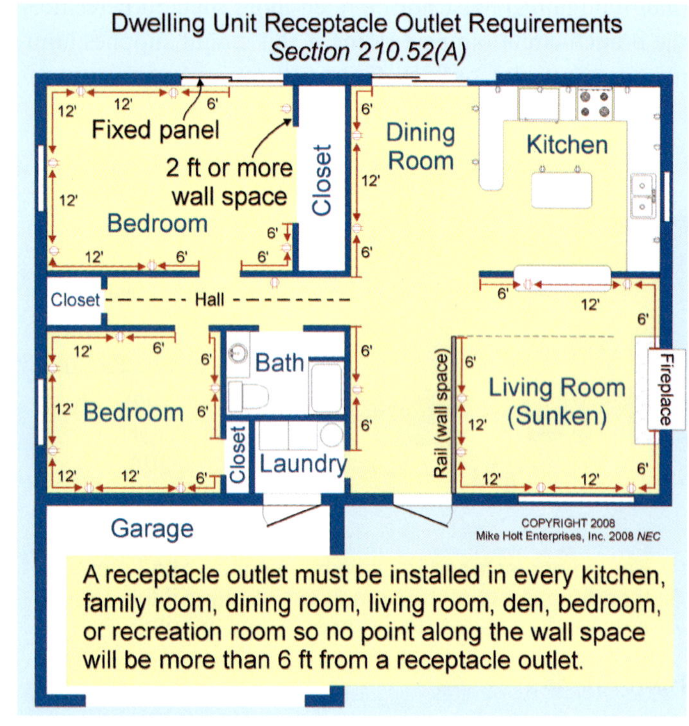

Figure 210–52

(1) Receptacle Placement. A receptacle outlet must be installed so that no point along the wall space is more than 6 ft, measured horizontally along the floor line, from a receptacle outlet.

Author's Comment: The purpose of this rule is to ensure that a general-purpose receptacle is conveniently located to reduce the chance that an extension cord will be used.

(2) Definition of Wall Space. Figure 210–53

(1) Any space 2 ft or more in width, unbroken along the floor line by doorways, fireplaces, and similar openings.

(2) The space occupied by fixed panels in exterior walls.

(3) The space occupied by fixed room dividers, such as freestanding bar-type counters or guard rails.

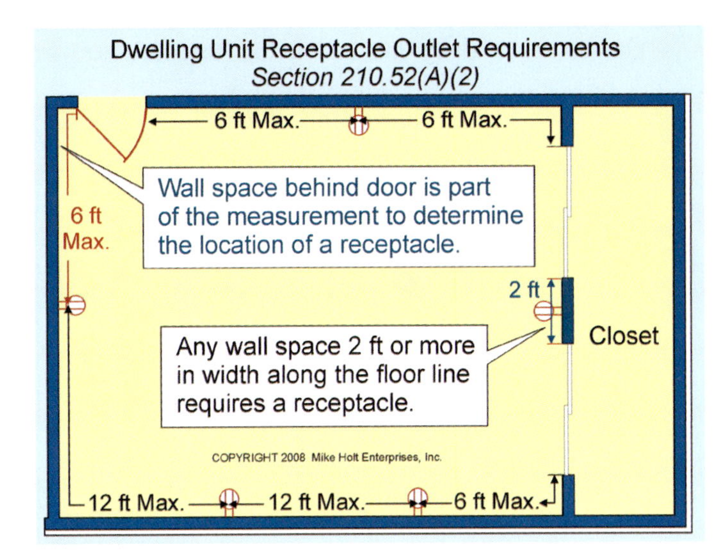

Figure 210–53

(3) Floor Receptacle Outlets. Floor receptacle outlets are not counted as the required receptacle wall outlet if they are located more than 18 in. from the wall. Figure 210–54

Figure 210–54

(B) Small-Appliance Circuits.

(1) Receptacle Outlets. The two or more 20A, 120V small-appliance branch circuits serving the kitchen, pantry, breakfast room, and dining room area of a dwelling unit [210.11(C)(1)] must serve all wall, floor and countertop receptacle outlets [210.52(C)], and the receptacle outlet for refrigeration equipment. Figure 210–55

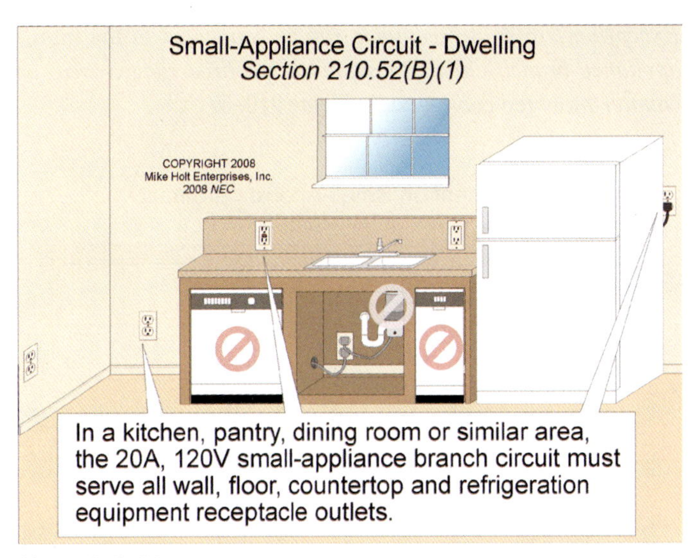

Figure 210–55

Exception No. 2: The receptacle outlet for refrigeration equipment can be supplied from an individual branch circuit rated 15A or greater. Figure 210–56

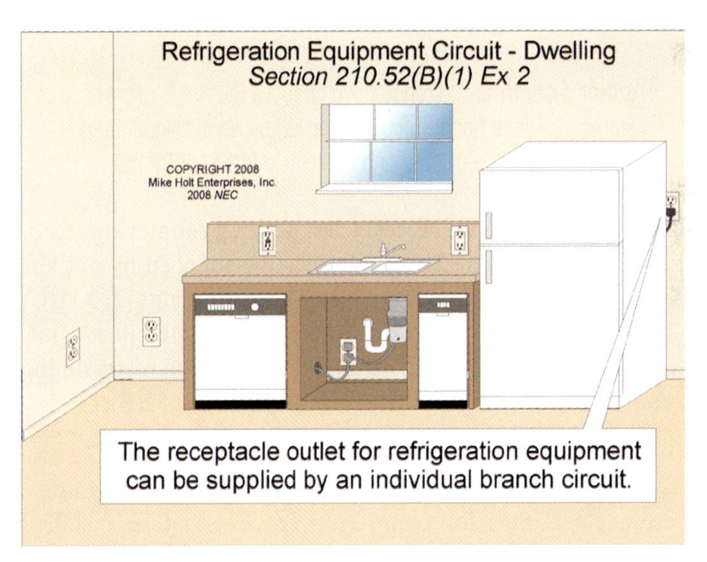

Figure 210–56

(2) Not Supply Other Outlets. The 20A, 120V small-appliance circuits required by 210.11(C)(1) must not supply outlets for luminaires or appliances.

Exception No. 1: The 20A, 120V small-appliance branch circuit can be used to supply a receptacle for an electric clock.

Exception No. 2: A receptacle can be connected to the small-appliance branch circuit to supply a gas-fired range, oven, or counter-mounted cooking unit. **Figure 210–57**

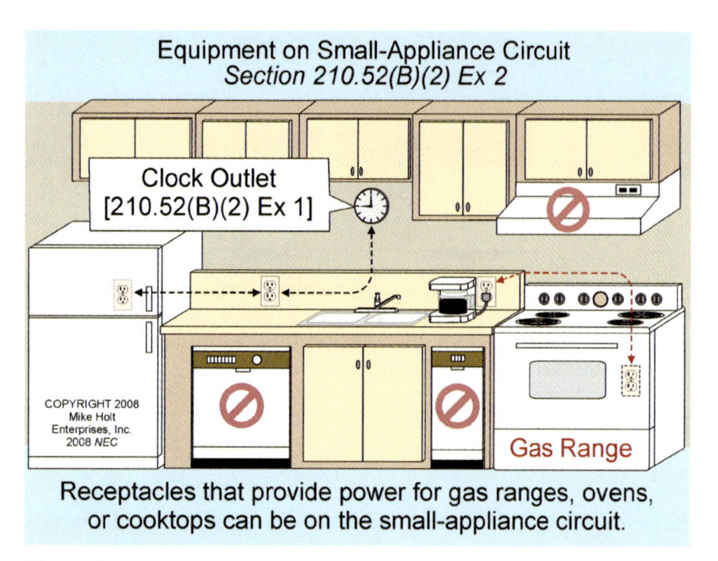

Figure 210–57

Author's Comment: A range hood or above the range microwave listed as a range hood must be supplied by an individual branch circuit if connected by cord and receptacle [422.16(B)(4)(5)].

(3) Kitchen Countertop Receptacles. Kitchen countertop receptacles, as required by 210.52(C), must be supplied by not less than two 20A, 120V small-appliance branch circuits [210.11(C)(1)]. Either or both of these circuits can supply receptacle outlets in the same kitchen, pantry, breakfast room, or dining room of the dwelling unit [210.11(C)(1) and 210.52(B)(1)].

(C) Countertop Receptacles. In kitchens, pantries, breakfast rooms, dining rooms and similar areas of dwelling units, receptacle outlets for countertop spaces must be installed according to (1) through (5) below.

Where a range, counter-mounted cooking unit, or sink is installed in an island or peninsular countertop, and the width of the counter behind the range, counter-mounted cooking unit, or sink is less than 12 in., the countertop space is considered to be two separate countertop spaces. **Figure 210–58**

Author's Comment: GFCI protection is required for all 15A and 20A, 125V receptacles that supply kitchen countertop surfaces [210.8(A)(6)].

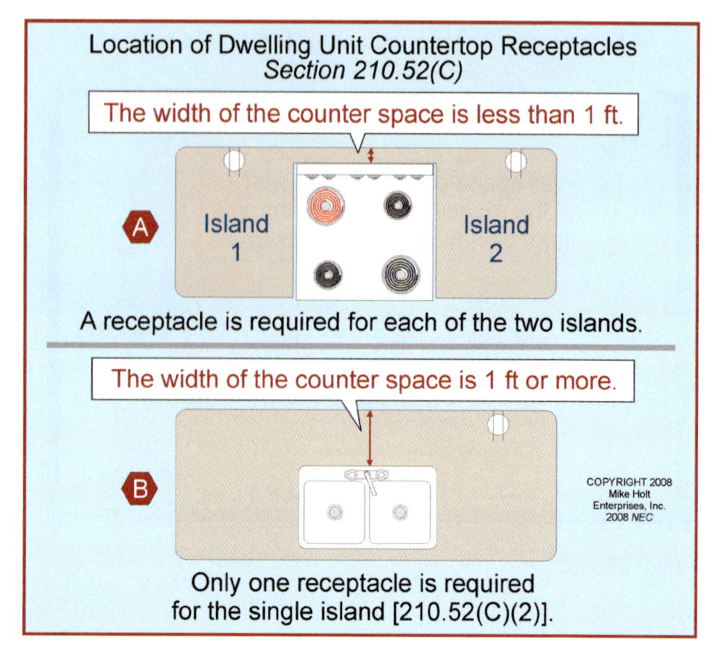

Figure 210–58

(1) Wall Countertop Spaces. A receptacle outlet must be installed for each kitchen and dining area countertop wall space 1 ft or wider, and receptacles must be placed so that no point along the countertop wall space is more than 2 ft, measured horizontally, from a receptacle outlet. **Figure 210–59**

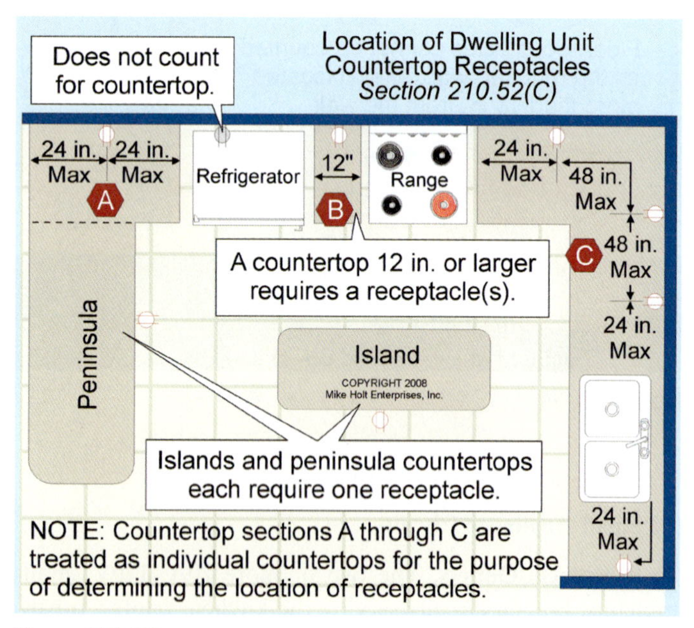

Figure 210–59

Exception: A receptacle outlet isn't required on a wall directly behind a range, counter-mounted cooking unit, or sink, in accordance with Figure 210.52(C)(1) in the NEC. Figure 210–60

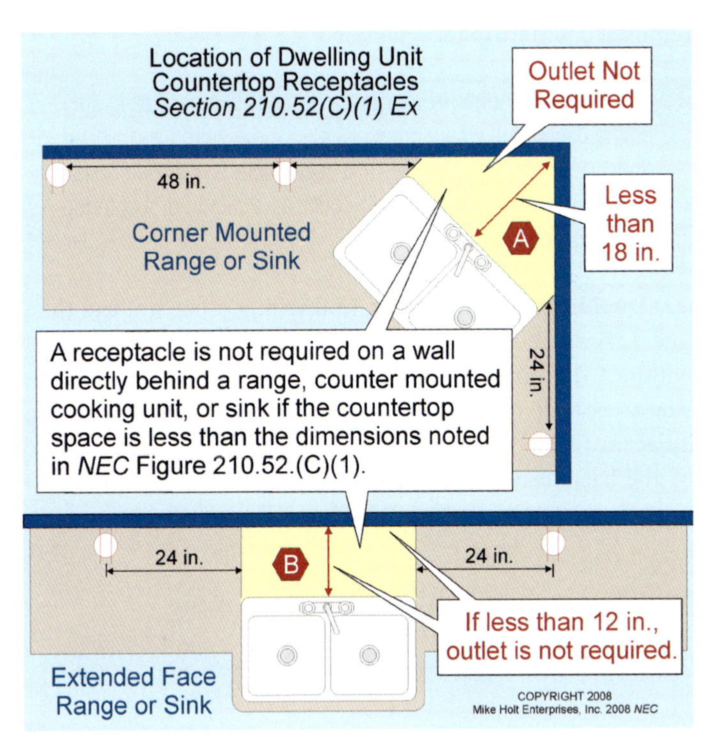

Figure 210–60

Author's Comment: If the countertop space behind a range or sink is larger than the dimensions noted in Figure 210.52(C)(1) of the *NEC*, then a GFCI-protected receptacle must be installed in that space. This is because, for all practical purposes, if there's sufficient space for an appliance, an appliance will be placed there.

(2) Island Countertop Spaces. At least one receptacle outlet must be installed at each island countertop space with a long dimension of 2 ft or more, and a short dimension of 1 ft or more.

(3) Peninsular Countertop Spaces. At least one receptacle outlet must be installed at each peninsular countertop with a long dimension of 2 ft or more, and a short dimension of 1 ft or more, measured from the connecting edge. Figure 210–61

Author's Comment: The *Code* doesn't require more than one receptacle outlet in an island or peninsular countertop space, regardless of the length of the countertop, unless the countertop is broken as described in 210.52(C)(4).

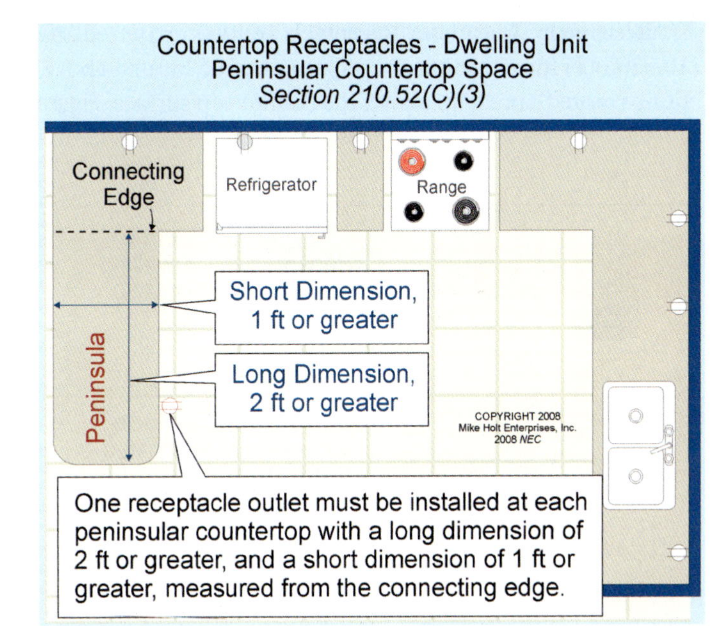

Figure 210–61

(4) Separate Countertop Spaces. When breaks occur in countertop spaces for rangetops, refrigerators, or sinks, each countertop space is considered as a separate countertop for determining receptacle placement. Figure 210–62

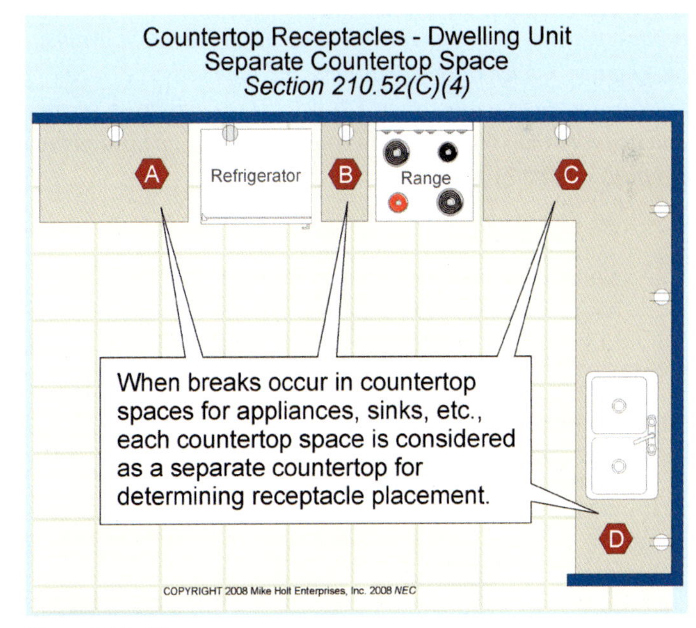

Figure 210–62

(5) Receptacle Location. Receptacle outlets required by 210.52(C)(1) for the countertop space must be located above, but not more than 20 in. above, the countertop surface. **Figure 210–63**

Figure 210–63

Exception: The receptacle outlet for the countertop space can be installed below the countertop only when wall space or a backsplash is not available, such as in an island or peninsular counter. Under these conditions, the required receptacle(s) must be located no more than 1 ft below the countertop surface and no more than 6 in. from the countertop edge, measured horizontally. **Figure 210–64**

Figure 210–64

Receptacle outlets rendered not readily accessible by appliances fastened in place, located in an appliance garage, behind sinks, or rangetops [210.52(C)(1) Ex], or supplying appliances that occupy dedicated space do not count as the required countertop receptacles.

Author's Comment: An "appliance garage" is an enclosed area on the countertop where an appliance can be stored and hidden from view when not in use. If a receptacle is installed inside an appliance garage, it doesn't count as a required countertop receptacle outlet.

(D) Dwelling Unit Bathrooms. In dwelling units, not less than one 15A or 20A, 125V receptacle outlet must be installed within 3 ft from the outside edge of each bathroom basin. The receptacle outlet must be located on a wall or partition adjacent to the basin counter surface, or on the side or face of the basin cabinet not more than 12 in. below the countertop [210.11(C)(3)]. **Figures 210–65** and **Figure 210–66**

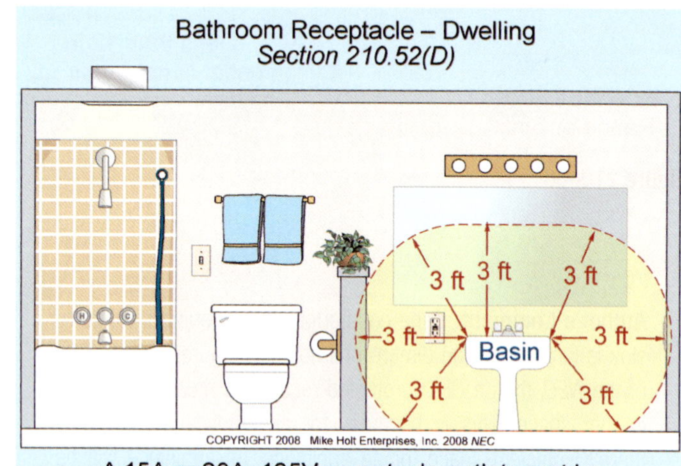

Figure 210–65

Author's Comments:

- One receptacle outlet can be located between two basins to meet the requirement, but only if the receptacle outlet is located within 3 ft of the outside edge of each basin. **Figure 210–67**
- The bathroom receptacles must be GFCI protected [210.8(A)(1)].

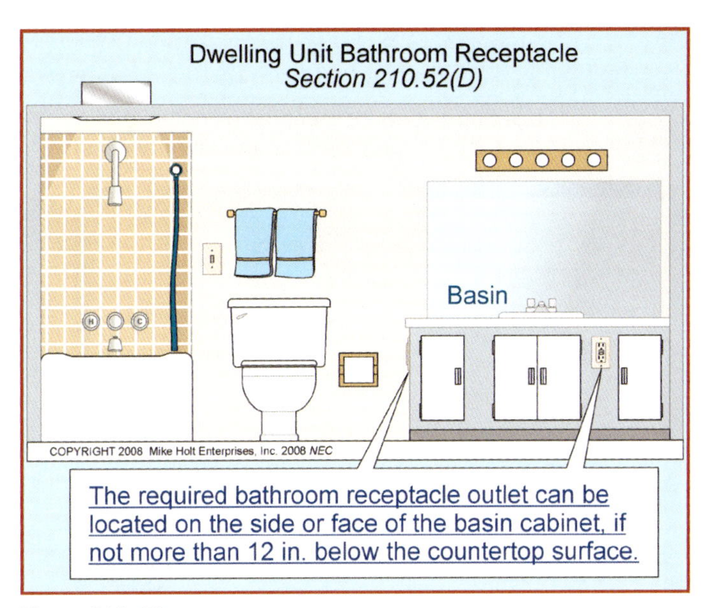

Figure 210–66

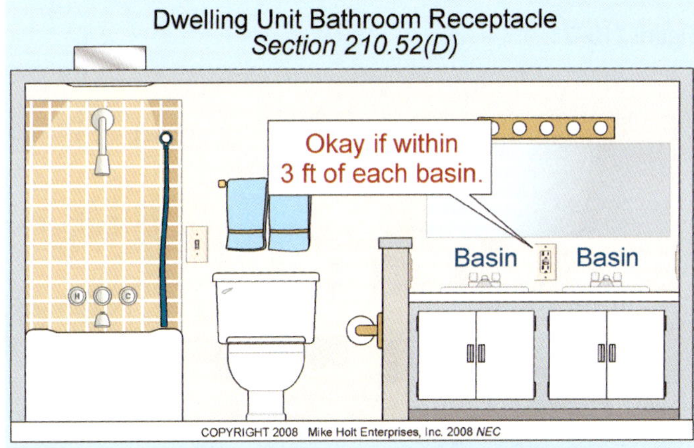

Figure 210–67

(E) Dwelling Unit Outdoor Receptacles.

(1) One- and Two-Family Dwellings. Two GFCI-protected 15A or 20A, 125V receptacle outlets that are accessible while standing at grade level must be installed outdoors for each dwelling unit, one at the front and one at the back, no more than 6½ ft above grade. Figure 210–68

(2) Multifamily Dwelling. Each dwelling unit of a multifamily dwelling that has an individual entrance at grade level must have at least one GFCI-protected 15A or 20A, 125V receptacle outlet accessible from grade level located not more than 6½ ft above grade. Figure 210–69

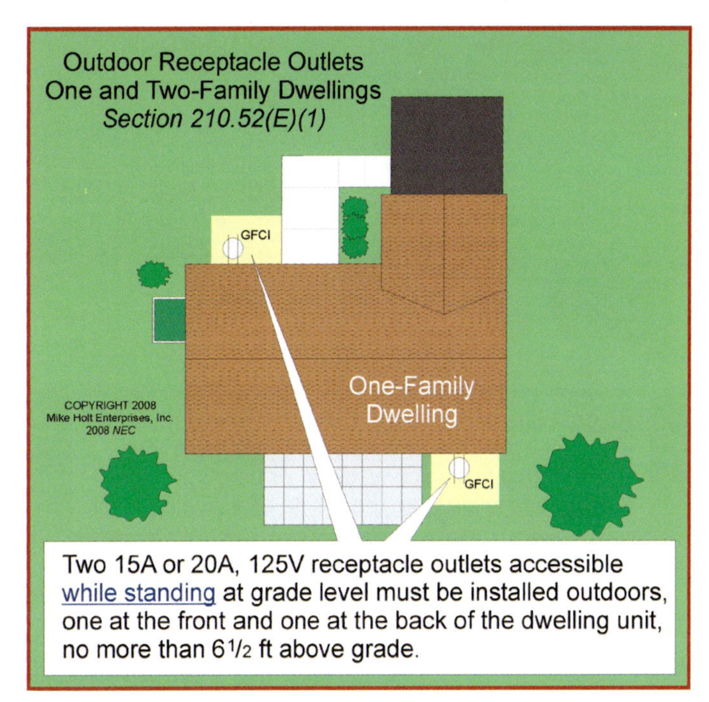

Figure 210–68

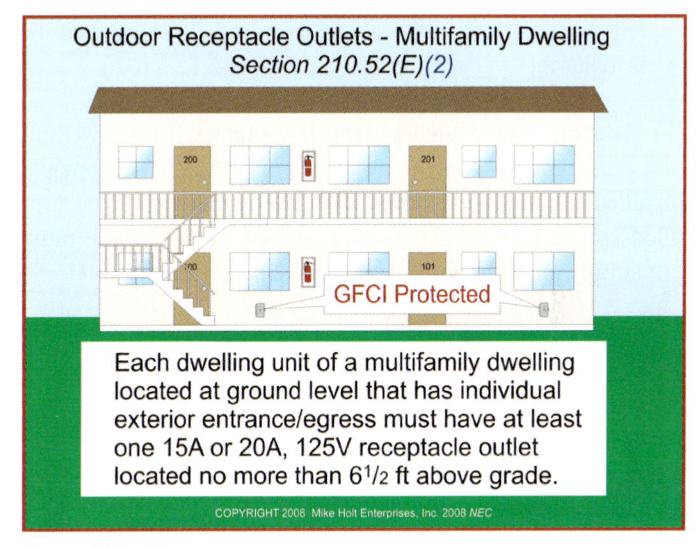

Figure 210–69

(3) Balconies, Decks, and Porches. At least one 15A or 20A, 125V receptacle must be installed within the perimeter and not more than 6½ ft above the balcony, deck, or porch surface that is accessible from the inside of a dwelling unit. Figure 210–70

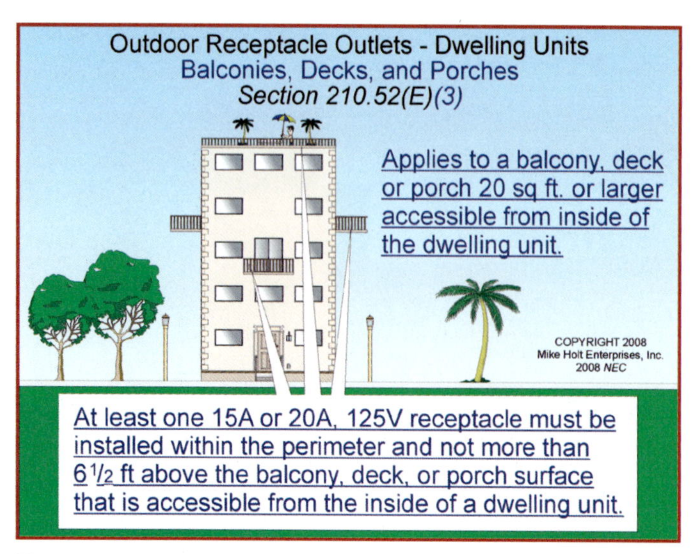

Figure 210–70

Exception: Balconies, decks or porches with a usable area of less than 20 sq ft are not required to have a receptacle installed.

Author's Comment: These receptacles must be GFCI protected [210.8(A)(3)].

(F) Dwelling Unit Laundry Area Receptacles. Each dwelling unit must have not less than one 15A or 20A, 125V receptacle installed in the laundry area. This receptacle(s) must be supplied by the 20A, 120V laundry branch circuit, which must not supply any other outlets [210.11(C)(2)]. Figure 210–71

Author's Comment: Receptacles located within 6 ft of a laundry room sink require GFCI protection [210.8(A)(7)].

Exception No. 1: A laundry receptacle outlet isn't required in a dwelling unit located in a multifamily building with laundry facilities available to all occupants.

(G) Dwelling Unit Garage and Basement Receptacles.

(1) Not less than one 15A or 20A, 125V receptacle outlet, in addition to any provided for a specific piece of equipment, must be installed in each basement, each attached garage, and each detached garage with electric power. Figure 210–72

Author's Comment: The garage and basement receptacles must be GFCI protected in accordance with 210.8(A)(2) for garages and 210.8(A)(5) for unfinished basements.

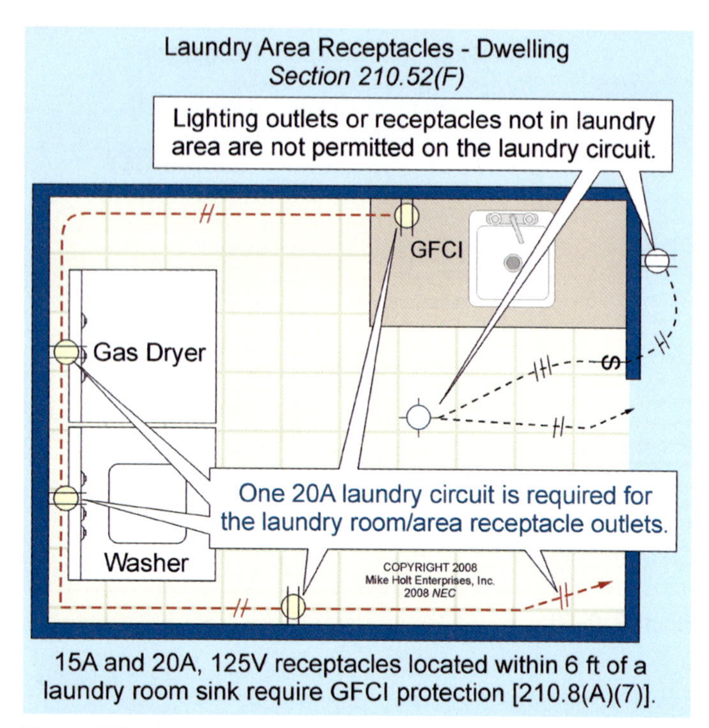

Figure 210–71

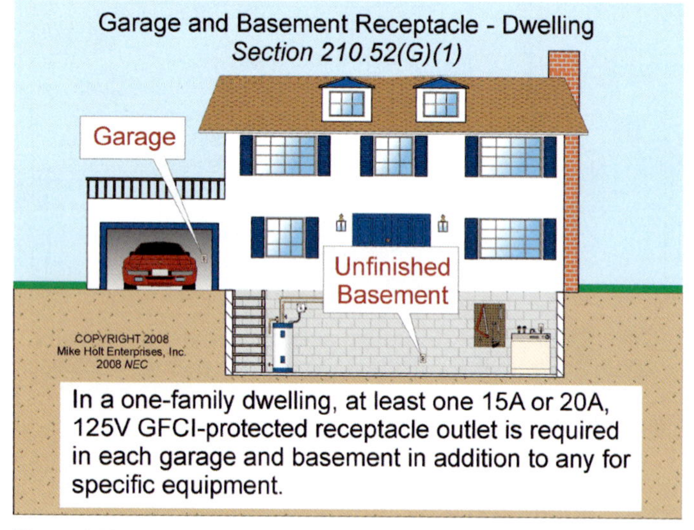

Figure 210–72

(2) Where a portion of the basement is finished into habitable rooms, each separate unfinished portion must have a 15A or 20A, 125V receptacle outlet installed. Figure 210–73

Author's Comment: The purpose of this requirement is to prevent an extension cord from a non-GFCI-protected receptacle from being used to supply power to loads in the unfinished portion of the basement.

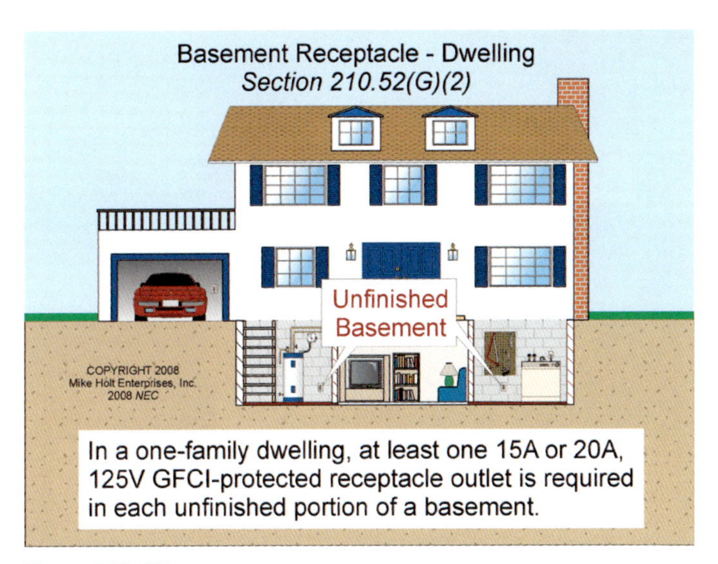

Figure 210–73

(H) Dwelling Unit Hallway Receptacles. One 15A or 20A, 125V receptacle outlet must be installed in each hallway that is at least 10 ft long, measured along the centerline of the hall without passing through a doorway. Figure 210–74

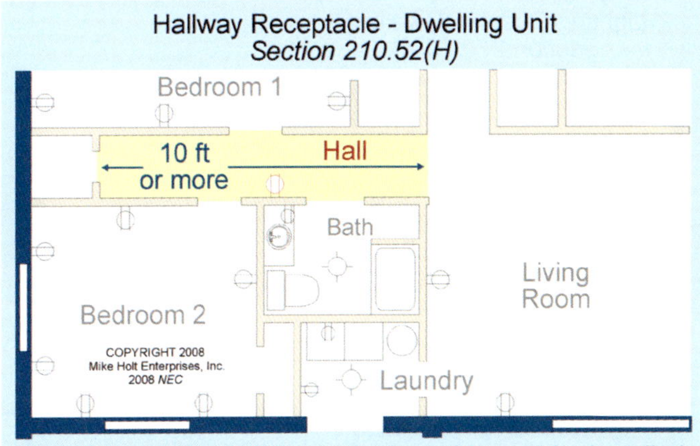

Figure 210–74

210.60 Receptacles in Guest Rooms, Guest Suites, Dormitories, and Similar Occupancies.

(A) General Requirements. Guest rooms or guest suites in hotels, motels, and sleeping rooms in dormitories and similar occupancies, must have receptacle outlets installed in accordance with all the requirements for a dwelling unit as described in 210.52(A) and 210.52(D). Guest rooms with permanent provisions for living, sleeping, cooking, and sanitation are considered dwelling units [Article 100, Dwelling unit], and must meet all of the rules for dwelling units.

(B) Receptacle Placement. The number of receptacle outlets required for guest rooms must not be less than that required for a dwelling unit, in accordance with 210.52(A). To eliminate the need for extension cords by guests for ironing, computers, refrigerators, etc., receptacles can be located to be convenient for permanent furniture layout, but not less than two receptacle outlets must be readily accessible.

Receptacle outlets behind a bed must be located so the bed will not make contact with the attachment plug, or the receptacle must be provided with a suitable guard. Figure 210–75

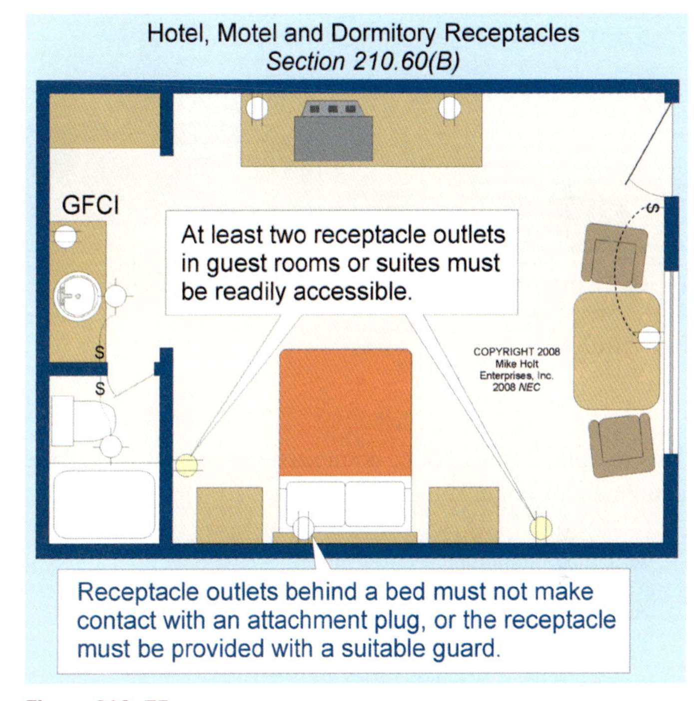

Figure 210–75

Author's Comment: See the definition of "Attachment Plug" in Article 100.

210.62 Show Windows.

At least one receptacle outlet must be installed within 18 in. of the top of a show window for each 12 linear ft or major fraction thereof measured horizontally at its maximum width.

Author's Comment: See the definition of "Show Window" in Article 100

210.63 Heating, Air-Conditioning, and Refrigeration (HACR) Equipment.

210.63 Heating, Air-Conditioning, and Refrigeration (HACR) Equipment. A 15A or 20A, 125V receptacle outlet must be installed at an accessible location for the servicing of heating, air-conditioning, and refrigeration equipment. The receptacle must be located within 25 ft of, and on the same level as, the heating, air-conditioning, and refrigeration equipment. Figure 210–76

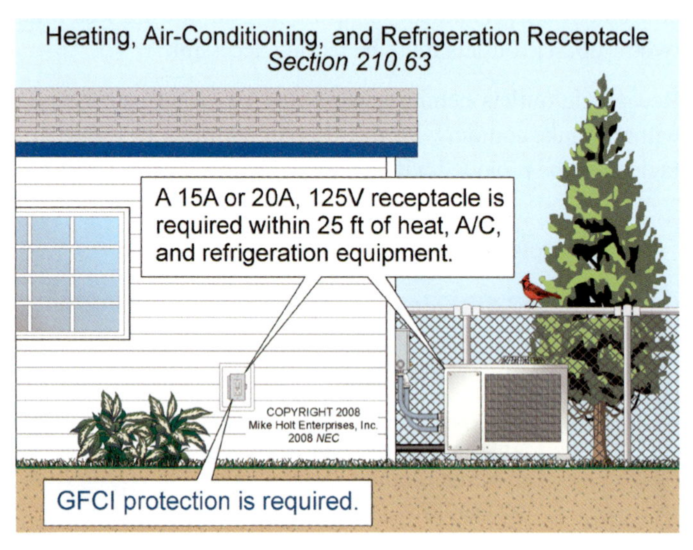

Figure 210–76

This receptacle must not be connected to the load side of the equipment disconnecting means.

Author's Comments:

- A receptacle outlet isn't required for ventilation equipment, because it's not heating, air-conditioning, or refrigeration equipment.
- The HACR receptacle must be GFCI protected if located outdoors [210.8(A)(3) and 210.8(B)(5)] or in the crawl space or unfinished basement of a dwelling unit [210.8(A)(4) and 210.8(A)(5)].
- The outdoor 15A or 20A, 125V receptacle outlet required for dwelling units [210.52(E)(1)] can be used to satisfy this requirement, if located on the front or back of the dwelling unit, and if it is within 25 ft of the HACR equipment. Figure 210–77

Exception: A receptacle outlet isn't required at one- and two-family dwellings for the service of evaporative coolers.

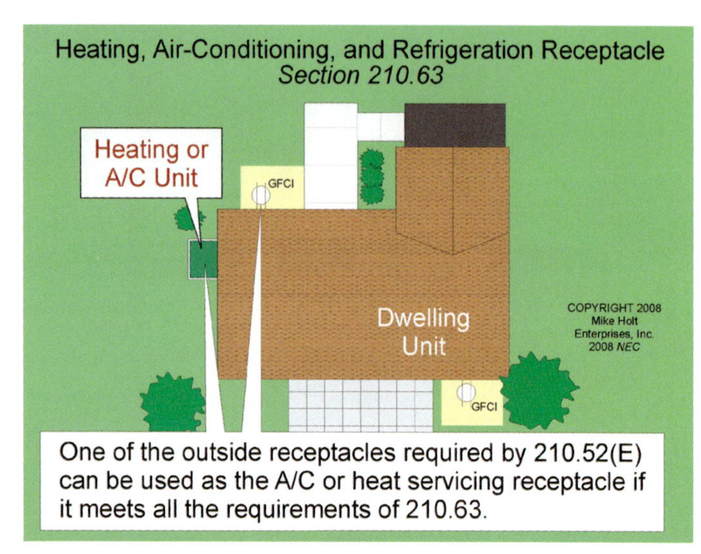

Figure 210–77

210.70 Lighting Outlet Requirements.

(A) Dwelling Unit Lighting Outlets. Lighting outlets must be installed in: Figure 210–78

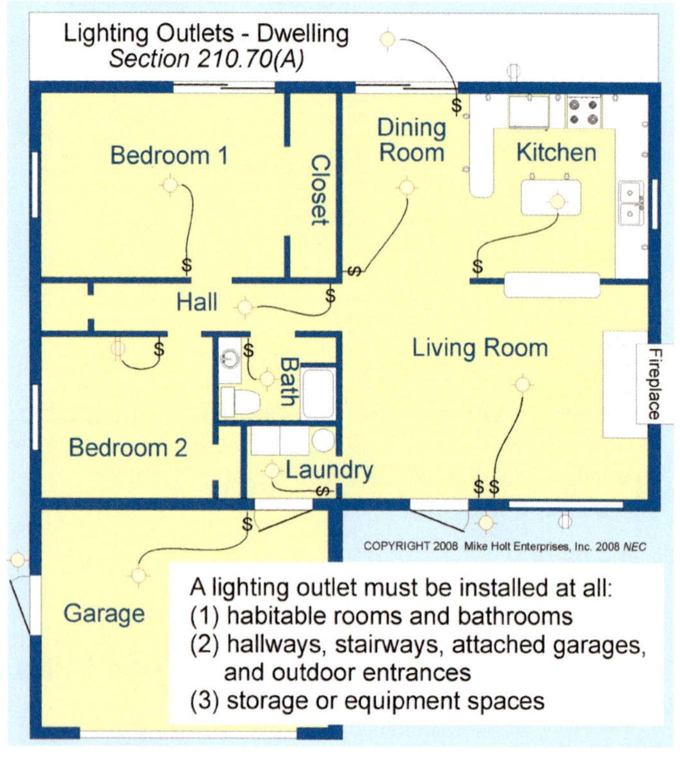

Figure 210–78

(1) Habitable Rooms. At least one wall switch-controlled lighting outlet must be installed in every habitable room and bathroom of a dwelling unit. Figure 210–79

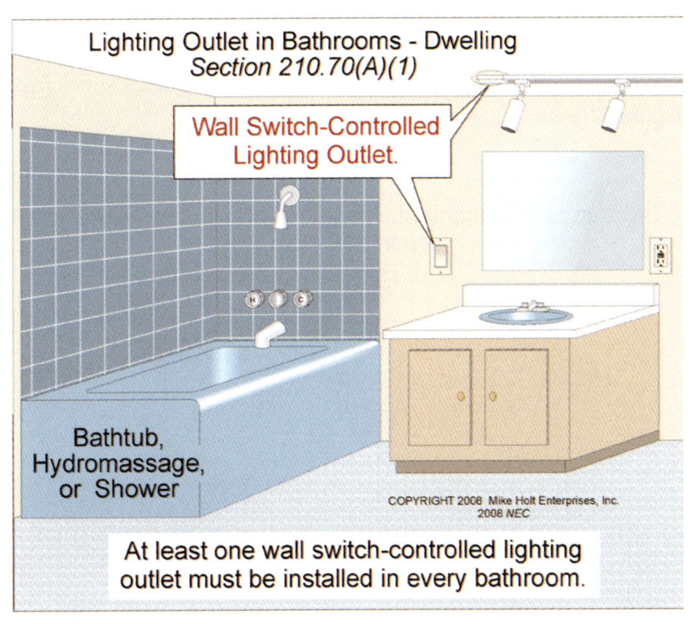

Figure 210–79

Author's Comment: See the definition of "Lighting Outlet" in Article 100.

Exception No. 1: In other than kitchens and bathrooms, a receptacle controlled by a wall switch can be used instead of a lighting outlet.

Exception No. 2: Lighting outlets can be controlled by occupancy sensors equipped with a manual override that permits the sensor to function as a wall switch. Figure 210–80

Author's Comment: The *Code* specifies the location of the wall switch-controlled lighting outlet, but it doesn't specify the switch location. Naturally, you wouldn't want to install a switch behind a door or other inconvenient location, but the *NEC* doesn't require you to relocate the switch to suit the swing of the door. When in doubt as to the best location to place a light switch, consult the job plans or ask the customer. Figure 210–81

(2) Other Areas.

(a) Hallways, Stairways and Garages. In dwelling units, not less than one wall switch-controlled lighting outlet must be installed in hallways, stairways, attached garages, and detached garages with electric power.

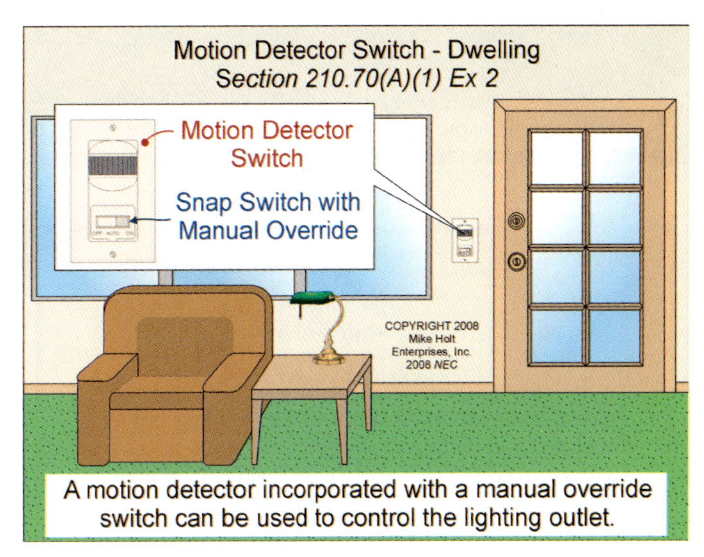

Figure 210–80

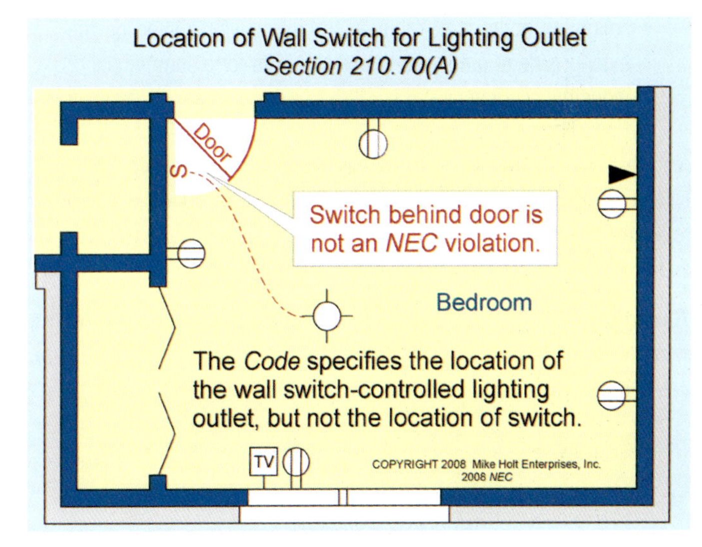

Figure 210–81

(b) Exterior Entrances. At least one wall switch-controlled lighting outlet must provide illumination on the exterior side of outdoor entrances or exits of dwelling units with grade-level access. Figure 210–82

Author's Comments:

- The *NEC* doesn't require a switch adjacent to each outdoor entrance or exit. The *Code* considers switch location a "design issue" which is beyond the purpose of the *NEC* [90.1(C)]. For this reason, proposals to mandate switch locations have been rejected.

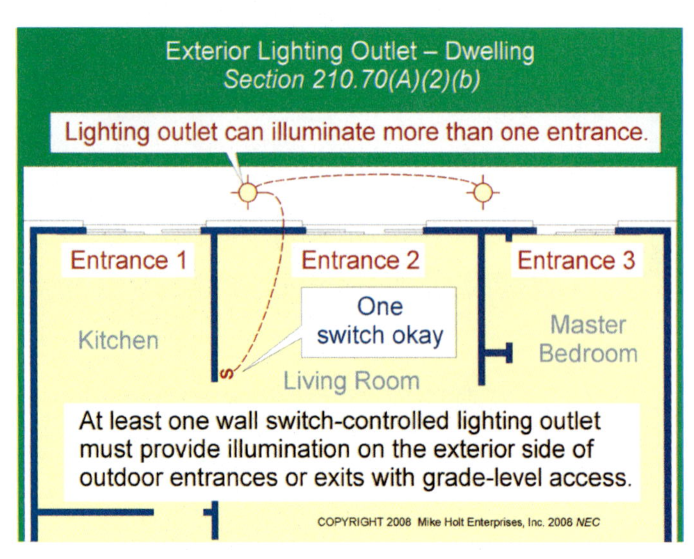

Figure 210–82

- A lighting outlet isn't required to provide illumination on the exterior side of outdoor entrances or exits for a commercial or industrial occupancy.

(c) Stairway. Where the stairway between floor levels has six risers or more, a wall switch must be located at each floor level and at each landing level that includes an entryway to control the illumination for the stairway.

Exception to (a), (b), and (c): Lighting outlets for hallways, stairways, and outdoor entrances can be switched by a remote, central, or automatic control device. Figure 210–83

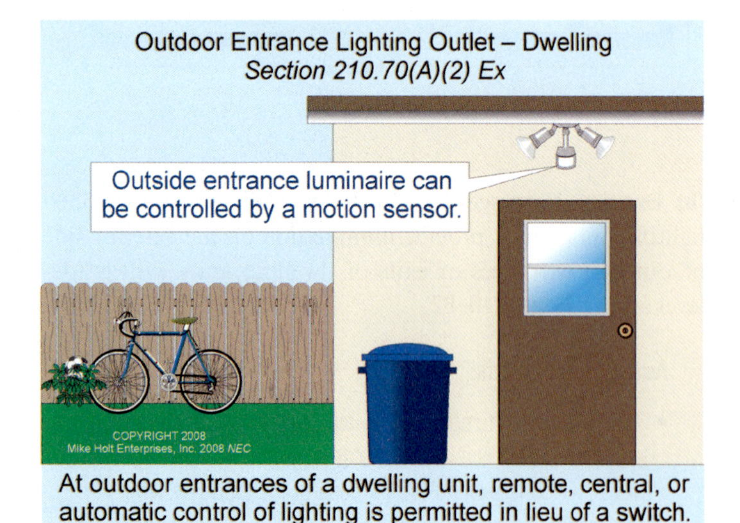

Figure 210–83

(3) Storage and Equipment Rooms. At least one lighting outlet that contains a switch or is controlled by a wall switch must be installed in attics, underfloor spaces, utility rooms, and basements used for storage or containing equipment that requires servicing. The switch must be located at the usual point of entry to these spaces, and the lighting outlet must be located at or near the equipment that requires servicing. Figure 210–84

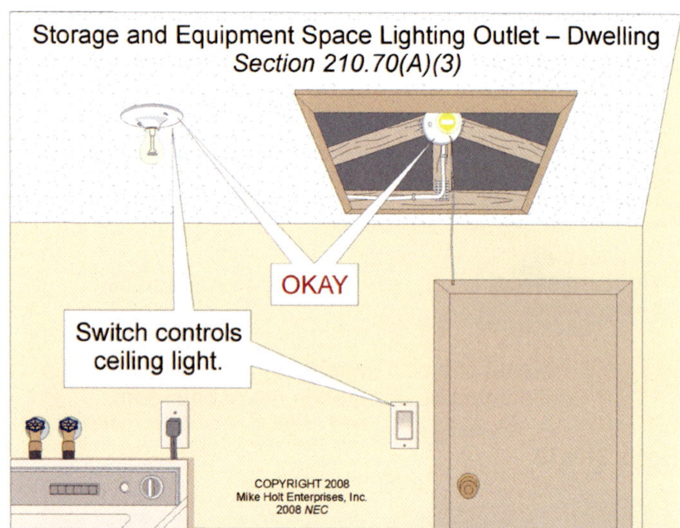

Figure 210–84

(B) Guest Rooms or Guest Suites. At least one wall switch-controlled lighting outlet must be installed in every habitable room and bathroom of a guest room or guest suite of hotels, motels, and similar occupancies.

Exception No. 1: In other than bathrooms and kitchens, a receptacle controlled by a wall switch is permitted in lieu of lighting outlets. Figure 210–85

Exception No. 2: Lighting outlets can be controlled by occupancy sensors equipped with a manual override that permits the sensor to function as a wall switch.

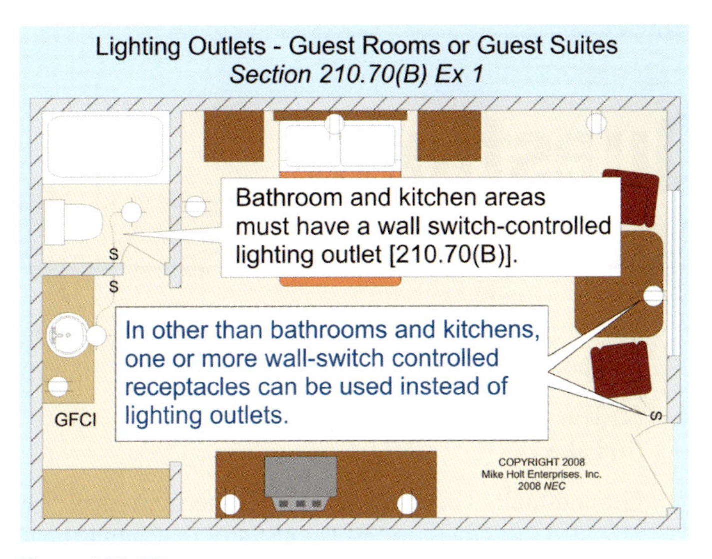

Figure 210–85

(C) Other Than Dwelling Units. At least one lighting outlet that contains a switch or is controlled by a wall switch must be installed in attics and underfloor spaces containing equipment that requires servicing. The switch must be located at the usual point of entry to these spaces, and the lighting outlet must be located at or near the equipment requiring servicing.

> **Author's Comment:** A 15A or 20A, 125V receptacle must be installed within 25 ft of HACR equipment [210.63].

ARTICLE 210 Practice Questions

ARTICLE 210. BRANCH CIRCUITS—PRACTICE QUESTIONS

1. A three-phase, 4-wire, _____ power system used to supply power to nonlinear loads may necessitate that the power system design allow for the possibility of high harmonic current on the neutral conductor.

   (a) wye-connected
   (b) delta-connected
   (c) wye/delta-connected
   (d) none of these

2. Multiwire branch circuits shall _____.

   (a) supply only line-to-neutral loads
   (b) not be permitted in dwelling units
   (c) have their conductors originate from different panelboards
   (d) none of these

3. In dwelling units, the voltage between conductors that supply the terminals of _____ shall not exceed 120V, nominal.

   (a) luminaires
   (b) cord-and-plug-connected loads of 1,440 VA or less
   (c) cord-and-plug-connected loads of more than ¼ hp
   (d) a and b

4. GFCI protection shall be provided for all 15A and 20A, 125V receptacles installed in a dwelling unit _____.

   (a) attic
   (b) garage
   (c) laundry room
   (d) all of these

5. All 15A and 20A, 125V receptacles installed in _____ of dwelling units shall have GFCI protection.

   (a) unfinished attics
   (b) finished attics
   (c) unfinished basements
   (d) finished basements

6. All 15A and 20A, 125V receptacles installed in dwelling unit boathouses shall have GFCI protection.

   (a) True
   (b) False

7. Ground-fault circuit-interrupter protection shall be provided for outlets not exceeding 240V that supply boat hoists installed in dwelling unit locations.

   (a) True
   (b) False

8. An individual 20A branch circuit can supply a single dwelling unit bathroom for receptacle outlet(s) and other equipment within the same bathroom.

   (a) True
   (b) False

9. Branch-circuit neutral conductors that supply a continuous load shall have an ampacity of not less than 125 percent of the continuous load.

   (a) True
   (b) False

10. A single receptacle installed on an individual branch circuit shall have an ampere rating not less than that of the branch circuit.

    (a) True
    (b) False

11. The total rating of utilization equipment fastened in place, shall not exceed _____ percent of the branch-circuit ampere rating where lighting units and cord-and-plug-connected utilization equipment are supplied.

    (a) 50
    (b) 75
    (c) 100
    (d) 125

12. Receptacle outlets installed for a specific appliance in a dwelling unit, such as laundry equipment, shall be located within _____ of the intended location of the appliance.

    (a) sight
    (b) 3 ft
    (c) 6 ft
    (d) none of these

13. In a dwelling unit, each wall space _____ or wider requires a receptacle.

    (a) 2 ft
    (b) 3 ft
    (c) 4 ft
    (d) 5 ft

14. In dwelling units, outdoor receptacles can be connected to one of the 20A small-appliance branch circuits.

    (a) True
    (b) False

15. Receptacles installed in a dwelling unit kitchen to serve countertop surfaces shall be supplied by not fewer than _____ small-appliance branch circuits.

    (a) one
    (b) two
    (c) three
    (d) four

16. In dwelling units, at least one receptacle outlet shall be installed at each peninsular countertop having a long dimension of _____ in. or greater, and a short dimension of _____ in. or greater.

    (a) 12, 24
    (b) 24, 12
    (c) 24, 48
    (d) 48, 24

17. Kitchen and dining room countertop receptacle outlets in dwelling units shall be installed above the countertop surface, and not more than ___ above the countertop.

    (a) 12 in.
    (b) 18 in.
    (c) 20 in.
    (d) 24 in.

18. There shall be a minimum of _____ receptacle(s) installed outdoors at a one-family or two-family dwelling unit.

    (a) one
    (b) two
    (c) three
    (d) four

19. For a one-family dwelling, at least one receptacle outlet shall be required in each _____.

    (a) basement
    (b) attached garage
    (c) detached garage with electric power
    (d) all of these

20. Guest rooms or guest suites provided with permanent provisions for _____ shall have receptacle outlets installed in accordance with all of the applicable requirements for a dwelling unit as specified in 210.52(A) and 210.52(D).

    (a) whirlpool tubs
    (b) bathing
    (c) cooking
    (d) internet access

21. At least one receptacle outlet shall be installed within 18 in. of the top of a show window for each _____, or major fraction thereof, of show-window area measured horizontally at its maximum width.

    (a) 10 ft
    (b) 12 ft
    (c) 18 ft
    (d) 24 ft

22. In rooms other than kitchens and bathrooms of dwelling units, one or more receptacles controlled by a wall switch shall be permitted in lieu of _____.

    (a) lighting outlets
    (b) luminaires
    (c) the receptacles required by 210.52(B) and (D)
    (d) all of these

23. When locating lighting outlets in dwelling units, a vehicle door in a garage shall be considered an outdoor entrance.

    (a) True
    (b) False

24. In a dwelling unit, at least one lighting outlet _____ shall be located at the point of entry to the attic, underfloor space, utility room, or basement where these spaces are used for storage or contain equipment requiring servicing.

    (a) that is unswitched
    (b) containing a switch
    (c) controlled by a wall switch
    (d) b or c

25. For attics and underfoot spaces containing heating, air-conditioning, and refrigeration equipment, at least one lighting outlet containing a switch or controlled by a wall switch shall be installed _____.

    (a) at the equipment requiring servicing
    (b) near the equipment requiring servicing
    (c) a or b
    (d) none of these

ARTICLE

215 Feeders

INTRODUCTION TO ARTICLE 215—FEEDERS

The next logical step up from the branch circuit is the feeder circuit. Consequently, Article 215 follows Article 210. This article covers the rules for the installation, minimum size, and ampacity of feeders.

This is a very short article, and that's puzzling at first glance. It might seem feeders should just be "heavier" branch circuits, and Article 215 should just be another Article 210 but with more stringent requirements. This, however, isn't the case at all.

If you go back and look at Article 210 again, you'll see it covers many types of branch circuits. It also devotes extensive space to dwelling-area branch circuits. Dwelling units don't have many feeders, but multifamily dwelling buildings will have at least one feeder for each dwelling unit.

Here's an object lesson in the value of Article 100. Go there now and review the definitions of branch circuit and feeder, and think about how much time you spend working with branch circuits and how little time you spend working with feeders. Once you've done that, you'll understand why Article 215 is so much shorter than Article 210.

215.1 Scope. Article 215 covers the installation, conductor sizing, and protection requirements for feeders.

> **Author's Comment:** See the definition of "Feeder" in Article 100. **Figure 215–1**

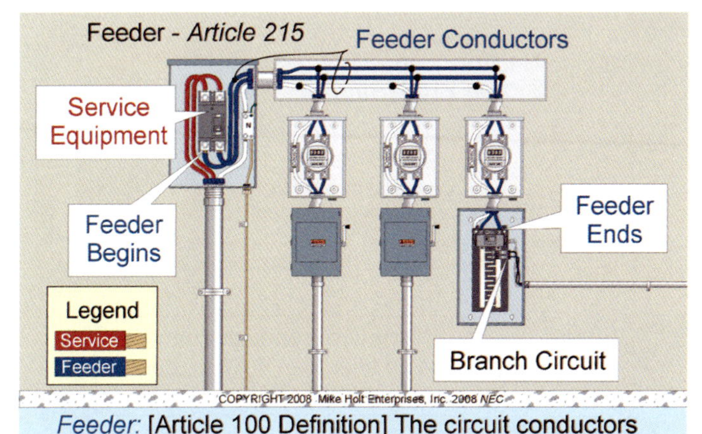

Figure 215–1

215.2 Minimum Rating.

(A) Feeder Conductor Size.

(1) Continuous and Noncontinuous Loads. The minimum feeder-circuit conductor ampacity, before the application of any adjustment and/or correction factors, must be no less than 125 percent of the continuous load, plus 100 percent of the noncontinuous load, based on the terminal temperature rating ampacities as listed in Table 310.16 [110.14(C)]. **Figure 215–2**

> **Author's Comment:** See 215.3 for the feeder overcurrent device sizing requirements for continuous and noncontinuous loads.

Exception No. 1: Where the assembly and the overcurrent device are both listed for operation at 100 percent of its rating, the conductors can be sized at 100 percent of the continuous load.

> **Author's Comment:** Equipment suitable for 100 percent continuous loading is rarely available in ratings under 400A.

Exception No. 2: Neutral conductors can be sized at 100 percent of the continuous and noncontinuous load. **Figure 215-3**

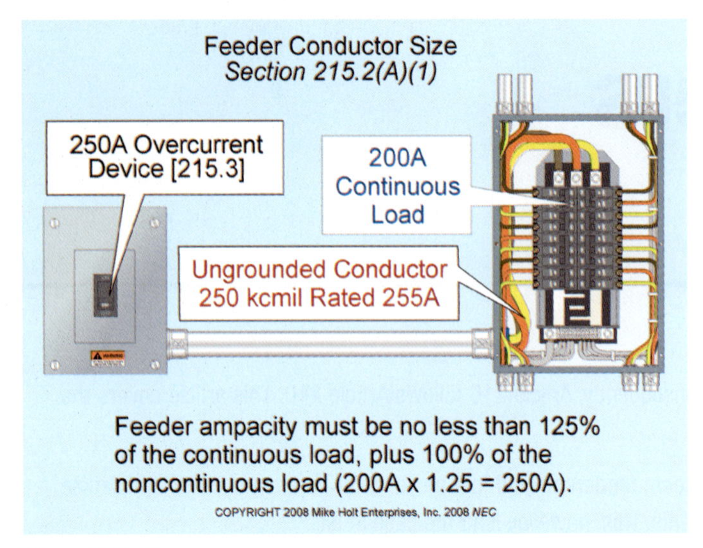

Figure 215–2

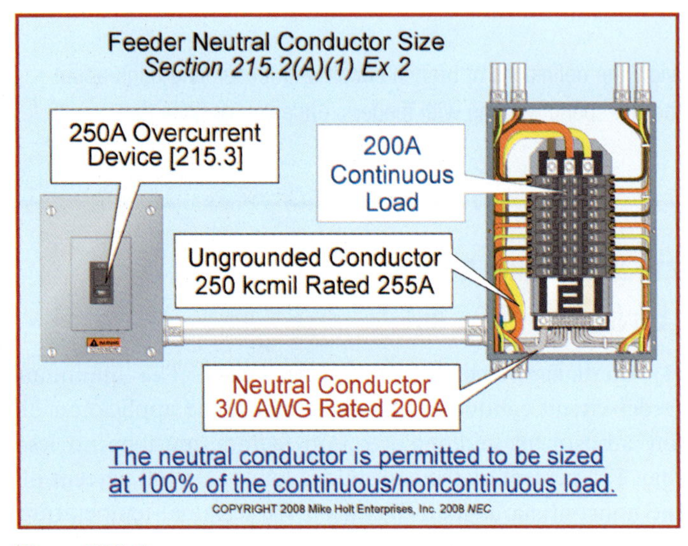

Figure 215–3

Question: *What size feeder conductors are required for a 200A continuous load if the terminals are rated 75°C?*

(a) 2/0 AWG ungrounded conductors and a 1/0 AWG neutral conductor

(b) 3/0 AWG ungrounded conductors and a 1/0 AWG neutral conductor

(c) 4/0 AWG ungrounded conductors and a 1/0 AWG neutral conductor

(d) 250 kcmil ungrounded conductors and a 3/0 AWG neutral conductor

Answer: *(d) 250 kcmil AWG ungrounded conductors and a 3/0 AWG neutral conductor*

Since the load is 200A continuous, the feeder conductors must have an ampacity of not less than 250A (200A x 1.25). The neutral conductor is sized to the 200A continuous load according to the 75°C column of Table 310.16. According to the 75°C column of Table 310.16, 250 kcmil has an ampere rating of 255A, and 3/0 has an ampacity of 200A.

The feeder neutral conductor must be sized to carry the maximum unbalanced load, in accordance with 220.61, and must not be smaller than the size listed in 250.122, based on the rating of the feeder overcurrent device.

Question: *What size neutral conductor is required for a feeder consisting of 250 kcmil ungrounded conductors and one neutral conductor protected by a 250A overcurrent device, where the unbalanced load is only 50A, with 75°C terminals?* **Figure 215–4**

(a) 6 AWG *(b) 4 AWG* *(c) 1/0 AWG* *(d) 3/0 AWG*

Answer: *(b) 4 AWG [based on Table 250.122]*

Table 310.16 and 220.61 permit an 8 AWG neutral conductor, rated 50A at 75°C to carry the 50A unbalanced load, but the neutral conductor is not permitted to be smaller than 4 AWG, as listed in Table 250.122 based on the 250A overcurrent device.

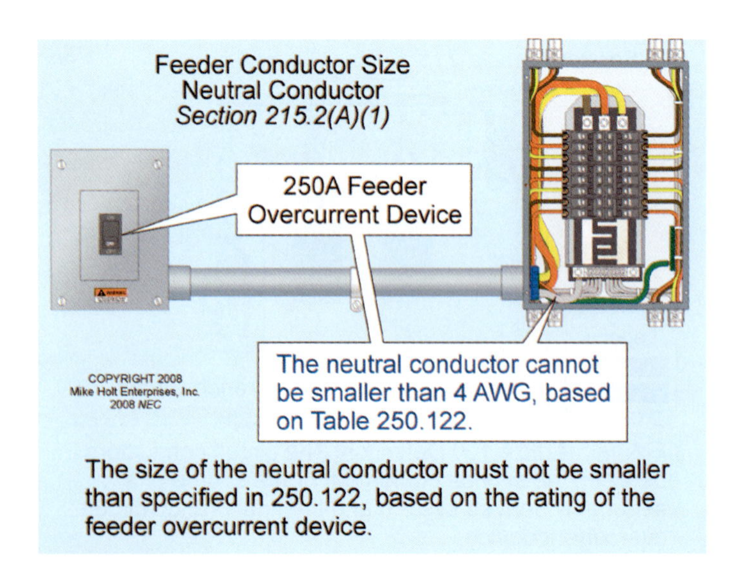

Figure 215–4

(2) Ampacity Relative to Service Conductors. The feeder conductor ampacity must not be less than that of the service conductors where the feeder conductors carry the total load supplied by service conductors with an ampacity of 55A or less.

(3) Dwelling Unit and Mobile Home Feeder Sizing. Feeder conductors for individual dwelling units or mobile homes need not be larger than service conductors sized to 310.15(B)(6).

> **FPN No. 2:** To provide reasonable efficiency of operation of electrical equipment, feeder conductors should be sized to prevent a voltage drop not to exceed 3 percent. In addition, the maximum total voltage drop on both feeders and branch circuits should not exceed 5 percent.

> **FPN No. 3:** See 210.19(A), FPN No. 4, for voltage drop for branch circuits.

215.3 Overcurrent Protection.

Feeder overcurrent devices must have an ampacity of not less than 125 percent of the continuous loads, plus 100 percent of the noncontinuous loads. Figure 215-5

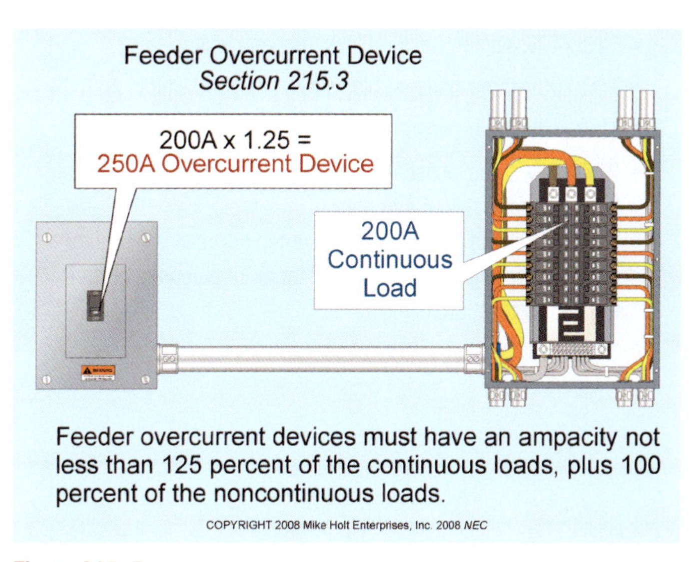

Figure 215–5

> **Author's Comment:** See 215.2(A)(1) for feeder conductor sizing requirements.

Exception: Where the assembly and the overcurrent device are both listed for operation at 100 percent of its rating, the overcurrent device can be sized at 100 percent of the continuous load.

> **Author's Comment:** Equipment suitable for 100 percent continuous loading is rarely available in ratings under 400A.

215.6 Equipment Grounding Conductor.

Feeder circuits must include or provide an equipment grounding conductor of a type listed in 250.118, and it must terminate in a manner so that branch-circuit equipment grounding conductors can be connected to it, and installed in accordance with 250.134. Figure 215-6

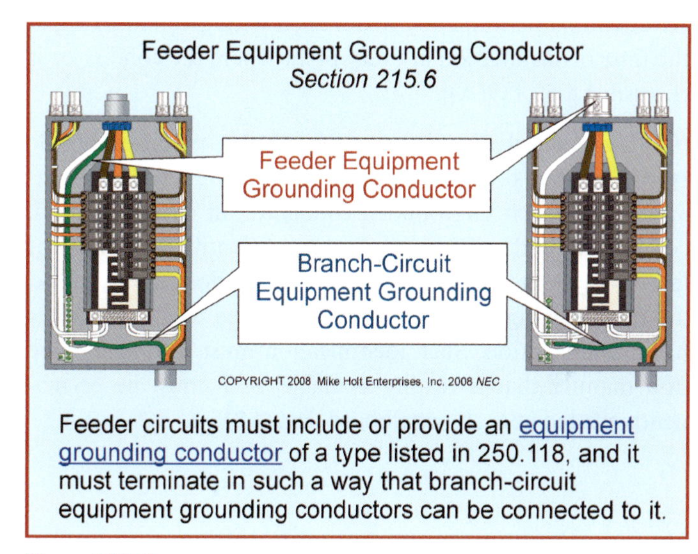

Figure 215–6

215.10 Ground-Fault Protection of Equipment.

Each feeder disconnect rated 1,000A or more supplied by a 4-wire, three-phase, 277/480V wye-connected system must be provided with ground-fault protection of equipment in accordance with 230.95 and 240.13.

> **Author's Comment:** See the definition of "Ground-Fault Protection of Equipment" in Article 100.

Exception No. 2. Equipment ground-fault protection isn't required if ground-fault protection of equipment is provided on the supply side of the feeder and on the load side of the transformer supplying the feeder.

> **Author's Comment:** Ground-fault protection of equipment isn't permitted for fire pumps [695.6(H)] and it's not required for emergency systems [700.26] or legally required standby systems [701.17].

215.12 Identification for Feeders.

(A) Neutral Conductor. The neutral conductor of a feeder must be identified in accordance with 200.6.

(B) Equipment Grounding Conductor. Equipment grounding conductors can be bare, and individually covered or insulated equipment grounding conductors sized 6 AWG and smaller must have a continuous outer finish either green or green with one or more yellow stripes [250.119].

Insulated equipment grounding conductors larger than 6 AWG can be permanently reidentified with green marking at the time of installation at every point where the conductor is accessible [250.119(A)].

(C) Ungrounded Conductors. Where the premises wiring system contains feeders supplied from more than one voltage system, each ungrounded conductor, at all termination, connection, and splice points, must be identified by phase or line and system. Identification can be by color coding, marking tape, tagging, or other means approved by the authority having jurisdiction. Such identification must be documented in a manner that is readily available, or it must be permanently posted at each panelboard. Figure 215-7

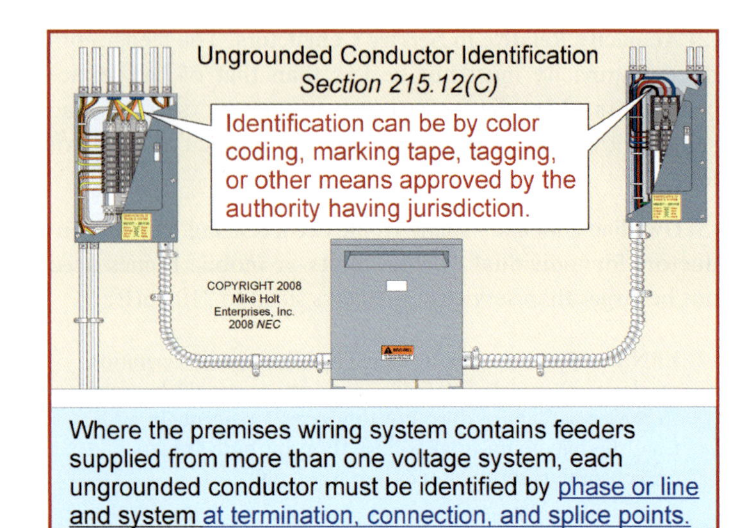

Figure 215–7

Author's Comment: Although the *NEC* doesn't require a specific color code for ungrounded conductors, electricians often use the following color system for power and lighting conductor identification:

- 120/240V, single-phase—black, red, and white
- 120/208V, three-phase—black, red, blue, and white
- 120/240V, three-phase—black, orange, blue, and white
- 277/480V, three-phase—brown, orange, yellow, and gray; or, brown, purple, yellow, and gray
- Conductors with insulation that is green or green with one or more yellow stripes can't be used for an ungrounded or neutral conductor [250.119].

ARTICLE 215 Practice Questions

ARTICLE 215. FEEDERS—PRACTICE QUESTIONS

1. The feeder conductor ampacity shall not be less than that of the service-entrance conductors where the feeder conductors carry the total load supplied by service conductors with an ampacity of _____ or less.

   (a) 30A
   (b) 55A
   (c) 60A
   (d) 100A

2. When a feeder supplies _____ in which equipment grounding conductors are required, the feeder shall include or provide an equipment grounding conductor.

   (a) an equipment disconnecting means
   (b) electrical systems
   (c) branch circuits
   (d) electric-discharge lighting equipment

3. Ground-fault protection of equipment shall be required for a feeder disconnect if _____.

   (a) the disconnect is rated 800A, 208V
   (b) the disconnect is rated 800A, 480V
   (c) the disconnect is rated 1,000A, 208V
   (d) the disconnect is rated 1,000A, 480V

4. Where a premises wiring system contains feeders supplied from more than one nominal voltage system, each ungrounded conductor of a feeder shall be identified by phase and system by _____, or other approved means.

   (a) color coding
   (b) marking tape
   (c) tagging
   (d) any of these

Branch-Circuit, Feeder, and Service Calculations

INTRODUCTION TO ARTICLE 220—BRANCH-CIRCUIT, FEEDER, AND SERVICE CALCULATIONS

This article provides the requirements for calculating branch circuits, feeders, and services, and for determining the number of receptacles on a nondwelling unit circuit, as well as the number of branch circuits required. It consists of five parts:

Part I describes the layout of Article 220 and provides a table of where other types of load calculations can be found in the *NEC*. Part II provides requirements for branch-circuit calculations and for specific types of branch circuits. Part III provides requirements for feeder and service calculations, just as the title says. Part IV provides some shortcut calculations you can use in place of the more complicated calculations provided in Parts II and III—if your installation meets certain requirements. Part IV covers just what it says, Farm Load Calculations.

Because often times the standard method (Part III) or the optional method (Part IV) can be used, you need to remember that there can be two right answers when doing the calculations.

Study Article 220 carefully. If something doesn't make sense at first, make a note of it and take a short break from your studies. Then go back to that item and read through the explanation, using the illustrations to help you understand. In addition to the illustrations, take some time to review the examples in Annex D of the *NEC*. The more practice you have with these calculations, the easier they will come to you. What you learn here will really stick if you also consider the why, not just the how.

PART I. GENERAL

220.1 Scope. This article contains the requirements necessary for calculating branch circuits, feeders, and services. In addition, this article can be used to determine the number of receptacles on a circuit and the number of general-purpose branch circuits required.

220.3 Application of Other Articles. Other articles contain calculations that are in addition to, or modify, those contained within Article 220. Take a moment to review the following additional calculation requirements found in these sections:

- Air-Conditioning and Refrigeration Equipment, 440.6, 440.21, 440.22, 440.31, 440.32, and 440.62
- Appliances, 422.10 and 422.11
- Branch Circuits, 210.19 and 210.20(A)
- Computers (Data Processing Equipment), 645.4 and 645.5(A)
- Conductors, 310.15
- Feeders, 215.2(B) and 215.3
- Fire Pumps, 695.7
- Fixed Electric Space-Heating Equipment, 424.3(B)
- Marinas, 555.12, 555.19(A)(4), and 555.19(B)
- Mobile Homes and Manufactured Homes, 550.12 and 550.18
- Motors, 430.6(A), 430.22(A), 430.24, 430.52, and 430.62
- Overcurrent Protection, 240.4 and 240.15
- Refrigeration (Hermetic), 440.6 and Part IV
- Recreational Vehicle Parks, 551.73(A)
- Electronic Equipment, 647.4(D)
- Services, 230.42(A) and 230.79
- Signs, 600.5
- Transformers, 450.3

220.5 Calculations.

(A) Voltage Used for Calculations. Unless other voltages are specified, branch-circuit, feeder, and service loads must be calculated on nominal system voltages, such as 120V, 120/240V, 120/208V, 240V, 277/480V, or 480V. Figure 220–1

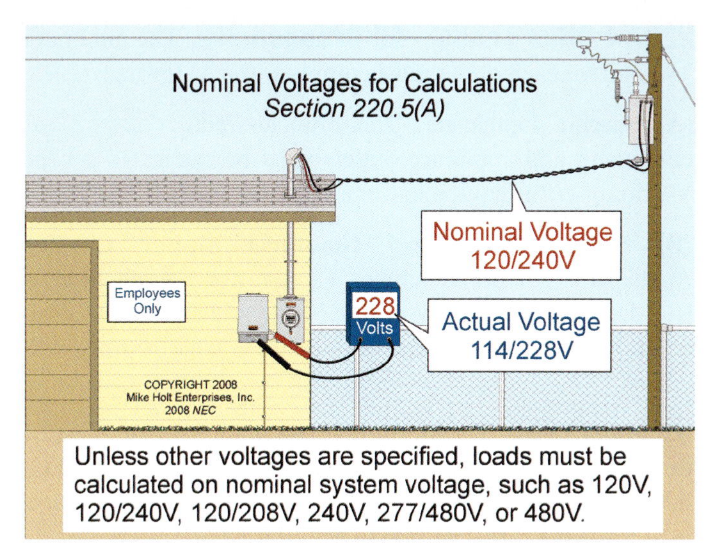

Figure 220–1

Author's Comment: A nominal value is assigned to a circuit for the purpose of convenient circuit identification. The actual voltage at which a circuit operates can vary from the nominal within a range that permits satisfactory operation of equipment [Article 100].

(B) Fractions of an Ampere (Rounding Amperes). Calculations that result in a fraction of less than one-half of an ampere can be dropped.

Author's Comment: When do you round—after each calculation, or at the final calculation? The *NEC* isn't specific on this issue, but rounding at each calculation can lead to accumulated errors that could be an issue with the authority having jurisdiction.

Question: *According to 424.3(B), the branch-circuit conductors and overcurrent device for electric space-heating equipment must be sized at no less than 125 percent of the total load. What size conductor is required to supply a 9 kW (37.50A), 240V, single-phase fixed space heater with a 3A blower motor, if equipment terminals are rated 75°C?* **Figure 220–2**

(a) 10 AWG *(b) 8 AWG* *(c) 6 AWG* *(d) 4 AWG*

Answer: *(c) 6 AWG*

Step 1: Determine the load for the heater:

I = VA/E
I = 9,000 VA/240V
I = 37.50A

Step 2 Conductor size at 125% of the load:

Conductor Size = (37.50A + 3A) x 1.25
Conductor Size = 50.63A, round up to 51A

If we rounded down, then 8 AWG rated 50A at 75°C could be used, but since we must round up, 6 AWG rated 65A at 75°C is required.

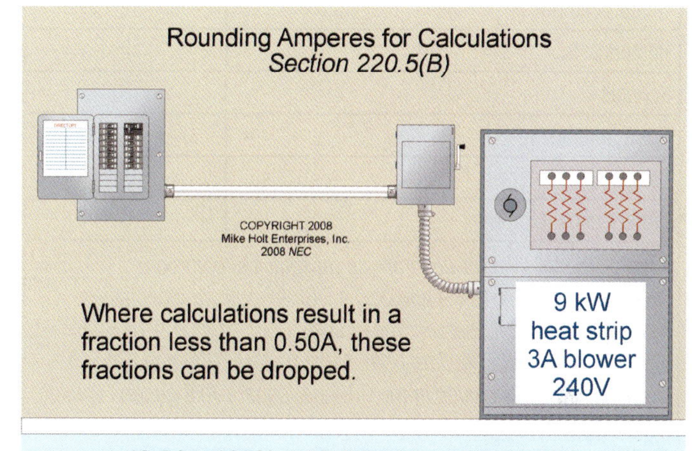

Figure 220–2

PART II. BRANCH-CIRCUIT LOAD CALCULATIONS

220.12 General Lighting. The general lighting load specified in Table 220.12 must be calculated from the outside dimensions of the building or area involved. **Figure 220-3**

Table 220.12 General Lighting VA per Square Foot

Occupancy	VA per Sq. Ft
Armories and auditoriums	1
Assembly halls and auditoriums	1
Banks	3½[b]
Barber shops and beauty parlors	3
Churches	1
Clubs	2
Courtrooms	2
Dwelling units	3[a]

Table 220.12 General Lighting VA per Square Foot	
Garages—commercial (storage)	½
Halls, corridors, closets, stairways	½
Hospitals	2
Hotels and motels without cooking facilities	2
Industrial commercial (loft buildings)	2
Lodge rooms	1½
Office buildings	3½[b]
Restaurants	2
Schools	3
Storage spaces	¼
Stores	3
Warehouses (storage)	¼
Note a: The VA load for general-use receptacles, bathroom receptacles [220.14(J)(1) and 210.11(C)(3)], outside receptacles, as well as garage, basement receptacles [220.14(J)(2), 210.52(E) and (G)], and lighting outlets [220.14(J)(3), 210.70(A) and (B)] in a dwelling unit are included in the 3 VA per-square-foot general lighting [220.14(J)].	
Note b: The receptacle calculated load for banks and office buildings is the largest calculation of either (1) or (2) [220.14(K)].	

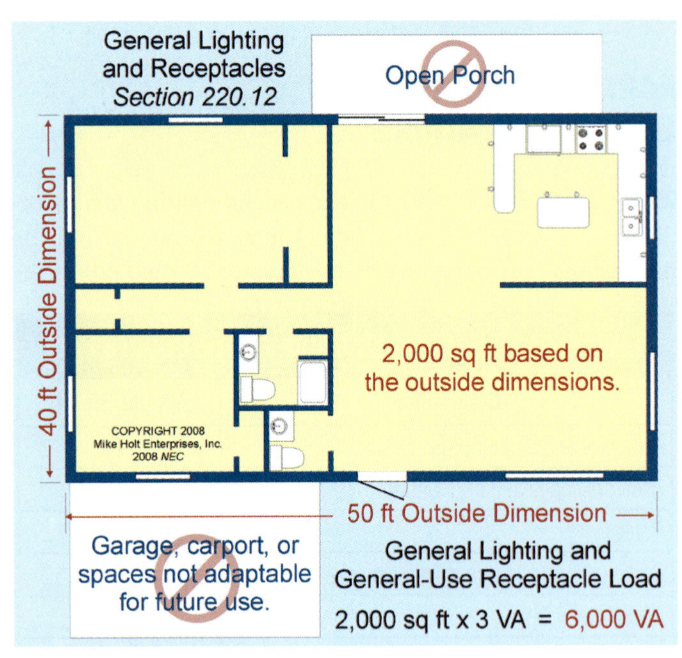

Figure 220–3

220.14 Other Loads—All Occupancies. The minimum VA load for each outlet must comply with (A) through (L).

(A) Specific Equipment. The branch-circuit VA load for equipment and appliance outlets must be calculated on the VA rating of the equipment or appliance.

(B) Electric Dryers and Household Electric Cooking Appliances. The branch-circuit VA load for household electric dryers must comply with 220.54, and household electric ranges and other cooking appliances must comply with 220.55.

(C) Motor Loads. The motor branch-circuit VA load must be determined by multiplying the motor full-load current (FLC) listed in Table 430.248 or 430.250 by the motor table voltage, in accordance with 430.22 [430.6(A)(1)].

(D) Luminaires. The branch-circuit VA load for recessed luminaires must be calculated based on the maximum VA rating for which the luminaires are rated.

(E) Heavy-Duty Lampholders. The branch-circuit VA load for heavy-duty lampholders must be calculated at a minimum of 600 VA.

(F) Sign Outlet. Each commercial occupancy that is accessible to pedestrians must have at least one 20A sign outlet [600.5(A)], which must have a minimum branch-circuit load of 1,200 VA. Figure 220–4

Figure 220–4

(G) Show Windows. The branch-circuit VA load for show-window lighting must be calculated in accordance with (1) or (2):

(1) 180 VA per outlet in accordance with 220.14(L), or

(2) 200 VA per linear foot of show-window lighting [220.43]. Figure 220–5

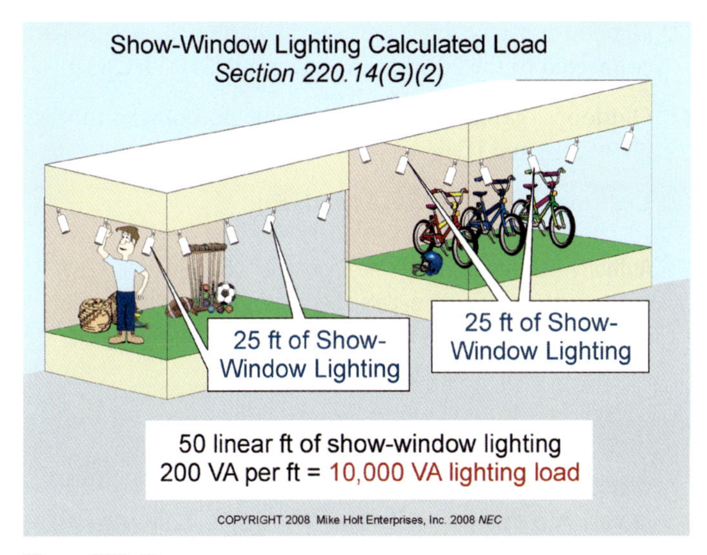

Figure 220–5

(H) Fixed Multioutlet Assemblies. Fixed multioutlet assemblies in commercial occupancies used in other than dwelling units, or in the guest rooms of hotels or motels, must be calculated in accordance with (1) or (2). Figure 220–6

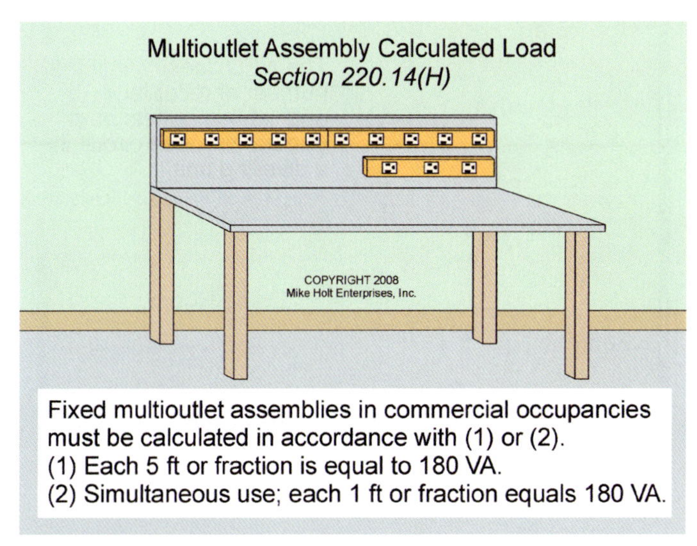

Figure 220–6

(1) Where appliances are unlikely to be used simultaneously, each 5 ft or fraction of 5 ft of multioutlet assembly is considered as one outlet of 180 VA.

(2) Where appliances are likely to be used simultaneously, each 1 ft or fraction of a foot of multioutlet assembly is considered as one outlet of 180 VA.

Author's Comments:

- See the definition of "Multioutlet Assembly" in Article 100.
- The feeder or service calculated load for fixed multioutlet assemblies can be calculated in accordance with the demand factors contained in 220.44.

(I) Receptacle Outlets. Except as covered in 200.14(J) and (K), each 15A or 20A, 125V general-use receptacle outlet is considered as a 180 VA per mounting strap. Figure 220–7

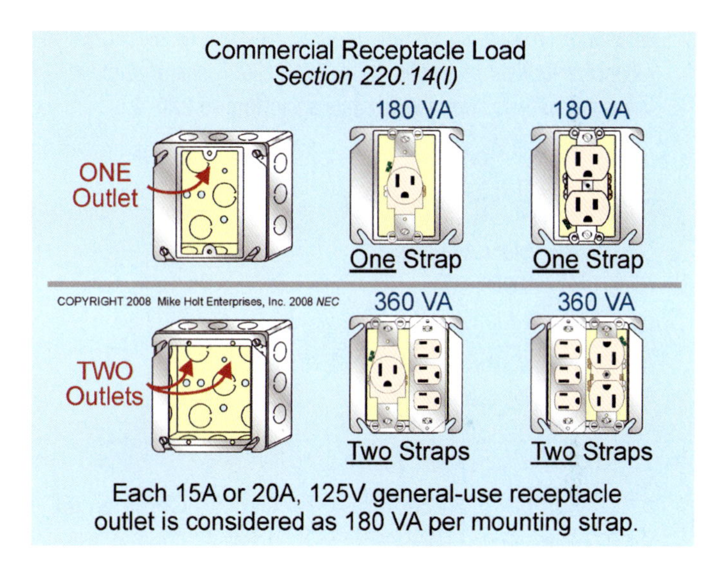

Figure 220–7

A single device consisting of four or more receptacles is considered as 90 VA per receptacle (360 VA for a quad receptacle). Figure 220–8

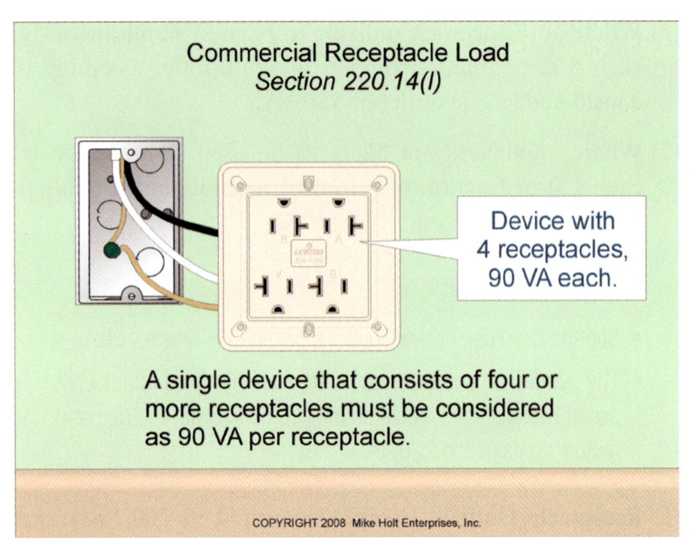

Figure 220–8

Question: *What is the maximum number of 15A or 20A, 125V receptacle outlets permitted on a 20A, 120V general-purpose branch circuit in a commercial occupancy?* **Figure 220–9**

(a) 4 *(b) 6* *(c) 10* *(d) 13*

Answer: *(d) 13*

Circuit VA = Volts x Amperes
Circuit VA = 120V x 20A
Circuit VA = 2,400 VA

Number of Receptacles = 2,400 VA/180 VA
Number of Receptacles = 13

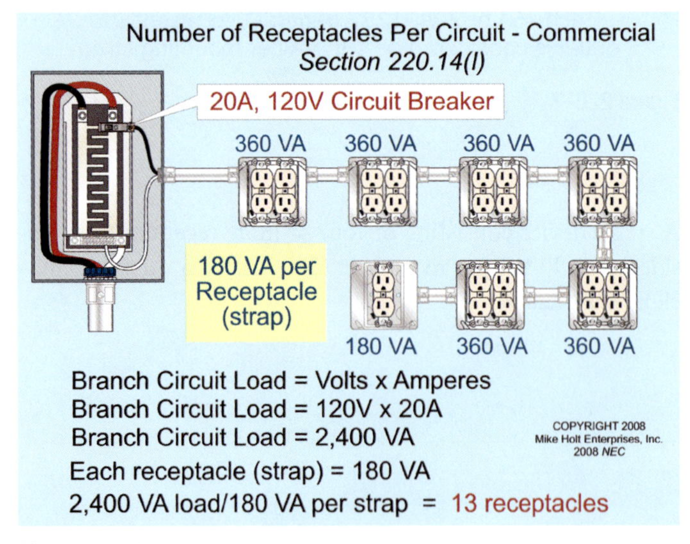

Figure 220–9

Author's Comment: According to the *NEC* Handbook, published by the NFPA, general-purpose receptacles aren't considered a continuous load.

(J) Residential Receptacle Load. In one-family, two-family, and multifamily dwellings, and in guest rooms of hotels and motels, the outlets specified in (1), (2), and (3) are included in the general lighting load calculations of 220.12.

(1) General-use receptacle outlets, including the receptacles connected to the 20A bathroom circuit [210.11(C)(3)].

(2) Outdoor, garage, and basement receptacle outlets [210.52(E) and (G)].

(3) Lighting outlets [210.70(A) and (B)].

Author's Comment: There's no VA load for 15A and 20A, 125V general-use receptacle outlets, because the loads for these devices are part of the 3 VA per-square-foot for general lighting as listed in Table 220.12 for dwelling units.

Question: *What is the maximum number of 15A or 20A, 125V receptacle outlets permitted on a 15A or 20A, 120V general-purpose branch circuit in a dwelling unit?* **Figure 220–10**

(a) 4 *(b) 6* *(c) 8* *(d) No limit*

Answer: *(d) No limit*

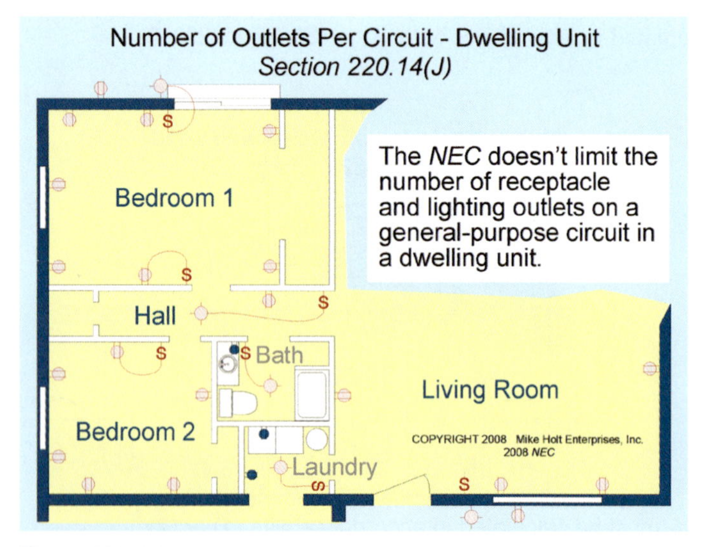

Figure 220–10

Author's Comment: The *NEC* doesn't limit the number of receptacle outlets on a general-purpose branch circuit in a dwelling unit. See the *NEC Handbook* for more information.

CAUTION: *There might be a local Code requirement that limits the number of receptacle outlets on a general-purpose branch circuit.*

Author's Comment: Although there's no limit on the number of receptacle outlets on dwelling general-purpose branch circuits, the *NEC* does require a minimum number of circuits to be installed for general-purpose receptacles and lighting outlets [210.11(A)]. In addition, the receptacle and lighting loads must be evenly distributed among the required circuits [210.11(B)].

(K) Banks and Office Buildings. The receptacle calculated load for banks and office buildings is the largest calculation of either (1) or (2).

(1) Determine the receptacle calculated load at 180 VA per receptacle yoke [220.14(I)], then apply the demand factor from Table 220.44, or

(2) Determine the receptacle load at 1 VA per sq ft.

Bank or Office General Lighting and Receptacle —Example 1

Question: *What is the calculated receptacle load for an 18,000 sq ft bank with 160 15A, 125V receptacles?*

(a) 15,400 VA (b) 19,400 VA (c) 28,800 VA (d) 142 kVA

Answer: *(b) 19,400 VA*

220.14(K)(1) and 220.14(I)
160 Receptacles x 180 VA = 28,800 VA

First 10,000 at 100% =	*–10,000 VA x 1.00 =*	*10,000 VA*
Remainder at 50% =	*18,800 VA x 0.50 =*	*+ 9,400 VA*
Receptacle Calculated Load =		*19,400 VA*

220.14(K)(2)
18,000 x 1 VA per sq ft *18,000 VA (smaller, omit*

Bank or Office General Lighting and Receptacle —Example 2

Question: *What is the receptacle calculated load for an 18,000 sq ft bank with 140 15A, 125V receptacles?* **Figure 220–11**

(a) 15,000 VA (b) 18,000 VA (c) 23,000 VA (d) 31,000 VA

Answer: *(b) 18,000 VA*

220.14(K)(1) and 220.14(I)
140 Receptacles x 180 VA = 25,200 VA

First 10,000 at 100% =	*–10,000 VA x 1.00 =*	*10,000 VA*
Remainder at 50% =	*15,200 VA x 0.50 =*	*+ 7,600 VA*
Receptacle Calculated Load =		*17,600 VA (smaller, omit)*

220.14(K)(2)
18,000 x 1 VA per sq ft = *18,000 VA*

Figure 220–11

(L) Other Outlets. Receptacle and lighting outlets not covered in (A) through (K) must have the branch-circuit VA load calculated at 180 VA per outlet.

220.18 Maximum Load on a Branch Circuit.

(A) Motor Operated Loads. Branch circuits that supply motor loads must be sized not smaller than 125 percent of the motor FLC, in accordance with 430.6(A) and 430.22(A).

Question: *What is the minimum size branch-circuit conductor for a 2 hp, 230V motor, where the conductor terminals are rated 60°C?* **Figure 220–12**

(a) 14 AWG *(b) 12 AWG* *(c) 10 AWG* *(d) 8 AWG*

Answer: *(a) 14 AWG*

Step 1: Determine the motor full-load current: [Table 430.248]:

2 hp FLC = 12A

Step 2: Size the branch-circuit conductors at 125 percent of the FLC in accordance with Table 310.16 [430.22(A)]:

Branch-Circuit Conductors = 12A x 1.25

Branch-Circuit Conductors = 15A, 14 AWG rated 20A at 60°C.

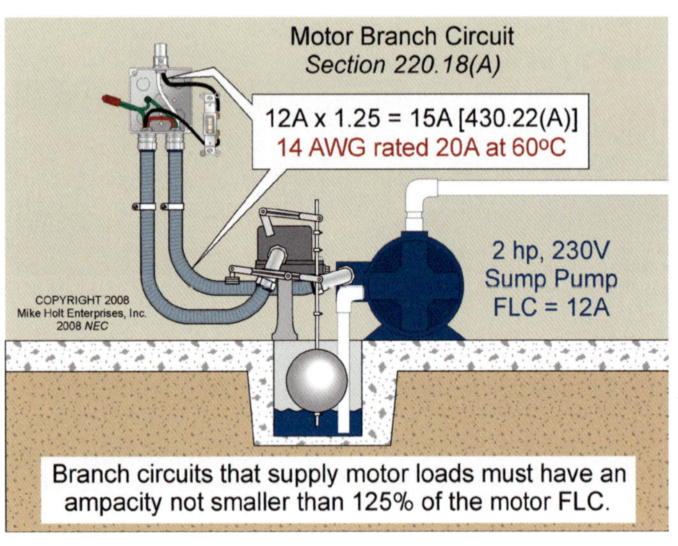

Figure 220–12

(B) Inductive Lighting Loads. Branch circuits that supply inductive luminaires, such as fluorescent and HID fixtures, must be sized to the ampere rating of the luminaire, not to the wattage of the lamps. **Figure 220–13**

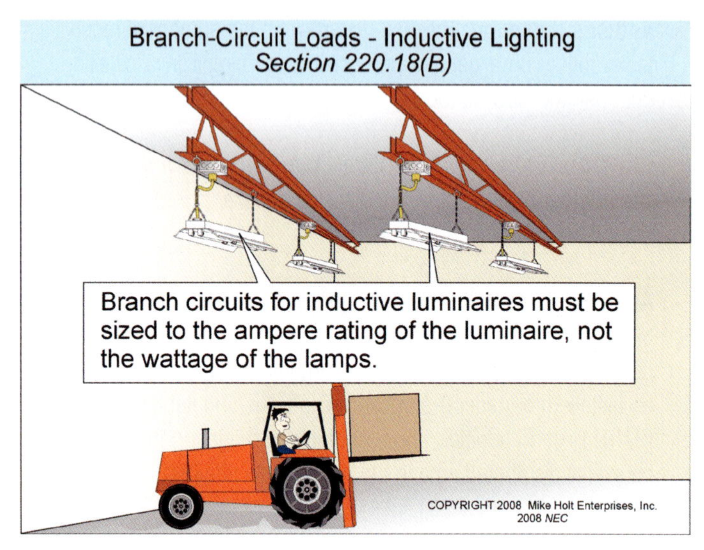

Figure 220–13

Question: *What is the maximum number of 1.34A fluorescent luminaires permitted on a 20A circuit if the luminaires operate for more than three hours?*

(a) 8 *(b) 11* *(c) 13* *(d) 15*

Answer: *(b) 11*

The maximum continuous load must not exceed 80 percent of the circuit rating [210.19(A)(1)].

Maximum Load = 20A x 0.80

Maximum Load = 16A

Luminaires on Circuit = 16A/1.34A

Luminaires on Circuit = 11.94 or 11 luminaires

Author's Comment: Because of power factor (inductive luminaires), the input VA of each luminaire is 162 VA (120V x 1.34A), which is greater than the 136W (34W x 4 lamps) of the lamps. This may seem complicated, but just remember—size all circuits that supply inductive loads to the ampere rating of the luminaire, not to the wattage of the lamps.

(C) Household Cooking Appliances. Branch-circuit conductors for household cooking appliances can be sized in accordance with Table 220.55; specifically, Note 4 for branch circuits.

Author's Comment: For ranges rated 8.75 kW or more, the minimum branch-circuit rating is 40A [210.19(A)(3)].

PART III. FEEDER AND SERVICE CALCULATIONS

220.40 General. The calculated load for a feeder or service must not be less than the sum of the branch-circuit loads, as determined by Part II of this article, as adjusted for the demand factors contained in Parts III, IV, or V.

FPN: See Examples D1(A) through D10 in Annex D.

220.42 General Lighting Demand Factors. The *Code* recognizes that not all luminaires will be on at the same time, and it permits the following demand factors to be applied to the general lighting load as determined in Table 220.42.

Table 220.42 General Lighting Demand Factors

Type of Occupancy	Lighting VA Load	Demand Factor
Dwelling Units	First 3,000 VA Next 117,000 VA Remainder at	100% 35% 25%
Hotels/motels without provision for cooking	First 20,000 VA Next 80,000 VA Remainder at	50% 40% 30%
Warehouses (storage)	First 12,500 VA Remainder	100% 50%
All others	Total VA	100%

Question: *What is the general lighting and receptacle calculated load, after demand factors, for a 40 x 50 ft (2,000 sq ft) dwelling unit?* Figure 220–14

(a) 2,050 VA (b) 3,050 VA (c) 4,050 VA (d) 5,050 VA

Answer: *(c) 4,050 VA*

General Lighting = 40 x 50 ft
General Lighting = 2,000 sq ft x 3 VA per sq ft
General Lighting = 6,000 VA

First 3,000 VA at 100% 3,000 VA x 1.00 = 3,000 VA
Next 117,000 VA at 35% 3,000 VA x 0.35 = + 1,050 VA

General Lighting and General-Use Receptacles Calculated Load = 4,050 VA

Author's Comment: For commercial occupancies, the VA load for receptacles [220.14(I)] and fixed multioutlet assemblies [220.14(H)] can be added to the general lighting load and subjected to the demand factors of Table 220.42 [220.44].

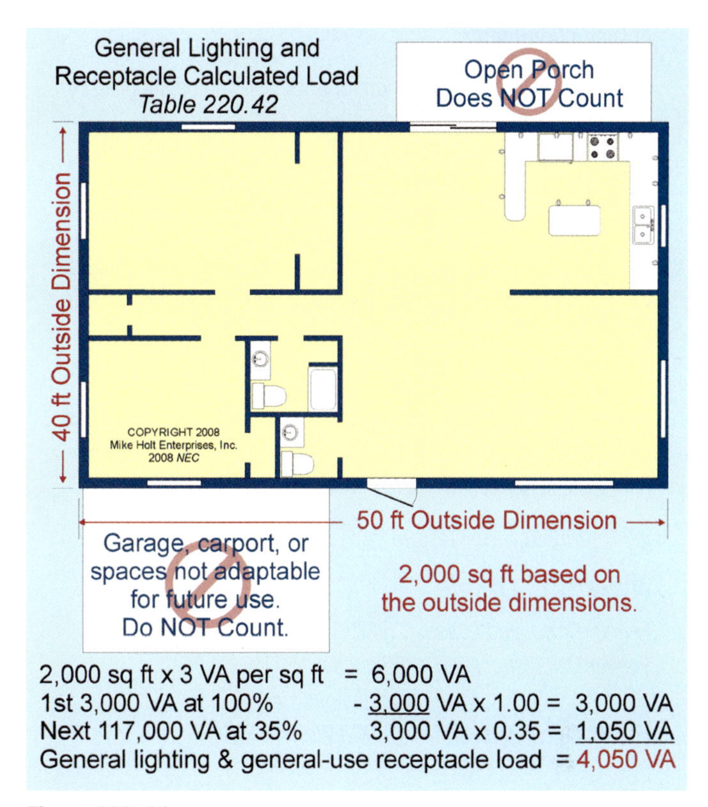

Figure 220–14

220.43 Commercial—Show Window and Track Lighting Load.

(A) Show Windows. The feeder/service VA load must not be less than 200 VA per-linear-foot.

(B) Track Lighting. The feeder/service VA load must not be less than 150 VA for every 2 ft of track lighting or fraction of that length. Figure 220–15

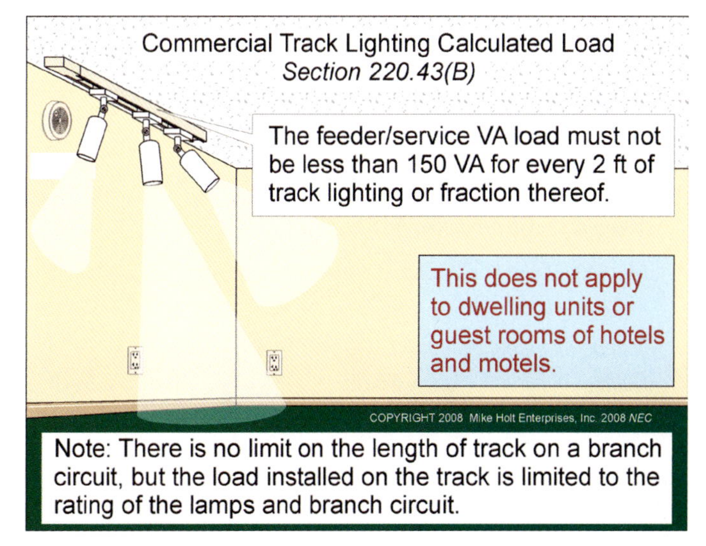

Figure 220–15

Author's Comments:

- There is no limit on the length of track that can be supplied by a single branch circuit.
- Where a feeder or service supplies continuous loads, the minimum feeder or service conductor size, before the application of any adjustment and/or correction factors, must have an allowable ampacity of at least 125 percent of the continuous load [215.2(A)(1) for feeders and 230.42(A) for services].

***Question:** What is the approximate feeder/service calculated load for conductor sizing for 150 ft of track lighting in a commercial occupancy?*

(a) 10,000 VA (b) 12,000 VA (c) 14,000 VA (d) 16,000 VA

***Answer:** (c) 14,000 VA*

Feeder Calculated Load = 150 ft/2 ft
Feeder Calculated Load = 75 units x 150 VA x 1.25 (continuous load)
Feeder Calculated Load = 14,063 VA

Author's Comment: This rule doesn't apply to branch circuits. Therefore, the maximum number of lampholders permitted on a track lighting system is based on the wattage rating of the lamps and the voltage and ampere rating of the circuit [410.151(B)]. Because lighting is a continuous load, the maximum load on a branch circuit must not exceed 80 percent of the circuit rating [210.19(A)(1)].

***Question:** How many 75W lampholders can be installed on a 20A, 120V track lighting circuit in a commercial occupancy if the track is 32 ft long?*

(a) 10 (b) 15 (c) 20 (d) 25

***Answer:** (d) 25*

Maximum Load Permitted on Circuit = 20A x 0.80
Maximum Load Permitted on Circuit = 16A

Maximum Load in VA = 120V x 16A
Maximum Load in VA = 1,920 VA

Number of Lampholders = 1,920 VA/75 VA
Number of Lampholders = 25.60

Author's Comments:

- There is no limit on the length of track lighting that can be supplied by a branch circuit.
- The total wattage of the lamps on the track is not permitted to exceed the rating of the track [410.151(B)].

220.44 Other than Dwelling Unit—Receptacle Load.

The feeder/service VA load for general-purpose receptacles [220.14(I)] and fixed multioutlet assemblies [220.14(H)] is determined by:

- Adding the receptacle and fixed multioutlet assembly VA load with the general lighting load [Table 220.12] and adjusting this VA value by the demand factors contained in Table 220.42, or
- Applying a 50 percent demand factor to that portion of the receptacle and fixed multioutlet receptacle load that exceeds 10 kVA.

***Question:** What is the calculated feeder/service VA load, after demand factors, for 150 general-purpose receptacles and 100 ft of fixed multioutlet assembly in a commercial occupancy? The appliances powered by the multioutlet assembly are not used simultaneously.* **Figure 220–16**

(a) 8,500 VA (b) 10,000 VA (c) 20,300 VA (d) 27,000 VA

***Answer:** (c) 20,300 VA*

Step 1: Determine the total connected load:

Receptacle Load = 150 receptacles x 180 VA
Receptacle Load = 27,000 VA [220.14(I)]

Multioutlet Load = 100 ft/5 ft
Multioutlet Load = 20 sections x 180 VA
Multioutlet Load = 3,600 VA [220.14(H)]

Step 2: Apply Table 220.44 demand factor:

Total Connected load = 30,600 VA

First 10,000 VA at 100% = 10,000 VA x 1.00 = 10,000 VA
Remainder at 50% = 20,600 VA x 0.50 = +10,300 VA
Receptacle Calculated Load = 20,300 VA

220.50 Motor Load.

The feeder/service load for motors must be sized not smaller than 125 percent of the largest motor load, plus the sum of the other motor loads. See 430.24 for example.

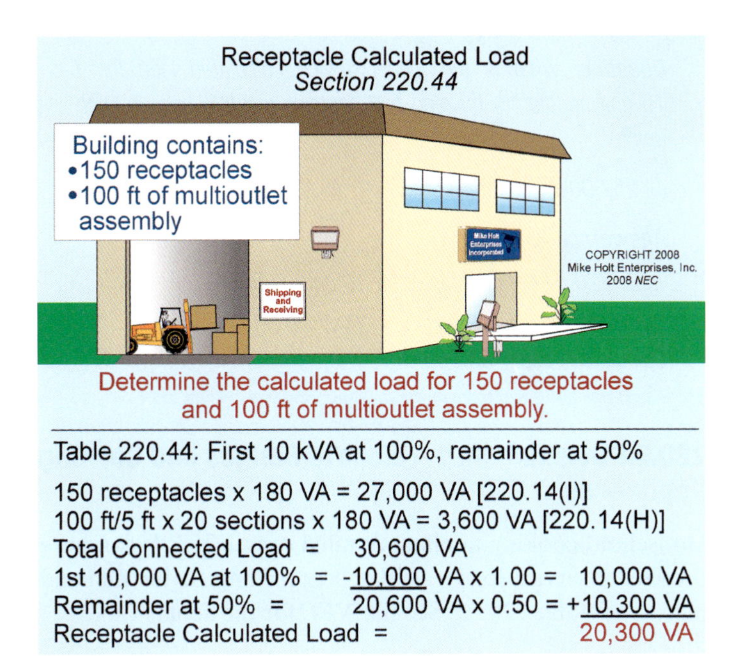

Figure 220–16

220.51 Fixed Electric Space-Heating Load.

The feeder/service load for fixed electric space-heating equipment must be calculated at 100 percent of the total connected load.

220.52 Dwelling Unit—Small-Appliance and Laundry Load.

(A) Small-Appliance Circuit Load. The feeder/service VA load for each 20A small-appliance circuit covered by 210.11(C)(1) is 1,500 VA, and this load can be subjected to the general lighting demand factors contained in Table 220.42.

Author's Comments:

- Each dwelling unit must have a minimum of two 20A, 120V small-appliance branch circuits for kitchen and dining room receptacles [210.11(C)(1)].
- The bathroom circuit covered by 210.11(C)(3) is not included in the service/feeder calculations.

(B) Laundry Circuit Load. The feeder/service VA load for each 20A laundry circuit covered by 210.11(C)(2) is 1,500 VA, and this load can be subjected to the general lighting demand factors contained in Table 220.42.

Author's Comment: A laundry circuit isn't required in each dwelling unit of a multifamily building if laundry facilities are provided on the premises for all building occupants [210.52(F) Ex 1].

220.53 Dwelling Unit—Appliance Load.

A demand factor of 75 percent can be applied to the total connected load of four or more appliances on the same feeder/service. This demand factor doesn't apply to electric space-heating equipment [220.51], electric clothes dryers [220.54], electric ranges [220.55], electric air-conditioning equipment [Article 440, Part IV], or motors [220.50].

Question: *What is the feeder/service appliance calculated load for a dwelling unit that contains a 1,000 VA disposal, a 1,500 VA dishwasher, and a 4,500 VA water heater?* **Figure 220–17**

(a) 3,000 VA (b) 4,500 VA (c) 6,000 VA (d) 7,000 VA

Answer: *(d) 7,000 VA*

No demand factor applies for three appliances.

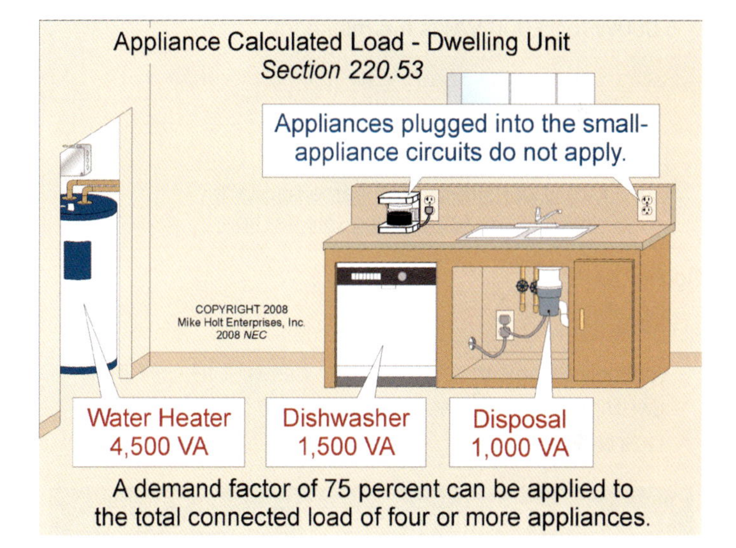

Figure 220–17

Question: *What is the feeder/service appliance calculated load, after demand factors, for a 12-unit multifamily dwelling if each unit contains a 1,000 VA disposal, a 1,500 VA dishwasher, and a 4,500 VA water heater?*

(a) 23,000 VA (b) 43,500 VA (c) 63,000 VA (d) 71,000 VA

Answer: *(c) 63,000 VA*

*Calculated Load = 7,000 VA x 12 units x 0.75**
Calculated Load = 63,000 VA

**Each dwelling unit has only three appliances, but the feeder supplies a total of 36 appliances (12 units x 3 appliances).*

220.54 Dwelling Unit—Electric Clothes Dryer Load.

The feeder/service load for electric clothes dryers located in a dwelling unit must not be less than 5,000W (5,000 VA), or the nameplate rating of the equipment if more than 5,000W (5,000 VA). Kilovolt-amperes (kVA) is considered equivalent to kilowatts (kW) for loads calculated in this section.

When a building contains five or more dryers, it's permissible to apply the demand factors listed in Table 220.54 to the total connected dryer load. Figure 220–18

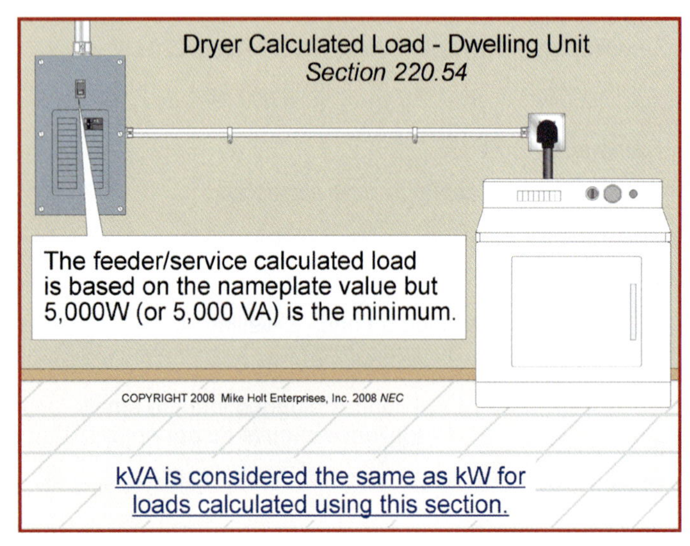

Figure 220–18

Author's Comment: A clothes dryer load isn't required if the dwelling unit doesn't have an electric clothes dryer circuit receptacle outlet.

Table 220.54 Dwelling Unit Dryer Demand Factors

Number of Dryers	Demand Factor(Percent)
1–4	100%
5	85%
6	75%
7	65%
8	60%
9	55%
10	50%
11	47%
12–23	47% minus 1% for each dryer exceeding 11
24–42	35% minus 0.50% for each dryer exceeding 23
43 and over	25%

Question: *What is the feeder/service calculated load for a 10-unit multifamily building that contains a 5 kW dryer in each unit?*

(a) 25,000W (b) 43,500W (c) 63,000W (d) 71,000W

Answer: *(a) 25,000W*

Table 220.54 demand factor for 10 units is 50%

Calculated Load = 10 units x 5,000W x 0.50
Calculated Load = 25,000W

220.55 Dwelling Unit—Electric Ranges and Cooking Appliances.

Household cooking appliances rated over 1.75 kW can have the feeder/service load calculated according to the demand factors of Table 220.55. See the *NEC* for the actual Table.

Note 1. For identically sized ranges individually rated more than 12 kW, the maximum demand in Column C must be increased 5 percent for each additional kilowatt of rating, or major fraction thereof, by which the rating of individual ranges exceeds 12 kW.

Question: *What is the feeder/service calculated load for three 15.60 kVA ranges?*

(a) 14 kVA (b) 15 kVA (c) 17 kVA (d) 21 kVA

Answer: *(c) 17 kVA (closest answer)*

Step 1: Determine the Column C demand load for 3 units: 14 kVA.

Step 2: Because the 15.60 kVA range exceeds 12 kVA by 3.60 kVA, increase the Column C demand load by 5% for each kVA or major fraction of kVA in excess of 12 kVA.

Step 3: Because 3.60 kVA is 3 kVA plus a major fraction of a kVA, increase the Column C value by 4 x 5% = 20%

Increase the Column C load (14 kVA) by 20%:
14 kVA x 1.20 = 16.80 kVA.

Note 2. For ranges individually rated more than 8.75 kW, but none exceeding 27 kW, and of different ratings, an average rating must be calculated by adding together the ratings of all ranges to obtain the total connected load (using 12 kW for any range rated less than 12 kW) and dividing this total by the number of ranges. Then the maximum demand in Column C must be increased 5 percent for each kilowatt, or major fraction thereof, by which this average value exceeds 12 kW.

Question: *What is the feeder/service calculated load for three ranges rated 9 kVA and three ranges rated 14 kVA?*

(a) 22 kVA (b) 36 kVA c) 42 kVA (d) 78 kVA

Answer: *(a) 22 kVA*

Step 1: Determine the total connected load:

9 kVA (minimum 12 kVA) 3 Ranges x 12 kVA = 36 kVA
14 kVA 3 Ranges x 14 kVA = + 42 kVA
Total Connected Load = 78 kVA

Step 2 Determine the average of range ratings:

78 kVA/6 units = 13 kVA average rating

Step 3: Demand load from Table 220.55 Column C:

6 ranges = 21 kVA

Step 4: Because the average of the ranges (13 kVA) exceeds 12 kVA by 1 kVA, increase the Column C demand load (21 kVA) by 5%:

Calculated Load = 21 kVA x 1.05
Calculated Load = 22.05 kVA

Note 4. It's permissible to compute the branch-circuit load for one range in accordance with Table 220.55.

Question: *What is the branch-circuit calculated load in amperes for a single 12 kW range connected on a 120/240V circuit?* **Figure 220–19**

(a) 20A (b) 33A (c) 41A (d) 50A

Answer: *(b) 33A*

Column C Calculated Load = 8 kW
Branch-Circuit Load in Amperes, I = P/E

P = 8,000W

E = 240V

I = 8,000W/240V
I = 33.33A

Note 4. The branch-circuit load for one wall-mounted oven or one counter-mounted cooking unit must be the nameplate rating of the appliance.

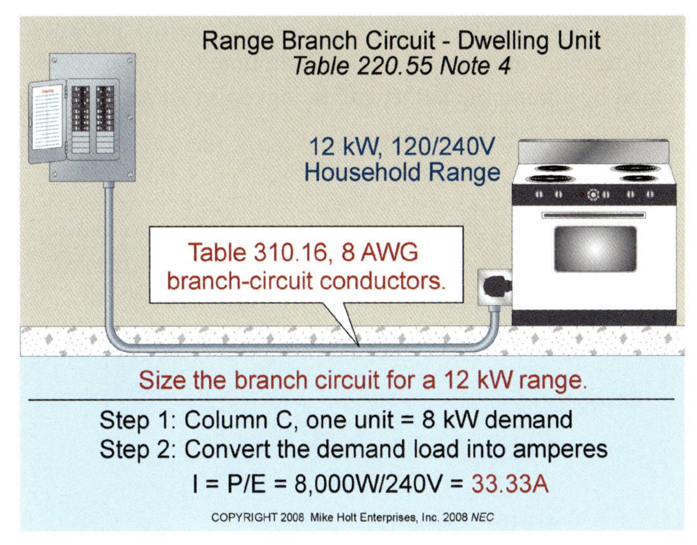

Figure 220–19

Question: *What size branch-circuit conductors are required for a 6 kW wall-mounted oven connected on a 120/240V circuit?* **Figure 220–20**

(a) 14 AWG (b) 12 AWG (c) 10 AWG (d) 8 AWG

Answer: *(c) 10 AWG*

Branch-circuit load in amperes, I = P/E

P = 6,000W
E = 240V

I = 6,000W/240V
I = 25A, 10 AWG rated 30A at 60°C, [Table 310.16 and 110.14(C)(1)]

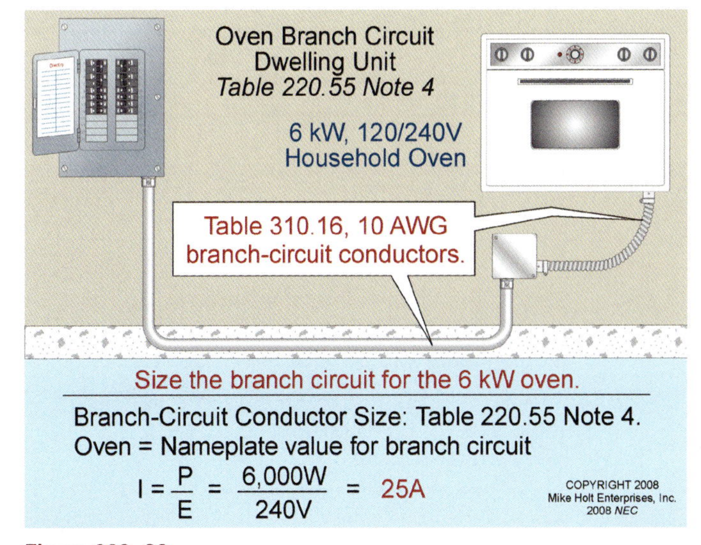

Figure 220–20

Note 4. The branch-circuit load for one counter-mounted cooking unit and up to two wall-mounted ovens is determined by adding the nameplate ratings together and treating that value as a single range.

Question: *What size branch circuit is required for one 6 kW counter-mounted cooking unit and two 3 kW wall-mounted ovens connected on a 120/240V circuit?* **Figure 220–21**

(a) 14 AWG *(b) 12 AWG* *(c) 10 AWG* *(d) 8 AWG*

Answer: *(d) 8 AWG*

Step 1: Determine the total connected load:

Total Connected Load = (6 kW + 3 kW + 3 kW).
Total Connected Load = 12 kW

Step 2: Determine the calculated VA load as a single 12 kW range:

Table 220.55 Column C = 8 kW

Step 3: Determine the branch-circuit load in amperes, I = P/E:

P = 8,000W
E = 240V

I = 8,000W/240V
I = 33.33A, 8 AWG rated 40A at 60°C [Table 310.16 and 110.14(C)(1)]

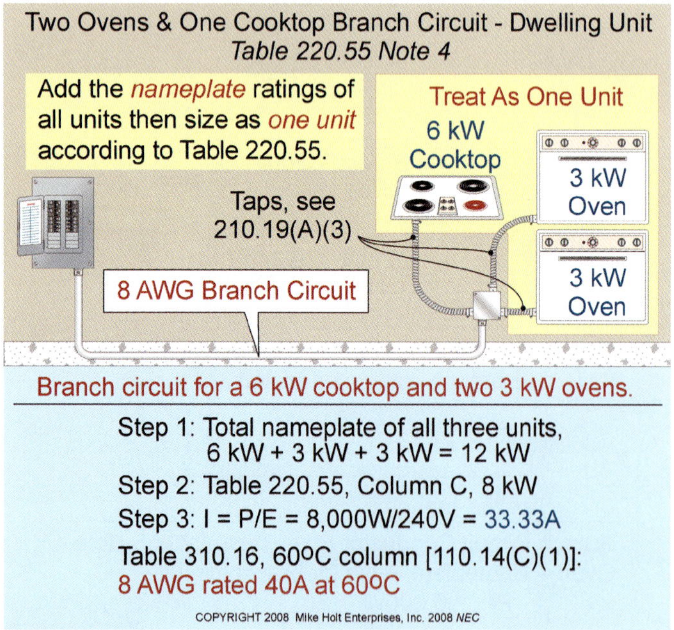

Two Ovens & One Cooktop Branch Circuit - Dwelling Unit
Table 220.55 Note 4
Add the nameplate ratings of all units then size as one unit according to Table 220.55.
Treat As One Unit
6 kW Cooktop
3 kW Oven
3 kW Oven
Taps, see 210.19(A)(3)
8 AWG Branch Circuit
Branch circuit for a 6 kW cooktop and two 3 kW ovens.
Step 1: Total nameplate of all three units, 6 kW + 3 kW + 3 kW = 12 kW
Step 2: Table 220.55, Column C, 8 kW
Step 3: I = P/E = 8,000W/240V = 33.33A
Table 310.16, 60ºC column [110.14(C)(1)]: 8 AWG rated 40A at 60ºC

Figure 220–21

Author's Comment: For ranges rated 8.75 kW or more, the minimum branch-circuit rating is 40A [210.19(A)(3)].

220.56 Commercial—Kitchen Equipment Load.

Table 220.56 can be used to calculate the feeder/service load for thermostat-controlled or intermittently used commercial electric cooking equipment, such as dishwasher booster heaters, water heaters, and other kitchen loads. The kitchen equipment feeder/service calculated load must not be less than the sum of the two largest kitchen equipment loads. Table 220.56 demand factors don't apply to space-heating, ventilating, or air-conditioning equipment.

Question: *What is the feeder/service calculated load for one 15 kW booster water heater, one 15 kW water heater, one 3 kW oven, and one 2 kW deep fryer in a commercial kitchen?* **Figure 220–22**

(a) 15 kW *(b) 20 kW* *(c) 26 kW* *(d) 30 kW*

Answer: *(d) 30 kW*

Step 1: Determine the total connected load:

Total Connected Load = 15 kW + 15 kW + 3 kW + 2 kW
Total Connected Load = 35 kW

Step 2: Determine the feeder/service calculated load:

35 kW x 0.80 = 28 kW, but it must not be less than the sum of the two largest appliances, or 30 kW.

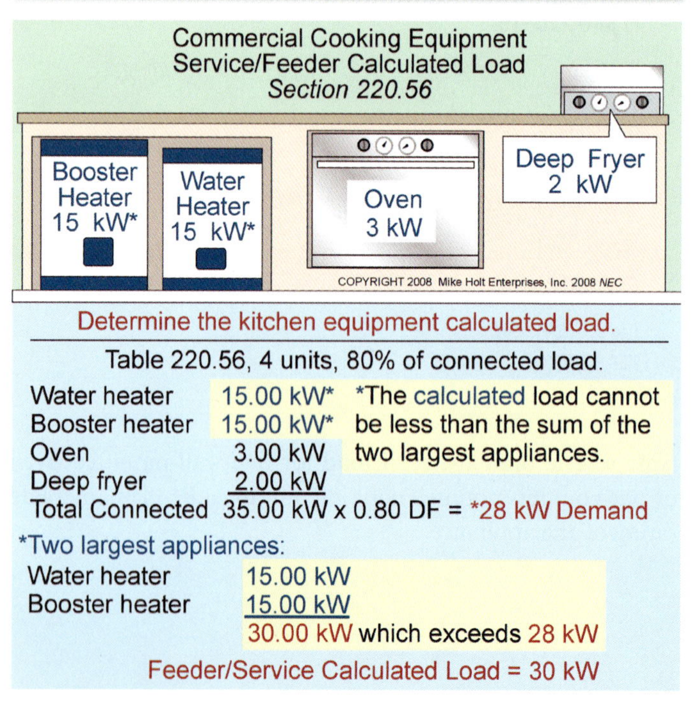

Figure 220–22

220.60 Noncoincident Loads. Where it's unlikely that two or more loads will be used at the same time, only the largest load(s) must be used to determine the feeder/service VA calculated load. Figure 220–23

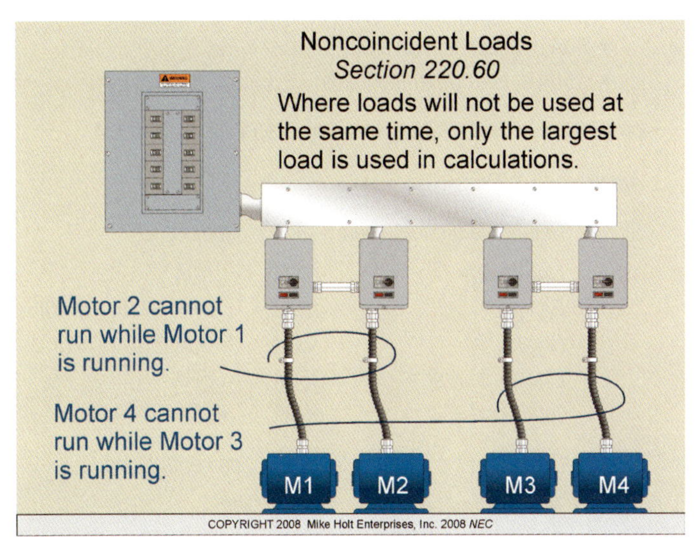

Figure 220–23

Question: *What is the feeder/service calculated load for a 5 hp, 230V air conditioner having a rated load current of 28A versus three electric space heaters, each rated 3 kW?* Figure 220–24

(a) 5,000W (b) 6,000W (c) 7,500W (d) 9,000W

Answer: *(d) 9,000W*

Air-Conditioning Load = 230V x 28A
Air-Conditioning Load = 6,440 VA (omit, smaller than 9,000W)

Electric Space Heating Load = 9,000W

220.61 Feeder/Service Neutral Unbalanced Load.

(A) Basic Calculation. The calculated neutral load for feeders/services is the maximum calculated load between the neutral conductor and any one ungrounded conductor. Line-to-line loads don't place any load on the neutral conductor, therefore they aren't considered.

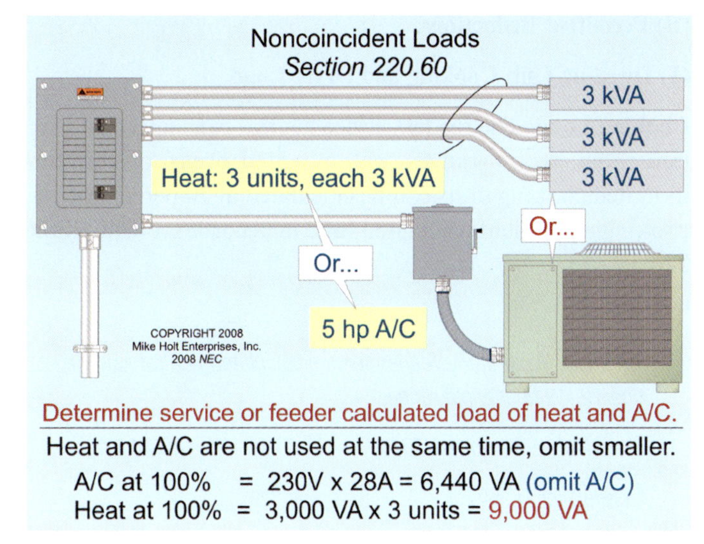

Figure 220–24

Question: *What is the minimum neutral conductor size for a 200A feeder, of which 100A is line-to-line loads with a maximum unbalanced neutral load of 100A?* Figure 220–25

(a) 3/0 AWG (b) 1/0 AWG (c) 1 AWG (d) 3 AWG

Answer: *(d) 3 AWG*

200A total load less 100A line-to-line loads = 100A neutral load

Table 310.16, 75°C column, 3 AWG rated 100A

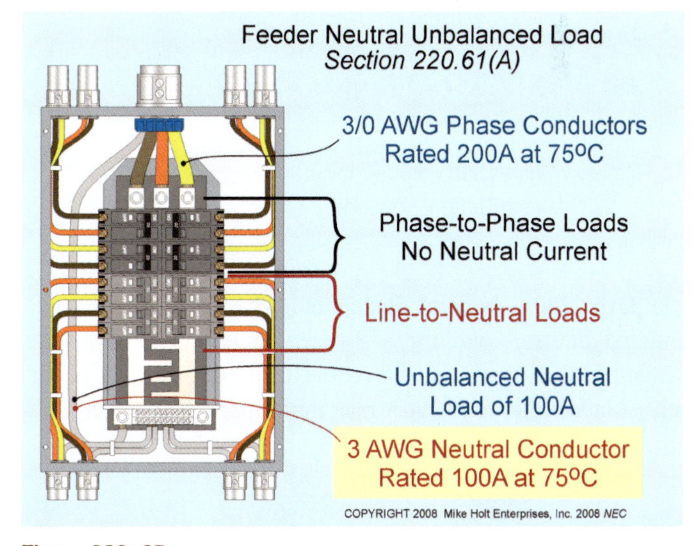

Figure 220–25

(B) Permitted Reductions.

(1) Dwelling Unit Cooking and Dryer Load.

Cooking Load. The feeder/service neutral calculated load for household electric ranges, wall-mounted ovens, or counter-mounted cooking units can be calculated at 70 percent of the cooking equipment calculated load in accordance with Table 220.55.

Question: *What is the feeder/service calculated neutral load for nine 12 kW household ranges?*

(a) 13 kW (b) 14.70 kW (c) 16.80 kW (d) 24 kW

Answer: *(c) 16.80 kW*

Step 1: Table 220.55 Column C = 24 kW

Step 2: Neutral Load = 24 kW x 0.70
Neutral Load = 16.80 kW

Dryer Load. The feeder/service neutral calculated load for household electric dryers can be calculated at 70 percent of the dryer calculated load in accordance with Table 220.54.

Question: *A 10-unit multifamily building has a 5 kW electric clothes dryer in each unit. What is the feeder/service neutral load for these dryers?*

(a) 17.50 kW (b) 23.50 kW (c) 33 kW (d) 41 kW

Answer: *(a) 17.50 kW*

Step 1: Table 220.54 = 10 units x 5 kW x 0.50
Table 220.54 = 25 kW

Step 2: Neutral Load = 25 kW x 0.70
Neutral Load = 17.50 kW

(2) Over 200A Neutral Reduction. The feeder/service calculated neutral load for a 3-wire, single-phase, or 4-wire, three-phase system, can be reduced for that portion of the unbalanced load over 200A by a multiplier of 70 percent.

Question: *What is the feeder/service neutral calculated load for the following? The voltage system is 120/240V, single phase.* **Figure 220–26**

- *100A of line-to-line loads*
- *100A of household ranges*
- *50A of household dryers*
- *350A of line-to-neutral loads*

(a) 200A (b) 379A (c) 455A (d) 600A

Answer: *(b) 379A*

Step 1: Determine the total feeder/service neutral load:

Line-to-line	100A	0A
Ranges	100A	70A (100A x 0.70)
Dryers	50A	35A (50A x 0.70)
Line-to-neutral	+ 350A	+ 350A
Total Load =	600A	455A

Step 2: Determine the demand feeder/service neutral load:

Total Neutral Load	455A	
First 200A at 100%	– 200A x 1.00 =	200A
Remainder at 70%	255A x 0.70 =	179A
Total Demand Neutral Load =		379A

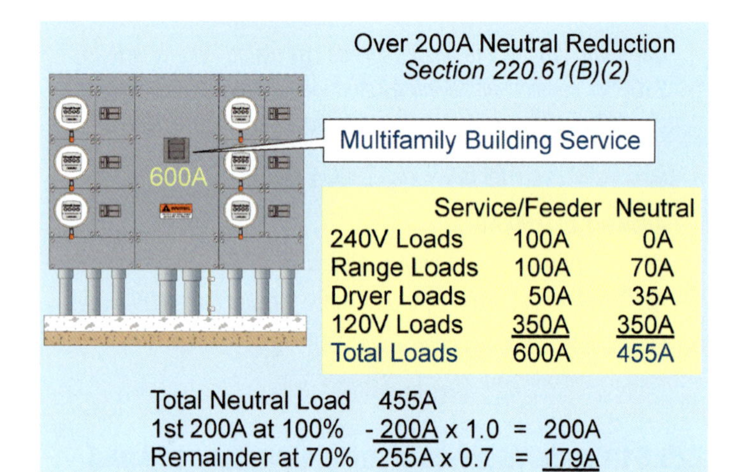

Figure 220–26

(C) Prohibited Reductions. No reduction of the neutral conductor calculated feeder/service load is permitted for the following:

(1) 3-Wire Circuits from 4-Wire Wye-Connected Systems. The feeder/service neutral calculated load must not be reduced for 3-wire circuits that consist of two ungrounded conductors and a neutral conductor supplied from a 4-wire, three-phase, 120/208V or 277/480V wye-connected system. This is because the neutral load on the 3-wire circuit will carry approximately the same amount of line-to-neutral current as the ungrounded conductors [310.15(B)(4)(c)].

Question: *What is the current on the neutral conductor of a 3-wire feeder supplied from a 4-wire, three-phase, 120/208V or 277/480V wye-connected system? The ungrounded conductors carry 200A of line-to-neutral loads.* **Figure 220–27**

(a) 200A (b) 379A (c) 455A (d) 600A

Answer: *(a) 200A*

$I_n = \sqrt{(I_{Line1}{}^2 + I_{Line2}{}^2) - (I_{Line1} \times I_{Line2})}$

$I_n = \sqrt{200A^2 + 200A^2 - (200A \times 200A)}$

$I_n = \sqrt{80{,}000 - 40{,}000}$

$I_n = \sqrt{40{,}000}$

$I_n = 200A$

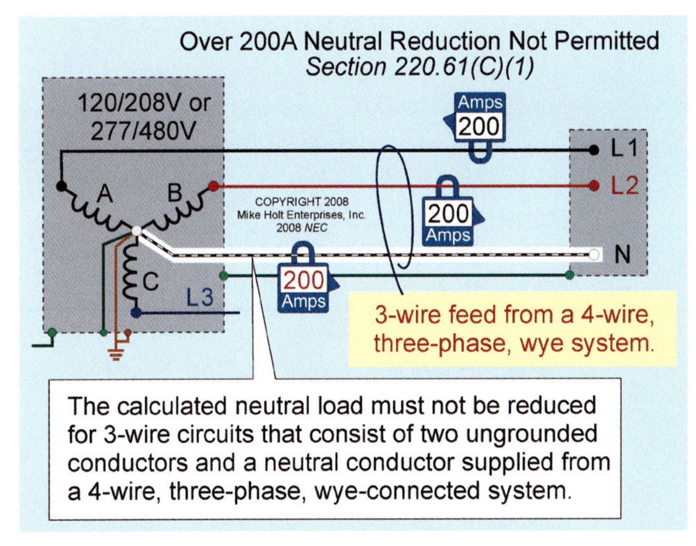

Figure 220–27

(2) Nonlinear Loads. The feeder/service neutral calculated load must not be reduced for nonlinear loads supplied from a 4-wire, three-phase, 120/208V or 277/480V wye-connected system.

Question: *What is the feeder/service neutral calculated load for the following?*

- *200A of line-to-line loads*
- *200A of line-to-neutral nonlinear loads*
- *200A of line-to-neutral linear loads*

(a) 200A (b) 400A (c) 500A (d) 600A

Answer: *(d) 400A*

Caution: The current on the neutral conductor for nonlinear loads can be as much as twice the maximum neutral load.

	Feeder/Service	Neutral Load
Line-to-Line Loads	200A	0A
Nonlinear Line-to-Neutral Loads	200A	200A
Linear Line-to-Neutral Loads	200A	200A
Total Calculated Load	600A	400A

PART IV. OPTIONAL CALCULATIONS FOR COMPUTING FEEDER AND SERVICE LOADS

220.82 Dwelling Unit—Optional Load Calculation.

(A) Feeder/Service Load. The 3-wire feeder/service load for a dwelling unit can be calculated by adding the calculated loads from 220.82(B) and (C). The feeder/service neutral calculated load must be determined in accordance with 220.61.

(B) General Loads. The feeder/service calculated load must not be less than 100 percent of the first 10 kVA, plus 40 percent of the remainder of the following:

(1) General Lighting. The general lighting load is based on 3 VA per sq ft for general lighting and general-use receptacles. The floor area is calculated from the outside dimensions of the dwelling unit, not including open porches, garages, or unused or unfinished spaces not adaptable for future use.

(2) Small-Appliance and Laundry Circuits. A load of 1,500 VA for each 20A small-appliance and laundry branch circuit [220.11(C)(1) and (2)]. Since a minimum of two small-appliance circuits and a laundry circuit are required, the minimum load for calculation purposes is 4,500 VA.

(3) Appliances. The nameplate rating of the following:

a. Appliances fastened in place, permanently connected or located to be on a specific circuit.

b. Ranges, wall-mounted ovens, counter-mounted cooking units

c. Clothes dryers

d. Water heaters

(4) Motor VA. The VA nameplate rating of all motors not part of an appliance.

(C) Air-Conditioning and Heating Equipment. The larger of (1) through (6):

(1) Air-Conditioning Equipment. 100 percent of the nameplate rating(s).

(2) Heat-Pump Compressor without Supplemental Heating. 100 percent of the heat-pump nameplate rating.

(3) Heat-Pump Compressor and Supplemental Heating. 100 percent of the nameplate rating of the heat-pump and 65 percent of the supplemental electric heating. If the heat-pump compressor is prevented from operating at the same time as the supplementary heat, it can be omitted in the calculation.

(4) Space-Heating Units (three or fewer units). 65 percent of the space-heating nameplate rating.

(5) Space-Heating Units (four or more units). 40 percent of the space-heating nameplate rating.

(6) Thermal Storage Heating. 100 percent of the thermal storage heating nameplate rating.

Author's Comment: One form of thermal storage heating involves heating bricks or water at night when the electric rates are lower. Then during the day, the building uses the thermally stored heat.

Question: *Using the optional calculation method, what size 3-wire, single-phase, 120/240V feeder/service ungrounded conductors are required for a 1,500 sq ft dwelling unit that contains the following loads?*

- *Dishwasher 1,200 VA*
- *Disposal 900 VA*
- *Cooktop 6,000 VA*
- *Oven 3,000 VA*
- *Dryer 4,000 VA*
- *Water heater 4,500 VA*
- *Heat pump compressor having a rating of 5 hp, with supplemental electric heat having a rating of 7 kW.*

(a) 100A (b) 110A (c) 125A (d) 150A

Answer: *(c) 125A*

Step 1: Determine the total feeder/service calculated load:

Lighting, receptacles, and appliance calculated load [220.82(B)]

Small appliance	1,500 VA x 2 =	3,000 VA
Laundry	1,500 VA x 1 =	1,500 VA
General lighting	1,500 sq ft x 3 VA/sq ft =	4,500 VA
Dishwasher	1,200 VA x 1 =	1,200 VA
Disposal	900 VA x 1 =	900 VA
Cooktop	6,000 VA x 1 =	6,000 VA
Oven	3,000 VA x 1 =	3,000 VA
Dryer	4,000 VA x 1 =	4,000 VA
Water heater	4,500 VA x 1 =	+ 4,500 VA
		28,600 VA

First 10,000 VA at 100% 10,000 VA x 1.00 = 10,000 VA
Remainder at 40% 18,600 VA x 0.40 = + 7,440 VA
220.82(B) Calculated Load = 17,440 VA

Largest of A/C or Heat [220.82(C)]
Heat pump 5 hp compressor at 100%
230V x 28A = 6,440 VA

Supplemental heat at 65% =
7,000 VA x 0.65 = + 4,550 VA

Total Calculated Load 220.82(B) and (C)
17,440 VA + 6,440 VA + 4,550 VA = 28,430 VA

Step 2: Determine the feeder/service calculated load in amperes:

I = VA/E
I = 28,430 VA/240V
I = 119A, 2 AWG [215.2(A)(3) and 310.15(B)(6)]

220.84 Multifamily—Optional Load Calculation.

(A) Feeder or Service Load. The feeder/service calculated load for a building with three or more dwelling units equipped with electric cooking equipment, and either electric space heating or air-conditioning, can be in accordance with the demand factors of Table 220.84, based on the number of dwelling units. The feeder/service neutral calculated load must be determined in accordance with 220.61.

(B) House Loads. House loads are calculated in accordance with Part III of Article 220, and then added to the Table 220.84 calculated load.

Author's Comment: House loads are those not directly associated with the individual dwelling units of a multifamily dwelling. Some examples of house loads are landscape and parking lot lighting, common area lighting, common laundry facilities, common pool and recreation areas, etc.

(C) Connected Loads. The connected loads from all of the dwelling units are added together, and then the Table 220.84 demand factors are applied to determine the calculated load.

(1) 3 VA per sq ft for general lighting and general-use receptacles.

(2) 1,500 VA for each 20A small-appliance circuit as required by 210.11(C)(1) (minimum of 2 circuits per dwelling unit), and 1,500 VA for each 20A laundry circuit as required by 210.11(C)(2).

Author's Comment: A laundry circuit isn't required in an individual unit of a multifamily dwelling if common laundry facilities are provided [210.52(F) Ex 1].

(3) Appliances. The nameplate rating of the following:

a. Appliances fastened in place, permanently connected or located to be on a specific circuit.

b. Ranges, wall-mounted ovens, counter-mounted cooking units.

c. Clothes dryers not connected to the required laundry circuit specified in 210.11(C)(2).

d. Water heaters.

(4) The nameplate rating of all motors not part of an appliance.

(5) The larger of the air-conditioning load or fixed electric space-heating load.

Question: *What size 4-wire, three-phase, 120/208V service is required for a multifamily building with twenty 1,500 sq ft dwelling units, where each unit contains the following loads?*

- *Dishwasher* — *1,200 VA*
- *Water heater* — *4,500 VA*
- *Disposal* — *900 VA*
- *Dryer* — *4,000 VA*
- *Cooktop* — *6,000 VA*
- *Oven* — *3,000 VA*
- *Heat* — *7,000 VA*
- *A/C, 5 hp compressor* — *6,440 VA*

(a) 400A *(b) 600A* *(c) 800A* *(d) 1,200A*

Answer: *(c) 800A [240.4 and 240.6(A)]*

Step 1: Determine the dwelling unit connected load:

General lighting	*1,500 sq ft x 3 VA/sq ft =*	*4,500 VA*
Small appliance	*1500 VA x 2 =*	*3,000 VA*
Laundry	*1,500 VA x 1 =*	*1,500 VA*
Dishwasher	*1,200 VA x 1 =*	*1,200 VA*
Water heater	*4,500 VA x 1 =*	*4,500 VA*
Disposal	*900 VA x 1 =*	*900 VA*
Dryer	*4,000 VA x 1 =*	*4,000 VA*
Cooktop	*6,000 VA x 1 =*	*6,000 VA*
Oven	*3,000 VA x 1 =*	*3,000 VA*
A/C 5 hp (omit)	*0 VA x 1 =*	*0 VA*
Heat =		*+ 7,000 VA*
Total Dwelling Unit Load =		*35,600 VA*

Step 2: Determine the calculated load for the multifamily building:

Demand Factor for 20 units = 0.38 [Tables 220.84]

35,600 VA x 20 x 0.38 = 270,560 VA

Step 3: Determine feeder/service conductor size:

$I = VA/(E \times \sqrt{3})$

$I = 270{,}560\ VA/(208V \times 1.732)$

$I = 751A$

I of Each Conductor Parallel Set = 751A/2 conductors

I of Each Conductor Parallel Set = 376A

Conductor = 500 kcmil, rated 380A at 75°C x 2

Conductor = 760A, Table 310.16

220.85 Optional Calculation—Two Dwelling Units. Where two dwelling units are supplied by a single feeder, and where the standard calculated load in accordance with Part II of this article exceeds that for three identical units calculated in accordance with 220.84, the lesser of the two calculated loads may be used.

220.87 Determining Existing Loads. The calculation of a feeder or service load for existing installations can be based on:

(1) The maximum demand data for one year.

Exception: Where the maximum demand data for one year is not available, the maximum power demand over a 15-minute period continuously recorded over a minimum 30-day period using a recording ammeter or power meter connected to the highest loaded phase, based on the initial loading at the start of the recording is permitted. The recording must be taken when the building or space is occupied based on the larger of the heating or cooling equipment load.

ARTICLE 220 Practice Questions

ARTICLE 220. BRANCH-CIRCUIT, FEEDER, AND SERVICE CALCULATIONS—PRACTICE QUESTIONS

1. Where fixed multioutlet assemblies are used in other than dwelling units or the guest rooms of hotels or motels, each _____ or fraction thereof of each separate and continuous length shall be considered as 180 VA where appliances are unlikely to be used simultaneously.

   (a) 5 ft
   (b) 5½ ft
   (c) 6 ft
   (d) 6½ ft

2. The 3 VA per-square-foot general lighting load for dwelling units includes general-use receptacles and lighting outlets.

   (a) True
   (b) False

3. Feeder and service loads for fixed electric space heating shall be calculated at _____ percent of the total connected load.

   (a) 80
   (b) 100
   (c) 125
   (d) 200

4. When sizing a feeder for the fixed appliance loads in dwelling units, a demand factor of 75 percent of the total nameplate ratings can be applied if there are _____ or more appliances fastened in place on the same feeder.

   (a) two
   (b) three
   (c) four
   (d) five

5. To determine the feeder calculated load for ten 3 kW household cooking appliances, use _____ of Table 220.55.

   (a) Column A
   (b) Column B
   (c) Column C
   (d) none of these

6. The demand factors of Table 220.56 apply to space heating, ventilating, or air-conditioning equipment.

   (a) True
   (b) False

7. The maximum unbalanced feeder load for household electric ranges, wall-mounted ovens, and counter-mounted cooking units shall be considered as _____ percent of the load on the ungrounded conductors.

   (a) 50
   (b) 70
   (c) 85
   (d) 115

8. A demand factor of _____ percent applies to a multifamily dwelling with ten units if the optional calculation method is used.

   (a) 43
   (b) 50
   (c) 60
   (d) 75

ARTICLE

225 Outside Branch Circuits and Feeders

INTRODUCTION TO ARTICLE 225—OUTSIDE BRANCH CIRCUITS AND FEEDERS

This article covers the installation requirements for equipment including conductors located outdoors on or between buildings, poles, and other structures on the premises. It has two parts:

Part I provides a listing of other articles that may provide additional requirements, addresses some general concerns, and briefly covers conductor sizing. It also addresses conductor support, attachment, and clearances.

Part II addresses how many supplies (branch circuits or feeders) are permitted to a building or structure and how to disconnect them. This includes such rules as where to locate the disconnecting means and how to group them.

PART I. GENERAL REQUIREMENTS

225.1 Scope. Article 225 contains the installation requirements for outside branch circuits and feeders run on or between buildings, structures, or poles. Figure 225–1

Figure 225–1

Author's Comment: Review the following definitions in Article 100:

- Branch Circuit
- Building
- Feeder
- Structure

225.2 Other Articles. Other articles containing important requirements include:

- Branch Circuits, Article 210
- Class 1, Class 2, and Class 3 Remote-Control, Signaling, and Power-Limited Circuits, Article 725
- Communications Circuits, Article 800
- Community Antenna Television and Radio Distribution Systems, Article 820
- Conductors for General Wiring, Article 310
- Electric Signs and Outline Lighting, Article 600
- Feeders, Article 215
- Floating Buildings, Article 553
- Grounding and Bonding, Article 250
- Marinas and Boatyards, Article 555
- Radio and Television Equipment, Article 810
- Services, Article 230
- Solar Photovoltaic Systems, Article 690
- Swimming Pools, Fountains, and Similar Installations, Article 680

225.6 Minimum Size of Conductors.

(A) Overhead Spans.

(1) Conductor Size. Conductors 10 AWG and larger are permitted for overhead spans up to 50 ft. For spans over 50 ft, the minimum size conductor is 8 AWG, unless supported by a messenger wire. Figure 225–2

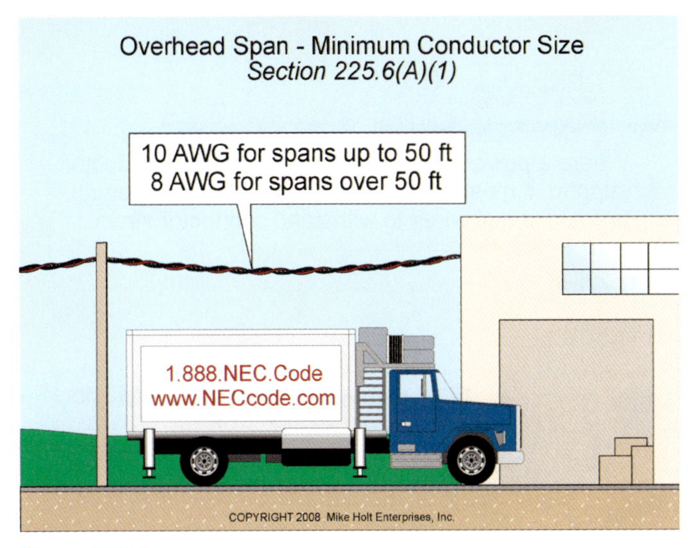

Figure 225–2

(B) Festoon Lighting. Overhead conductors for festoon lighting must not be smaller than 12 AWG, unless messenger wires support the conductors. The overhead conductors must be supported by messenger wire, with strain insulators, whenever the spans exceed 40 ft. Figure 225–3

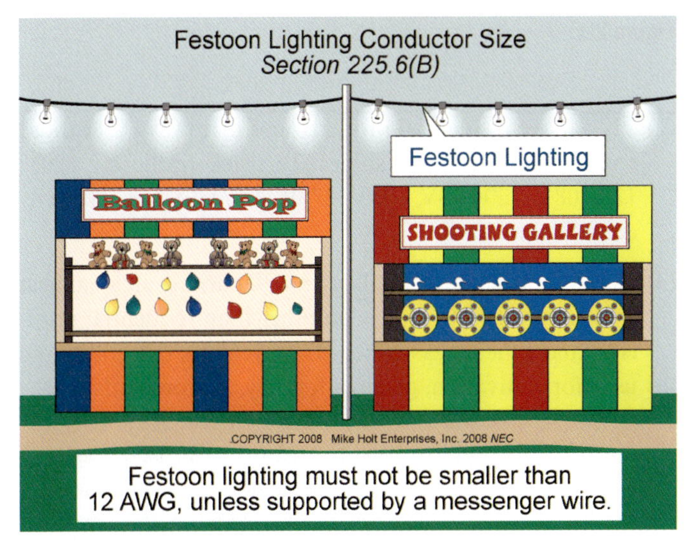

Figure 225–3

Author's Comment: Festoon lighting is a string of outdoor lights suspended between two points [Article 100]. It's commonly used at carnivals, circuses, fairs, and Christmas tree lots [525.20(C)].

225.7 Luminaires Installed Outdoors.

(C) 277V to Ground Circuits. Lighting equipment on a 277V or 480V branch circuit must not be located within 3 ft of windows that open, platforms, fire escapes, and the like. Figure 225–4

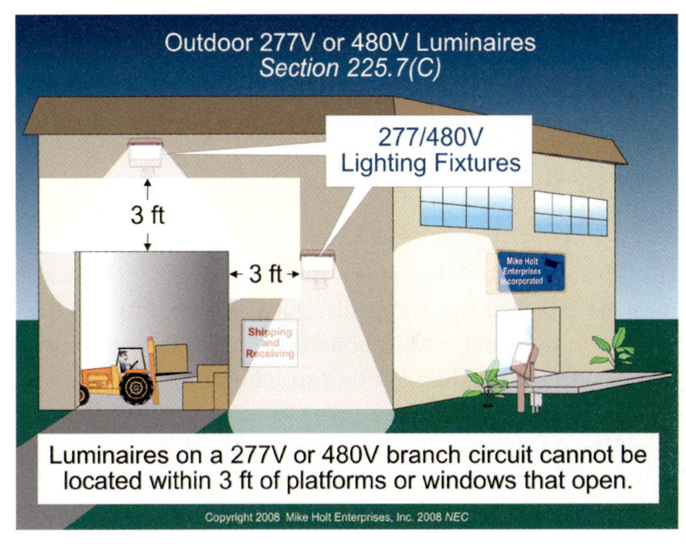

Figure 225–4

Author's Comment: See 210.6(C) for the types of luminaires permitted on 277V or 480V branch circuits.

225.15 Supports Over Buildings. Conductor spans over a building must be securely supported by substantial structures. Where practicable, such supports must be independent of the building [230.29].

225.16 Attachment.

(A) Point of Attachment. The point of attachment for overhead conductors must not be less than 10 ft above the finished grade, and it must be located so the minimum conductor clearance required by 225.18 can be maintained.

CAUTION: *Conductors might need to have the point of attachment raised so the overhead conductors will comply with the clearances required by 225.19 from building openings and other building areas.* Figure 225–5

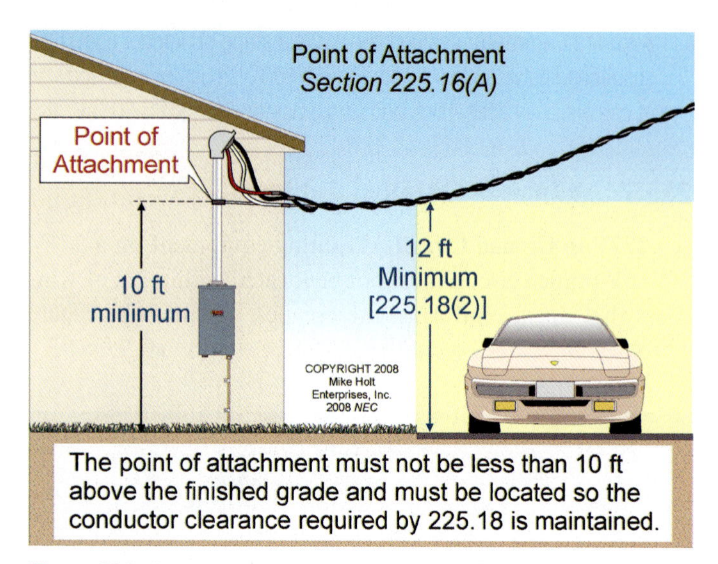

Figure 225–5

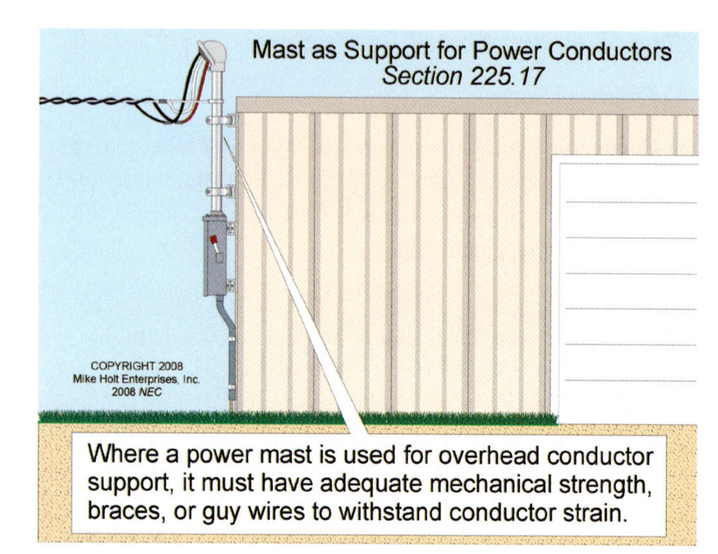

Figure 225–6

(B) Means of Attachment to Buildings. Open conductors must be attached to fittings identified for use with conductors, or to noncombustible, nonabsorbent insulators securely attached to the building or other structure.

Author's Comment: The point of attachment of the overhead conductor spans to a building or other structure must provide the minimum clearances as specified in 225.18 and 225.19. In no case can this point of attachment be less than 10 ft above the finished grade.

225.17 Masts as Support.

Where a mast is used for overhead conductor support, it must have adequate mechanical strength, braces, or guy wires to withstand the strain caused by the conductors. Only branch-circuit or feeder conductors can be attached to the mast. Figure 225–6

Author's Comment: Aerial cables and antennas for radio and TV equipment must not be attached to the feeder or branch-circuit mast [810.12]. In addition, 800.133(B) prohibits communications cables from being attached to raceways, including a mast for power conductors. Figure 225–7

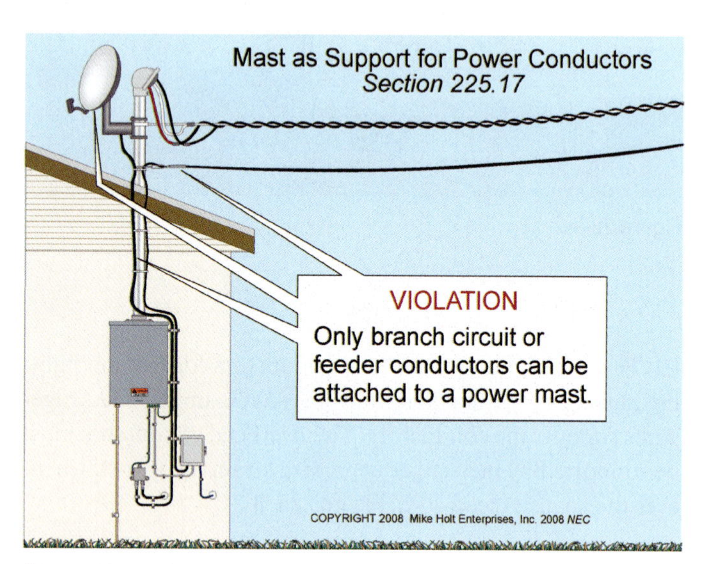

Figure 225–7

225.18 Clearance for Overhead Conductors.

Overhead conductor spans must maintain the vertical clearances as follows:

(1) 10 ft above finished grade, sidewalks, platforms, or projections from which they might be accessible to pedestrians for 120V, 120/208V, 120/240V, or 240V circuits.

(2) 12 ft above residential property and driveways, and those commercial areas not subject to truck traffic for 120V, 120/208V, 120/240V, 240V, 277V, 277/480V, or 480V circuits. Figure 225–8

(4) 18 ft over public streets, alleys, roads, parking areas subject to truck traffic, driveways on other than residential property, and other areas traversed by vehicles (such as those used for cultivation, grazing, forestry, and orchards).

Author's Comment: Overhead conductors located above pools, outdoor spas, outdoor hot tubs, diving structures, observation stands, towers, or platforms must be installed in accordance with the clearance requirements in 680.8.

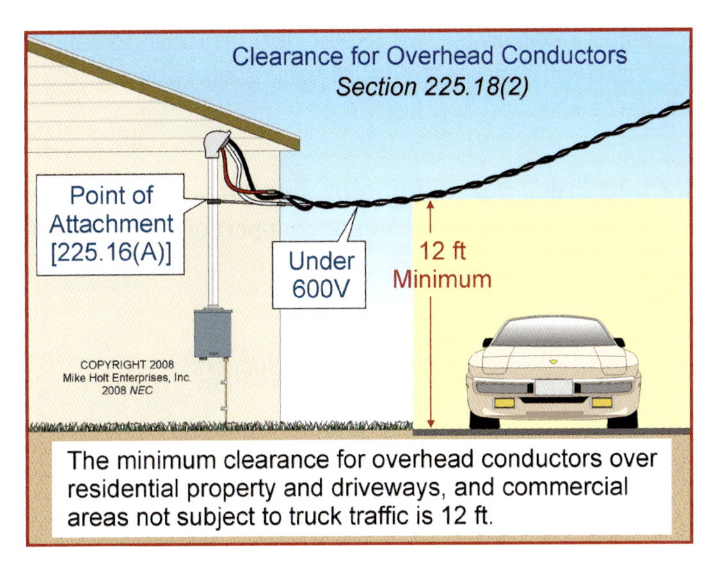

Figure 225–8

225.19 Clearances from Buildings

(A) Above Roofs. Overhead conductors must maintain a vertical clearance of 8 ft above the surface of a roof that must be maintained for a distance of at least 3 ft from the edge of the roof. Figure 225–9

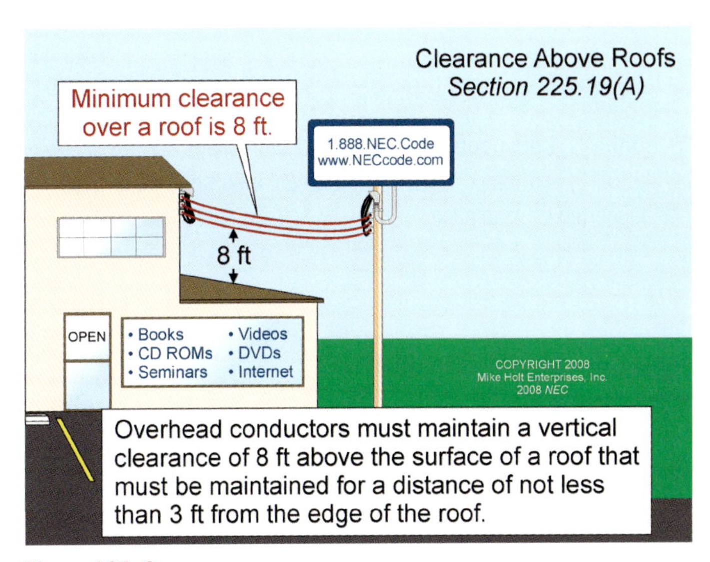

Figure 225–9

Exception No. 2: The overhead conductor clearances from the roof can be reduced from 8 ft to 3 ft if the slope of the roof meets or exceeds 4 in. vertical rise for every 12 in. of horizontal run.

Exception No. 3: For 120/208V or 120/240V circuits, the conductor clearance over the roof overhang can be reduced from 8 ft to 18 in., if no more than 6 ft of conductor passes over no more than 4 ft of roof. Figure 225–10

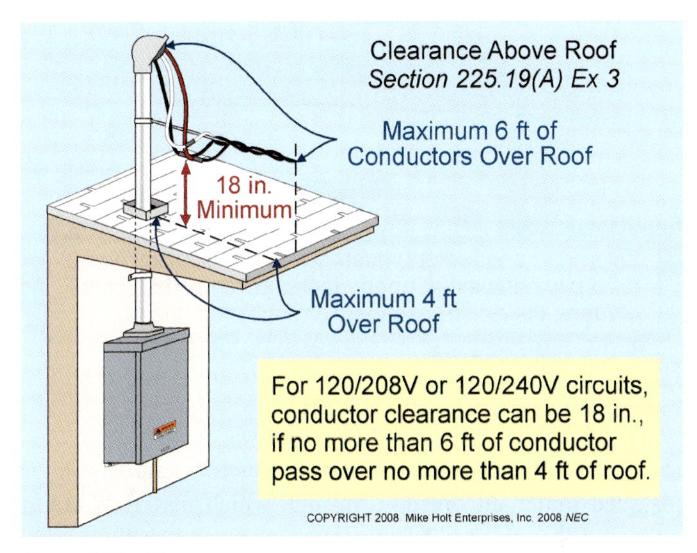

Figure 225–10

Exception No. 4: The 3 ft clearance from the roof edge doesn't apply when the point of attachment is on the side of the building below the roof.

(B) From Other Structures. Overhead conductors must maintain a clearance of at least 3 ft from signs, chimneys, radio and television antennas, tanks, and other nonbuilding or nonbridge structures.

(D) Final Span Clearance.

(1) Clearance from Windows. Overhead conductors must maintain a clearance of 3 ft from windows that open, doors, porches, balconies, ladders, stairs, fire escapes, or similar locations. Figure 225–11

Exception: Overhead conductors run above a window aren't required to maintain the 3 ft distance from the window.

(2) Vertical Clearance. Overhead conductors must maintain a vertical clearance of at least 10 ft above platforms, projections, or surfaces from which they might be reached. This vertical clearance must be maintained for 3 ft, measured horizontally from the platforms, projections, or surfaces from which they might be reached.

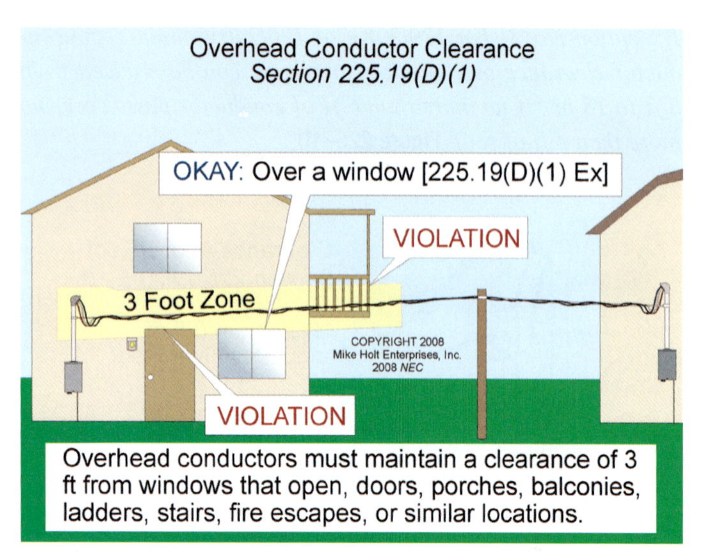

Figure 225–11

(3) Below Openings. Overhead conductors must not be installed under an opening through which materials might pass, and they must not be installed where they will obstruct an entrance to building openings. Figure 225–12

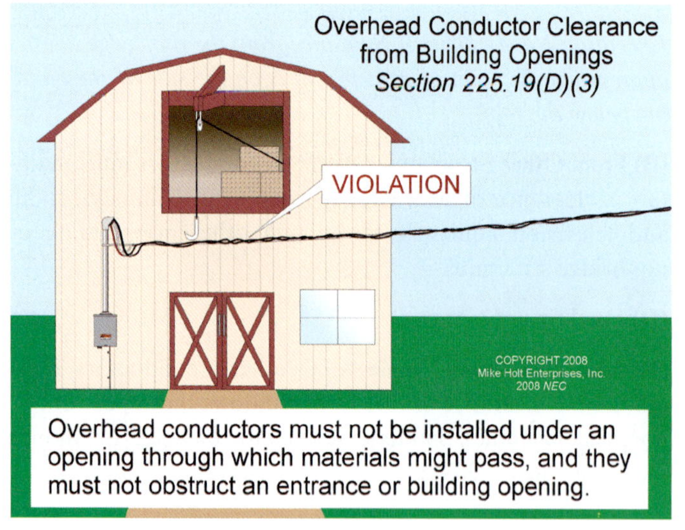

Figure 225–12

225.22 Raceways on Exterior Surfaces of Buildings or Other Structures. Raceways on exterior surfaces of buildings or other structures must be arranged to drain, and in wet locations must be raintight.

> **Author's Comment:** A "Wet Location" is an area subject to saturation with water and unprotected locations exposed to weather [Article 100].

225.26 Trees for Conductor Support. Trees or other vegetation must not be used for the support of overhead conductor spans. Figure 225–13

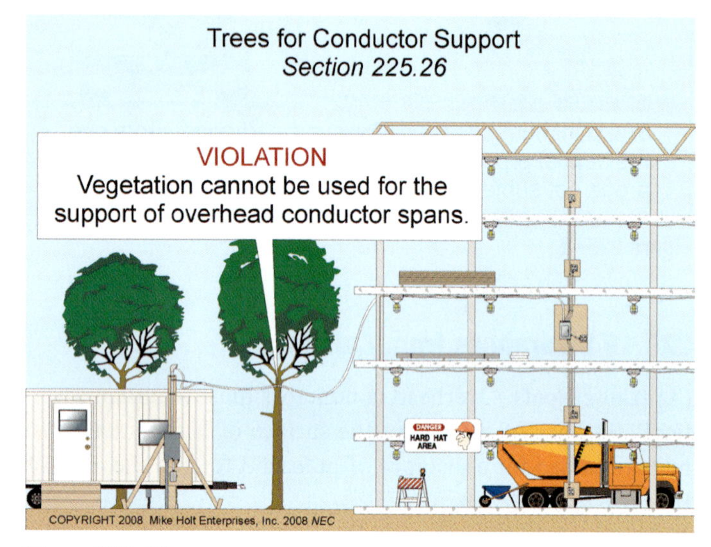

Figure 225–13

> **Author's Comment:** Overhead conductor spans for services [230.10] and temporary wiring [590.4(J)] aren't permitted to be supported by vegetation.

PART II. MORE THAN ONE BUILDING OR STRUCTURE

225.30 Number of Supplies. Where more than one building or structure is on the same property, each building or structure must be served by no more than one feeder or branch circuit, except as permitted in (A) through (E).

> **Author's Comment:** Article 100 defines "Structure" as, "That which is built or constructed."

A multiwire branch circuit is a single circuit.

(A) Special Conditions. Additional supplies are permitted for:

(1) Fire pumps

(2) Emergency systems

(3) Legally required standby systems

(4) Optional standby systems

(5) Parallel power production systems

(6) Systems designed for connection to multiple sources of supply for the purpose of enhanced reliability.

Author's Comment: To minimize the possibility of accidental interruption, the disconnecting means for the fire pump or standby power must be located remotely away from the normal power disconnect [225.34(B)].

(B) Special Occupancies. By special permission, additional supplies are permitted for:

(1) Multiple-occupancy buildings where there's no available space for supply equipment accessible to all occupants, or

(2) A building or structure so large that two or more supplies are necessary.

(C) Capacity Requirements. Additional supplies are permitted for a building or structure where the capacity requirements exceed 2,000A.

(D) Different Characteristics. Additional supplies are permitted for different voltages, frequencies, or uses, such as control of outside lighting from multiple locations.

(E) Documented Switching Procedures. Additional supplies are permitted where documented safe switching procedures are established and maintained for disconnection.

225.31 Disconnecting Means. A disconnect is required for all conductors that enter or pass through a building or structure.

225.32 Disconnect Location. The disconnecting means for a building or structure must be installed at a readily accessible location either outside or inside nearest the point of entrance of the conductors. Figure 225–14

Supply conductors are considered outside of a building or other structure where they are encased or installed under not less than 2 in. of concrete or brick [230.6]. Figure 225–15

Exception No. 1: Where documented safe switching procedures are established and maintained, the building/structure disconnecting means can be located elsewhere on the premises, if monitored by qualified persons.

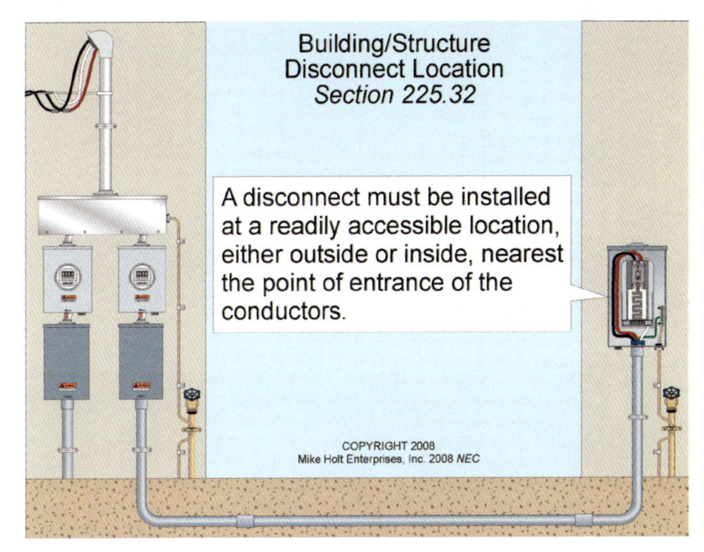

Figure 225–14

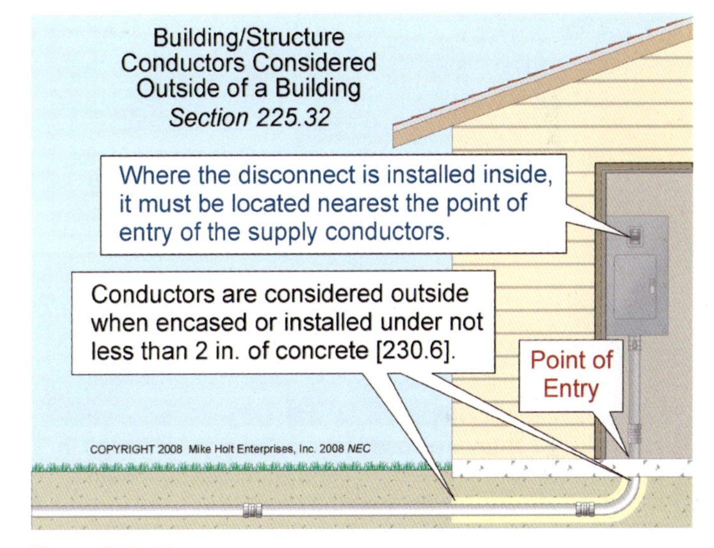

Figure 225–15

Author's Comment: A "Qualified Person" is one who has skills and knowledge related to the construction and operation of the electrical equipment and installation, and has received safety training to recognize and avoid the hazards involved with electrical systems [Article 100].

Exception No. 3: A disconnecting means isn't required within sight of poles that support luminaires. Figure 225–16

Exception No. 4: The disconnecting means for a sign isn't required to be readily accessible if installed in accordance with the requirements for signs. Figure 225–17

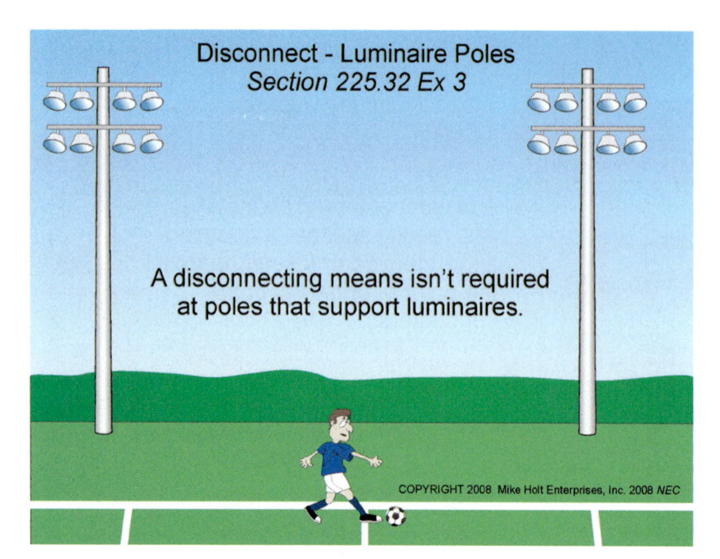

Figure 225–16

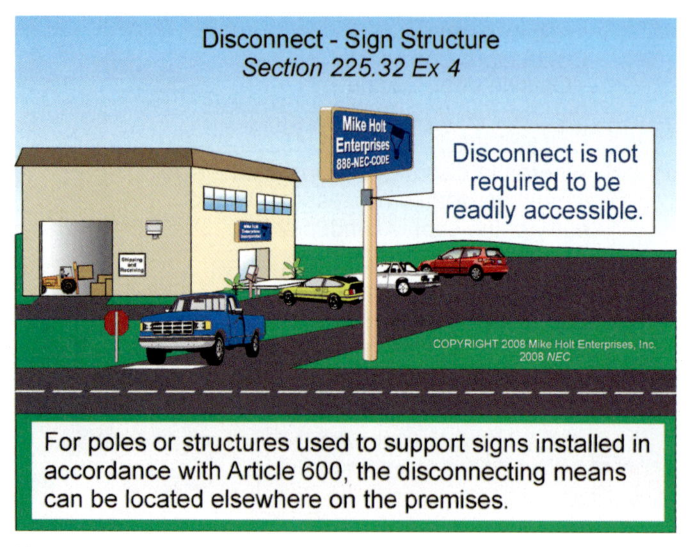

Figure 225–17

Author's Comment: Each sign must be controlled by an externally operable switch or circuit breaker that opens all ungrounded conductors to the sign. The sign disconnecting means must be within sight of the sign, or the disconnecting means must be capable of being locked in the open position [600.6(A)].

225.33 Maximum Number of Disconnects.

(A) General. The building or structure disconnecting means can consist of no more than six switches or six circuit breakers in a single enclosure, or separate enclosures for each supply permitted by 225.30. Figure 225–18

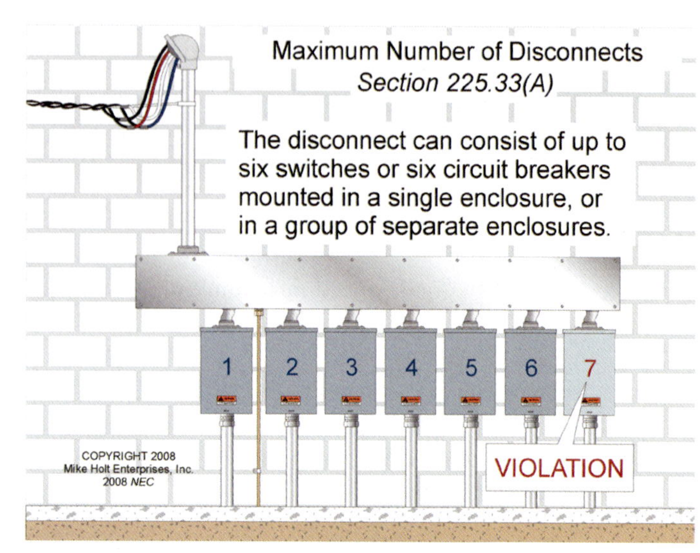

Figure 225–18

225.34 Grouping of Disconnects.

(A) General. The building or structure disconnecting means must be grouped in one location, and they must be marked to indicate the loads they serve [110.22].

(B) Additional Disconnects. To minimize the possibility of accidental interruption of the critical power systems, the disconnecting means for a fire pump or for standby power is to be located remotely away from the normal power disconnect.

225.35 Access to Occupants.

In a multiple-occupancy building, each occupant must have access to the disconnecting means for their occupancy.

Exception: The occupant's disconnecting means can be accessible to only building management, if electrical maintenance under continuous supervision is provided by the building management.

225.36 Identified as Suitable for Service Equipment.

The building or structure disconnecting means must be identified as "suitable for use as service equipment." Figure 225–19

Exception: A snap switch or a set of 3-way or 4-way snap switches can be used as the disconnecting means for garages and outbuildings on residential property, without having a "service equipment" rating.

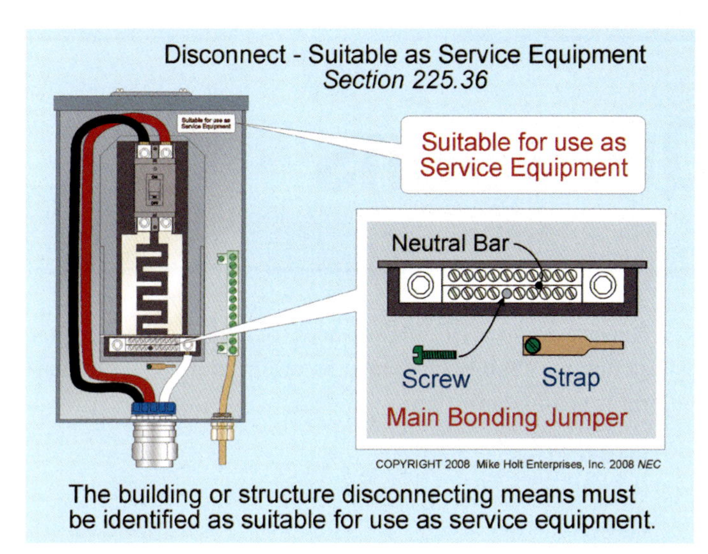

Figure 225–19

225.37 Identification of Multiple Feeders. Where a building or structure is fed by more than one supply, a permanent plaque or directory must be installed at each feeder disconnect location denoting all other feeders or branch circuits that supply that building or structure, and the area served by each.

225.38 Disconnect Construction.

(A) Manual or Power-Operated Circuit Breakers. The building or structure disconnecting means can consist of either a manually or power-operated switch or circuit breaker capable of being operated manually.

Author's Comment: A shunt-trip pushbutton can be used to open a power-operated circuit breaker. The circuit breaker is the disconnecting means, not the pushbutton.

225.39 Rating of Disconnecting Means. A single disconnecting means for a building or structure must have an ampere rating not less than the calculated load as determined by Article 220. Where the disconnecting means consists of more than one switch or circuit breaker, the combined ratings of the circuit breakers must not be less than the calculated load as determined by Article 220. In addition, the disconnecting means must not be rated lower than:

(A) One-Circuit Installation. For installations consisting of a single branch circuit, the disconnecting means must have a rating of not less than 15A.

(B) Two-Circuit Installation. For installations consisting of two 2-wire branch circuits, the feeder disconnecting means must have a rating of not less than 30A.

(C) One-Family Dwelling. For a one-family dwelling, the feeder disconnecting means must have a rating of not less than 100A, 3-wire.

(D) All Others. For all other installations, the feeder or branch-circuit disconnecting means must have a rating of not less than 60A.

ARTICLE 225 Practice Questions

ARTICLE 225. OUTSIDE BRANCH CIRCUITS AND FEEDERS—PRACTICE QUESTIONS

1. The point of attachment of overhead premises wiring to a building shall in no case be less than _____ above finished grade.

   (a) 8 ft
   (b) 10 ft
   (c) 12 ft
   (d) 15 ft

2. The minimum clearance for overhead feeder conductors not exceeding 600V that pass over commercial areas subject to truck traffic is _____.

   (a) 10 ft
   (b) 12 ft
   (c) 15 ft
   (d) 18 ft

3. The requirement for maintaining a 3 ft vertical clearance from the edge of the roof shall not apply to the final feeder conductor span where the conductors are attached to _____.

   (a) a building pole
   (b) the side of a building
   (c) an antenna
   (d) the base of a building

4. Raceways on exterior surfaces of buildings or other structures shall be arranged to drain, and in _____ locations shall be raintight.

   (a) damp
   (b) wet
   (c) dry
   (d) all of these

5. The disconnecting means for a building supplied by a feeder shall be installed at a(n) _____ location.

   (a) accessible
   (b) readily accessible
   (c) outdoor
   (d) indoor

6. The two to six disconnects for a disconnecting means for a building supplied by a feeder shall be _____.

   (a) the same size
   (b) grouped
   (c) in the same enclosure
   (d) none of these

7. The disconnecting means for a building supplied by a feeder shall plainly indicate whether it is in the _____ position.

   (a) open or closed
   (b) correct
   (c) up or down
   (d) none of these

Services

INTRODUCTION TO ARTICLE 230—SERVICES

This article covers the installation requirements for service conductors and service equipment. The requirements for service conductors differ from those for other conductors. For one thing, service conductors for one structure can't pass through the interior of another structure [230.3], and you apply different rules depending on whether a service conductor is inside or outside a structure. When are they "outside" as opposed to "inside?" The answer may seem obvious, but isn't.

It's usually good to start a service installation by determining which conductors are actually parts of the service. What you determine here will determine how you do the rest of the job. To identify a service conductor, you must first determine whether you're dealing with the service (line side of the service point) or a premises wiring system (load side of the service point).

Let's review the following definitions in Article 100 to understand when the requirements of Article 230 apply:

- Service Point—The point of connection between the serving utility and the premises wiring.
- Service Conductors—The conductors from the service point to the service equipment (the service disconnecting means, not the meter). Service-entrance conductors may be either overhead (service drop) or underground (service lateral).
- Service Equipment—The necessary equipment, usually consisting of circuit breakers or switches and fuses and their accessories, connected to the load end of service conductors at a building or other structure (or an otherwise designated area), and intended to constitute the main control and cutoff of the electrical supply. Service equipment doesn't include metering components, such as the meter and meter socket enclosure [230.66].

After reviewing these definitions, you should understand that service conductors originate at the serving utility (service point) and terminate on the line side of the service disconnecting means (service equipment). Conductors and equipment on the load side of service equipment are considered feeder conductors or branch circuits, and must be installed in accordance with Articles 210 and 215. They must also comply with Article 225 if they are the supply to a building or structure. Feeder conductors include: Figure 230–1

- Secondary conductors from customer-owned transformers,
- Conductors from generators, UPS systems, or photovoltaic systems, and
- Conductors to remote buildings or structures

Article 230 consists of seven parts:

- PART I. GENERAL
- PART II. OVERHEAD SERVICE-DROP CONDUCTORS
- PART III. UNDERGROUND SERVICE-LATERAL CONDUCTORS
- PART IV. SERVICE-ENTRANCE CONDUCTORS
- PART V. SERVICE EQUIPMENT
- PART VI. DISCONNECTING MEANS
- PART VII. OVERCURRENT PROTECTION

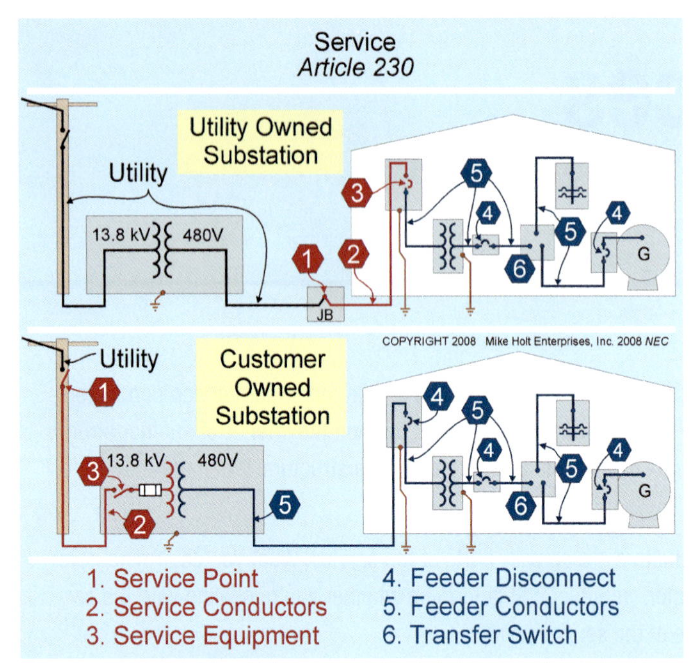

Figure 230–1

PART I. GENERAL

230.1 Scope. Article 230 covers the installation requirements for service conductors and service equipment.

230.2 Number of Services. A building or structure can only be served by one service drop or service lateral, except as permitted by (A) through (D). Figure 230–2

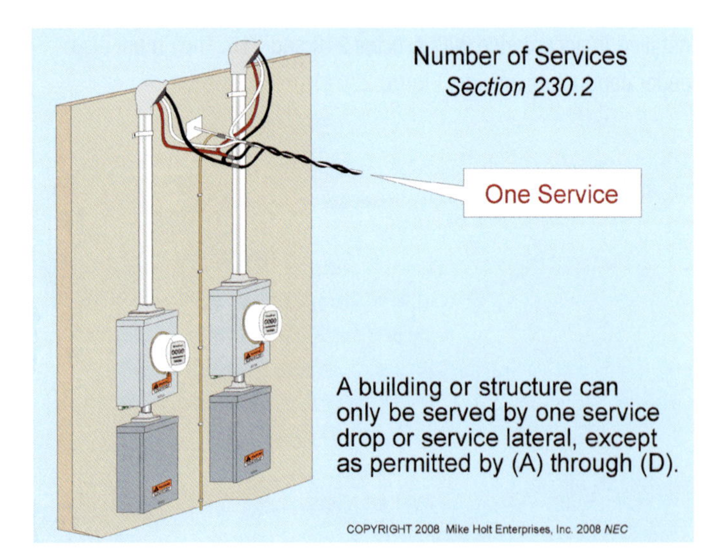

Figure 230–2

Service laterals 1/0 AWG and larger run to the same location and connected together at their supply end, but not connected together at their load end, are considered to be a single service.

(A) Special Conditions. Additional services are permitted for the following:

(1) Fire pumps

(2) Emergency systems

(3) Legally required standby systems

Author's Comment: A separate service for emergency and legally required systems is permitted only when approved by the authority having jurisdiction [700.12(D) and 701.11(D)].

(4) Optional standby power

(5) Parallel power production systems

(6) Systems designed for connection to multiple sources of supply for the purpose of enhanced reliability.

Author's Comment: To minimize the possibility of accidental interruption, the disconnecting means for the fire pump, emergency system or standby power system must be located remotely away from the normal power disconnect [230.72(B)].

(B) Special Occupancies. By special permission, additional services are permitted for:

(1) Multiple-occupancy buildings where there's no available space for supply equipment accessible to all occupants, or

(2) A building or other structure so large that two or more supplies are necessary.

(C) Capacity Requirements. Additional services are permitted:

(1) Where the capacity requirements exceed 2,000A, or

(2) Where the load requirements of a single-phase installation exceeds the utility's capacity, or

(3) By special permission.

Author's Comment: Special permission is defined in Article 100 as "the written consent of the authority having jurisdiction."

(D) Different Characteristics. Additional services are permitted for different voltages, frequencies, or phases, or for different uses, such as for different electricity rate schedules.

(E) Identification of Multiple Services. Where a building or structure is supplied by more than one service, or a combination of feeders and services, a permanent plaque or directory must be installed at each service and feeder disconnect location to denote all other services and feeders supplying that building or structure, and the area served by each. Figure 230–3

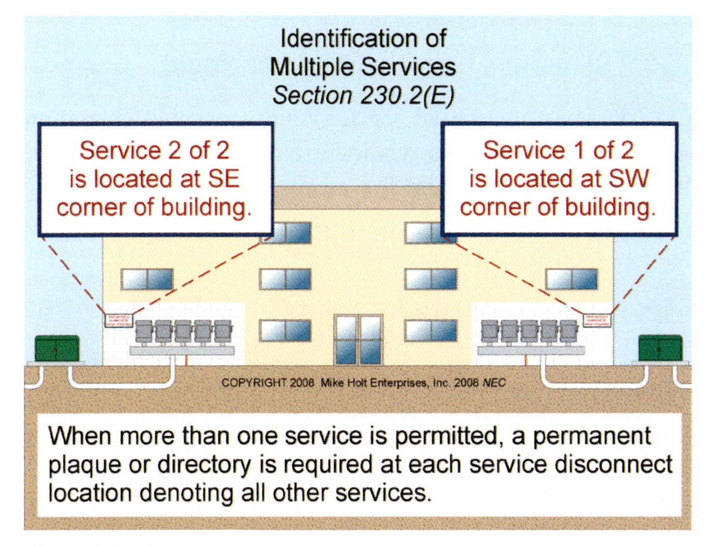

Figure 230–3

230.3 Not to Pass Through a Building or Structure.

Service conductors must not pass through the interior of another building or other structure.

230.6 Conductors Considered Outside a Building.

Conductors are considered outside of a building when they are installed:

(1) Under not less than 2 in. of concrete beneath a building or structure. Figure 230–4

(2) Within a building or structure in a raceway encased in not less than 2 in. of concrete or brick.

(3) Installed in a vault that meets the construction requirements of Article 450, Part III.

(4) In a raceway under not less than 18 in. of earth beneath a building or structure.

230.7 Service Conductors Separate from Other Conductors. Service conductors must not be installed in the same raceway or cable with feeder or branch-circuit conductors. Figure 230–5

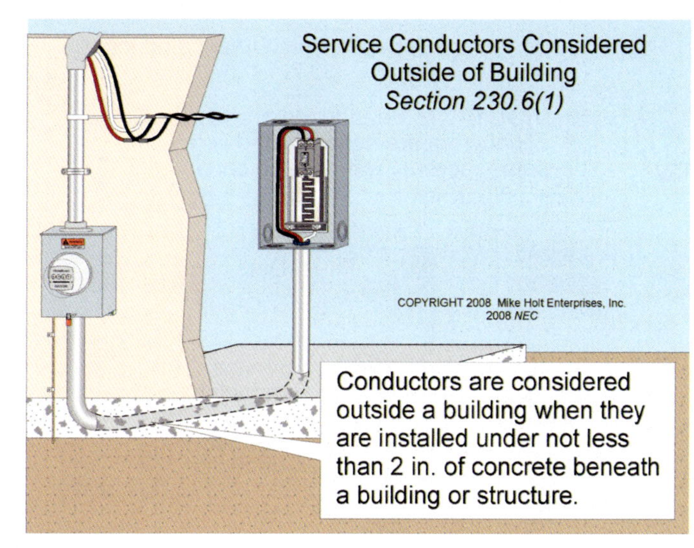

Figure 230–4

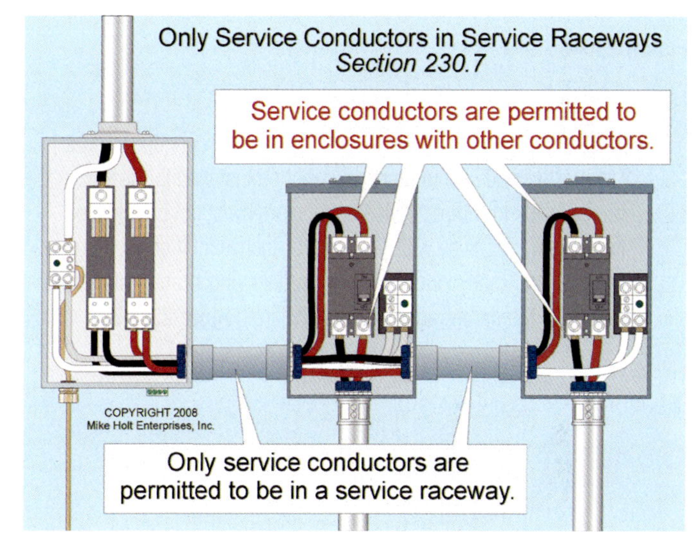

Figure 230–5

WARNING: *Overcurrent protection for the feeder or branch circuit conductors can be bypassed if service conductors are mixed with feeder or branch circuit conductors in the same raceway and a fault occurs between the service and feeder or branch circuit conductors.* Figure 230–6

Author's Comments:

- This rule doesn't prohibit the mixing of service, feeder, and branch-circuit conductors in the same service equipment enclosure.

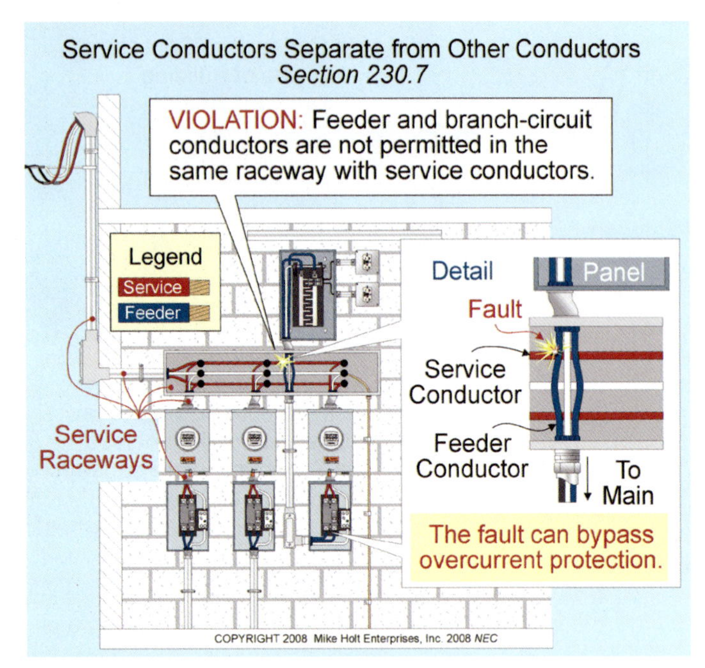

Figure 230–6

- This requirement may be the root of the misconception that "line" and "load" conductors must not be installed in the same raceway. It's true that service conductors must not be installed in the same raceway with feeder or branch-circuit conductors, but line and load conductors for feeders and branch circuits can be in the same raceway or enclosure. Figure 230–7

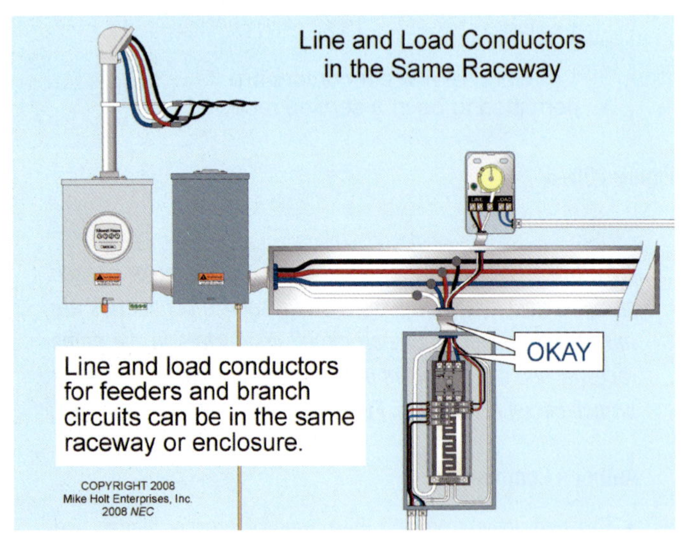

Figure 230–7

230.8 Raceway Seals. Underground raceways (used or unused) must be sealed or plugged to prevent moisture from contacting energized live parts [300.5(G)].

> **Author's Comment:** Sealing can be accomplished with the use of a putty-like material called duct seal, or a fitting identified for the purpose. A seal of the type required in Chapter 5 for hazardous (classified) locations isn't required.

230.9 Clearance from Building Openings.

(A) Clearance. Overhead service conductors must maintain a clearance of 3 ft from windows that open, doors, porches, balconies, ladders, stairs, fire escapes, or similar locations. Figure 230–8

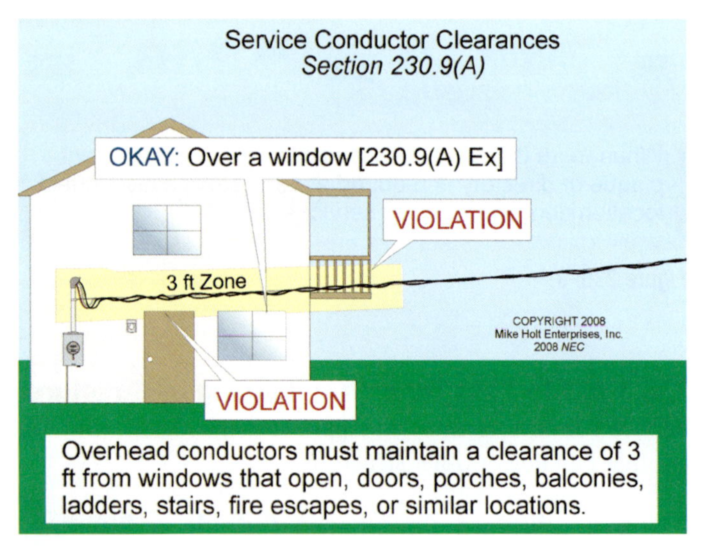

Figure 230–8

Exception: Overhead conductors run above a window aren't required to maintain the 3 ft distance.

(B) Vertical Clearance. Overhead service conductors must maintain a vertical clearance not less than 10 ft above platforms, projections, or surfaces from which they might be reached [230.24(B)]. This vertical clearance must be maintained for 3 ft, measured horizontally from the platform, projections, or surfaces from which people might reach them.

(C) Below Openings. Service conductors must not be installed under an opening through which materials might pass, and they must not be installed where they will obstruct entrance to building openings. Figure 230–9

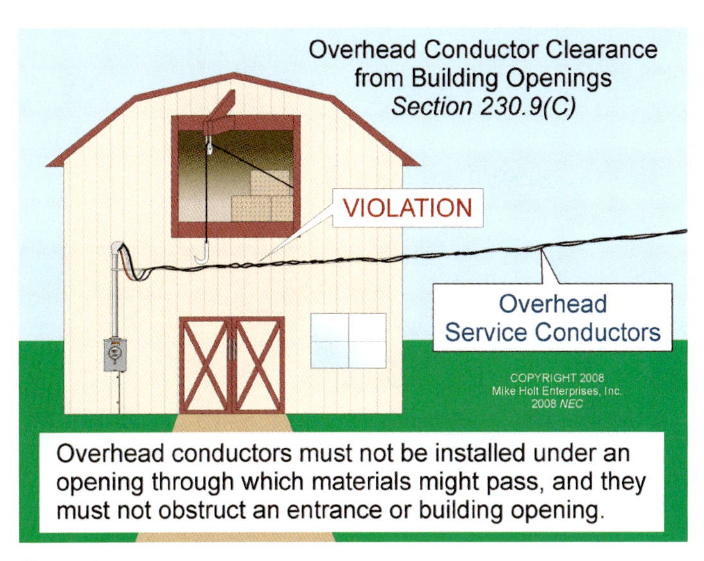

Figure 230–9

230.10 Vegetation as Support. Trees or other vegetation must not be used for the support of overhead service conductor spans. Figure 230–10

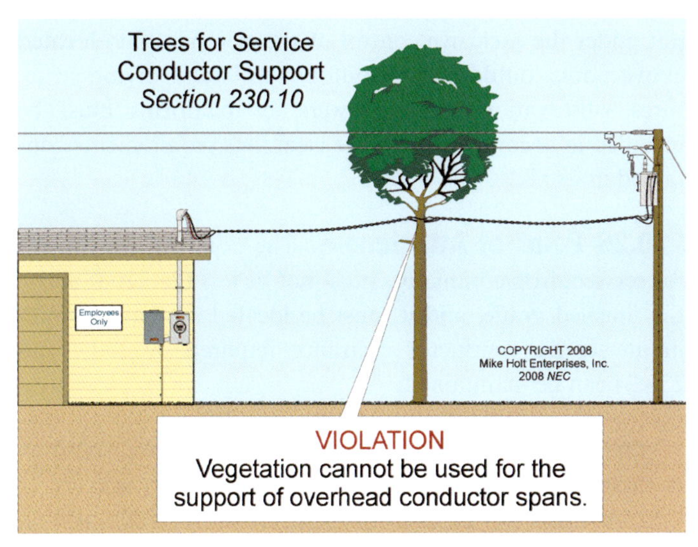

Figure 230–10

Author's Comment: Service-drop conductors installed by the electric utility must comply with the *National Electrical Safety Code* (NESC), not the *National Electrical Code* [90.2(B)(5)]. Overhead service conductors that are not under exclusive control of the electric utility must be installed in accordance with the *NEC*.

PART II. OVERHEAD SERVICE-DROP CONDUCTORS

230.23 Size and Rating.

(A) Ampacity of Service-Drop Conductors. Service-drop conductors must have adequate mechanical strength, and they must have sufficient ampacity to carry the load as calculated in accordance with Article 220.

(B) Ungrounded Conductor Size. Service-drop conductors must not be smaller than 8 AWG copper or 6 AWG aluminum.

Exception: Service-drop conductors can be as small as 12 AWG for limited-load installations.

(C) Neutral Conductor Size. The neutral service-drop conductor must be sized to carry the maximum unbalanced load, in accordance with 220.61, and it must not be sized smaller than required by 250.24(C).

WARNING: *In all cases, the service neutral conductor size must not be smaller than required by 250.24(C) to ensure that it has sufficiently low impedance and current-carrying capacity to safely carry fault current in order to facilitate the operation of the overcurrent device.*

Question: *What size neutral conductor is required for a structure with a 400A service supplied with 500 kcmil conductors if the maximum line-to-neutral load is no more than 100A?*

(a) 3 AWG (b) 2 AWG (c) 1 AWG (d) 1/0 AWG

Answer: *(d) 1/0 AWG*

According to Table 310.16, 3 AWG rated 100A at 75°C [110.14(C)] is sufficient to carry 100A of neutral current. However, the service neutral conductor must be sized not smaller than 1/0 AWG, in accordance with Table 250.66, based on the area of the service conductor [250.24(C)].

230.24 Vertical Clearance for Service-Drop Conductors. Service-drop conductor spans must maintain the vertical clearances as follows:

(A) Above Roofs. A minimum of 8 ft above the surface of a roof for a minimum distance of 3 ft in all directions from the edge of the roof.

Exception No. 2: If the slope of the roof exceeds 4 in. of vertical rise for every 12 in. of horizontal run, 120/208V or 120/240V conductor clearances can be reduced to 3 ft over the roof.

Exception No. 3: If no more than 6 ft of conductors pass over no more than 4 ft of roof, 120/208V or 120/240V conductor clearances over the roof overhang can be reduced to 18 in. **Figure 230–11**

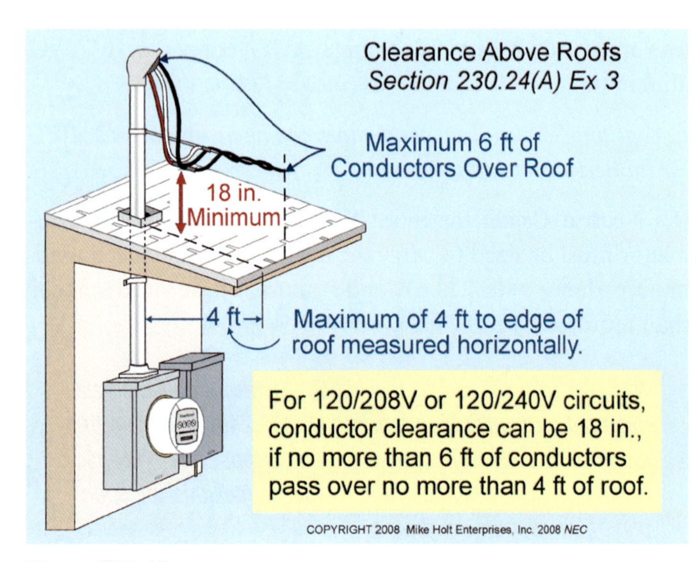

Figure 230–11

Exception No. 4: The 3 ft vertical clearance that extends from the roof doesn't apply when the point of attachment is on the side of the building (below the roof).

(B) Vertical Clearance for Service-Drop Conductors. Overhead conductor spans must maintain the following vertical clearances: **Figure 230–12**

(1) 10 ft above finished grade, sidewalks, or platforms or projections from which they might be accessible to pedestrians for 120/208V or 120/240V circuits.

(2) 12 ft above residential property and driveways, and those commercial areas not subject to truck traffic for 120/208V, 120/240V, or 277/480V circuits.

(4) 18 ft over public streets, alleys, roads, parking areas subject to truck traffic, driveways on other than residential property, and other areas traversed by vehicles, such as those used for cultivation, grazing, forestry, and orchards.

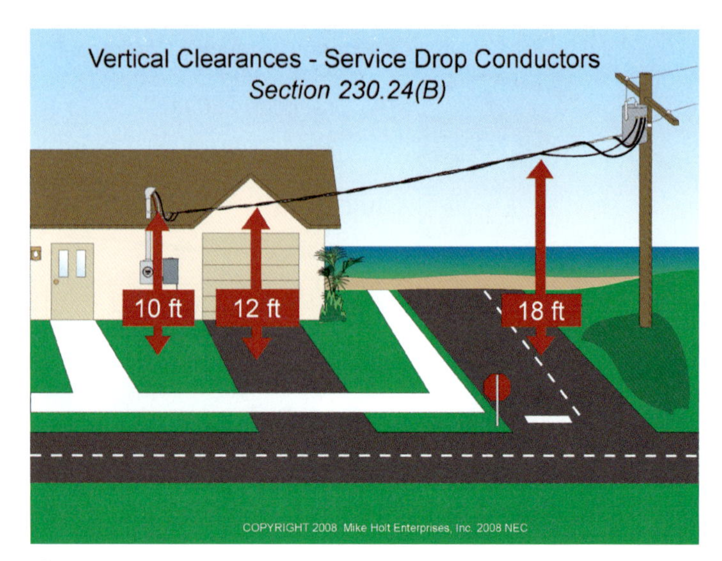

Figure 230–12

Author's Comment: Department of Transportation (DOT) type right-of-ways in rural areas are often used by slow-moving and tall farming machinery to avoid impeding road traffic.

(D) Swimming Pools. Overhead service conductors that are not under the exclusive control of the electric utility located above pools, outdoor spas, outdoor hot tubs, diving structures, observation stands, towers, or platforms must be installed in accordance with the clearance requirements contained in 680.8.

230.26 Point of Attachment.

The point of attachment for service-drop conductors must not be less than 10 ft above the finished grade, and it must be located so that the minimum service conductor clearance required by 230.9 and 230.24 can be maintained.

CAUTION: *The point of attachment for conductors might need to be raised so the overhead conductors will comply with the clearances from building openings required by 230.9 and from other areas by 230.24.* **Figure 230–13**

230.27 Means of Attachment.

Multiconductor cables used for service drops must be attached to buildings or other structures by fittings identified for use with service conductors.

Open conductors must be attached to fittings identified for use with service conductors or to noncombustible, nonabsorbent insulators securely attached to the building or other structure.

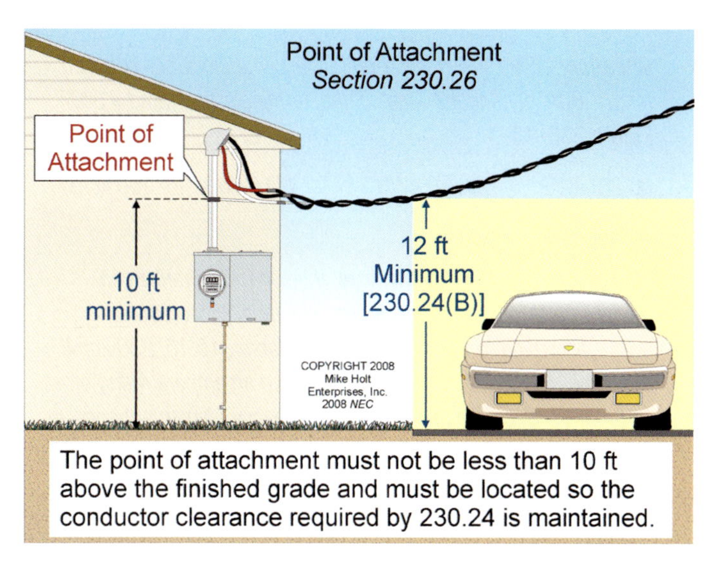

Figure 230–13

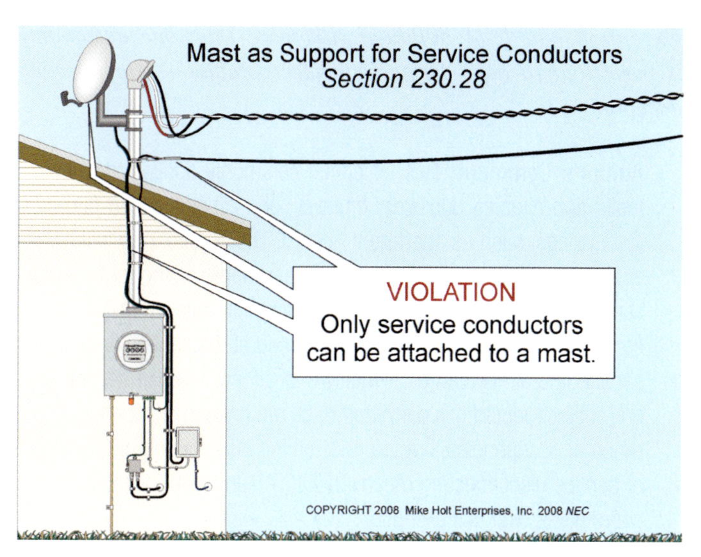

Figure 230–14

230.28 Service Masts Used as Supports. The service mast used as overhead conductor support must have adequate mechanical strength, or braces or guy wires to support it, to withstand the strain caused by the service-drop conductors.

Author's Comment: Some local codes require a minimum 2 in. rigid metal conduit for the service mast. In addition, many electric utilities contain specific requirements for the installation of the service mast.

Only electric utility service-drop conductors can be attached to a service mast.

Author's Comment: 810.12 and 820.44(C) specify that aerial cables for radio, TV, or CATV must not be attached to the service mast, and 810.12 prohibits antennas from being attached to the service mast. In addition, 800.133(B) and 830.133(B) prohibit broadband communications cables from being attached to raceways, including a service mast. Figure 230–14

PART III. UNDERGROUND SERVICE-LATERAL CONDUCTORS

Author's Comment: Underground service-lateral conductors installed by the electric utility must comply with the *National Electrical Safety Code* (NESC), not the *National Electrical Code* [90.2(B)(5)]. Underground conductors that are not under the exclusive control of the electric utility must be installed in accordance with the *NEC*.

230.31 Service-Lateral Conductor Size and Rating.

(A) General. Service-lateral conductors must have adequate mechanical strength, and they must have sufficient ampacity to carry the load as calculated in accordance with Article 220.

(B) Ungrounded Conductor Size. Service-lateral conductors must not be smaller than 8 AWG copper or 6 AWG aluminum.

Exception: Service-lateral conductors can be as small as 12 AWG for limited-load installations.

(C) Neutral Conductor Size. The neutral service-lateral conductor must be sized to carry the maximum unbalanced load in accordance with 220.61, and it must not be sized smaller than required by 250.24(C).

Author's Comment: 250.24(C) requires the service neutral conductor to be sized according to Table 250.66.

230.32 Protection Against Damage. Where installed underground, service lateral conductors must be installed so as to have minimum cover in accordance with Table 300.5.

PART IV. SERVICE-ENTRANCE CONDUCTORS

230.40 Number of Service-Entrance Conductor Sets.

Each service drop or service lateral can supply only one set of service-entrance conductors.

Exception No. 1: A building with more than one occupancy is permitted to have service-entrance conductors run to each occupancy.

Author's Comment: This exception commonly applies to strip malls and even to duplexes, triplexes, and other multiple-family buildings, such as apartment houses. The general rule is that each service drop or service lateral can supply only one set of service-entrance conductors. This exception allows metering equipment to be located at or near the end of the service drop or service lateral, instead of at multiple locations. A set of service-entrance conductors is permitted to be run to each occupancy or group of occupancies, but the rules on the location and grouping of service disconnecting means in 230.71 and 230.72 must be coordinated with this exception.

Exception No. 2: Service-entrance conductors can supply two to six service disconnecting means as permitted in 230.71(A).

Exception No. 3: A single-family dwelling unit with a separate structure can have one set of service-entrance conductors run to each structure from a single service drop or lateral.

Exception No. 5: One set of service-entrance conductors connected to the supply side of the normal service disconnecting means can supply standby power systems, fire pump equipment, and fire and sprinkler alarms [230.82(5) and (6)].

230.42 Size and Rating.

(A) Load Calculations. Service-entrance conductors must have sufficient ampacity for the loads to be served in accordance with Parts III, IV, or V of Article 220.

(1) Continuous and Noncontinuous Loads. The ampacity of the service-entrance conductors, before the application of any adjustment or correction factors, must be sized no smaller than 125 percent of the continuous loads, plus 100 percent of the noncontinuous loads, based on the terminal temperature rating ampacities as listed in Table 310.16, before any ampacity adjustment [110.14(C)].

Author's Comment: See 215.3 for the sizing requirements of feeder overcurrent devices for continuous and noncontinuous loads.

Question: *What size service-entrance conductors are required for a 200A continuous load, if the terminals are rated 75°C?* **Figure 230–15**

(a) 2/0 AWG (b) 3/0 AWG (c) 4/0 AWG (d) 250 kcmil

Answer: *(d) 250 kcmil*

Since the load is 200A continuous, the service-entrance conductors must have an ampacity not less than 250A (200A x 1.25). According to the 75°C column of Table 310.16, 250 kcmil conductors are suitable because they have an ampere rating of 255A at 75°C, before any conductor ampacity adjustment and/or correction.

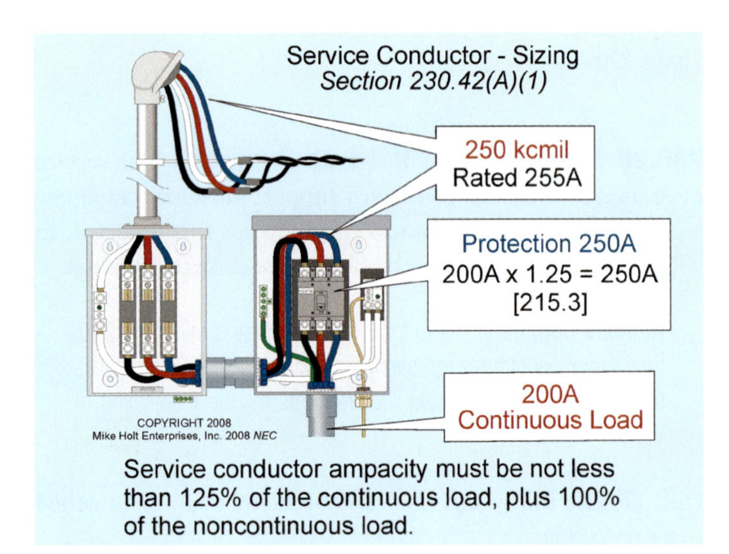

Figure 230–15

(C) Neutral Conductor Size. The service neutral conductor must be sized to carry the maximum unbalanced load in accordance with 220.61, and must not be sized smaller than required by 250.24(C).

WARNING: *In all cases the service neutral conductor size must not be smaller than required by 250.24(C) to ensure that it has sufficiently low impedance and current-carrying capacity to safely carry fault current in order to facilitate the operation of the overcurrent device.*

230.43 Wiring Methods. Service-entrance conductors must be installed with one of the following wiring methods:

(1) Open wiring on insulators

(3) Rigid metal conduit

(4) Intermediate metal conduit

(5) Electrical metallic tubing

(6) Electrical nonmetallic tubing

(7) Service-entrance cables

(8) Wireways

(9) Busways

(11) PVC conduit

(13) Type MC cable

(15) Flexible metal conduit or liquidtight flexible metal conduit not longer than 6 ft

(16) Liquidtight flexible nonmetallic conduit

230.44 Cable Trays. Cable trays used to support service-entrance conductors must only contain service-entrance conductors.

230.46 Spliced Conductors. Service-entrance conductors can be spliced or tapped in accordance with 110.14, 300.5(E), 300.13, and 300.15. Figure 230–16

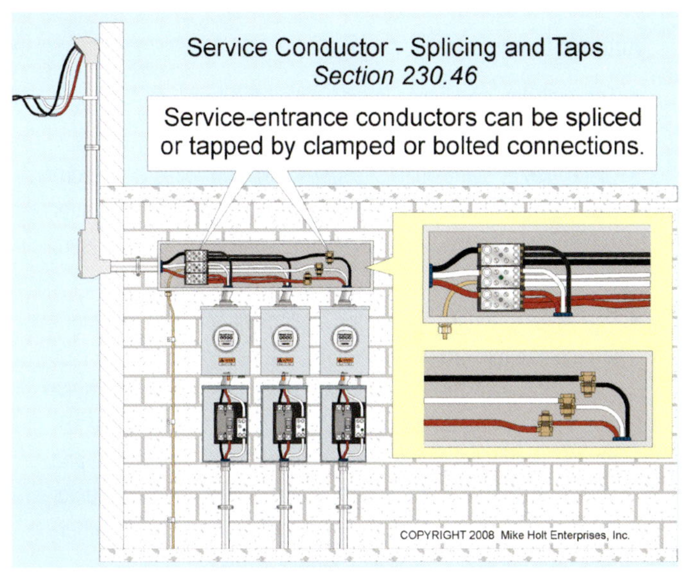

Figure 230–16

230.50 Protection Against Physical Damage.

(A) Underground Service-Entrance Conductors. Underground service-entrance conductors must be protected against physical damage in accordance with 300.5.

(B) All Other Service-Entrance Conductors. Service-entrance conductors that are not installed underground must be protected against physical damage as follows:

(1) Service Cables. Service cables that are subject to physical damage must be protected by one of the following: Figure 230–17

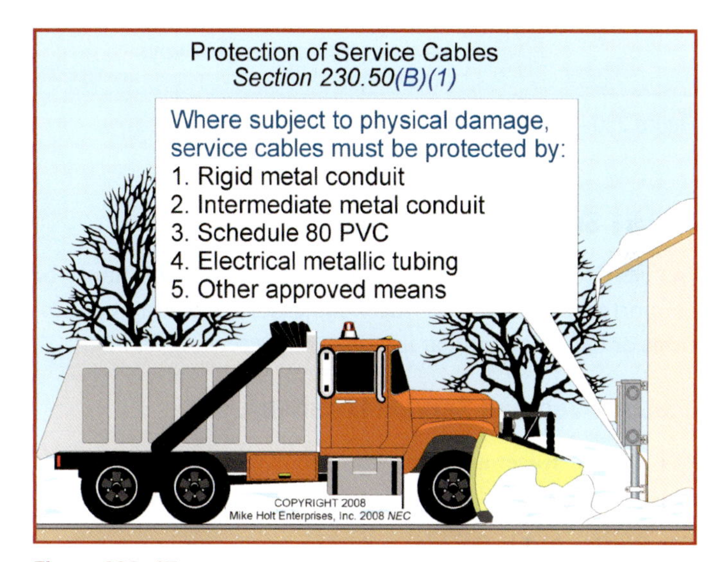

Figure 230–17

(1) Rigid metal conduit

(2) Intermediate metal conduit

(3) Schedule 80 PVC conduit

> **Author's Comment:** If the authority having jurisdiction determines the raceway isn't subject to physical damage, Schedule 40 PVC conduit can be used. Figure 230–18

(4) Electrical metallic tubing

(5) Other means approved by the authority having jurisdiction

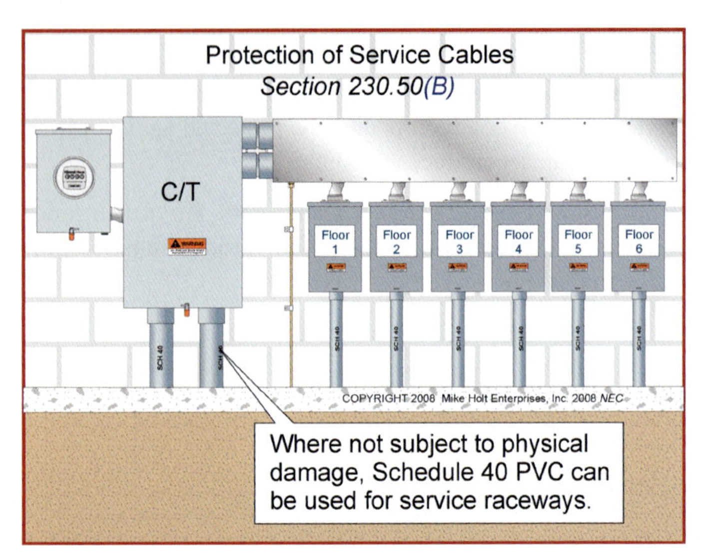

Figure 230–18

230.51 Service Cable Supports.

(A) Service Cable Supports. Service-entrance cable must be supported within 1 ft of the weatherhead, raceway connections or enclosure, and at intervals not exceeding 30 in.

230.54 Overhead Service Locations.

(A) Service Raceway. Raceways for overhead service drops must have a weatherhead.

(B) Service Cable. Service cables must be equipped with a weatherhead.

Exception: SE cable can be formed into a gooseneck and taped with self-sealing weather-resistant thermoplastic.

(C) Above the Point of Attachment. Service heads and goosenecks must be located above the point of attachment. See 230.26.

Exception: Where it's impractical to locate the service head above the point of attachment, it must be located within 2 ft of the point of attachment.

(E) Opposite Polarity Through Separately Bushed Holes. Service heads must provide a bushed opening, and ungrounded conductors must be in separate openings.

(F) Drip Loops. Drip loop conductors must be below the service head or below the termination of the service-entrance cable sheath.

(G) Arranged so Water Will Not Enter. Service drops and service-entrance conductors must be arranged to prevent water from entering service equipment.

230.56 High-Leg Identification. On a 4-wire, delta-connected, three-phase system, where the midpoint of one phase winding is grounded (high-leg system), the conductor with the higher phase voltage-to-ground (208V) must be durably and permanently marked by an outer finish that is orange in color, or by other effective means. Such identification must be placed at each point on the system where a connection is made if the neutral conductor is present [110.15, 215.8, and 230.56]. Figure 230-19

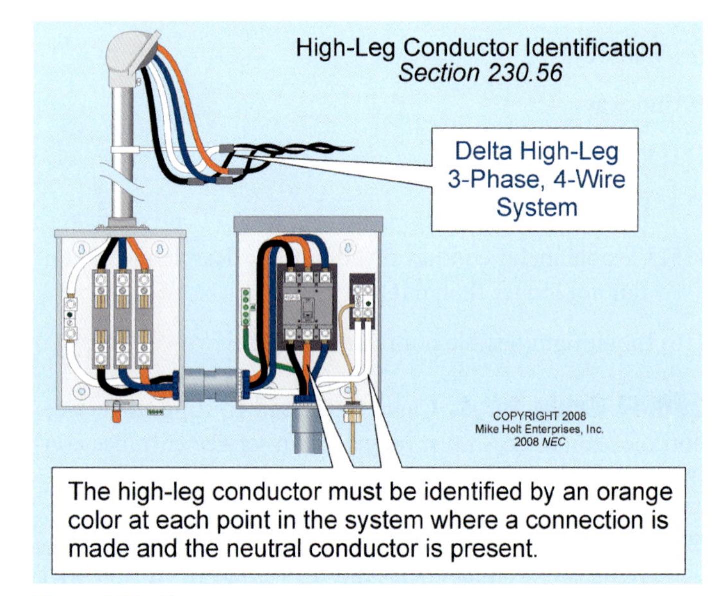

Figure 230–19

Author's Comments:

- The high-leg conductor is also called the "wild leg," "stinger leg," or "bastard leg."
- Since 1975, panelboards supplied by a 4-wire, delta-connected, three-phase system must have the high-leg conductor (208V) terminate to the "B" (center) phase of a panelboard [408.3(E)].
- The ANSI standard for meter equipment requires the high-leg conductor (208V to neutral) to terminate on the "C" (right) phase of the meter socket enclosure. This is because the demand meter needs 120V and it gets this from the "B" phase. Hopefully, the utility lineman is not color blind and doesn't inadvertently cross the "orange" high-leg conductor (208V) with the red (120V) service conductor at the weatherhead. It's happened before... Figure 230–20

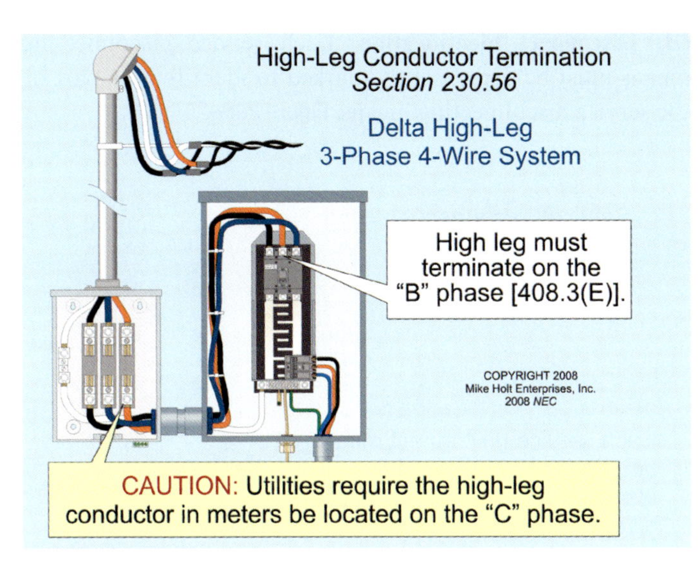

Figure 230–20

PART V. SERVICE EQUIPMENT—GENERAL

230.66 Identified as Suitable for Service Equipment.

The service disconnecting means must be identified as suitable for use as service equipment.

Author's Comment: "Suitable for use as service equipment" means the service disconnecting means is supplied with a main bonding jumper so a neutral-to-case connection can be made, as required in 250.24(C) and 250.142(A). Figure 230–21

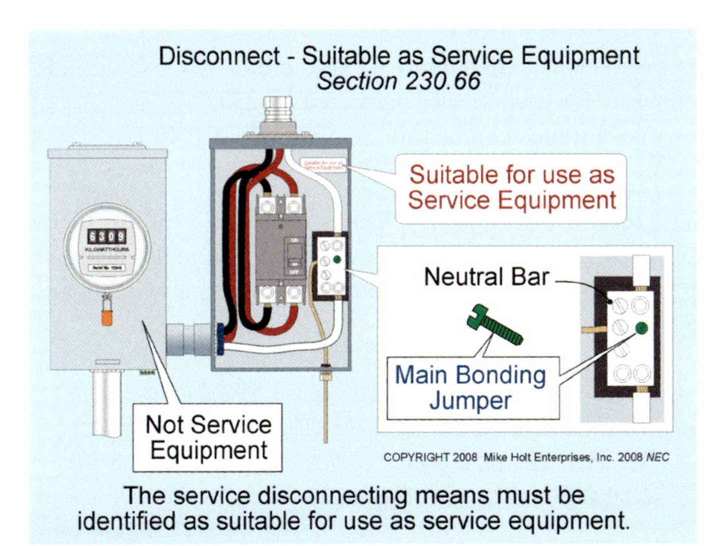

Figure 230–21

PART VI. SERVICE EQUIPMENT—DISCONNECTING MEANS

230.70 General. The service disconnecting means must open all service-entrance conductors from the building or structure premises wiring.

(A) Location.

(1) Readily Accessible. The service disconnecting means must be placed at a readily accessible location either outside the building or structure, or inside nearest the point of service conductor entry. Figure 230–22

WARNING: *Because service-entrance conductors don't have short-circuit or ground-fault protection, they must be limited in length when installed inside a building. Some local jurisdictions have a specific requirement as to the maximum length permitted within a building.*

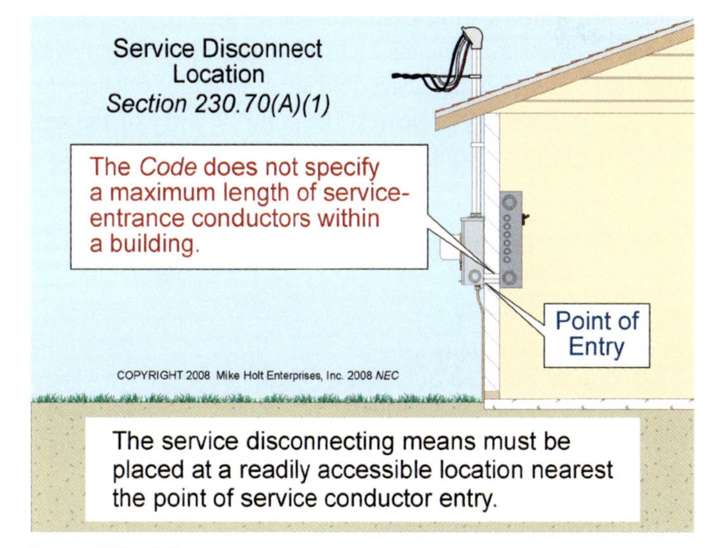

Figure 230–22

(2) Bathrooms. The service disconnecting means is not permitted to be installed in a bathroom. Figure 230–23

Author's Comment: Overcurrent devices must not be located in the bathrooms of dwelling units, or guest rooms or guest suites of hotels or motels [240.24(E)].

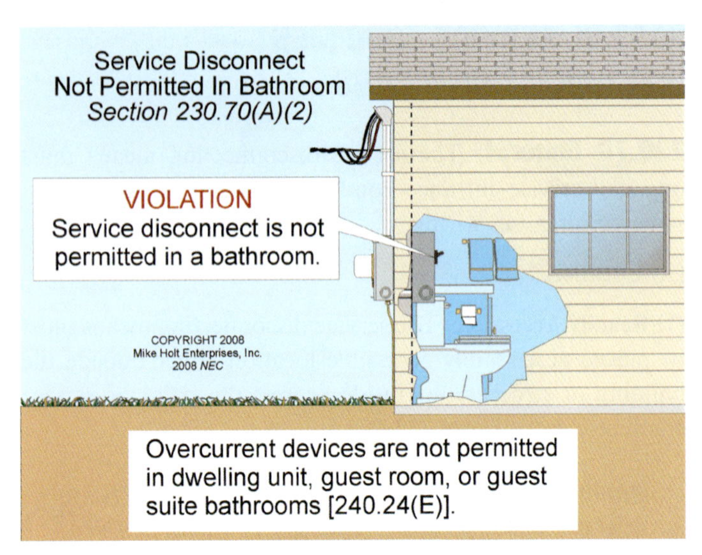

Figure 230–23

(3) Remote Control. Where a remote-control device (such as a pushbutton for a shunt-trip breaker) is used to actuate the service disconnecting means, the service disconnecting means must still be at a readily accessible location either outside the building or structure, or nearest the point of entry of the service conductors as required by 230.70(A)(1). Figure 230–24

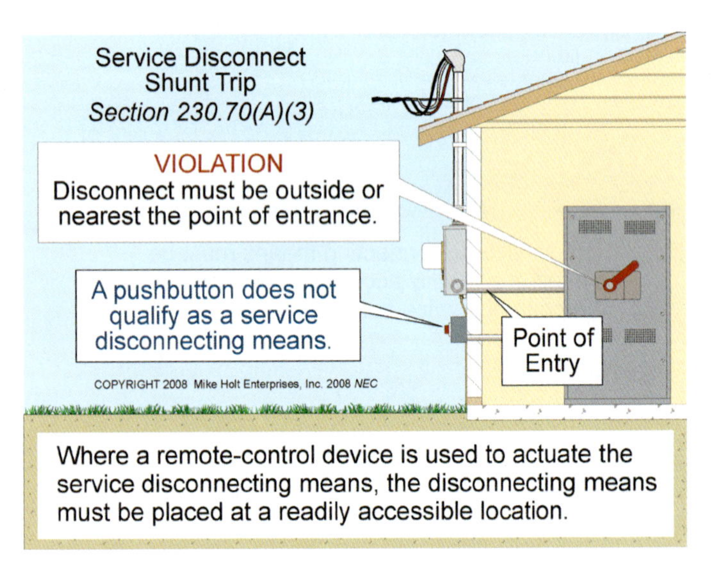

Figure 230–24

Author's Comments:

- See the definition of "Remote Control" in Article 100.
- The service disconnecting means must consist of either a manually operated switch, or a power-operated switch or circuit breaker also capable of being operated manually [230.76].

(B) Disconnect Identification. Each service disconnecting means must be permanently marked to identify it as part of the service disconnecting means. Figure 230–25

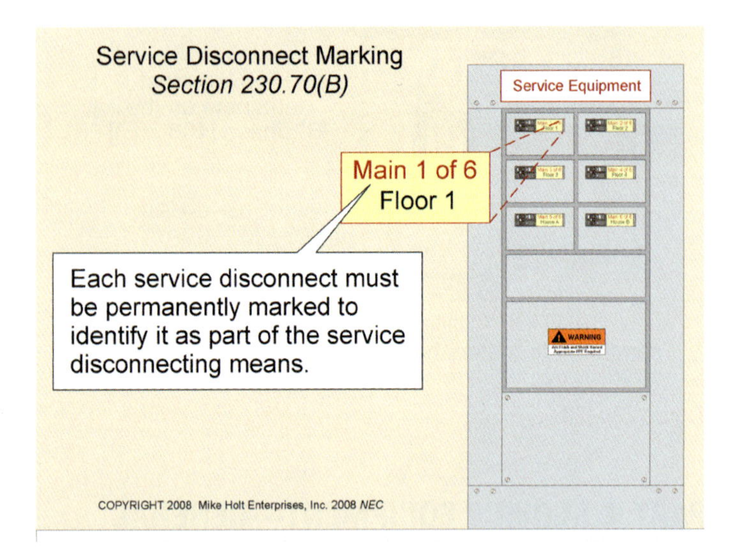

Figure 230–25

Author's Comment: When a building or structure has multiple services and/or feeders, a plaque is required at each service or feeder disconnect location to show the location of the other service or feeder disconnect locations. See 230.2(E).

230.71 Number of Disconnects.

(A) Maximum. There must be no more than six service disconnects for each service permitted by 230.2, or each set of service-entrance conductors permitted by 230.40 Ex 1, 3, 4, or 5.

The service disconnecting means can consist of up to six switches or six circuit breakers mounted in a single enclosure, in a group of separate enclosures, or in or on a switchboard. Figure 230–26

CAUTION: *The rule is six disconnecting means for each service, not for each building. If the building has two services, then there can be a total of twelve service disconnects (six disconnects per service).* Figure 230–27

230.72 Grouping of Disconnects.

(A) Two to Six Disconnects. The service disconnecting means for each service must be grouped.

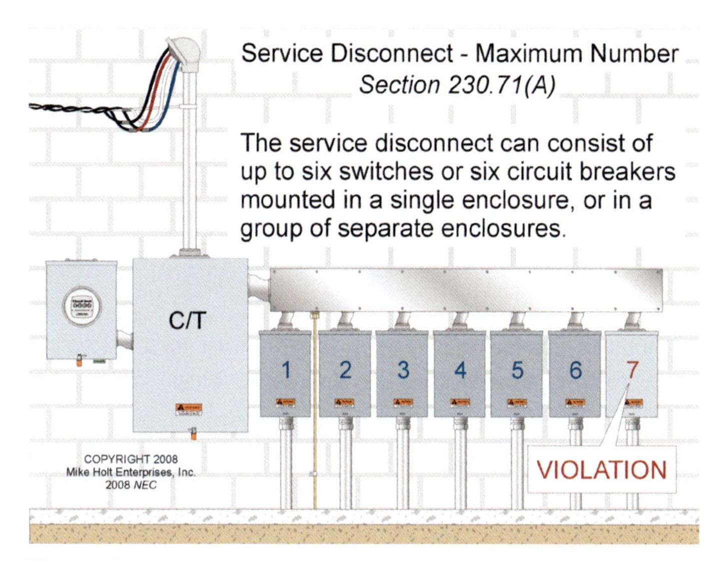

Figure 230–26

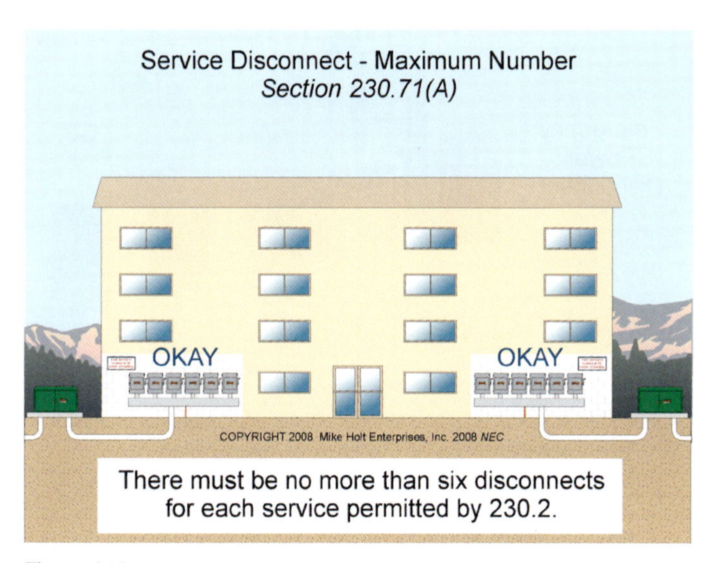

Figure 230–27

(B) Additional Service Disconnecting Means. To minimize the possibility of accidental interruption of power, the disconnecting means for fire pumps [Article 695], emergency [Article 700], legally required standby [Article 701], or optional standby [Article 702] systems must be located remote from the one to six service disconnects for normal service.

> **Author's Comment:** Because emergency systems are just as important as fire pumps and standby systems, they need to have the same safety precautions to prevent unintended interruption of the supply of electricity.

(C) Access to Occupants. In a multiple-occupancy building, each occupant must have access to their service disconnecting means.

Exception: In multiple-occupancy buildings where electrical maintenance is provided by continuous building management, the service disconnecting means can be accessible only to building management personnel.

230.76 Manual or Power Operated. The service disconnecting means can consist of: Figure 230–28

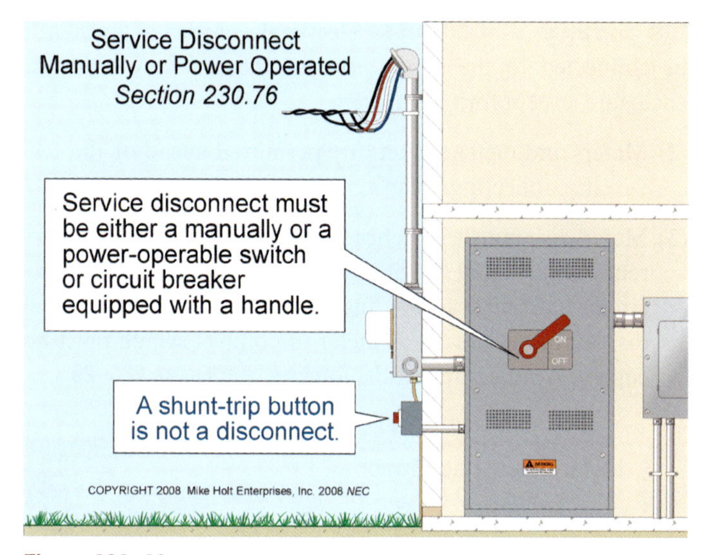

Figure 230–28

(1) A manually operable switch or circuit breaker equipped with a handle or suitable operating means.

(2) A power-operated switch or circuit breaker, provided it can be opened by hand in the event of a power supply failure.

230.79 Rating of Disconnect. The service disconnecting means for a building or structure must have an ampere rating of not less than the calculated load according to Article 220, and in no case less than:

(A) One-Circuit Installation. For installations consisting of a single branch circuit, the disconnecting means must have a rating not less than 15A.

(B) Two-Circuit Installation. For installations consisting of two 2-wire branch circuits, the disconnecting means must have a rating not less than 30A.

(C) One-Family Dwelling. For a one-family dwelling, the disconnecting means must have a rating not less than 100A, 3-wire.

(D) All Others. For all other installations, the disconnecting means must have a rating not less than 60A.

> **Author's Comment:** A shunt-trip button does not qualify as a service disconnect because it does not meet any of the above requirements.

230.82 Equipment Connected to the Supply Side of the Service Disconnect.

Electrical equipment must not be connected to the supply side of the service disconnect enclosure, except for:

(2) Meters and meter sockets are permitted ahead of the service disconnecting means.

(3) Meter disconnect switches that have a short-circuit current rating equal to or more than the available short-circuit current can be installed ahead of the service disconnecting means. A meter disconnect switch must be capable of interrupting the load served. Figure 230–29

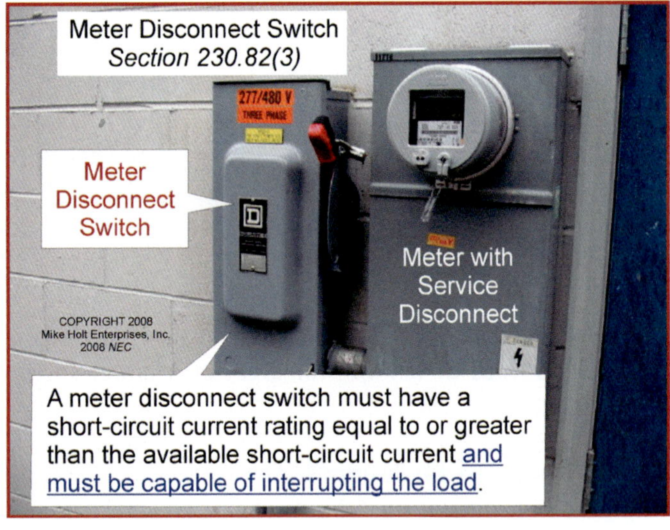

Figure 230–29

> **Author's Comment:** Electric utilities often require a meter disconnect switch for 277/480V services to enhance safety for utility personnel when they install or remove a meter.

(4) A Type 1 surge protective device can be installed ahead of the service disconnecting means.

> **Author's Comment:** A Type 1 surge protective device is listed to be permanently connected on the line side of service equipment [285.23].

(5) Tap conductors for legally required [701] and optional standby [702] power systems, fire pump equipment, fire and sprinkler alarms, and load (energy) management devices.

> **Author's Comment:** Emergency standby power must not be supplied by the connection ahead of service equipment [700.12]. Figure 230–30

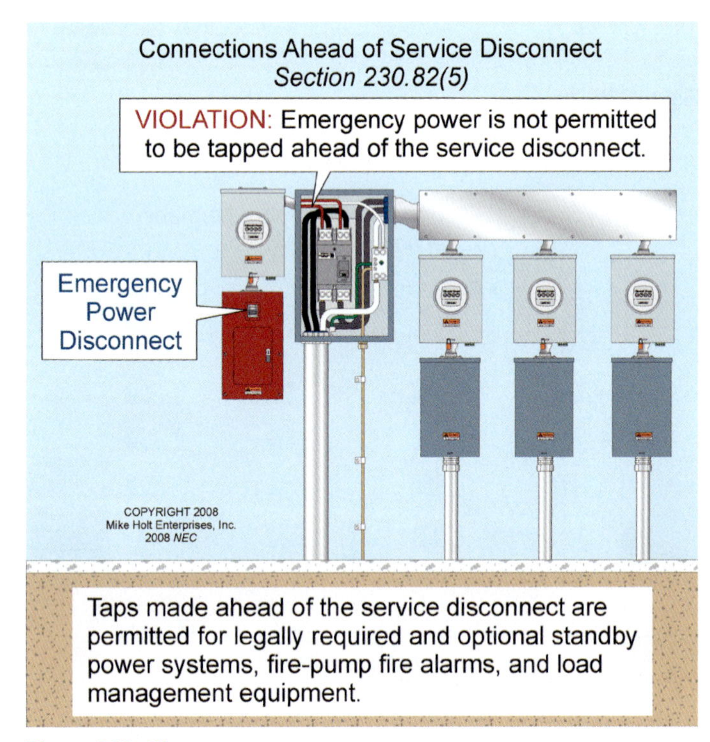

Figure 230–30

(6) Solar photovoltaic systems, fuel cell systems, or interconnected electric power production sources are permitted ahead of the service disconnecting means.

(7) Control circuits for power-operable service disconnecting means, if suitable overcurrent protection and disconnecting means are provided, are permitted ahead of the service disconnecting means.

PART VII. SERVICE EQUIPMENT OVERCURRENT PROTECTION

Author's Comment: The *NEC* doesn't require service conductors to be provided with short-circuit or ground-fault protection, but the feeder overcurrent device provides overload protection for the service conductors.

230.90 Overload Protection Required. Each ungrounded service conductor must have overload protection at the point where the service conductors terminate [240.21(D)]. Figure 230–31

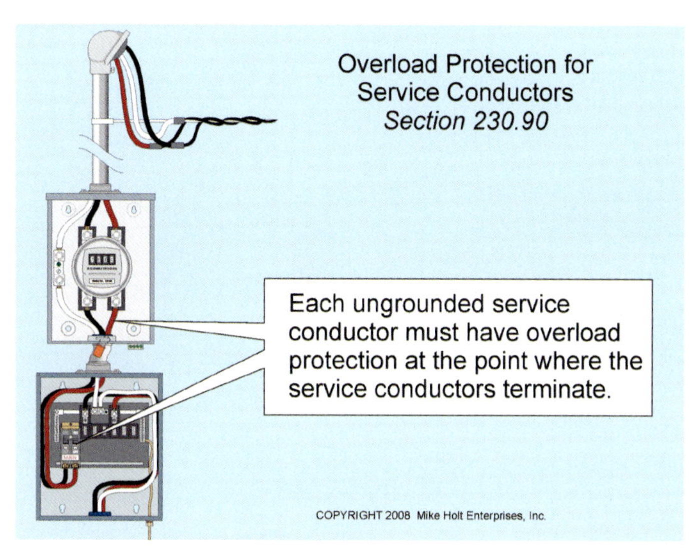

Figure 230–31

(A) Overcurrent Device Rating. The rating of the overcurrent device must not be more than the ampacity of the conductors.

Exception No. 2: Where the ampacity of the ungrounded conductors doesn't correspond with the standard rating of overcurrent devices as listed in 240.6(A), the next higher overcurrent device can be used, if it doesn't exceed 800A [240.4(B)].

Example: *Two sets of parallel 500 kcmil THHN conductors (each rated 380A at 75°C) can be protected by an 800A overcurrent device.* Figure 230–32

Exception No. 3: The combined ratings of two to six service disconnecting means can exceed the ampacity of the service conductors provided the calculated load, in accordance with Article 220, doesn't exceed the ampacity of the service conductors. Figure 230–33

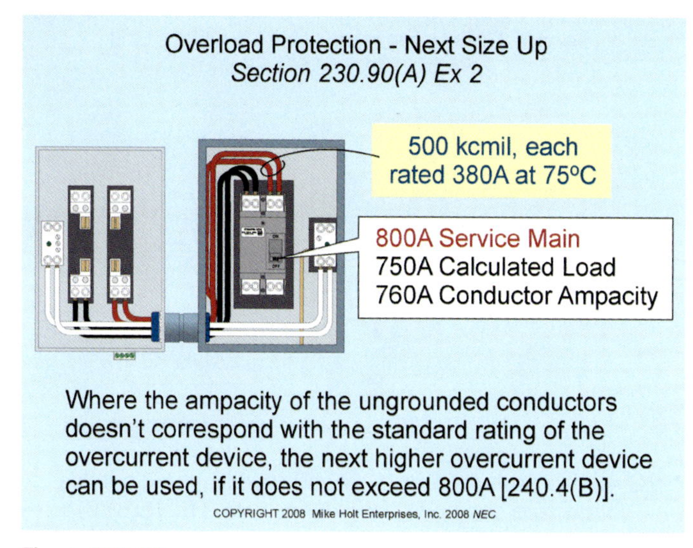

Figure 230–32

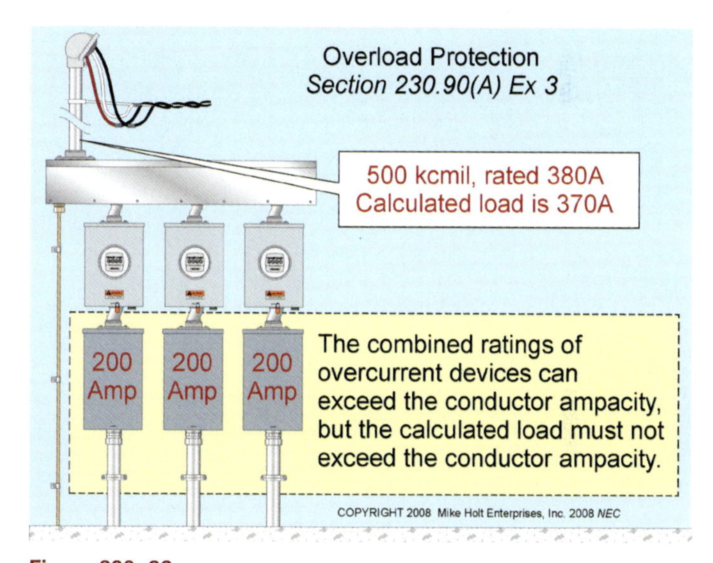

Figure 230–33

Exception No. 5: Overload protection for 3-wire, single-phase, 120/240V dwelling unit service conductors can be in accordance with 310.15(B)(6). Figure 230–34

230.95 Ground-Fault Protection of Equipment. Ground-fault protection of equipment is required for each service disconnecting means rated 1,000A or more that is supplied by a 4-wire, three-phase, 277/480V wye-connected system.

The rating of the service disconnecting means is considered to be the rating of the largest fuse that can be installed or the highest continuous current trip setting of a circuit breaker.

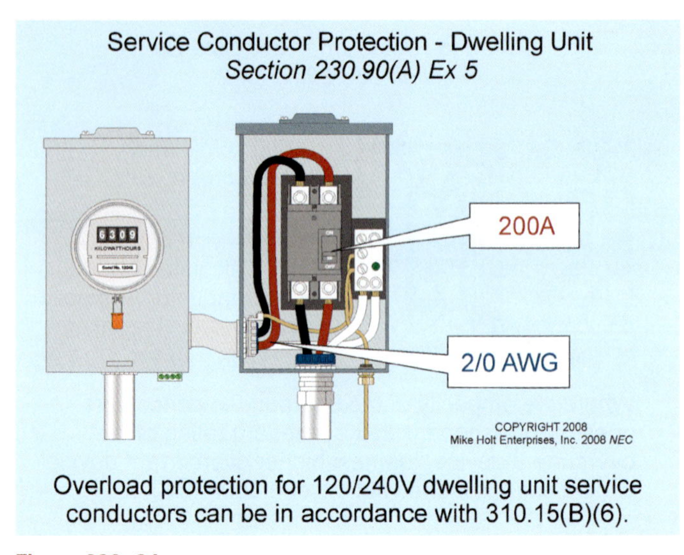

Figure 230–34

Author's Comments:

- Ground-fault protection of equipment isn't permitted for fire pumps [695.6(H)] and it's not required for emergency systems [700.26] or legally required standby systems [701.17].
- Article 100 defines "Ground-Fault Protection of Equipment" as a system intended to provide protection of equipment from ground faults by opening the overcurrent device at current levels less than those required to protect conductors from damage. This type of protective system isn't intended to protect people, only connected equipment. See 215.10 for similar requirements for feeders.

ARTICLE 230 Practice Questions

ARTICLE 230. SERVICES—PRACTICE QUESTIONS

1. Additional services shall be permitted for a single building or other structure sufficiently large to make two or more services necessary if permitted by _____.

   (a) the registered design professional
   (b) special permission
   (c) the engineer of record
   (d) master electricians

2. Service conductors supplying a building or other structure shall not _____ of another building or other structure.

   (a) be installed on the exterior walls
   (b) pass through the interior
   (c) a and b
   (d) none of these

3. Where a service raceway enters a building or structure from a(n) _____, it shall be sealed in accordance with 300.5(G).

   (a) transformer vault
   (b) underground distribution system
   (c) cable tray
   (d) overhead rack

4. Service conductors shall not be installed beneath openings through which materials may be moved and shall not be installed where they will obstruct entrance to these buildings' openings.

   (a) True
   (b) False

5. Service-drop conductors shall have _____.

   (a) sufficient ampacity to carry the load
   (b) adequate mechanical strength
   (c) a or b
   (d) a and b

6. If a set of 120/240V service-drop conductors terminates at a through-the-roof raceway or approved support, with less than 6 ft of these conductors passing over the roof overhang, the minimum clearance above the roof for these service conductors shall be _____.

   (a) 12 in.
   (b) 18 in.
   (c) 2 ft
   (d) 5 ft

7. The minimum clearance for service-drop conductors not exceeding 600V that pass over commercial areas subject to truck traffic is _____.

   (a) 10 ft
   (b) 12 ft
   (c) 15 ft
   (d) 18 ft

8. Where raceway-type service masts are used, all raceway fittings shall be _____ for use with service masts.

   (a) identified
   (b) approved
   (c) of a heavy-duty type
   (d) listed

9. Service-lateral conductors that supply power to limited loads of a single branch circuit shall not be smaller than _____.

   (a) 14 AWG copper
   (b) 14 AWG aluminum
   (c) 12 AWG copper
   (d) 12 AWG aluminum

10. Underground service conductors shall be protected against physical damage.

    (a) True
    (b) False

11. Overhead service cables shall be equipped with a _____.

    (a) raceway
    (b) service head
    (c) cover
    (d) all of these

12. To prevent moisture from entering service equipment, service-entrance conductors shall _____.

    (a) be connected to service-drop conductors below the level of the service head
    (b) have drip loops formed on the service-entrance conductors
    (c) a or b
    (d) a and b

13. A service disconnecting means shall be installed at a(n) _____ location.

    (a) dry
    (b) readily accessible
    (c) outdoor
    (d) indoor

14. Each service disconnecting means shall be permanently _____ to identify it as a service disconnect.

    (a) identified
    (b) positioned
    (c) marked
    (d) none of these

15. The additional service disconnecting means for fire pumps, emergency systems, legally required standby, or optional standby services, shall be installed remote from the one to six service disconnecting means for normal service to minimize the possibility of _____ interruption of supply.

    (a) intentional
    (b) accidental
    (c) simultaneous
    (d) prolonged

16. When the service disconnecting means is a power-operated switch or circuit breaker, it shall be able to be opened by hand in the event of a _____.

    (a) ground fault
    (b) short circuit
    (c) power surge
    (d) power supply failure

17. A meter disconnect switch located ahead of service equipment must have a short-circuit current rating equal to or greater than the available short-circuit current and be capable of interrupting the load served.

    (a) True
    (b) False

18. As defined by 230.95, the rating of the service disconnect shall be considered to be the rating of the largest _____ that can be installed or the highest continuous current trip setting for which the actual overcurrent device installed in a circuit breaker is rated or can be adjusted.

    (a) fuse
    (b) circuit
    (c) conductor
    (d) all of these

ARTICLE

240 Overcurrent Protection

INTRODUCTION TO ARTICLE 240—OVERCURRENT PROTECTION

This article provides the requirements for selecting and installing overcurrent devices. Reviewing the basic concept of overcurrent protection will help you avoid confusion as we move forward. Overcurrent exists when current exceeds the rating of conductors or equipment. This can be due to overload, short circuit, or ground fault [Article 100].

Overload. An overload is a condition where equipment or conductors carry current exceeding their rated ampacity [Article 100]. An example of an overload is plugging two 12.50A (1,500W) hair dryers into a 20A branch circuit.

Ground Fault. A ground fault is an unintentional, electrically conducting connection between an ungrounded conductor of an electrical circuit and the normally noncurrent-carrying conductors, metallic enclosures, metallic raceways, metallic equipment, or earth [250.2]. During the period of a ground fault, dangerous voltages will be present on metal parts until the circuit overcurrent device opens.

Short Circuit. A short circuit is the unintentional electrical connection between any two normally current-carrying conductors of an electrical circuit, either line-to-line or line-to-neutral.

Overcurrent devices protect conductors and equipment. But it's important to note they protect conductors differently than equipment.

An overcurrent device protects conductors by opening when the current reaches a value that will cause an excessive temperature rise in conductors. The overcurrent device's interrupting rating must be sufficient for the maximum possible fault current available on the line-side terminals of the equipment [110.9]. Using a water analogy, current rises like water in a tank—at a certain level, the overcurrent device shuts off the faucet. Think in terms of normal operating conditions that just get too far out of the normal range.

An overcurrent device protects equipment by opening when it detects a short circuit or ground fault. Every piece of electrical equipment must have a short-circuit current rating that permits the overcurrent device (for that equipment) to clear short circuits or ground faults without extensive damage to the electrical components of the circuit [110.10]. Short circuits and ground faults aren't normal operating conditions, so the overcurrent devices for equipment will have different characteristics than overcurrent devices for conductors.

PART I. GENERAL

240.1 Scope. Article 240 covers the general requirements for overcurrent protection and the installation requirements of overcurrent devices. Figure 240–1

Author's Comment: Overcurrent is a condition where the current exceeds the rating of conductors or equipment due to overload, short circuit, or ground fault [Article 100]. Figure 240–2

FPN: An overcurrent device protects the circuit by opening the device when the current reaches a value that will cause excessive or dangerous temperature rise (overheating) in conductors. Overcurrent devices must have an interrupting rating sufficient for the maximum possible fault current available on the line-side terminals of the equipment [110.9]. Electrical equipment must have a short-circuit current rating that permits the circuit's overcurrent device to clear short circuits or ground faults without extensive damage to the circuit's electrical components [110.10].

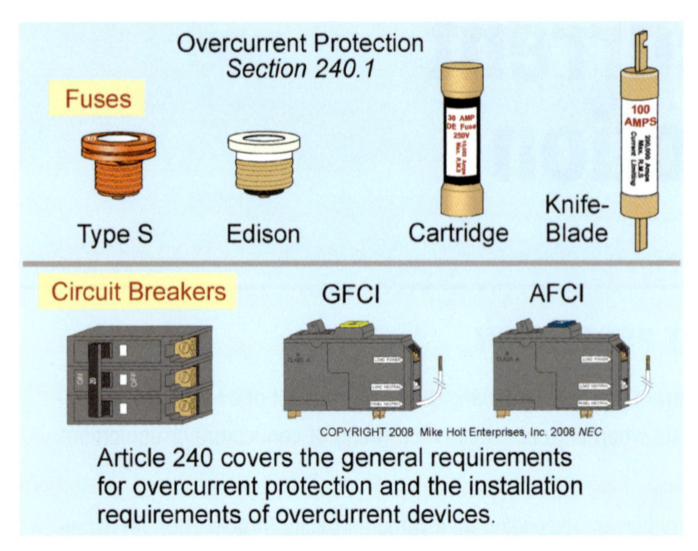

Article 240 covers the general requirements for overcurrent protection and the installation requirements of overcurrent devices.

Figure 240–1

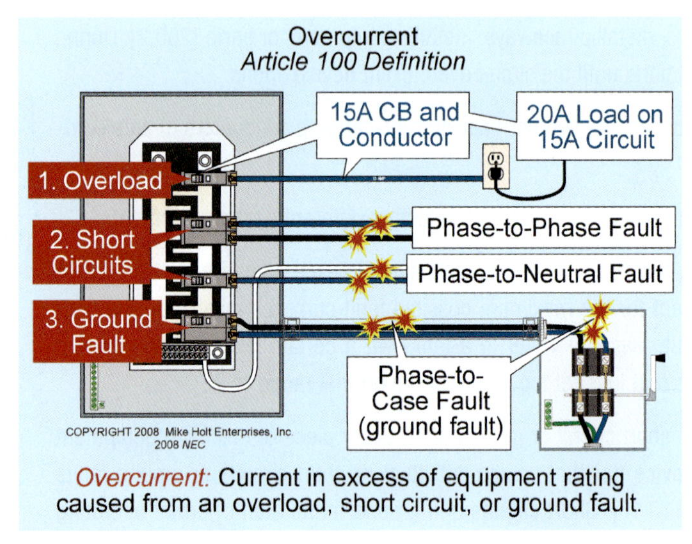

Overcurrent: Current in excess of equipment rating caused from an overload, short circuit, or ground fault.

Figure 240–2

240.2 Definitions.

Current-Limiting Overcurrent Device. An overcurrent device (typically a fast-acting fuse) that reduces the fault current to a magnitude substantially less than that obtainable in the same circuit if the current-limiting device was not used. See 240.40 and 240.60(B). Figure 240–3

Author's Comment: A current-limiting fuse is a type of fuse designed for operations related to short circuits only. When a fuse operates in its current-limiting range, it will begin to melt in less than a quarter cycle, and it will open a bolted short circuit in less than half of a cycle. This type of fuse limits the instantaneous peak let-through current to a value substantially less

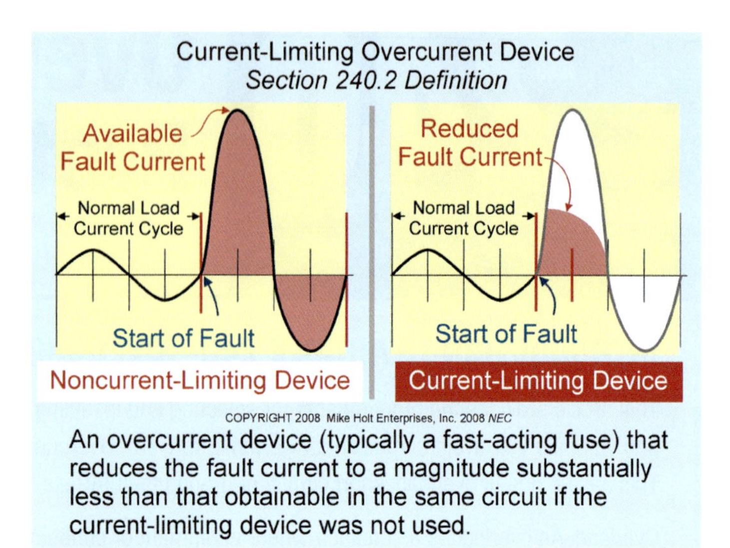

An overcurrent device (typically a fast-acting fuse) that reduces the fault current to a magnitude substantially less than that obtainable in the same circuit if the current-limiting device was not used.

Figure 240–3

than what will occur in the same circuit if the fuse is replaced with a solid conductor of equal impedance. If the available short-circuit current exceeds the equipment/conductor short-circuit current rating, then the thermal and magnetic forces can cause the equipment circuit conductors, as well as the circuit equipment grounding conductors, to vaporize. The only solutions to the problem of excessive available fault current are to:

- Install equipment with a higher short-circuit rating, or
- Protect the components of the circuit by a current-limiting overcurrent device such as a fast-clearing fuse, which can reduce the let-through energy.

A breaker or a fuse does limit current, but it may not be listed as a current-limiting device. A thermal-magnetic circuit breaker typically clears fault current in less than three to five cycles when subjected to a short circuit or ground fault of 20 times its rating. A standard fuse will clear the same fault in less than one cycle and a current-limiting fuse in less than half of a cycle.

Tap Conductors. A conductor, other than a service conductor, that has overcurrent protection rated more than the ampacity of a conductor. See 240.21(A) and 240.21(B) for details. Figure 240–4

240.3 Protection of Equipment.

The following equipment and their conductors are protected against overcurrent in accordance with the article that covers the type of equipment:

- Air-Conditioning and Refrigeration Equipment, 440.22

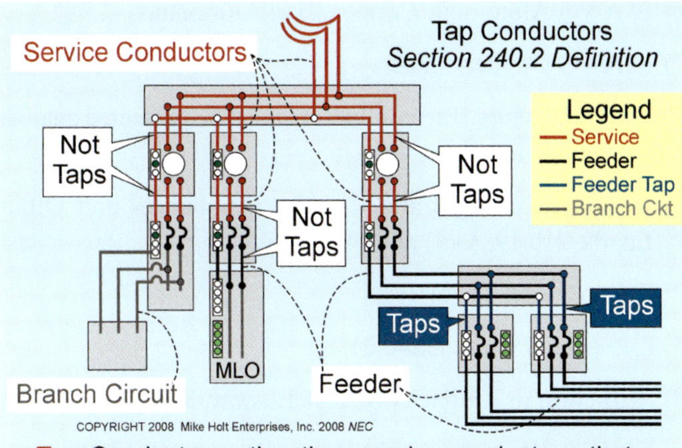

Tap: Conductors, other than service conductors, that have overcurrent protection ahead of the point of supply that exceeds the value permitted for similar conductors.

Figure 240–4

- Appliances, Article 422
- Audio Circuits, 640.9
- Branch Circuits, 210.20
- Class 1, 2, and 3 Circuits, Article 725
- Feeder Conductors, 215.3
- Flexible Cords, 240.5(B)(1)
- Fire Alarms, Article 760
- Fire Pumps, Article 695
- Fixed Electric Space-Heating Equipment, 424.3(B)
- Fixture Wire, 240.5(B)(2)
- Panelboards, 408.36
- Service Conductors, 230.90(A)
- Transformers, 450.3

240.4 Protection of Conductors.

Except as permitted by (A) through (G), conductors must be protected against overcurrent in accordance with their ampacity after ampacity adjustment, as specified in 310.15.

(A) Power Loss Hazard. Conductor overload protection is not required, but short-circuit protection is required where the interruption of the circuit will create a hazard; such as in a material-handling electromagnet circuit or fire pump circuit.

(B) Overcurrent Devices Rated 800A or Less. The next higher standard rating of overcurrent device (above the ampacity of the ungrounded conductors being protected) is permitted, provided all of the following conditions are met:

(1) The conductors don't supply multioutlet receptacle branch circuits.

(2) The ampacity of a conductor, after ampacity adjustment and/or correction, doesn't correspond with the standard rating of a fuse or circuit breaker in 240.6(A).

(3) The overcurrent device rating doesn't exceed 800A.

Example: *A 400A overcurrent device can protect 500 kcmil conductors, where each conductor has an ampacity of 380A at 75°C, in accordance with Table 310.16.* **Figure 240–5**

Author's Comment: This "next size up" rule doesn't apply to feeder tap conductors [240.21(B)], or transformer secondary conductors [240.21(C)].

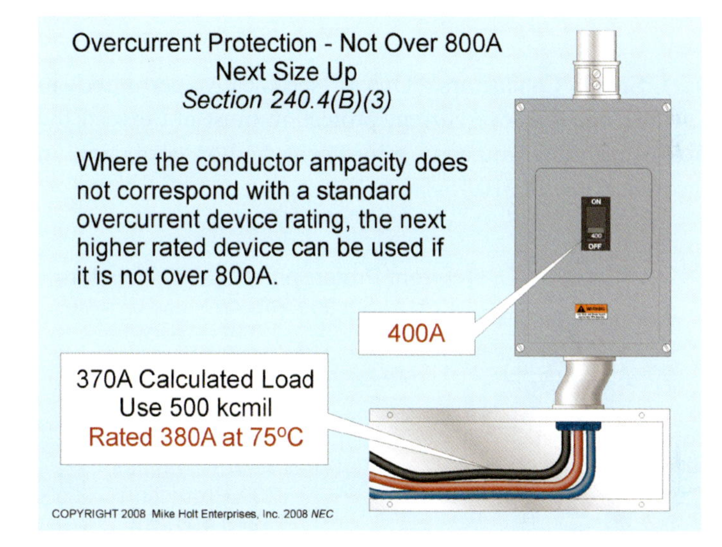

Figure 240–5

(C) Overcurrent Devices Rated Over 800A. If the circuit's overcurrent device exceeds 800A, the conductor ampacity, after ampacity adjustment and/or correction, must have a rating of not less than the rating of the overcurrent device.

Example: *A 1,200A overcurrent device can protect three sets of 600 kcmil conductors per phase, where each conductor has an ampacity of 420A at 75°C, in accordance with Table 310.16.* **Figure 240–6**

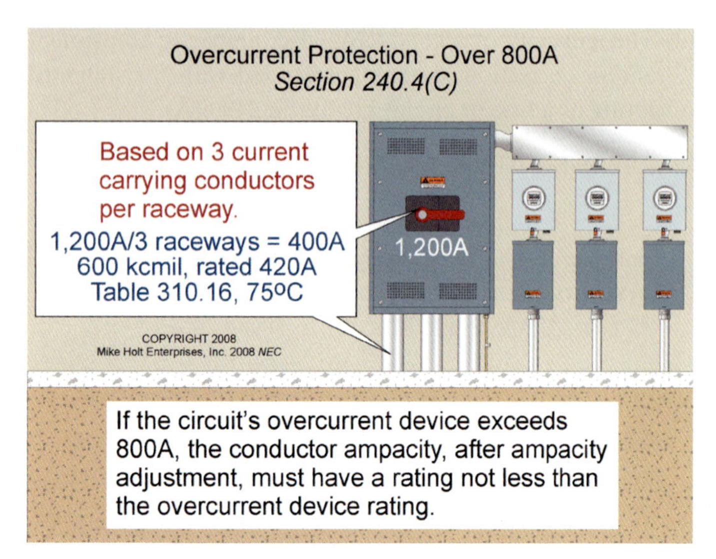

Figure 240–6

(D) Small Conductors. Unless specifically permitted in 240.4(E) or (G), overcurrent protection must not exceed the following after ampacity adjustment and/or correction in accordance with 310.15: Figure 240–7

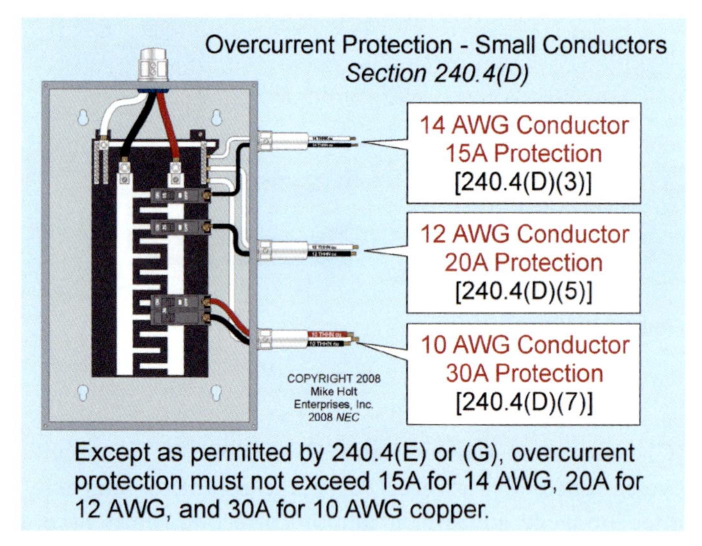

Figure 240–7

(1) 18 AWG Copper—7A

(2) 16 AWG Copper—10A

(3) 14 AWG Copper—15A

(4) 12 AWG Aluminum/Copper-Clad Aluminum—15A

(5) 12 AWG Copper—20A

(6) 10 AWG Aluminum/Copper-Clad Aluminum—25A

(7) 10 AWG Copper—30A

(E) Tap Conductors. Tap conductors must be protected against overcurrent as follows:

(1) Household Ranges and Cooking Appliances and Other Loads, 210.19(A)(3) and (4)

(2) Fixture Wire, 240.5(B)(2)

(3) Location in Circuit, 240.21

(4) Reduction in Ampacity Size of Busway, 368.17(B)

(5) Feeder or Branch Circuits (busway taps), 368.17(C)

(6) Single Motor Taps, 430.53(D)

(F) Transformer Secondary Conductors. The primary overcurrent device sized in accordance with 450.3(B) is considered suitable to protect the secondary conductors of a 2-wire (single voltage) system, provided the primary overcurrent device doesn't exceed the value determined by multiplying the secondary conductor ampacity by the secondary-to-primary transformer voltage ratio.

Question: *What is the minimum secondary conductor size required for a 2-wire, 480V to 120V transformer rated 1.50 kVA?* Figure 240–8

(a) 16 AWG (b) 14 AWG (c) 12 AWG (d) 10 AWG

Answer: *(b) 14 AWG*

Primary Current = VA/E
VA = 1,500 VA
E = 480V

Primary Current = 1,500 VA/480V
Primary Current = 3.13A

Primary Protection [450.3(B)] = 3.13A x 1.67
Primary Protection = 5.22A or 5A Fuse

Secondary Current = 1,500 VA/120V
Secondary Current = 12.50A

Secondary Conductor = 14 AWG, rated 20A at 60C, Table 310.16

The 5A primary overcurrent device can be used to protect 14 AWG secondary conductors because it doesn't exceed the value determined by multiplying the secondary conductor ampacity by the secondary-to-primary transformer voltage ratio (5A = 20A x 120V/480V).

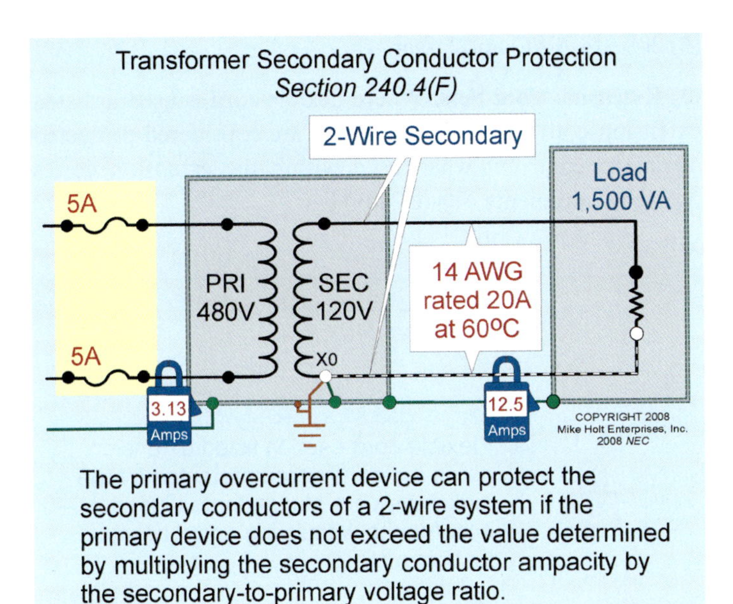

Figure 240–8

(G) Overcurrent Protection for Specific Applications. Overcurrent protection for specific equipment and conductors must comply with the requirements referenced in Table 240.4(G).

Air-Conditioning and Refrigeration [Article 440]. Air-conditioning and refrigeration equipment, and their circuit conductors, must be protected against overcurrent in accordance with 440.22.

> **Author's Comment:** Typically, the branch-circuit ampacity and protection size is marked on the equipment nameplate [440.4(A)].

Question: *What size branch-circuit overcurrent device is required for an air conditioner (18A) when the nameplate indicates the minimum circuit ampacity is 23A, with maximum overcurrent protection of 40A?* **Figure 240–9**

(a) 12 AWG, 40A protection *(b) 12 AWG, 50A protection*
(c) 12 AWG, 60A protection *(d) 12 AWG, 70A protection*

Answer: *(a) 12 AWG, 40A protection*

> **Author's Comment:** Air-conditioning and refrigeration nameplate values are calculated by the manufacturer according to the following:

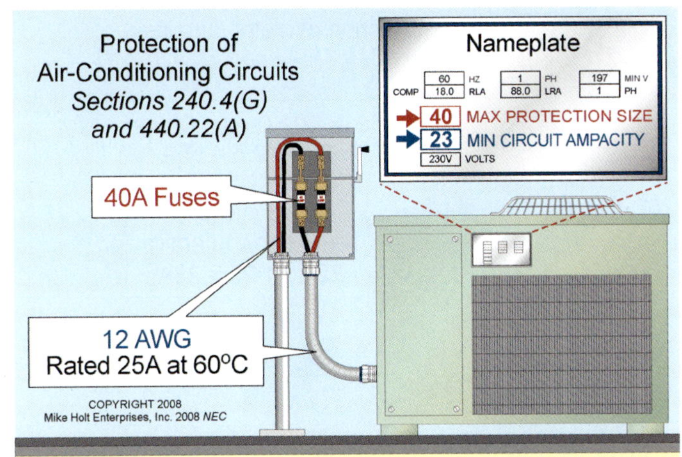

Figure 240–9

- Branch-Circuit Conductor Size [440.32]
  18A x 1.25 = 22.50A, 12 AWG rated 25A at 60°C
- Branch-Circuit Protection Size [440.22(A)]
  18A x 2.25 = 40.50A, 40A maximum overcurrent protection size {240.6(A)]
- Motors [Article 430]. Motor circuit conductors must be protected against short circuits and ground faults in accordance with 430.52 and 430.62 [430.51].

If the nameplate calls for fuses, fuses must be used to comply with the manufacturer's instructions [110.3(B)].

Question: *What size branch-circuit conductor and overcurrent device (circuit breaker) is required for a 7½ hp, 230V, three-phase motor?* **Figure 240–10**

(a) 10 AWG, 50A breaker *(b) 10 AWG, 60A breaker*
(c) a or b *(d) none of these*

Answer: *(c) 10 AWG, 50A or 60A breaker*

Step 1: Determine the branch-circuit conductor size [Table 310.16, 430.22, and Table 430.250]:

FLC = 28A [Table 430-250]

22A x 1.25 = 28A, 10 AWG, rated 30A at 60°C

Step 2: Determine the branch-circuit protection size [240.6(A), 430.52(C)(1) Ex 1, Table 430.250].

Inverse Time Breaker: 22A x 2.50 = 55A
Next size up = 60A

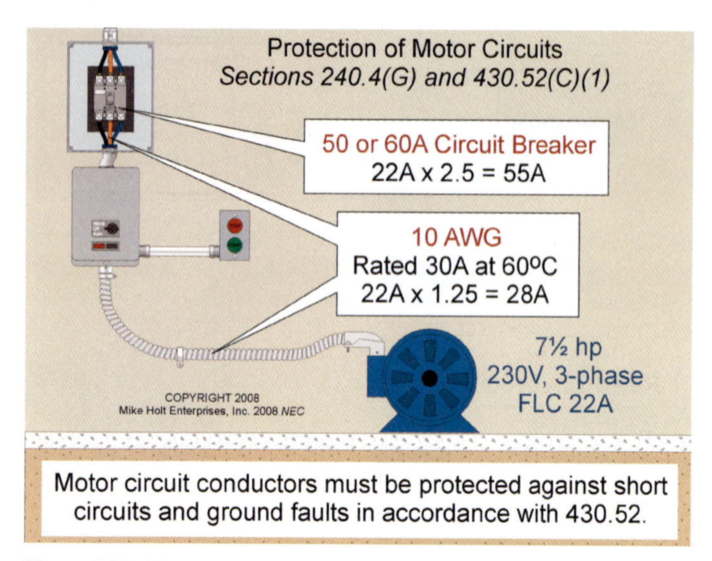

Figure 240–10

Motor Control [Article 430]. Motor control circuit conductors must be sized and protected in accordance with 430.72.

Remote-Control, Signaling, and Power-Limited Circuits [Article 725]. Remote-control, signaling, and power-limited circuit conductors must be protected against overcurrent in accordance with 725.43.

240.5 Protection of Flexible Cords and Fixture Wires.

(A) Ampacities. Flexible cord must be protected by an overcurrent device in accordance with its ampacity as specified in Table 400.5(A) or Table 400.5(B). Fixture wire must be protected against overcurrent in accordance with its ampacity as specified in Table 402.5. Supplementary overcurrent protection, as discussed in 240.10, is permitted to provide this protection.

(B) Branch-Circuit Overcurrent Protection.

(1) Cords for Listed Appliances or Luminaires. Where flexible cord is used with a specific listed appliance or luminaire, the conductors are considered protected against overcurrent when used within the appliance or luminaire listing requirements.

> **Author's Comment:** The *NEC* only applies to premises wiring, not to the supply cords of listed appliances and luminaires.

(2) Fixture Wire. Fixture wires can be tapped to the following circuits:

(1) 20A–18 AWG, up to 50 ft of run length

(2) 20A–16 AWG, up to 100 ft of run length

(3) 20A–14 AWG and larger

(3) Extension Cord Sets. Where flexible cord is used in listed extension cord sets, the conductors are considered protected against overcurrent when used within the extension cord's listing requirements. Figure 240–11

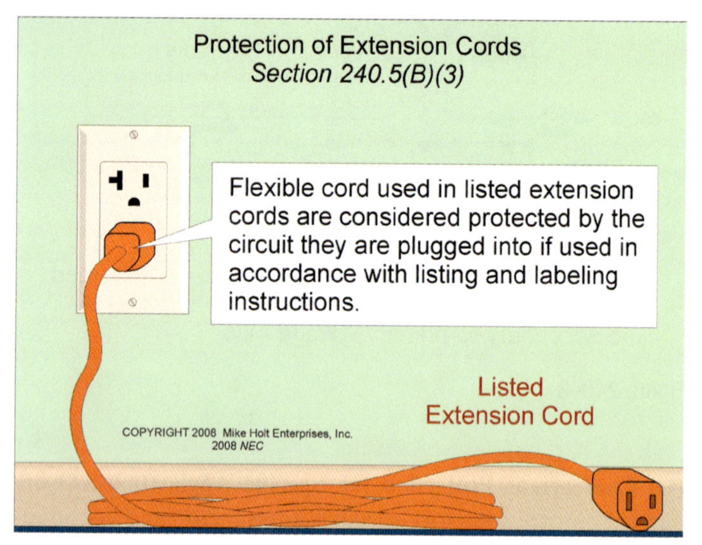

Figure 240–11

240.6 Standard Ampere Ratings.

(A) Fuses and Fixed-Trip Circuit Breakers. The standard ratings in amperes for fuses and inverse time breakers are: 15, 20, 25, 30, 35, 40, 45, 50, 60, 70, 80, 90, 100, 110, 125, 150, 175, 200, 225, 250, 300, 350, 400, 450, 500, 600, 700, 800, 1,000, 1,200, 1,600, 2,000, 2,500, 3,000, 4,000, 5,000 and 6,000. Figure 240–12

Additional standard ampere ratings for fuses include 1, 3, 6, 10, and 601.

> **Author's Comment:** Fuses rated less than 15A are sometimes required for the protection of fractional horsepower motor circuits [430.52], motor control circuits [430.72], small transformers [450.3(B)], and remote-control circuit conductors [725.43].

(B) Adjustable Circuit Breakers. The ampere rating of an adjustable circuit breaker is equal to its maximum long-time pickup current setting.

(C) Restricted Access, Adjustable-Trip Circuit Breakers. The ampere rating of adjustable-trip circuit breakers that have restricted access to the adjusting means is equal to their adjusted long-time pickup current settings.

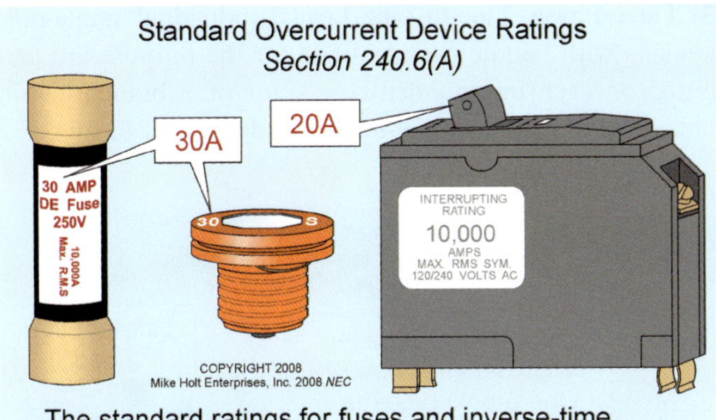

Figure 240–12

240.10 Supplementary Overcurrent Protection.

Supplementary overcurrent devices (usually an internal fuse), often used in luminaires, appliances, and equipment for internal circuits and components, must not be used as the required branch-circuit overcurrent device. Figure 240–13

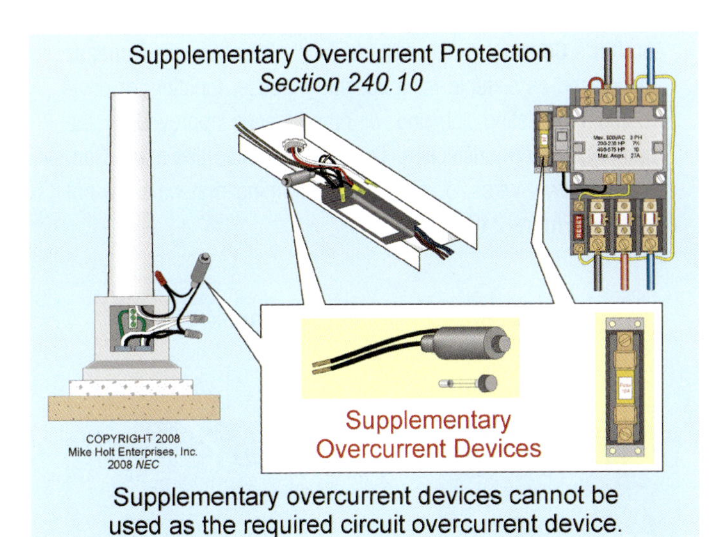

Figure 240–13

A supplementary overcurrent device isn't required to be readily accessible [240.24(A)(2)].

Author's Comment: Article 100 defines a "Supplementary Overcurrent Device" as a device intended to provide limited overcurrent protection for specific applications and utilization equipment. This limited protection is in addition to the protection provided in the required branch circuit by the branch-circuit overcurrent device.

240.13 Ground-Fault Protection of Equipment.

Service equipment and feeder circuits rated 1,000A or more, supplied from a 4-wire, three-phase, 277/480V wye-connected system must be protected against ground-faults in accordance with 230.95 [215.10 and 230.95].

The requirement for ground-fault protection of equipment doesn't apply to:

(1) Continuous industrial processes where a nonorderly shutdown will introduce additional or increased hazards.

(2) Installations where ground-fault protection of equipment is already provided.

(3) Fire pumps [695.6(H)].

Author's Comments:

- Article 100 defines "Ground-Fault Protection of Equipment" as a system intended to provide protection of equipment from ground faults by opening the overcurrent device at current levels less than those required to protect conductors from damage. This type of protective system isn't intended to protect people, only connected equipment. See 215.10 and 230.95 for similar requirements for feeders and services.
- Ground-fault protection of equipment isn't required for emergency power systems [700.26] or legally required standby power systems [701.17].

240.15 Ungrounded Conductors.

(B) Circuit Breaker as an Overcurrent Device. Circuit breakers must automatically open all ungrounded conductors of the circuit, except as follows:

(1) Multiwire Branch Circuits. Single-pole breakers with identified handle ties are permitted for a multiwire branch circuit that only supplies line-to-neutral loads.

Author's Comments:

- Multiwire branch circuits must be provided with a means to disconnect simultaneously all ungrounded conductors at the point where the branch circuit originates [210.4(B)]. This can be accomplished by single-pole circuit breakers with handle ties identified for the purpose or a 2- or 3-pole breaker with common internal trip. Figure 240–14

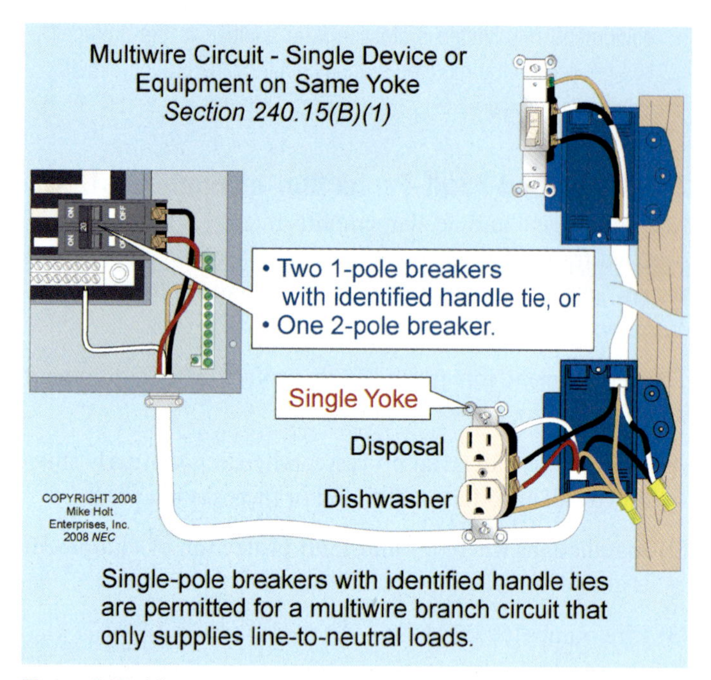

Figure 240–14

- Single-pole AFCI or GFCI circuit breakers are not suitable for protecting multiwire branch circuits. AFCI or GFCI circuit breakers for multiwire branch circuits must be of the 2-pole type.

(2) Single-Phase, Line-to-Line Loads. Individual single-pole circuit breakers with handle ties identified for the purpose are permitted on each ungrounded conductor of a branch circuit that supplies single-phase, line-to-line loads. Figure 240–15

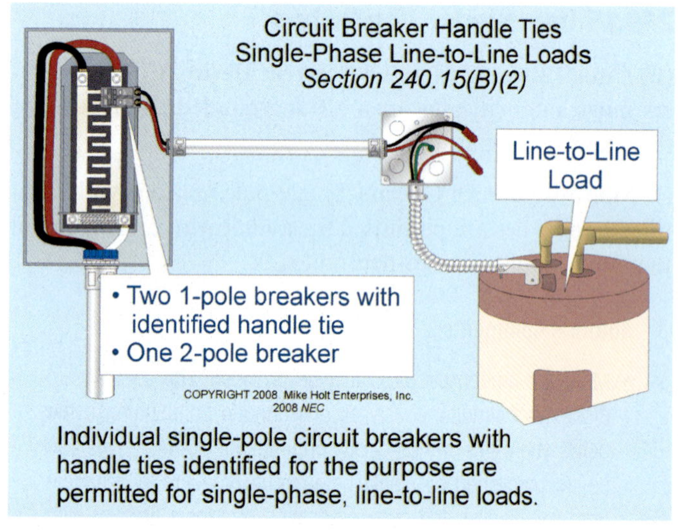

Figure 240–15

(3) Three-Phase, Line-to-Line Loads. Individual single-pole breakers with handle ties identified for the purpose are permitted on each ungrounded conductor of a branch circuit that serves three-phase, line-to-line loads. Figure 240–16

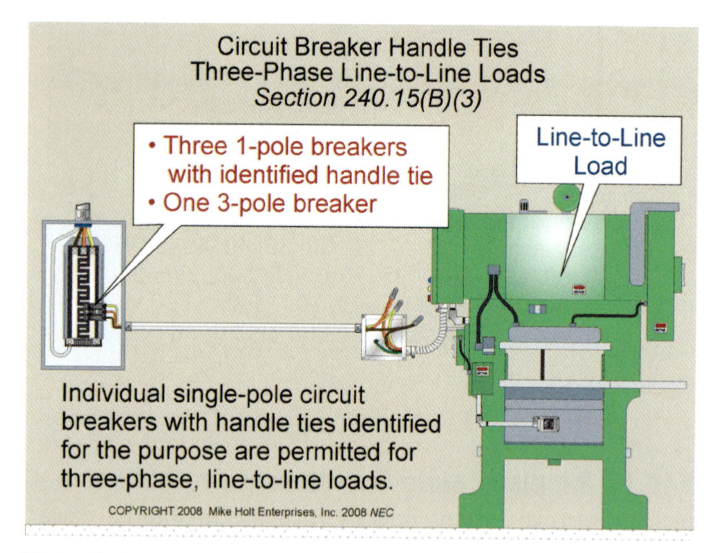

Figure 240–16

Author's Comment: According to Article 100 "Identified" means recognized as suitable for a specific purpose, function, or environment by listing, labeling, or other means approved by the authority having jurisdiction. This means handle ties made from nails, screws, wires, or other nonconforming materials are not suitable. Figure 240–17

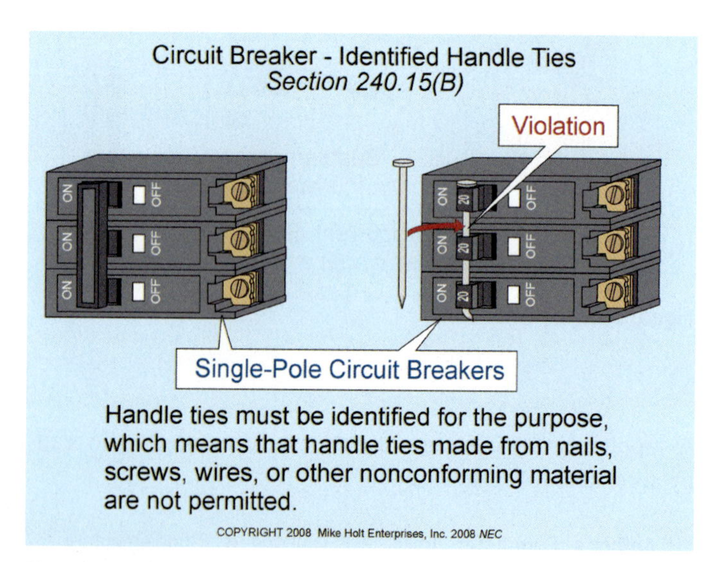

Figure 240–17

PART II. LOCATION

240.21 Overcurrent Protection Location in Circuit.

Except as permitted by (A) through (G), overcurrent devices must be placed at the point where the branch circuit or feeder conductors receive their power. Taps and transformer secondary conductors are not permitted to supply another conductor (tapping a tap is not permitted). Figure 240–18

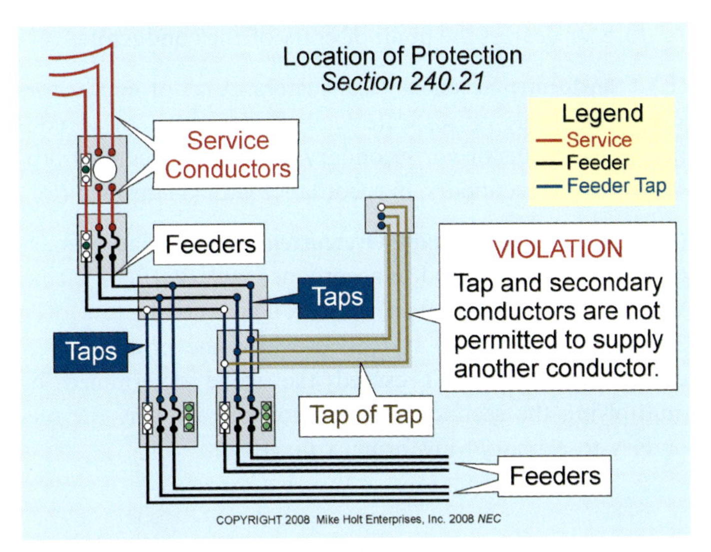

Figure 240–18

(A) Branch-Circuit Taps. Branch-circuit taps are permitted in accordance with 210.19.

(B) Feeder Taps. Conductors can be tapped to a feeder as specified 240.21(B)(1) through (B)(5).

(1) 10-Foot Feeder Tap. Feeder tap conductors up to 10 ft long are permitted without overcurrent protection at the tap location if installed as follows: Figure 240–19

(1) The ampacity of the tap conductor must not be less than:

a. The calculated load in accordance with Article 220, and

b. The rating of the device or overcurrent device supplied by the tap conductors.

(2) The tap conductors must not extend beyond the equipment they supply.

(3) The tap conductors must be installed in a raceway if they leave the enclosure.

(4) The tap conductors must have an ampacity not less than 10 percent of the ampacity of the overcurrent device that protects the feeder.

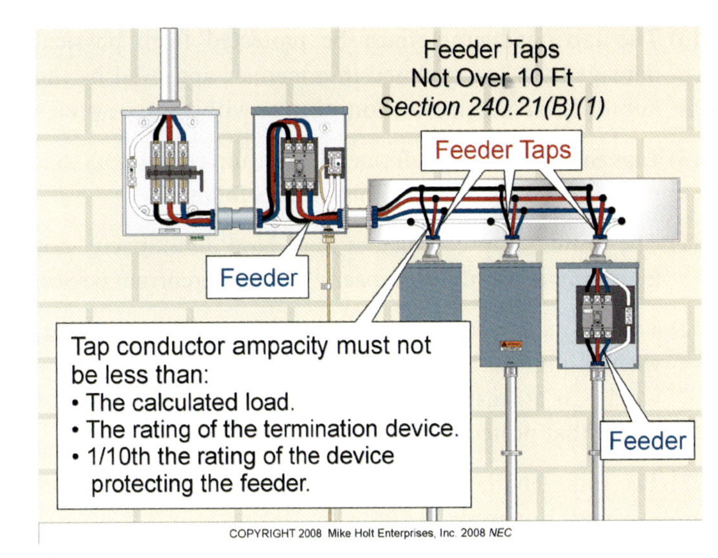

Figure 240–19

FPN: See 408.36 for the overcurrent protection requirements for panelboards.

(2) 25-Foot Feeder Tap. Feeder tap conductors up to 25 ft long are permitted without overcurrent protection at the tap location if installed as follows: Figure 240–20

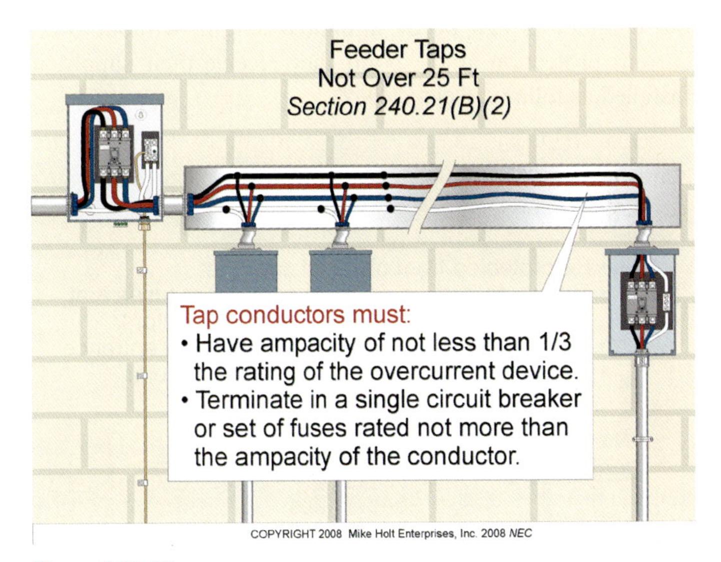

Figure 240–20

(1) The ampacity of the tap conductors must not be less than one-third the ampacity of the overcurrent device that protects the feeder.

(2) The tap conductors terminate in a single circuit breaker, or set of fuses rated no more than the tap conductor ampacity in accordance with 310.15 [Table 310.16].

(3) The tap conductors must be protected from physical damage by being enclosed in a manner approved by the authority having jurisdiction, such as within a raceway.

(3) Taps Supplying a Transformer. Feeder tap conductors that supply a transformer must be installed as follows:

(1) The primary tap conductors must have an ampacity not less than one-third the ampacity of the overcurrent device.

(2) The secondary conductors must have an ampacity that, when multiplied by the ratio of the primary-to-secondary voltage, is at least one-third the rating of the overcurrent device that protects the feeder conductors.

(3) The total length of the primary and secondary conductors must not exceed 25 ft.

(4) Primary and secondary conductors must be protected from physical damage by being enclosed in a manner approved by the authority having jurisdiction, such as within a raceway.

(5) Secondary conductors must terminate in a single circuit breaker, or set of fuses rated no more than the tap conductor ampacity in accordance with 310.15 [Table 310.16].

(5) Outside Feeder Taps of Unlimited Length. Outside feeder tap conductors can be of unlimited length, without overcurrent protection at the point they receive their supply, if installed as follows: Figure 240–21

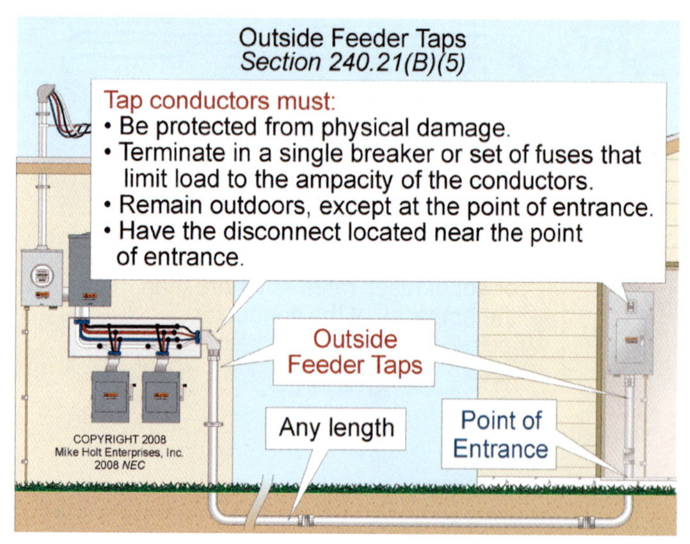

Figure 240–21

(1) The tap conductors must be suitably protected from physical damage in a raceway or manner approved by the authority having jurisdiction.

(2) The tap conductors must terminate at a single circuit breaker or a single set of fuses that limits the load to the ampacity of the conductors.

(3) The overcurrent device for the tap conductors must be an integral part of the disconnecting means, or it must be located immediately adjacent to it.

(4) The disconnecting means must be located at a readily accessible location, either outside the building or structure, or nearest the point of entry of the conductors.

(C) Transformer Secondary Conductors. A set of conductors supplying single or separate loads is permitted to be connected to a transformer secondary, without overcurrent protection at the secondary, in accordance with (1) through (6):

(1) Protection by Primary Overcurrent Device. The primary overcurrent device sized in accordance with 450.3(B) is considered suitable to protect the secondary conductors of a 2-wire (single voltage) system, provided the primary overcurrent device doesn't exceed the value determined by multiplying the secondary conductor ampacity by the secondary-to-primary transformer voltage ratio.

Question: *What is the minimum size secondary conductor required for a 2-wire, 480V to 120V transformer rated 1.50 kVA?* Figure 240–22

(a) 16 AWG *(b) 14 AWG* *(c) 12 AWG* *(d) 10 AWG*

Answer: *(b) 14 AWG*

Primary Current = VA/E

VA = 1,500 VA
E = 480V

Primary Current = 1,500 VA/480V
Primary Current = 3.13A

Primary Protection [450.3(B)] = 3.13A x 1.67
Primary Protection [450.3(B)] = 5.22A or 5A Fuse

Secondary Current = 1,500 VA/120V
Secondary Current = 12.50A

Secondary Conductor = 14 AWG, rated 20A at 60°C, Table 310.16

The 5A primary overcurrent device can be used to protect 14 AWG secondary conductors because it doesn't exceed the value determined by multiplying the secondary conductor ampacity by the secondary-to-primary transformer voltage ratio (5A = 20A x 120V/480V).

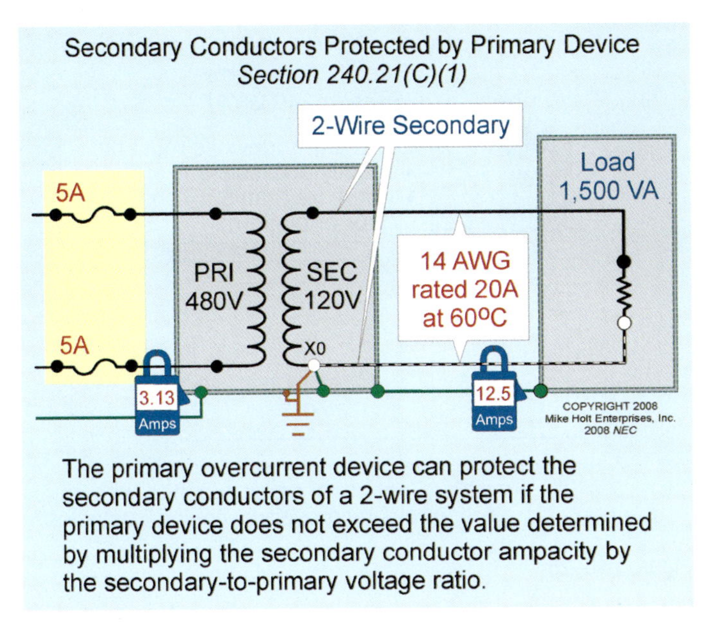

Figure 240–22

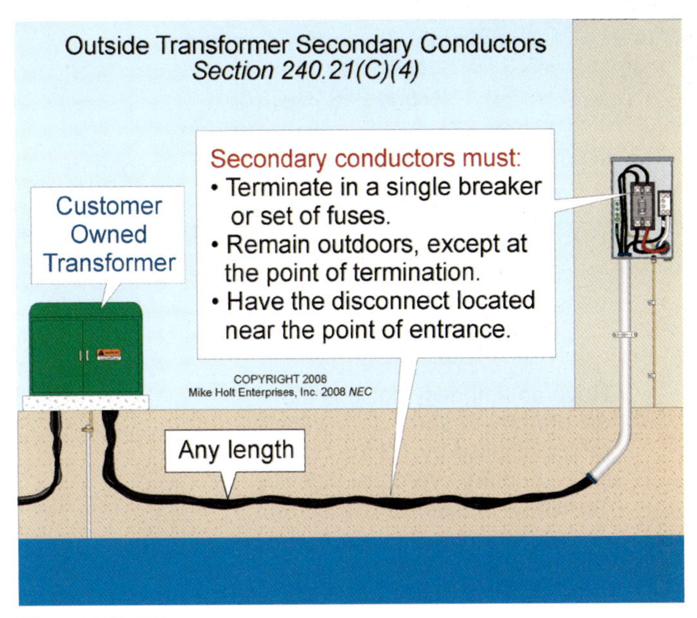

Figure 240–23

(2) 10 Ft Secondary Conductors. Secondary conductors can be run up to 10 ft without overcurrent protection if installed as follows:

(1) The ampacity of the secondary conductor must not be less than:

a. The calculated load in accordance with Article 220,

b. The rating of the device supplied by the secondary conductors or the overcurrent device at the termination of the secondary conductors, and

c. Not less than one-tenth the rating of the overcurrent device protecting the primary of the transformer, multiplied by the primary-to-secondary transformer voltage ratio.

FPN: See 408.36 for the overcurrent protection requirements for panelboards.

(2) The secondary conductors must not extend beyond the switchboard, panelboard, disconnecting means, or control devices they supply.

(3) The secondary conductors must be enclosed in a raceway.

(4) Outside Secondary Conductors of Unlimited Length. Outside secondary conductors can be of unlimited length, without overcurrent protection at the point they receive their supply, if they are installed as follows: Figure 240–23

(1) The conductors must be suitably protected from physical damage in a raceway or manner approved by the authority having jurisdiction.

(2) The conductors must terminate at a single circuit breaker or a single set of fuses that limit the load to the ampacity of the conductors.

(3) The overcurrent device for the ungrounded conductors must be an integral part of a disconnecting means or it must be located immediately adjacent thereto.

(4) The disconnecting means must be located at a readily accessible location that complies with one of the following:

a. Outside of a building or structure.

b. Inside, nearest the point of entrance of the conductors.

c. Where installed in accordance with 230.6, nearest the point of entrance of the conductors.

(5) Secondary Conductors from a Feeder Tapped Transformer. Transformer secondary conductors must be installed in accordance with 240.21(B)(3).

(6) 25-Foot Secondary Conductor. Secondary conductors can be run up to 25 ft without overcurrent protection if they comply with all of the following: Figure 240–24

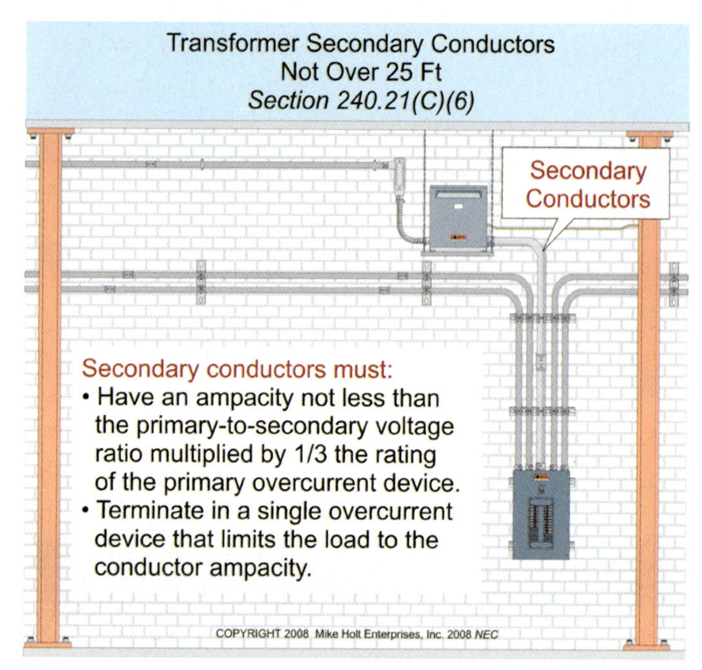

Figure 240–24

(1) The secondary conductors have an ampacity not less than the value of the primary-to-secondary voltage ratio multiplied by one-third of the rating of the overcurrent device that protects the primary of the transformer.

(2) Secondary conductors terminate in a single circuit breaker or set of fuses rated no more than the tap conductor ampacity in accordance with 310.15 [Table 310.16].

(3) The secondary conductors must be protected from physical damage by being enclosed in a manner approved by the authority having jurisdiction, such as within a raceway.

(D) Service Conductors. Service conductors must be protected against overload by the service disconnect overcurrent device in accordance with 230.91.

240.24 Location of Overcurrent Devices.

(A) Readily Accessible. Circuit breakers and fuses must be readily accessible, and they must be installed so the center of the grip of the operating handle of the fuse switch or circuit breaker, when in its highest position, isn't more than 6 ft 7 in. above the floor or working platform, unless the installation is for: Figure 240–25

(1) Busways, as provided in 368.17(C).

(2) Supplementary overcurrent devices [240.10]. Figure 240–26

(3) For overcurrent devices, as described in 225.40 and 230.92.

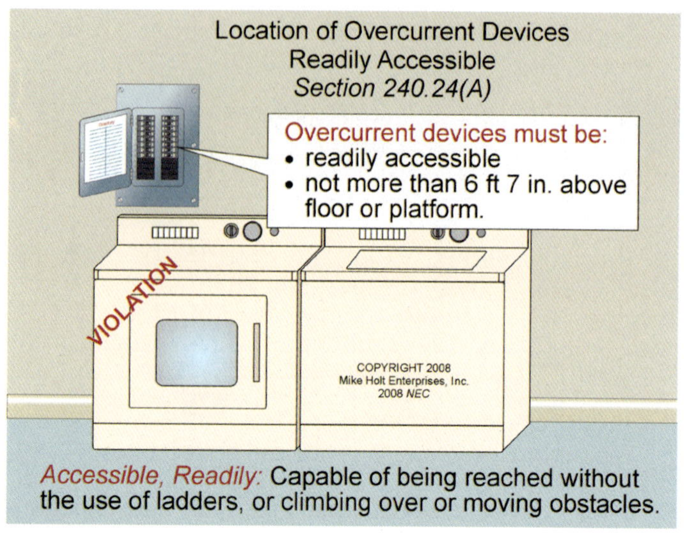

Figure 240–25

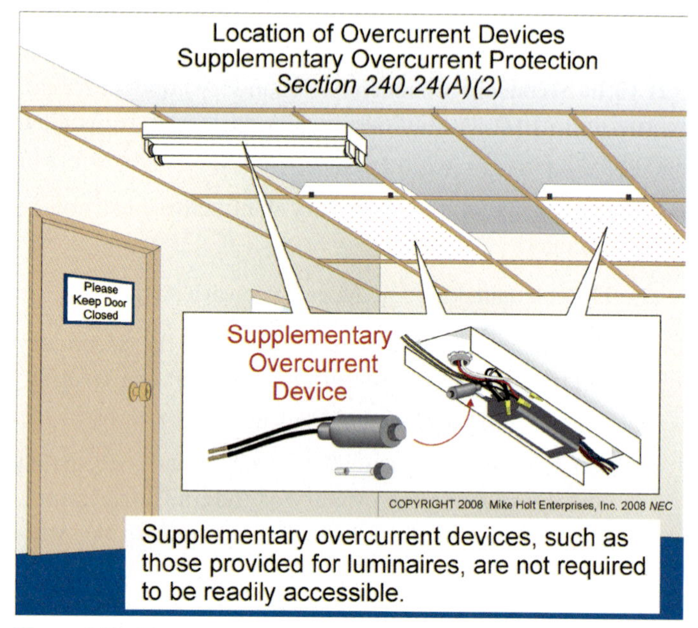

Figure 240–26

(4) Overcurrent devices located next to equipment can be mounted above 6 ft 7 in., if accessible by portable means [404.8(A) Ex 2]. Figure 240–27

(C) Not Exposed to Physical Damage. Overcurrent devices must not be exposed to physical damage. Figure 240–28

> **FPN:** Electrical equipment must be suitable for the environment, and consideration must be given to the presence of corrosive gases, fumes, vapors, liquids, or chemicals that have a deteriorating effect on conductors or equipment [110.11]. Figure 240–29

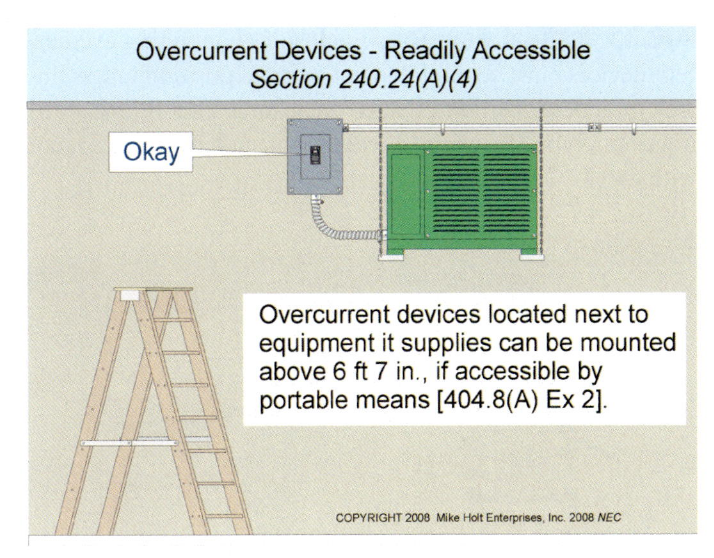

Figure 240–27

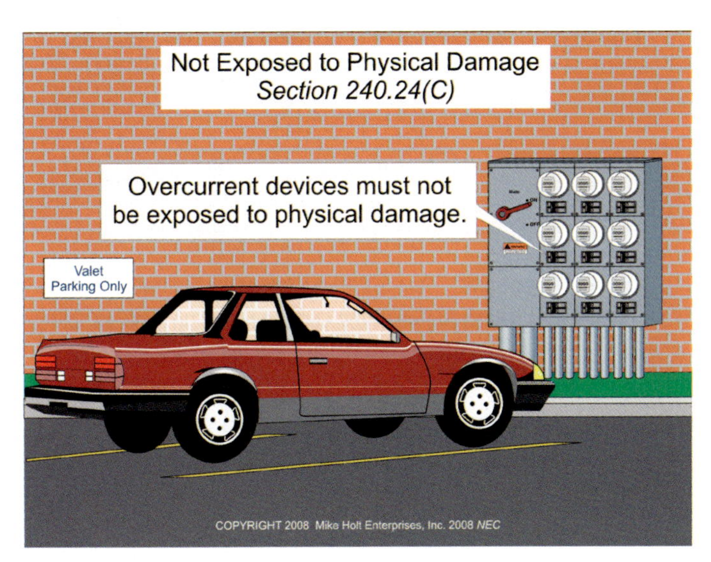

Figure 240–28

(D) Not in Vicinity of Easily Ignitible Material. Overcurrent devices must not be located near easily ignitible material, such as in clothes closets. Figure 240–30

> **Author's Comment:** The purpose of keeping overcurrent devices away from easily ignitible material is to prevent fires, not to keep them out of clothes closets.

(E) Not in Bathrooms. Overcurrent devices must not be located in the bathrooms of dwelling units, or guest rooms or guest suites of hotels or motels. Figure 240–31

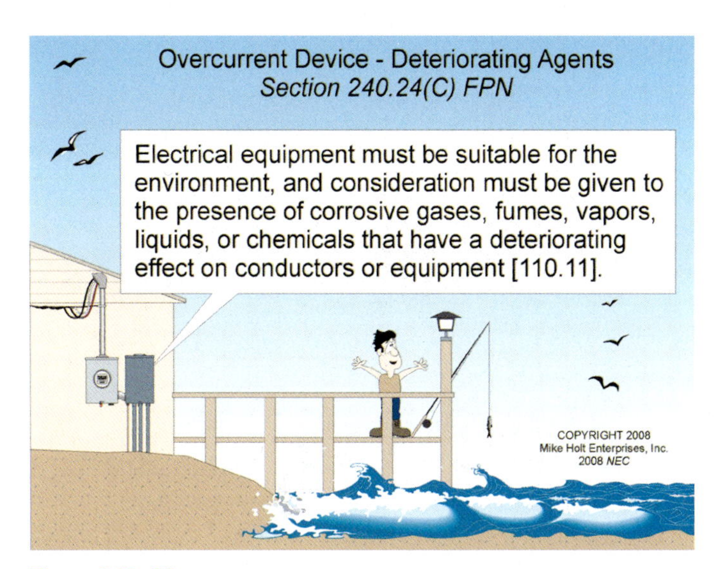

Figure 240–29

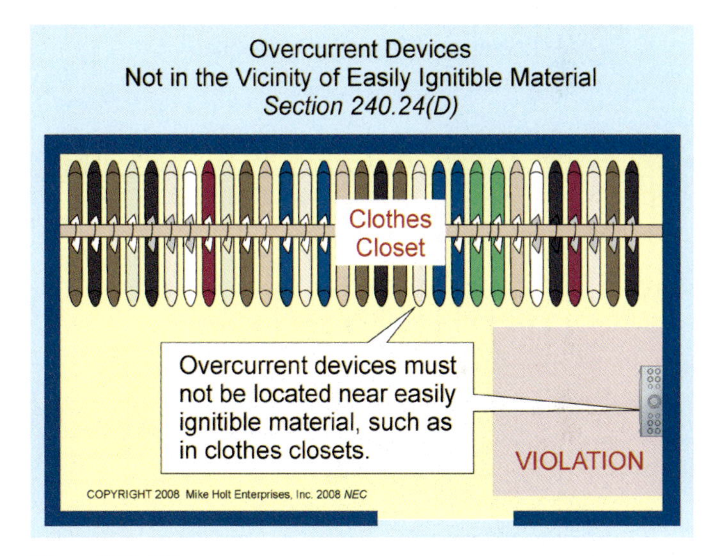

Figure 240–30

> **Author's Comment:** The service disconnecting means must not be located in a bathroom, even in commercial or industrial facilities [230.70(A)(2)].

(F) Over Steps. Overcurrent devices must not be located over the steps of a stairway. Figure 240–32

> **Author's Comment:** Clearly, it's difficult for electricians to safely work on electrical equipment that is located on uneven surfaces such as over stairways.

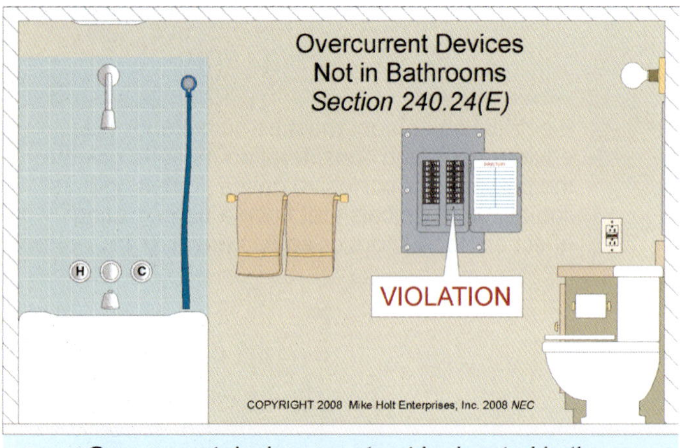

Figure 240–31

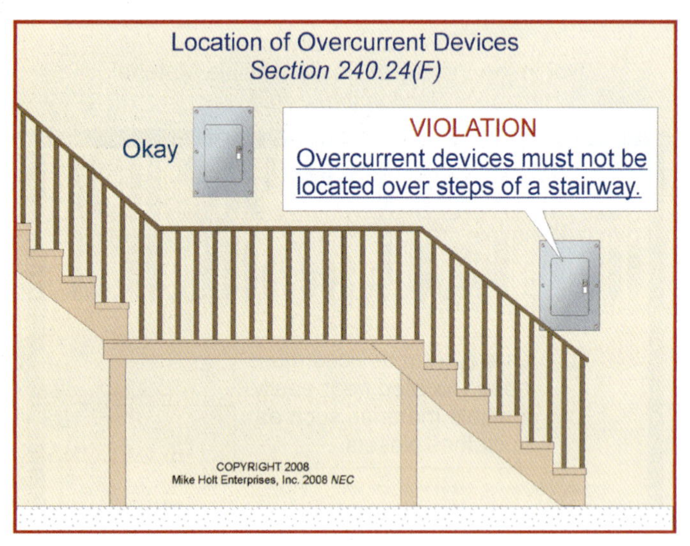

Figure 240–32

PART III. ENCLOSURES

240.32 Damp or Wet Locations. In damp or wet locations, enclosures containing overcurrent devices must prevent moisture or water from entering or accumulating within the enclosure. When the enclosure is surface mounted in a wet location, it must be mounted with not less than ¼ in. of air space between it and the mounting surface. See 312.2.

240.33 Vertical Position. Enclosures containing overcurrent devices must be mounted in a vertical position unless this isn't practical. Circuit-breaker enclosures can be mounted horizontally if the circuit breaker is installed in accordance with 240.81. **Figure 240–33**

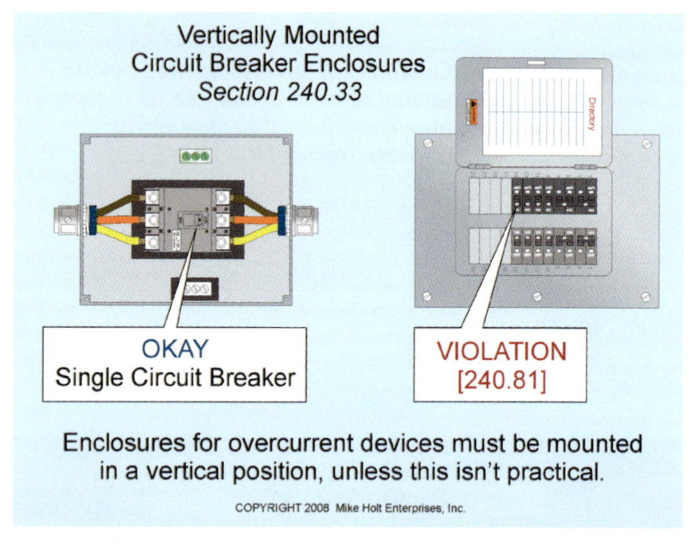

Figure 240–33

Author's Comment: Section 240.81 specifies that where circuit-breaker handles are operated vertically, the "up" position of the handle must be in the "on" position. So, in effect, an enclosure that contains one row of circuit breakers can be mounted horizontally, but an enclosure that contains a panelboard with multiple circuit breakers on opposite sides of each other will have to be mounted vertically.

PART V. PLUG FUSES, FUSEHOLDERS, AND ADAPTERS

240.51 Edison-Base Fuse.

(A) Classification. Edison-base fuses are classified to operate at not more than 125V and have an ampere rating of not more than 30A.

(B) Replacement Only. Edison-base fuses are permitted only for replacement in an existing installation where there's no evidence of tampering or overfusing.

240.53 Type S Fuses.

(A) Classification. Type S fuses operate at not more than 125V and have ampere ratings of 15A, 20A, and 30A. Figure 240–34

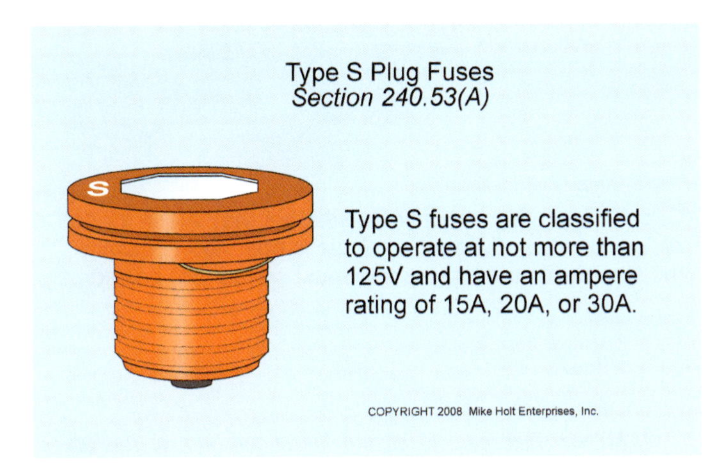

Figure 240–34

(B) Not Interchangeable. Type S fuses are made so different ampere ratings aren't interchangeable.

240.54 Type S Fuses, Adapters, and Fuseholders.

(A) Type S Adapters. Type S adapters are designed to fit Edison-base fuseholders.

(B) Prevent Edison-Base Fuses. Type S fuseholders and adapters are designed for Type S fuses only.

(C) Nonremovable Adapters. Type S adapters are designed so they cannot be removed once installed.

PART VI. CARTRIDGE FUSES AND FUSEHOLDERS

Author's Comment: There are two basic designs of cartridge fuses, the ferrule type with a maximum rating of 60A and the knife-blade type rated over 60A. The fuse length and diameter varies with the voltage and current rating. Figure 240–35

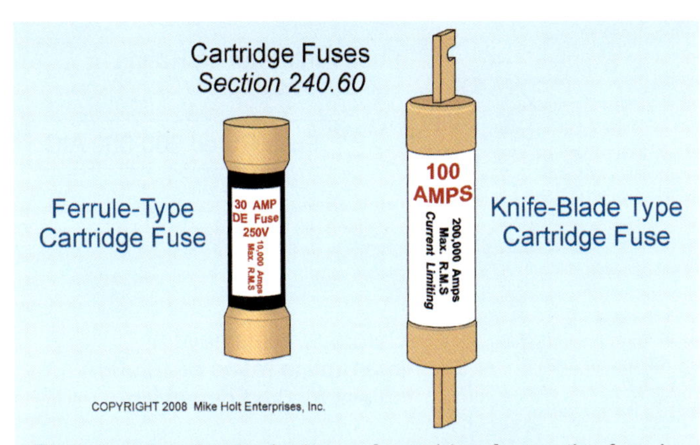

There are two basic designs of cartridge fuses, the ferrule type (maximum 60A) and the knife-blade type. The size of the fuse varies with the fuse voltage and current rating.

Figure 240–35

240.60 General.

(A) Maximum Voltage—300V Type. Cartridge fuses and fuseholders of the 300V type can only be used for:

- Circuits not exceeding 300V between conductors.
- Circuits not exceeding 300V from any ungrounded conductor to the neutral point.

(B) Noninterchangeable Fuseholders. Fuseholders must be designed to make it difficult to interchange fuses of any given class for different voltages and current ratings.

Fuseholders for current-limiting fuses must be designed so only current-limiting fuses can be inserted.

Author's Comment: A current-limiting fuse is a fast-clearing overcurrent device that reduces the fault current to a magnitude substantially lower than that obtainable in the same circuit if the current-limiting device is not used [240.2].

(C) Marking. Cartridge fuses have an interrupting rating of 10,000A, unless marked otherwise.

WARNING: *Fuses must have an interrupting rating sufficient for the short-circuit current available at the line terminals of the equipment. Using a fuse with an inadequate interrupting current rating can cause equipment to be destroyed from a line-to-line or ground fault, and result in death or serious injury. See 110.9 for more details.* Figure 240–36

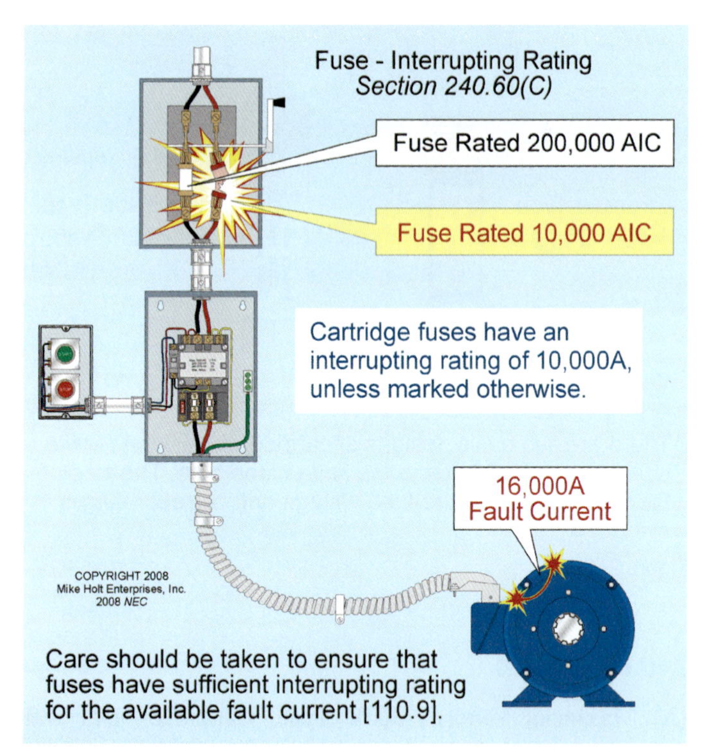

Figure 240–36

240.61 Classification. Cartridge fuses and fuseholders are classified according to voltage and amperage ranges. Fuses rated 600V, nominal, or less are permitted for voltages at or below their ratings.

PART VII. CIRCUIT BREAKERS

240.80 Method of Operation. Circuit breakers must be capable of being opened and closed by hand. Nonmanual means of operating a circuit breaker, such as electrical shunt trip or pneumatic operation, are permitted as long as the circuit breaker can also be manually operated.

240.81 Indicating. Circuit breakers must clearly indicate whether they are in the open "off" or closed "on" position. When the handle of a circuit breaker is operated vertically, the "up" position of the handle must be the "on" position. See 240.33 and 404.6(C). Figure 240–37

240.83 Markings.

(C) Interrupting Rating. Circuit breakers have an interrupting rating of 5,000A unless marked otherwise.

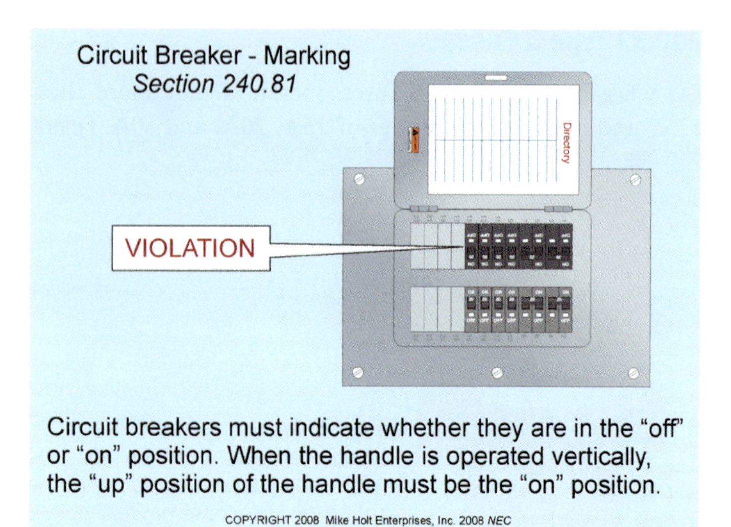

Figure 240–37

WARNING: *Take care to ensure the circuit breaker has an interrupting rating sufficient for the short-circuit current available at the line terminals of the equipment. Using a circuit breaker with inadequate interrupting current rating can cause equipment to be destroyed from a line-to-line or ground fault, and result in death or serious injury. See 110.9 for more details.* Figure 240–38

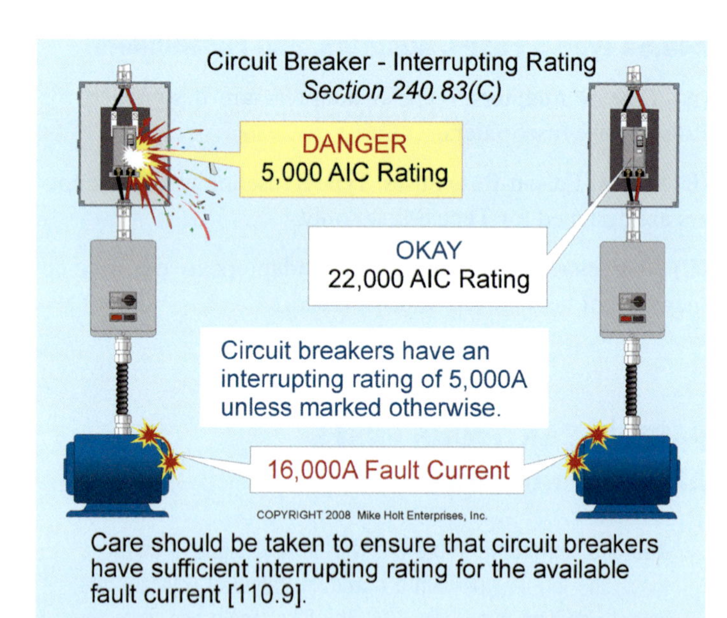

Figure 240–38

(D) Used as Switches. Circuit breakers used to switch 120V or 277V fluorescent lighting circuits must be listed and marked SWD or HID. Circuit breakers used to switch high-intensity discharge lighting circuits must be listed and marked HID. Figure 240–39

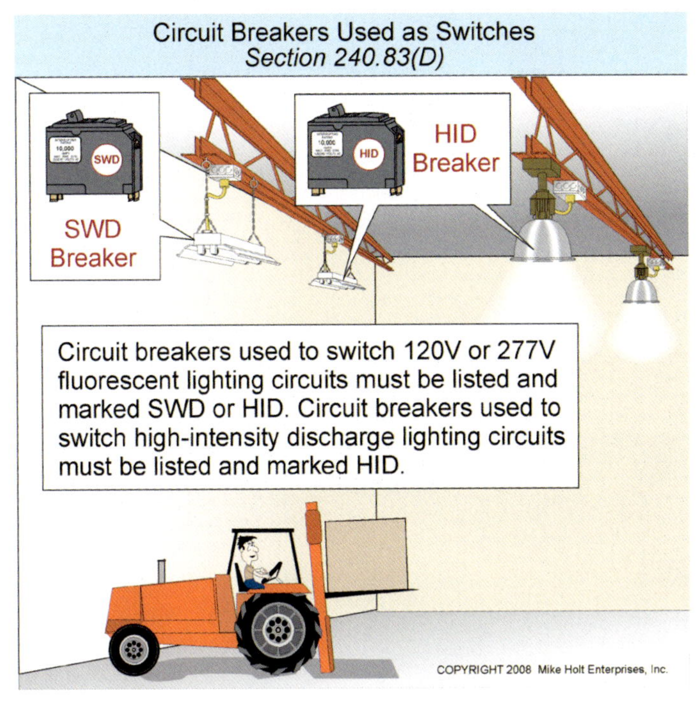

Figure 240–39

Author's Comments:

- This rule applies only when the circuit breaker is used as the switch. If a general-use snap switch or contactor is used to control the lighting, this rule does not apply.
- UL 489, *Standard for Molded Case Circuit Breakers*, permits "HID" breakers to be rated up to 50A, whereas an "SWD" breaker can only be rated up to 20A. The tests for "HID" breakers include an endurance test at 75 percent power factor, whereas "SWD" breakers are endurance-tested at 100 percent power factor. The contacts and the spring of an "HID" breaker are of a heavier-duty material to dissipate the increased heat caused by the increased current flow in the circuit, because the "HID" luminaire takes a minute or two to ignite the lamp.

(E) Voltage Markings. Circuit breakers must be marked with a voltage rating that corresponds with their interrupting rating. See 240.85.

240.85 Applications

Straight Voltage Rating. A circuit breaker with a straight voltage rating, such as 240V or 480V, is permitted on a circuit where the nominal voltage between any two conductors (line-to-neutral or line-to-line) doesn't exceed the circuit breaker's voltage rating. Figure 240–40

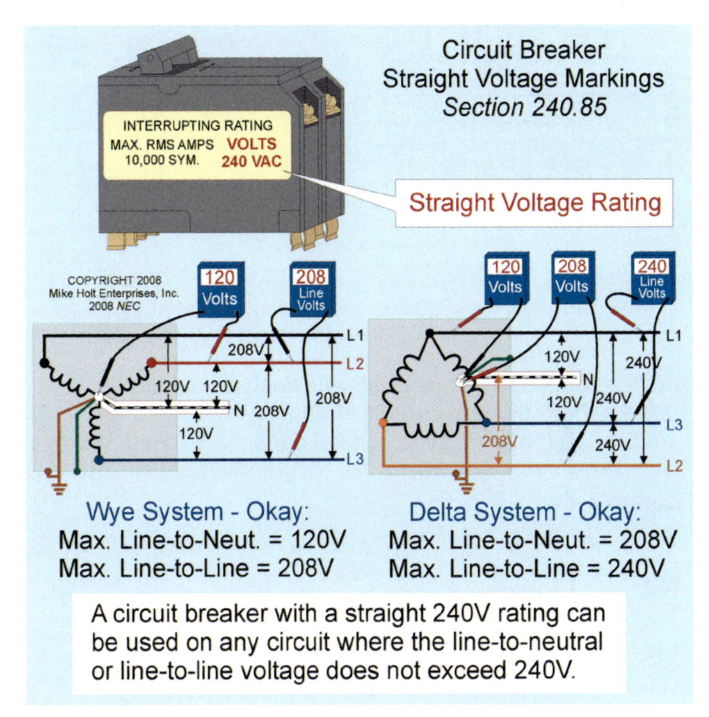

Figure 240–40

Slash Voltage Rating. A circuit breaker with a slash rating, such as 120/240V or 277/480V, is permitted on a solidly grounded system where the nominal voltage of any one conductor to ground doesn't exceed the lower of the two values, and the nominal voltage between any two conductors doesn't exceed the higher value.

CAUTION: *A 120/240V slash circuit breaker must not be used on the high leg of a solidly grounded 4-wire, three-phase, 120/240V delta-connected system, because the line-to-ground voltage of the high leg is 208V, which exceeds the 120V line-to-ground voltage rating of the breaker.* **Figure 240–41**

FPN: When installing circuit breakers on corner-grounded delta systems, consideration needs to be given to the circuit breakers' individual pole-interrupting capability.

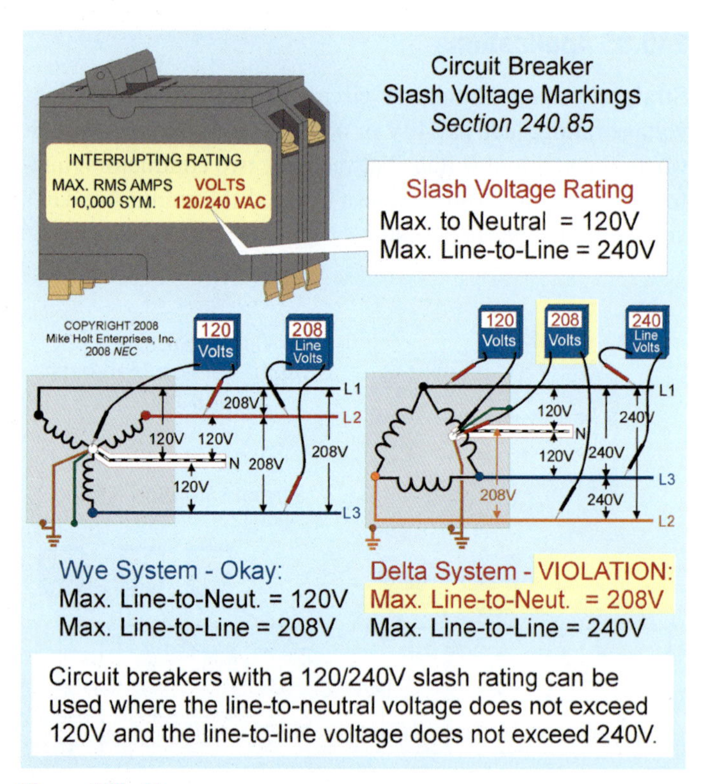

Figure 240–41

ARTICLE 240 Practice Questions

ARTICLE 240. OVERCURRENT PROTECTION—PRACTICE QUESTIONS

1. A device that, when interrupting currents in its current-limiting range, reduces the current flowing in the faulted circuit to a magnitude substantially less than that obtainable in the same circuit if the device were replaced with a solid conductor having comparable impedance, is a(n) _____ protective device.

   (a) short-circuit
   (b) overload
   (c) ground-fault
   (d) current-limiting

2. If the circuit's overcurrent device exceeds _____, the conductor ampacity must have a rating not less than the rating of the overcurrent device.

   (a) 800A
   (b) 1,000A
   (c) 1,200A
   (d) 2,000A

3. Flexible cords used in listed extension cord sets shall be considered protected against overcurrent when used in(within) _____.

   (a) indoor installations
   (b) unclassified locations
   (c) the extension cord's listing requirements
   (d) 50 ft of the branch-circuit panelboard

4. Supplementary overcurrent protection _____.

   (a) shall not be used in luminaires
   (b) may be used as a substitute for a branch-circuit overcurrent device
   (c) may be used to protect internal circuits of equipment
   (d) shall be readily accessible

5. Circuit breakers shall _____ all ungrounded conductors of the circuit.

   (a) open
   (b) close
   (c) isolate
   (d) inhibit

6. Tap conductors not over 25 ft long shall be permitted, providing the _____.

   (a) ampacity of the tap conductors is not less than one-third the rating of the overcurrent device protecting the feeder conductors being tapped
   (b) tap conductors terminate in a single circuit breaker or set of fuses that limit the load to the ampacity of the tap conductors
   (c) tap conductors are suitably protected from physical damage
   (d) all of these

7. Outside secondary conductors can be of unlimited length without overcurrent protection at the point they receive their supply if:

   (a) The conductors are suitably protected from physical damage.
   (b) The conductors terminate at a single circuit breaker or a single set of fuses that limits the load to the ampacity of the conductors.
   (c) a and b
   (d) none of these

8. Overcurrent devices shall not be located _____.

   (a) where exposed to physical damage
   (b) near easily ignitible materials, such as in clothes closets
   (c) in bathrooms of dwelling units
   (d) all of these

9. Plug fuses of the Edison-base type shall have a maximum rating of _____.

(a) 20A
(b) 30A
(c) 40A
(d) 50A

10. Type _____ fuse adapters shall be designed so that once inserted in a fuseholder they cannot be removed.

(a) A
(b) E
(c) S
(d) P

11. Cartridge fuses and fuseholders shall be classified according to _____ ranges.

(a) voltage
(b) amperage
(c) a or b
(d) a and b

12. Where the circuit breaker handles are operated vertically, the "up" position of the handle shall be the _____.

(a) "on" position
(b) "off" position
(c) tripped position
(d) any of these

13. Circuit breakers used to switch high-intensity discharge lighting circuits shall be listed and marked as _____.

(a) SWD
(b) HID
(c) a or b
(d) a and b

14. A circuit breaker with a _____ rating, such as 120/240V or 277/480V can be used on a solidly grounded circuit where the nominal voltage of any conductor to ground does not exceed the lower of the two values, and the nominal voltage between any two conductors does not exceed the higher value.

(a) straight
(b) slash
(c) high
(d) low

Grounding and Bonding

INTRODUCTION TO ARTICLE 250—GROUNDING AND BONDING

No other article can match Article 250 for misapplication, violation, and misinterpretation. People often insist on building electrical installations that violate this article.

Article 250 covers the grounding requirements for providing a low-impedance path to the earth to reduce overvoltage from lightning, and the bonding requirements for a low-impedance fault current path necessary to facilitate the operation of overcurrent devices in the event of a ground fault.

Over the past four *Code* cycles, this article was extensively revised to organize it better and make it easier to understand and implement. It's arranged in a logical manner, so it's a good idea to just read through Article 250 to get a big picture view—after you review the definitions. Next, study the article closely so you understand the details. The illustrations will help you understand the key points.

PART I. GENERAL

250.1 Scope. Article 250 contains the following grounding and bonding requirements:

(1) What systems and equipment are required to be grounded

(3) Location of grounding connections

(4) Types of electrodes and sizes of grounding and bonding conductors

(5) Methods of grounding and bonding

250.2 Definitions.

Bonding Jumper, System. The conductor, screw, or strap that bonds the metal parts of a separately derived system to the system neutral point according to 250.30(A)(1). Figure 250–1

> **Author's Comment:** The system bonding jumper provides a low-impedance fault current path to the power supply to facilitate the clearing of a ground fault by opening the circuit overcurrent device. For more information, see 250.4(A)(5), 250.28, and 250.30(A)(1).

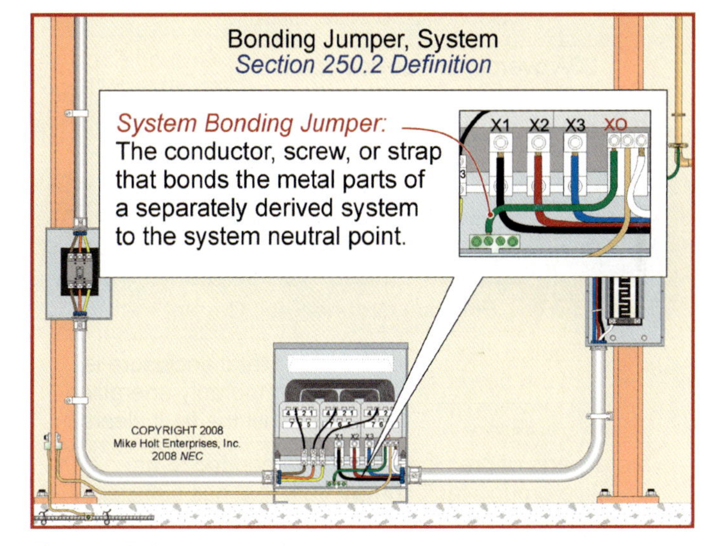

Figure 250–1

Effective Ground-Fault Current Path. An intentionally constructed low-impedance conductive path designed to carry fault current from the point of a ground fault on a wiring system to the electrical supply source. Figure 250–2

The effective ground-fault current path is intended to help remove dangerous voltage from a ground fault by opening the circuit overcurrent device. Figure 250–3

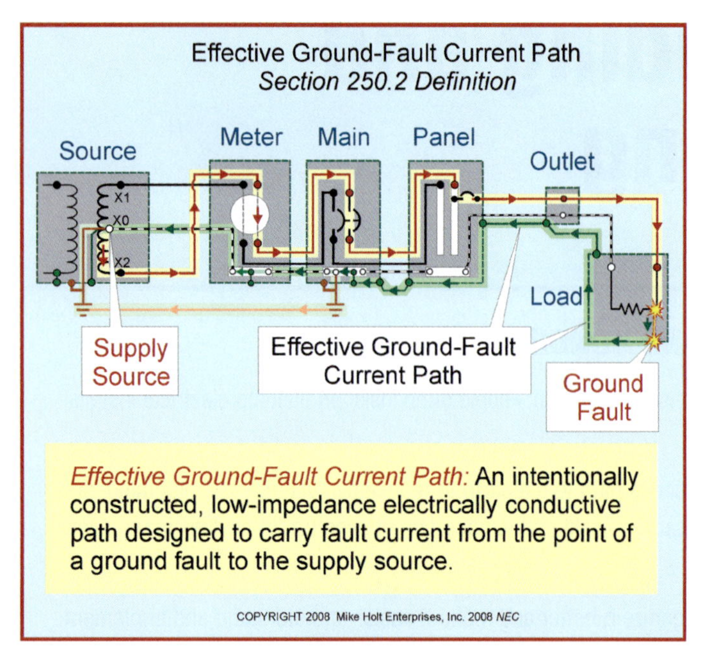

Figure 250–2

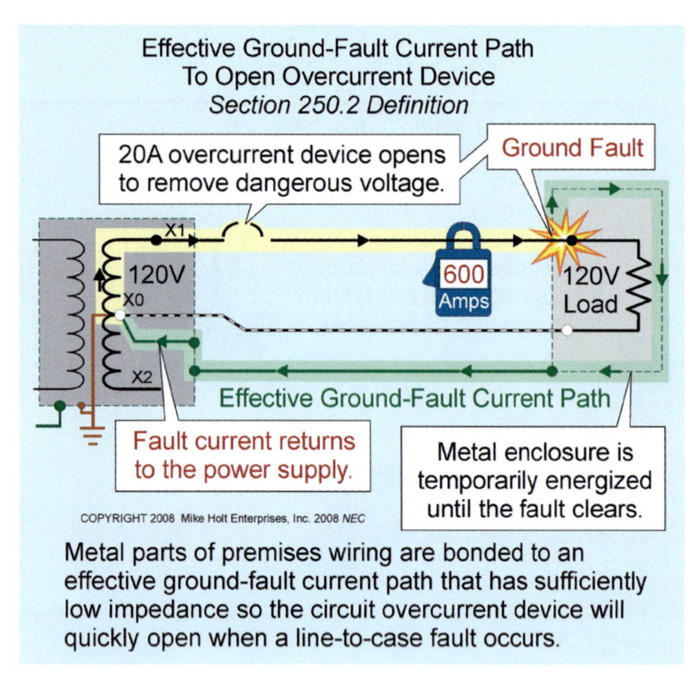

Figure 250–3

Ground Fault. An unintentional connection between an ungrounded conductor and the metal parts of enclosures, raceways, or equipment. Figure 250–4

Ground-Fault Current Path. An electrically conductive path from a ground fault to the electrical supply source.

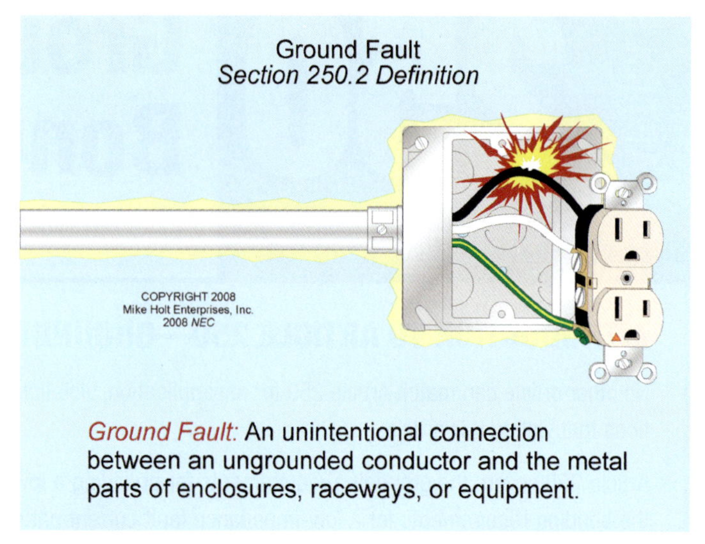

Figure 250–4

FPN: The ground-fault current path could be metal raceways, cable sheaths, electrical equipment, or other electrically conductive materials, such as metallic water or gas piping, steel-framing members, metal ducting, reinforcing steel, or the shields of communications cables. Figure 250–5

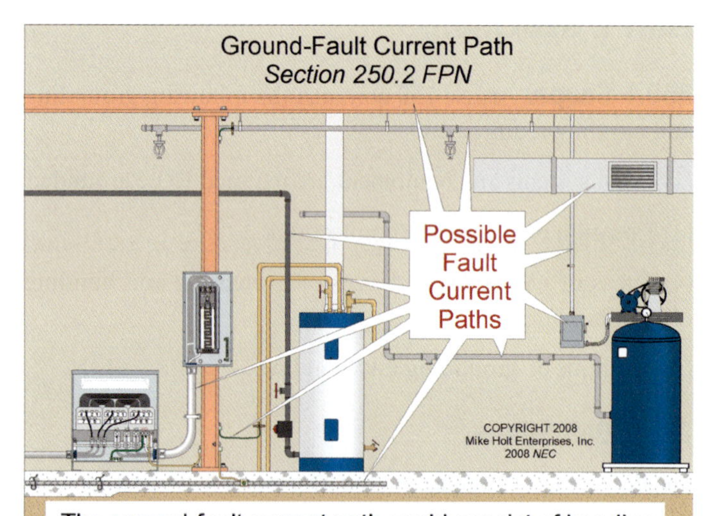

Figure 250–5

Author's Comment: The difference between an "effective ground-fault current path" and a "fault current path" is the effective ground-fault current path is "intentionally" constructed to provide a low-impedance fault current path to the electrical supply source for the purpose of clearing a ground fault. A ground-fault current path is all of the available conductive paths over which fault current flows on its return to the electrical supply source during a ground fault.

250.4 General Requirements for Grounding and Bonding.

(A) Solidly Grounded Systems.

(1) Electrical System Grounding. Electrical power systems, such as the secondary winding of a transformer are grounded (connected to the earth) to limit the voltage caused by lightning, line surges, or unintentional contact by higher-voltage lines. Figure 250–6

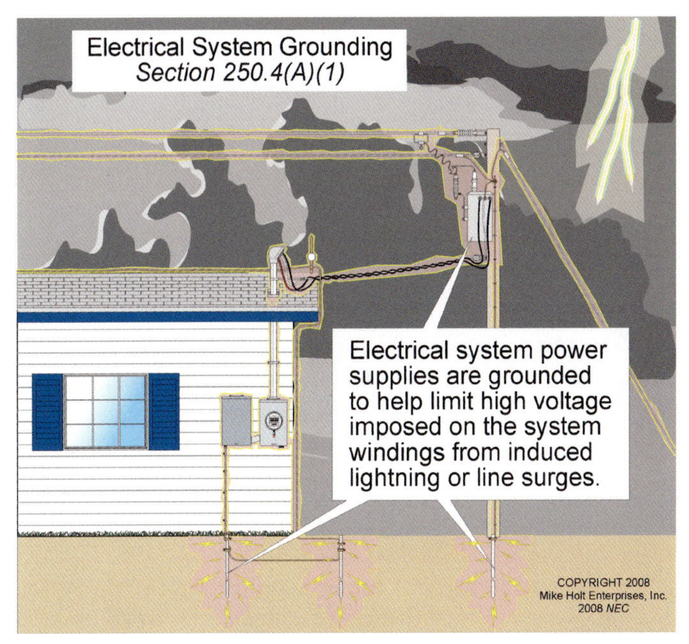

Figure 250–6

Author's Comment: System grounding helps reduce fires in buildings as well as voltage stress on electrical insulation, thereby ensuring longer insulation life for motors, transformers, and other system components. Figure 250–7

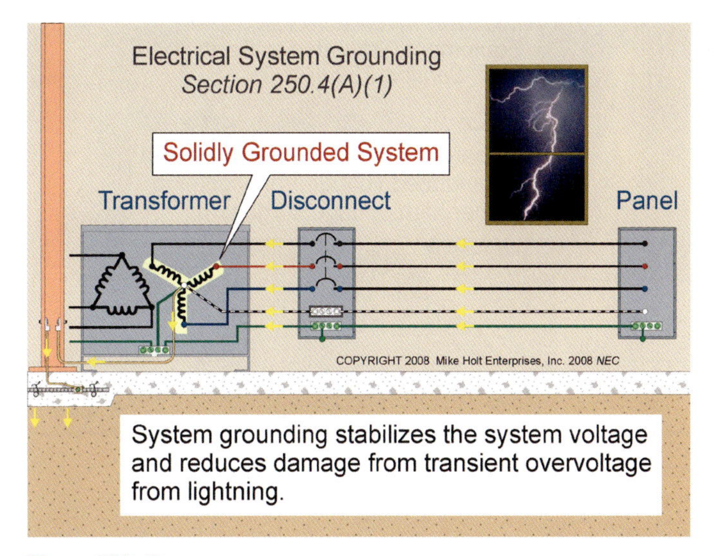

Figure 250–7

FPN: An important consideration for limiting the imposed voltage is the routing of bonding and grounding conductors so that they are not any longer than necessary to complete the connection without disturbing the permanent parts of the installation and so that unnecessary bends and loops are avoided. Figure 250–8

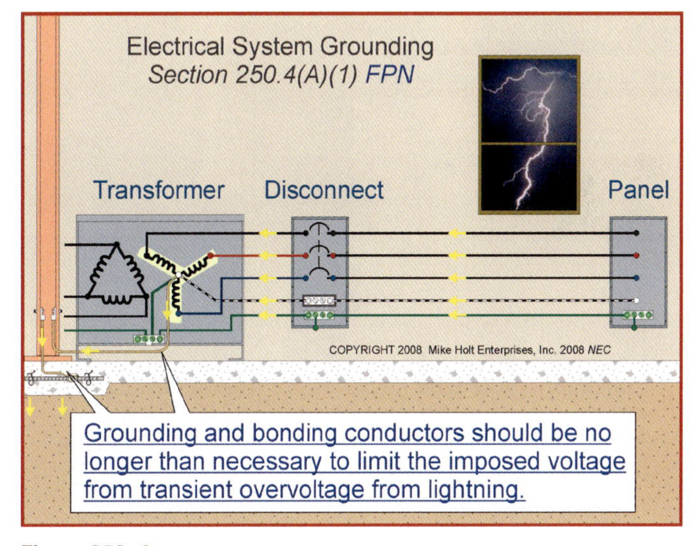

Figure 250–8

(2) Equipment Grounding. Metal parts of electrical equipment are grounded (connected to the earth) to reduce induced voltage on metal parts from exterior lightning so as to prevent fires from an arc within the building or structure. Figure 250–9

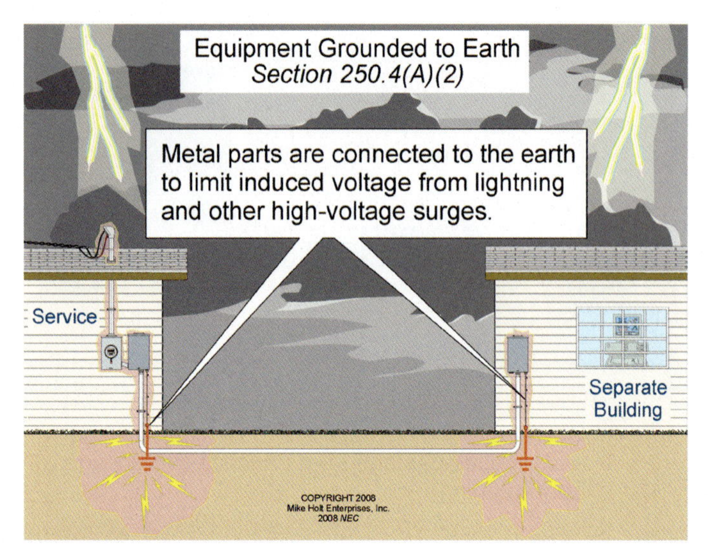

Figure 250–9

DANGER: *Failure to ground the metal parts can result in high-voltage on metal parts from an indirect lightning strike to seek a path to the earth within the building—possibly resulting in a fire and or electric shock.* **Figure 250-10**

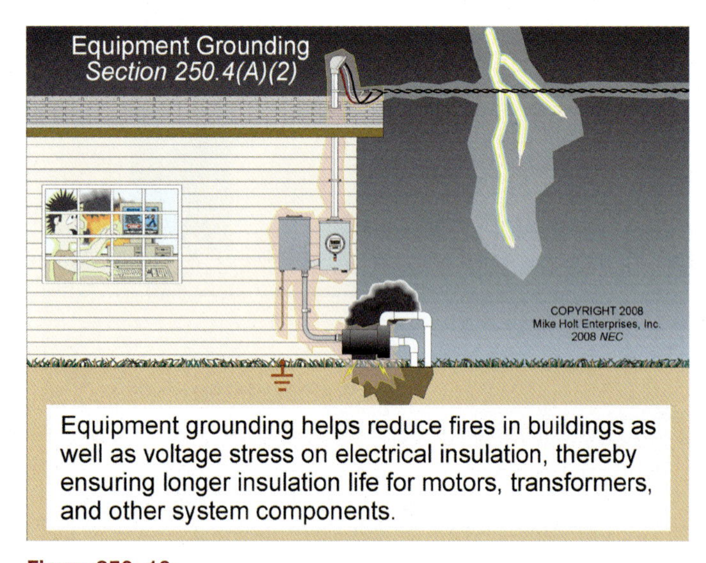

Figure 250–10

Author's Comment: Grounding metal parts helps drain off static electricity charges before flashover potential is reached. Static grounding is often used in areas where the discharge (arcing) of the voltage buildup (static) can cause dangerous or undesirable conditions [500.4 FPN No. 3].

DANGER: *Because the contact resistance of an electrode to the earth is so high, very little fault current returns to the power supply if the earth is the only fault current return path. Result—the circuit overcurrent device will not open and clear the ground fault, and all metal parts associated with the electrical installation, metal piping, and structural building steel will become and remain energized* **Figure 250–11**

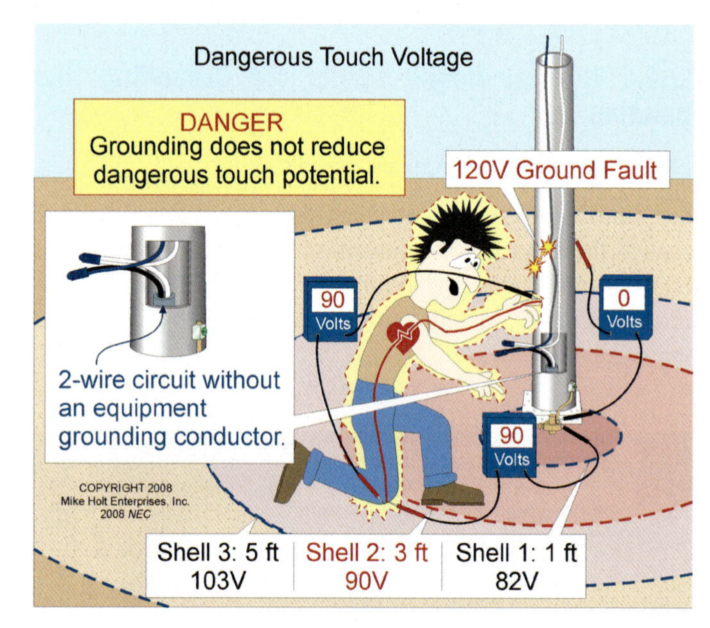

Figure 250–11

(3) Equipment Bonding. Metal parts of electrical raceways, cables, enclosures, and equipment must be connected to the supply source via an equipment grounding conductor of a type recognized in 250.118. **Figures 250–12** and **250–13**

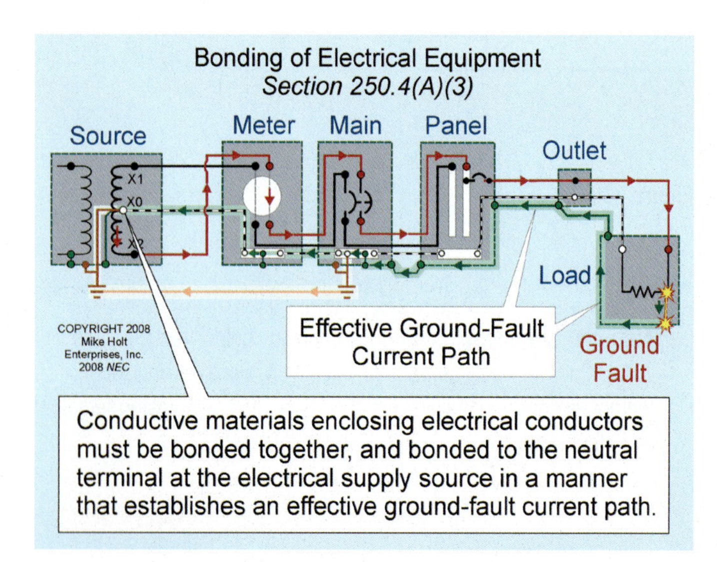

Figure 250–12

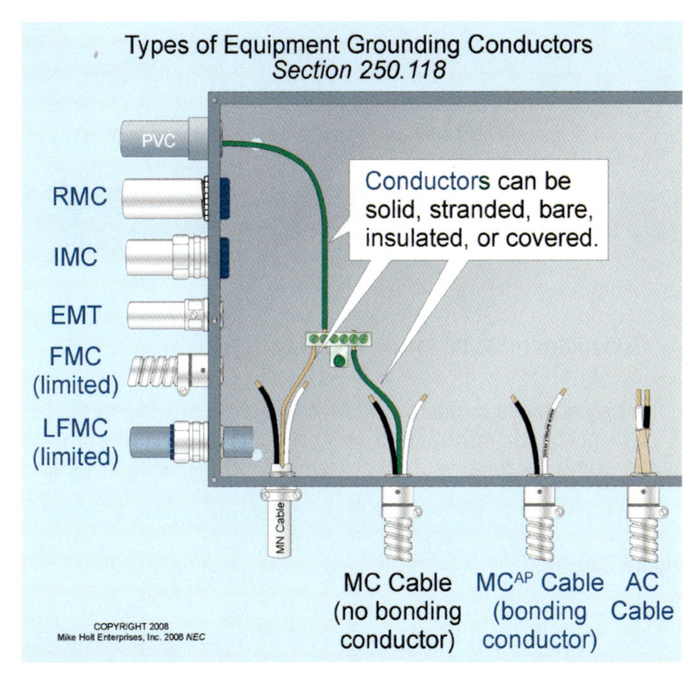

Figure 250–13

Author's Comment: To quickly remove dangerous touch voltage on metal parts from a ground fault, the fault current path must have sufficiently low impedance to the source so that fault current will quickly rise to a level that will open the branch-circuit overcurrent device. **Figure 250–14**

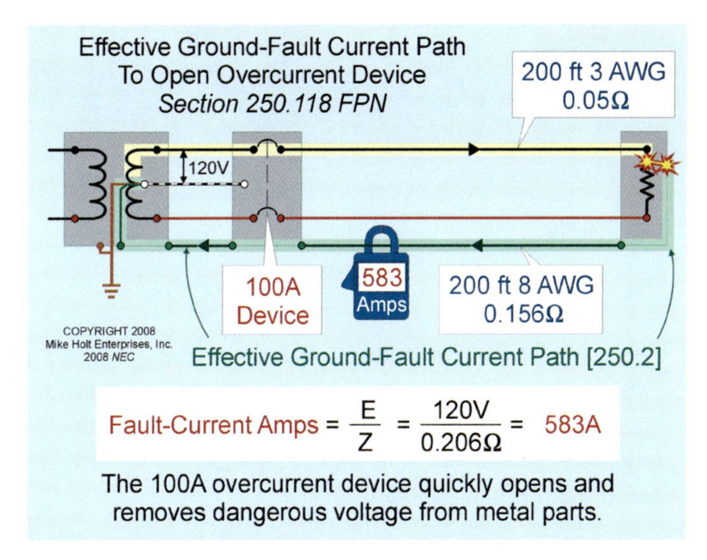

Figure 250–14

Author's Comment: The time it takes for an overcurrent device to open is inversely proportional to the magnitude of the fault current. This means the higher the ground-fault current value, the less time it will take for the overcurrent device to open and clear the fault. For example, a 20A circuit with an overload of 40A (two times the 20A rating) takes 25 to 150 seconds to open the overcurrent device. At 100A (five times the 20A rating) the 20A breaker trips in 5 to 20 seconds. **Figure 250–15**

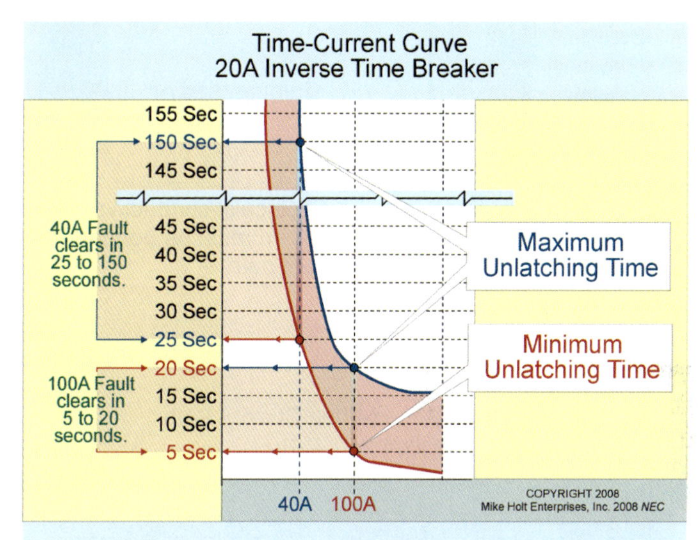

Figure 250–15

(4) Bonding Conductive Materials. Electrically conductive materials such as metal water piping systems, metal sprinkler piping, metal gas piping, and other metal-piping systems, as well as exposed structural steel members likely to become energized, must be connected to the supply source via an equipment grounding conductor of a type recognized in 250.118. **Figure 250–16**

Author's Comment: The phrase "likely to become energized" is subject to interpretation by the authority having jurisdiction.

(5) Effective Ground-Fault Current Path. Metal parts of electrical raceways, cables, enclosures, or equipment must be bonded together and to the supply system in a manner that creates a low-impedance path for ground-fault current that facilitates the operation of the circuit overcurrent device. **Figure 250–17**

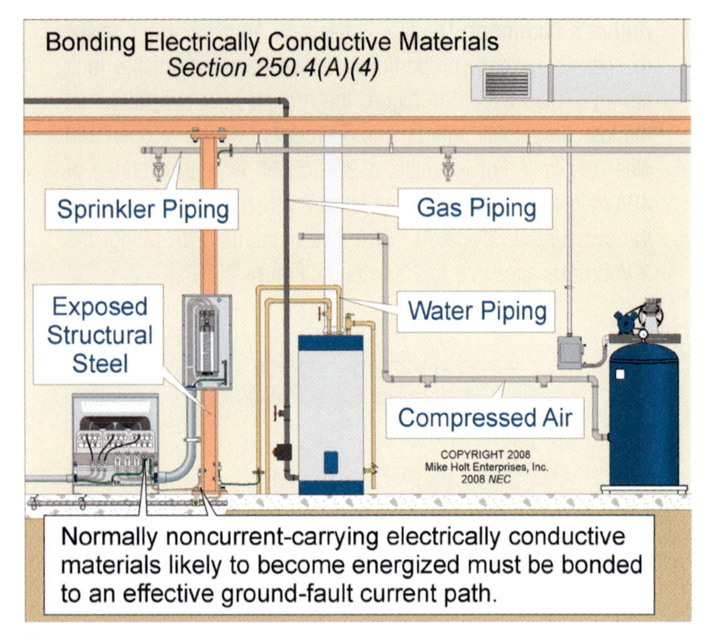

Figure 250–16

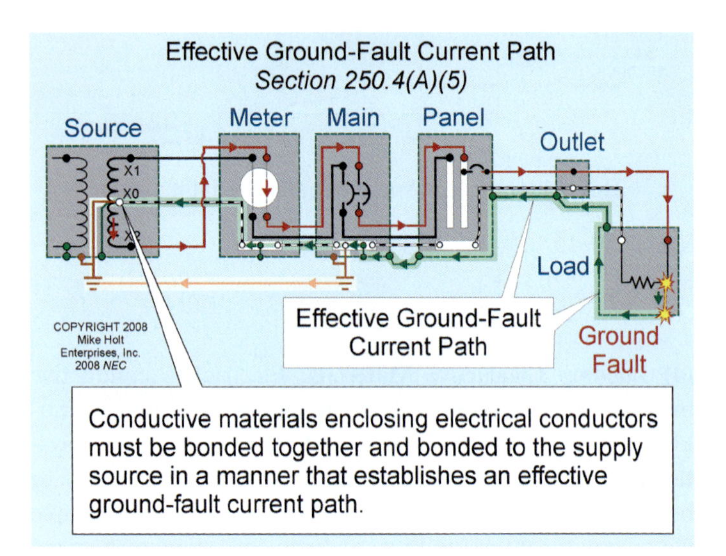

Figure 250–17

Author's Comment: To assure a low-impedance ground-fault current path, all circuit conductors must be grouped together in the same raceway, cable, or trench [300.3(B), 300.5(I), and 300.20(A)]. Figure 250–18

Because the earth is not suitable to serve as the required effective ground-fault current path, an equipment grounding conductor is required to be run with all circuits. Figure 250–19

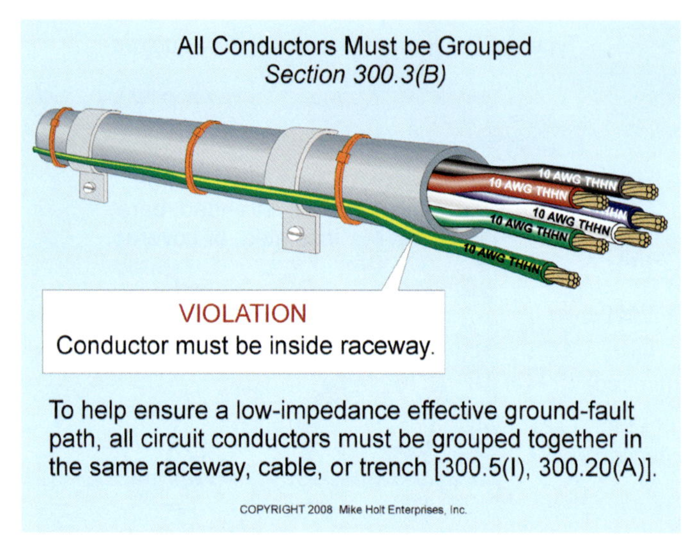

Figure 250–18

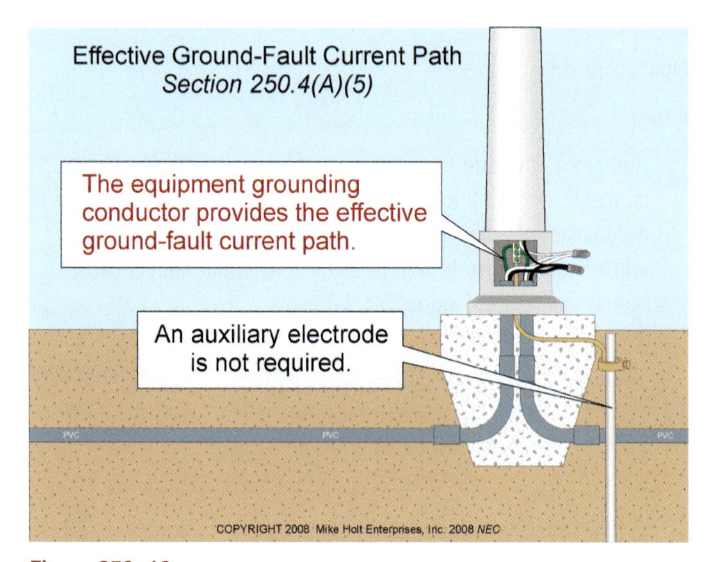

Figure 250–19

Question: *What is the maximum fault current that can flow through the earth to the power supply from a 120V ground fault to metal parts of a light pole that is grounded (connected to the earth) via a ground rod having a contact resistance to the earth of 25 ohms?* Figure 250–20

(a) 4.80A *(b) 20A* *(c) 40A* *(d) 100A*

Answer: *(a) 4.80A*

$I = E/R$

$I = 120V/25\ ohms$

$I = 4.80A$

Figure 250–20

DANGER: *Because the contact resistance of an electrode to the earth is so high, very little fault current returns to the power supply if the earth is the only fault current return path. Result—the circuit overcurrent device will not open and all metal parts associated with the electrical installation, metal piping, and structural building steel will become and remain energized.* **Figure 250–21**

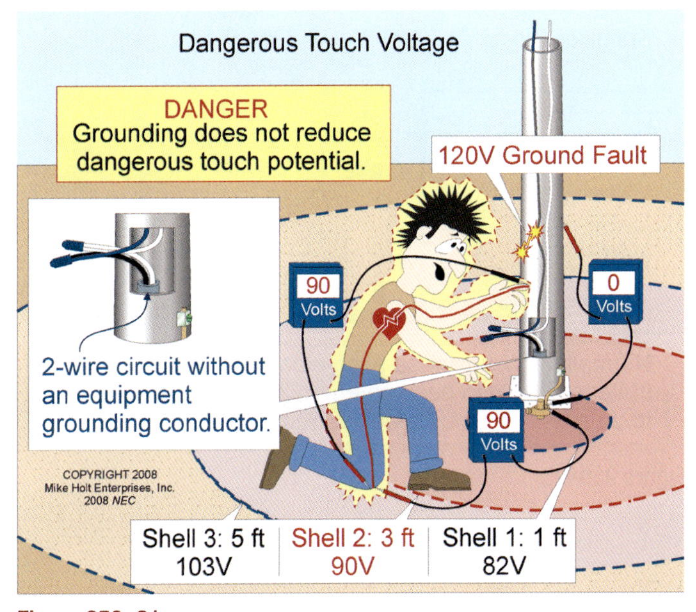

Figure 250–21

EARTH SHELLS

According to ANSI/IEEE 142, *Recommended Practice for Grounding of Industrial and Commercial Power Systems* (Green Book) [4.1.1], the resistance of the soil outward from a ground rod is equal to the sum of the series resistances of the earth shells. The shell nearest the rod has the highest resistance and each successive shell has progressively larger areas and progressively lower resistances. Don't be concerned if you don't understand this statement; just review the table below. **Figure 250–22**

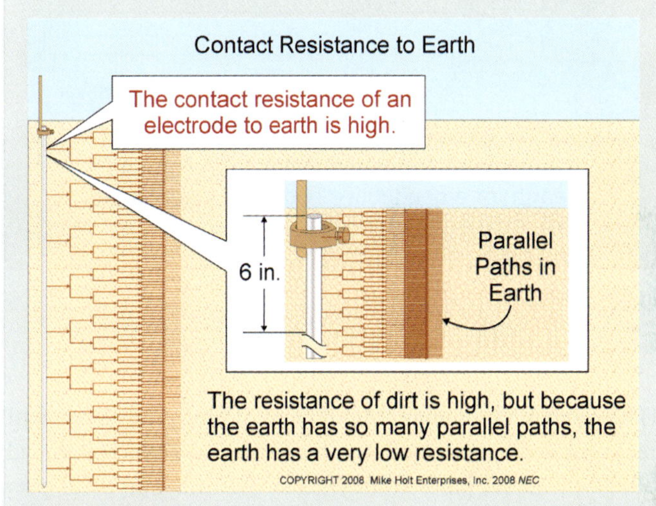

Figure 250–22

Distance from Rod	Soil Contact Resistance
1 ft (Shell 1)	68% of total contact resistance
3 ft (Shells 1 and 2)	75% of total contact resistance
5 ft (Shells 1, 2, and 3)	86% of total contact resistance

Since voltage is directly proportional to resistance, the voltage gradient of the earth around an energized ground rod will be as follows, assuming a 120V ground fault:

Distance from Rod	Soil Contact Resistance	Voltage Gradient
1 ft (Shell 1)	68%	82V
3 ft (Shells 1 and 2)	75%	90V
5 ft (Shells 1, 2, and 3)	86%	103V

(B) Ungrounded Systems.

Author's Comment: Ungrounded systems are those systems with no connection to the ground or a conductive body that extends the ground connection [Article 100]. **Figure 250–23**

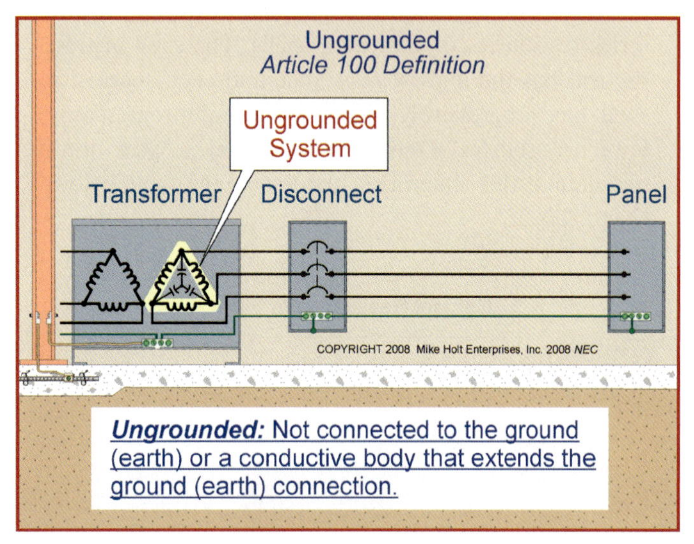

Figure 250–23

(1) Equipment Grounding. Metal parts of electrical equipment are grounded (connected to the earth) to reduce induced voltage on metal parts from exterior lightning so as to prevent fires from an arc within the building or structure. **Figure 250–24**

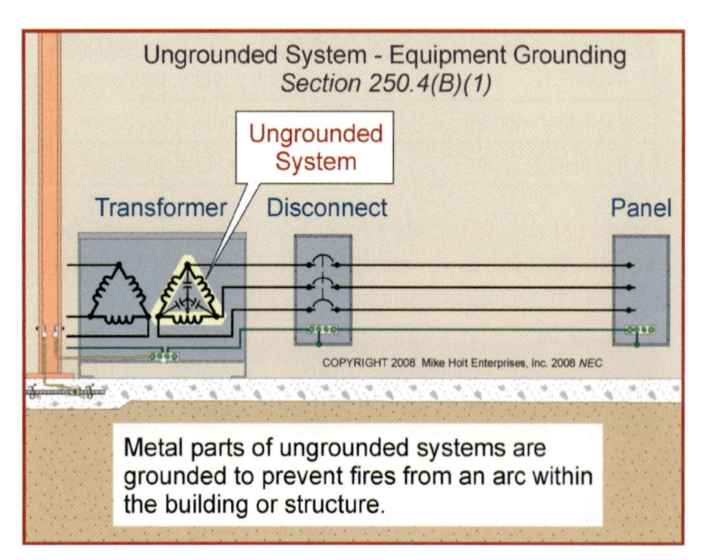

Figure 250–24

Author's Comment: Grounding metal parts helps drain off static electricity charges before an electric arc takes place (flashover potential). Static grounding is often used in areas where the discharge (arcing) of the voltage buildup (static) can cause dangerous or undesirable conditions [500.4 FPN No. 3].

CAUTION: *Connecting metal parts to the earth (grounding) serves no purpose in electrical shock protection.*

(2) Equipment Bonding. Metal parts of electrical raceways, cables, enclosures, or equipment must be bonded together in a manner that creates a low-impedance path for ground-fault current to facilitate the operation of the circuit overcurrent device.

The fault current path must be capable of safely carrying the maximum ground-fault current likely to be imposed on it from any point on the wiring system where a ground fault may occur to the electrical supply source.

(3) Bonding Conductive Materials. Conductive materials such as metal water piping systems, metal sprinkler piping, metal gas piping, and other metal-piping systems, as well as exposed structural steel members likely to become energized must be bonded together in a manner that creates a low-impedance fault current path that is capable of carrying the maximum fault current likely to be imposed on it. **Figure 250–25**

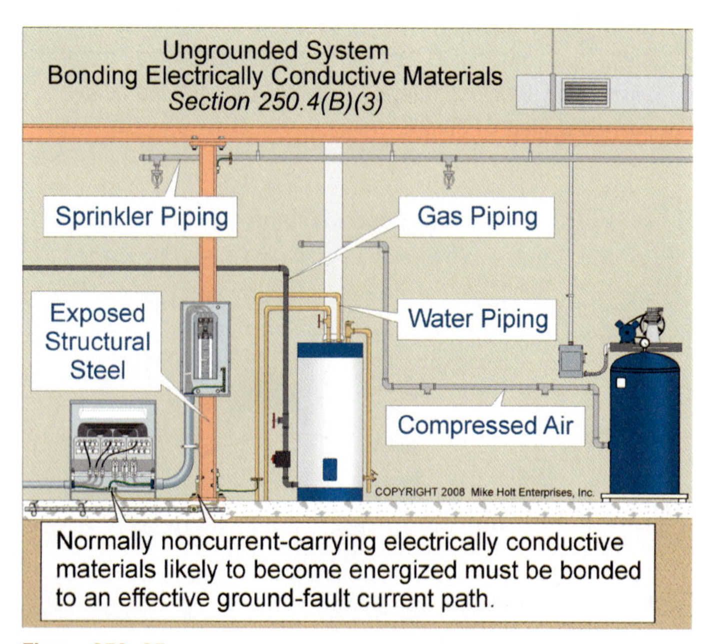

Figure 250–25

Author's Comment: The phrase "likely to become energized" is subject to interpretation by the authority having jurisdiction.

(4) Fault Current Path. Electrical equipment, wiring, and other electrically conductive material likely to become energized must be installed in a manner that creates a low-impedance fault current path to facilitate the operation of overcurrent devices should a second ground fault from a different phase occur. Figure 250–26

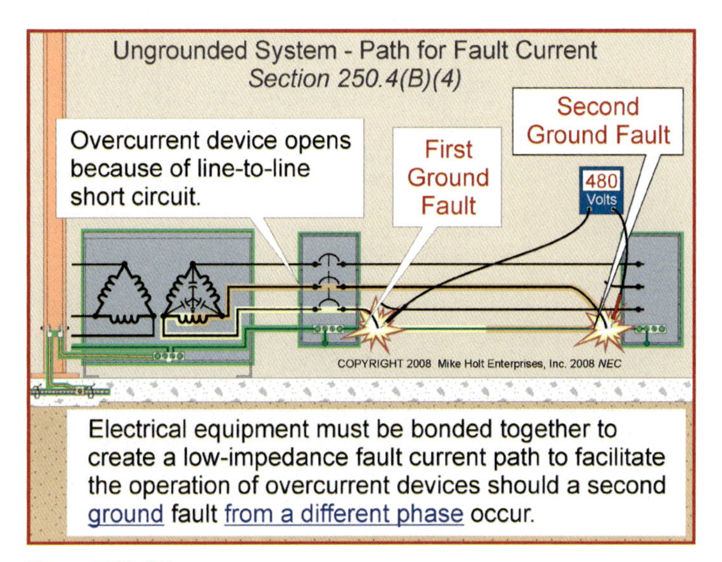

Figure 250–26

Author's Comment: A single ground fault cannot be cleared on an ungrounded system because there's no low-impedance fault current path to the power source. The first ground fault simply grounds the previously ungrounded system. However, a second ground fault on a different phase results in a line-to-line short circuit between the two ground faults. The conductive path, between the ground faults, provides the low-impedance fault current path necessary so the overcurrent device will open.

250.6 Objectionable Current.

(A) Preventing Objectionable Current. To prevent a fire, electric shock, or improper operation of circuit overcurrent devices or electronic equipment, electrical systems and equipment must be installed in a manner that prevents objectionable neutral current from flowing on metal parts.

OBJECTIONABLE CURRENT

Objectionable neutral current occurs because of improper neutral-to-case connections or wiring errors that violate 250.142(B).

Improper Neutral-to-Case Connection [250.142]

Panelboards. Objectionable neutral current will flow when the neutral conductor is connected to the metal case of a panelboard that is not used as service equipment. Figure 250–27

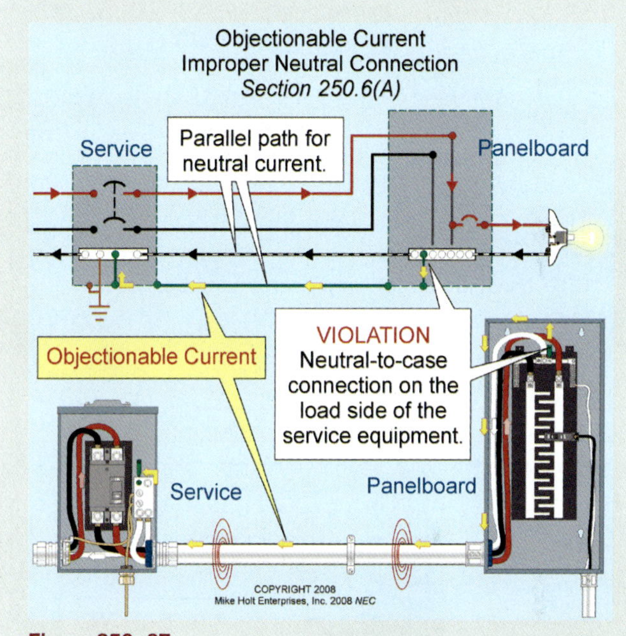

Figure 250–27

Separately Derived Systems. Objectionable neutral current will flow on conductive metal parts and conductors if the neutral conductor is connected to the circuit equipment grounding conductor on the load side of the system bonding jumper for separately derived system. Figures 250–28 and 250–29

Disconnects. Objectionable neutral current will flow when the neutral conductor is connected to the metal case of a disconnecting means not part of service equipment. Figure 250–30

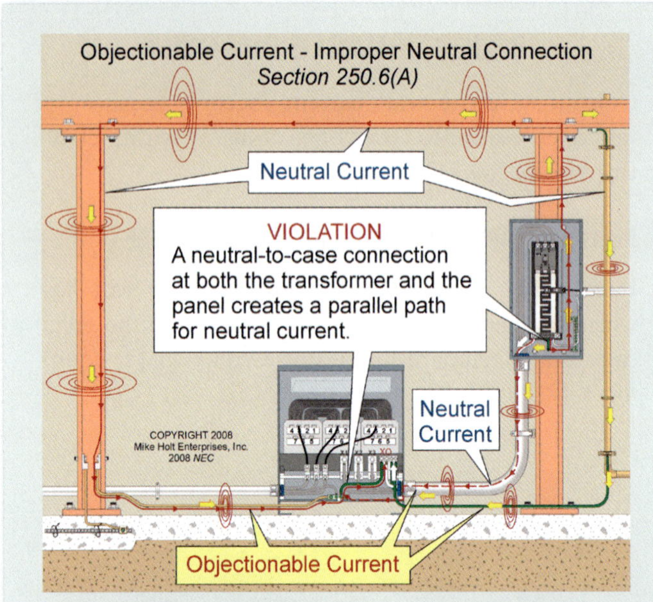

Figure 250–28

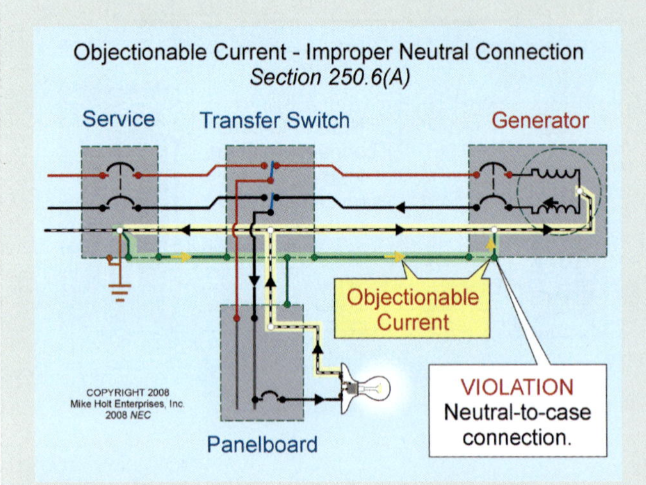

Figure 250–29

Wiring Errors. Objectionable neutral current will flow when the neutral conductor from one system is connected to a circuit of a different system. Figure 250–31

Objectionable neutral current will flow on metal parts when the circuit equipment grounding conductor is used as a neutral conductor:

- A 230V time-clock motor is replaced with a 115V time-clock motor, and the circuit equipment grounding conductor is used for neutral return current.

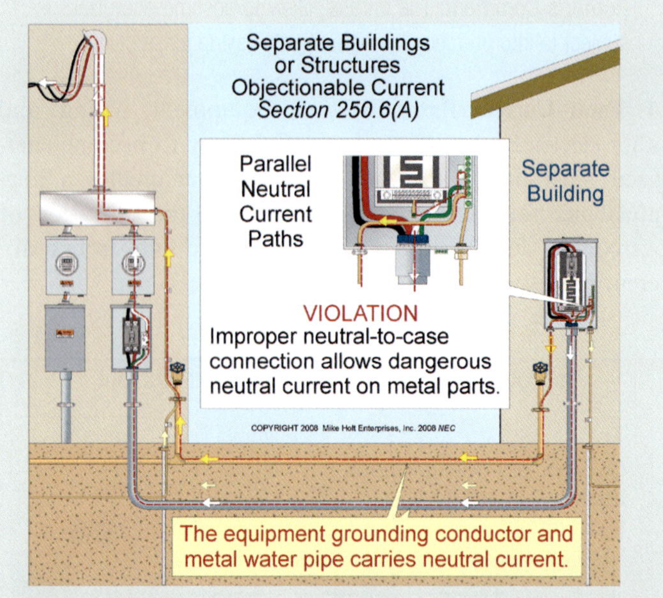

Figure 250–30

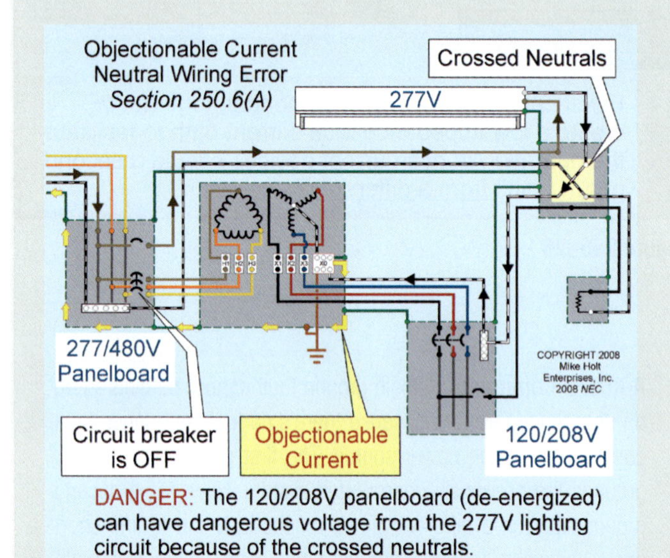

Figure 250–31

- A 115V water filter is wired to a 240V well-pump motor circuit, and the circuit equipment grounding conductor is used for neutral return current. Figure 250–32
- Where the circuit equipment grounding conductor is used for neutral return current. Figure 250–33

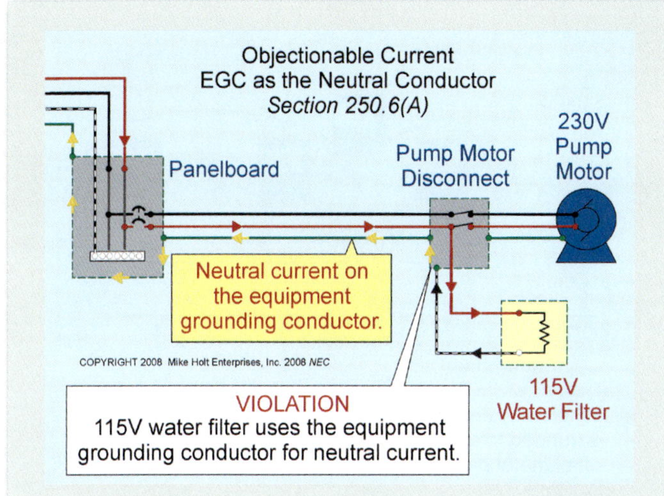

Figure 250–32

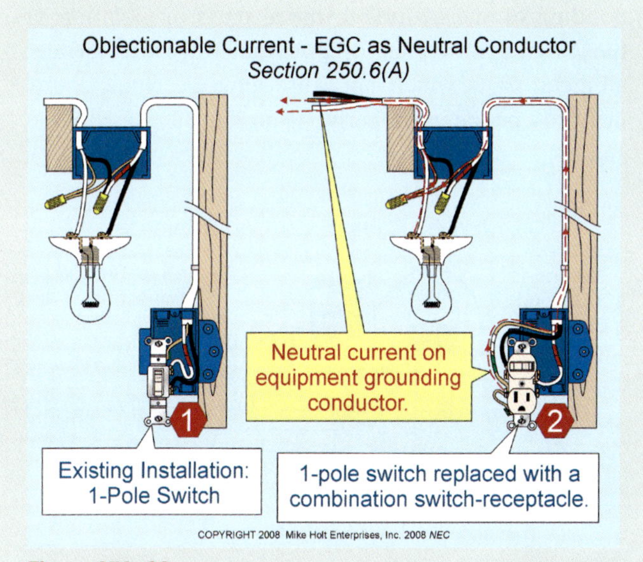

Figure 250–33

Dangers of Objectionable Current

Objectionable neutral current on metal parts can cause electric shock, fires, and improper operation of electronic equipment and overcurrent devices such as GFPs, GFCIs, and AFCIs.

Shock Hazard. When objectionable neutral current flows on metal parts, electric shock and even death can occur from elevated voltage on the metal parts. Figure 250–34 and 250–35

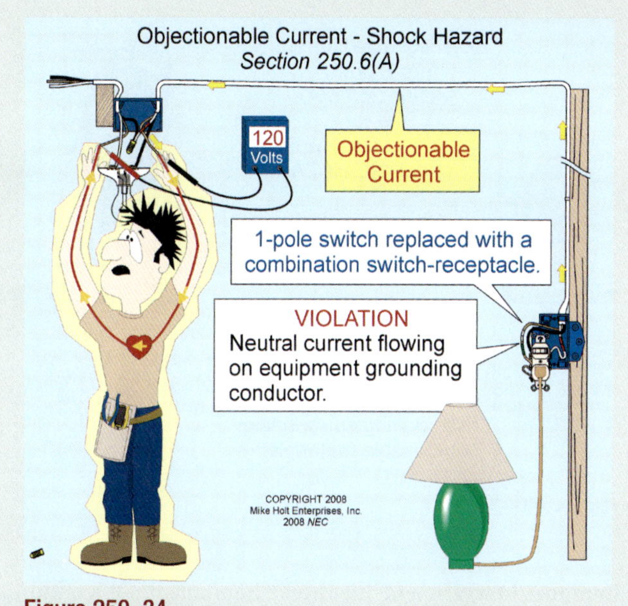

Figure 250–34

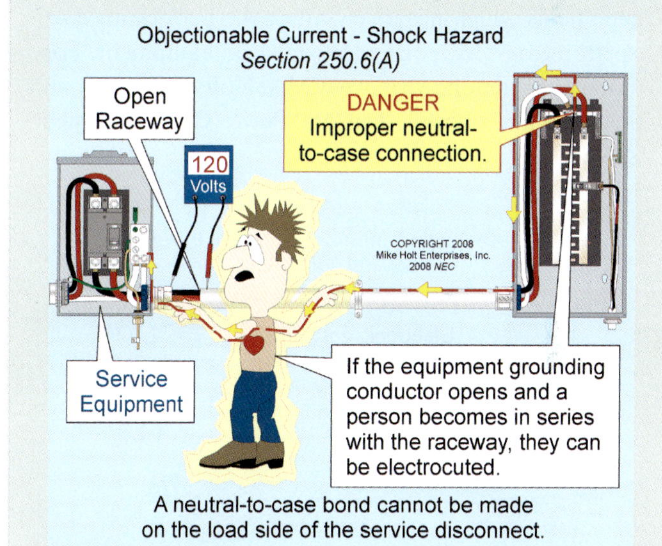

Figure 250–35

Fire Hazard. When objectionable neutral current flows on metal parts, a fire can ignite adjacent combustible material. Heat is generated whenever current flows, particularly over high-resistance parts. In addition, arcing at loose connections is especially dangerous in areas containing easily ignitible and explosive gases, vapors, or dust. Figure 250–36

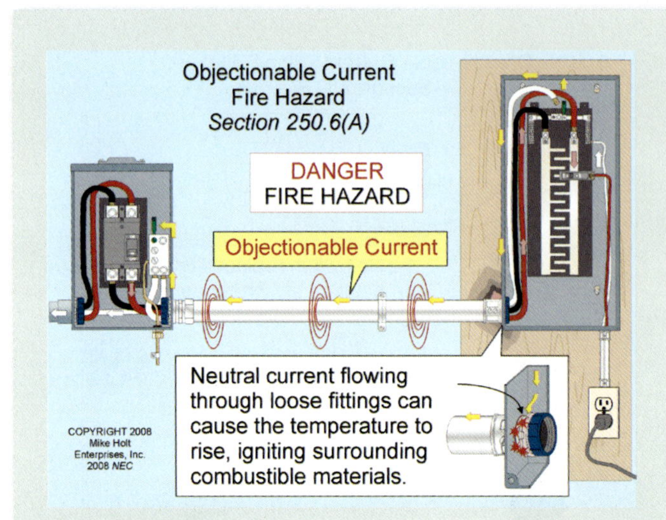

Figure 250–36

Improper Operation of Electronic Equipment. Objectionable neutral current flowing on metal parts of electrical equipment and building parts can cause electromagnetic fields which negatively affect the performance of electronic devices, particularly medical equipment. For more information, visit www.MikeHolt.com, click on the "Technical Link," then on "Power Quality." Figure 250–37

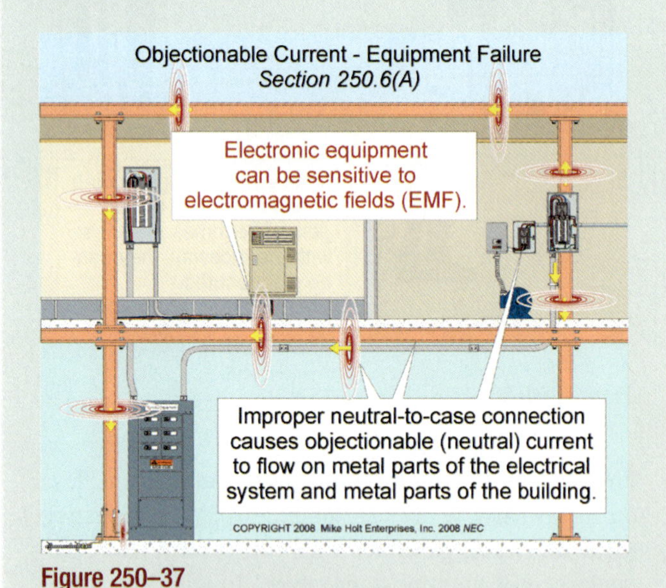

Figure 250–37

When a system is properly grounded and bonded, the voltage of all metal part to the earth and to each other will be zero. Figure 250–38 When objectionable neutral current travels on metal parts because of improper bonding of the neutral to metal parts in violation of the *NEC*, a difference of potential will exist between all metal parts, which can cause some electronic equipment to operate improperly. Figure 250–39

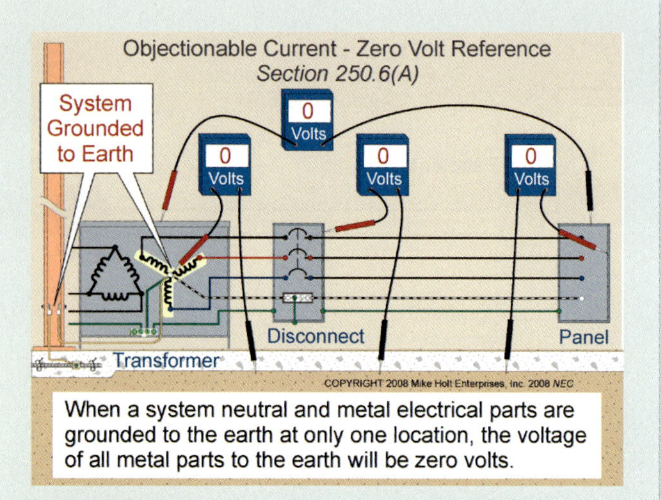

Figure 250–38

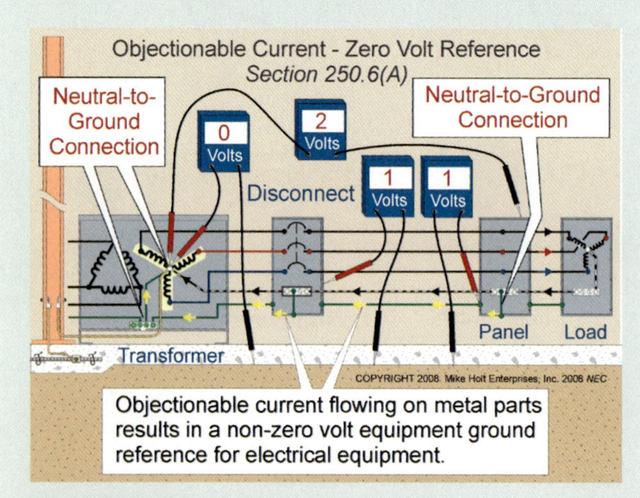

Figure 250–39

Operation of Overcurrent Devices. When objectionable neutral current travels on metal parts, tripping of electronic overcurrent devices equipped with ground-fault protection can occur because some neutral current flows on the circuit equipment grounding conductor instead of the neutral conductor.

250.8 Termination of Grounding and Bonding Conductors.

(A) Permitted Methods. Grounding and bonding conductors must terminate in one of the following methods:

(1) Listed pressure connectors

(2) Terminal bars

(3) Pressure connectors listed for direct burial or concrete encasement [250.70]

(4) Exothermic welding

(5) Machine screws that engage at least two threads or are secured with a nut **Figure 250–40**

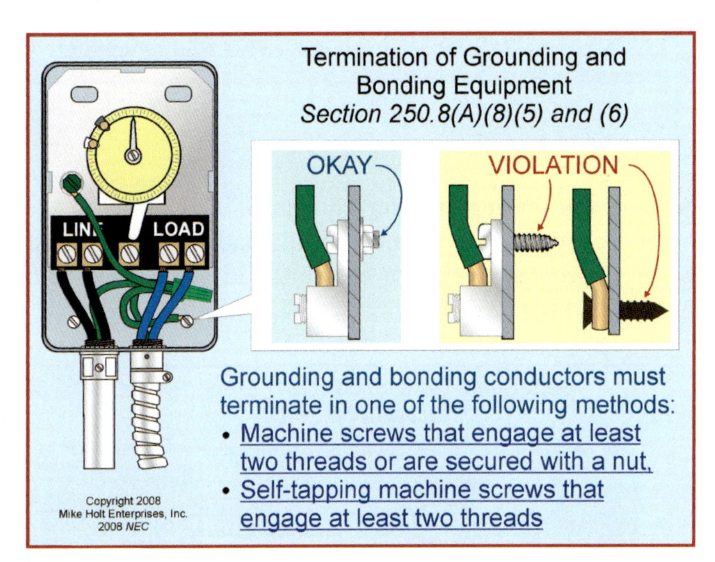

Figure 250–40

(6) Self-tapping machine screws that engage at least two threads

(7) Connections that are part of a listed assembly

(8) Other listed means

250.10 Protection of Fittings.

Grounding and bonding fittings must be protected from physical damage by:

(1) Locating the fittings so they aren't likely to be damaged.

(2) Enclosing the fittings in metal, wood, or equivalent protective covering.

> **Author's Comment:** Grounding and bonding fittings can be buried or encased in concrete if they are installed in accordance with 250.53(G), 250.68(A) Ex 1, and 250.70.

250.12 Clean Surfaces.

Nonconductive coatings, such as paint, must be removed to ensure good electrical continuity, or the termination fittings must be designed so as to make such removal unnecessary [250.53(A) and 250.96(A)].

> **Author's Comment:** Tarnish on copper water pipe need not be removed before making a termination.

PART II. SYSTEM GROUNDING AND BONDING

250.20 Systems Required to be Grounded.

(A) Systems Below 50 Volts. Systems operating below 50V are not required to be grounded (connected to the earth) unless the transformer's primary supply is from: **Figure 250–41**

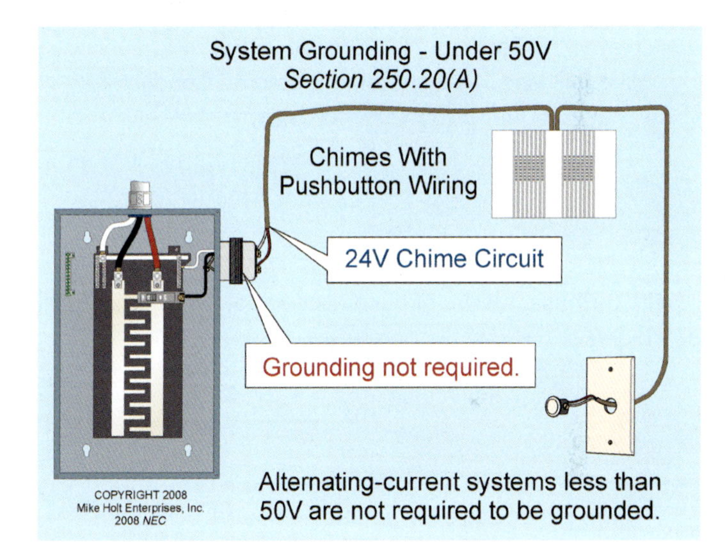

Figure 250–41

(1) A 277V or 480V system

(2) An ungrounded system

(B) Systems Over 50 Volts. The following systems must be connected (grounded) to the earth:

(1) Single-phase systems where the neutral conductor is used as a circuit conductor. **Figure 250–42**

(2) Three-phase, wye-connected systems where the neutral conductor is used as a circuit conductor. **Figure 250–43A**

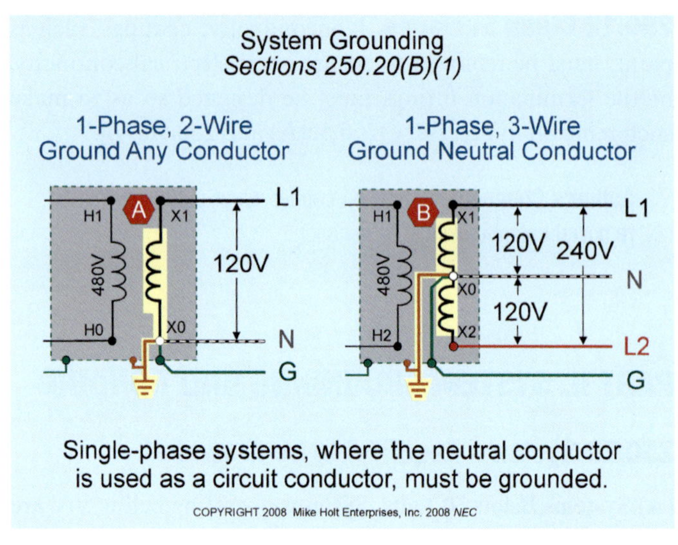

Figure 250–42

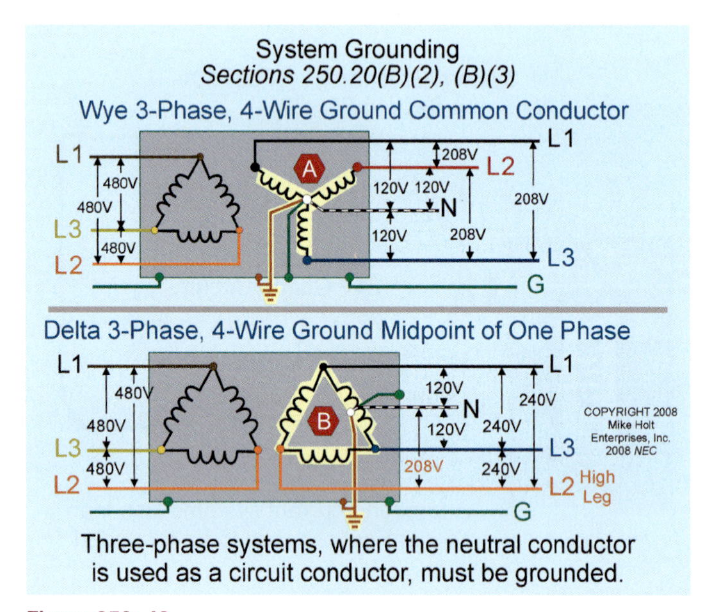

Figure 250–43

(3) Three-phase, high-leg delta-connected systems where the neutral conductor is used as a circuit conductor. **Figure 250–43B**

(D) Separately Derived Systems. Separately derived systems as covered in 250.20(B) must be grounded (connected to the earth) in accordance with 250.30(A).

Author's Comment: A separately derived system is a wiring system whose power is derived from a source where there's no direct electrical connection to the supply conductors of another system. This includes most transformers because the primary circuit conductors don't have any direct electrical connection to the secondary circuit conductors [Article 100]. **Figure 250–44**

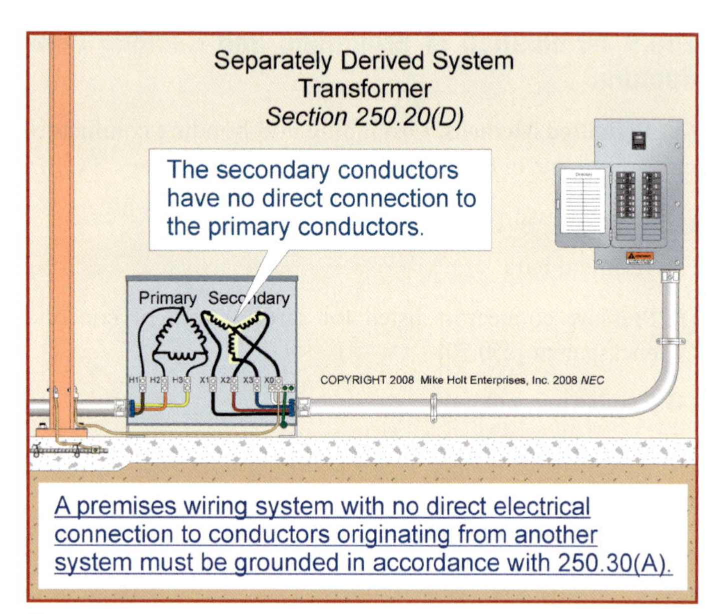

Figure 250–44

A generator having transfer equipment that switches the neutral conductor or has no neutral conductor at all must be grounded (connected to the earth) in accordance with 250.30(A). **Figure 250–45**

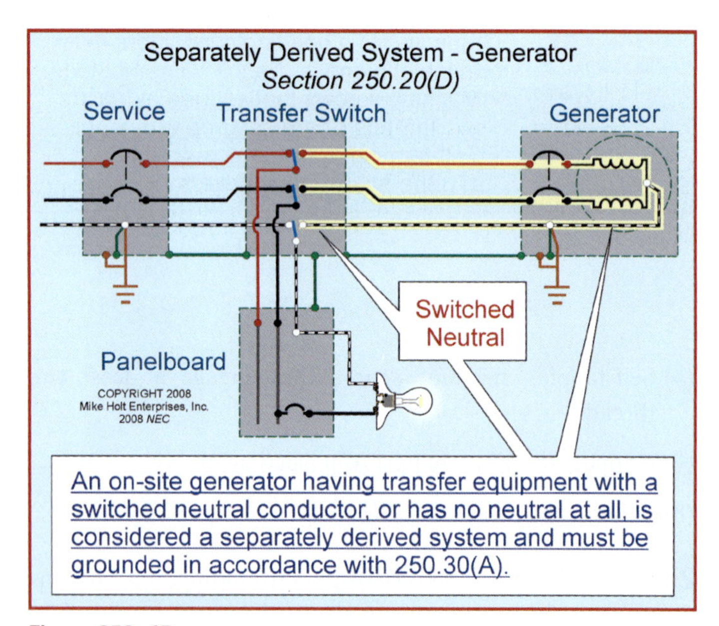

Figure 250–45

FPN No. 1: An alternate ac power source such as an on-site generator is not a separately derived system if the neutral conductor is solidly interconnected to a service-supplied system neutral conductor. An example would be a generator provided with a transfer switch that includes a neutral conductor that is not switched. **Figure 250–46**

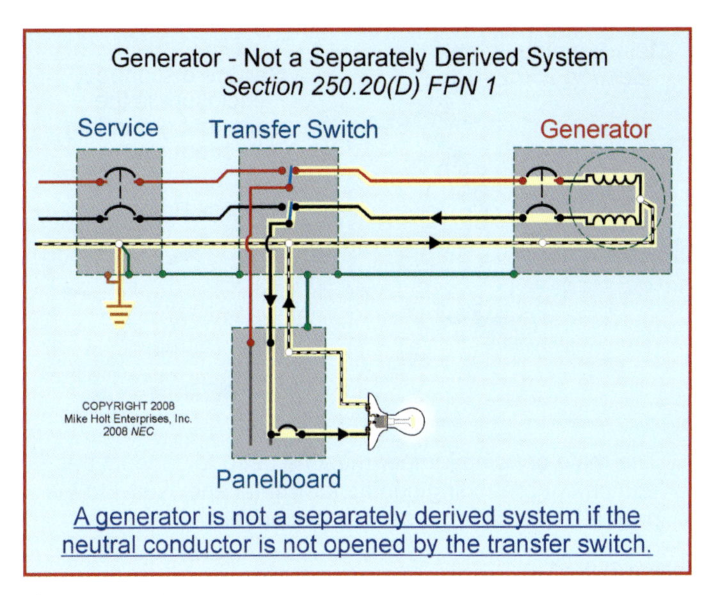

Figure 250–46

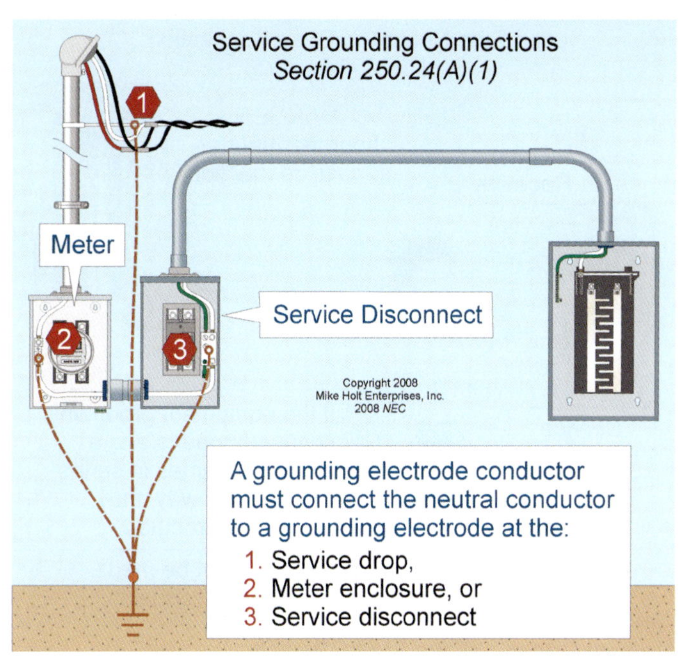

Figure 250–47

250.24 Service Equipment—Grounding and Bonding.

(A) Grounded System. Service equipment supplied from a grounded system must have the neutral conductor terminate in accordance with (1) through (5).

(1) Grounding Location. A grounding electrode conductor must connect the service neutral conductor to the grounding electrode at any accessible location, from the load end of the service drop or service lateral, up to and including the service disconnecting means. Figure 250–47

Author's Comment: Some inspectors require the service neutral conductor to be grounded (connected to the earth) from the meter socket enclosure, while other inspectors insist that the service neutral conductor be grounded (connected to the earth) only from the service disconnect.

(4) Grounding Termination. When the service neutral conductor is connected to the service disconnecting means [250.24(B)] by a wire or busbar [250.28], the grounding electrode conductor is permitted to terminate to either the neutral terminal or the equipment grounding terminal within the service disconnect.

(5) Neutral-to-Case Connection. A neutral-to-case connection is not permitted on the load side of service equipment, except as permitted by 250.142(B). Figure 250–48

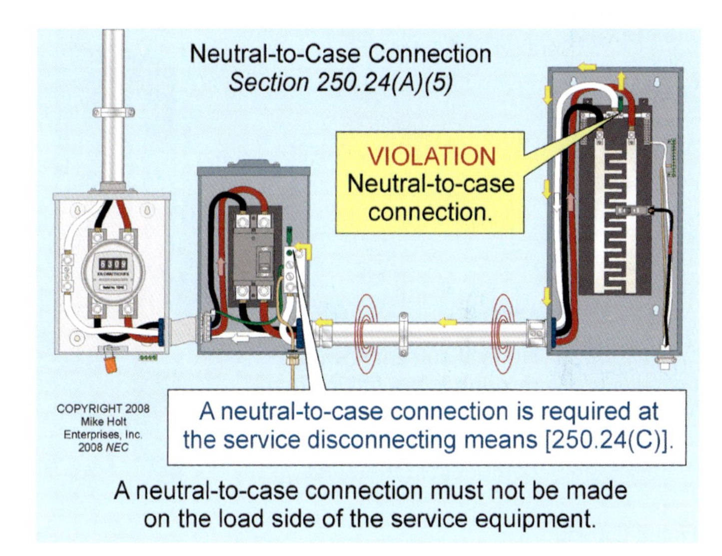

Figure 250–48

Author's Comment: If a neutral-to-case connection is made on the load side of service equipment, dangerous objectionable neutral current will flow on conductive metal parts of electrical equipment [250.6(A)]. Objectionable neutral current on metal parts of electrical equipment can cause electric shock and even death from ventricular fibrillation, as well as a fire. Figures 250–49 and 250–50

(B) Bonding. A main bonding jumper [250.28] must be installed for the purpose of connecting the neutral conductor to the metal parts of the service disconnecting means. Figure 250–51

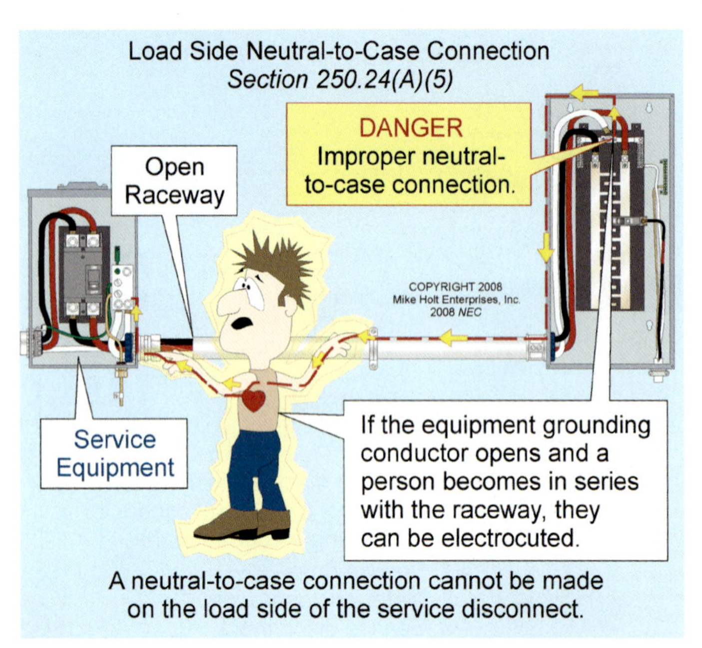

Figure 250–49

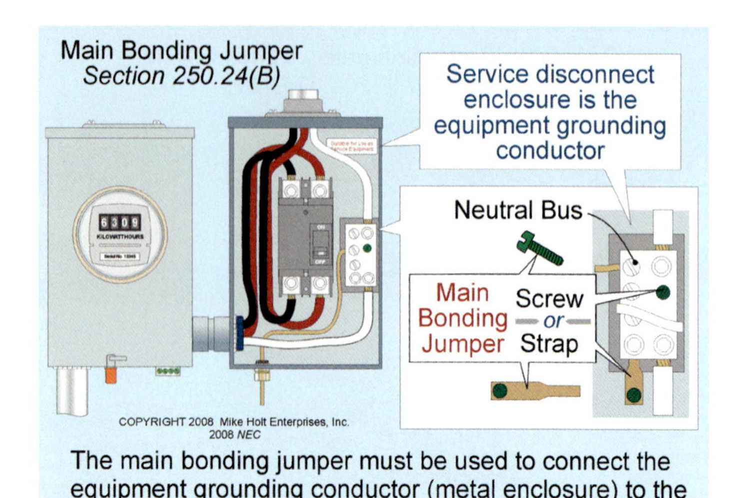

Figure 250–51

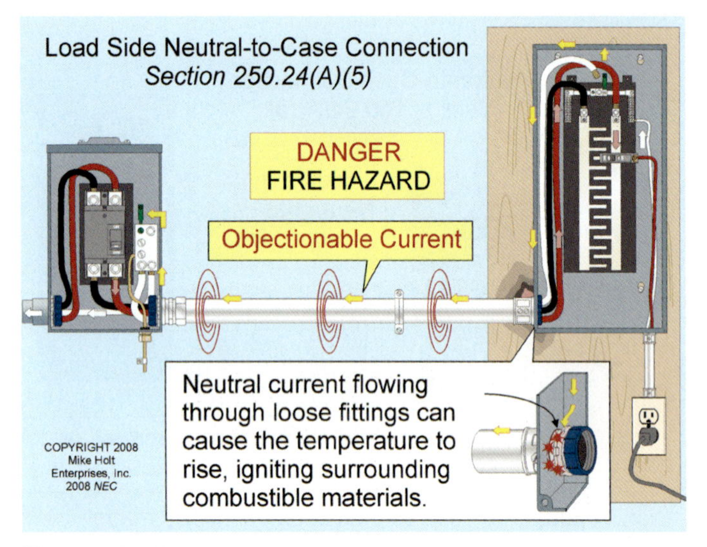

Figure 250–50

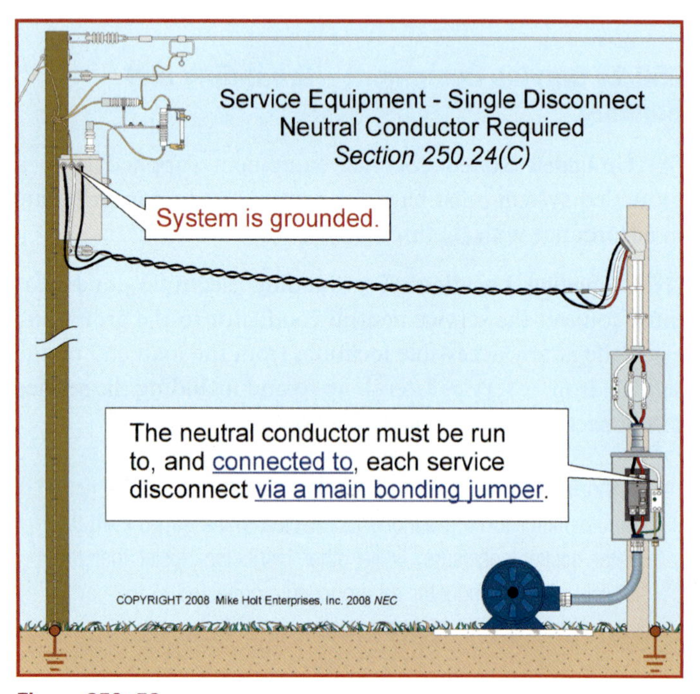

Figure 250–52

(C) Neutral Conductor. A service neutral conductor from the electric utility must terminate to each service disconnecting means via a main bonding jumper [250.24(B)] that is installed between the service neutral conductor and each service disconnecting means enclosure. Figures 250–52 and 250–53

Author's Comment: The service neutral conductor provides the effective ground-fault current path to the power supply to ensure that dangerous voltage from a ground fault will be quickly removed by opening the overcurrent device [250.4(A)(3) and 250.4(A)(5)]. Figure 250–54

DANGER: *Dangerous voltage from a ground fault will not be removed from metal parts, metal piping, and structural steel if the service disconnecting means enclosure is not connected to the service neutral conductor. This is because the contact resistance of a grounding electrode to the earth is so great that insufficient fault current returns to the power supply if the earth is the only fault current return path to open the circuit overcurrent device.* Figure 250–55

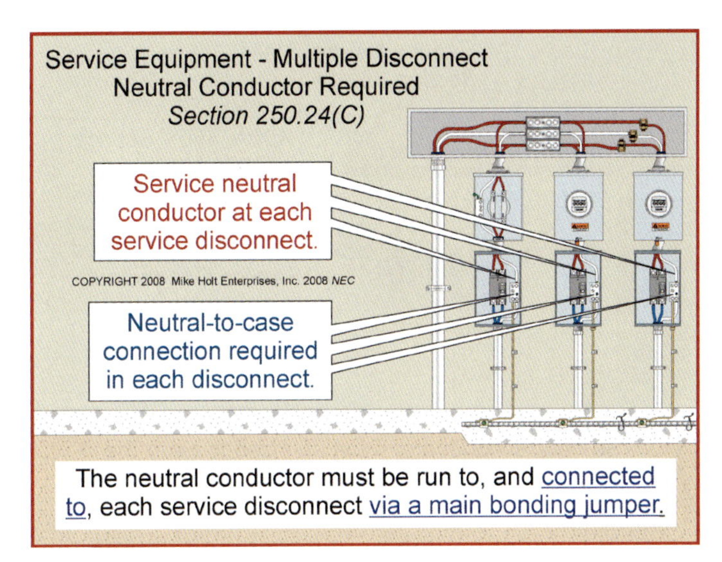

Figure 250–53

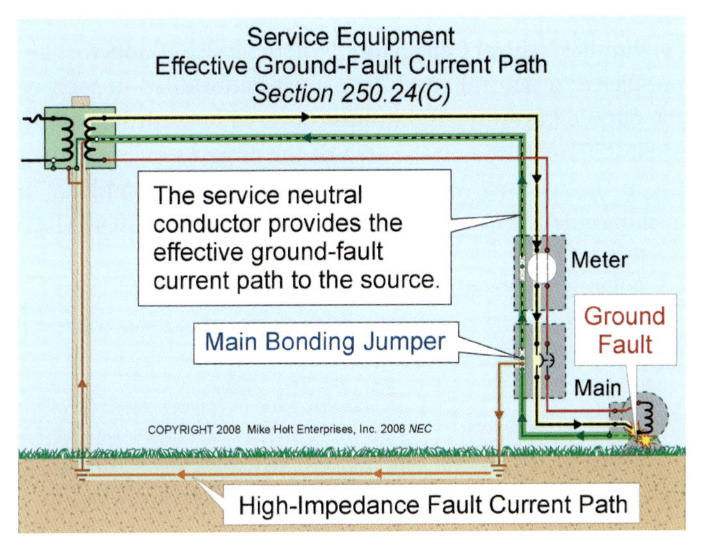

Figure 250–54

Author's Comment: For example, if the neutral conductor is opened, dangerous voltage will be present on metal parts under normal conditions, providing the potential for electric shock. If the earth's ground resistance is 25 ohms and the load's resistance is 25 ohms, the voltage drop across each of these resistors will be half of the voltage source. Since the neutral is connected to the service disconnect, all metal parts will be elevated to 60V above the earth's potential for a 120/240V system. Figure 250–56

To determine the actual voltage on the metal parts from an open service neutral conductor, you need to do some complex math calculations. Visit www.MikeHolt.com and go to the "Free Stuff" link to download a spreadsheet for this purpose.

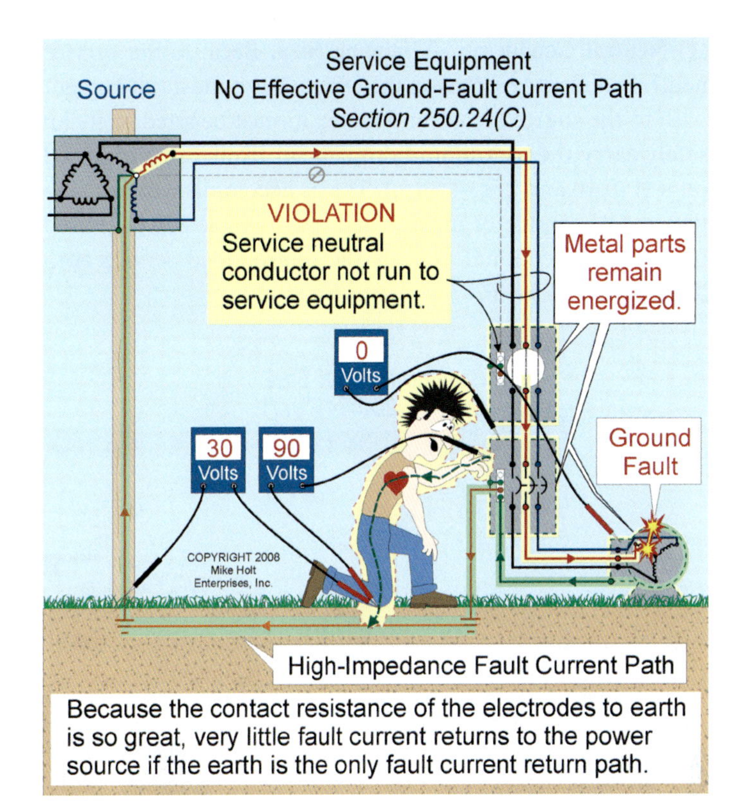

Figure 250–55

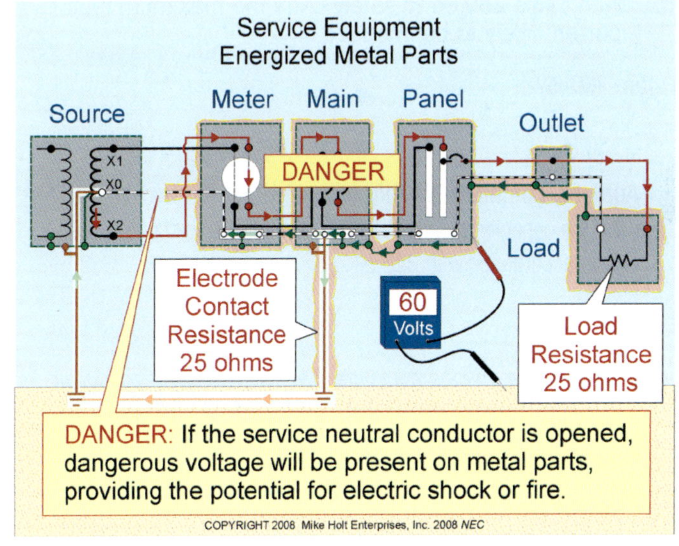

Figure 250–56

(1) Neutral Conductor—Minimum Size. Because the service neutral conductor serves as the effective ground-fault current path to the source for ground faults, it must be sized so it can safely carry the maximum fault current likely to be imposed on it [110.10 and 250.4(A)(5)]. This is accomplished by sizing the neutral conductor in accordance with Table 250.66, based on the cross-sectional area of the ungrounded service conductor. Figure 250–57

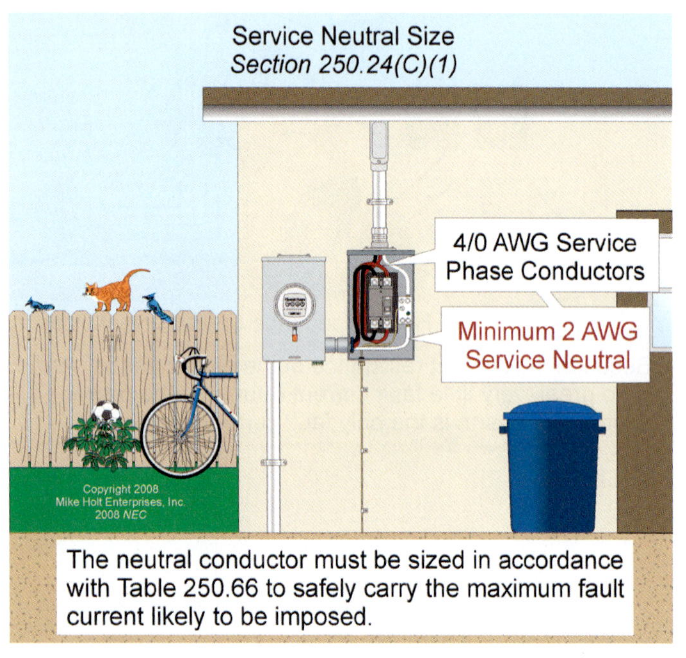

Figure 250–57

Author's Comment: In addition, the neutral conductors must have the capacity to carry the maximum unbalanced neutral current in accordance with 220.61.

Question: *What is the minimum size service neutral conductor required for a 480V, three-phase service where the ungrounded service conductors are sized at 500 kcmil and the maximum unbalanced load is 100A?* **Figure 250–58**

(a) 3 AWG (b) 2 AWG (c) 1 AWG (d) 1/0 AWG

Answer: *(d) 1/0 AWG [Table 250.66]*

The unbalanced load requires a 3 AWG service neutral conductor, which is rated 100A at 75°C in accordance with Table 310.16 [220.61]. However, the service neutral conductor must not be smaller than 1/0 AWG to ensure it will accommodate the maximum fault current likely to be imposed on it [Table 250.66].

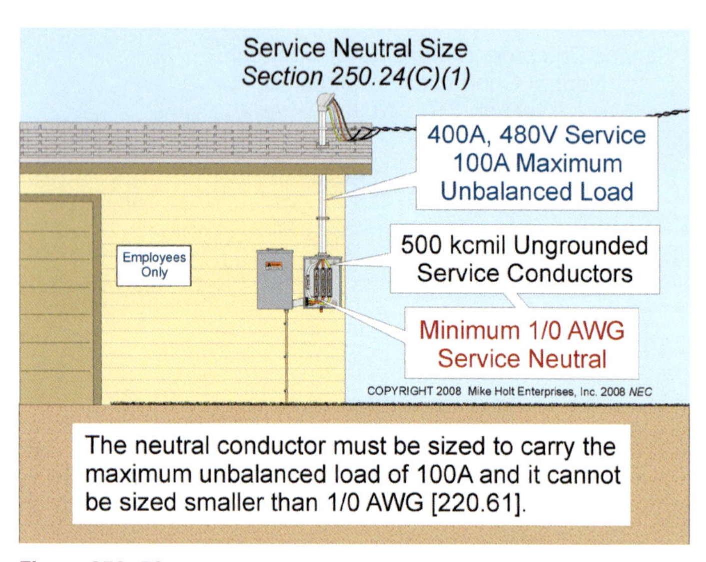

Figure 250–58

(2) Parallel Neutral Conductors. Where service conductors are paralleled, a neutral conductor must be installed in each of the parallel raceways and it must be sized in accordance with Table 250.66, based on the area of the largest service conductor in the raceway. In no case can the neutral conductor in each parallel set be sized smaller than 1/0 AWG [310.4(A)].

Author's Comment: In addition, the neutral conductors must have the capacity to carry the maximum unbalanced neutral current in accordance with 220.61.

Question: *What is the minimum size service neutral conductor required for a 480V, three-phase service installed in two raceways where the ungrounded service conductors in each of the raceways are 350 kcmil and the maximum unbalanced load is 100A?* **Figure 250–59**

(a) 3 AWG (b) 2 AWG (c) 1 AWG (d) 1/0 AWG

Answer: *(d) 1/0 AWG per raceway [Table 250.66 and 310.4]*

The unbalanced load of 50A in each raceway requires an 8 AWG service neutral conductor, which is rated 50A at 75°C in accordance with Table 310.16 [220.61]. However, the smallest service neutral conductor permitted to be run in parallel in each raceway must not be smaller than 1/0 AWG [Table 250.66].

(D) Grounding Electrode Conductor. A grounding electrode conductor, sized in accordance with 250.66 based on the area of the ungrounded service conductor, must connect the metal parts of service equipment enclosures to a grounding electrode in accordance with Part III of Article 250.

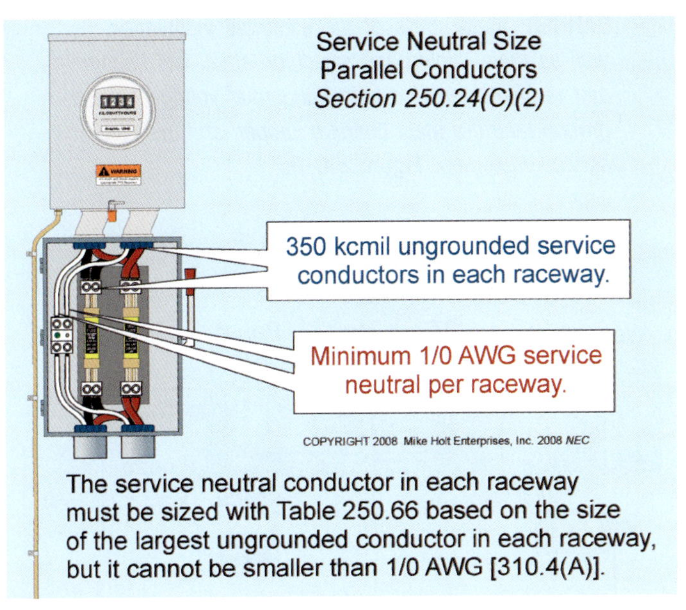

Figure 250–59

Question: *What is the minimum size grounding electrode conductor for a 400A service where the ungrounded service conductors are sized at 500 kcmil?* **Figure 250–60**

(a) 3 AWG (b) 2 AWG (c) 1 AWG (d) 1/0 AWG

Answer: *(d) 1/0 AWG [Table 250.66]*

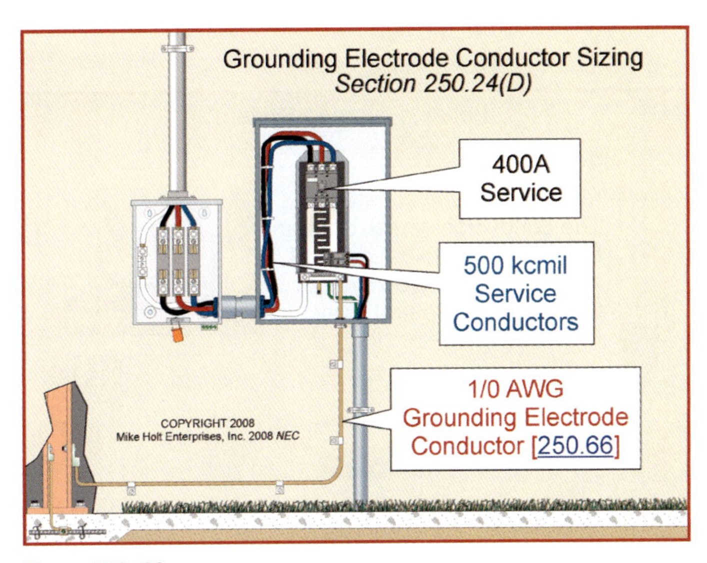

Figure 250–60

Author's Comment: Where the grounding electrode conductor is connected to a ground rod, that portion of the conductor that is the sole connection to the ground rod isn't required to be larger than 6 AWG copper [250.66(A)], **Figure 250–61**. Where the grounding electrode conductor is connected to a concrete-encased electrode, that portion of the conductor that is the sole connection to the concrete-encased electrode isn't required to be larger than 4 AWG copper [250.66(B)]. **Figure 250–62**

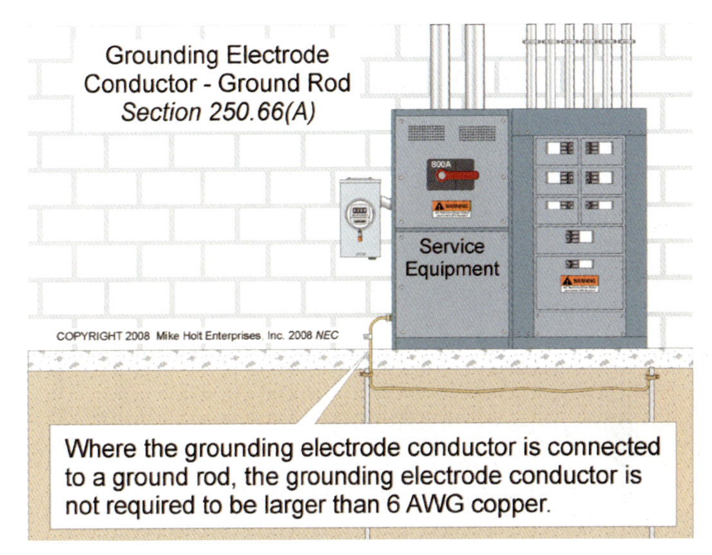

Figure 250–61

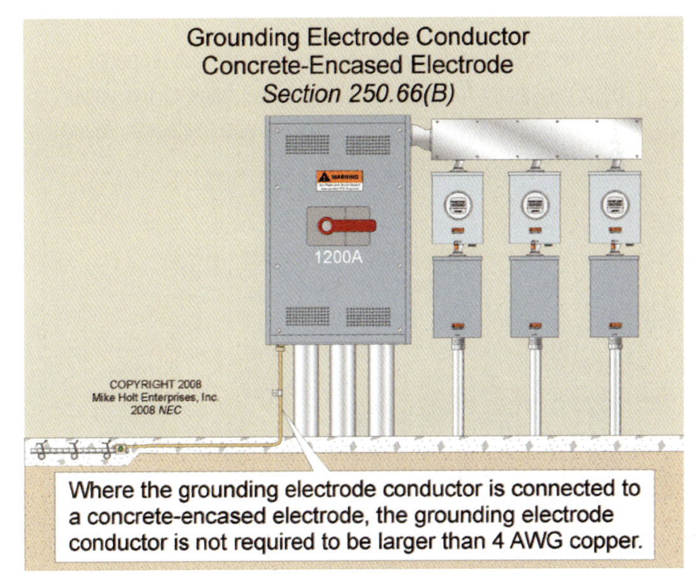

Figure 250–62

250.28 Main Bonding Jumper and System Bonding Jumper.

Main and system bonding jumpers must be installed as follows:

Author's Comments:

- **Main Bonding Jumper.** At service equipment, a main bonding jumper must be installed to electrically connect the neutral conductor to the service disconnect enclosure [250.24(B)]. **Figure 250–63**

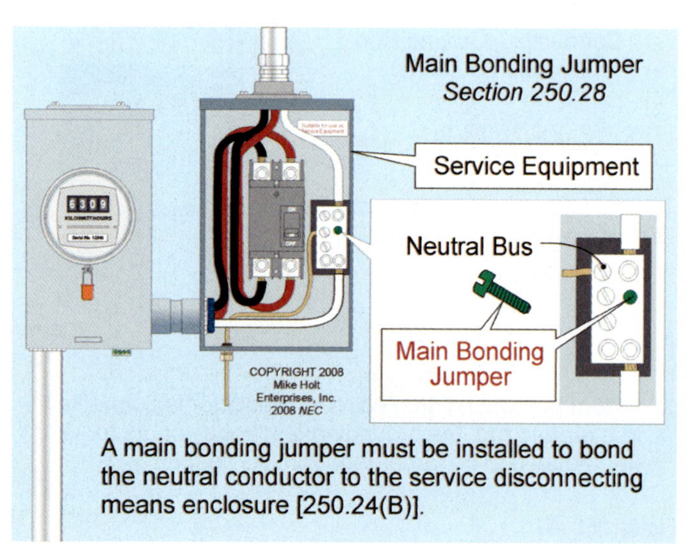

Figure 250–63

The main bonding jumper provides the low-impedance path necessary for fault current to travel back to the power supply to open the circuit overcurrent device to clear a ground fault [250.24(C)]. **Figure 250–64**

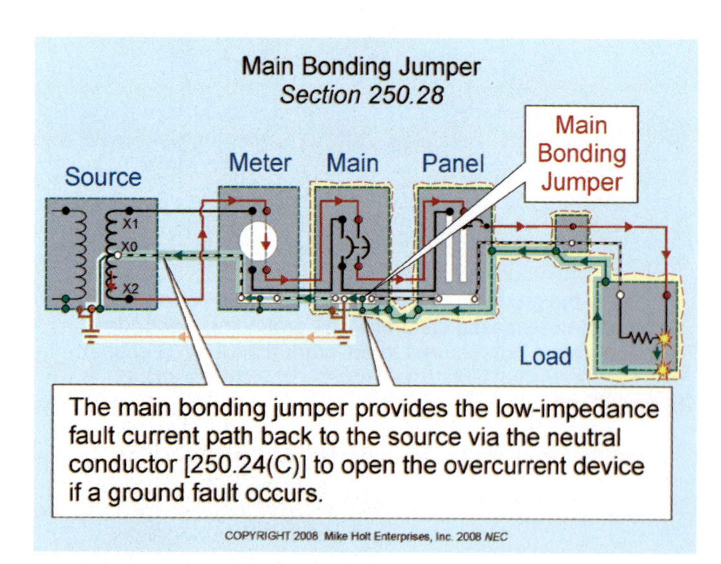

Figure 250–64

DANGER: *Metal parts of the electrical installation, as well as metal piping and structural steel, will become and remain energized with dangerous voltage from a ground fault if a main bonding jumper isn't installed at service equipment.* **Figure 250–65**

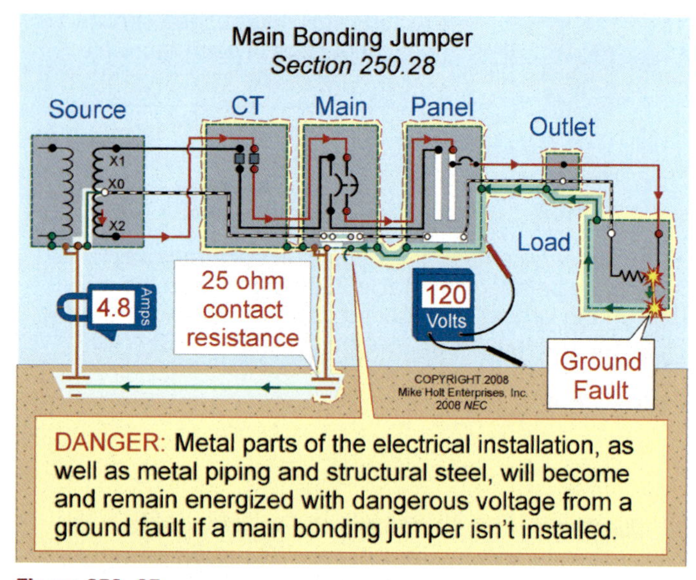

Figure 250–65

- **System Bonding Jumper.** A system bonding jumper is installed between the neutral terminal of a separately derived system and the circuit equipment grounding conductor of the secondary system [250.2 and 250.30(A)(1)]. **Figure 250–66**

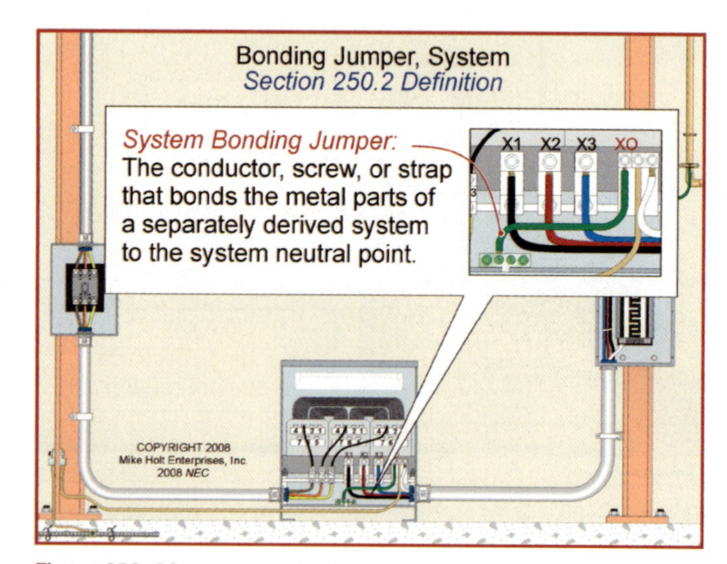

Figure 250–66

DANGER: *Metal parts of the electrical installation as well as metal piping and structural steel, will remain energized with dangerous voltage from a ground fault if a system bonding jumper isn't installed at a separately derived system.* **Figure 250–67**

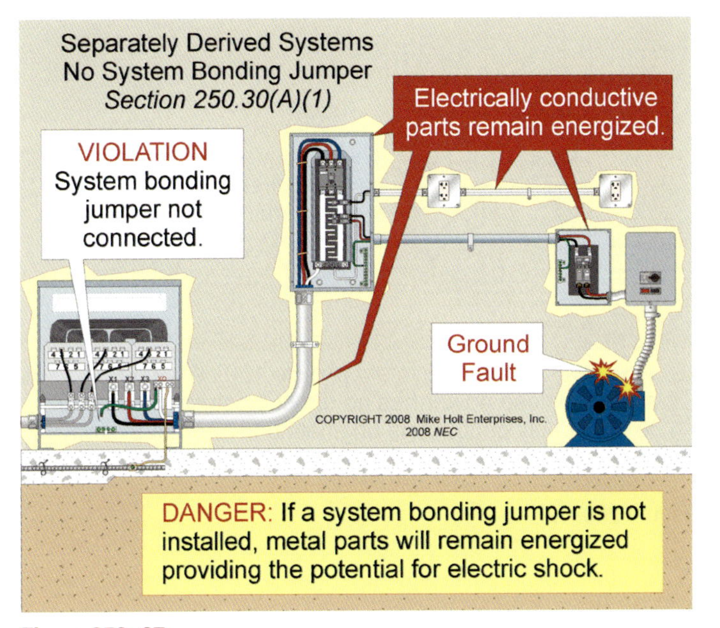

Figure 250–67

(A) Material. The bonding jumper can be a wire, bus, or screw.

(B) Construction. If the bonding jumper is a screw, it must be identified with a green finish visible with the screw installed.

(C) Attachment. Main and system bonding jumpers must terminate by one of the following means according to 250.8(A):

- Listed pressure connectors
- Terminal bars
- Pressure connectors listed as grounding and bonding equipment
- Exothermic welding
- Machine screw-type fasteners that engage not less than two threads or are secured with a nut
- Thread-forming machine screws that engage not less than two threads in the enclosure
- Connections that are part of a listed assembly
- Other listed means.

(D) Size. Main and system bonding jumpers must be sized not smaller than the sizes shown in Table 250.66. Where the service or secondary conductors have a total area larger than 1,100 kcmil copper or 1,750 kcmil aluminum, the bonding jumper must have an area not less than 12½ percent of the total conductor area of the largest ungrounded conductor. **Figures 250–68, 250–69,** and **250–70**

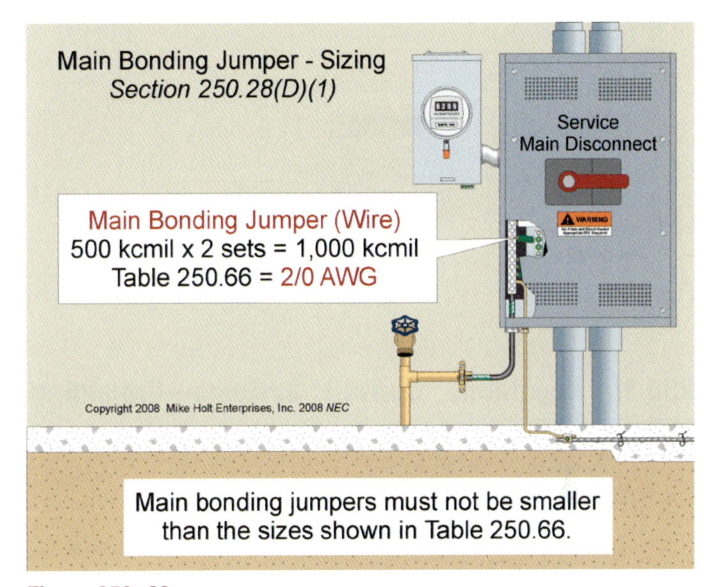

Figure 250–68

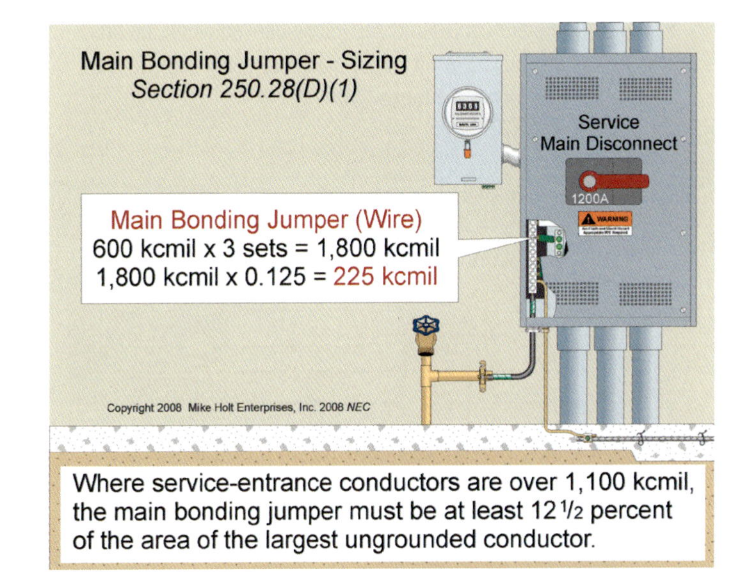

Figure 250–69

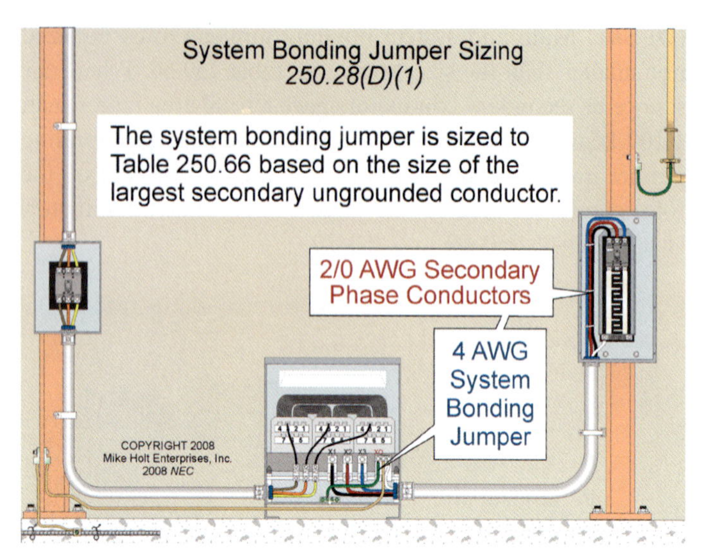

Figure 250–70

250.30 Separately Derived Systems—Grounding and Bonding.

Author's Comments:

- According to Article 100, a separately derived system is a wiring system whose power is derived from a source where there's no direct electrical connection to the supply conductors of another system.
- Transformers are considered separately derived when the primary conductors have no direct electrical connection to the secondary conductors. Figure 250–71

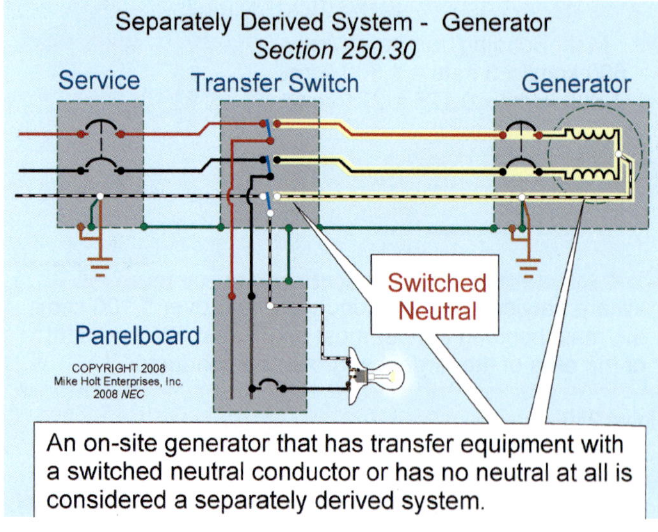

Figure 250–71

- Generators that supply transfer equipment that switches the neutral conductor are an example of a separately derived system. Figure 250–72

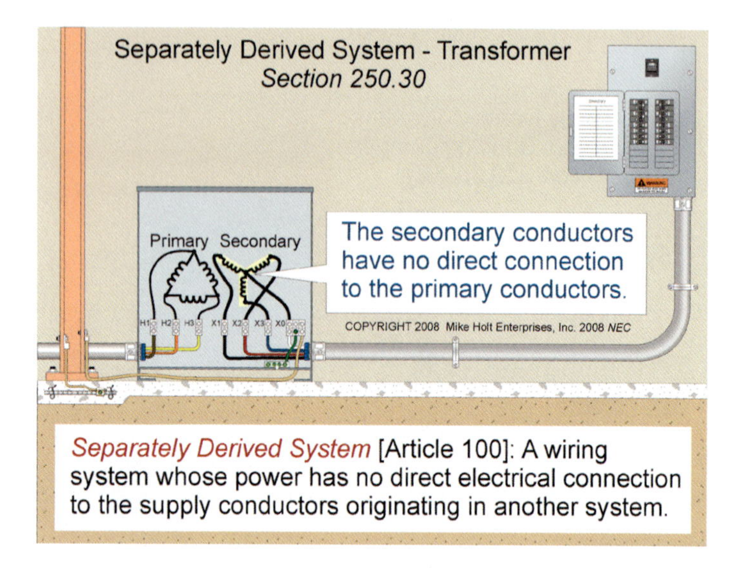

Figure 250–72

(A) Grounded Systems. Separately derived systems must be grounded (connected to the earth) and bonded in accordance with (1) through (8). A neutral-to-case connection must not be on the load side of the system bonding jumper, except as permitted by 250.142(B).

CAUTION: *Dangerous objectionable neutral current will flow on conductive metal parts of electrical equipment as well as metal piping and structural steel, in violation of 250.6(A), if more than one system bonding jumper is installed, or if it's not located where the grounding electrode conductor terminates to the neutral conductor.* Figure 250–73

(1) System Bonding Jumper. A system bonding jumper must be installed at the same location where the grounding electrode conductor terminates to the neutral terminal of the separately derived system; either at the separately derived system or the system disconnecting means, but not at both locations [250.30(A)(3)]. Figure 250–74

Author's Comment: A system bonding jumper is a conductor, screw, or strap that bonds the metal parts of a separately derived system to a system neutral point [250.2] and it's sized to Table 250.66 in accordance with 250.28(D).

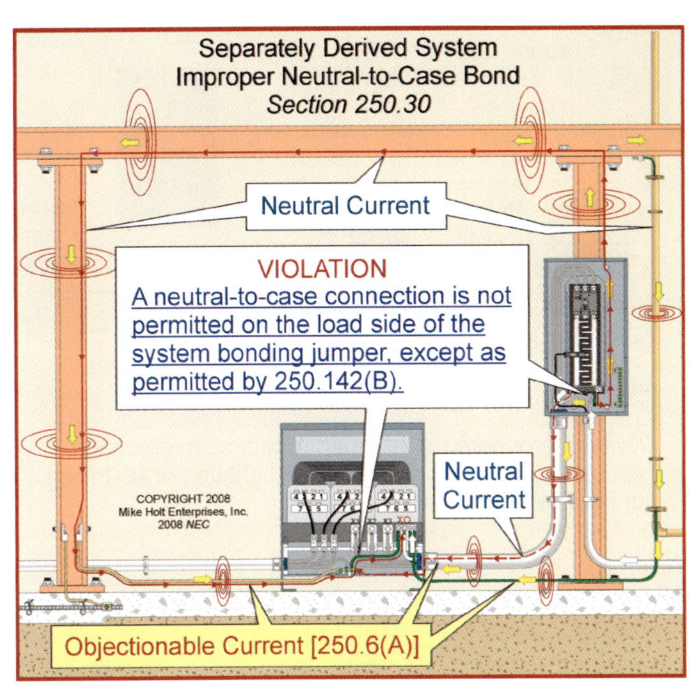

Figure 250–73

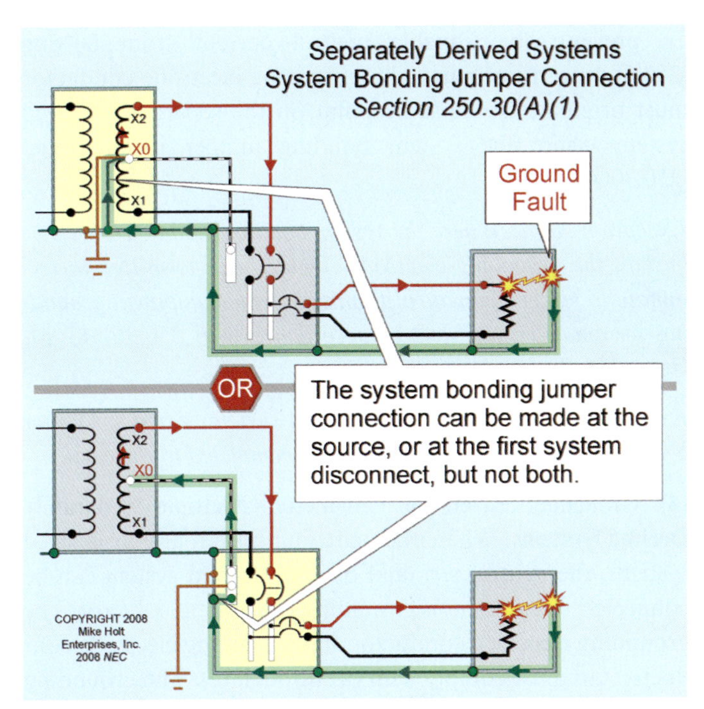

Figure 250–74

During a ground fault, metal parts of electrical equipment, as well as metal piping and structural steel, will become and remain energized providing the potential for electric shock and fire if the system bonding jumper is not installed. **Figure 250–75**

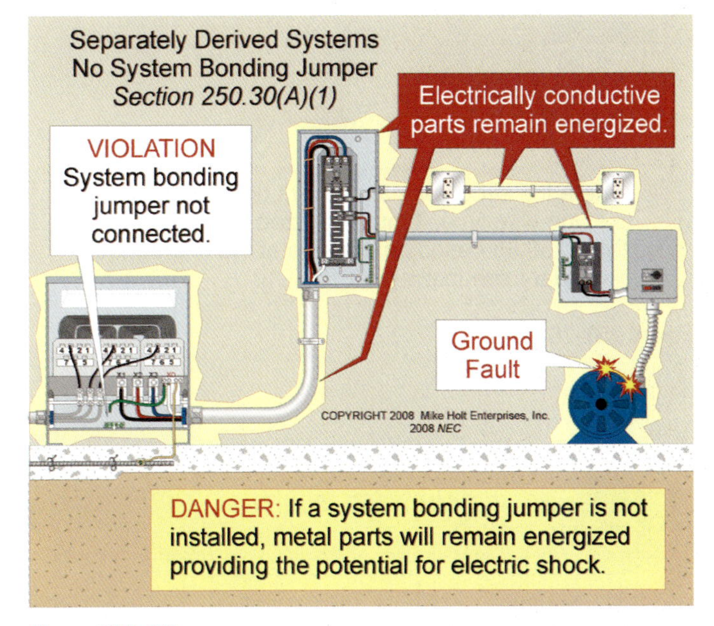

Figure 250–75

(2) Equipment Bonding Jumper Size. An equipment bonding jumper must be run to the secondary system disconnecting means. Where the secondary equipment grounding conductor is of the wire type, it must be sized in accordance with Table 250.66, based on the area of the largest ungrounded secondary conductor in the raceway or cable.

Question: *What size equipment bonding jumper is required for flexible metal conduit containing 300 kcmil secondary conductors?* **Figure 250–76**

(a) 3 AWG *(b) 2 AWG* *(c) 1 AWG* *(d) 1/0 AWG*

Answer: *(b) 2 AWG [Table 250.66]*

(3) Grounding—Single Separately Derived System. A grounding electrode conductor must connect the neutral terminal of a separately derived system to a grounding electrode of a type identified in 250.30(A)(7). The grounding electrode conductor must be sized in accordance with 250.66, based on the area of the ungrounded secondary conductor. **Figure 250–77**

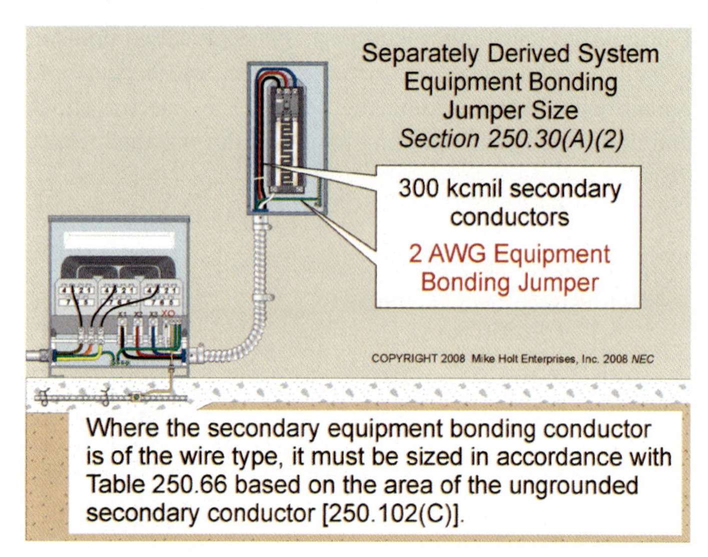

Figure 250–76

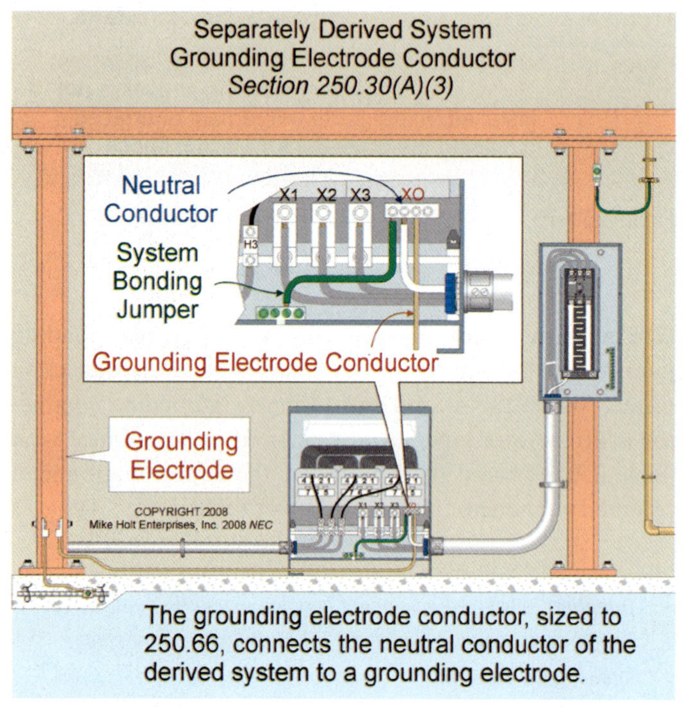

Figure 250–77

Author's Comments:

- System grounding is intended to reduce overvoltage caused by induction from indirect lightning, or restriking/intermittent ground faults. Induced voltage imposed from lightning can be reduced by short grounding conductors and eliminating unnecessary bends and loops [250.4(A)(1) FPN]. Figure 250–78

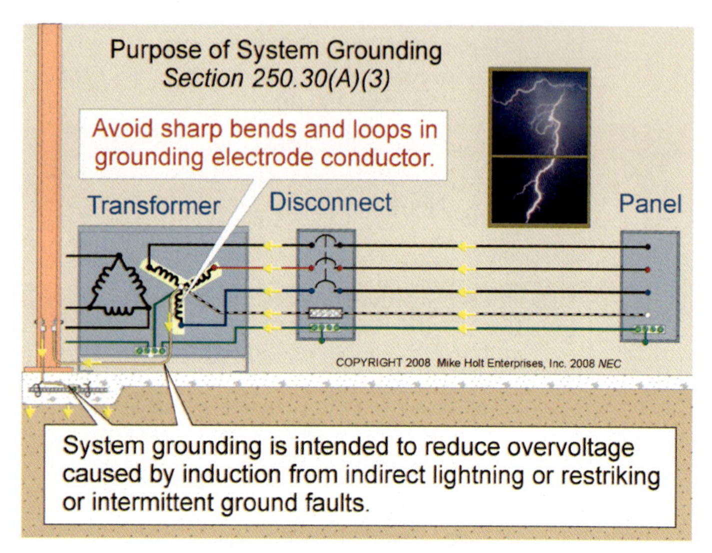

Figure 250–78

- System grounding also helps reduce fires in buildings as well as voltage stress on electrical insulation, thereby ensuring longer insulation life for motors, transformers, and other system components.

To prevent objectionable neutral current from flowing [250.6] onto metal parts, the grounding electrode conductor must originate at the same point on the separately derived system where the system bonding jumper is connected [250.30(A)(1)].

Exception No. 1: Where the system bonding jumper is a wire or busbar, the grounding electrode conductor is permitted to terminate to either the neutral terminal or the equipment grounding terminal, bar, or bus in accordance with 250.30(A)(1). Figure 250–79

Exception No. 3: Separately derived systems rated 1 kVA or less are not required to be grounded (connected to the earth).

(4) Grounding Electrode Conductor, Multiple Separately Derived Systems. Where there are multiple separately derived systems, the neutral terminal of each derived system can be connected to a common grounding electrode conductor. The grounding electrode conductor and grounding electrode conductor tap must comply with (a) through (c). The grounding electrode conductor and its taps must terminate at the same point on the separately derived system where the system bonding jumper is connected. Figure 250–80

Exception No. 1: Where the system bonding jumper is a wire or busbar, the grounding electrode conductor tap can terminate to either the neutral terminal or the equipment grounding terminal, bar, or bus in accordance with 250.30(A)(1).

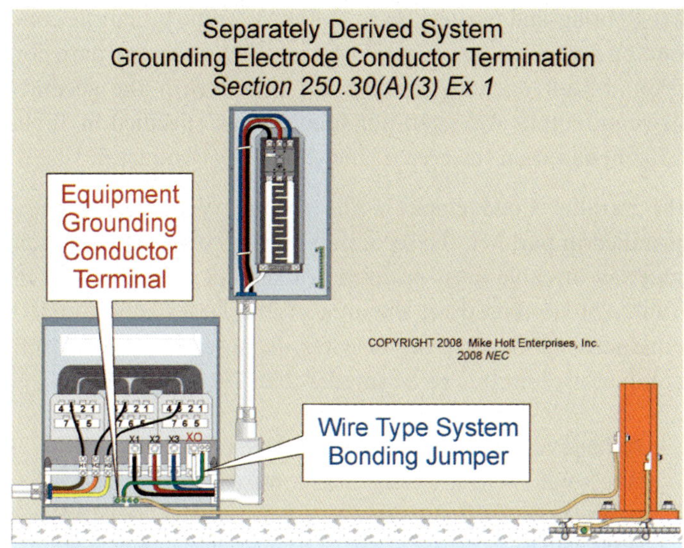

Figure 250–79

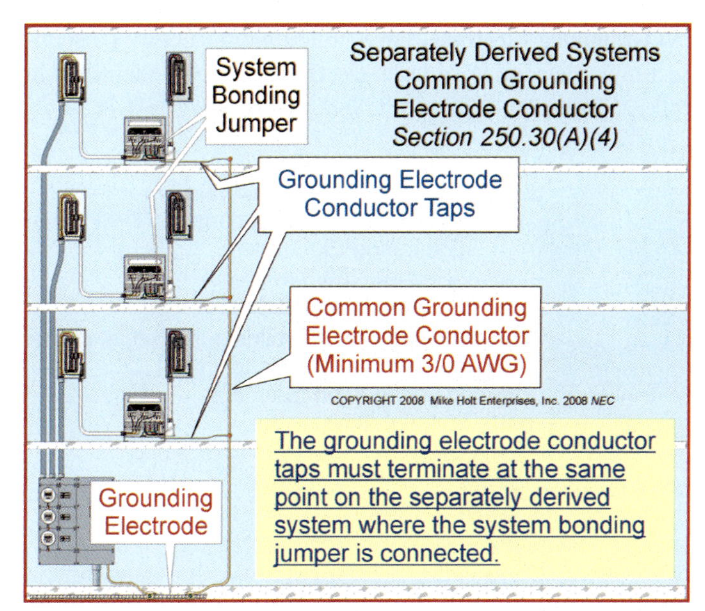

Figure 250–80

Exception No. 2: Separately derived systems rated 1 kVA or less are not required to be grounded (connected to the earth).

(a) Common Grounding Electrode Conductor. The common grounding electrode conductor must not be smaller than 3/0 AWG copper.

(b) Tap Conductor Size. Grounding electrode conductor taps must be sized in accordance with Table 250.66, based on the area of the largest ungrounded secondary conductor.

(c) Connections. Grounding electrode conductor tap connections must be made at an accessible location by:

(1) Listed connector.

(2) Listed connections to aluminum or copper busbars not less than ¼ x 2 in. Where aluminum busbars are used, the installation must comply with 250.64(A).

(3) Exothermic welding.

Grounding electrode conductor taps must be connected to the common grounding electrode conductor so the common grounding electrode conductor isn't spliced.

(5) Installation. The grounding electrode conductor must comply with the following:

- Be of copper where within 18 in. of earth [250.64(A)].
- Securely fastened to the surface on which it's carried [250.64(B)].
- Adequately protected if exposed to physical damage [250.64(B)].
- Metal enclosures enclosing a grounding electrode conductor must be made electrically continuous from the point of attachment to cabinets or equipment to the grounding electrode [250.64(E)].

(6) Structural Steel and Metal Piping. To ensure dangerous voltage from a ground fault is removed quickly, structural steel and metal piping in the area served by a separately derived system must be connected to the neutral conductor at the separately derived system in accordance with 250.104(D).

(7) Grounding Electrode. The grounding electrode must be as close as possible, and preferably in the same area where the system bonding jumper is located and be one of the following: **Figure 250–81**

(1) Metal water pipe electrode, within 5 ft of entry to the building [250.52(A)(1)].

(2) Metal building frame electrode [250.52(A)(2)].

Exception No. 1: Where the electrodes listed in (1) or (2) are not available, one of the following must be used:

- *A concrete-encased electrode encased by not less than 2 in. of concrete, located horizontally near the bottom or vertically, and within that portion of concrete foundation or footing that is in direct contact with the earth, consisting of at least 20 ft of electrically conductive steel reinforcing bars or rods of not less than ½ in. diameter [250.52(A)(3)].*

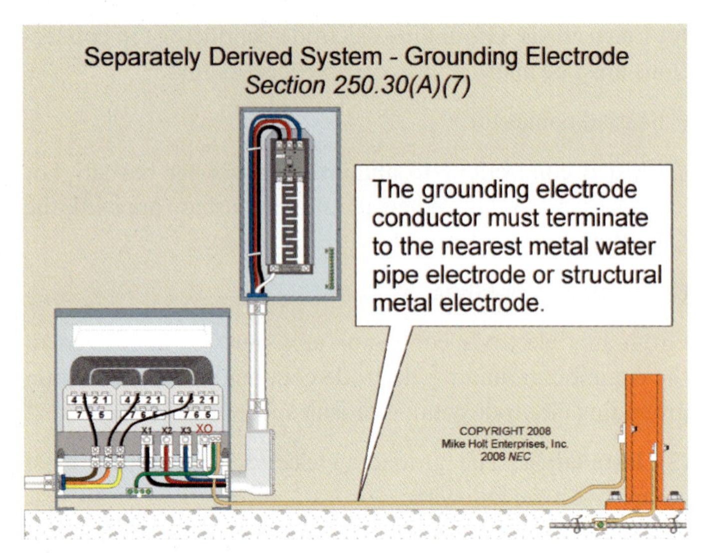

Figure 250–81

- *A ground ring electrode encircling the building or structure, buried not less than 30 in. below grade, consisting of at least 20 ft of bare copper conductor not smaller than 2 AWG [250.52(A)(4) and 250.53(F)].*
- *A ground rod electrode having not less than 8 ft of contact with the soil meeting the requirements of 250.56 [250.52(A)(5) and 250.53(G)].*
- *Other metal underground systems, piping systems, or underground tanks [250.52(A)(8)].*

(8) System Bonding at Disconnect. Where the system bonding jumper is installed at the disconnecting means instead of at the source, the following requirements apply: Figure 250–82

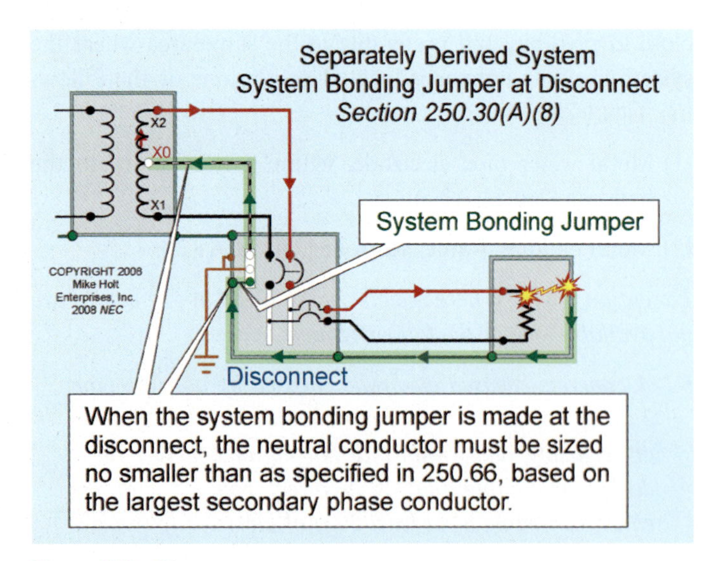

Figure 250–82

(a) Routing and Sizing. Because the secondary neutral conductor serves as the effective ground-fault current path for ground-fault current, it must be routed with the secondary conductors and sized not smaller than specified in Table 250.66, based on the area of the secondary conductor.

(b) Parallel Conductors. If the secondary conductors are installed in parallel, the secondary neutral conductor in each raceway or cable must be sized not smaller than specified in Table 250.66, based on the area of the largest ungrounded conductor in the raceway or cable. In no case is the neutral conductor permitted to be smaller than 1/0 AWG [310.4].

> **Author's Comment:** Where the system bonding jumper is installed at the disconnecting means instead of at the source, an equipment bonding conductor must connect the metal parts of the separately derived system to the neutral conductor at the disconnecting means in accordance with 250.30(A)(2).

250.32 Buildings or Structures Supplied by a Feeder or Branch Circuit.

(A) Grounding Electrode. Each building or structure's disconnect must be connected to an electrode of a type identified in 250.52. Figure 250–83

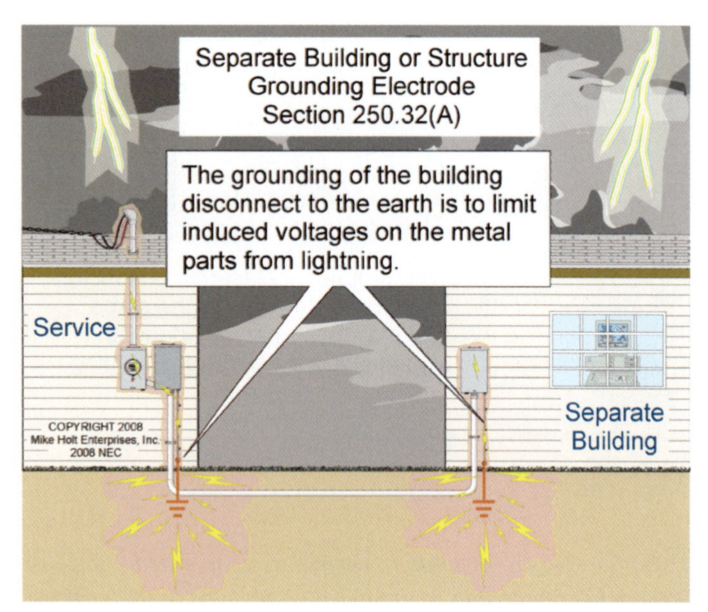

Figure 250–83

Author's Comments:

- The grounding of the building or structure disconnecting means to the earth is intended to help in limiting induced voltages on the metal parts from nearby lightning strikes [250.4(A)(1)].
- The *Code* prohibits the use of the earth to serve as an effective ground-fault current path [250.4(A)(5) and 250.4(B)(4)].

Exception: A grounding electrode isn't required where the building or structure is served with a 2-wire, 3-wire, or 4-wire multiwire branch circuit. **Figure 250–84**

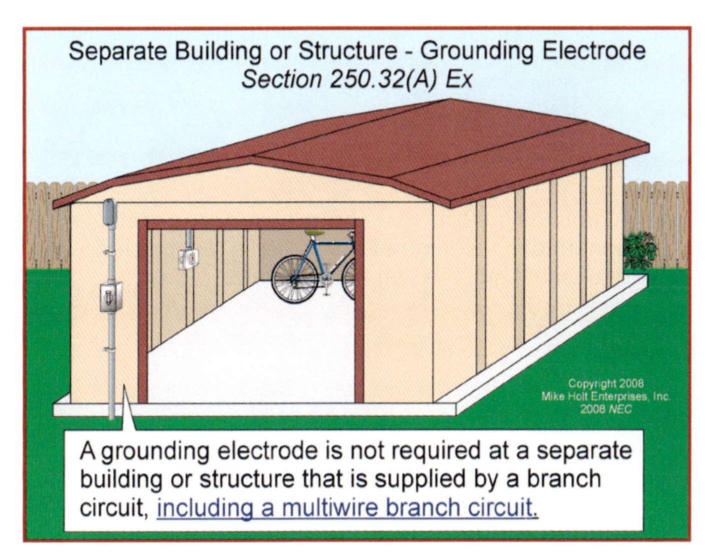

Figure 250–84

(B) Equipment Grounding Conductor. To quickly clear a ground fault and remove dangerous voltage from metal parts, the building or structure disconnecting means must be connected to the circuit equipment grounding conductor of a type described in 250.118. Where the supply circuit equipment grounding conductor is of the wire type, it must be sized in accordance with 250.122, based on the rating of the overcurrent device. **Figure 250–85**

CAUTION: *To prevent dangerous objectionable neutral current from flowing onto metal parts [250.6(A)], the supply circuit neutral conductor is not permitted to be connected to the remote building or structure disconnecting means [250.142(B)].* **Figure 250–86**

Exception: For existing premises, when an equipment grounding conductor was not run to the building or structure disconnecting means, the building or structure disconnecting means

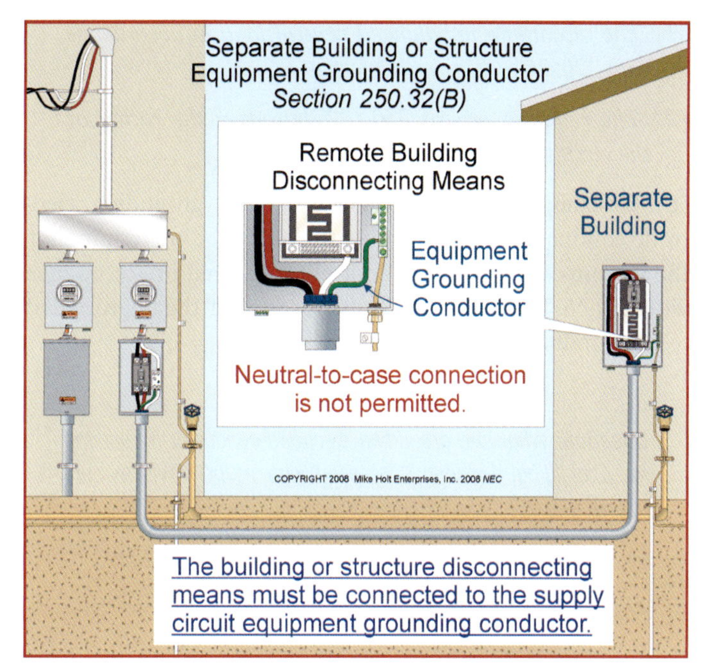

Figure 250–85

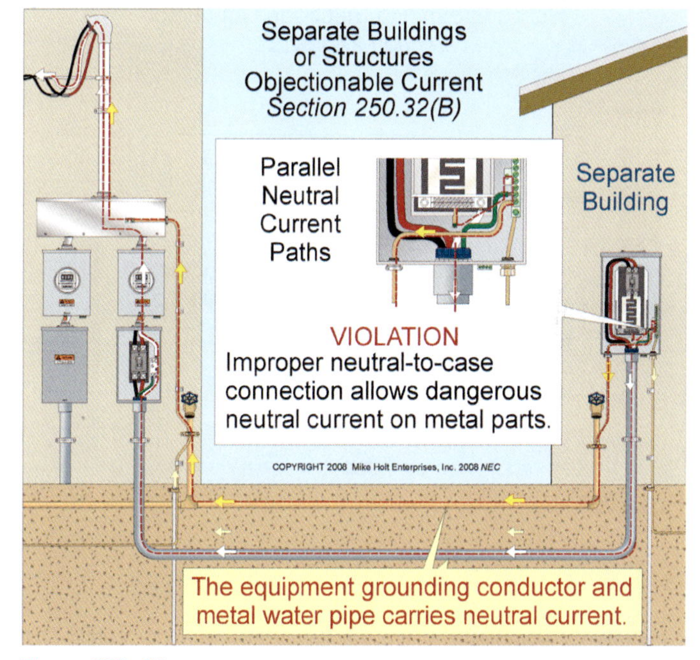

Figure 250–86

can remain connected to the neutral conductor where there are no continuous metallic paths between buildings and structures, ground-fault protection of equipment isn't installed on the supply side of the circuit, and the neutral conductor is sized no smaller than the larger of:

(1) The maximum unbalanced neutral load in accordance with 220.61.

(2) The requirements of 250.122, based on the rating of the circuit overcurrent device

(E) Grounding Electrode Conductor. The grounding electrode conductor must terminate to the grounding terminal of the disconnecting means, and it must be sized in accordance with 250.66, based on the conductor area of the ungrounded feeder conductor.

Question: *What size grounding electrode conductor is required for a building disconnect supplied with a 3/0 AWG feeder?* Figure 250–87

(a) 4 AWG *(b) 3 AWG* *(c) 2 AWG* *(d) 1 AWG*

Answer: *(a) 4 AWG [Table 250.66]*

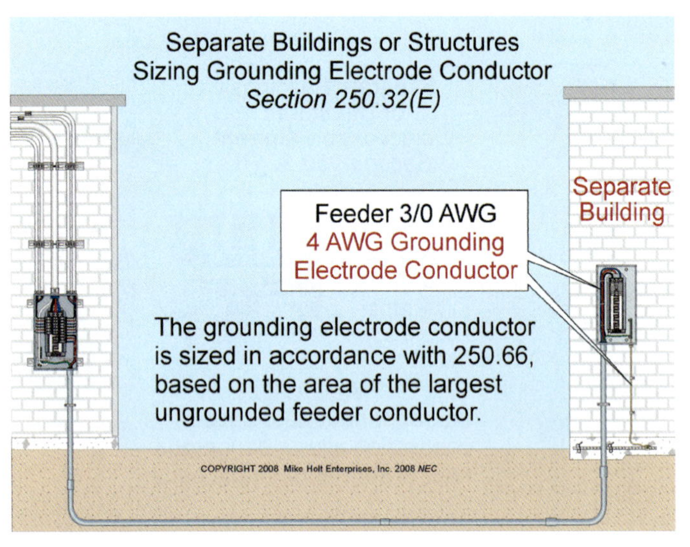

Figure 250–87

Author's Comment: Where the grounding electrode conductor is connected to a ground rod, that portion of the conductor that is the sole connection to the ground rod isn't required to be larger than 6 AWG copper [250.66(A)]. Where the grounding electrode conductor is connected to a concrete-encased electrode, that portion of the conductor that is the sole connection to the concrete-encased electrode isn't required to be larger than 4 AWG copper [250.66(B)].

250.34 Generators—Portable and Vehicle-Mounted.

(A) Portable Generators. The frame of a portable generator isn't required to be grounded (connected to the earth) if: Figure 250–88

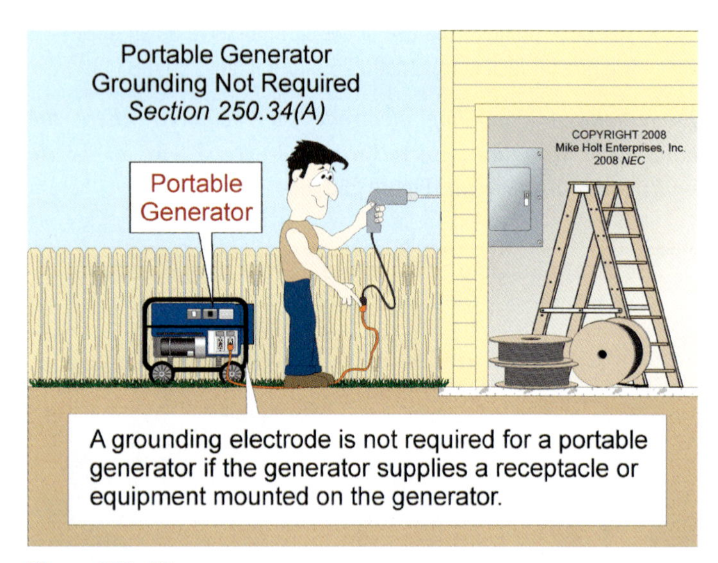

Figure 250–88

(1) The generator only supplies equipment or receptacles mounted on the generator, and

(2) The metal parts of the generator and the receptacle grounding terminal are connected to the generator frame.

(B) Vehicle-Mounted Generators. The frame of a vehicle-mounted generator isn't required to be grounded (connected to the earth) if: Figure 250–89

(1) The generator frame is bonded to the vehicle frame,

(2) The generator only supplies equipment or receptacles mounted on the vehicle or generator, and

(3) The metal parts of the generator and the receptacle grounding terminal are connected to the generator frame.

(C) Separately Derived Portable or Vehicle-Mounted Generator. A portable or vehicle-mounted generator used as a separately derived system to supply equipment or receptacles mounted on the vehicle or generator must have the neutral conductor connected to the generator frame.

FPN: A portable or vehicle-mounted generator supplying fixed wiring of premises must be grounded (connected to the earth) and bonded in accordance with 250.30 for separately derived systems and 250.35 for nonseparately derived systems.

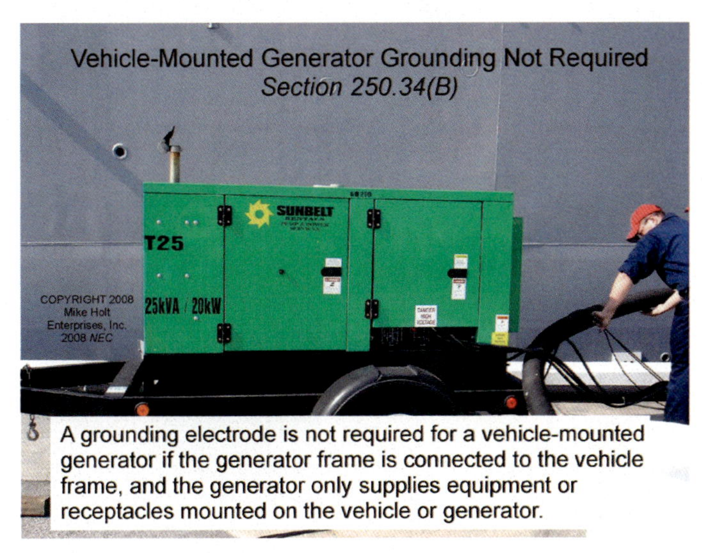

Figure 250–89

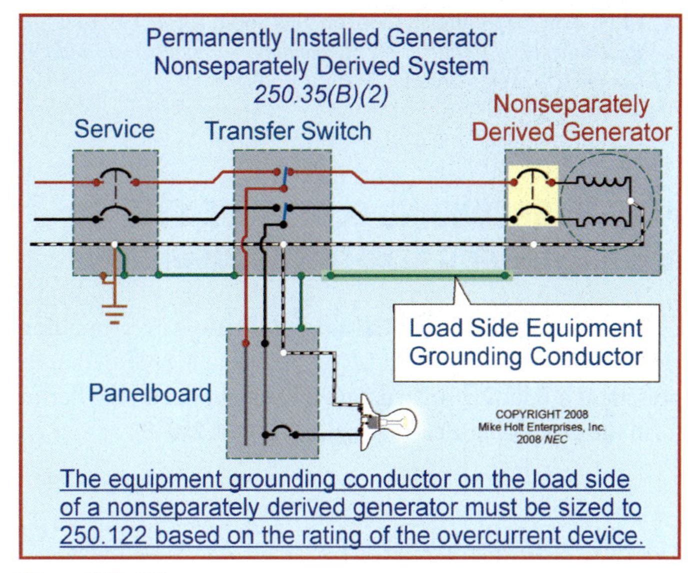

Figure 250–90

250.35 Permanently Installed Generators.

(A) Separately Derived System. Where the generator is installed as a separately derived system, the system must be grounded (connected to the earth) and bonded in accordance with 250.30.

(B) Nonseparately Derived System. An equipment bonding jumper must be installed according to (1) or (2) as follows.

(1) Supply Side of Disconnect. An equipment bonding jumper must be installed from the generator equipment grounding terminal to the equipment grounding terminal of the generator disconnecting means. The bonding jumper must be of the wire type and be sized in accordance with Table 250.66, based on the cross-sectional area of the ungrounded generator conductors [250.102(C)].

(2) Load Side. The circuit equipment grounding conductor on the load side of each generator overcurrent device is required [250.86], and where of the wire type, it must be sized in accordance with 250.122, based on the rating of the overcurrent device [250.102(D)]. Figure 250–90

> **Author's Comment:** The frame of a nonseparately derived system generator is not required to be connected to a grounding electrode.

250.36 High-Impedance Grounded Systems.

High-impedance grounded systems are only permitted for three-phase systems where all of the following conditions are met:

(1) Conditions of maintenance and supervision ensure that only qualified persons service the installation.

(2) Ground detectors are installed on the system [250.21(B)].

(3) Line-to-neutral loads aren't served.

> **Author's Comment:** High-impedance grounded systems are generally referred to as "high-resistance grounded systems" in the industry.

(A) Grounding Impedance Location. To limit fault current to a very low value, high-impedance grounded systems have a resistor installed between the neutral point of the derived system and the grounding electrode conductor. Figure 250–91

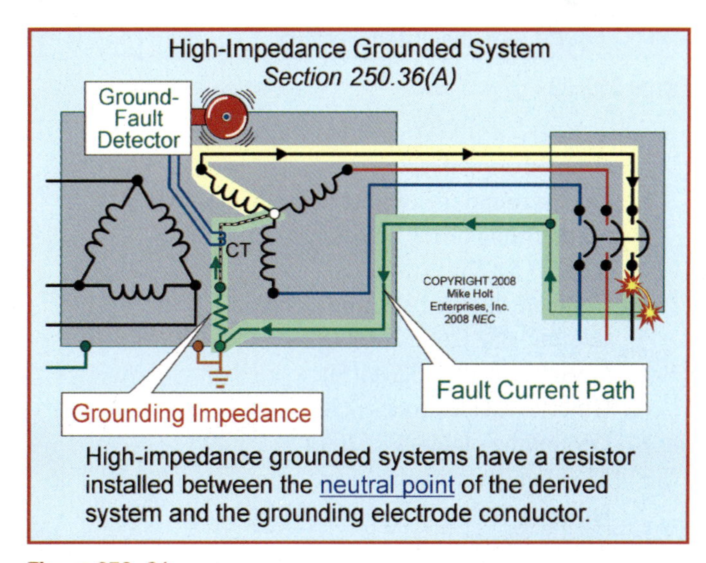

Figure 250–91

FPN: For more information on this topic see IEEE 142, *Recommended Practice for Grounding of Industrial and Commercial Power Systems* (Green Book).

PART III. GROUNDING ELECTRODE SYSTEM AND GROUNDING ELECTRODE CONDUCTOR

250.50 Grounding Electrode System. Any grounding electrode described in 250.52(A)(1) through (A)(8) that is present at a building or structure must be bonded together to form the grounding electrode system. Figure 250–92

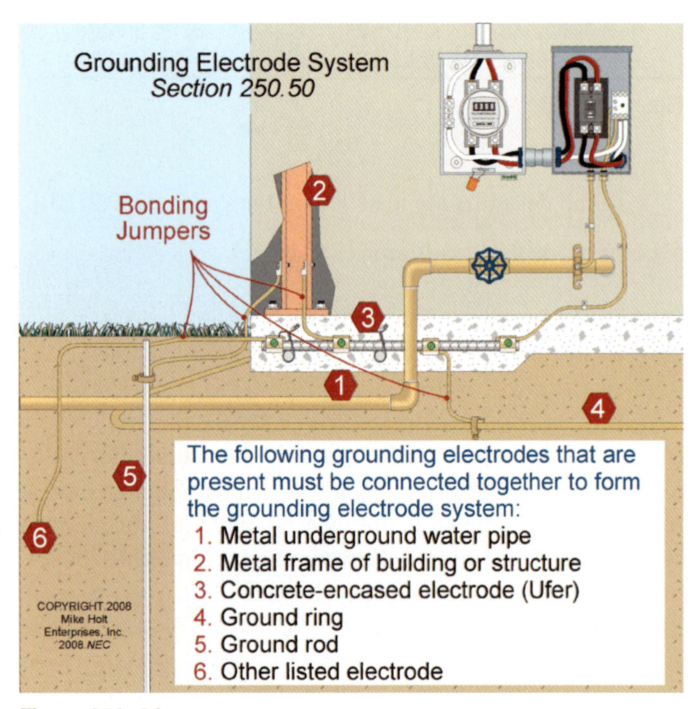

Figure 250–92

- Underground metal water pipe [250.52(A)(1)]
- Metal frame of the building or structure [250.52(A)(2)]
- Concrete-encased electrode [250.52(A)(3)]
- Ground ring [250.52(A)(4)]
- Ground rod [250.52(A)(5)]
- Other listed electrodes [250.52(A)(6)]
- Grounding plate [250.52(A)(7)]
- Metal underground systems, piping systems, or underground tanks [250.52(A)(8)].

Exception: Concrete-encased electrodes are not required for existing buildings or structures where the conductive steel reinforcing bars aren't accessible without chipping up the concrete. Figure 250–93

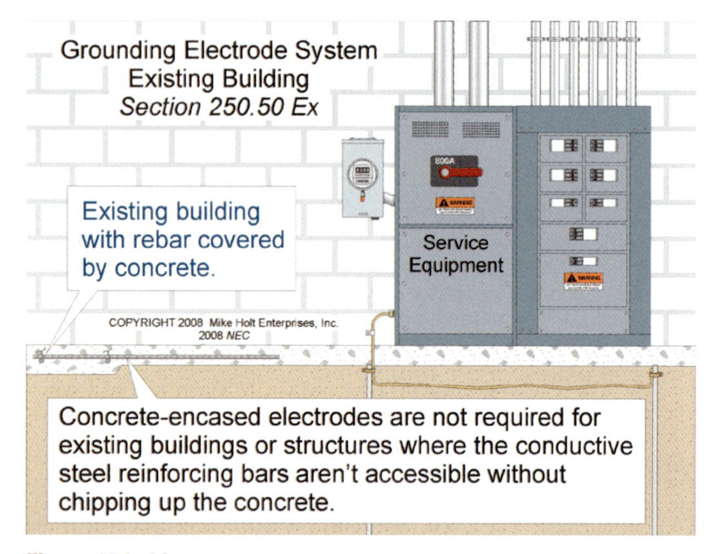

Figure 250–93

250.52 Grounding Electrode Types.

(A) Electrodes Permitted for Grounding.

(1) Underground Metal Water Pipe Electrode. Underground metal water pipe in direct contact with earth for 10 ft or more can serve as a grounding electrode. Figure 250–94

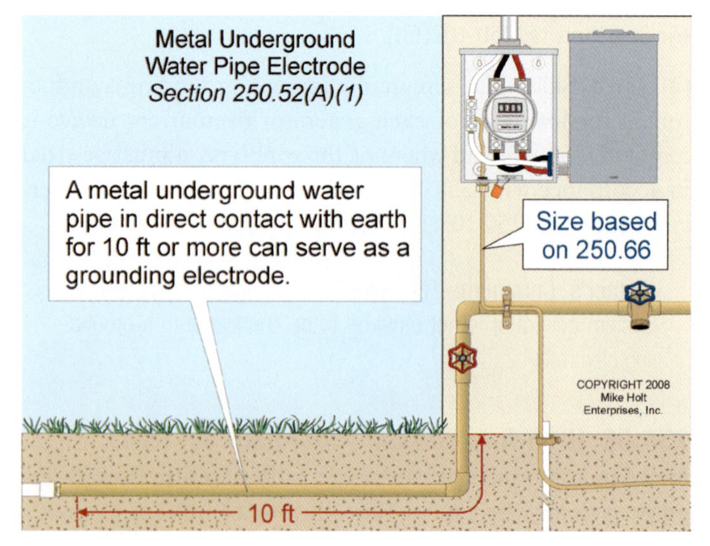

Figure 250–94

Author's Comment: The grounding electrode conductor to the water pipe electrode must be sized in accordance with Table 250.66.

An underground metal water pipe electrode that may be interrupted, such as with a water meter, must be made electrically continuous with a bonding jumper sized according to 250.66, based on the area of the ungrounded service conductors [250.68(B)].

Interior metal water piping located more than 5 ft from the point of entrance to the building or structure can't be used to interconnect electrodes that are part of the grounding electrode system.

Exception: In industrial, institutional, and commercial buildings where conditions of maintenance and supervision ensure only qualified persons service the installation, the entire length of the metal water piping system can be used for grounding purposes, provided the entire length, other than short sections passing through walls, floors, or ceilings, is exposed.

Author's Comment: Controversy about using metal underground water supply piping as a grounding electrode has existed since the early 1900s. The water industry believes that neutral current flowing on water piping corrodes the metal. For more information, contact the American Water Works Association about their report *Effects of Electrical Grounding on Pipe Integrity and Shock Hazard*, Catalog No. 90702, 1-800-926-7337. Figure 250–95

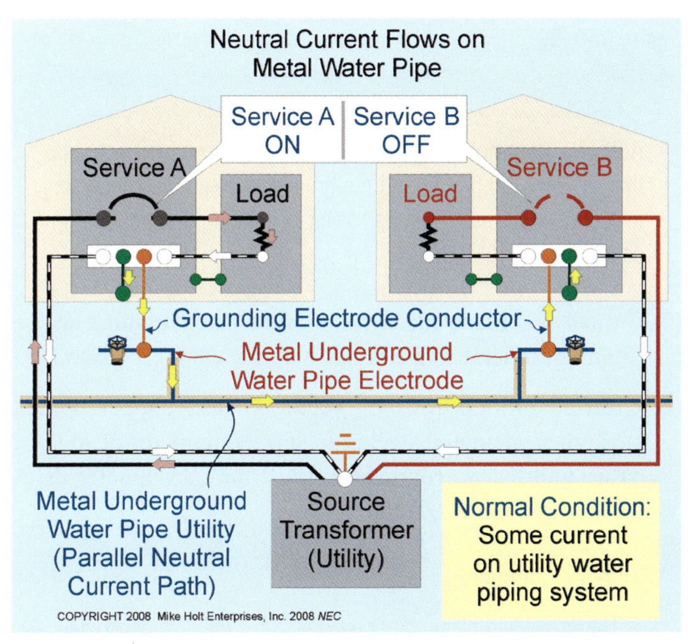

Figure 250–95

(2) Metal Building Frame Electrode. The metal frame of a building or structure can serve as a grounding electrode when it meets one of the following conditions:

(1) A single structural metal member (10 ft or more) is in direct contact with the earth or encased in concrete that is in direct contact with the earth,

(2) The structural metal frame is connected to a concrete-encased electrode as provided in 250.52(A)(3) or a ground ring as provided in 250.52(A)(4), Figure 250–96

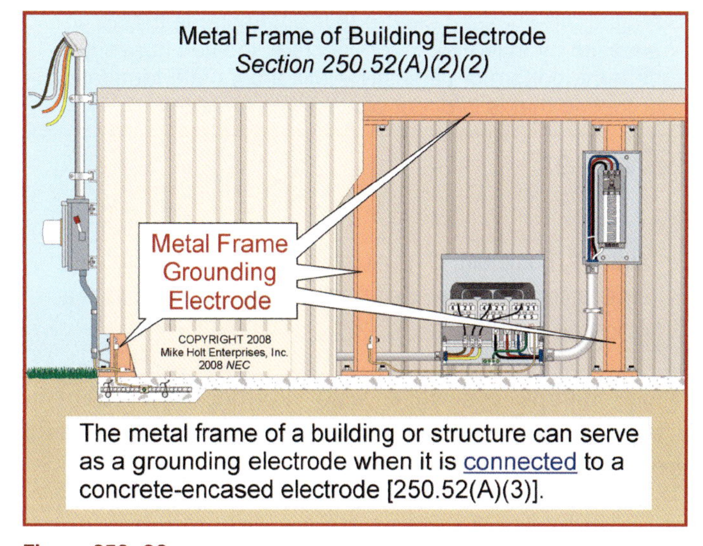

Figure 250–96

(3) The structural metal frame is connected to a rod electrode [250.52(A)(5)] that has a contact resistance of 25 ohms or less [250.56], or

(4) A single structural metal member is connected to the earth by other approved means.

(3) Concrete-Encased Grounding Electrode. A concrete-encased electrode is an electrode that is encased by at least 2 in. of concrete, located horizontally near the bottom or vertically within a concrete foundation or footing that is in direct contact with the earth consisting of one of the following: Figure 250–97

- Twenty feet of one or more bare, zinc-galvanized, or otherwise electrically conductive steel reinforcing bars mechanically connected together by steel tie wires not less than ½ in. in diameter, or
- Twenty feet of bare copper conductor not smaller than 4 AWG.

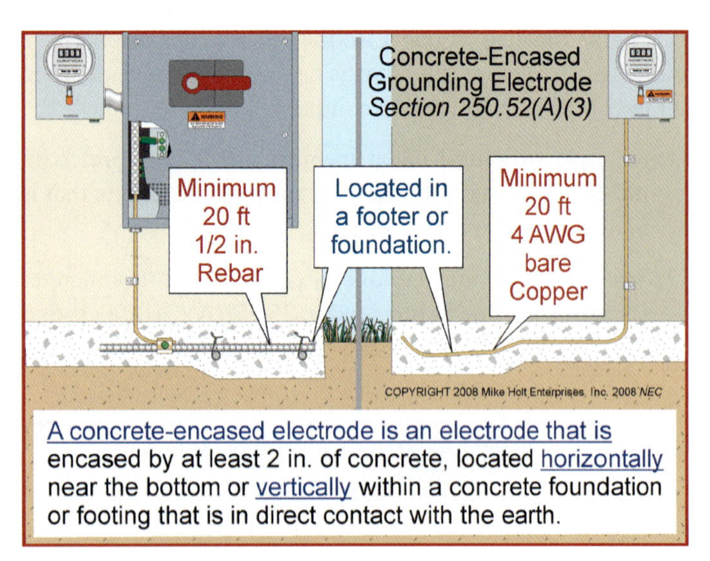

Figure 250–97

Author's Comment: If a moisture/vapor barrier is installed under a concrete footer, then the steel rebar can't be part of a concrete-encased electrode.

Where multiple concrete-encased electrodes are present at a building or structure, only one is required to serve as a grounding electrode. **Figure 250–98**

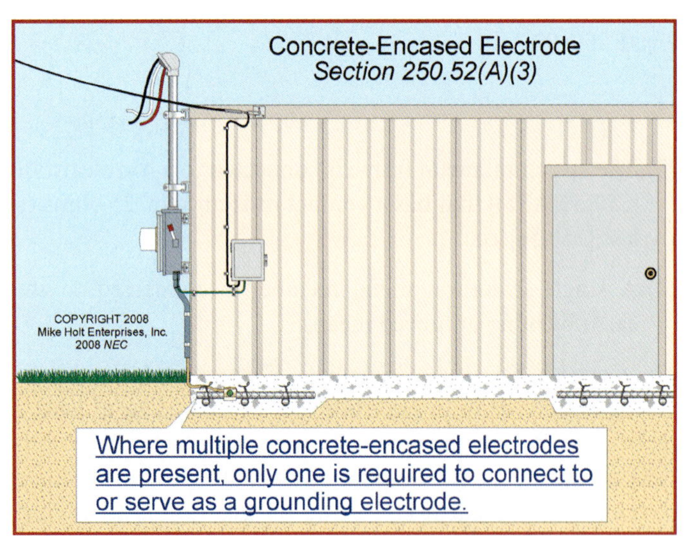

Figure 250–98

Author's Comments:

- The grounding electrode conductor to a concrete-encased grounding electrode isn't required to be larger than 4 AWG copper [250.66(B)].
- The concrete-encased grounding electrode is also called a "Ufer Ground," named after a consultant working for the U.S. Army during World War II. The technique Mr. Ufer came up with was necessary because the site needing grounding had no underground water table and little rainfall. The desert site was a series of bomb storage vaults in the area of Flagstaff, Arizona. This type of grounding electrode generally offers the lowest ground resistance for the cost.

(4) Ground Ring Electrode. A bare copper conductor, not smaller than 2 AWG buried in the earth encircling a building or structure, can serve as a grounding electrode. **Figure 250–99**

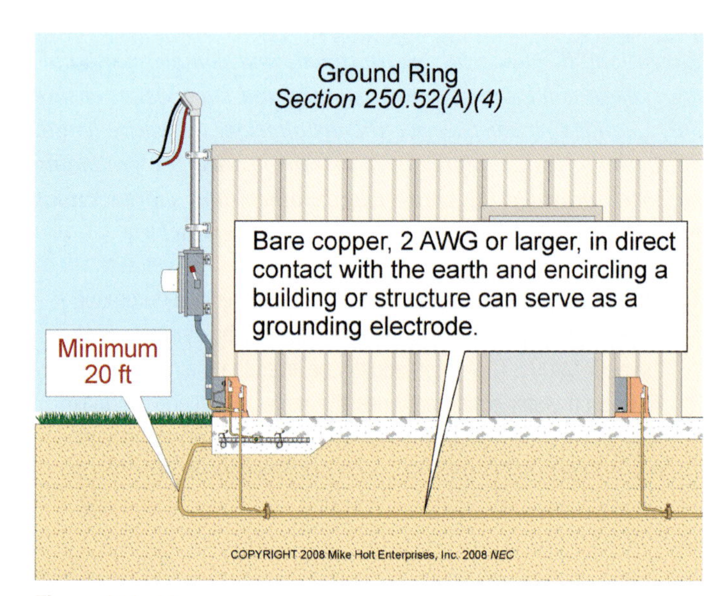

Figure 250–99

Author's Comment: The ground ring must be buried not less than 30 in. [250.53(F)], and the grounding electrode conductor to a ground ring isn't required to be larger than the ground ring conductor size [250.66(C)].

(5) Ground Rod Electrode. Ground rod electrodes must not be less than 8 ft in length in contact with the earth [250.53(G)].

(b) Unlisted ground rods of stainless steel, copper coated steel, or zinc coated steel must have a diameter of at least ⅝ in. and listed ground rods must have a diameter of at least ½ in. **Figure 250–100**

Author's Comments:

- The grounding electrode conductor, if it's the sole connection to the ground rod, is not required to be larger than 6 AWG copper [250.66(A)].

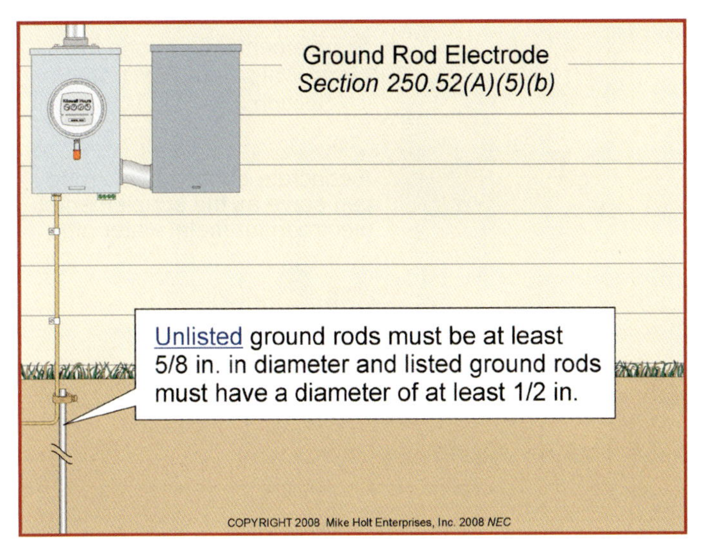

Figure 250–100

- The diameter of a ground rod has an insignificant effect on the contact resistance of a ground rod to the earth. However, larger diameter ground rods (¾ in. and 1 in.) are sometimes installed where mechanical strength is desired, or to compensate for the loss of the electrode's metal due to corrosion.

(6) Listed Electrode. Other listed grounding electrodes.

(7) Ground Plate Electrode. A buried iron or steel plate with not less than ¼ in. of thickness, or a copper metal plate not less than 0.06 in. of thickness, with an exposed surface area not less than 2 sq ft.

(8) Metal Underground Systems Electrode. Metal underground piping systems, underground tanks, and underground metal well casings can serve as a grounding electrode.

Author's Comment: The grounding electrode conductor to the metal underground system must be sized according to Table 250.66.

(B) Not Permitted for Use as Grounding Electrode.

(1) Underground metal gas piping systems. Figure 250–101

250.53 Grounding Electrode Installation Requirements.

(A) Rod Electrodes. Where practicable, rod electrodes must be embedded below permanent moisture level.

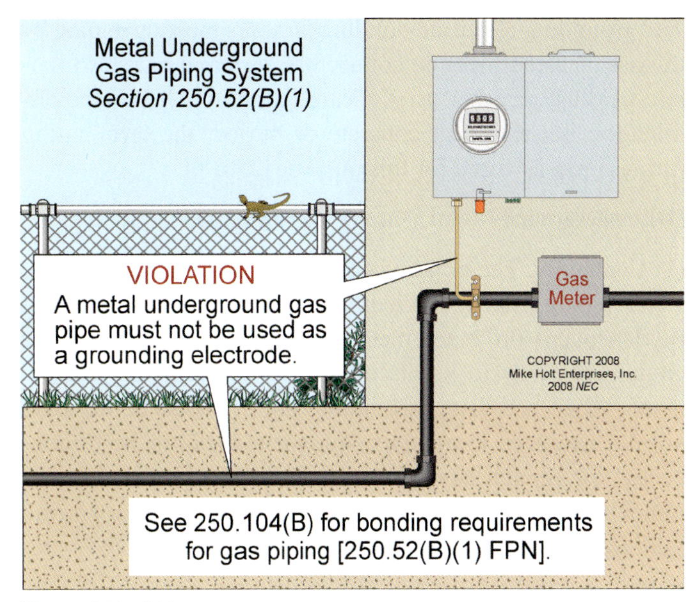

Figure 250–101

(B) Electrode Spacing. Electrodes for power systems must not be less than 6 ft from any other electrode of another grounding system. Two or more grounding electrodes that are bonded together are considered a single grounding electrode system.

(C) Grounding Electrode Bonding Jumper. Grounding electrode bonding jumpers must be copper when within 18 in. of earth [250.64(A)], be securely fastened to the surface, and be protected if exposed to physical damage [250.64(B)]. The bonding jumper to each electrode must be sized according to 250.66. Figure 250–102

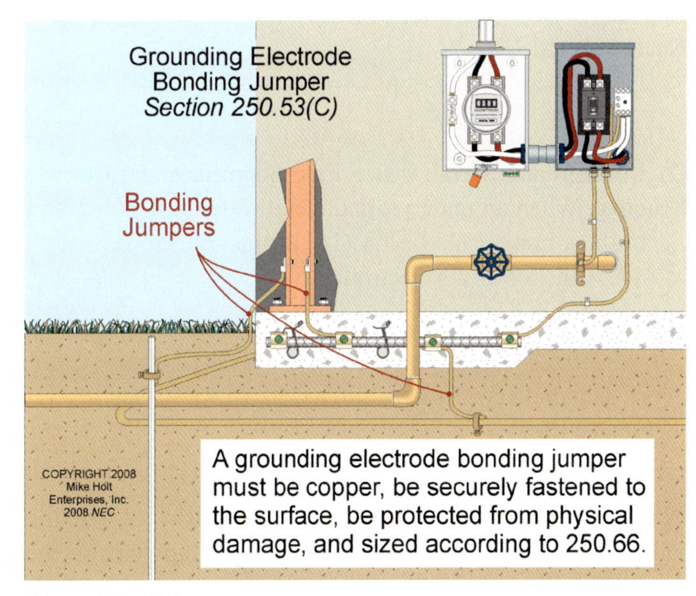

Figure 250–102

The grounding electrode bonding jumpers must terminate by the use of listed pressure connectors, terminal bars, exothermic welding, or other listed means [250.8(A)]. When the termination is encased in concrete or buried, the termination fittings must be listed for this purpose [250.70].

(D) Underground Metal Water Pipe Electrode.

(1) Continuity. The bonding connection to the interior metal water piping system, as required by 250.104(A), must not be dependent on water meters, filtering devices, or similar equipment likely to be disconnected for repairs or replacement. When necessary, a bonding jumper must be installed around insulated joints and equipment likely to be disconnected for repairs or replacement to assist in clearing and removing dangerous voltage on metal parts due to a ground fault [250.68(B)]. Figure 250–103

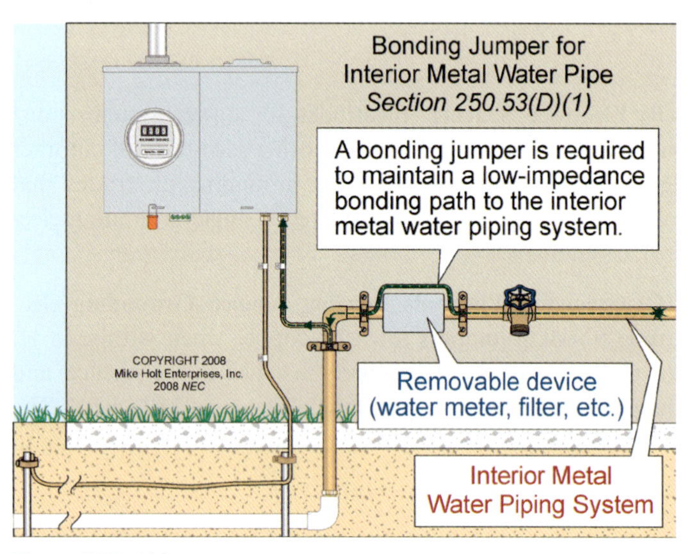

Figure 250–103

(2) Underground Metal Water Pipe Supplemental Electrode Required. When an underground metal water pipe grounding electrode is present [250.52(A)(1)], it must be supplemented by one of the following electrodes:

- Metal frame of the building or structure electrode [250.52(A)(2)]
- Concrete-encased electrode [250.52(A)(3)] Figure 250–104
- Ground ring electrode [250.52(A)(4)]

Where none of the above electrodes are present, one of the following electrodes must be installed to supplement the water pipe electrode:

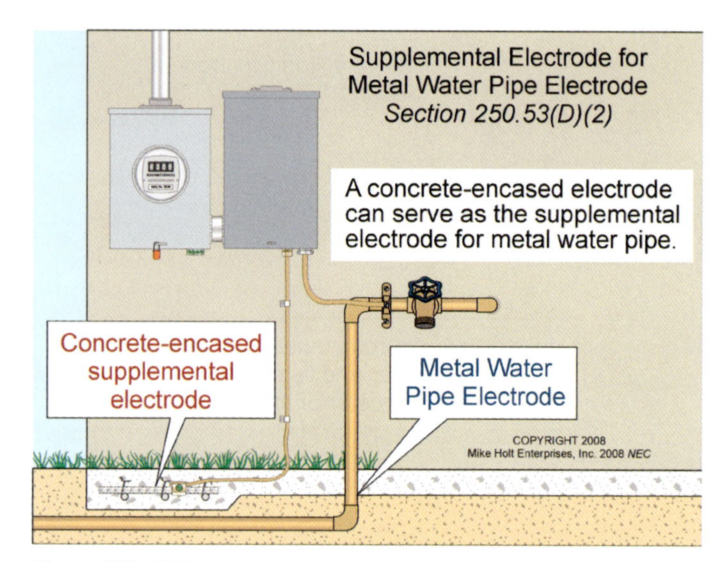

Figure 250–104

- Ground rod electrode meeting the requirements of 250.56 [250.52(A)(5)]
- Other listed electrodes [250.52(A)(6)]
- Metal underground systems, piping systems, or underground tanks [250.52(A)(8)]

The termination of the supplemental grounding electrode conductor must be to one of the following locations: Figure 250–105

- Grounding electrode conductor
- Service neutral conductor
- Metal service raceway
- Service equipment enclosure

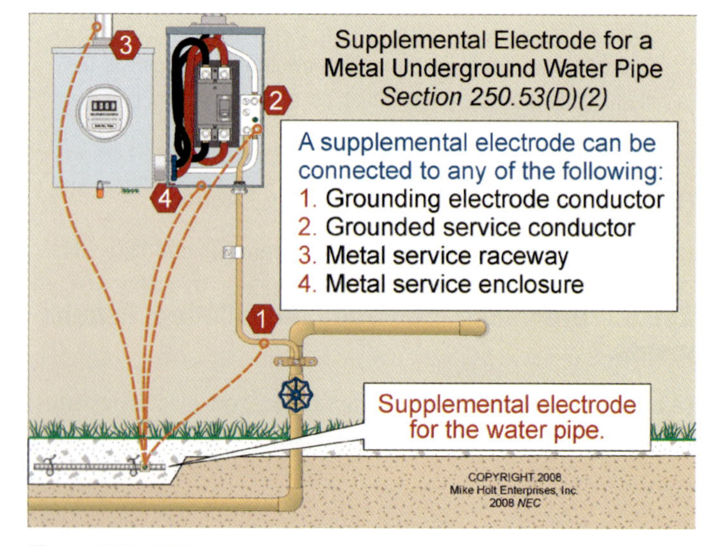

Figure 250–105

(E) Supplemental Ground Rod Electrode. The grounding electrode conductor to a ground rod that serves as a supplemental electrode isn't required to be larger than 6 AWG copper.

(F) Ground Ring. A ground ring encircling the building or structure, consisting of at least 20 ft of bare copper conductor not smaller than 2 AWG, must be buried not less than 30 in. [250.52(A)(4)]. Figure 250–106

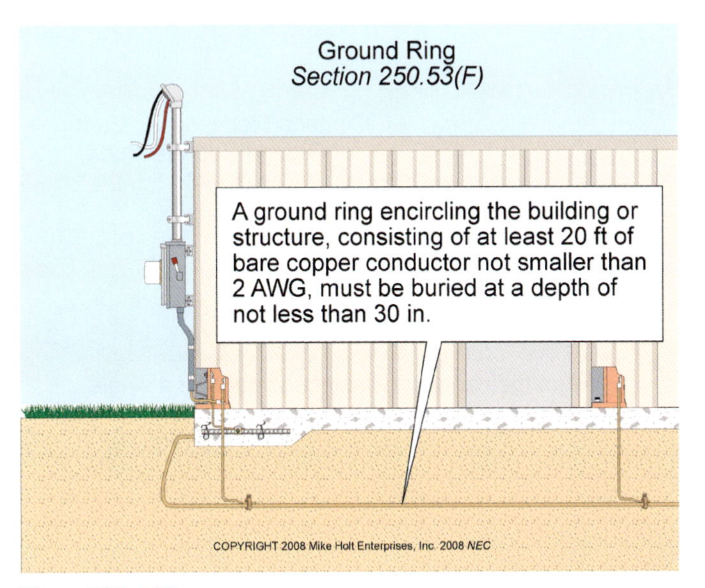

Figure 250–106

(G) Ground Rod Electrodes. Ground rod electrodes must be installed so that not less than 8 ft of length is in contact with the soil. Where rock bottom is encountered, the ground rod must be driven at an angle not to exceed 45 degrees from vertical. If rock bottom is encountered at an angle up to 45 degrees from vertical, the ground rod can be buried in a minimum 30 in. deep trench. Figure 250–107

The upper end of the ground rod must be flush with or underground unless the grounding electrode conductor attachment is protected against physical damage as specified in 250.10.

> **Author's Comment:** When the grounding electrode attachment fitting is located underground, it must be listed for direct soil burial [250.68(A) Ex 1 and 250.70].

250.54 Auxiliary Grounding Electrodes.

Auxiliary electrodes can be connected to the circuit equipment grounding conductor. They are not required to be bonded to the building or structure grounding electrode system, the grounding conductor to the electrode is not required to be sized to 250.66, and its contact resistance to the earth is not required to comply with the 25 ohm requirement of 250.56. Figures 250–108 and 250–109

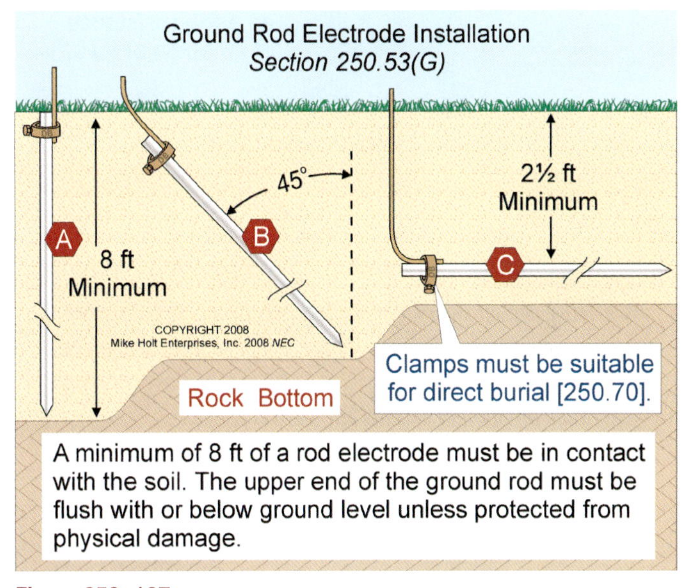

Figure 250–107

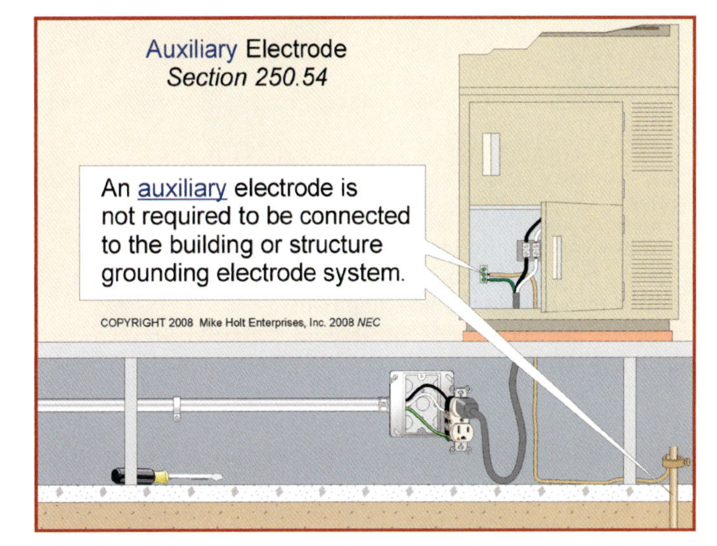

Figure 250–108

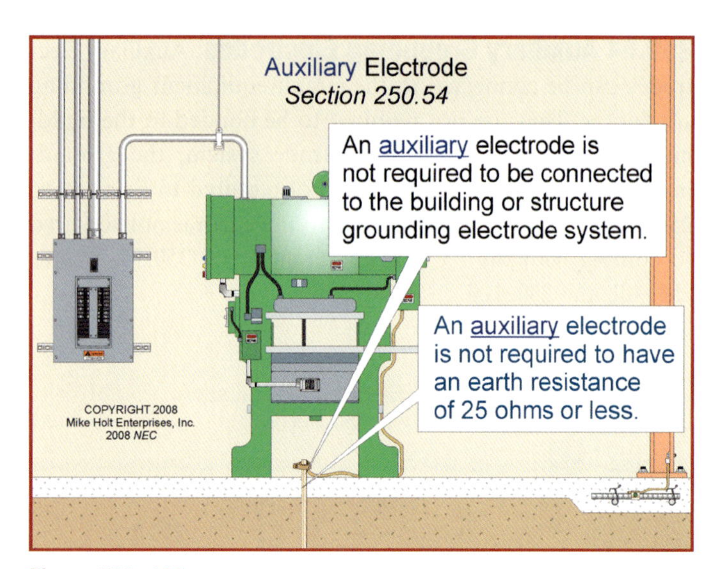

Figure 250–109

CAUTION: *An auxiliary electrode typically serves no useful purpose, and in some cases it may actually cause equipment failures by providing a path for lightning to travel through electronic equipment.* Figure 250–110

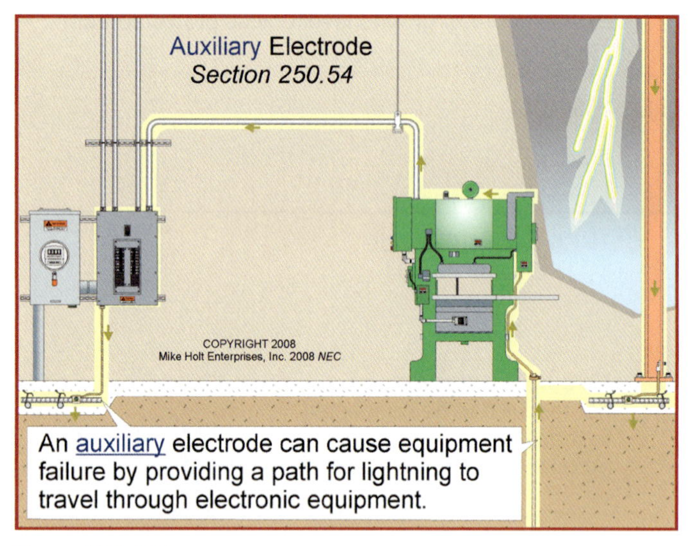

Figure 250–110

The earth must not be used as the effective ground-fault current path required by 250.4(A)(5).

DANGER: *Because the contact resistance of an electrode to the earth is so great, very little fault current returns to the power supply if the earth is the only fault current return path. Result—the circuit overcurrent device will not open and clear the ground fault, and all metal parts associated with the electrical installation, metal piping, and structural building steel will become and remain energized.*

250.56 Contact Resistance of Ground Rod to the Earth. When the grounding electrode system consists of a single ground rod having a contact resistance to the earth of over 25 ohms, it must be augmented with an additional electrode located not less than 6 ft away. Figure 250–111

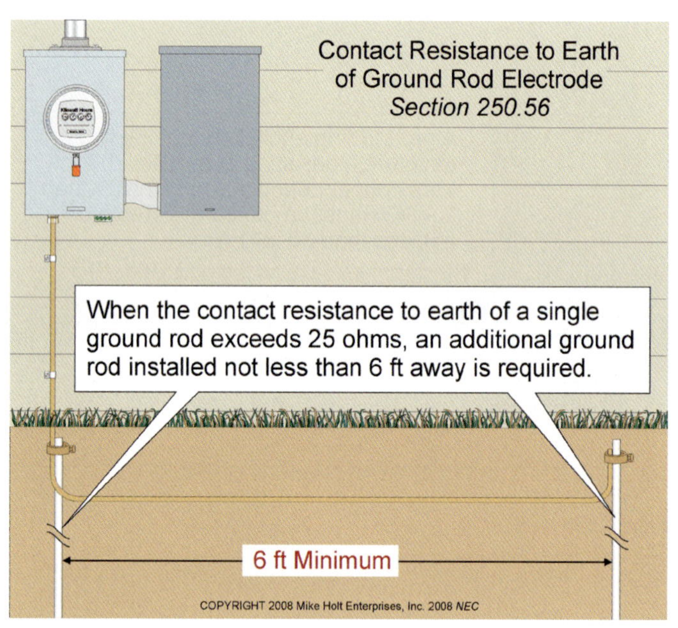

Figure 250–111

Author's Comment: If the contact resistance of two ground rods to the earth exceeds 25 ohms, no additional electrodes are required.

MEASURING THE GROUND RESISTANCE

A ground resistance clamp meter, or a three-point fall of potential ground resistance meter, can be used to measure the contact resistance of a grounding electrode to the earth.

Ground Clamp Meter. The ground resistance clamp meter measures the contact resistance of the grounding system to the earth by injecting a high-frequency signal via the service neutral conductor to the utility ground, and then measuring the strength of the return signal through the earth to the grounding electrode being measured. Figure 250–112

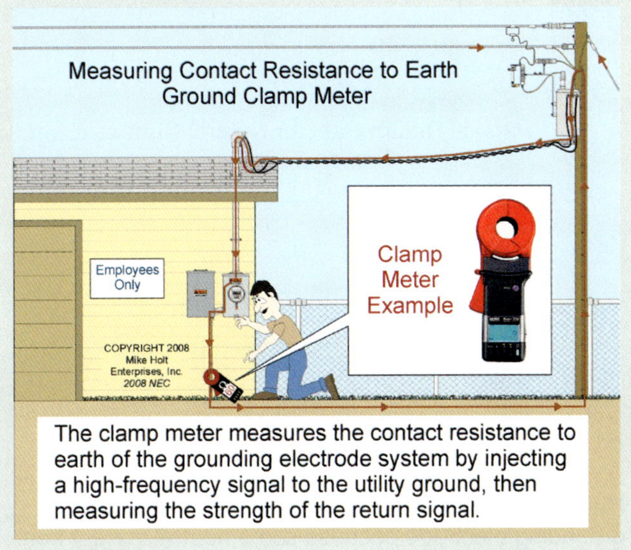

Figure 250–112

Fall of Potential Ground Resistance Meter. The three-point fall of potential ground resistance meter determines the contact resistance of a single grounding electrode to the earth by using Ohm's Law: R=E/I.

This meter divides the voltage difference between the electrode to be measured and a driven potential test stake (P) by the current flowing between the electrode to be measured and a driven current test stake (C). The test stakes are typically made of ¼ in. diameter steel rods, 24 in. long, driven two-thirds of their length into earth.

The distance and alignment between the potential and current test stakes, and the electrode, is extremely important to the validity of the earth contact resistance measurements. For an 8 ft ground rod, the accepted practice is to space the current test stake (C) 80 ft from the electrode to be measured.

The potential test stake (P) is positioned in a straight line between the electrode to be measured and the current test stake (C). The potential test stake should be located at approximately 62 percent of the distance the current test stake is located from the electrode. Since the current test stake (C) for an 8 ft ground rod is located 80 ft from the grounding electrode, the potential test stake (P) will be about 50 ft from the electrode to be measured.

Question: *If the voltage between the ground rod and the potential test stake (P) is 3V and the current between the ground rod and the current test stake (C) is 0.20A, then the earth contact resistance of the electrode to the earth will be _____.* Figure 250–113

(a) 5 ohms (b) 10 ohms (c) 15 ohms (d) 25 ohms

Answer: *(c) 15 ohms*

Resistance = Voltage/Current

E (Voltage) = 3V
I (Current) = 0.20A

R = E/I

Resistance = 3V/0.20A
Resistance = 15 ohms

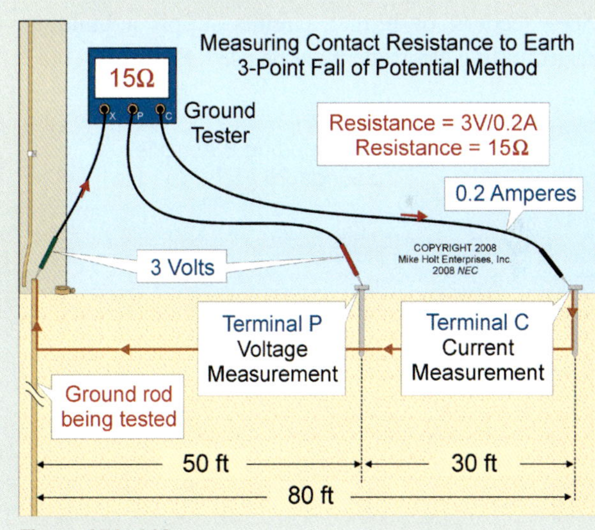

Figure 250–113

Author's Comment: The three-point fall of potential meter can only be used to measure the contact resistance of one electrode to the earth at a time, and this electrode must be independent and not connected to any part of the electrical system. The contact resistance of two electrodes bonded together must not be measured until they have been separated. The contact resistance of two separate electrodes to the earth is calculated as if they are two resistors connected in parallel.

Soil Resistivity

The earth's ground resistance is directly impacted by soil resistivity, which varies throughout the world. Soil resistivity is influenced by electrolytes, which consist of moisture, minerals, and dissolved salts. Because soil resistivity changes with moisture content, the resistance of any grounding system varies with the seasons of the year. Since moisture is stable at greater distances below the surface of the earth, grounding systems are generally more effective if the grounding electrode can reach the water table. In addition, having the grounding electrode below the frost line helps to ensure less deviation in the system's contact resistance to the earth year round.

The contact resistance to the earth can be lowered by chemically treating the earth around the grounding electrodes with electrolytes designed for this purpose.

250.58 Common Grounding Electrode. Where separate services, feeders, or branch circuits supply a building, the same grounding electrode must be used. **Figure 250–114**

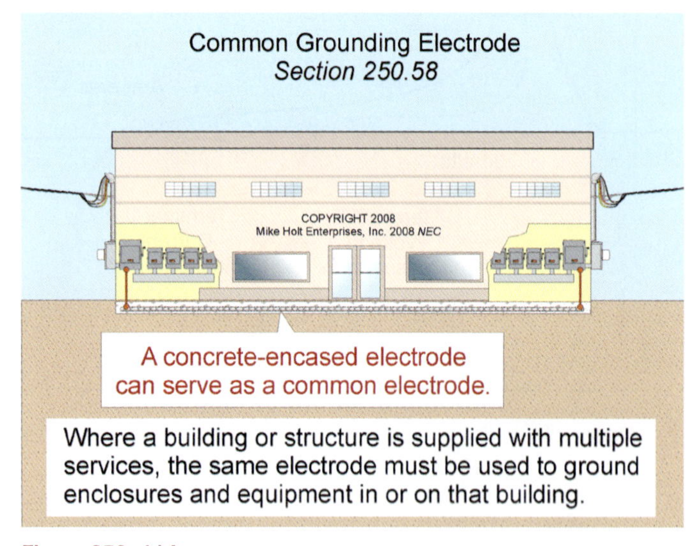

Figure 250–114

Author's Comment: Metal parts of the electrical installation are grounded (connected to the earth) to reduce induced voltage on the metal parts from lightning so as to prevent fires from a surface arc within the building or structure. Grounding electrical equipment doesn't serve the purpose of providing a low-impedance fault current path to open the circuit overcurrent device in the event of a ground fault.

CAUTION: *Potentially dangerous objectionable neutral current flows on the metal parts when multiple service disconnecting means are connected to the same electrode. This is because neutral current from each service can return to the utility via the common grounding electrode and its conductors. This is especially a problem if a service neutral conductor is opened.* **Figure 250–115**

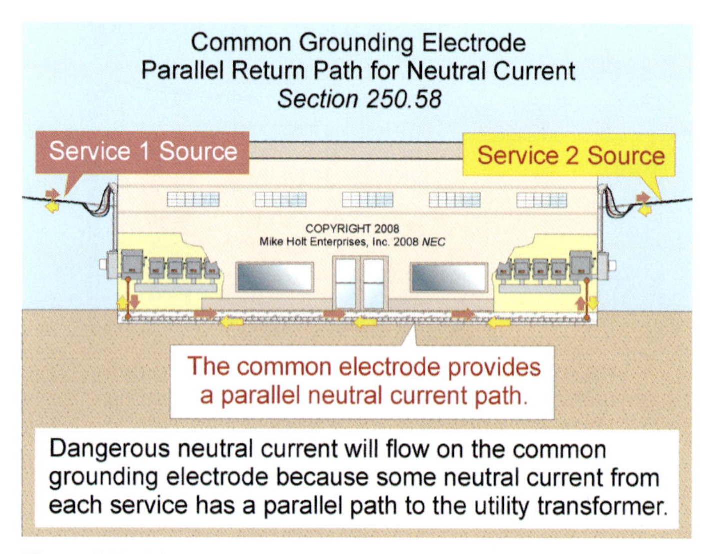

Figure 250–115

250.60 Lightning Protection Electrode. The grounding electrode for a lightning protection system must not be used as the required grounding electrode system for the buildings or structures. **Figure 250–116**

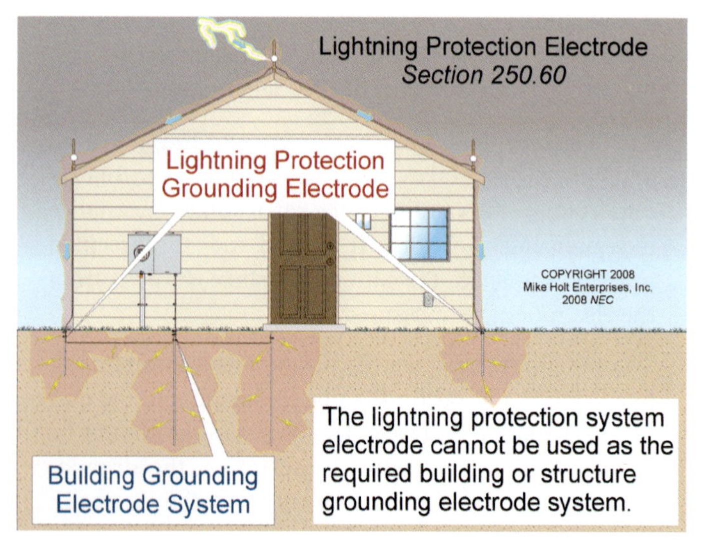

Figure 250–116

Author's Comment: Where a lightning protection system is installed, the lightning protection system must be bonded to the building or structure grounding electrode system [250.106]. Figure 250–117

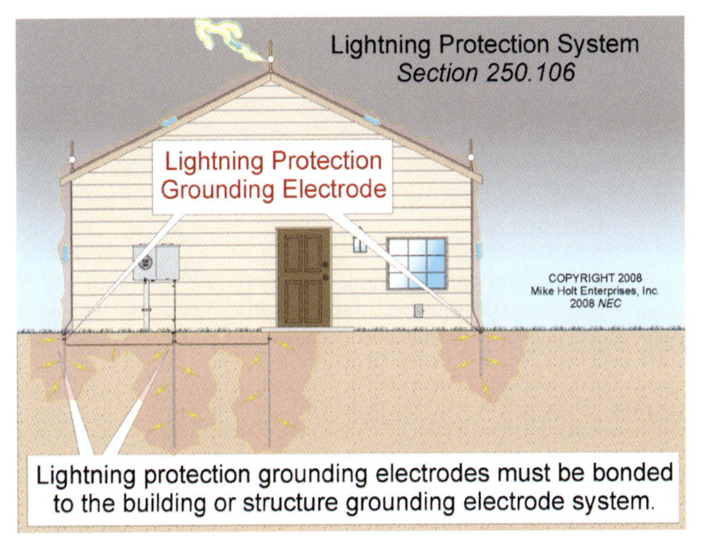

Figure 250–117

250.62 Grounding Electrode Conductor.

The grounding electrode conductor must be solid or stranded, insulated or bare, and it must be copper if within 18 in. of the earth [250.64(A)]. Figure 250–118

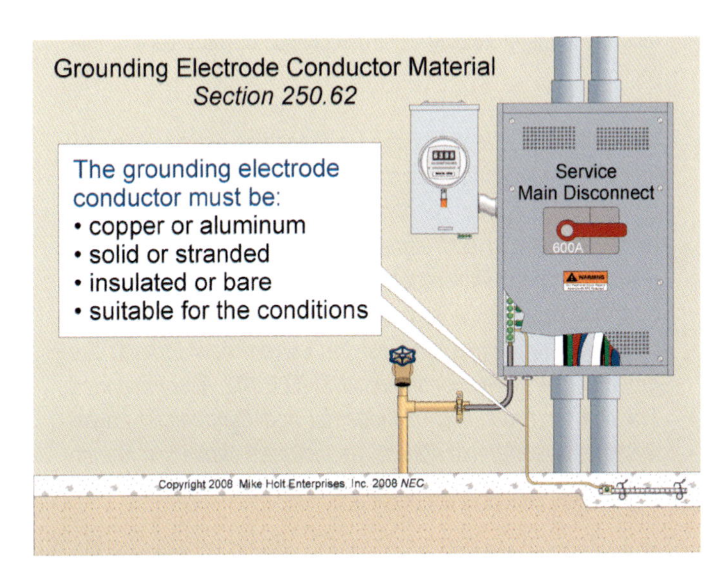

Figure 250–118

250.64 Grounding Electrode Conductor Installation.

Grounding electrode conductors must be installed as specified in (A) through (F).

(A) Aluminum Conductors. Aluminum grounding electrode conductors must not be within 18 in. of earth.

(B) Conductor Protection. Where run exposed, grounding electrode conductors must be protected where subject to physical damage, and grounding electrode conductors 6 AWG copper and larger can be run exposed along the surface of the building if securely fastened and not subject to physical damage.

Grounding electrode conductors sized 8 AWG must be installed in rigid metal conduit, intermediate metal conduit, PVC conduit, or electrical metallic tubing.

Author's Comment: A ferrous metal raceway containing a grounding electrode conductor must be made electrically continuous by bonding each end of that type of raceway to the grounding electrode conductor [250.64(E)]. So it's best to use PVC conduit.

(C) Continuous Run. The grounding electrode conductor is not permitted to be spliced except as permitted in (1) or (2): Figure 250–119

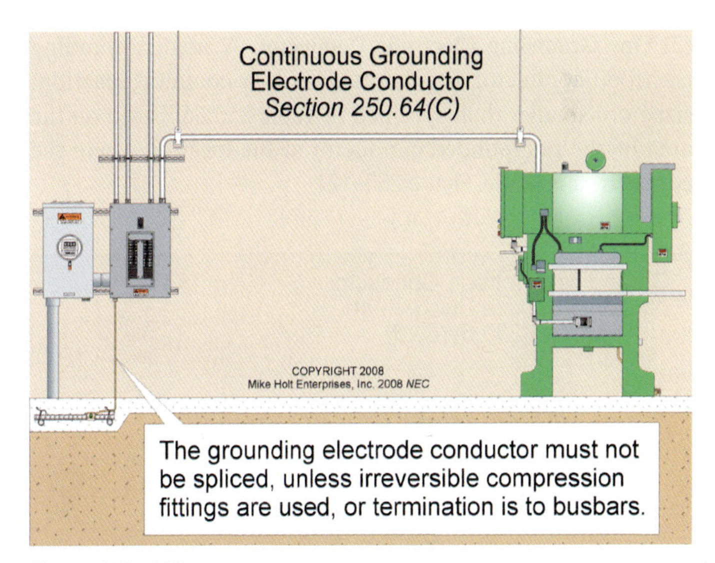

Figure 250–119

(1) Irreversible compression-type connectors listed for grounding, or by exothermic welding.

(2) Sections of busbars connected together.

(D) Grounding Electrode Conductor for Multiple Service Disconnects.

(2) Grounding Electrode Conductor from Each Disconnect. A grounding electrode conductor is permitted from each service disconnecting means sized not smaller than specified in Table 250.66, based on the area of the ungrounded conductor for each service disconnecting means. Figure 250–120

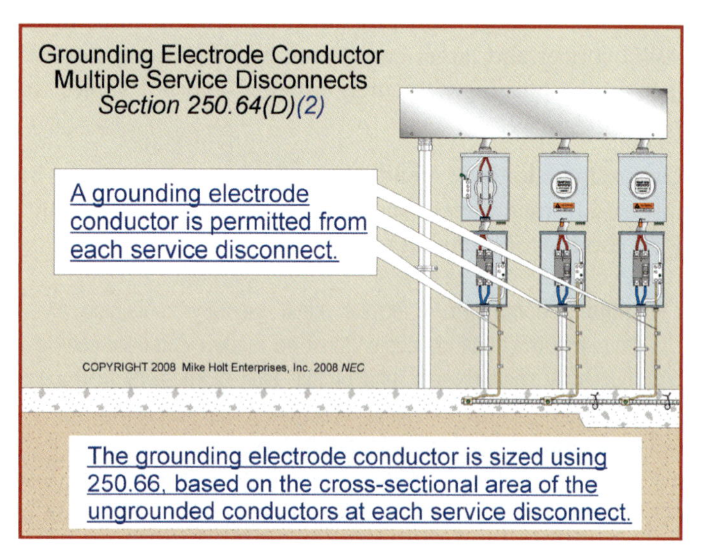

Figure 250–120

(3) One Grounding Electrode Conductor. A single grounding electrode conductor is permitted from a common location, sized not smaller than specified in Table 250.66, based on the area of the ungrounded conductor at the location where the connection is made. Figure 250–121

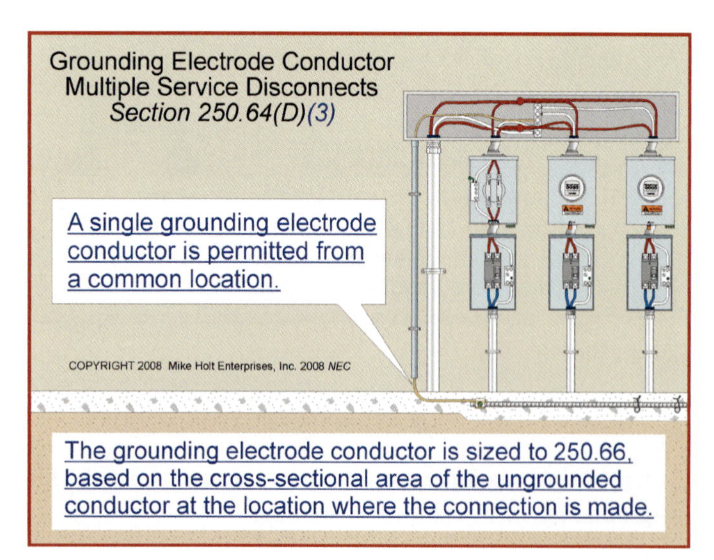

Figure 250–121

(E) Ferrous Metal Enclosures Containing Grounding Electrode Conductors. To prevent inductive choking of grounding electrode conductors; ferrous raceways and enclosures containing grounding electrode conductors must have each end of the raceway or enclosure bonded to the grounding electrode conductor in accordance with 250.92(B). Figure 250–122

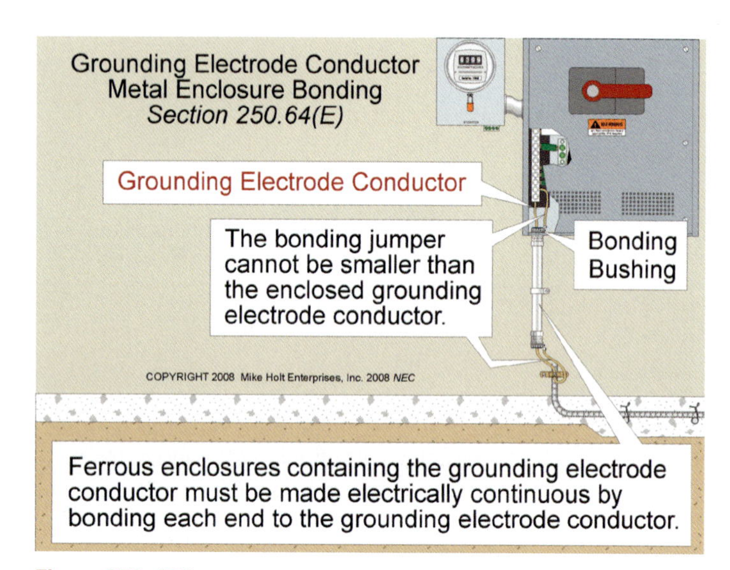

Figure 250–122

Author's Comment: Nonferrous metal raceways, such as aluminum rigid metal conduit, enclosing the grounding electrode conductor aren't required to meet the "bonding each end of the raceway to the grounding electrode conductor" provisions of this section.

CAUTION: *The effectiveness of a grounding electrode is significantly reduced if a ferrous metal raceway containing a grounding electrode conductor isn't bonded to the ferrous metal raceway at both ends. This is because a single conductor carrying high-frequency induced lightning current in a ferrous raceway causes the raceway to act as an inductor, which severely limits (chokes) the current flow through the grounding electrode conductor. ANSI/IEEE 142, Recommended Practice for Grounding of Industrial and Commercial Power Systems (Green Book) states: "An inductive choke can reduce the current flow by 97 percent."*

Author's Comment: To save a lot of time and effort, run the grounding electrode conductor exposed if it's not subject to physical damage [250.64(B)], or enclose it in PVC conduit suitable for the application [352.10(F)].

(F) Termination to Grounding Electrode.

(1) Single Grounding Electrode Conductor. A single grounding electrode conductor is permitted to terminate to any grounding electrode of the grounding electrode system. Figure 250–123

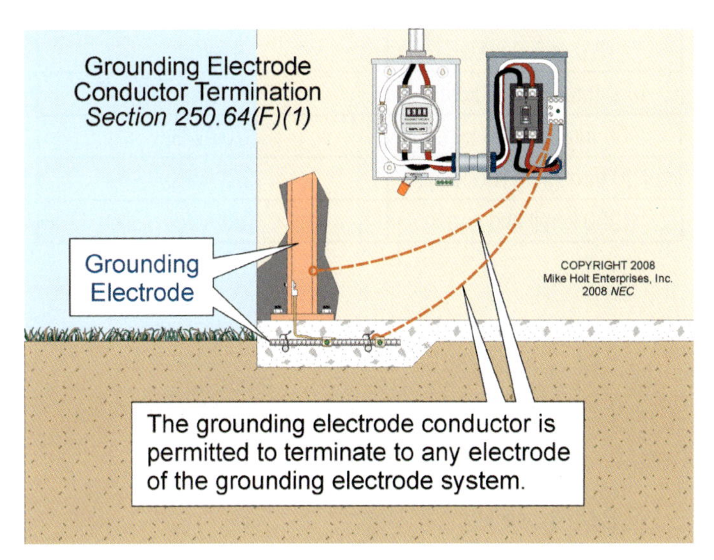

Figure 250–123

(2) Multiple Grounding Electrode Conductors. When multiple grounding electrode conductors are installed [250.64(D)(2)], each grounding electrode conductor is permitted to terminate to any grounding electrode of the grounding electrode system. Figure 250–124

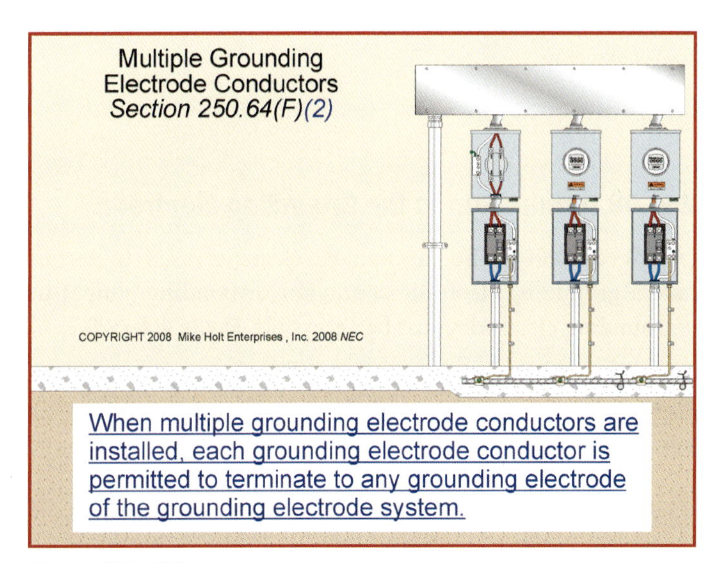

Figure 250–124

(3) Termination to Busbar. A grounding electrode conductor and grounding electrode bonding jumpers are permitted to terminate to a busbar sized not less than ¼ in. × 2 in. that is securely fastened at an accessible location. The terminations to the busbar must be made by a listed connector or by exothermic welding. Figure 250–125

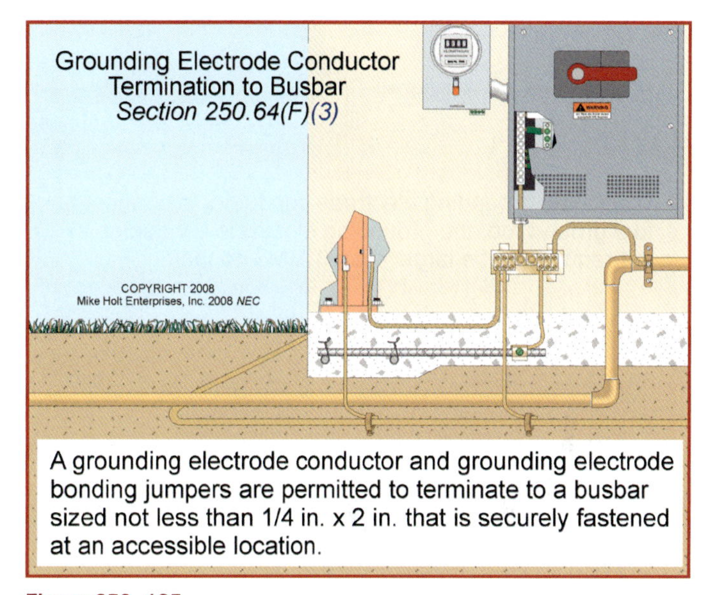

Figure 250–125

250.66 Sizing Grounding Electrode Conductor.

Except as permitted in (A) through (C), a grounding electrode conductor must be sized in accordance with Table 250.66.

(A) Ground Rod. Where the grounding electrode conductor is connected to a ground rod as permitted in 250.52(A)(5), that portion of the grounding electrode conductor that is the sole connection to the ground rod isn't required to be larger than 6 AWG copper. Figure 250–126

(B) Concrete-Encased Grounding Electrode. Where the grounding electrode conductor is connected to a concrete-encased electrode, that portion of the grounding electrode conductor that is the sole connection to the concrete-encased electrode isn't required to be larger than 4 AWG copper. Figure 250–127

(C) Ground Ring. Where the grounding electrode conductor is connected to a ground ring, that portion of the conductor that is the sole connection to the ground ring isn't required to be larger than the conductor used for the ground ring.

Author's Comment: A ground ring encircling the building or structure in direct contact with earth must consist of at least 20 ft of bare copper conductor not smaller than 2 AWG [250.52(A)(4)]. See 250.53(F) for the installation requirements for a ground ring.

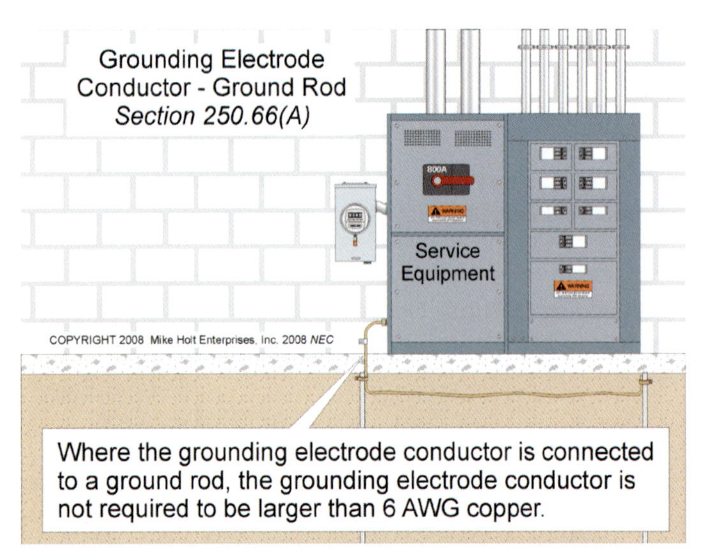

Figure 250–126

Table 250.66 Sizing Grounding Electrode Conductor	
Conductor or Area of Parallel Conductors	Copper Grounding Electrode Conductor
12 through 2 AWG	8 AWG
1 or 1/0 AWG	6 AWG
2/0 or 3/0 AWG	4 AWG
4/0 through 350 kcmil	2 AWG
400 through 600 kcmil	1/0 AWG
700 through 1,100 kcmil	2/0 AWG
1,200 kcmil and larger	3/0 AWG

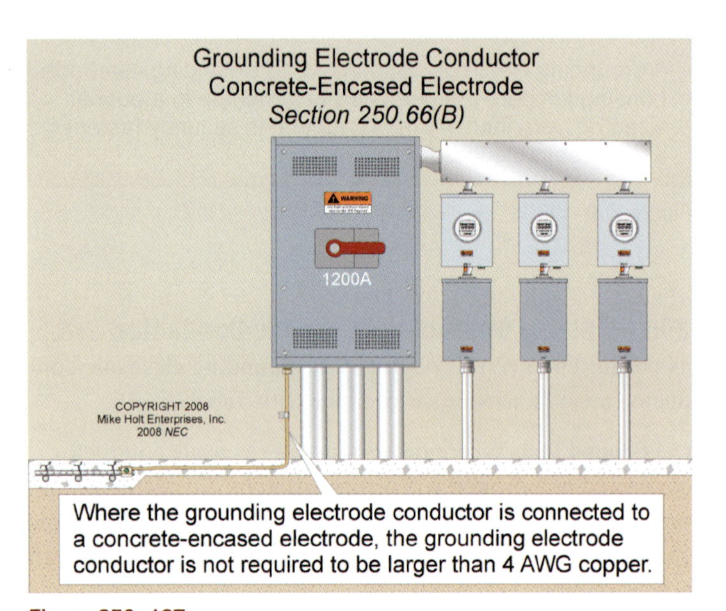

Figure 250–127

Author's Comment: Table 250.66 is used to size the grounding electrode conductor when the conditions of 250.66(A), (B), or (C) do not apply. Figure 250–128

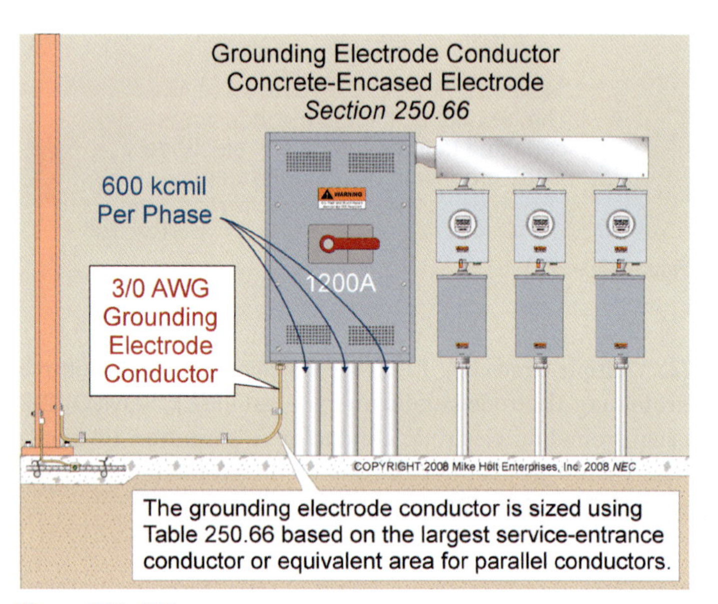

Figure 250–128

250.68 Termination to the Grounding Electrode.

(A) Accessibility. The mechanical elements used to terminate a grounding electrode conductor or bonding jumper to a grounding electrode must be accessible. Figure 250–129

Exception No. 1: The termination is not required to be accessible if the termination to the electrode is encased in concrete or buried in the earth. Figure 250–130

Author's Comment: Where the grounding electrode attachment fitting is encased in concrete or buried in the earth, it must be listed for direct soil burial or concrete encasement [250.70].

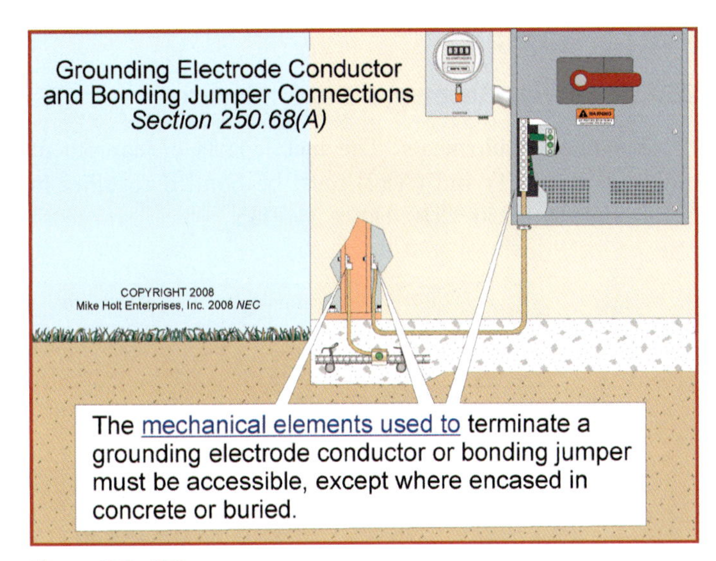

Figure 250–129

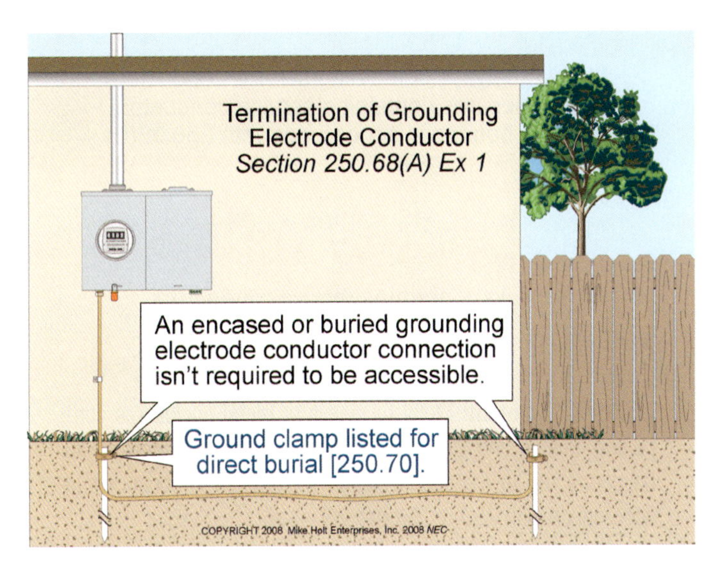

Figure 250–130

Exception No. 2: Exothermic or irreversible compression connections, together with the mechanical means used to attach to fireproofed structural metal, are not required to be accessible.

(B) Integrity of Underground Metal Water Pipe Electrode. A bonding jumper must be installed around insulated joints and equipment likely to be disconnected for repairs or replacement for an underground metal water piping system used as a grounding electrode. The bonding jumper must be of sufficient length to allow the removal of such equipment while retaining the integrity of the grounding path. Figure 250–131

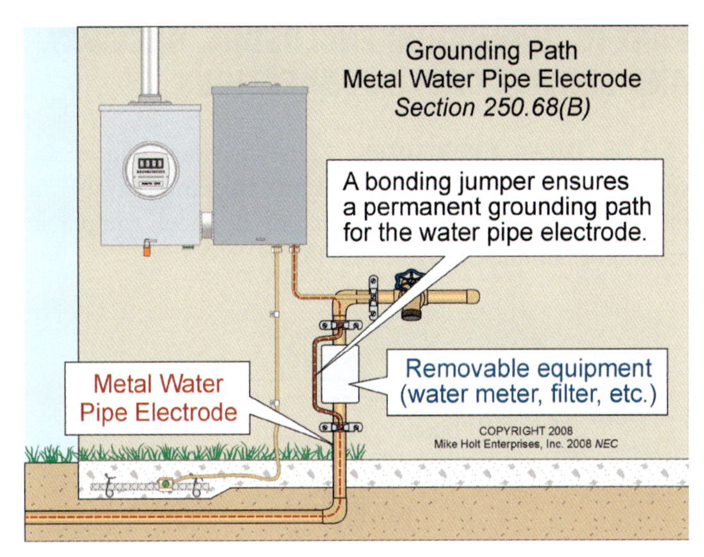

Figure 250–131

250.70 Grounding Electrode Conductor Termination Fittings.

The grounding electrode conductor must terminate to the grounding electrode by exothermic welding, listed lugs, listed pressure connectors, listed clamps, or other listed means. In addition, fittings terminating to a grounding electrode must be listed for the materials of the grounding electrode.

When the termination to a grounding electrode is encased in concrete or buried in the earth, the termination fitting must be listed for direct soil burial or concrete encasement. No more than one conductor can terminate on a single clamp or fitting unless the clamp or fitting is listed for multiple connections. Figure 250–132

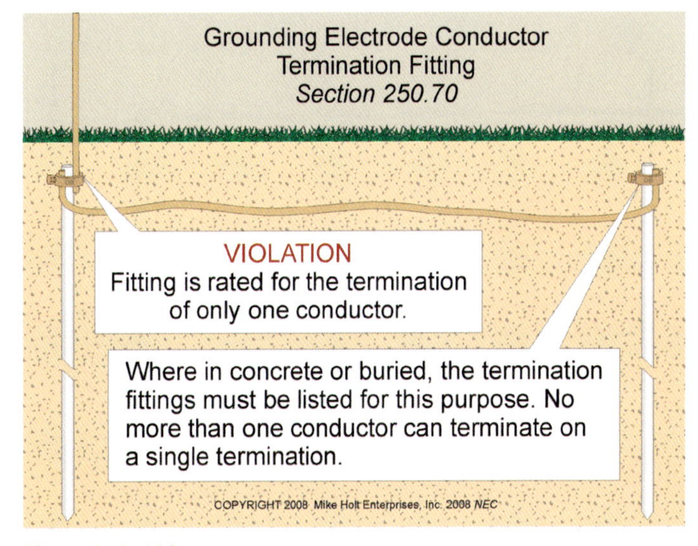

Figure 250–132

PART IV. GROUNDING ENCLOSURE, RACEWAY, AND SERVICE CABLE CONNECTIONS

250.86 Other Enclosures. Metal raceways and enclosures containing electrical conductors operating at 50V or more [250.20(A)] must be connected to the circuit equipment grounding conductor. Figure 250–133

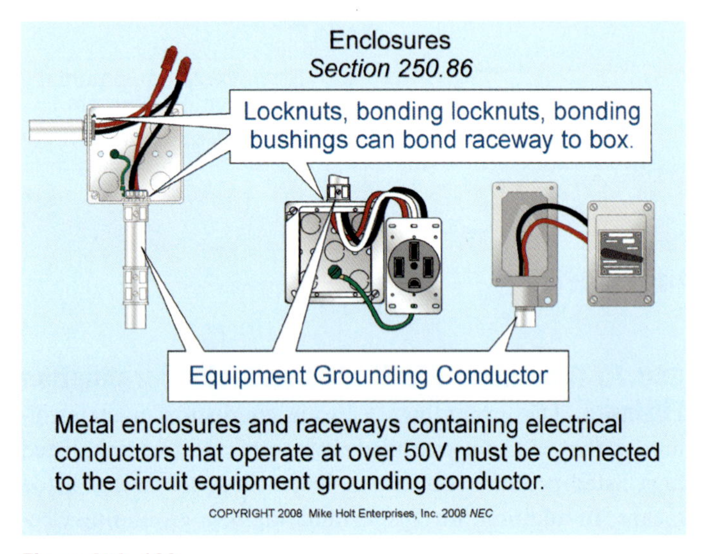

Figure 250–133

Exception No. 2: Short sections of metal raceways used for the support or physical protection of cables aren't required to be connected to the circuit equipment grounding conductor. Figure 250–134

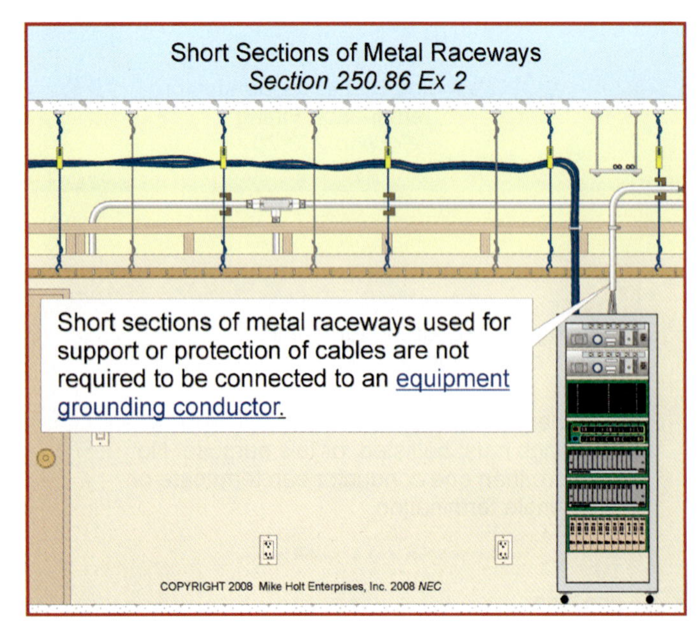

Figure 250–134

PART V. BONDING

250.92 Service Raceways and Enclosures.

(A) Bonding Requirements. The metal parts of equipment indicated in (A)(1) and (A)(2) must be bonded together in accordance with 250.92(B). Figure 250–135

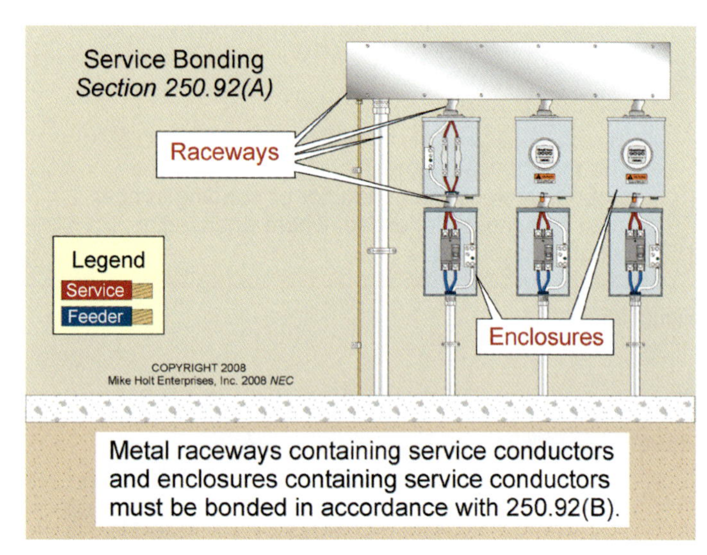

Figure 250–135

(1) Metal raceways containing service conductors.

(2) Metal enclosures containing service conductors.

Author's Comment: Metal raceways or metal enclosures containing feeder and branch-circuit conductors are required to be connected to the circuit equipment grounding conductor in accordance with 250.86. Figure 250–136

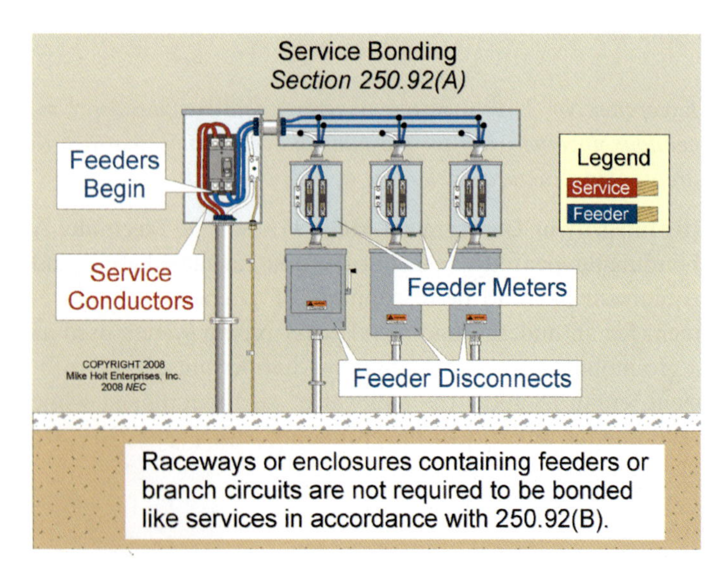

Figure 250–136

(B) Methods of Bonding. Metal raceways and metal enclosures containing service conductors must be bonded by one of the following methods:

(1) Neutral Conductor. By bonding the metal parts to the service neutral conductor. Figure 250–137

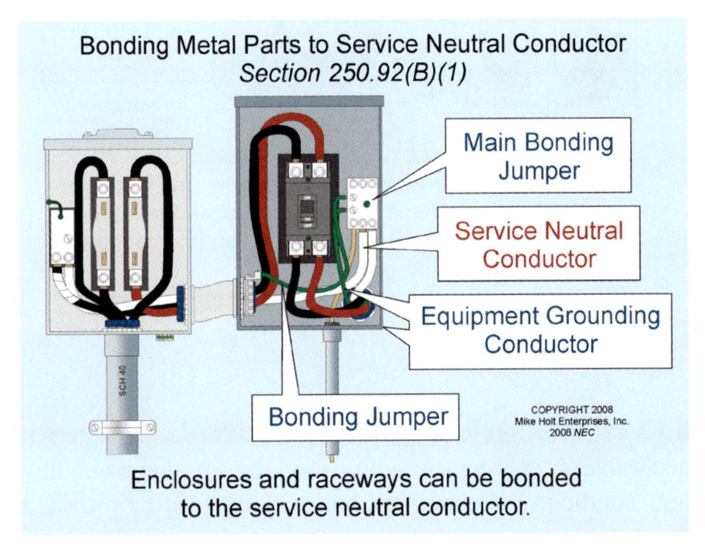

Figure 250–137

Author's Comments:

- A main bonding jumper is required to bond the service disconnect to the service neutral conductor [250.24(B) and 250.28].
- At service equipment, the service neutral conductor provides the effective ground-fault current path to the power supply [250.24(C)]; therefore, an equipment grounding conductor isn't required to be installed within PVC conduit containing service-entrance conductors [250.142(A)(1) and 352.60 Ex 2]. Figure 250–138

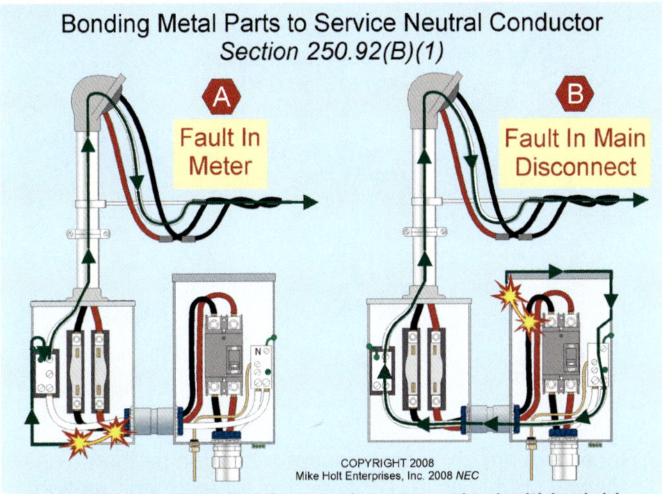

Figure 250–138

(2) Threaded Fittings or Entries. By using threaded couplings or threaded entries made up wrenchtight. Figure 250–139

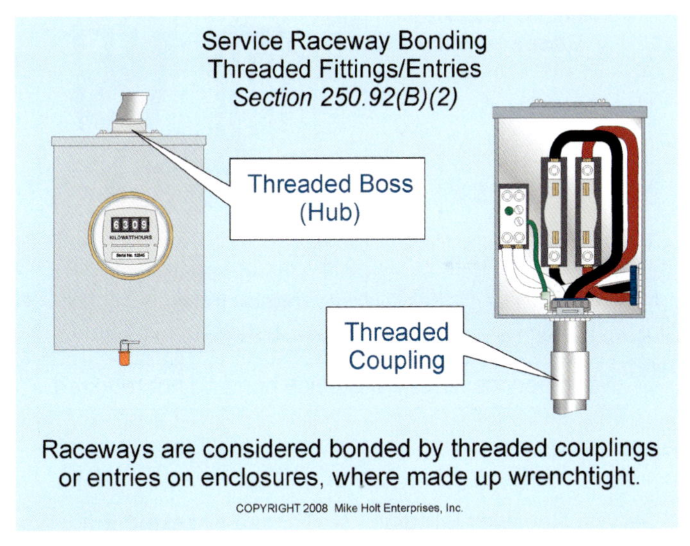

Figure 250–139

(3) Threadless Fittings. By using threadless raceway couplings and connectors made up tight. Figure 250–140

(4) Bonding Fittings. When a metal service raceway terminates to an enclosure with a ringed knockout, a listed bonding wedge or bushing with a bonding jumper must be used to bond one end of the service raceway to the service neutral conductor. The bonding jumper used for this purpose must be sized in accordance with Table 250.66, based on the area of the service conductors within the raceway [250.92(B)(4) and 250.102(C)]. Figure 250–141

Author's Comments:

- When a metal raceway containing service conductors terminates to an enclosure without a ringed knockout, a bonding-type locknut can be used. Figure 250–142
- A bonding locknut differs from a standard locknut in that it has a bonding screw with a sharp point that drives into the metal enclosure to ensure a solid connection.
- Bonding one end of a service raceway to the service neutral provides the low-impedance fault current path to the source. Figure 250–143

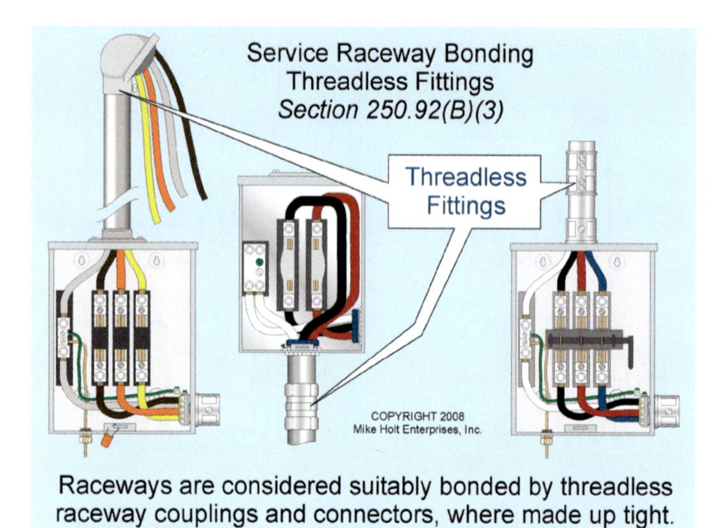

Figure 250–140

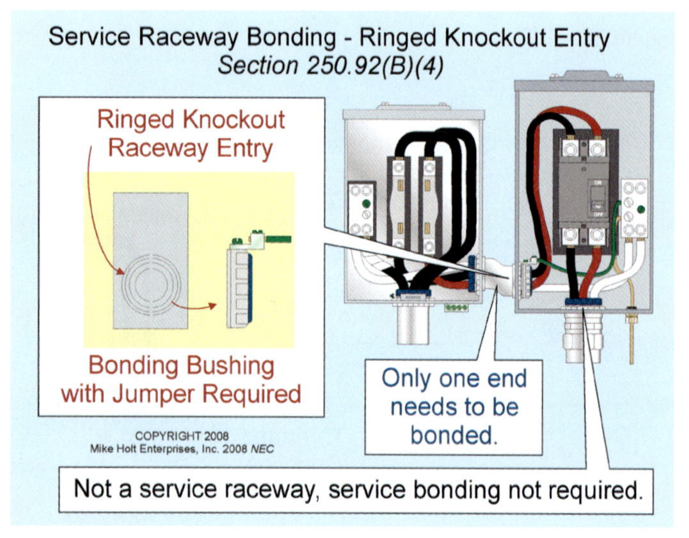

Figure 250–141

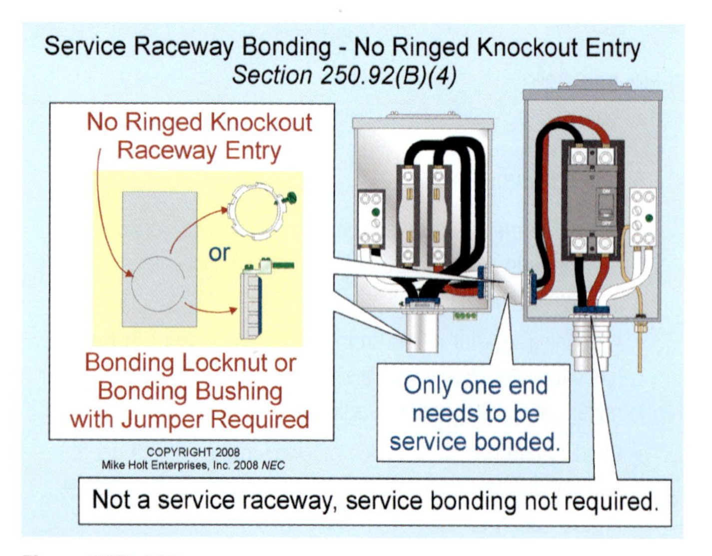

Figure 250–142

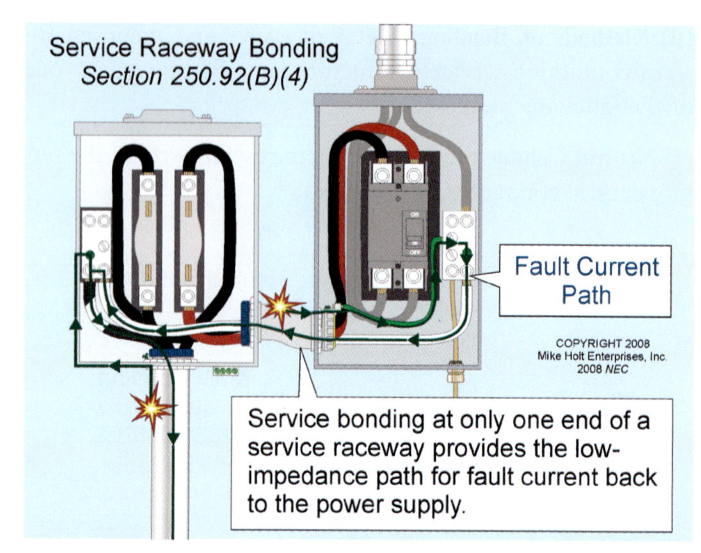

Figure 250–143

250.94 Intersystem Bonding Terminal. An external accessible intersystem bonding terminal for the grounding and bonding of communications systems must be provided at service equipment and disconnecting means for buildings or structures supplied by a feeder. The intersystem bonding terminal must not interfere with the opening of any equipment enclosure and be one of the following: Figure 250–144

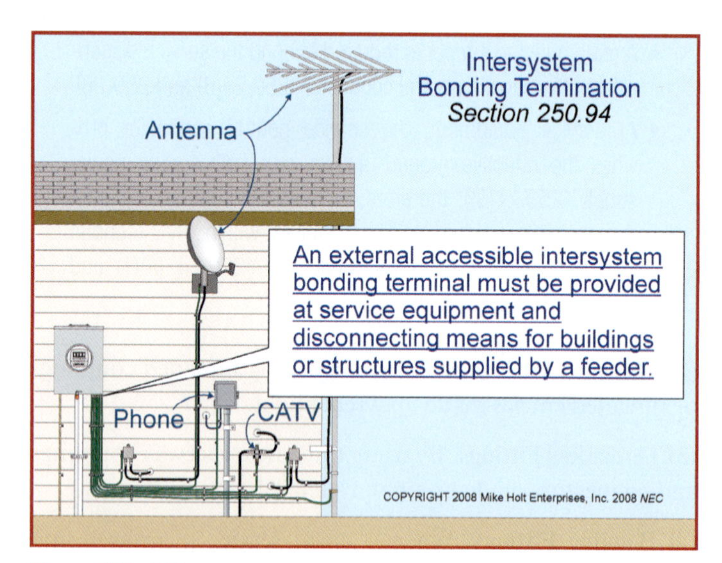

Figure 250–144

(1) Terminals listed for grounding and bonding attached to a meter socket enclosure. Figure 250–145

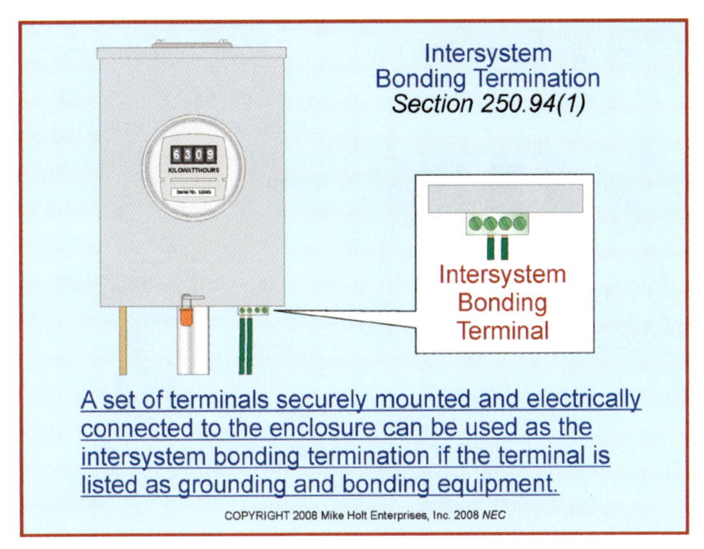

Figure 250–145

(2) Bonding bar connected to the service equipment enclosure or metal service raceway with a minimum 6 AWG copper conductor.

(3) Bonding bar connected to the grounding electrode conductor with a minimum 6 AWG copper conductor. Figure 250–146

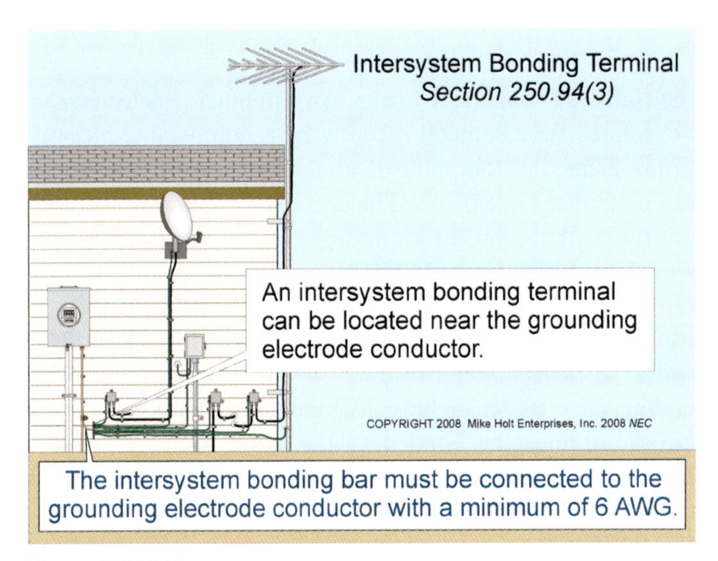

Figure 250–146

Author's Comment: According to Article 100, an intersystem bonding terminal is a device that provides a means to connect communications systems grounding and bonding conductors to the building grounding electrode system.

Exception: At existing buildings or structures, an external accessible means for bonding communications systems together can be by the use of:

(1) Nonflexible metallic raceway,

(2) Grounding electrode conductor, or

(3) Connection approved by the authority having jurisdiction.

FPN No. 2: Communications systems must be bonded to the intersystem bonding terminal in accordance with the following: Figures 250–147 and 250–148

- Antennas/Satellite Dishes, 810.15 and 810.21
- CATV, 820.100
- Telephone Circuits, 800.100

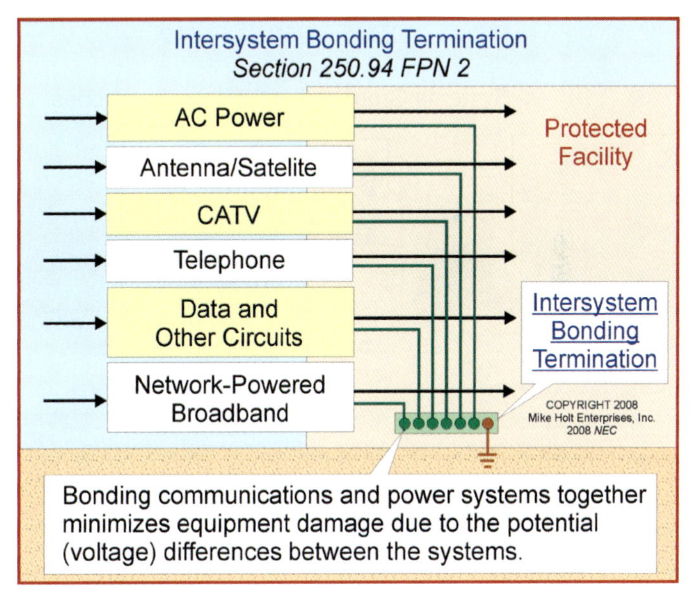

Figure 250–147

Author's Comment: All external communications systems must be connected to the intersystem bonding terminal to minimize the damage to communications systems from induced potential (voltage) differences between the systems from a lightning event. Figure 250–149

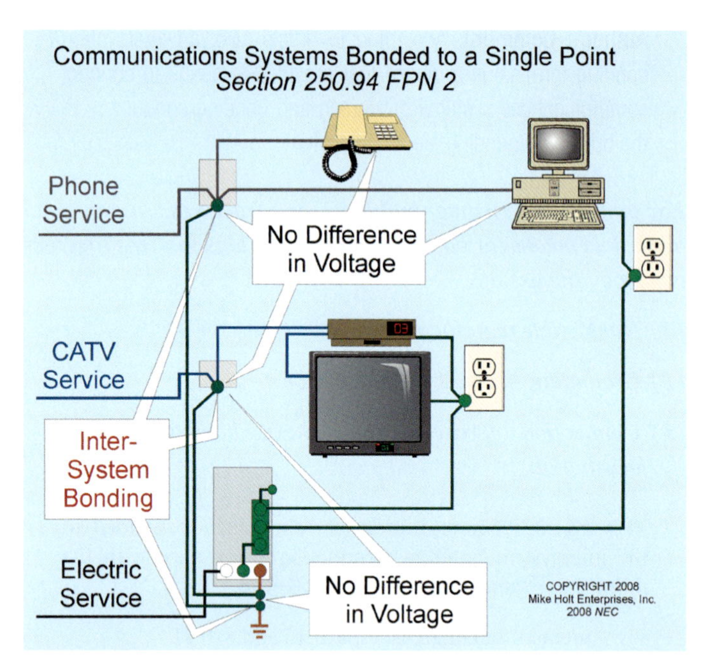

Figure 250–148

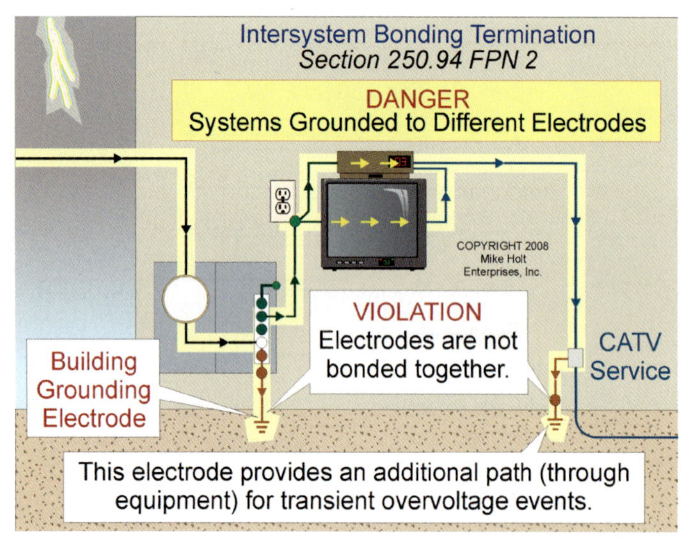

Figure 250–149

250.96 Bonding Other Enclosures.

(A) Maintaining Effective Ground-Fault Current Path. Metal parts intended to serve as equipment grounding conductors including raceways, cables, equipment, and enclosures must be bonded together to ensure they have the capacity to conduct safely any fault current likely to be imposed on them [110.10, 250.4(A)(5), and Note to Table 250.122]. Figure 250–150

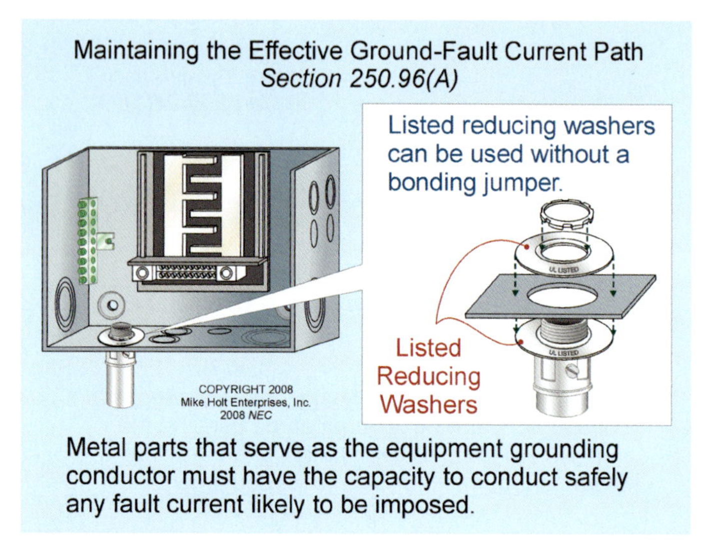

Figure 250–150

Nonconductive coatings such as paint, lacquer, and enamel on equipment must be removed to ensure an effective ground-fault current path, or the termination fittings must be designed so as to make such removal unnecessary [250.12].

> **Author's Comment:** The practice of driving a locknut tight with a screwdriver and pliers is considered sufficient in removing paint and other nonconductive finishes to ensure an effective ground-fault current path.

(B) Isolated Grounding Circuits. An equipment enclosure can be isolated from a metal raceway by a nonmetallic raceway fitting located at the point of attachment of the raceway to the equipment enclosure. The metal raceway must contain an insulated equipment grounding conductor in accordance with 250.146(D). Figure 250–151

250.97 Bonding Metal Parts Containing 277V and 480V Circuits.

Metal raceways or cables containing 277V and/or 480V feeder or branch circuits terminating at ringed knockouts must be bonded to the metal enclosure with a bonding jumper sized in accordance with 250.122, based on the rating of the circuit overcurrent device [250.102(D)]. Figure 250–152

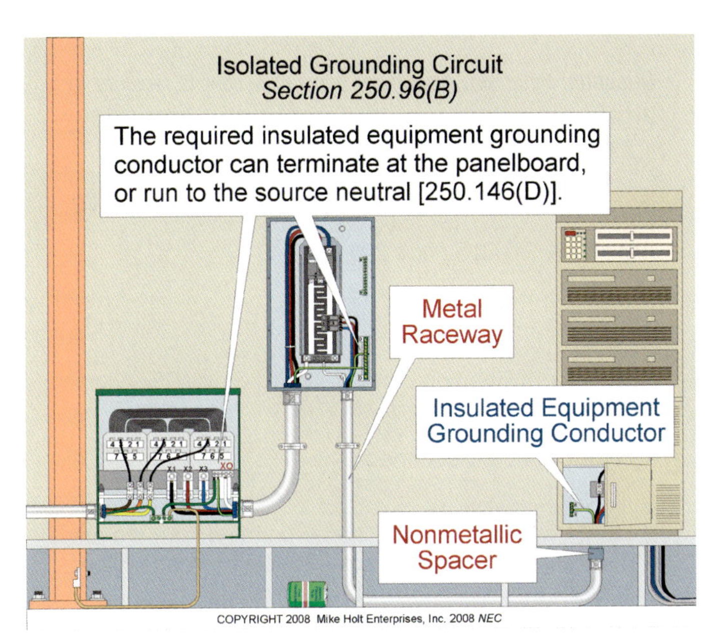

Figure 250–151

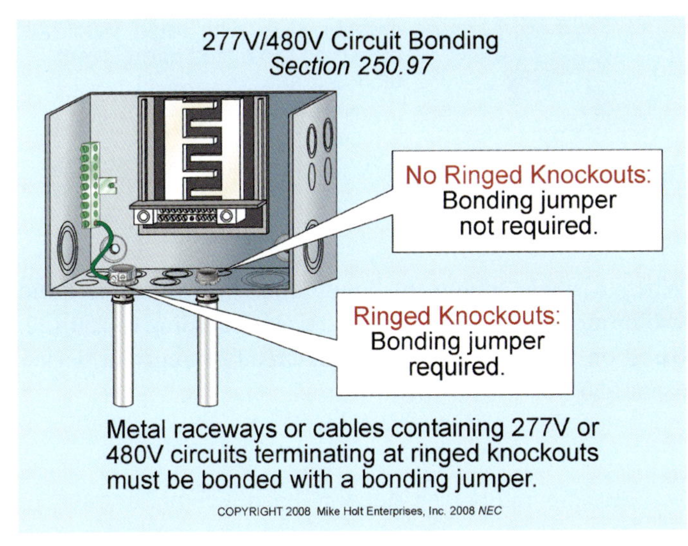

Figure 250–152

Author's Comments:

- Bonding jumpers for raceways and cables containing 277V or 480V circuits are required at ringed knockout terminations to ensure the ground-fault current path has the capacity to safely conduct the maximum ground-fault current likely to be imposed [110.10, 250.4(A)(5), and 250.96(A)].
- Ringed knockouts aren't listed to withstand the heat generated by a 277V ground fault, which generates five times as much heat as a 120V ground fault. **Figure 250–153**

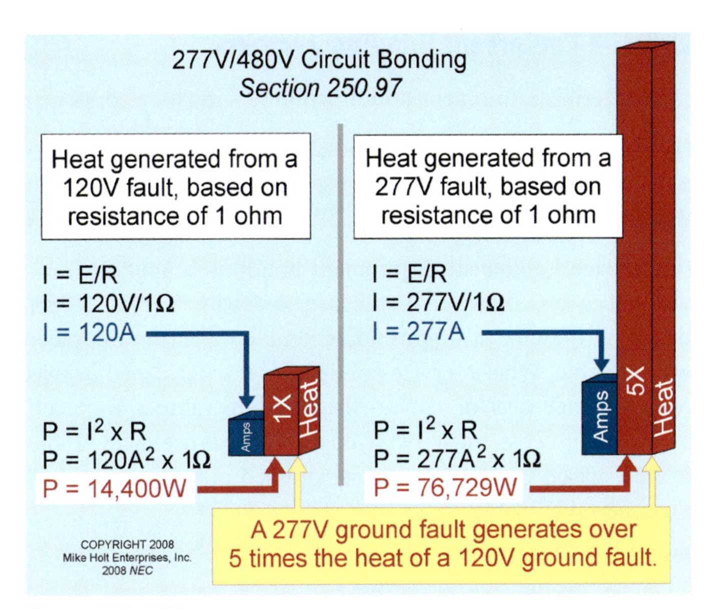

Figure 250–153

Exception: A bonding jumper isn't required where ringed knockouts aren't encountered, knockouts are totally punched out, or where the box is listed to provide a reliable bonding connection. **Figure 250–154**

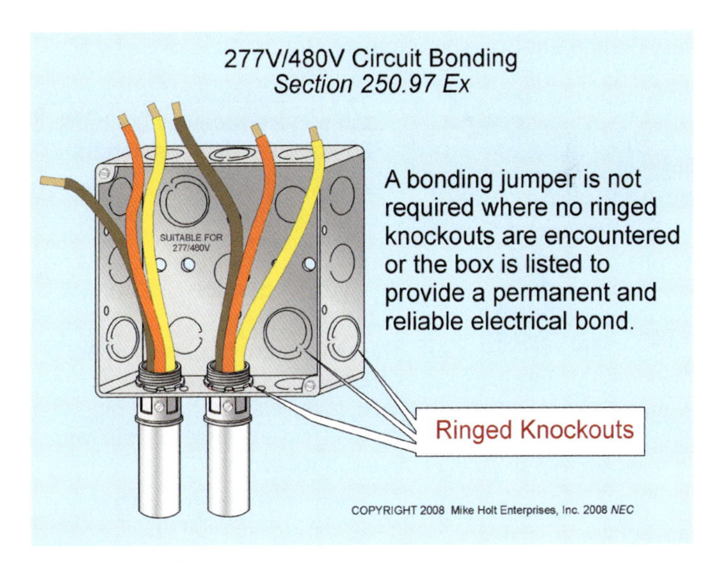

Figure 250–154

250.102 Equipment Bonding Jumpers.

(A) Material. Equipment bonding jumpers must be copper.

(B) Termination. Equipment bonding jumpers must terminate by listed pressure connectors, terminal bars, exothermic welding, or other listed means [250.8(A)].

(C) Service Equipment. Equipment bonding jumpers for service raceways are sized in accordance with Table 250.66, based on the largest ungrounded conductor within the raceway or cable. Where service conductors are paralleled in two or more raceways or cables, the bonding jumper for each raceway or cable must be sized in accordance with Table 250.66, based on the largest ungrounded conductors within the raceway or cable.

Question: *What size equipment bonding jumper is required for a metal raceway containing 700 kcmil service conductors?* **Figure 250–155**

(a) 1 AWG *(b) 1/0 AWG* *(c) 2/0 AWG* *(d) 3/0 AWG*

Answer: *(c) 2/0 AWG [Table 250.66]*\

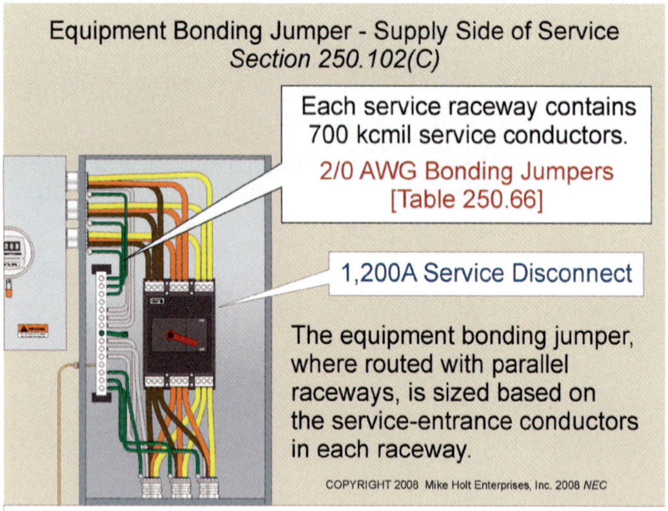

Figure 250–155

(D) Feeder and Branch Circuits. Equipment bonding jumpers for feeders and branch circuits are sized in accordance with 250.122, based on the rating of the circuit overcurrent device.

Question: *What size equipment bonding jumper is required for a metal raceway where the circuit conductors are protected by a 1,200A overcurrent device?* **Figure 250–156**

(a) 1 AWG *(b) 1/0 AWG* *(c) 2/0 AWG* *(d) 3/0 AWG*

Answer: *(d) 3/0 AWG [Table 250.122]*

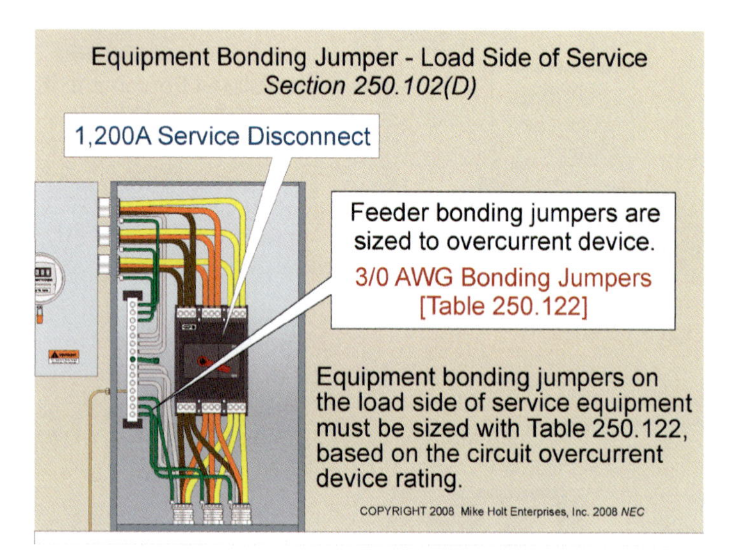

Figure 250–156

Where a single equipment bonding jumper is used to bond two or more raceways, it must be sized according to 250.122, based on the rating of the largest circuit overcurrent device. **Figure 250–157**

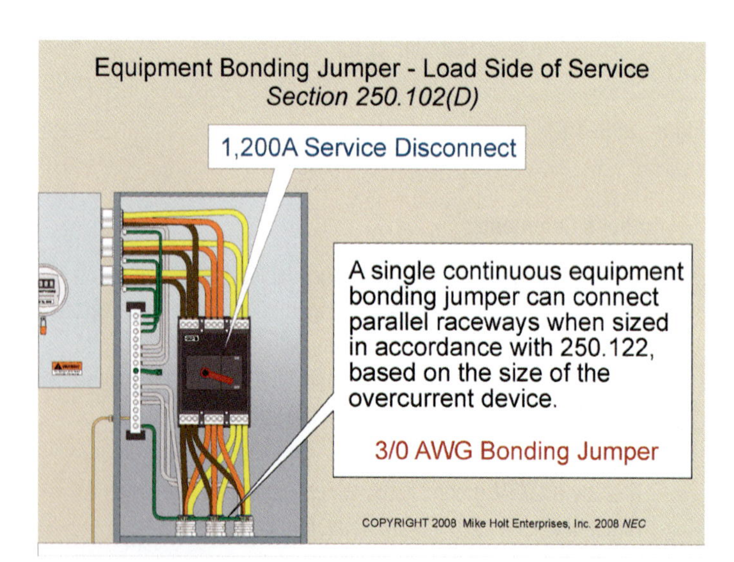

Figure 250–157

(E) Installation. Where the equipment bonding jumper is installed outside of a raceway, its length must not exceed 6 ft and it must be routed with the raceway. Figure 250–158

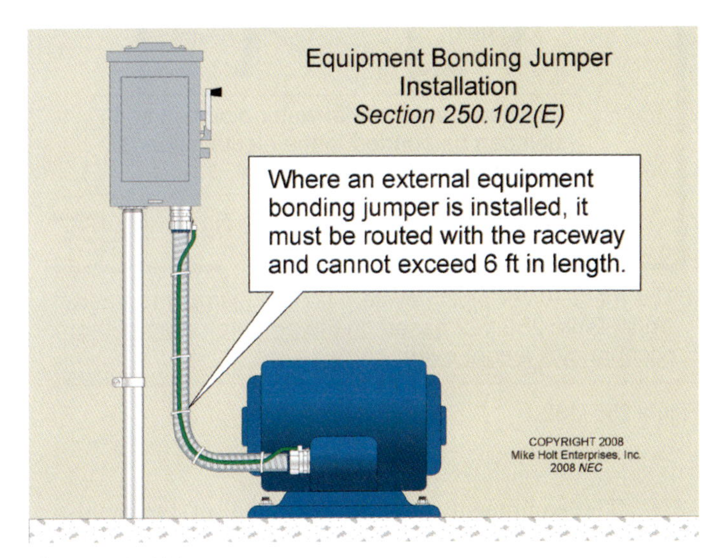

Figure 250–158

250.104 Bonding of Piping Systems and Exposed Structural Metal.

Author's Comment: To remove dangerous voltage on metal parts from a ground fault, electrically conductive metal water piping systems, metal sprinkler piping, metal gas piping, as well as exposed structural steel members likely to become energized, must be connected to an effective ground-fault current path [250.4(A)(4)].

(A) Metal Water Piping System. The metal water piping system must be bonded as required in (A)(1), (A)(2), or (A)(3). The bonding jumper must be copper where within 18 in. of earth [250.64(A)], securely fastened to the surface on which it's mounted [250.64(B)], and adequately protected if exposed to physical damage [250.64(B)]. In addition, all points of attachment must be accessible.

Author's Comment: Bonding isn't required for isolated sections of metal water piping connected to a nonmetallic water piping system. Figure 250-159

Figure 250–159

(1) Building or Structure Supplied by a Service. The metal water piping system, including the metal sprinkler water piping system of a building or structure supplied with service conductors must be bonded to the: Figure 250–160

- Service equipment enclosure,
- Service neutral conductor,
- Grounding electrode conductor of sufficient size, or
- Grounding electrode system.

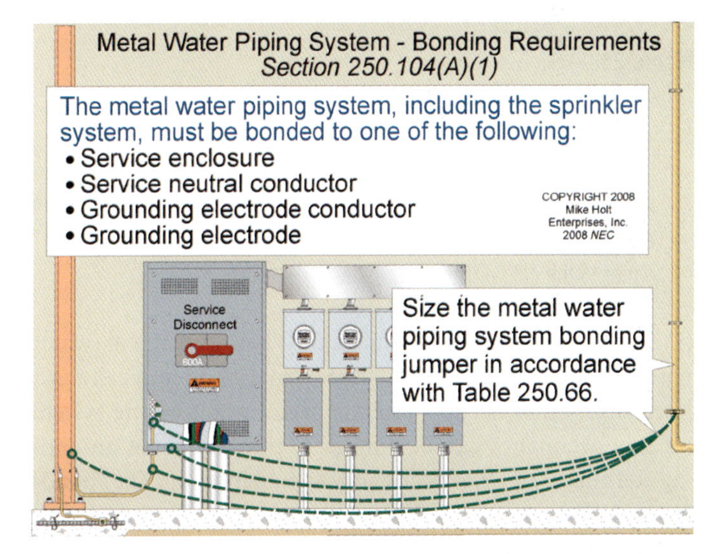

Figure 250–160

The metal water piping system bonding jumper must be sized according to Table 250.66, based on the cross-sectional area of the ungrounded service conductors.

Question: *What size bonding jumper is required for the metal water piping system if the 300 kcmil service conductors are paralleled in two raceways?* **Figure 250–161**

(a) 6 AWG (b) 4 AWG (c) 2 AWG (d) 1/0 AWG

Answer: *(d) 1/0 AWG, based on 600 kcmil conductors, in accordance with Table 250.66*

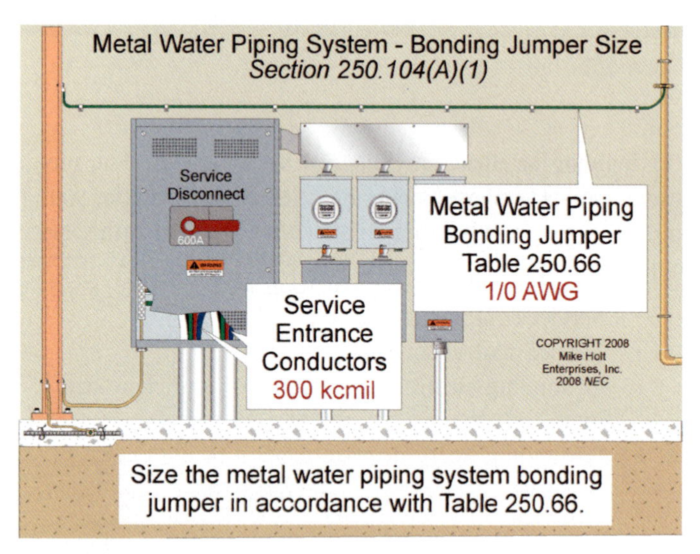

Figure 250–161

Author's Comment: Where hot and cold metal water pipes are electrically connected, only one bonding jumper is required, either to the cold or hot water pipe.

(2) Multiple Occupancy Building. When the metal water piping system in an individual occupancy is metallically isolated from other occupancies, the metal water piping system for that occupancy can be bonded to the equipment grounding terminal of the occupancy's panelboard. The bonding jumper must be sized in accordance with Table 250.122, based on the ampere rating of the occupancy's feeder overcurrent device. **Figure 250–162**

(3) Building or Structure Supplied by a Feeder. The metal water piping system of a building or structure supplied by a feeder must be bonded to:

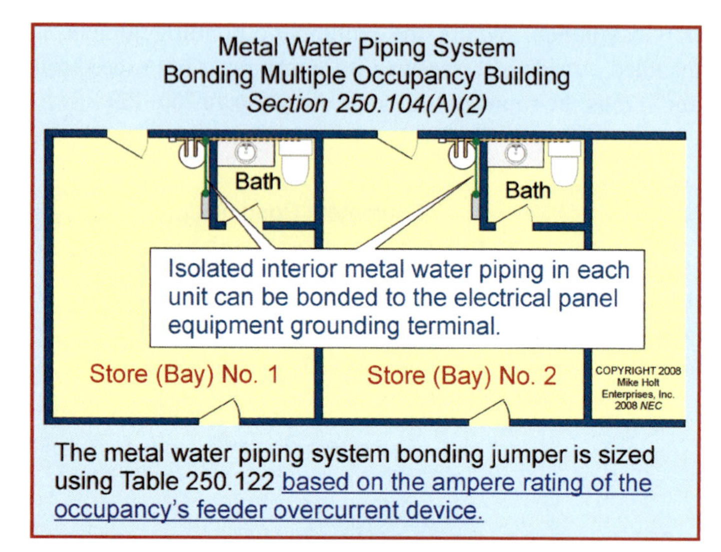

Figure 250–162

- The equipment grounding terminal of the building disconnect enclosure,
- The feeder equipment grounding conductor, or
- The grounding electrode system.

The bonding jumper must be sized according to Table 250.66, based on the cross-sectional area of the ungrounded feeder conductor.

(B) Other Metal Piping Systems. Metal piping systems such as gas or air that are likely to become energized can be bonded to the equipment grounding conductor for the circuit that may energize the piping. **Figure 250–163**

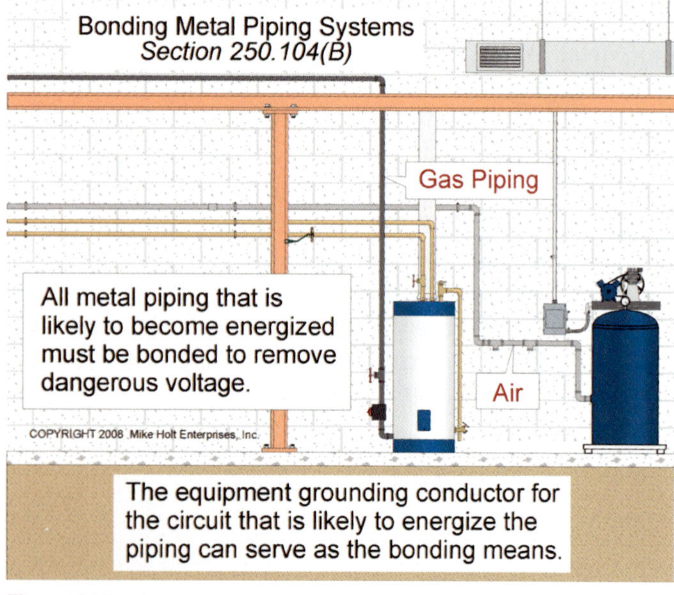

Figure 250–163

(C) Structural Metal. Exposed structural metal that forms a metal building frame that is likely to become energized must be bonded to the: Figure 250–164

- Service equipment enclosure,
- Service neutral conductor,
- Grounding electrode conductor of sufficient size, or
- Grounding electrode system.

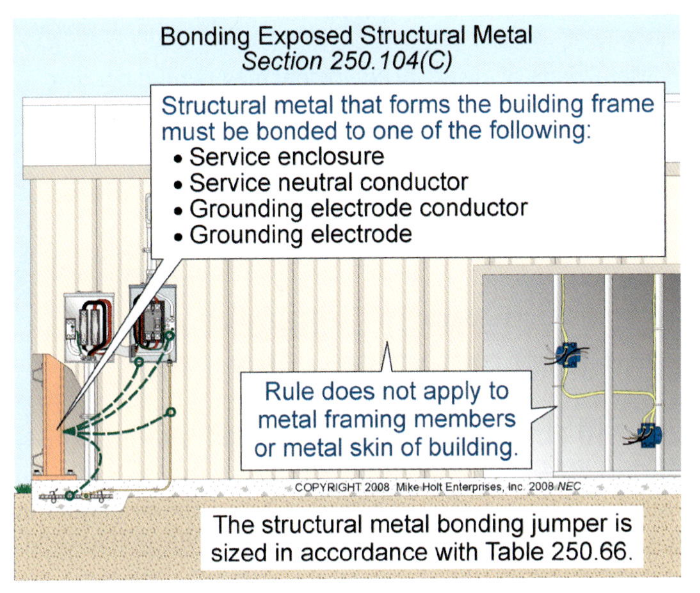

Figure 250–164

Author's Comment: This rule doesn't require the bonding of sheet metal framing members (studs) or the metal skin of a wood frame building.

The bonding jumper must be sized in accordance with Table 250.66, based on the area of ungrounded supply conductors. The bonding jumper must be copper where within 18 in. of earth [250.64(A)], securely fastened to the surface on which it's carried [250.64(B)], and adequately protected if exposed to physical damage [250.64(B)]. In addition, all points of attachment must be accessible.

(D) Separately Derived Systems. Metal water piping systems and structural metal that forms a building frame must be bonded as required in (D)(1) through (D)(3).

(1) Metal Water Pipe. The nearest available point of the metal water piping system in the area served by a separately derived system must be bonded to the neutral point of the separately derived system where the grounding electrode conductor is connected. Figure 250–165

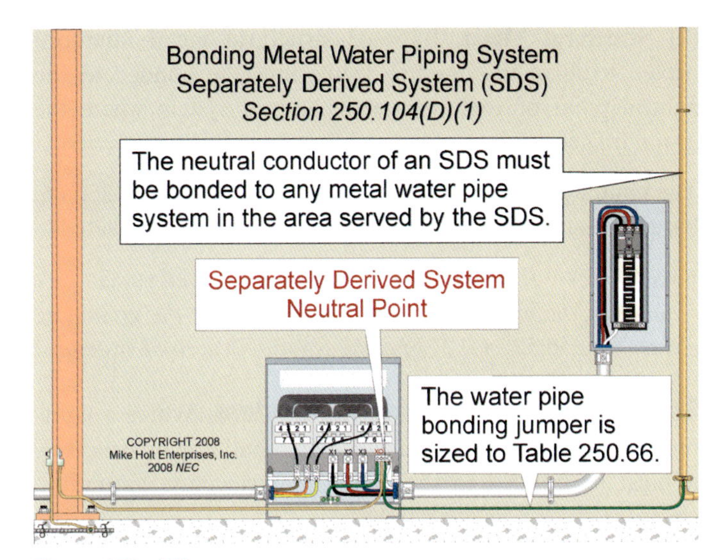

Figure 250–165

The bonding jumper must be sized in accordance with Table 250.66, based on the area of the ungrounded secondary conductors.

Exception No. 2: The metal water piping system is permitted to be bonded to the structural metal building frame if it serves as the grounding electrode [250.52(A)(1)] for the separately derived system. Figure 250–166

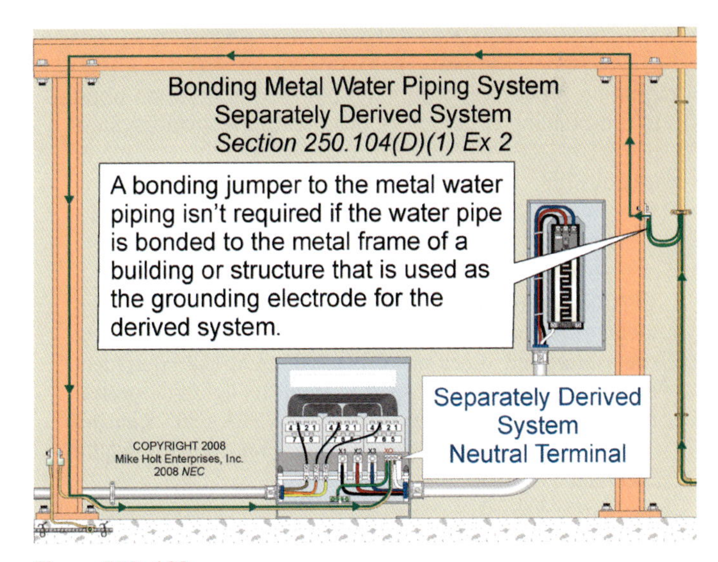

Figure 250–166

(2) Structural Metal. Exposed structural metal interconnected to form the building frame must be bonded to the neutral point of each separately derived system where the grounding electrode conductor is connected.

The bonding jumper must be sized according to Table 250.66, based on the area of the ungrounded secondary conductors.

Exception No. 1: Bonding to the separately derived system isn't required if the metal structural frame serves as the grounding electrode [250.52(A)(2)] for the separately derived system.

250.106 Lightning Protection System. Where a lightning protection system is installed on a building or structure, it must be bonded to the building or structure grounding electrode system. Figure 250–167

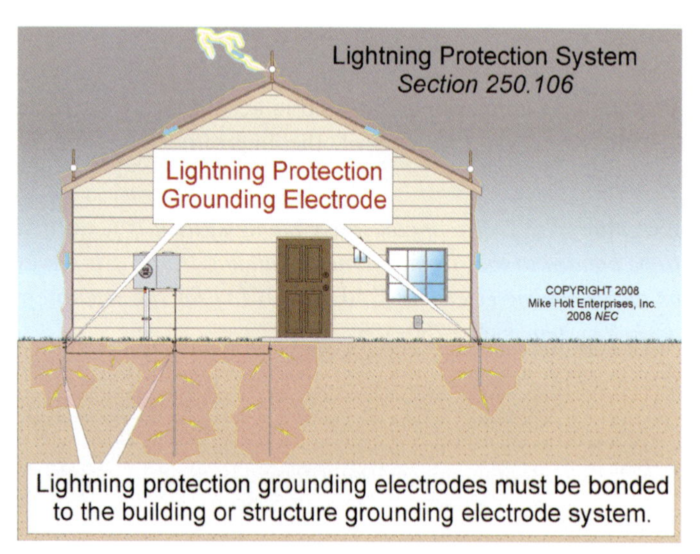

Figure 250–167

FPN No. 2: To minimize the likelihood of arcing between metal parts because of induced voltage, metal raceways, enclosures, and other metal parts of electrical equipment may require bonding or spacing from the lightning protection conductors in accordance with NFPA 780, *Standard for the Installation of Lightning Protection Systems.* Figure 250–168

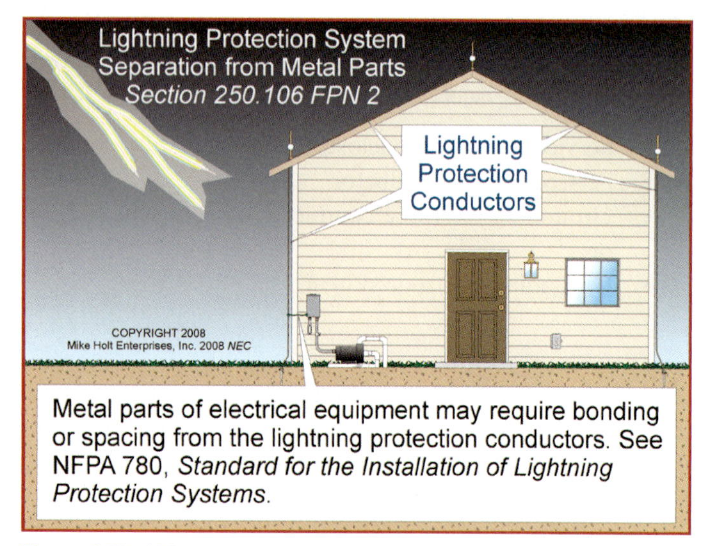

Figure 250–168

PART VI. EQUIPMENT GROUNDING AND EQUIPMENT GROUNDING CONDUCTORS

250.110 Fixed Equipment Connected by Permanent Wiring Methods—General. Exposed metal parts of fixed equipment likely to become energized must be connected to the circuit equipment grounding conductor where the equipment is:

(1) Within 8 ft vertically or 5 ft horizontally of the earth or a grounded metal object

(2) Located in a wet or damp location

(3) In electrical contact with metal

(4) In a hazardous (classified) location [Articles 500 through 517]

(5) Supplied by a wiring method that provides an equipment grounding conductor

(6) Supplied by a 277V or 480V circuit

Exception No. 3: Listed equipment distinctively marked as double-insulated isn't required to be connected to the circuit equipment grounding conductor.

250.112 Fastened in Place or Connected by Permanent Wiring Methods (Fixed). Except as permitted in 250.112(I), exposed metal parts of equipment and enclosures must be connected to the circuit equipment grounding conductor. Figure 250–169

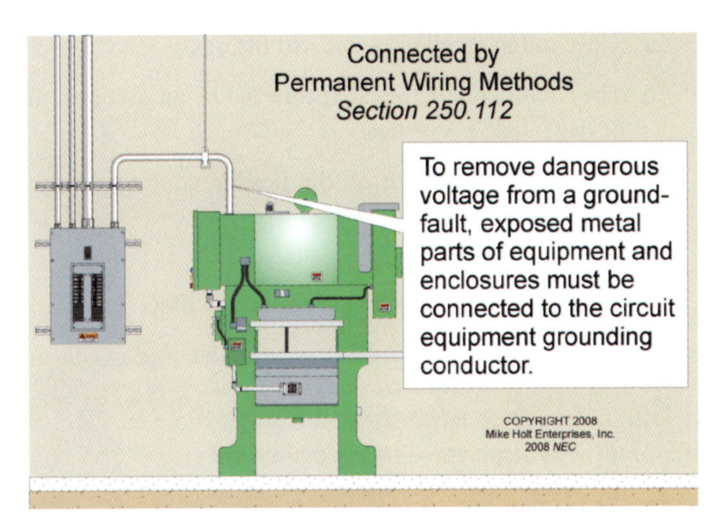

Figure 250–169

(I) Remote-Control, Signaling, and Fire Alarm Circuits. Equipment supplied by circuits operating at 50V or less is not required to be connected to the circuit equipment grounding conductor. Figure 250–170

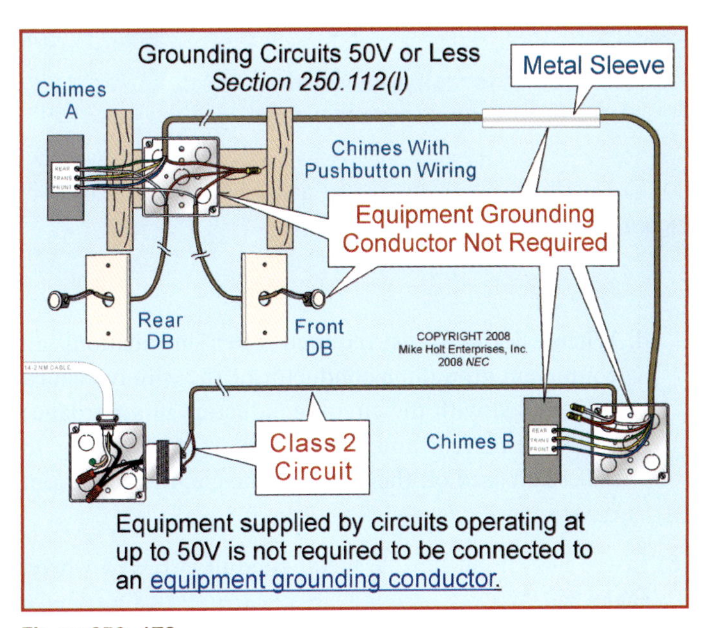

Figure 250–170

Author's Comment: Class 1 power-limited circuits, Class 2, and Class 3 remote-control and signaling circuits, and fire alarm circuits operating at 50V or less don't need to have any metal parts connected to an equipment grounding conductor.

250.114 Cord-and-Plug-Connected Equipment.

Metal parts of cord-and-plug-connected equipment must be connected to the circuit equipment grounding conductor.

Exception: Listed equipment distinctively marked as double-insulated isn't required to be connected to the circuit equipment grounding conductor.

250.118 Types of Equipment Grounding Conductors.

An equipment grounding conductor can be any one or a combination of the following: Figure 250–171

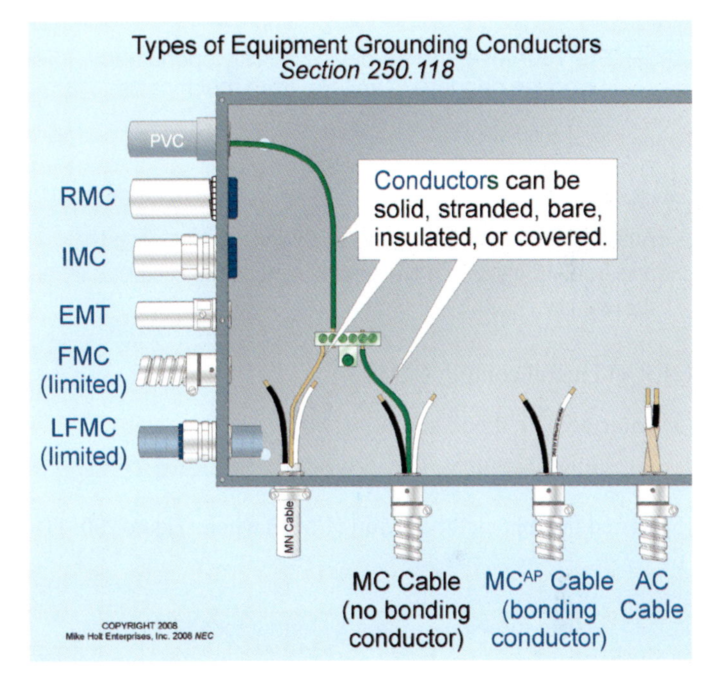

Figure 250–171

FPN: The equipment grounding conductor is intended to serve as the effective ground-fault current path. See 250.2.

Author's Comment: The effective ground-fault path is an intentionally constructed low-impedance conductive path designed to carry fault current from the point of a ground fault on a wiring system to the electrical supply source. Its purpose is to quickly remove dangerous voltage from a ground fault by opening the circuit overcurrent device [250.2]. Figure 250–172

(1) A bare or insulated copper or aluminum conductor sized in accordance with 250.122.

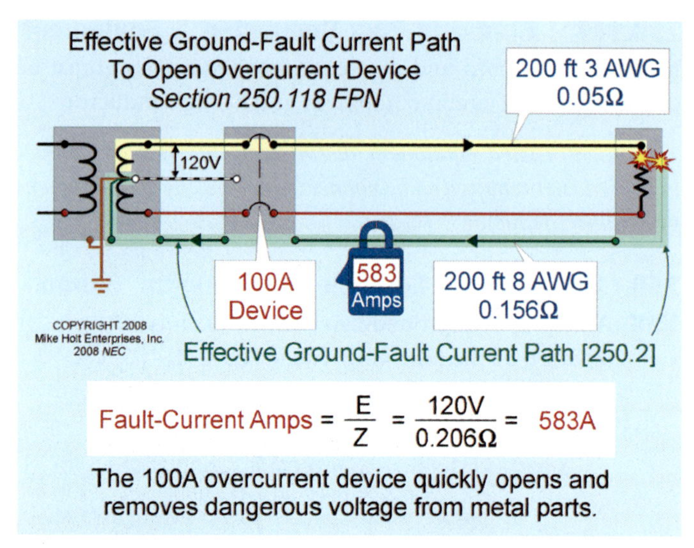

Figure 250–172

Author's Comment: Examples include PVC conduit, Type NM, and Type MC cable with an equipment grounding conductor of the wire type.

(2) Rigid metal conduit (RMC).

(3) Intermediate metal conduit (IMC).

(4) Electrical metallic tubing (EMT).

(5) Listed flexible metal conduit (FMC) where: Figure 250–173

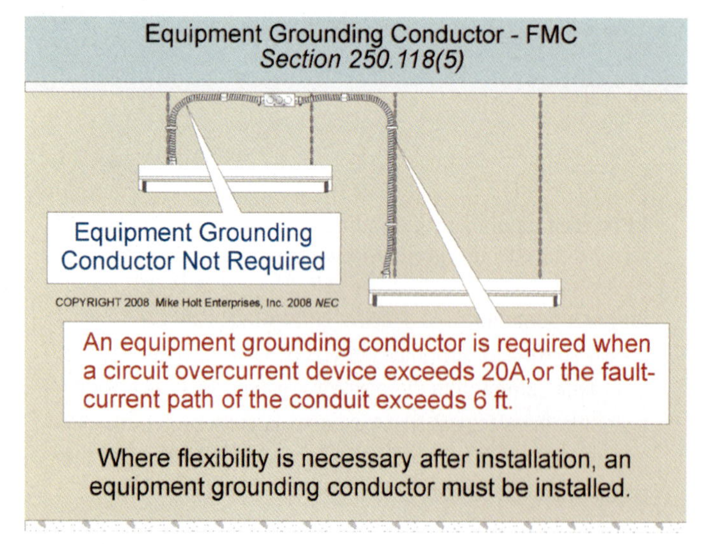

Figure 250–173

a. The conduit terminates in listed fittings.

b. The circuit conductors are protected by an overcurrent device rated 20A or less.

c. The combined length of the flexible conduit doesn't exceed 6 ft. Figure 250–174

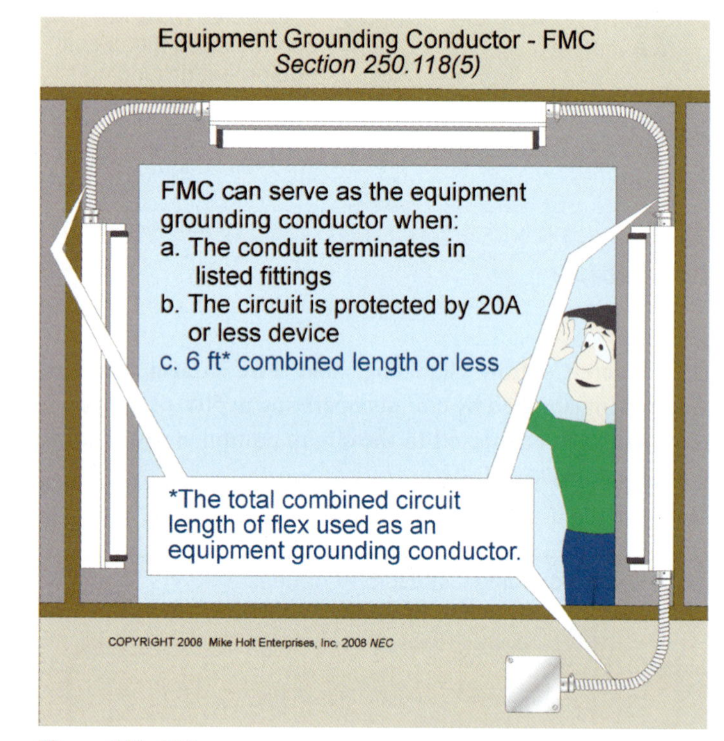

Figure 250–174

d. Where flexibility is required after installation, an equipment grounding conductor of the wire type must be installed with the circuit conductors in accordance with 250.102(E), and it must be sized according to 250.122, based on the rating of the circuit overcurrent device.

(6) Listed liquidtight flexible metal conduit (LFMC) where: Figure 250–175

a. The conduit terminates in listed fittings.

b. For 3⁄8 in. through ½ in., the circuit conductors are protected by an overcurrent device rated 20A or less.

c. For ¾ through 1¼ in., the circuit conductors are protected by an overcurrent device rated 60A or less.

d. The combined length of the flexible conduit doesn't exceed 6 ft.

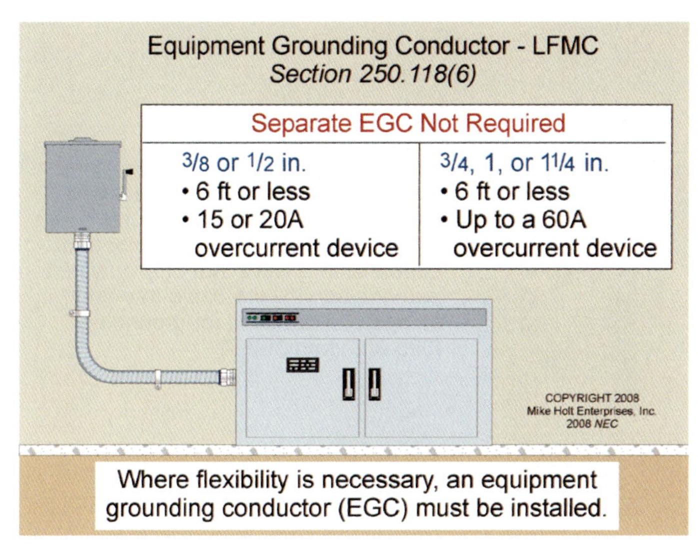

Figure 250–175

e. Where flexibility is required after installation, an equipment grounding conductor of the wire type must be installed with the circuit conductors in accordance with 250.102(E), and it must be sized in accordance with 250.122, based on the rating of the circuit overcurrent device.

(8) The sheath of Type AC cable containing an aluminum bonding strip. Figure 250–176

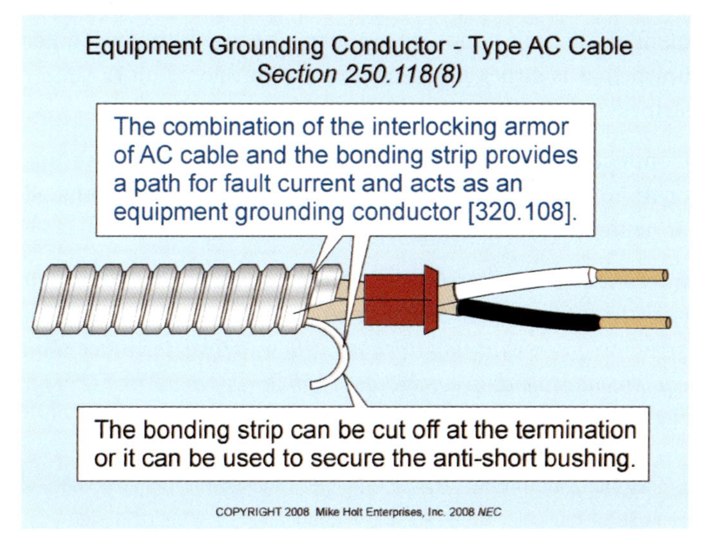

Figure 250–176

Author's Comments:

- The internal aluminum bonding strip isn't an equipment grounding conductor, but it allows the interlocked armor to serve as an equipment grounding conductor because it reduces the impedance of the armored spirals to ensure that a ground fault will be cleared. It's the aluminum bonding strip in combination with the cable armor that creates the circuit equipment grounding conductor. Once the bonding strip exits the cable, it can be cut off because it no longer serves any purpose.
- The effective ground-fault current path must be maintained by the use of fittings specifically listed for Type AC cable [320.40]. See 300.12, 300.15, and 320.100.

(9) The copper sheath of Type MI cable.

(10) The sheath of Type MC cable where listed and identified for grounding:

a. Interlocked Type MC cable containing a bare aluminum bonding strip in direct contact with the interlocked metal armor that is listed and identified for grounding. Figure 250–177

b. Smooth or corrugated-tube Type MC cable without an equipment grounding conductor.

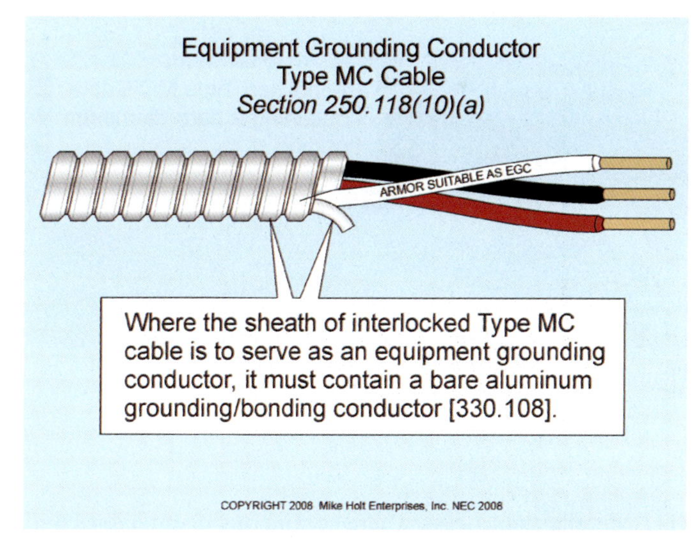

Figure 250–177

Author's Comment: MC$^{AP®}$ cable is Type MC cable constructed with THHN copper insulated circuit conductors and interlocked armor that is listed and identified for grounding. After considering potential cable applications, Southwire decided to name the product MC^{AP} cable to reflect the all-purpose aspect of the product. The "AP" in MC^{AP} stands for All-Purpose, which means that MC^{AP} cable can be used in both MC and AC cable applications. The dual use is permitted because the armor of MC^{AP} cable is listed and identified as a suitable equipment grounding path, unlike conventional MC cable.

An aluminum grounding/bonding conductor in direct contact with the interlocked armor throughout the entire cable length allows the armor of MC^{AP} cable to serve as an equipment grounding conductor.

Once the bare aluminum grounding/bonding conductor exits the cable, it can be cut off because it no longer serves any purpose. The effective ground-fault current path must be maintained by the use of fittings specifically listed for Type MC^{AP} cable [330.40]. See 300.12, 300.15, and 330.100. Figure 250–178

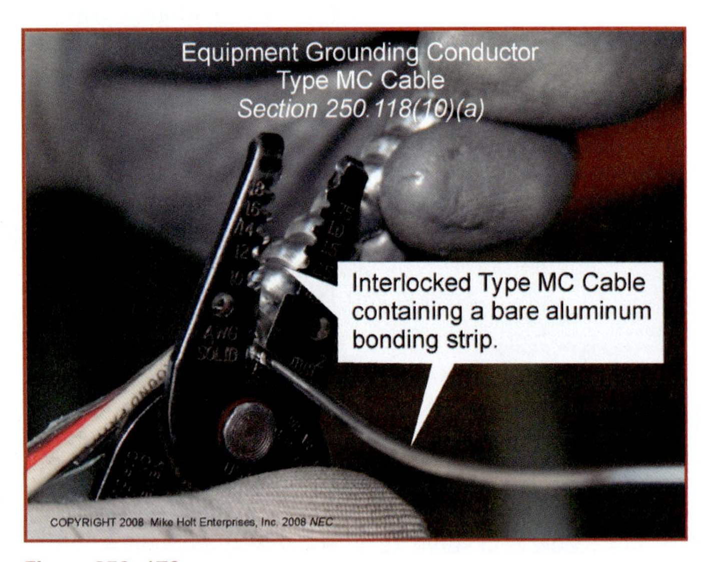

Figure 250–178

(11) Metallic cable trays where continuous maintenance and supervision ensure only qualified persons will service the cable tray [392.3(C)], with cable tray and fittings identified for grounding and the cable tray, fittings, and raceways are bonded using bolted mechanical connectors or bonding jumpers sized and installed in accordance with 250.102 [392.7]. Figure 250–179

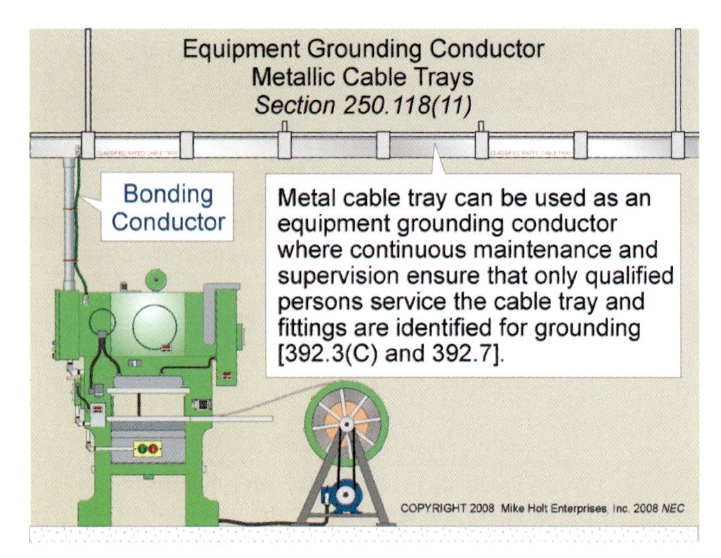

Figure 250–179

(13) Listed electrically continuous metal raceways, such as metal wireways [Article 376] or strut-type channel raceways [384.60].

(14) Surface metal raceways listed for grounding [Article 386].

250.119 Identification of Equipment Grounding Conductors.

Unless required to be insulated, equipment grounding conductors can be bare, covered or insulated. Insulated equipment grounding conductors must have a continuous outer finish that is either green or green with one or more yellow stripes.

Conductors with insulation that is green, or green with one or more yellow stripes must not be used for an ungrounded or neutral conductor. Figure 250–180

Exception: Class 2 or Class 3 cables supplying equipment that operates at less than 50V can use conductors with insulated green or green with one or more yellow stripes for other than equipment grounding conductors [250.20(A) and 250.112(I)]. Figure 250–181

Author's Comment: The *NEC* neither requires nor prohibits the use of the color green for the identification of grounding electrode conductors. Figure 250–182

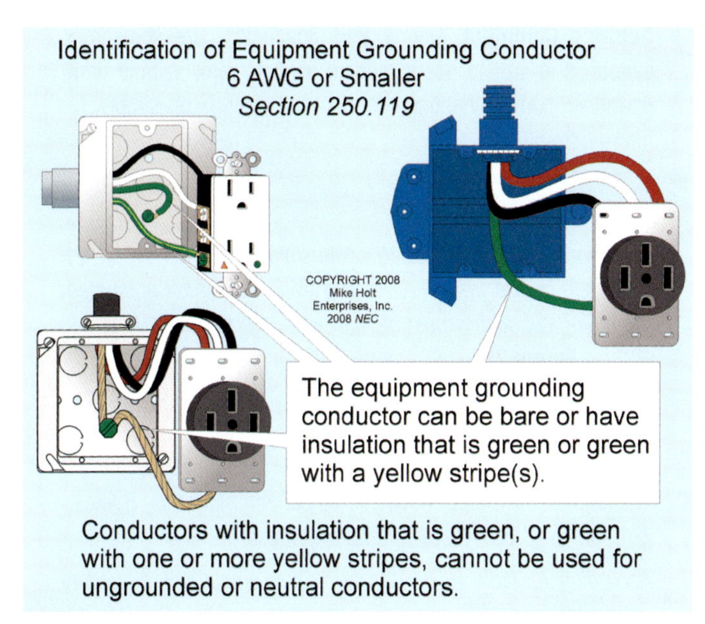

Figure 250–180

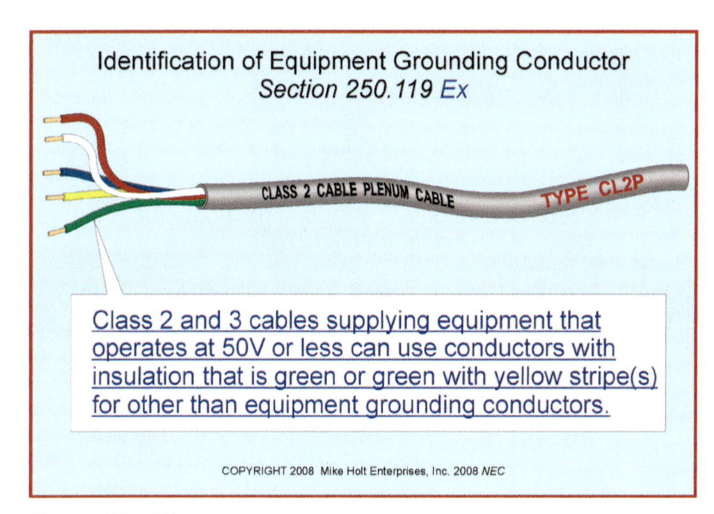

Figure 250–181

(A) Conductors Larger Than 6 AWG.

(1) Identified Where Accessible. Insulated equipment grounding conductors larger than 6 AWG can be permanently reidentified at the time of installation at every point where the conductor is accessible.

Exception: Identification of equipment grounding conductors larger than 6 AWG in conduit bodies is not required.

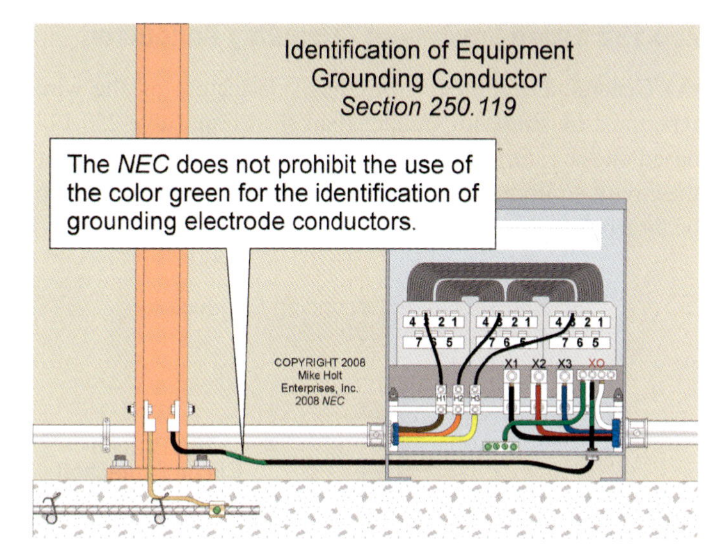

Figure 250–182

(2) Identification Method. Equipment grounding conductor identification must encircle the conductor by: Figure 250–183

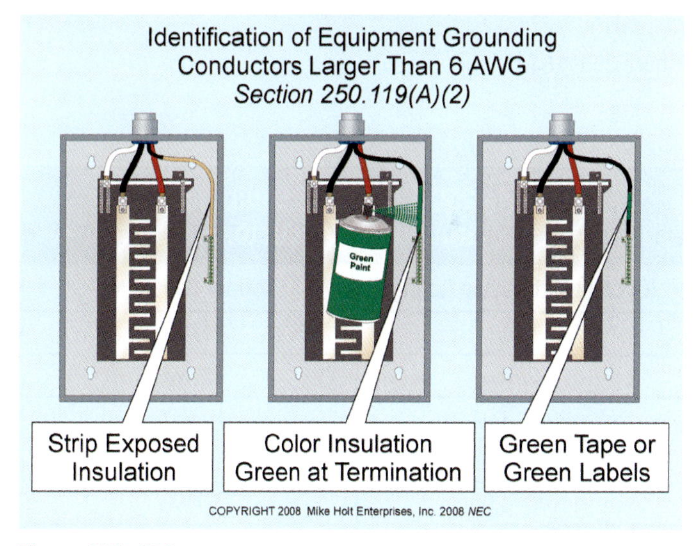

Figure 250–183

a. Removing insulation at termination

b. Coloring the insulation green at termination

c. Marking the insulation at termination with green tape or green adhesive labels

250.122 Sizing Equipment Grounding Conductor.

(A) General. Equipment grounding conductors of the wire type must be sized not smaller than shown in Table 250.122 based on the rating of the circuit overcurrent device; however the circuit equipment grounding conductor is not required to be larger than the circuit conductors. Figure 250–184

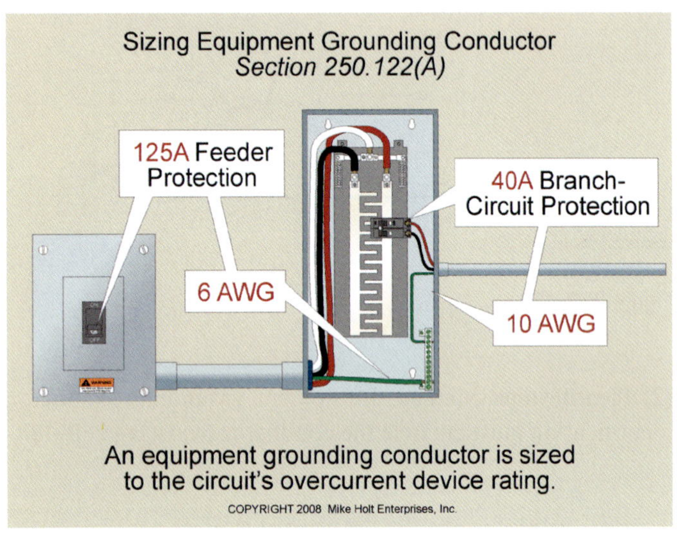

Figure 250–184

Table 250.122 Sizing Equipment Grounding Conductor	
Overcurrent Device Rating	**Copper Conductor**
15A	14 AWG
20A	12 AWG
30A—60A	10 AWG
70A—100A	8 AWG
110A—200A	6 AWG
225A—300A	4 AWG
350A—400A	3 AWG
450A—500A	2 AWG
600A	1 AWG
700A—800A	1/0 AWG
1,000A	2/0 AWG
1,200A	3/0 AWG

(B) Increased in Size. When ungrounded circuit conductors are increased in size, the circuit equipment grounding conductor must be proportionately increased in size according to the circular mil area of the ungrounded conductors.

Author's Comment: Ungrounded conductors are sometimes increased in size to accommodate for conductor voltage drop, harmonic current heating, short-circuit rating, or simply for future capacity.

Question: *If the ungrounded conductors for a 40A circuit are increased in size from 8 AWG to 6 AWG, the circuit equipment grounding conductor must be increased in size from 10 AWG to _____.* Figure 250–185

(a) 10 AWG (b) 8 AWG (c) 6 AWG (d) 4 AWG

Answer: *(b) 8 AWG*

The circular mil area of 6 AWG is 59 percent more than 8 AWG (26,240 cmil/16,510 cmil) [Chapter 9, Table 8].

According to Table 250.122, the circuit equipment grounding conductor for a 40A overcurrent device will be 10 AWG (10,380 cmil), but the circuit equipment grounding conductor for this circuit must be increased in size by a multiplier of 1.59.

Conductor Size = 10,380 cmil x 1.59
Conductor Size = 16,504 cmil
Conductor Size = 8 AWG, Chapter 9, Table 8

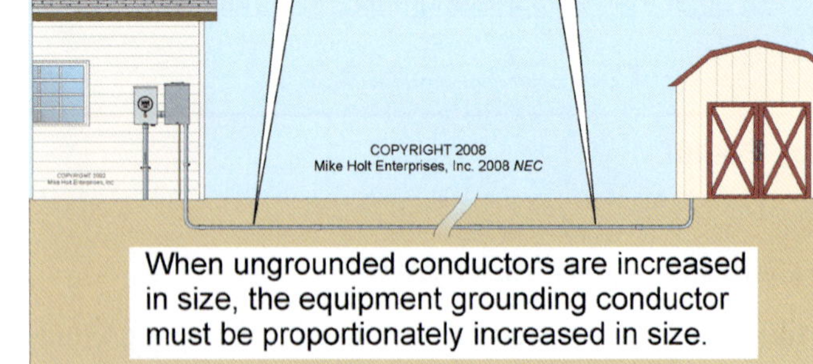

Figure 250–185

(C) Multiple Circuits. When multiple circuits are installed in the same raceway, cable, or cable tray, only one equipment grounding conductor is required for the multiple circuits, sized in accordance with 250.122, based on the rating of the largest circuit overcurrent device. Figures 250–186 and 250–187

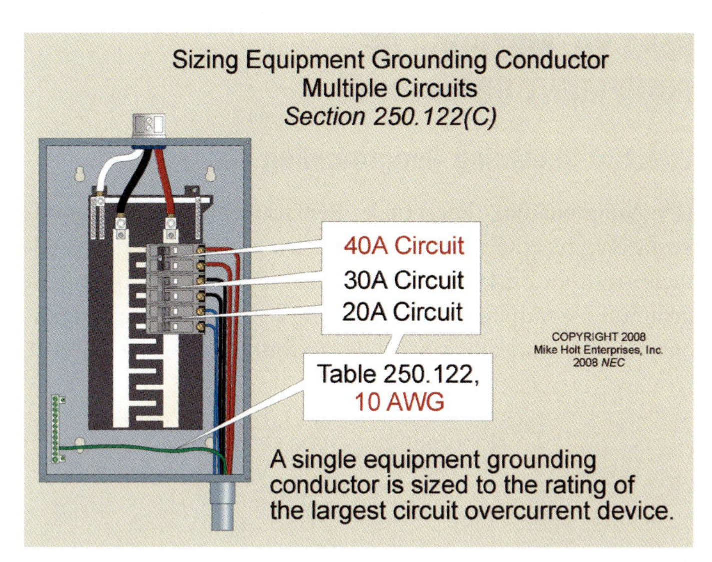

Figure 250–186

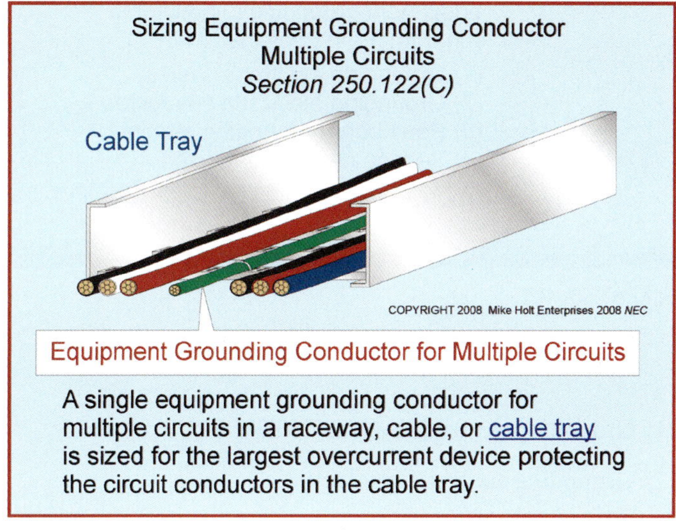

Figure 250–187

Author's Comment: Single conductors used as equipment grounding conductors in cable trays must be sized 4 AWG or larger [392.3(B)(1)(c)].

(D) Motor Branch Circuits.

(1) General. The equipment grounding conductor, where of the wire type, must be sized in accordance with Table 250.122, based on the rating of the motor circuit branch-circuit short-circuit and ground-fault overcurrent device, but this conductor is not required to be larger than the circuit conductors [250.122(A)].

Question: *What size equipment grounding conductor is required for a 2 hp, 230V, single-phase motor?* **Figure 250–188**

(a) 14 AWG (b) 12 AWG (c) 10 AWG (d) 8 AWG

Answer: *(a) 14 AWG*

Step 1: Determine the branch-circuit conductor size [Table 310.16 and 430.22(A)]

2 hp, 230V Motor FLC = 12A [Table 430.248]

12A x 1.25 = 15A, 14 AWG, rated 20A at 75°C [Table 310.16]

Step 2: Determine the branch-circuit protection [240.6(A), 430.52(C)(1), and Table 430.248]

12A x 2.50 = 30A

Step 3: The circuit equipment grounding conductor must be sized to the 30A overcurrent device—10 AWG [Table 250.122], but it's not required to be sized larger than the circuit conductors—14 AWG.

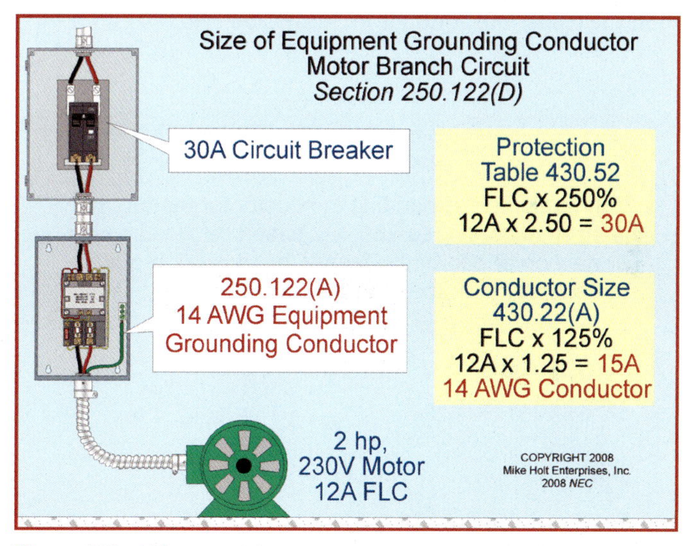

Figure 250–188

(F) Parallel Runs. When circuit conductors are run in parallel [310.4], an equipment grounding conductor must be installed with each parallel conductor set and it must be sized in accordance with Table 250.122, based on the rating of the circuit overcurrent device, but this conductor is not required to be larger than the circuit conductors [250.112(A)]. **Figure 250–189**

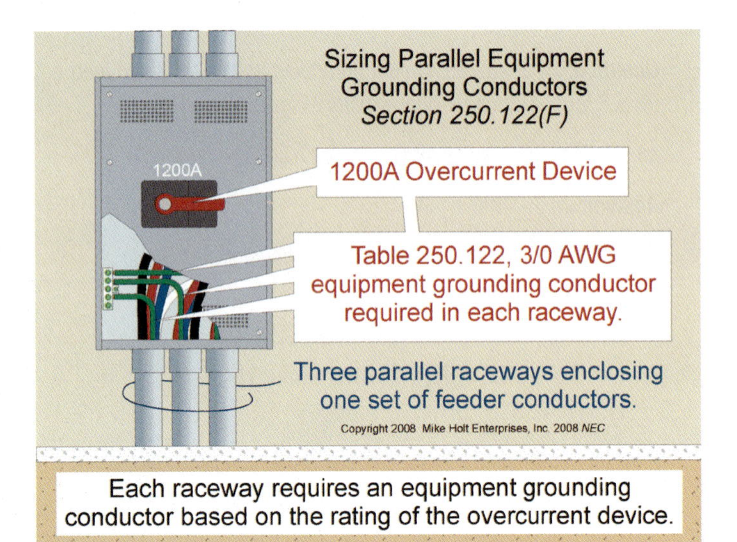

Figure 250–189

(G) Feeder Tap Conductors. Equipment grounding conductors for feeder taps must be sized in accordance with Table 250.122, based on the ampere rating of the overcurrent device ahead of the feeder, but in no case is it required to be larger than the feeder tap conductors. Figure 250–190

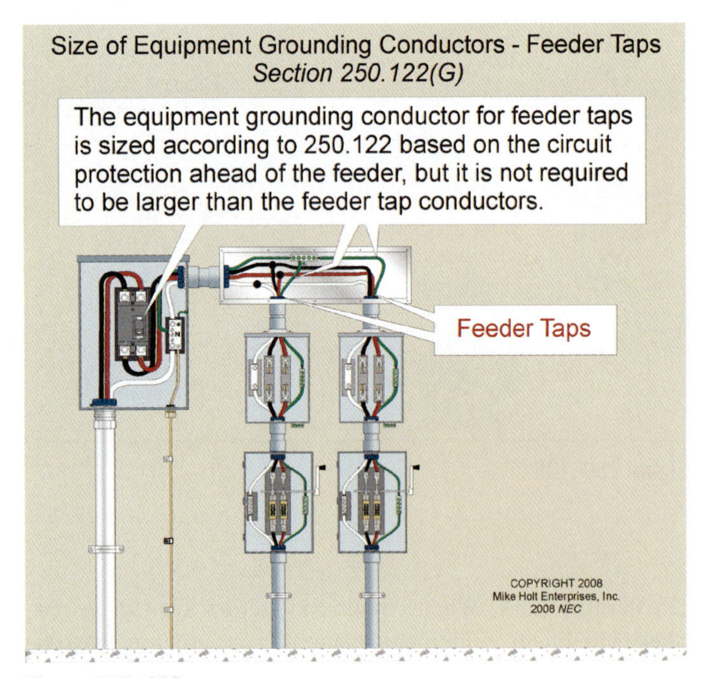

Figure 250–190

PART VII. METHODS OF EQUIPMENT GROUNDING

250.130 Replacing Nongrounding Receptacles.

(C) Nongrounding Receptacle Replacement. Where a nongrounding receptacle is replaced with a grounding-type receptacle from an outlet box that doesn't contain an equipment grounding conductor, the grounding contacts of the receptacle must be connected to one of the following: Figure 250–191

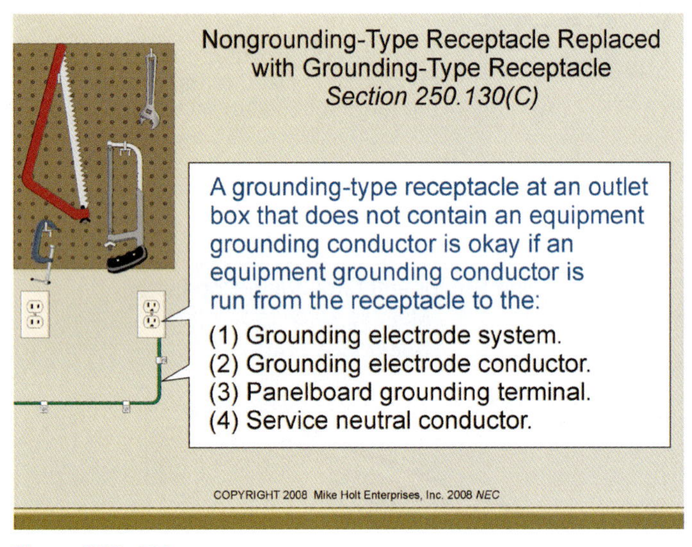

Figure 250–191

(1) Grounding electrode system [250.50]

(2) Grounding electrode conductor

(3) Panelboard equipment grounding terminal

(4) Service neutral conductor

> **FPN:** A grounding-type receptacle can replace a nongrounding type receptacle, without having the grounding terminal connected to an equipment grounding conductor, if the receptacle is GFCI protected and marked in accordance with 406.3(D)(3). Figure 250–192

250.134 Equipment Fastened in Place or Connected by Wiring Methods. Unless connected to the neutral conductor at services or separately derived systems as permitted or required by 250.142, metal parts of equipment, raceways, and enclosures must be connected to an equipment grounding conductor by one of the following methods:

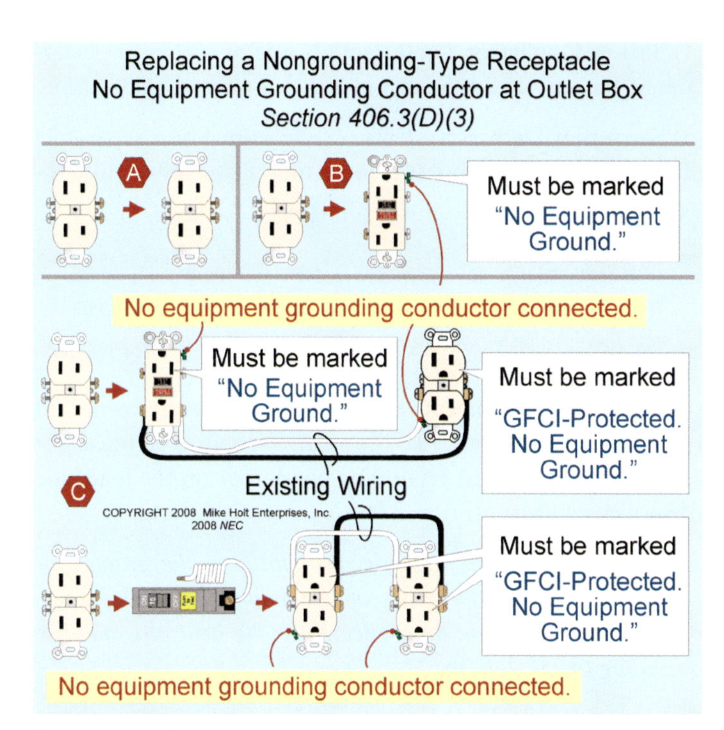

Figure 250–192

(A) Equipment Grounding Conductor Types. By connecting to equipment grounding conductors identified in 250.118.

(B) With Circuit Conductors. Where an equipment grounding conductor of the wire type is installed, it must be installed in the same raceway, cable tray, trench, cable, or cord with the circuit conductors in accordance with 300.3(B), except as permitted by 250.102(E). Figures 250–193 and 250–194

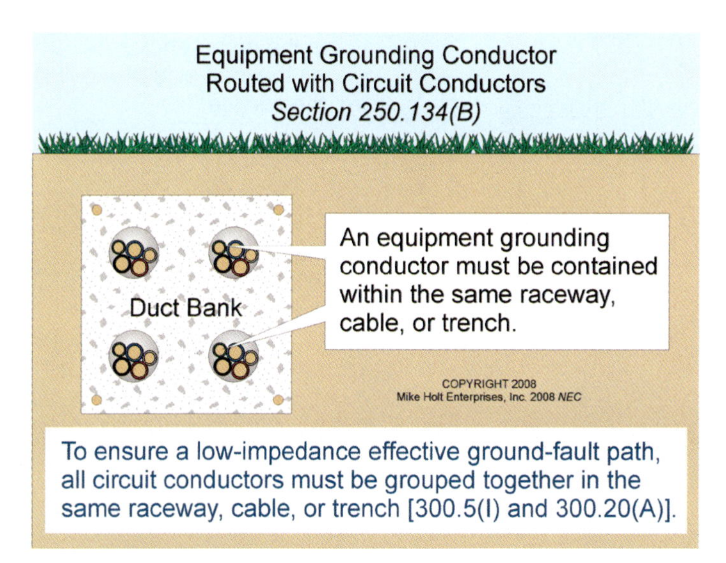

Figure 250–193

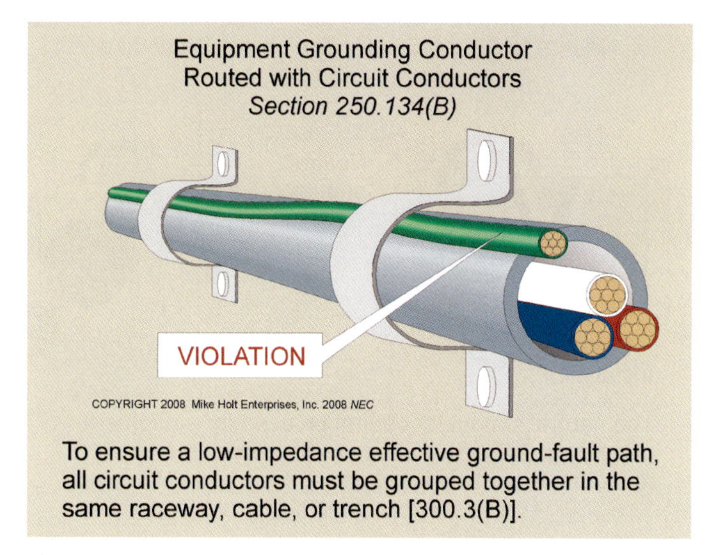

Figure 250–194

250.136 Equipment Considered Grounded.

(A) Equipment Secured to Grounded Metal Supports. The structural metal frame of a building must not be used as the required equipment grounding conductor.

250.138 Cord-and-Plug-Connected Equipment.

(A) Equipment Grounding Conductor. Metal parts of cord-and-plug-connected equipment must be connected to an equipment grounding conductor that terminates to a grounding-type attachment plug.

250.140 Ranges, Ovens, and Clothes Dryers. The frames of electric ranges, wall-mounted ovens, counter-mounted cooking units, clothes dryers, and outlet boxes that are part of the circuit for these appliances must be connected to the equipment grounding conductor [250.134(A)]. Figure 250–195

CAUTION: *Ranges, dryers, and ovens have their metal cases connected to the neutral conductor at the factory. This neutral-to-case connection must be removed when these appliances are installed in new construction, and a 4-wire cord and receptacle must be used [250.142(B)].*

Exception: For existing installations where an equipment grounding conductor isn't present in the outlet box, the frames of electric ranges, wall-mounted ovens, counter-mounted cooking units, clothes dryers, and outlet boxes that are part of the circuit for these appliances may be connected to the neutral conductor. Figure 250–196

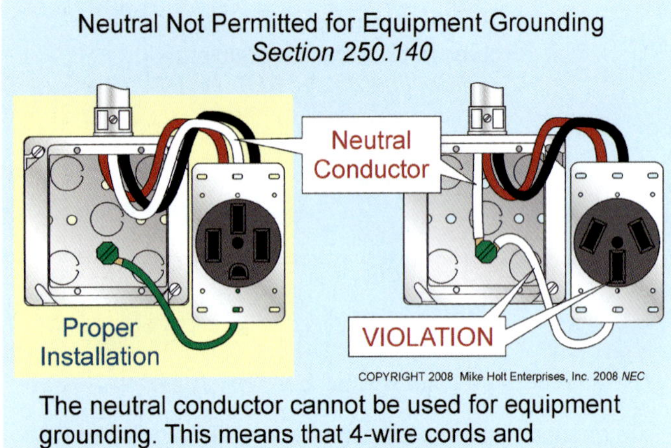

The neutral conductor cannot be used for equipment grounding. This means that 4-wire cords and receptacles are required for ranges, dryers, and ovens. See 250.140 Ex for existing installations.

Figure 250–195

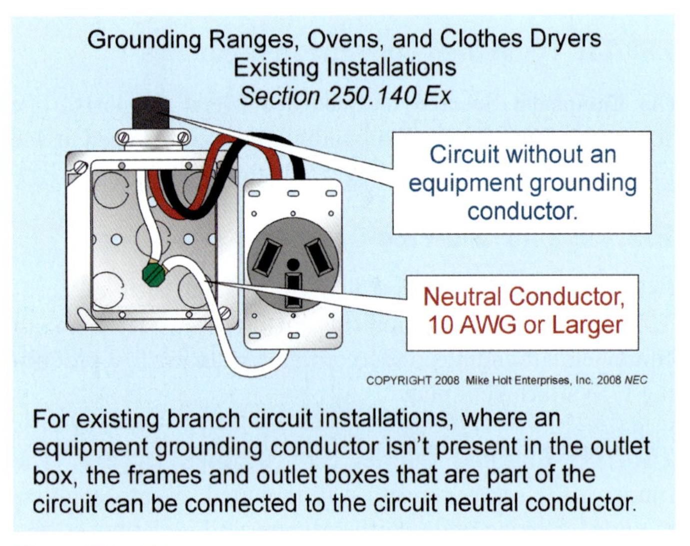

For existing branch circuit installations, where an equipment grounding conductor isn't present in the outlet box, the frames and outlet boxes that are part of the circuit can be connected to the circuit neutral conductor.

Figure 250–196

250.142 Use of Neutral Conductor for Equipment Grounding.

Author's Comment: To remove dangerous voltage on metal parts from a ground fault, the metal parts of electrical raceways, cables, enclosures, and equipment must be connected to an equipment grounding conductor of a type recognized in 250.118 in accordance with 250.4(A)(3).

(A) Supply Side Equipment. The neutral conductor can be used as the circuit equipment grounding conductor for metal parts of equipment, raceways, and enclosures at the following locations:

(1) Service Equipment. On the supply side or within the enclosure of the service disconnect in accordance with 250.24(B).

(3) Separately Derived Systems. At the source of a separately derived system or within the enclosure of the system disconnecting means in accordance with 250.30(A)(1).

DANGER: *Failure to install the system bonding jumper as required by 250.30(A)(1) creates a condition where dangerous touch voltage from a ground fault will not be removed.*

(B) Load Side Equipment. Except for service equipment and separately derived systems, the neutral conductor must not serve as an equipment grounding conductor on the load side of service equipment.

Exception No. 1: In existing installations, the frames of ranges, wall-mounted ovens, counter-mounted cooking units, and clothes dryers can be connected to the neutral conductor according to 250.140 Ex.

Exception No. 2: The neutral conductor can be connected to meter socket enclosures on the load side of the service disconnecting means if: Figure 250–197

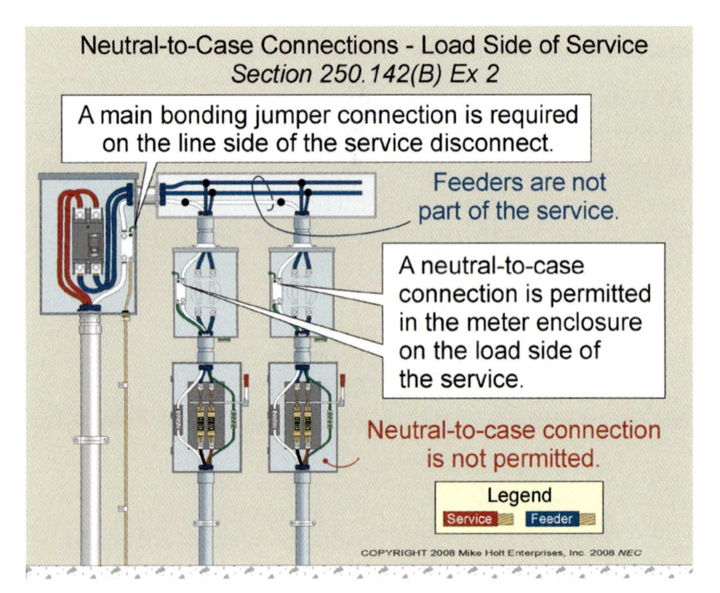

Figure 250–197

(1) Ground-fault protection is not provided on service equipment,

(2) Meter socket enclosures are immediately adjacent to the service disconnecting means, and

(3) The neutral conductor is sized in accordance with 250.122, based on the ampere rating of the occupancy's feeder overcurrent device.

250.146 Connecting Receptacle Grounding Terminal to Metal Enclosure. An equipment bonding jumper sized in accordance with 250.122, based on the rating of the circuit overcurrent device, must connect the grounding terminal of a receptacle to a metal box, except as permitted for (A) through (D). Figure 250–198

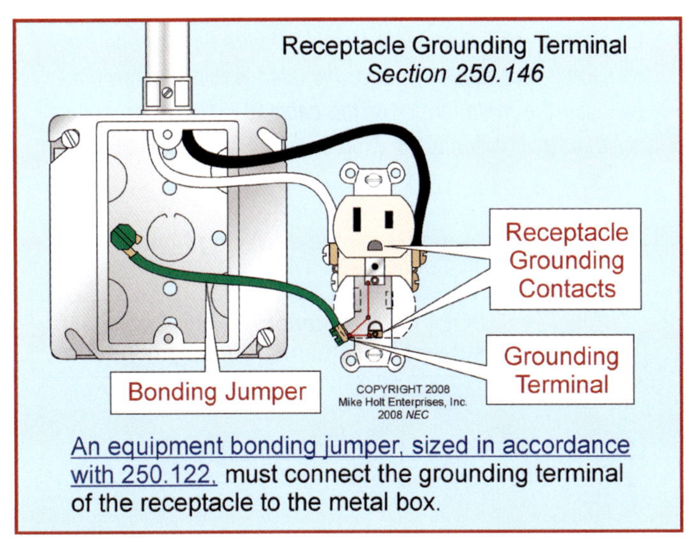

Figure 250–198

Author's Comment: The *NEC* does not restrict the position of the receptacle grounding terminal; it can be up, down, or sideways. *Code* proposals to specify the mounting position of receptacles have always been rejected. Figure 250–199

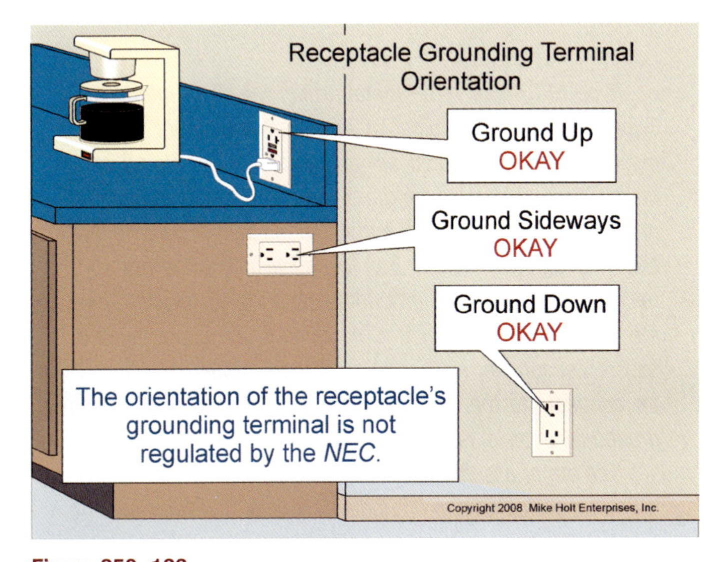

Figure 250–199

(A) Surface-Mounted Box. An equipment bonding jumper from a receptacle to a metal box that is surface mounted is not required where there is direct metal-to-metal contact between the device yoke and the metal box. To ensure a suitable bonding path between the device yoke and a metal box, at least one of the insulating retaining washers on the yoke screw must be removed. Figure 250–200

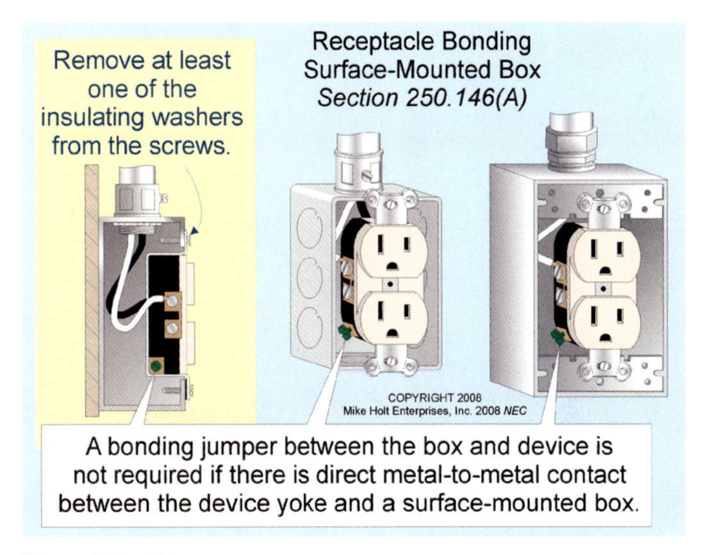

Figure 250–200

An equipment bonding jumper is not required for receptacles attached to listed exposed work covers when the receptacle is attached to the cover with at least two fasteners that have a thread locking or screw locking means, and the cover mounting holes are located on a flat non-raised portion of the cover. Figure 250–201

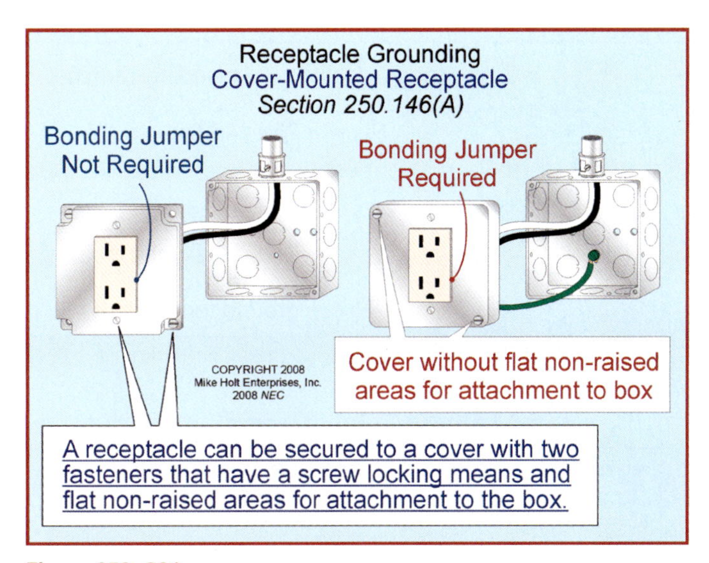

Figure 250–201

(B) Self-Grounding Receptacles. Receptacle yokes listed as self-grounding are designed to establish the bonding path between the device yoke and a metal box via the two metal mounting screws. Figure 250–202

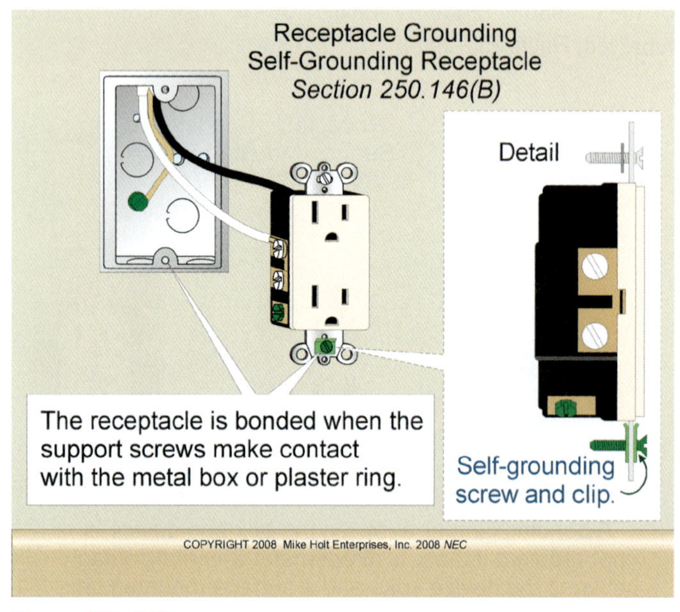

Figure 250–202

(C) Floor Boxes. Listed floor boxes are designed to establish the bonding path between the device yoke and a metal box.

(D) Isolated Ground Receptacles. Where installed for the reduction of electrical noise, the grounding terminal of an isolated ground receptacle must be connected to an insulated equipment grounding conductor run with the circuit conductors. Figure 250–203

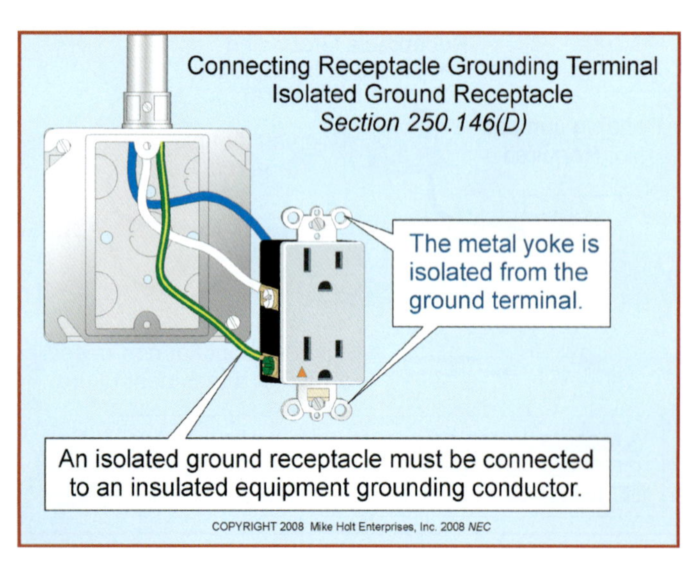

Figure 250–203

The circuit equipment grounding conductor is permitted to pass through panelboards [408.40 Ex], boxes, wireways, or other enclosures [250.148 Ex] without a connection to the enclosure as long as it terminates at an equipment grounding conductor terminal of the derived system or service.

CAUTION:

Type AC Cable—Type AC cable containing an insulated equipment grounding conductor of the wire type can be used to supply receptacles having insulated grounding terminals because the metal armor of the cable is listed as an equipment grounding conductor [250.118(8)]. Figure 250–204

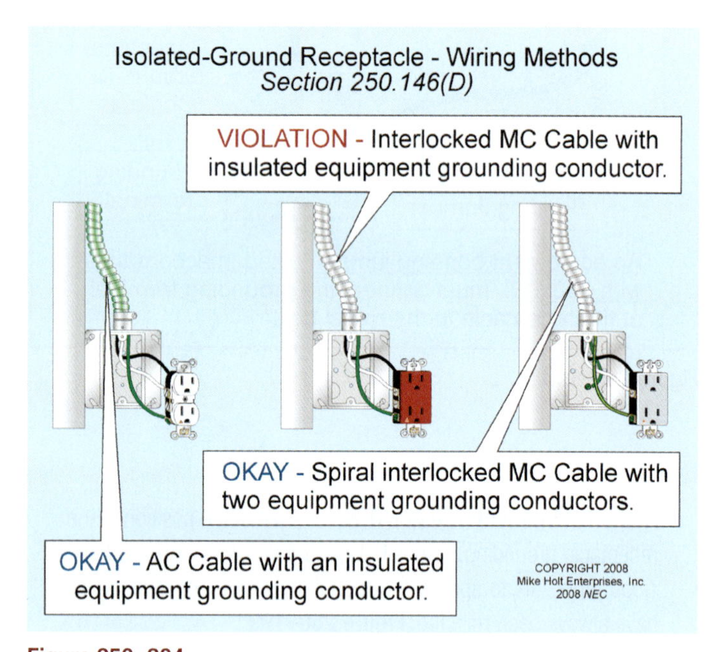

Figure 250–204

Type MC Cable—The metal armor sheath of interlocked Type MC cable containing an insulated equipment grounding conductor isn't listed as an equipment grounding conductor. Therefore, this wiring method with a single equipment grounding conductor can't supply an isolated ground receptacle installed in a metal box (because the box is not connected to an equipment grounding conductor). However, Type MC cable with two insulated equipment grounding conductors is acceptable, since one equipment grounding conductor connects to the metal box and the other to the isolated ground receptacle. See Figure 250–204

The armor assembly of interlocked Type MC^AP cable with a 10 AWG bare aluminum grounding/bonding conductor running just below the metal armor is listed to serve as an equipment grounding conductor in accordance with 250.118(10)(a).

Nonmetallic Boxes—Because the grounding terminal of an isolated ground receptacle is insulated from the metal mounting yoke, a metal faceplate must not be used when an isolated ground receptacle is installed in a nonmetallic box. The reason is that the metal faceplate is not connected to an equipment grounding conductor [406.2(D)(2)]. **Figure 250–205**

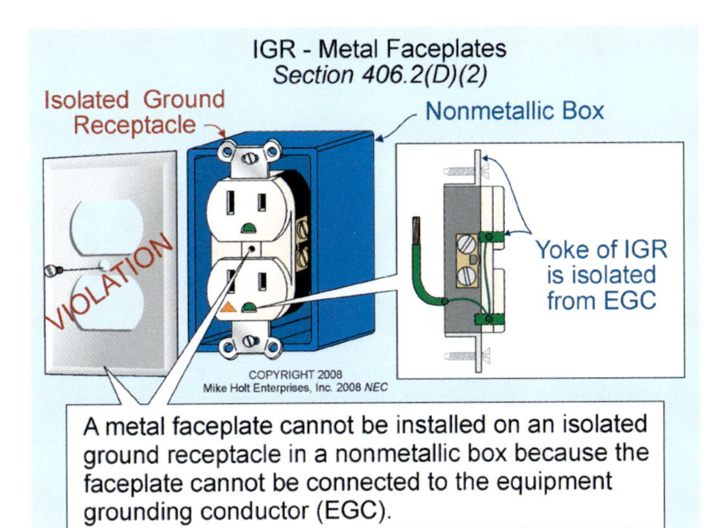

Figure 250–205

Author's Comment: When should an isolated ground receptacle be installed and how should the isolated ground system be designed? These questions are design issues and must not be answered based on the *NEC* alone [90.1(C)]. In most cases, using isolated ground receptacles is a waste of money. For example, IEEE 1100, *Powering and Grounding Electronic Equipment* (Emerald Book) states: "The results from the use of the isolated-ground method range from no observable effects, the desired effects, or worse noise conditions than when standard equipment bonding configurations are used to serve electronic load equipment [8.5.3.2]."

In reality, few electrical installations truly require an isolated-ground system. For those systems that can benefit from an isolated-ground system, engineering opinions differ as to what is a proper design. Making matters worse—of those properly designed, few are correctly installed and even fewer are properly maintained. For more information on how to properly ground electronic equipment, go to: www.MikeHolt.com, click on the "Technical" link, and then visit the "Power Quality" page.

250.148 Continuity and Attachment of Equipment Grounding Conductors in Boxes.

Where circuit conductors are spliced or terminated on equipment within a metal box, the equipment grounding conductor associated with those circuits must be connected to the box in accordance with the following: **Figure 250–206**

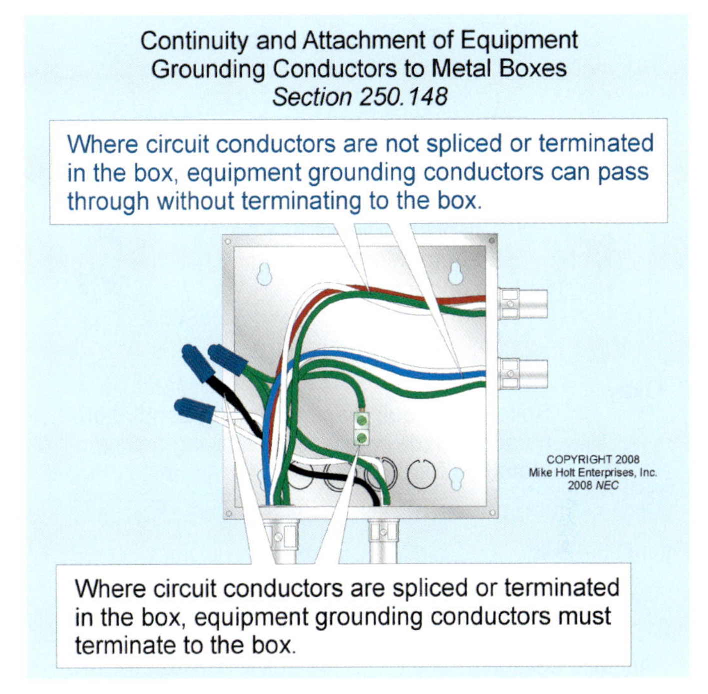

Figure 250–206

Exception: The circuit equipment grounding conductor for an isolated ground receptacle installed in accordance with 250.146(D) isn't required to terminate to a metal box. **Figure 250–207**

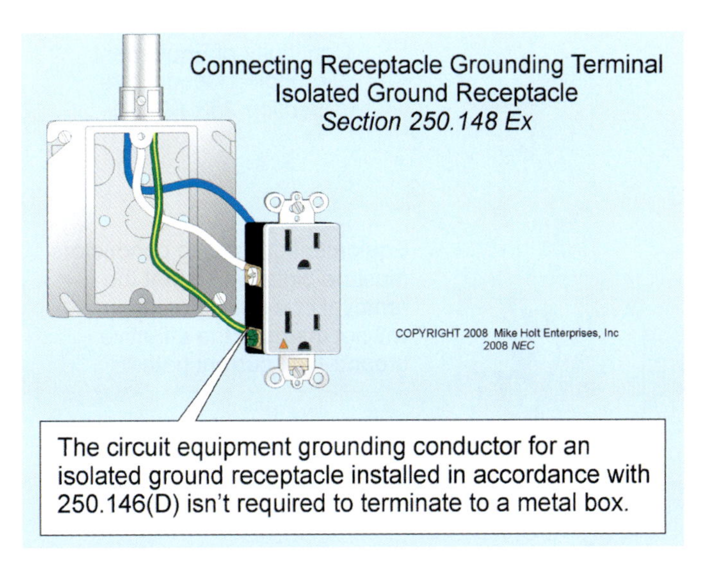

Figure 250–207

(A) Splicing. Equipment grounding conductors must be spliced together with a device listed for the purpose [110.14(B)]. Figure 250–208

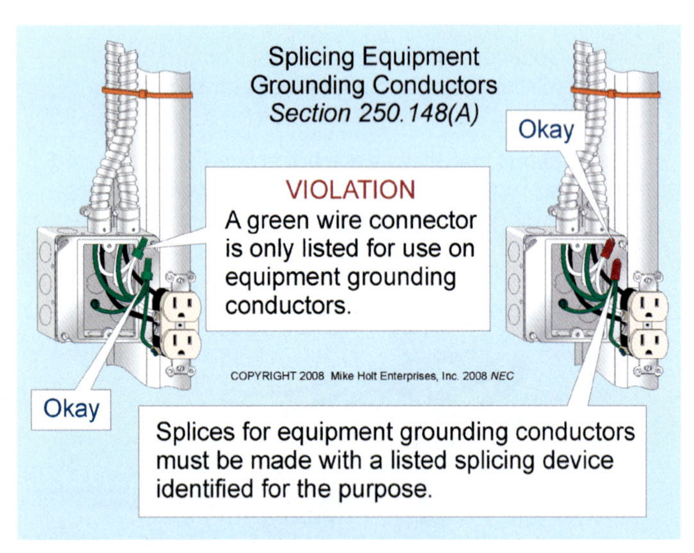

Figure 250–208

Author's Comment: Wire connectors of any color can be used with equipment grounding conductor splices, but green wire connectors can only be used with equipment grounding conductors.

(B) Equipment Grounding Continuity. Equipment grounding conductors must terminate in a manner that the disconnection or the removal of a receptacle, luminaire, or other device will not interrupt the grounding continuity. Figure 250–209

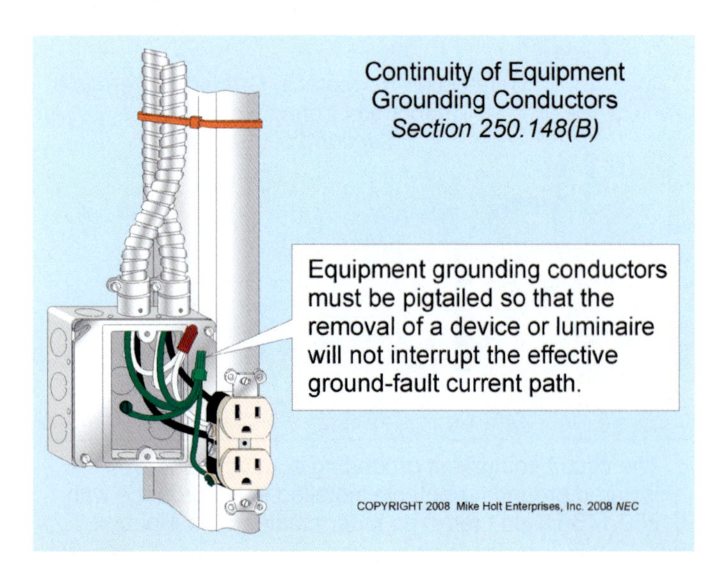

Figure 250–209

(C) Metal Boxes. Equipment grounding conductors within metal boxes must be connected to the metal box with a grounding screw that is not used for any other purpose, equipment fitting listed for grounding, or a listed grounding device such as a ground clip. Figure 250–210

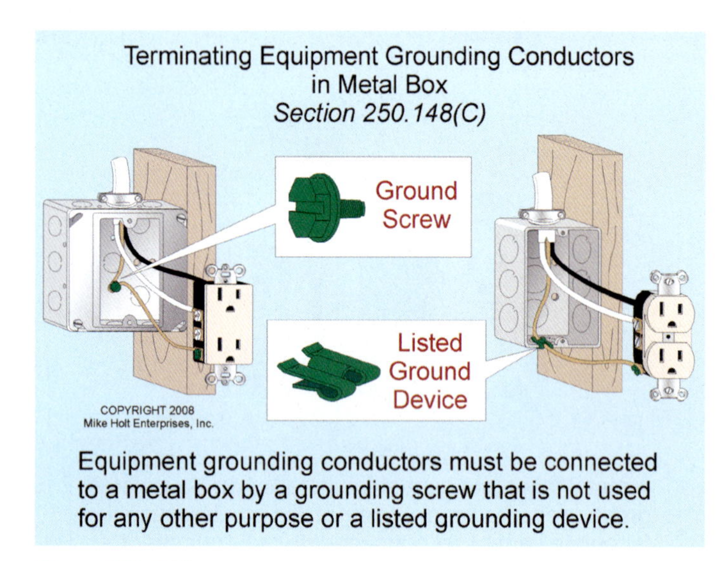

Figure 250–210

Author's Comment: Equipment grounding conductors aren't permitted to terminate to a screw that secures a plaster ring. Figure 250–211

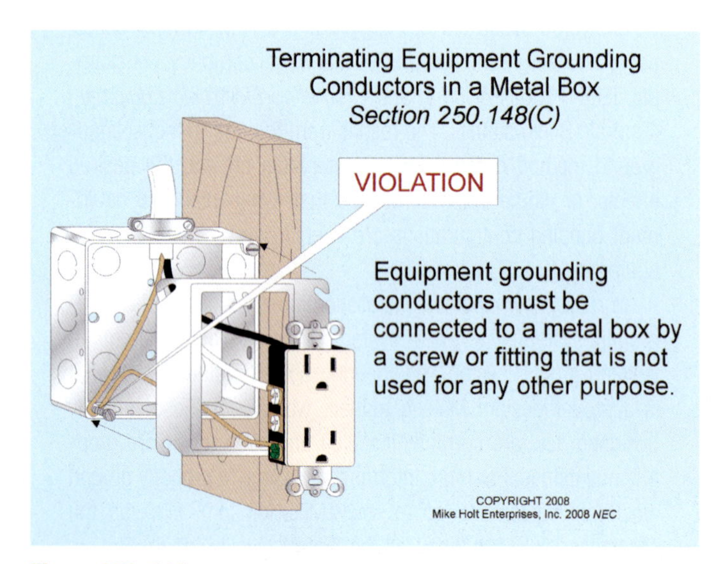

Figure 250–211

ARTICLE 250 Practice Questions

ARTICLE 250. GROUNDING AND BONDING—PRACTICE QUESTIONS

PART I. GENERAL

1. A(n) _____ is an unintentional, electrically conducting connection between an ungrounded normally current-carrying conductor of an electrical circuit, and the normally noncurrent-carrying conductors, metallic enclosures, metallic raceways, metallic equipment, or earth.

   (a) grounded conductor
   (b) ground fault
   (c) equipment ground
   (d) bonding jumper

2. Grounded electrical systems shall be connected to earth in a manner that will _____.

   (a) limit voltages due to lightning, line surges, or unintentional contact with higher-voltage lines
   (b) stabilize the voltage-to-ground during normal operation
   (c) facilitate overcurrent device operation in case of ground faults
   (d) a and b

3. In grounded systems, normally noncurrent-carrying electrically conductive materials that are likely to become energized shall be _____ in a manner that establishes an effective ground-fault current path.

   (a) connected together
   (b) connected to the electrical supply source
   (c) connected to the closest grounded conductor
   (d) a and b

4. For grounded systems, the earth is considered an effective ground-fault current path.

   (a) True
   (b) False

5. Electrically conductive materials that are likely to _____ in ungrounded systems shall be connected together and to the supply system grounded equipment in a manner that creates a low-impedance path for ground-fault current that is capable of carrying the maximum fault current likely to be imposed on it.

   (a) become energized
   (b) require service
   (c) be removed
   (d) be coated with paint or nonconductive materials

6. Temporary currents resulting from accidental conditions, such as ground faults, are not considered to be objectionable currents.

   (a) True
   (b) False

7. _____ on equipment to be grounded shall be removed from contact surfaces to ensure good electrical continuity.

   (a) Paint
   (b) Lacquer
   (c) Enamel
   (d) any of these

PART II. SYSTEM GROUNDING AND BONDING

1. AC systems of 50V to 1,000V that supply premises wiring systems shall be grounded where the system is three-phase, 4-wire, wye-connected with the neutral conductor used as a circuit conductor.

   (a) True
   (b) False

2. _____ AC systems operating at 480V shall have ground detectors installed on the system.

   (a) Grounded
   (b) Solidly grounded
   (c) Effectively grounded
   (d) Ungrounded

3. For a grounded system, an unspliced _____ shall be used to connect the equipment grounding conductor(s) and the service disconnect enclosure to the grounded conductor of the system within the enclosure for each service disconnect.

(a) grounding electrode
(b) main bonding jumper
(c) busbar
(d) insulated copper conductor

4. When service-entrance conductors exceed 1,100 kcmil for copper, the required grounded conductor for the service shall be sized not less than _____ percent of the area of the largest ungrounded service-entrance conductor.

(a) 9
(b) 11
(c) 12½
(d) 15

5. A main bonding jumper shall be a _____ or similar suitable conductor.

(a) wire
(b) bus
(c) screw
(d) any of these

6. Main bonding jumpers and system bonding jumpers shall not be smaller than the sizes shown in _____.

(a) Table 250.66
(b) Table 250.122
(c) Table 310.16
(d) Chapter 9, Table 8

7. A grounded conductor shall not be connected to normally noncurrent-carrying metal parts of equipment on the _____ of the point of grounding of the separately derived system except as otherwise permitted in Article 250.

(a) supply
(b) grounded side
(c) high voltage side
(d) load side

8. The connection of the system bonding jumper for a separately derived system shall be made _____ on the separately derived system from the source to the first system disconnecting means or overcurrent device.

(a) in at least two locations
(b) in every location that the grounded conductor is present
(c) at any single point
(d) none of these

9. The grounding electrode conductor for a single separately derived system is used to connect the grounded conductor of the derived system to the grounding electrode.

(a) True
(b) False

10. Each tap conductor to a common grounding electrode conductor for multiple separately derived systems shall be sized in accordance with _____ based on the derived phase conductors of the separately derived system it serves.

(a) 250.66
(b) 250.118
(c) 250.122
(d) 310.15

11. In an area served by a separately derived system, the _____ shall be connected to the grounded conductor of the separately derived system.

(a) structural steel
(b) metal piping
(c) metal building skin
(d) a and b

12. A grounding electrode shall be required if a building or structure is supplied by a feeder.

(a) True
(b) False

13. The size of the grounding electrode conductor for a building or structure supplied by a feeder shall not be smaller than that identified in _____, based on the largest ungrounded supply conductor.

(a) 250.66
(b) 250.122
(c) Table 310.16
(d) not specified

14. High-impedance grounded neutral systems shall be permitted for three-phase ac systems of 480 volts to 1,000 volts where _____.

    (a) the conditions of maintenance ensure that only qualified persons service the installation
    (b) ground detectors are installed on the system
    (c) line-to-neutral loads are not served
    (d) all of these

PART III. GROUNDING ELECTRODE SYSTEM AND GROUNDING ELECTRODE CONDUCTOR

1. For a metal underground water pipe to be used as a grounding electrode, it shall be in direct contact with the earth for _____.

   (a) 5 ft
   (b) 10 ft or more
   (c) less than 10 ft
   (d) 20 ft or more

2. An electrode encased by at least 2 in. of concrete, located horizontally near the bottom or vertically and within that portion of a concrete foundation or footing that is in direct contact with the earth, shall be permitted as a grounding electrode when it consists of _____.

   (a) at least 20 ft of ½ in. or larger steel reinforcing bars or rods
   (b) at least 20 ft of bare copper conductor of 4 AWG or larger
   (c) a or b
   (d) none of these

3. A ground ring encircling the building or structure can be used as a grounding electrode when _____.

   (a) the ring is in direct contact with the earth
   (b) the ring consists of at least 20 ft of bare conductor
   (c) the bare copper conductor is not smaller than 2 AWG
   (d) all of these

4. Local metal underground systems or structures such as _____ are permitted to serve as grounding electrodes.

   (a) piping systems
   (b) underground tanks
   (c) underground metal well casings
   (d) all of these

5. Two or more grounding electrodes bonded together are considered a single grounding electrode system.

   (a) True
   (b) False

6. When a ground ring is used as a grounding electrode, it shall be buried at a depth below the earth's surface of not less than _____.

   (a) 18 in.
   (b) 24 in.
   (c) 30 in.
   (d) 8 ft

7. Where rock bottom is encountered when driving a ground rod at an angle up to 45 degrees, the electrode can be buried in a trench that is at least _____ deep.

   (a) 18 in.
   (b) 30 in.
   (c) 4 ft
   (d) 8 ft

8. Where the resistance-to-ground of 25 ohms or less is not achieved for a single rod electrode, _____.

   (a) other means besides electrodes shall be used in order to provide grounding
   (b) the single rod electrode shall be augmented by one additional electrode
   (c) no additional electrodes are required
   (d) none of these

9. Grounding electrode conductors shall be made of _____ wire.

   (a) solid
   (b) stranded
   (c) insulated or bare
   (d) any of these

10. Grounding electrode conductors smaller than _____ shall be in rigid metal conduit, IMC, PVC conduit, electrical metallic tubing, or cable armor.

    (a) 10 AWG
    (b) 8 AWG
    (c) 6 AWG
    (d) 4 AWG

11. A service consisting of 12 AWG service-entrance conductors requires a grounding electrode conductor sized no less than _____.

    (a) 10 AWG
    (b) 8 AWG
    (c) 6 AWG
    (d) 4 AWG

12. In an ac system, the size of the grounding electrode conductor to a concrete-encased electrode shall not be required to be larger than a(n) _____ copper conductor.

    (a) 10 AWG
    (b) 8 AWG
    (c) 6 AWG
    (d) 4 AWG

13. Exothermic or irreversible compression connections, together with the mechanical means used to attach to fireproofed structural metal, shall not be required to be accessible.

    (a) True
    (b) False

PART IV. ENCLOSURE, RACEWAY, AND SERVICE CABLE CONNECTIONS

1. Metal enclosures and raceways for other than service conductors shall be connected to the grounded conductor.

    (a) True
    (b) False

PART V. BONDING

1. Service equipment, service raceways, and service conductor enclosures shall be bonded _____.

    (a) to the grounded service conductor
    (b) by threaded raceways into enclosures, couplings, hubs, conduit bodies, etc.
    (c) by listed bonding devices with bonding jumpers
    (d) any of these

2. A means external to enclosures for connecting intersystem _____ conductors shall be provided at service equipment and disconnecting means of other buildings or structures.

    (a) bonding
    (b) grounding
    (c) secondary
    (d) a and b

3. Equipment bonding jumpers shall be of copper or other corrosion-resistant material.

    (a) True
    (b) False

4. The equipment bonding jumper on the supply side of services shall be sized according to the _____.

    (a) overcurrent device rating
    (b) service-entrance conductor size
    (c) service-drop size
    (d) load to be served

5. A service is supplied by three metal raceways, each containing 600 kcmil ungrounded conductors. Determine the bonding jumper size for each service raceway.

    (a) 1/0 AWG
    (b) 3/0 AWG
    (c) 250 kcmil
    (d) 500 kcmil

6. Metal water piping system(s) shall be bonded to the _____.

    (a) grounded conductor at the service
    (b) service equipment enclosure
    (c) equipment grounding bar or bus at any panelboard within a single occupancy building
    (d) a or b

7. Where isolated metal water piping systems are installed in a multi-occupancy building, the water pipes can be bonded with bonding jumpers sized per Table 250.122, based on the size of _____.

    (a) service-entrance conductors
    (b) feeder conductors
    (c) rating of the service equipment overcurrent device
    (d) rating of overcurrent device supplying the occupancy.

8. Metal gas piping shall be considered bonded by the equipment grounding conductor of the circuit that is likely to energize the piping.

   (a) True
   (b) False

PART VI. EQUIPMENT GROUNDING AND EQUIPMENT GROUNDING CONDUCTORS

1. Listed FMC and LFMC shall contain an equipment grounding conductor if the raceway is installed for the reason of _____.

   (a) physical protection
   (b) flexibility after installation
   (c) protection from moisture
   (d) communications systems

2. The equipment grounding conductor shall not be required to be larger than the circuit conductors.

   (a) True
   (b) False

3. Equipment grounding conductors for motor branch circuits shall be sized in accordance with Table 250.122, based on the rating of the _____ device.

   (a) motor overload
   (b) motor over-temperature
   (c) motor short-circuit and ground-fault protective
   (d) feeder overcurrent protection

4. The terminal of a wiring device for the connection of the equipment grounding conductor shall be identified by a green-colored, _____.

   (a) not readily removable terminal screw with a hexagonal head
   (b) hexagonal, not readily removable terminal nut
   (c) pressure wire connector
   (d) any of these

PART VII. METHODS OF EQUIPMENT GROUNDING

1. Metal parts of cord-and-plug-connected equipment shall be connected to an equipment grounding conductor that terminates to a grounding-type attachment plug.

   (a) True
   (b) False

2. A(n) _____ shall be used to connect the grounding terminal of a grounding-type receptacle to a grounded box.

   (a) equipment bonding jumper
   (b) grounded conductor jumper
   (c) a or b
   (d) a and b

3. Receptacle yokes designed and _____ as self-grounding can establish the grounding circuit between the device yoke and a grounded outlet box.

   (a) approved
   (b) advertised
   (c) listed
   (d) installed

4. The arrangement of grounding connections shall be such that the disconnection or the removal of a receptacle, luminaire, or other device does not interrupt the grounding continuity.

   (a) True
   (b) False

ARTICLE

285 Surge Protective Devices (SPDs)

INTRODUCTION TO ARTICLE 285—SURGE PROTECTIVE DEVICES (SPDs)

This article covers the general requirements, installation requirements, and connection requirements for surge protective devices rated 1kV or less that are permanently installed on premises wiring systems. The *NEC* does not require surge protective devices to be installed, but if they are, they must comply with this article.

Surge protective devices are designed to reduce transient voltages present on premises power distribution wiring and load-side equipment, particularly electronic equipment such as computers, telecommunications equipment, security systems, and electronic appliances.

PART I. GENERAL

285.1 Scope. Article 285 covers the installation and connection requirements for permanently installed surge protective devices, which are intended to limit transient voltages on circuit conductors by diverting or limiting surge current. Figure 285–1

Figure 285–1

FPN No. 1: Surge arresters rated less than 1,000V are known as Type 1 surge protective devices.

285.3 Uses Not Permitted. A surge protective device must not be used in:

(1) Circuits that exceed 1,000V.

(2) Ungrounded systems, impedance grounded systems, or corner-grounded delta systems, unless listed specifically for use on these systems.

(3) Where the voltage rating of the surge protective device is less than the maximum continuous phase-to-ground voltage available at the point of connection.

285.4 Number Required. Where used, the surge protective device must be connected to each ungrounded conductor of the circuit.

285.5 Listing. Surge protective devices must be listed.

> **Author's Comment:** According to UL 1449, *Standard for Surge Protective Devices*, these units are intended to limit the maximum amplitude of transient voltage surges on power lines to specified values. They aren't intended to function as lightning arresters. The adequacy of the voltage suppression level to protect connected equipment from voltage surges has not been evaluated.

285.6 Short-Circuit Current Rating. Surge protective devices must be marked with their short-circuit current rating, and they must not be installed where the available fault current exceeds that rating. This short-circuit current marking requirement doesn't apply to receptacles containing surge protective device protection.

WARNING: *Surge protective devices of the series type are susceptible to failure at high fault currents. A hazardous condition is present if the surge protective device short-circuit current rating is less than the available fault current.*

PART II. INSTALLATION

285.11 Location. Surge protective devices can be located indoors or outdoors.

285.12 Routing of Conductors. Surge protective device conductors must not be any longer than necessary, and unnecessary bends must be avoided.

PART III. CONNECTING SURGE PROTECTIVE DEVICES

285.23 Type 1 SPD—Line Side of Service Equipment.

(A) Installation. Type 1 surge protective devices can be connected as follows: Figure 285–2

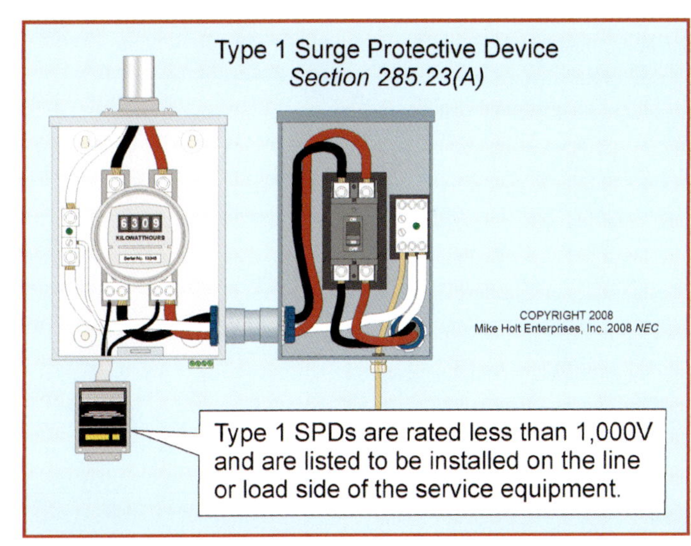

Figure 285–2

(1) Supply side of service equipment [230.82(4)]

(2) Load side of service equipment in accordance with 285.24.

(B) Grounding. Type 1 surge protective devices located at service equipment must be connected to one of the following:

(1) Service neutral conductor,

(2) Grounding electrode conductor,

(3) Grounding electrode for the service, or

(4) Equipment grounding terminal in the service equipment.

Author's Comment: Only one conductor can be connected to a terminal, unless the terminal is identified for multiple conductors [110.14(A)]. Figure 285–3

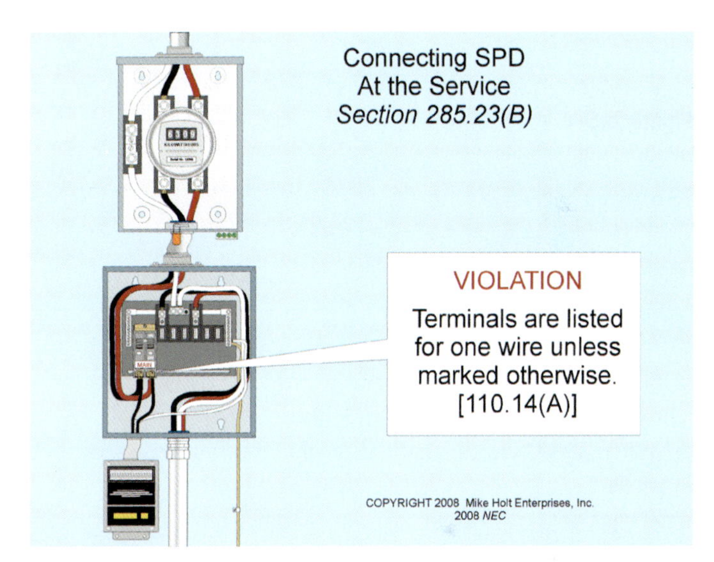

Figure 285–3

285.24 Type 2 SPD—Feeder Circuits.

(A) Service Equipment. Type 2 surge protective devices can be connected to the load side of service equipment. Figure 285–4

Author's Comment: Type 2 surge protective devices are listed for installation only on the load side of service equipment because they are not listed to accommodate lightning-induced surges beyond their capacity.

(B) Feeder-Supplied Buildings or Structures. Type 2 surge protective devices can be connected anywhere on the load side of the building or structure overcurrent device.

(C) Separately Derived Systems. Type 2 surge protective devices can be connected anywhere on the premises wiring of the separately derived system.

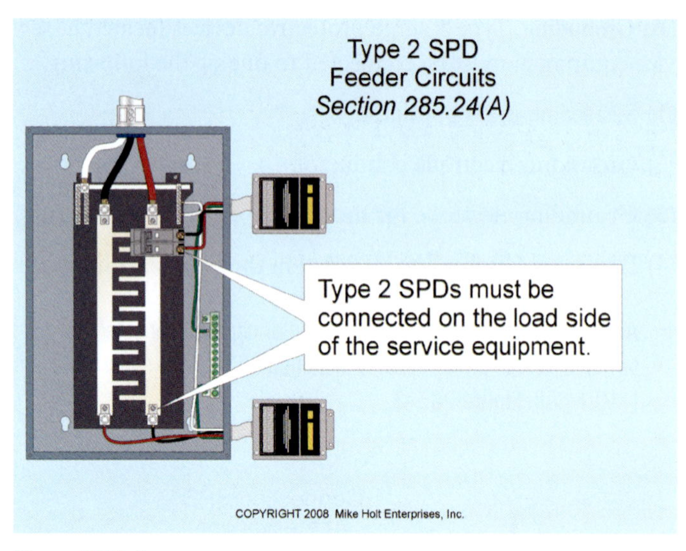

Figure 285–4

285.25 Type 3 SPD—Branch Circuits. Type 3 surge protective devices can be installed anywhere on the load side of a branch-circuit overcurrent device up to the equipment served, provided the connection is a minimum 30 ft of conductor distance from the service or separately derived system disconnect.

ARTICLE 285 Practice Questions

ARTICLE 285. SURGE PROTECTION DEVICES—PRACTICE QUESTIONS

1. Article 285 covers surge protective devices rated over 1 kV.

   (a) True
   (b) False

2. Surge protective devices shall be listed.

   (a) True
   (b) False

3. The conductors used to connect the surge protective device to ground shall not be any longer than _____ and shall avoid unnecessary bends.

   (a) 6 in.
   (b) 12 in.
   (c) 18 in.
   (d) necessary

CHAPTER 2

Notes

WIRING METHODS AND MATERIALS

INTRODUCTION TO CHAPTER 3—WIRING METHODS AND MATERIALS

Chapter 2 provided the general rules for wiring and protection of conductors, and was primarily concerned with the correct sizing of circuits and the means of protecting them. This differs from the purpose of Chapter 3, which is to correctly install the conductors that make up those circuits.

Chapter 2 was a bit of an uphill climb, because many rules had a kind of abstract quality to them. Chapter 3, on the other hand, gets very specific about conductors, cables, boxes, raceways, and fittings. It's also highly detailed about the installation and restrictions involved with wiring methods.

It's because of that detail that many people incorrectly apply the Chapter 3 wiring methods rules. Be sure to pay careful attention to the details, rather than making the mistake of glossing over something. This is especially true when it comes to applying the Tables.

The type of wiring method you'll use depends on several factors: *Code* requirements, the environment, need, and cost are among them.

Power quality is a major concern today. The cost of poor power quality runs into the millions of dollars each month in the United States alone. Grounding and bonding deficiencies (refer back to Article 250) constitute the number one cause of power quality problems. Violations of the Chapter 3 wiring methods rules constitute the number two cause of power quality problems. *Code* violations can also lead to fire, shock, and other hazards. This is particularly true of Chapter 3 violations.

Chapter 3 is really a modular assembly of articles, each detailing a specific area of an electrical installation. It starts with wiring methods [Article 300], covers conductors [Article 310], and then enclosures [Articles 312 and 314]. The next series [Articles 320 through 340] addresses specific types of cables, with Articles 342 through 390 covering specific types of raceways. We close with Article 392, a support system, and the last string [Articles 394 through 398] for open wiring.

Notice as you read through the various wiring methods that, for the most part, the section numbering remains the same in each article. This makes it very easy to locate specific requirements in a particular article. For example, the rules for securing and supporting can be found in 3xx.30 of each article. In addition to this, you'll find a "uses permitted" and "uses not permitted" section in nearly every article.

Wiring Method Articles

- **Article 300.** Wiring Methods. Article 300 contains the general requirements for all wiring methods included in the *NEC*, except for signaling and communications systems, which are covered in Chapters 7 and 8.

- **Article 310.** Conductors for General Wiring. This article contains the general requirements for conductors, such as insulation markings, ampacity ratings, and conductor use. Article 310 doesn't apply to conductors that are part of flexible cords, fixture wires, or conductors that are an integral part of equipment [90.6 and 300.1(B)].

- **Article 312.** Cabinets, Cutout Boxes, and Meter Socket Enclosures. Article 312 covers the installation and construction specifications for cabinets, cutout boxes, and meter socket enclosures.

- **Article 314.** Outlet, Device, Pull and Junction Boxes, Conduit Bodies, Fittings, and Handhole Enclosures. Installation requirements for outlet boxes, pull and junction boxes, as well as conduit bodies, and handhole enclosures are contained in this article.

Cable Articles

Articles 320 through 340 address specific types of cables. If you take the time to become familiar with the various types of cables, you'll:

- Understand what's available for doing the work.
- Recognize cable types that have special *Code* requirements.
- Avoid buying cable that you can't install due to *Code* requirements you can't meet with that particular wiring method.

Here's a brief overview of each one:

- **Article 320.** Armored Cable (Type AC). Armored cable is an assembly of insulated conductors, 14 AWG through 1 AWG, individually wrapped with waxed paper. The conductors are contained within a flexible spiral metal (steel or aluminum) sheath that interlocks at the edges. Armored cable looks like flexible metal conduit. Many electricians call this metal cable BX®.

- **Article 330.** Metal-Clad Cable (Type MC). Metal-clad cable encloses insulated conductors in a metal sheath of either corrugated or smooth copper or aluminum tubing, or spiral interlocked steel or aluminum. The physical characteristics of Type MC cable make it a versatile wiring method permitted in almost any location and for almost any application. The most commonly used Type MC cable is the interlocking type, which looks similar to armored cable or flexible metal conduit.

- **Article 334.** Nonmetallic-Sheathed Cable (Type NM). Nonmetallic-sheathed cable encloses two, three, or four insulated conductors, 14 AWG through 2 AWG, within a nonmetallic outer jacket. Because this cable is nonmetallic, it contains a separate equipment grounding conductor. Nonmetallic-sheathed cable is a common wiring method used for residential and commercial branch circuits. Many electricians call this plastic cable Romex®.

- **Article 336.** Power and Control Tray Cable (Type TC). Power and control tray cable is a factory assembly of two or more insulated conductors under a nonmetallic sheath for installation in cable trays, in raceways, or where supported by a messenger wire.

- **Article 338.** Service-Entrance Cable (Types SE and USE). Service-entrance cable can be a single-conductor or a multiconductor assembly within an overall nonmetallic covering. This cable is used primarily for services not over 600V, but is also permitted for feeders and branch circuits.

- **Article 340.** Underground Feeder and Branch-Circuit Cable (Type UF). Underground feeder cable is a moisture-, fungus-, and corrosion-resistant cable suitable for direct burial in the earth, and it comes in sizes 14 AWG through 4/0 AWG [340.104]. Multiconductor UF cable is covered in molded plastic that surrounds the insulated conductors.

Raceway Articles

Articles 342 through 390 address specific types of raceways. Refer to Article 100 for the definition of raceway. If you take the time to become familiar with the various types of raceways, you'll:

- Understand what's available for doing the work.
- Recognize raceway types that have special *Code* requirements.
- Avoid buying a raceway that you can't install due to *NEC* requirements you can't meet with that particular wiring method.

Here's a brief overview of each one:

- **Article 342.** Intermediate Metal Conduit (Type IMC). Intermediate metal conduit is a circular metal raceway with the same outside diameter as rigid metal conduit. The wall thickness of intermediate metal conduit is less than that of rigid metal conduit, so it has a greater interior cross-sectional area for holding conductors. Intermediate metal conduit is lighter and less expensive than rigid metal conduit, but it's permitted in all the same locations as rigid metal conduit. Intermediate metal conduit also uses a different steel alloy, which makes it stronger than rigid metal conduit, even though the walls are thinner.

- **Article 344.** Rigid Metal Conduit (Type RMC). Rigid metal conduit is similar to intermediate metal conduit, except the wall thickness is greater, so it has a smaller interior cross-sectional area. Rigid metal conduit is heavier than intermediate metal conduit and it's permitted to be installed in any location, just like intermediate metal conduit.

- **Article 348.** Flexible Metal Conduit (Type FMC). Flexible metal conduit is a raceway of circular cross section made of a helically wound, interlocked metal strip of either steel or aluminum. It's commonly called "Greenfield" or "Flex."

- **Article 350.** Liquidtight Flexible Metal Conduit (Type LFMC). Liquidtight flexible metal conduit is a raceway of circular cross section with an outer liquidtight, nonmetallic, sunlight-resistant jacket over an inner flexible metal core, with associated couplings, connectors, and fittings. It's listed for the installation of electric conductors. Liquidtight flexible metal conduit is commonly called Sealtite® or simply "liquidtight." Liquidtight flexible metal conduit is of similar construction to flexible metal conduit, but it has an outer thermoplastic covering.

- **Article 352.** Rigid Polyvinyl Chloride Conduit (Type PVC). Rigid polyvinyl chloride conduit is a nonmetallic raceway of circular cross section with integral or associated couplings, connectors, and fittings. It's listed for the installation of electrical conductors.

- **Article 353.** High-Density Polyethylene Conduit (Type HDPE). This article covers the use, installation, and construction specifications for high-density polyethylene (HDPE) conduit and associated fittings. It's lightweight and durable. It resists decomposition, oxidation, and hostile elements that cause damage to other materials. HDPE is mechanically and chemically resistant to a host of environmental conditions. Uses include communication, data, cable television, and general-purpose raceways.

- **Article 354.** Nonmetallic Underground Conduit with Conductors (Type NUCC). Nonmetallic underground conduit with conductors is a factory assembly of conductors or cables inside a nonmetallic, smooth wall conduit with a circular cross section. It can also be supplied on reels without damage or distortion and is of sufficient strength to withstand abuse, such as impact or crushing when handled and installed, without damage to the conduit or conductors.

- **Article 356.** Liquidtight Flexible Nonmetallic Conduit (Type LFNC). Liquidtight flexible nonmetallic conduit is a raceway of circular cross section with an outer liquidtight, nonmetallic, sunlight-resistant jacket over an inner flexible core, with associated couplings, connectors, and fittings. It's listed for the installation of electric conductors. LFNC is available in three types:
  - Type LFNC-A (orange). A smooth seamless inner core and cover bonded together with reinforcement layers inserted between the core and covers.
  - Type LFNC-B (gray). A smooth inner surface with integral reinforcement within the conduit wall.
  - Type LFNC-C (black). A corrugated internal and external surface without integral reinforcement within the conduit wall.

- **Article 358.** Electrical Metallic Tubing (EMT). Electrical metallic tubing is a nonthreaded thinwall raceway of circular cross section designed for the physical protection and routing of conductors and cables. Compared to rigid metal conduit and intermediate metal conduit, electrical metallic tubing is relatively easy to bend, cut, and ream. EMT isn't threaded, so all connectors and couplings are of the threadless type. Today, EMT is available in a range of colors, such as red and blue.

- **Article 362.** Electrical Nonmetallic Tubing (ENT). Electrical nonmetallic tubing is a pliable, corrugated, circular raceway made of PVC. It's often called "Smurf Pipe" or "Smurf Tube," because it was available only in blue when it originally came out at the time the children's cartoon characters "The Smurfs," were most popular.

- **Article 376.** Metal Wireways. This article covers the use, installation, and construction specifications for metal wireways and associated fittings. A metal wireway is a sheet metal trough with hinged or removable covers for housing and protecting electric conductors and cable, in which conductors are placed after the wireway has been installed as a complete system.

- **Article 378.** Nonmetallic Wireways. A nonmetallic wireway is a flame-retardant trough with hinged or removable covers for housing and protecting electric conductors and cable, in which conductors are placed after the wireway has been installed as a complete system.

- **Article 380.** Multioutlet Assemblies. A multioutlet assembly is a surface, flush, or freestanding raceway designed to hold conductors and receptacles. It's assembled in the field or at the factory.

- **Article 384.** Strut-Type Channel Raceway. Strut-type channel raceway is a metallic raceway intended to be mounted to the surface or suspended with associated accessories, in which conductors are placed after the raceway has been installed as a complete system.

- **Article 386.** Surface Metal Raceways. A surface metal raceway is a metallic raceway intended to be mounted to the surface with associated accessories, in which conductors are placed after the raceway has been installed as a complete system.

- **Article 388.** Surface Nonmetallic Raceways. A surface nonmetallic raceway is intended to be surface mounted with associated accessories. Conductors are placed inside after the raceway has been installed as a complete system.

Cable Tray

- **Article 392.** Cable Trays. A cable tray system is a unit or assembly of units or sections with associated fittings that form a structural system used to securely fasten or support cables and raceways. A cable tray isn't a raceway; it is a support system for raceways, cables, and enclosures.

ARTICLE

300 Wiring Methods

INTRODUCTION TO ARTICLE 300—WIRING METHODS

Article 300 contains the general requirements for all wiring methods included in the *NEC*. However, this article doesn't apply to communications systems, which is covered in Chapter 8, except when Article 300 is specifically referenced in Chapter 8.

This article is primarily concerned with how to install, route, splice, protect, and secure conductors and raceways. How well you conform to the requirements of Article 300 will generally be evident in the finished work, because many of the requirements tend to determine the appearance of the installation.

Because of this, it's often easy to spot Article 300 problems if you're looking for *Code* violations. For example, you can easily see when someone runs an equipment grounding conductor outside a raceway instead of grouping all conductors of a circuit together, as required by 300.3(B).

This is just one of the common points of confusion your studies here will clear up for you. To help achieve that end, be sure to carefully consider the accompanying illustrations, and also refer to Article 100 as needed.

PART I. GENERAL REQUIREMENTS

300.1 Scope.

(A) Wiring Installations. Article 300 contains the general requirements for power and lighting wiring methods.

Author's Comment: The requirements contained in Article 300 don't apply to the wiring methods for signaling and communications systems. However, the following sections in Chapter 7 and 8 refer back to Article 300 requirements.

- CATV, 820.3
- Class 2 and 3 Circuits, 725.3
- Communications Cables and Raceways, 800.133(A)(2)
- Fire Alarm Circuits, 760.3

(B) Integral Parts of Equipment. The requirements contained in Article 300 don't apply to the internal parts of electrical equipment. Figure 300–1

(C) Trade Sizes. Designators for raceway trade sizes are given in Table 300.1(C).

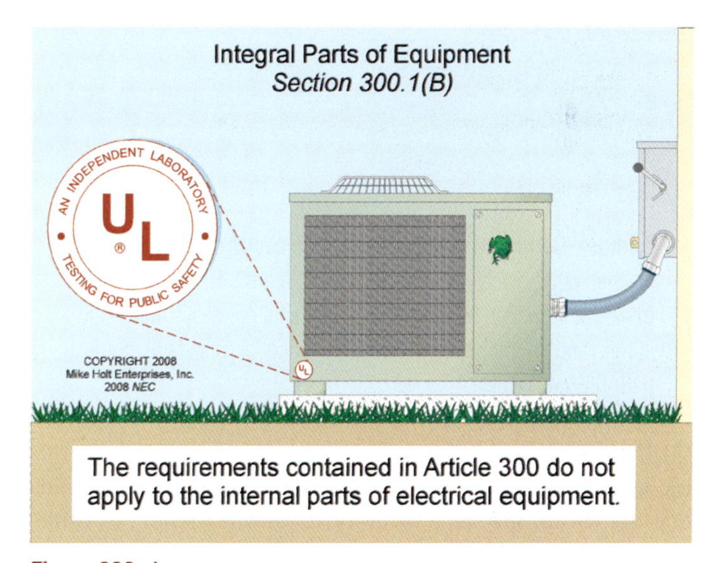

Figure 300–1

Author's Comment: Industry practice is to describe raceways using inch sizes, such as ½ in., 2 in., etc.; however, the proper reference (2005 *NEC* change) is to use "Trade Size ½," or "Trade Size 2." In this textbook we use the term "Trade Size."

300.3 Conductors.

(A) Conductors. Conductors must be installed within a Chapter 3 wiring method, such as a raceway, cable, or enclosure.

Exception: Overhead conductors can be installed in accordance with 225.6.

(B) Circuit Conductors Grouped Together. All conductors of a circuit must be installed in the same raceway, cable, trench, cord, or cable tray, except as permitted by (1) through (4). Figure 300–2

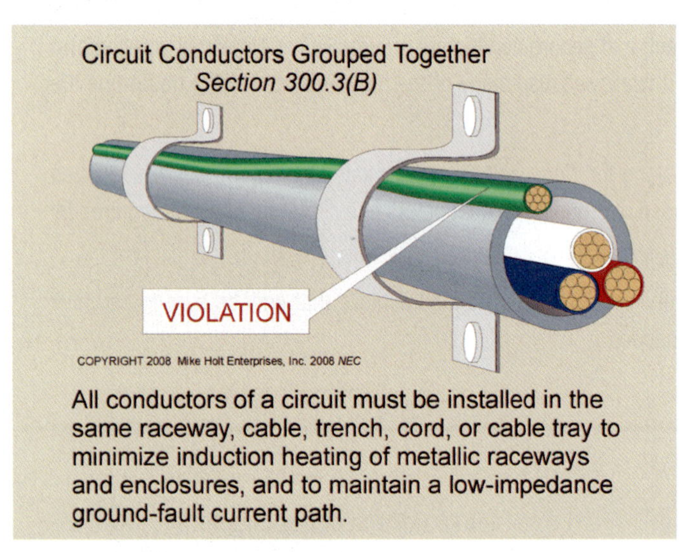

Figure 300–2

Author's Comment: All conductors of a circuit must be installed in the same raceway, cable, trench, cord, or cable tray to minimize induction heating of ferrous metal raceways and enclosures, and to maintain a low-impedance ground-fault current path [250.4(A)(3)].

(1) Paralleled Installations. Conductors can be run in parallel, in accordance with 310.4, and must have all circuit conductors within the same raceway, cable tray, trench, or cable. Figure 300–3

Exception: Parallel conductors run underground can be run in different raceways (Phase A in raceway 1, Phase B in raceway 2, etc.) if, in order to reduce or eliminate inductive heating, the raceway is nonmetallic or nonmagnetic and the installation complies with 300.20(B). See 300.3(B)(3) and 300.5(I) Ex 2.

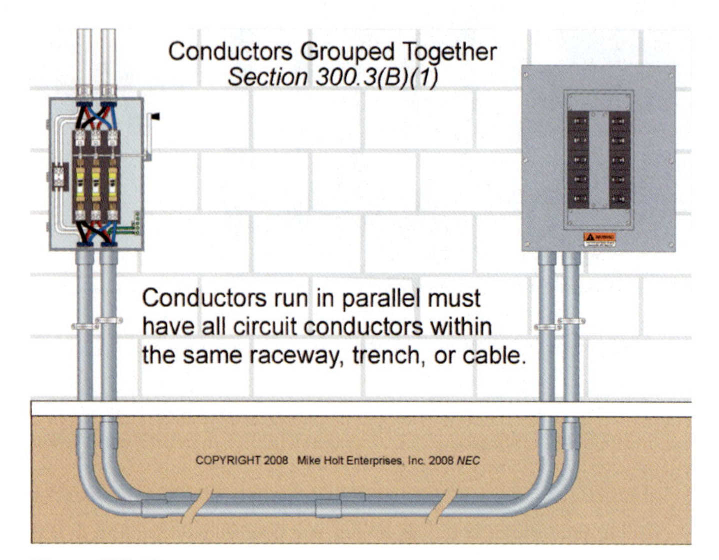

Figure 300–3

(2) Grounding and Bonding Conductors. Equipment grounding conductors can be installed outside of a raceway or cable assembly for certain existing installations. See 250.130(C). Equipment grounding jumpers can be located outside of a flexible raceway if the bonding jumper is installed in accordance with 250.102(E). Figure 300–4

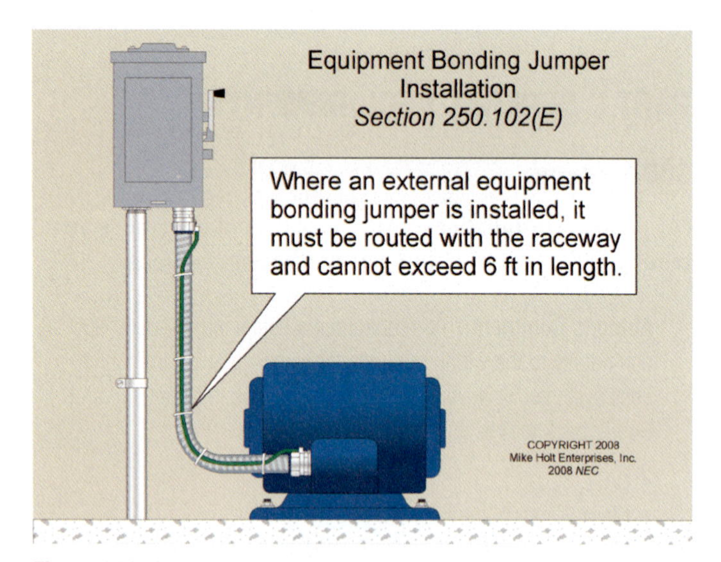

Figure 300–4

(3) Nonferrous Wiring Methods. Circuit conductors can be run in different raceways (Phase A in raceway 1, Phase B in raceway 2, etc.) if, in order to reduce or eliminate inductive heating, the raceway is nonmetallic or nonmagnetic and the installation complies with 300.20(B). See 300.3(B)(1) and 300.5(I) Ex 2.

(C) Conductors of Different Systems.

(1) Mixing. Power conductors of ac and dc systems rated 600V or less can occupy the same raceway, cable, or enclosure if all conductors have an insulation voltage rating not less than the maximum circuit voltage. Figure 300–5

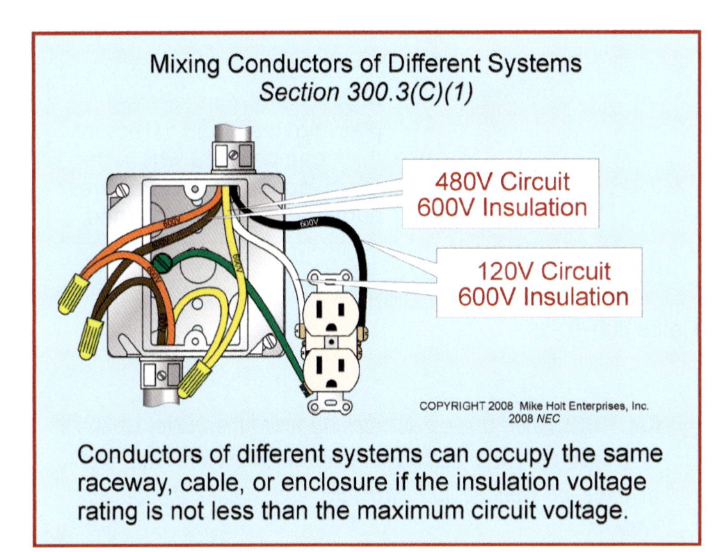

Figure 300–5

FPN: See 725.136(A) for Class 2 and Class 3 circuit conductors.

Author's Comments:

- Control, signal, and communications wiring must be separated from power and lighting circuits so the higher-voltage conductors don't accidentally energize the control, signal, or communications wiring:
  - CATV Coaxial Cable, 820.133(A)
  - Class 1, Class 2, and Class 3 Control Circuits, 725.48 and 725.136(A). Figure 300–6
  - Communications Circuits, 800.133(A)(1)(c)
  - Fire Alarm Circuits, 760.136(A)
  - Instrumentation Tray Cable, 727.5
  - Sound Circuits, 640.9(C)
- A Class 2 circuit that has been reclassified as a Class 1 circuit [725.130(A) Ex 2] can be run with associated power conductors [725.48(B)(1)] if all conductors have an insulation voltage rating not less than the maximum circuit voltage [300.3(C)(1)]. Figure 300–7

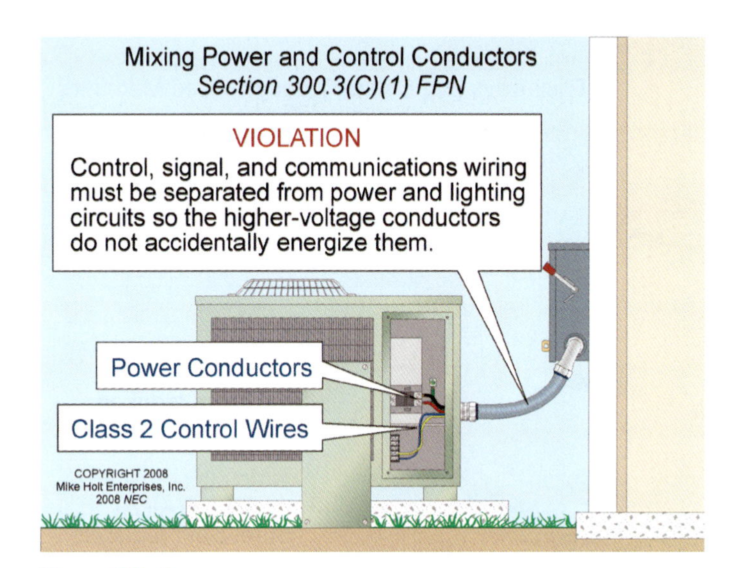

Figure 300–6

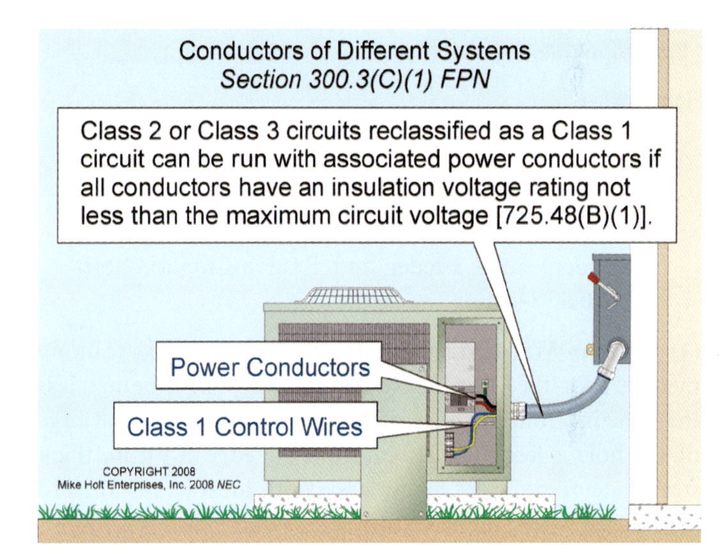

Figure 300–7

300.4 Protection Against Physical Damage. Conductors must be protected against physical damage [110.27(B)].

(A) Cables and Raceways Through Wood Members. When the following wiring methods are installed through wood members, they must comply with (1) and (2). Figure 300–8

- Armored Cable, Article 320
- Electrical Nonmetallic Tubing, Article 362
- Flexible Metal Conduit, Article 348
- Liquidtight Flexible Metal Conduit, Article 350
- Liquidtight Flexible Nonmetallic Conduit, Article 356
- Metal-Clad Cable, Article 330

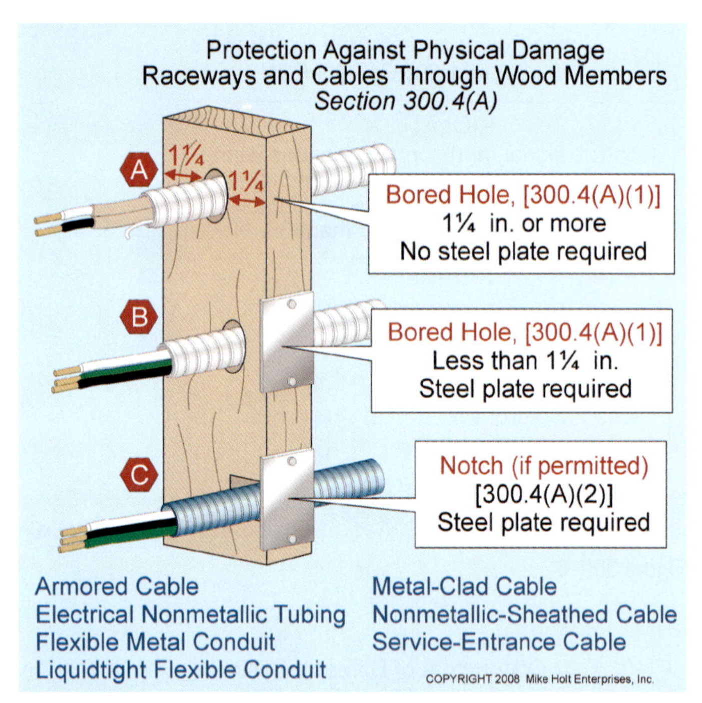

Figure 300–8

- Nonmetallic-Sheathed Cable, Article 334
- Service-Entrance Cable, Article 338
- Underground Feeder and Branch-Circuit Cable, Article 340

(1) Holes in Wood Members. Holes through wood framing members for the above cables or raceways must be not less than 1¼ in. from the edge of the wood member. If the edge of the hole is less than 1¼ in. from the edge, a 1⁄16 in. thick steel plate of sufficient length and width must be installed to protect the wiring method from screws and nails.

Exception No. 1: A steel plate isn't required to protect rigid metal conduit, intermediate metal conduit, PVC conduit, or electrical metallic tubing.

Exception No. 2: A listed and marked steel plate less than 1⁄16 in. thick that provides equal or better protection against nail or screw penetration is permitted. Figure 300–9

Author's Comment: Hardened steel plates thinner than 1⁄16 in. have been tested and found to provide better protection from screw and nail penetration than the thicker plates.

(2) Notches in Wood Members. Where notching of wood framing members for cables and raceways are permitted by the building code, a 1⁄16 in. thick steel plate of sufficient length and width must be installed to protect the wiring method laid in these wood notches from screws and nails.

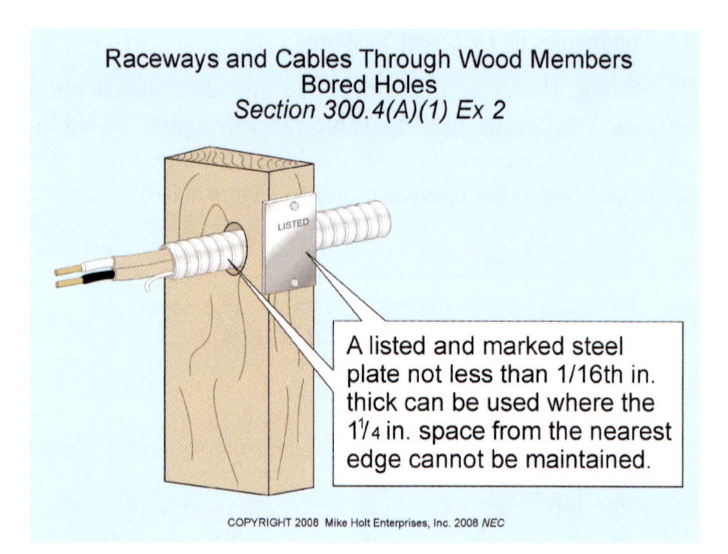

Figure 300–9

CAUTION: *When drilling or notching wood members, be sure to check with the building inspector to ensure you don't damage or weaken the structure and violate the building code.*

Exception No. 1: A steel plate isn't required to protect rigid metal conduit, intermediate metal conduit, PVC conduit, or electrical metallic tubing.

Exception No. 2: A listed and marked steel plate less than 1⁄16 in. thick that provides equal or better protection against nail or screw penetration is permitted. Figure 300–10

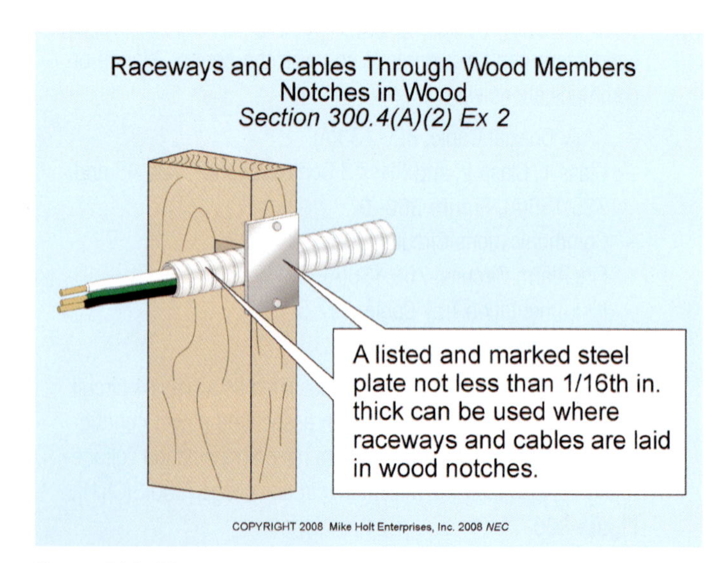

Figure 300–10

(B) Nonmetallic-Sheathed Cable and Electrical Nonmetallic Tubing Through Metal Framing Members.

(1) Nonmetallic-Sheathed Cable (NM). Where Type NM cables pass through factory or field openings in metal framing members, the cable must be protected by listed bushings or listed grommets that cover all metal edges. The protection fitting must be securely fastened in the opening before the installation of the cable. Figure 300–11

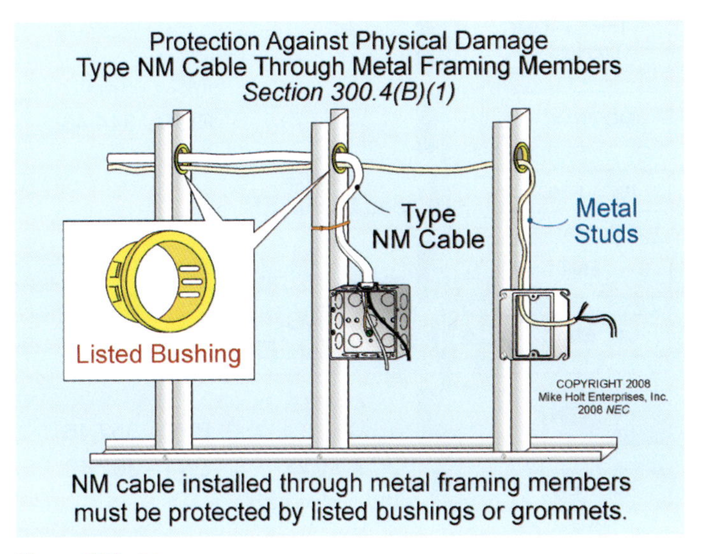

Figure 300–11

(2) Type NM Cable and Electrical Nonmetallic Tubing. Where nails or screws are likely to penetrate Type NM cable or electrical nonmetallic tubing, a steel sleeve, steel plate, or steel clip not less than 1/16 in. in thickness must be installed to protect the cable or tubing.

Exception: A listed and marked steel plate less than 1/16 in. thick that provides equal or better protection against nail or screw penetration is permitted.

(C) Behind Suspended Ceilings. Wiring methods, such as cables or raceways, installed behind panels designed to allow access must be supported in accordance with their applicable article. Figure 300–12

Author's Comment: This rule doesn't apply to control, signal, and communications cables, but similar requirements are contained in Chapters 6, 7, and 8 as follows:

- CATV Coaxial Cable, 820.21 and 820.24
- Communications Cable, 800.21
- Control and Signaling Cable, 725.21 and 725.24

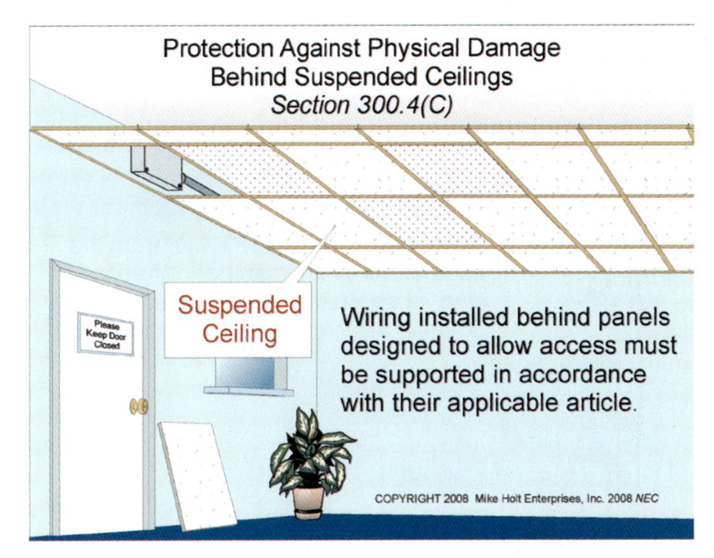

Figure 300–12

- Fire Alarm Cable, 760.7 and 760.8
- Optical Fiber Cable, 770.21 and 770.24
- Audio Cable, 640.6(B)

(D) Cables and Raceways Parallel to Framing Members and Furring Strips. Cables or raceways run parallel to framing members or furring strips must be protected where they are likely to be penetrated by nails or screws, by installing the wiring method so it isn't less than 1¼ in. from the nearest edge of the framing member or furring strip. If the edge of the framing member or furring strip is less than 1¼ in. away, a 1/16 in. thick steel plate of sufficient length and width must be installed to protect the wiring method from screws and nails. Figure 300–13

Author's Comment: This rule doesn't apply to control, signaling, and communications cables, but similar requirements are contained in Chapters 6, 7, and 8 as follows:

- CATV Coaxial Cable, 820.24
- Communications Cable, 800.24
- Control and Signaling Cable, 725.24
- Optical Fiber Cable, 770.24
- Fire Alarm Cable, 760.8
- Audio Cable, 640.6(B)

Exception No. 1: Protection isn't required for rigid metal conduit, intermediate metal conduit, PVC conduit, or electrical metallic tubing.

Exception No. 2: For concealed work in finished buildings, or finished panels for prefabricated buildings where such supporting is impracticable, the cables can be fished between access points.

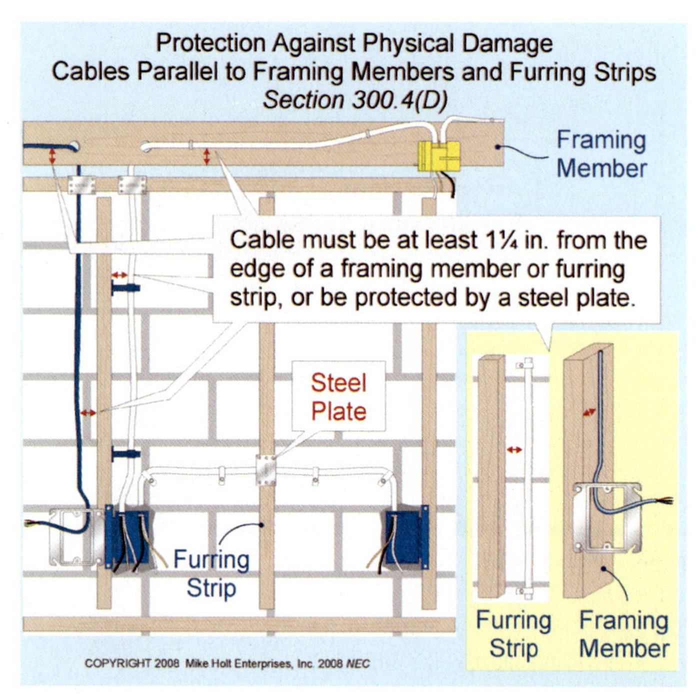

Figure 300–13

Exception No. 3: A listed and marked steel plate less than 1⁄16 in. thick that provides equal or better protection against nail or screw penetration is permitted.

(E) Wiring Under Roof Decking. Wiring under metal-corrugated sheet roof decking must not be less than 1½ in. from the nearest surface of the roof decking.

Exception: Spacing from roof decking doesn't apply to rigid metal conduit and intermediate metal conduit.

(F) Cables and Raceways Installed in Grooves. Cables and raceways installed in a groove must be protected by a 1⁄16 in. thick steel plate or sleeve, or by 1¼ in. of free space.

> **Author's Comment:** An example is Type NM cable installed in a groove cut into the Styrofoam-type insulation building block structure and then covered with wallboard.

Exception No. 1: Protection isn't required if the cable is installed in rigid metal conduit, intermediate metal conduit, PVC conduit, or electrical metallic tubing.

Exception No. 2: A listed and marked steel plate less than 1⁄16 in. thick that provides equal or better protection against nail or screw penetration is permitted.

(G) Insulating Fittings. Where raceways contain insulated conductors 4 AWG and larger that enter an enclosure, the conductors must be protected from abrasion during and after installation by a fitting that provides a smooth, rounded insulating surface, such as an insulating bushing. Figure 300–14

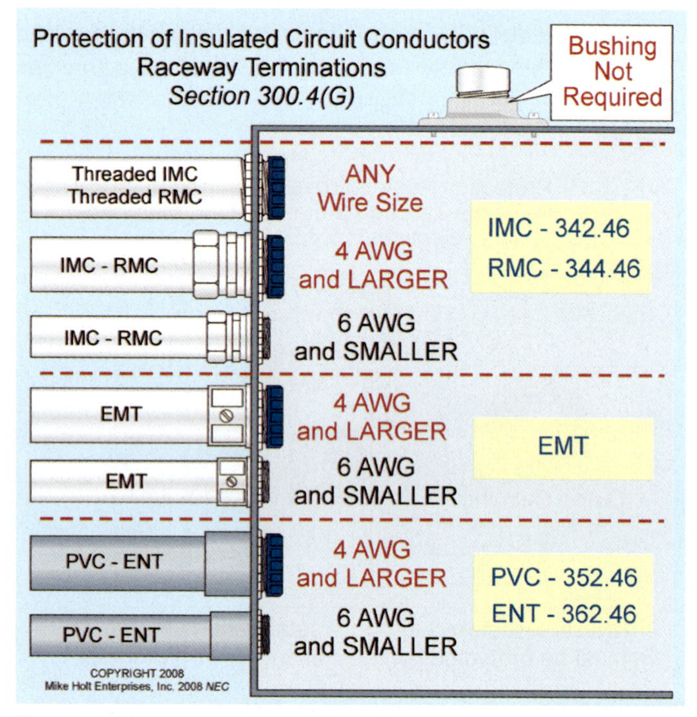

Figure 300–14

> **Author's Comment:** Where IMC or RMC conduit enters an enclosure without a connector, a bushing must be provided, regardless of the conductor size [342.46 and 344.46].

Exception: Insulating bushings aren't required where a raceway terminates in a threaded raceway entry that provides a smooth, rounded, or flared surface for the conductors. An example would be a meter hub fitting or a Meyer's hub-type fitting.

300.5 Underground Installations.

(A) Minimum Burial Depths. When cables or raceways are run underground, they must have a minimum "cover" in accordance with Table 300.5. Figure 300–15

> **Author's Comment:** Note 1 to Table 300.5 defines "Cover" as the distance from the top of the underground cable or raceway to the surface of finish grade. Figure 300–16

Underground Installations - Minimum Cover Depths
Table 300.5

	UF or USE Cables or Conductors	RMC or IMC	PVC not encased in concrete	Residential 15 & 20A GFCI Branch Circuits
Applications not listed below	24 in.	6 in.	18 in.	12 in.
Street Driveway Parking Lot	24 in.	24 in.	24 in.	24 in.
Driveways One - Two Family	18 in.	18 in.	18 in.	12 in.
Solid Rock With not less than 2 in. of concrete	Raceway Only			Raceway Only

Figure 300–15

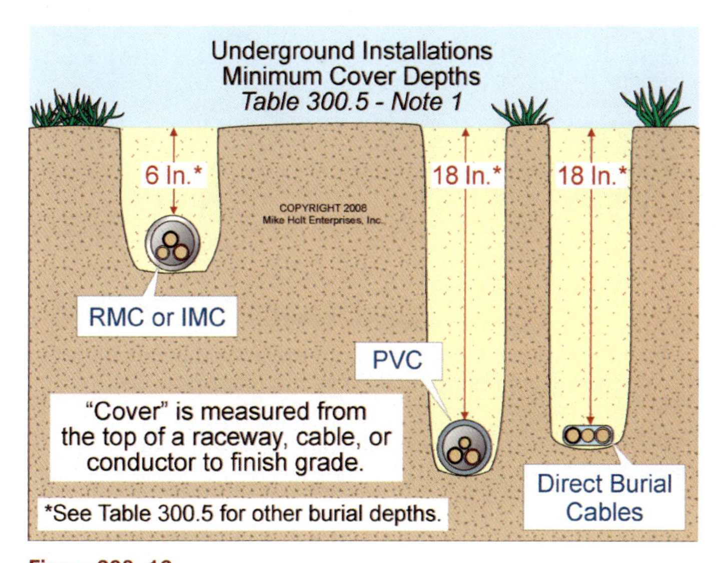

Figure 300–16

Table 300.5 Minimum Cover Requirements in Inches			
Location	**Buried Cables**	**Metal Raceway**	**Nonmetallic Raceway**
Under Building	0	0	0
Dwelling Unit	24/12*	6	18
Dwelling Unit Driveway	18/12*	6	18/12*
Under Roadway	24	24	24
Other Locations	24	6	18

**Residential branch circuits rated 120V or less with GFCI protection and maximum overcurrent protection of 20A. Note: This is a summary of NEC Table 300.5. See the table in the NEC for full details.*

Author's Comment: The cover requirements contained in 300.5 don't apply to the following signaling, communications and other power limited wiring: **Figure 300–17**

- CATV, 90.3
- Class 2 and 3 Circuits, 725.3
- Communications Cables and Raceways, 90.3
- Fire Alarm Circuits, 760.3
- Optical Fiber Cables and Raceways, 770.3

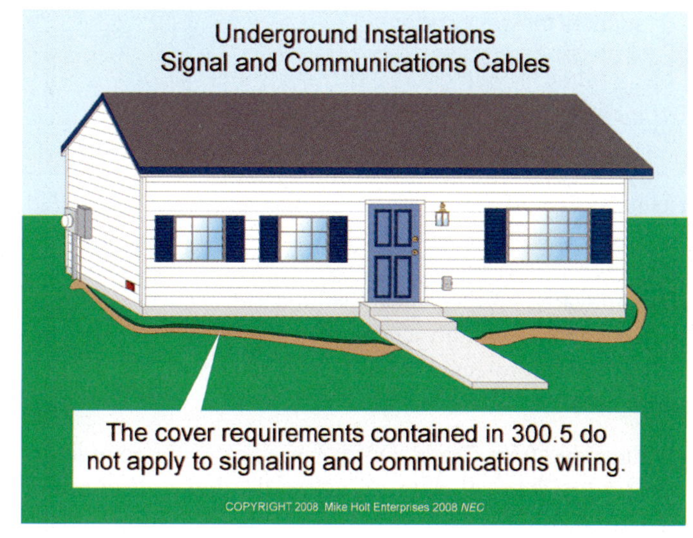

Figure 300–17

(B) Wet Locations. The interior of enclosures or raceways installed in an underground installation are considered to be a wet location. Cables and insulated conductors installed in underground enclosures or raceways must be listed for use in wet locations according to 310.8(C). Splices within an underground enclosure must be listed as suitable for wet locations [110.14(B)]. **Figure 300–18**

Author's Comment: The definition of a "Wet Location," as contained in Article 100, includes installations underground, in concrete slabs in direct contact with the earth, locations subject to saturation with water, and unprotected locations exposed to weather. Where raceways are installed in wet locations above grade, the interior of these raceways is also considered to be a wet location [300.9].

(C) Cables Under Buildings. Cables run under a building must be installed in a raceway that extends past the outside walls of the building.

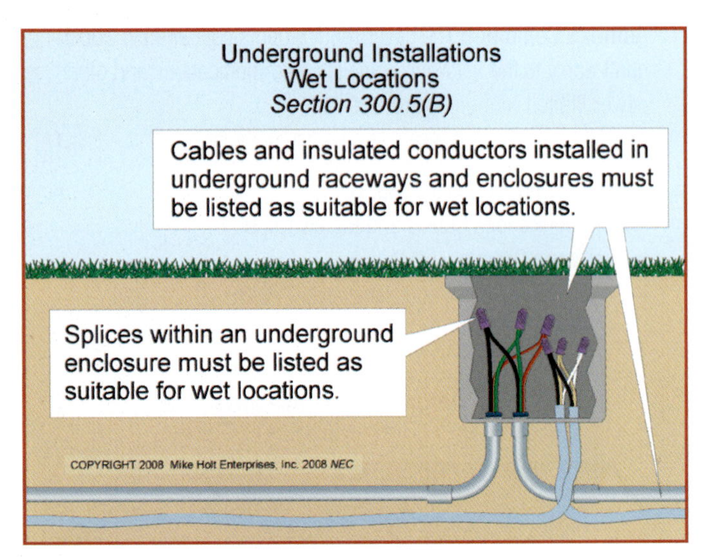

Figure 300–18

(D) Protecting Underground Cables and Conductors. Direct-buried conductors and cables such as Types MC, UF and USE must be protected from damage in accordance with (1) through (4).

(1) Emerging from Grade. Direct-buried cables or conductors that emerge from grade must be installed in an enclosure or raceway to protect against physical damage. Protection isn't required to extend more than 18 in. below grade, and protection above ground must extend to a height not less than 8 ft. Figure 300–19

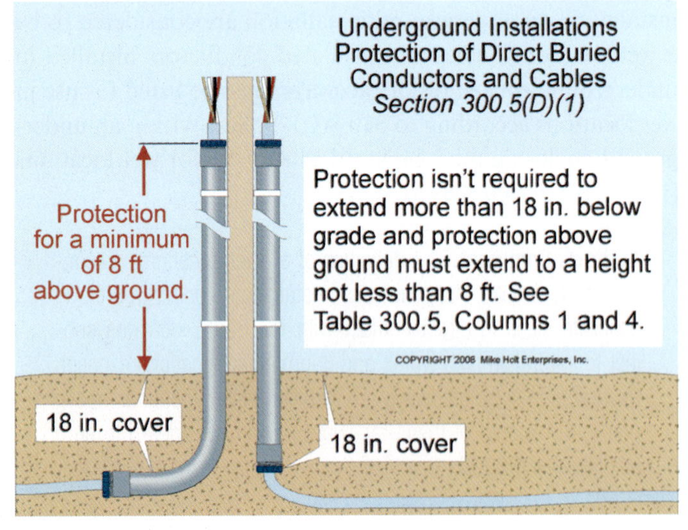

Figure 300–19

(2) Conductors Entering Buildings. Conductors that enter a building must be protected to the point of entrance.

(3) Service Lateral Conductors. Direct-buried service-lateral conductors that aren't under the exclusive control of the electric utility, and are buried 18 in. or more below grade, must have their location identified by a warning ribbon placed in the trench not less than 1 ft above the underground installation. Figure 300–20

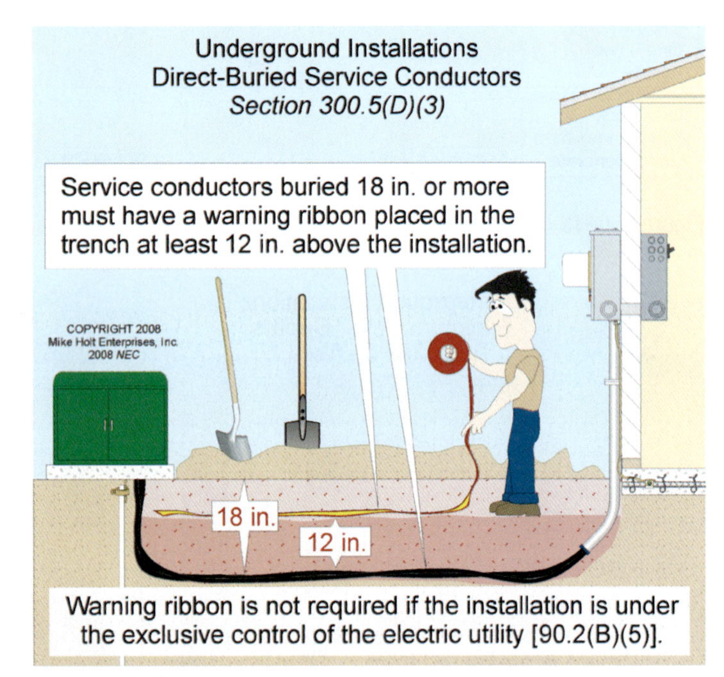

Figure 300–20

(4) Enclosure or Raceway Damage. Where direct-buried cables, enclosures, or raceways are subject to physical damage, the conductors must be installed in rigid metal conduit, intermediate metal conduit, or Schedule 80 PVC conduit.

(E) Underground Splices and Taps. Direct-buried conductors or cables can be spliced or tapped underground without a splice box [300.15(G)], if the splice or tap is made in accordance with 110.14(B). Figure 300–21

(F) Backfill. Backfill material for underground wiring must not damage the underground cable or raceway, or contribute to the corrosion of the metal raceway.

Author's Comment: Large rocks, chunks of concrete, steel rods, mesh, and other sharp-edged objects must not be used for backfill material, because they can damage the underground conductors, cables, or raceways.

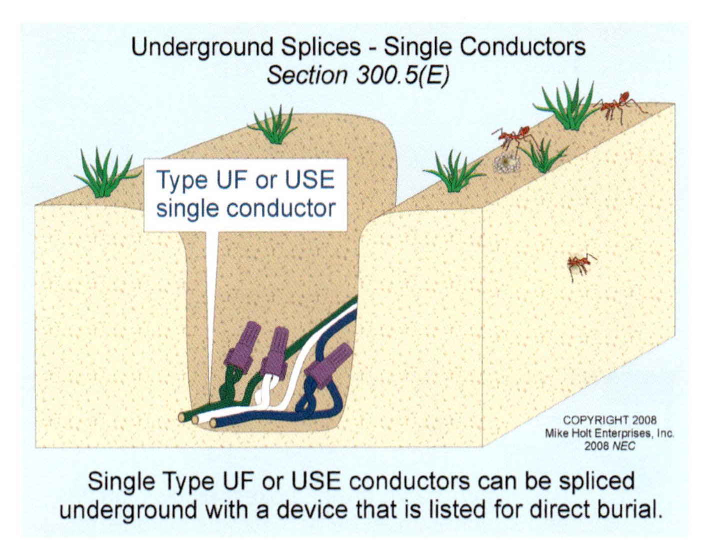

Figure 300–21

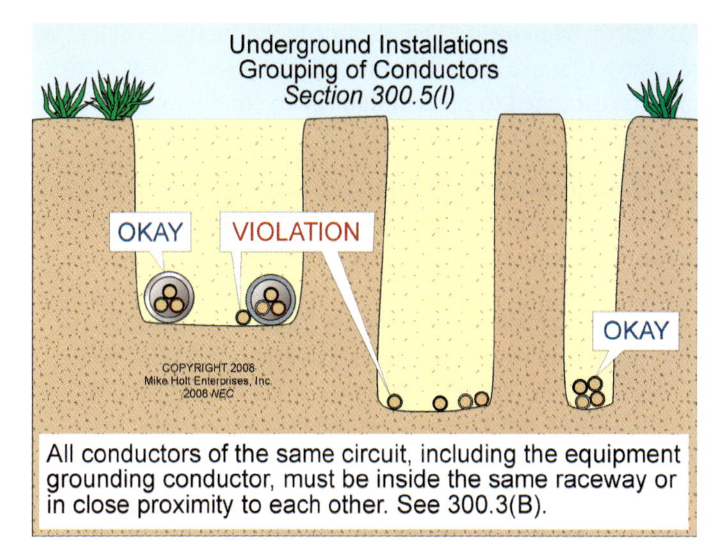

Figure 300–22

(G) Raceway Seals. Where moisture could enter a raceway and contact energized live parts, a seal must be installed at one or both ends of the raceway.

Author's Comment: This is a common problem for equipment located downhill from the supply, or in underground equipment rooms. See 230.8 for service raceway seals and 300.7(A) for different temperature area seals.

FPN: Hazardous explosive gases or vapors make it necessary to seal underground conduits or raceways that enter the building in accordance with 501.15.

Author's Comment: It isn't the intent of this FPN to imply that sealing fittings of the types required in hazardous (classified) locations be installed in unclassified locations, except as required in Chapter 5. This also doesn't imply that the sealing material provides a watertight seal, but only that it prevents moisture from entering the conduits or raceways.

(H) Bushing. Raceways that terminate underground must have a bushing or fitting at the end of the raceway to protect emerging cables or conductors.

(I) Conductors Grouped Together. All conductors of the same circuit, including the equipment grounding conductor, must be inside the same raceway, or in close proximity to each other. See 300.3(B). Figure 300–22

Exception No. 1: Conductors can be installed in parallel in accordance with 310.4.

Exception No. 2: Individual sets of parallel circuit conductors can be installed in underground PVC conduits, if inductive heating at raceway terminations is reduced by complying with 300.20(B) [300.3(B)(1) and 300.3(B)(3)]. Figure 300–23

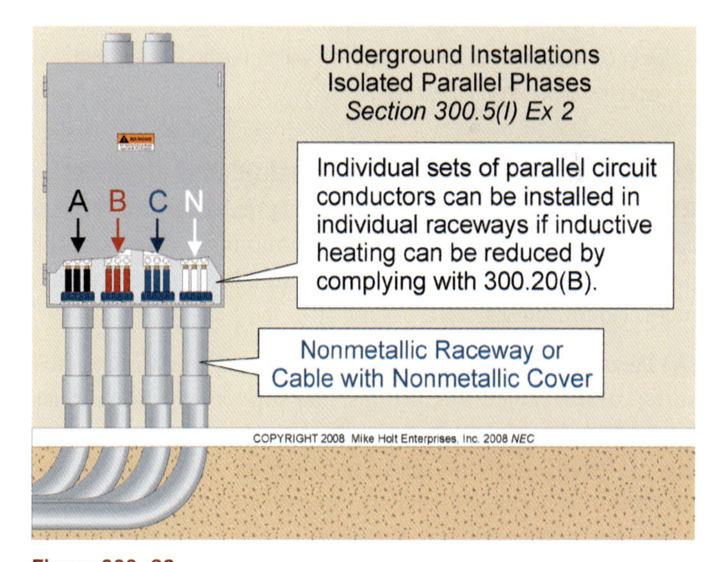

Figure 300–23

Author's Comment: Installing ungrounded and neutral conductors in different PVC conduits makes it easier to terminate larger parallel sets of conductors, but it will result in higher levels of electromagnetic fields (EMF), which can cause computer monitors to flicker in a distracting manner.

(J) Earth Movement. Direct-buried conductors, cables, or raceways that are subject to movement by settlement or frost must be arranged to prevent damage to conductors or equipment connected to the wiring.

(K) Directional Boring. Cables or raceways installed using directional boring equipment must be approved by the authority having jurisdiction for this purpose.

> **Author's Comment:** Directional boring technology uses a directional drill, which is steered continuously from point "A" to point "B." When the drill head comes out of the earth at point "B," it's replaced with a back-reamer and the duct or conduit being installed is attached to it. The size of the boring rig (hp, torque, and pull-back power) comes into play, along with the types of soil, in determining the type of raceways required. For telecom work, multiple poly innerducts are pulled in at one time. At major crossings, such as expressways, railroads, or rivers, outerduct may be installed to create a permanent sleeve for the innerducts.
>
> "Innerduct" and "outerduct" are terms usually associated with optical fiber cable installations, while "unitduct" comes with conductors factory installed. All of these come in various sizes. Galvanized rigid metal conduit, schedule 40 and schedule 80 PVC, HDPE conduit and nonmetallic underground conduit with conductors (NUCC) are common wiring methods used with directional boring installations.

300.6 Protection Against Corrosion and Deterioration.

Raceways, cable trays, cablebus, cable armor, boxes, cable sheathing, cabinets, elbows, couplings, fittings, supports, and support hardware must be suitable for the environment. Figure 300–24

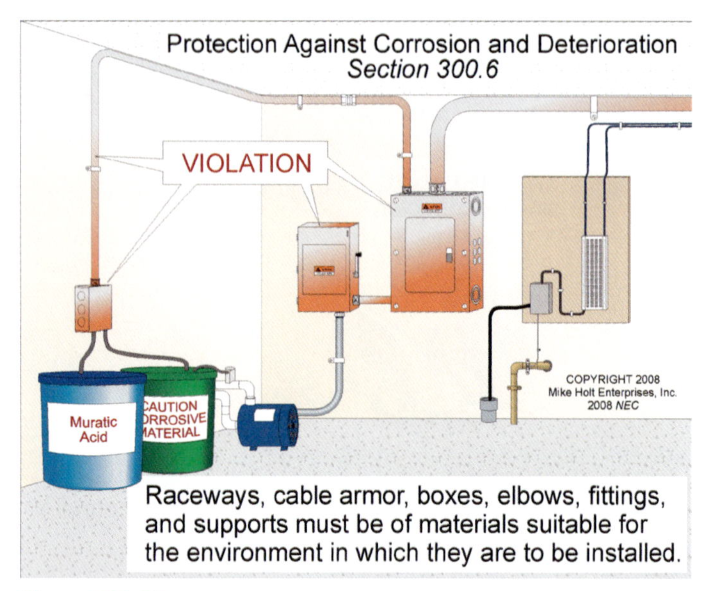

Figure 300–24

(A) Ferrous Metal Equipment. Ferrous metal raceways, enclosures, cables, cable trays, fittings, and support hardware must be protected against corrosion inside and outside by a coating of listed corrosion-resistant material. Where corrosion protection is necessary, such as underground and in wet locations, and the conduit is threaded in the field, the threads must be coated with an approved electrically conductive, corrosion-resistant compound, such as cold zinc.

> **Author's Comment:** Nonferrous metal raceways, such as aluminum rigid metal conduit, do not have to meet the provisions of this section.

(1) Protected from Corrosion Solely by Enamel. Where ferrous metal parts are protected from corrosion solely by enamel, they must not be used outdoors or in wet locations as described in 300.6(D).

(2) Organic Coatings on Boxes or Cabinets. Boxes or cabinets having a system of organic coatings marked "Raintight," "Rainproof," or "Outdoor Type," can be installed outdoors.

(3) In Concrete or in Direct Contact with the Earth. Ferrous metal raceways, cable armor, boxes, cable sheathing, cabinets, elbows, couplings, nipples, fittings, supports, and support hardware can be installed in concrete or in direct contact with the earth, or in areas subject to severe corrosive influences where made of material approved for the condition, or where provided with corrosion protection approved for the condition.

> **Author's Comment:** Galvanized steel electrical metallic tubing can be installed in concrete at grade level and in direct contact with earth, but supplementary corrosion protection is usually required (UL White Book, product category FJMX). Electrical metallic tubing can be installed in concrete above the ground floor slab generally without supplementary corrosion protection. Figure 300–25

(B) Aluminum Equipment. Aluminum raceways, cable trays, cablebus, cable armor, boxes, cable sheathing, cabinets, elbows, couplings, nipples, fittings, supports, and support hardware embedded or encased in concrete or in direct contact with the earth must be provided with supplementary corrosion protection.

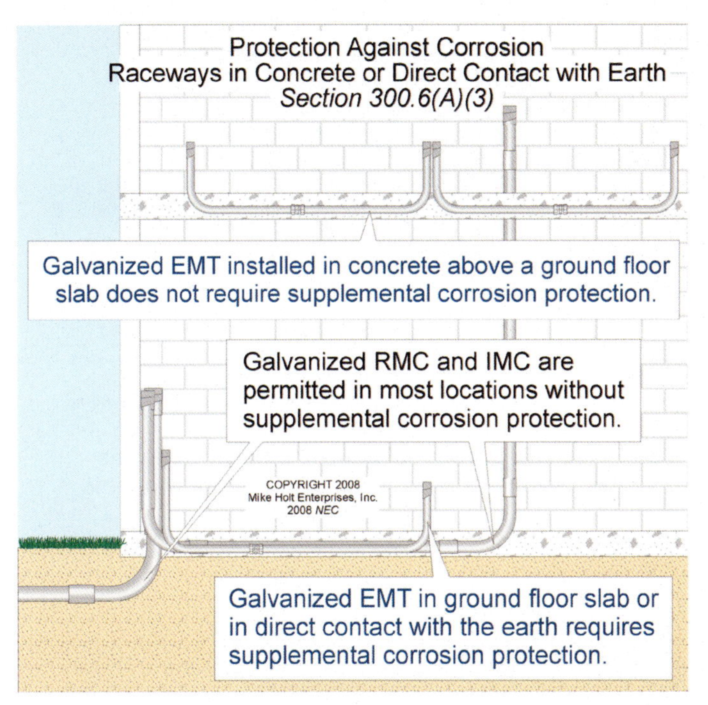

Figure 300–25

(C) Nonmetallic Equipment. Nonmetallic raceways, cable trays, cablebus, boxes, cables with a nonmetallic outer jacket and internal metal armor or jacket, cable sheathing, cabinets, elbows, couplings, nipples, fittings, supports, and support hardware must be made of material identified for the condition, and must comply with (1) and (2). Figure 300–26

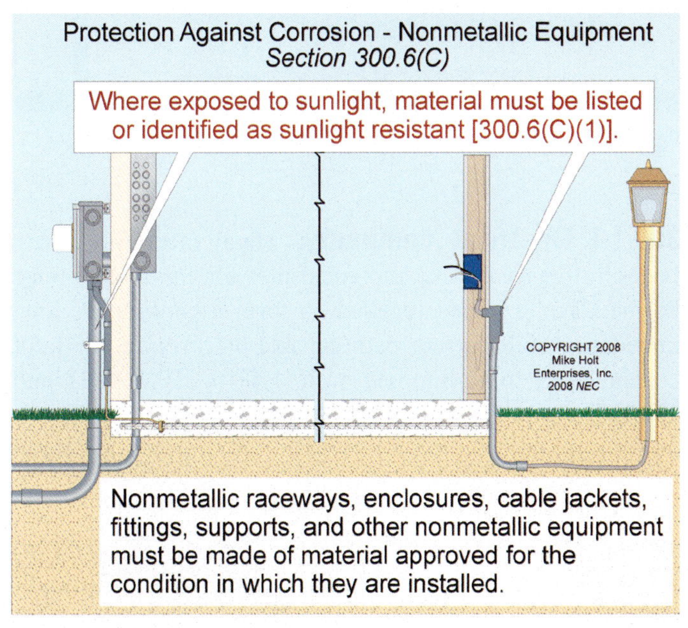

Figure 300–26

(1) Exposed to Sunlight. Where exposed to sunlight, the materials must be listed or identified as sunlight resistant.

(2) Chemical Exposure. Where subject to exposure to chemical solvents, vapors, splashing, or immersion, materials or coatings must either be inherently resistant to chemicals based upon their listing, or be identified for the specific chemical.

(D) Indoor Wet Locations. In portions of dairy processing facilities, laundries, canneries, and other indoor wet locations, and in locations where walls are frequently washed or where there are surfaces of absorbent materials, such as damp paper or wood, the entire wiring system, where installed exposed, including all boxes, fittings, raceways, and cables, must be mounted so there's at least ¼ in. of airspace between it and the wall or supporting surface.

Author's Comment: See the definitions of "Exposed" and "Location, Wet" in Article 100.

Exception: Nonmetallic raceways, boxes, and fittings are permitted without the airspace on a concrete, masonry, tile, or similar surface.

FPN: Areas where acids and alkali chemicals are handled and stored may present corrosive conditions, particularly when wet or damp. Severe corrosive conditions may also be present in portions of meatpacking plants, tanneries, glue houses, and some stables; in installations immediately adjacent to a seashore or swimming pool, spa, hot tub, and fountain areas; in areas where chemical deicers are used; and in storage cellars or rooms for hides, casings, fertilizer, salt, and bulk chemicals.

300.7 Raceways Exposed to Different Temperatures.

(A) Sealing. Where a raceway is subjected to different temperatures, and where condensation is known to be a problem, the raceway must be filled with a material approved by the authority having jurisdiction that will prevent the circulation of warm air to a colder section of the raceway. An explosionproof seal isn't required for this purpose. Figure 300–27

(B) Expansion Fittings. Raceways must be provided with expansion fittings where necessary to compensate for thermal expansion and contraction. Figure 300–28

FPN: Table 352.44 provides the expansion characteristics for PVC conduit. The expansion characteristics for metal raceways are determined by multiplying the values from Table 352.44 by 0.20, and the expansion characteristics for aluminum raceways is determined by multiplying the values from Table 352.44 by 0.40. Table 354.44 provides the expansion characteristics for reinforced thermosetting resin conduit (RTRC).

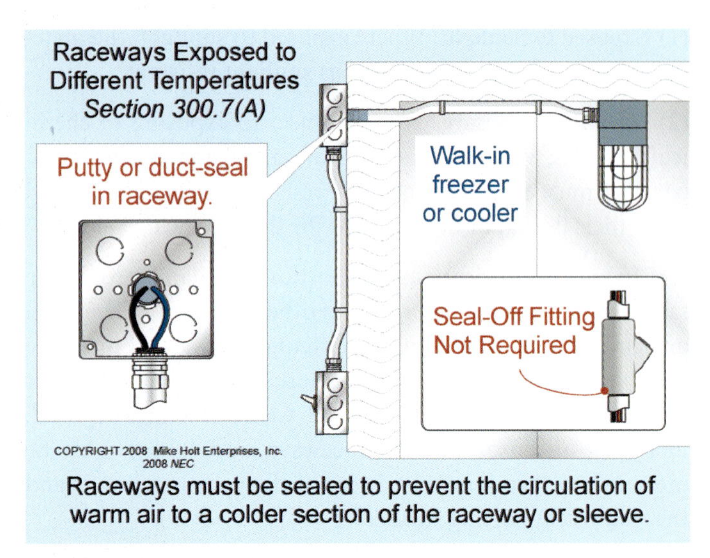

Figure 300–27

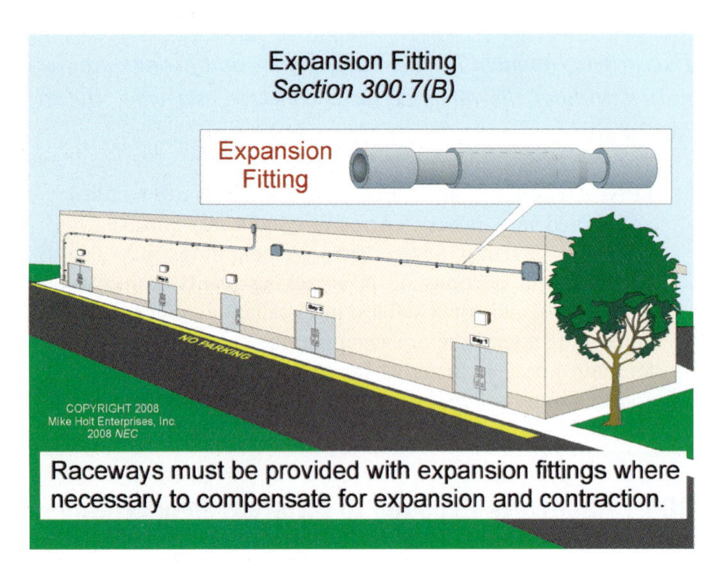

Figure 300–28

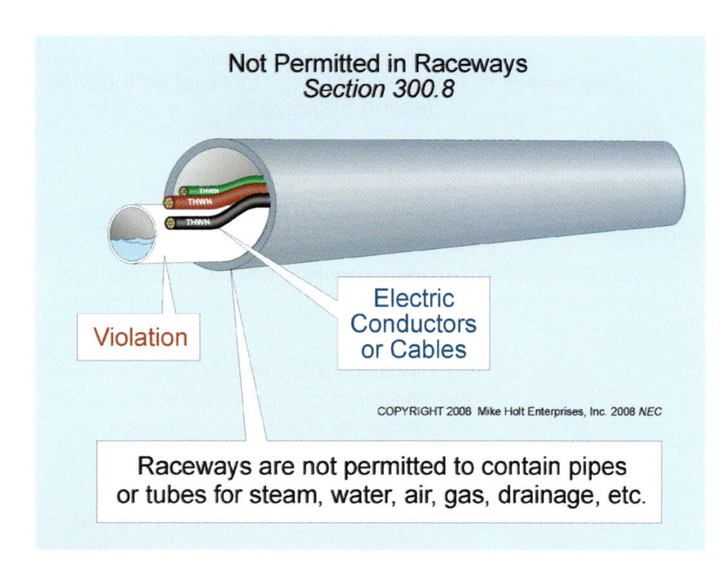

Figure 300–29

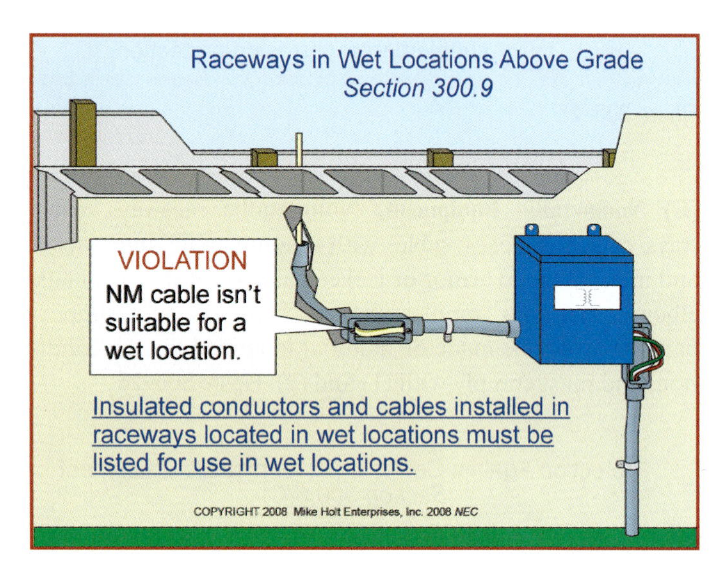

Figure 300–30

300.8 Not Permitted in Raceways. Raceways are designed for the exclusive use of electrical conductors and cables, and are not permitted to contain nonelectrical components, such as pipes or tubes for steam, water, air, gas, drainage, etc. Figure 300–29

300.9 Raceways in Wet Locations Above Grade. Insulated conductors and cables installed in raceways in above ground wet locations must be listed for use in wet locations according to 310.8(C). Figure 300–30

300.10 Electrical Continuity. Metal raceways, cables, boxes, fittings, cabinets, and enclosures for conductors must be metallically joined together to form a continuous, low-impedance fault current path capable of carrying any fault current likely to be imposed on it [110.10, 250.4(A)(3), and 250.122]. Figure 300–31

Metal raceways and cable assemblies must be mechanically secured to boxes, fittings, cabinets, and other enclosures.

Exception No. 1: Short lengths of metal raceways used for the support or protection of cables aren't required to be electrically continuous, nor are they required to be connected to an equipment grounding conductor of a type recognized in 250.118 [250.86 Ex 2 and 300.12 Ex]. Figure 300–32

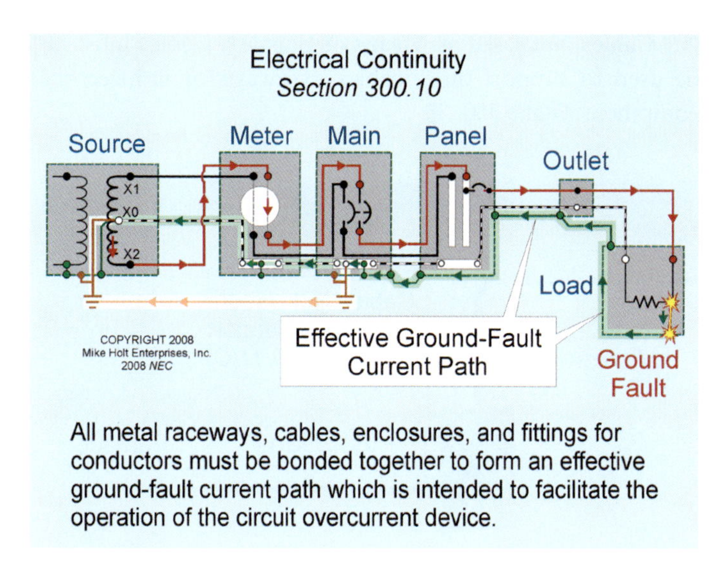

Figure 300–31

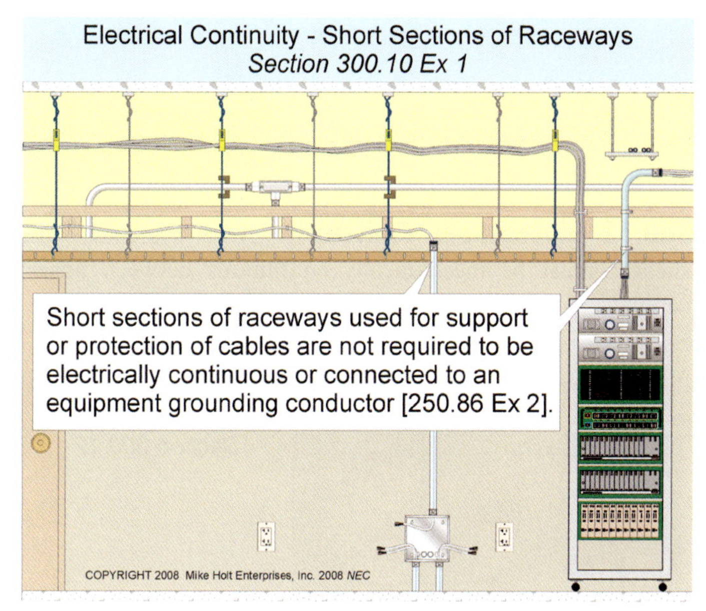

Figure 300–32

300.11 Securing and Supporting.

(A) Secured in Place. Raceways, cable assemblies, boxes, cabinets, and fittings must be securely fastened in place. The ceiling-support wires or ceiling grid must not be used to support raceways and cables (power, signaling, or communications). However, independent support wires that are secured at both ends and provide secure support are permitted. Figure 300–33

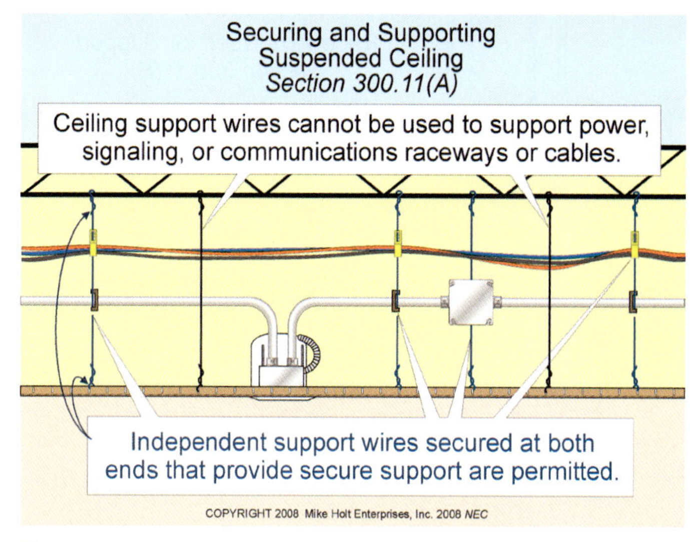

Figure 300–33

Author's Comment: Outlet boxes [314.23(D)] and luminaires can be secured to the suspended-ceiling grid if securely fastened to the ceiling-framing member by mechanical means such as bolts, screws, or rivets, or by the use of clips or other securing means identified for use with the type of ceiling-framing member(s) [410.36(B)].

(1) Fire-Rated Assembly. Electrical wiring within the cavity of a fire-rated floor-ceiling or roof-ceiling assembly can be supported by independent support wires attached to the ceiling assembly. The independent support wires must be distinguishable from the suspended-ceiling support wires by color, tagging, or other effective means.

(2) Nonfire-Rated Assembly. Wiring in a nonfire-rated floor-ceiling or roof-ceiling assembly is not permitted to be secured to, or supported by, the ceiling assembly, including the ceiling support wires. Independent support wires used for support can be attached to the nonfire-rated assembly.

Author's Comment: Support wires within nonfire-rated assemblies aren't required to be distinguishable from the suspended-ceiling framing support wires. Most suspended ceiling systems are not part of a fire-resistance rated assembly.

(B) Raceways Used for Support. Raceways must not be used as a means of support for other raceways, cables, or nonelectrical equipment, except as permitted in (1) through (3). Figure 300–34

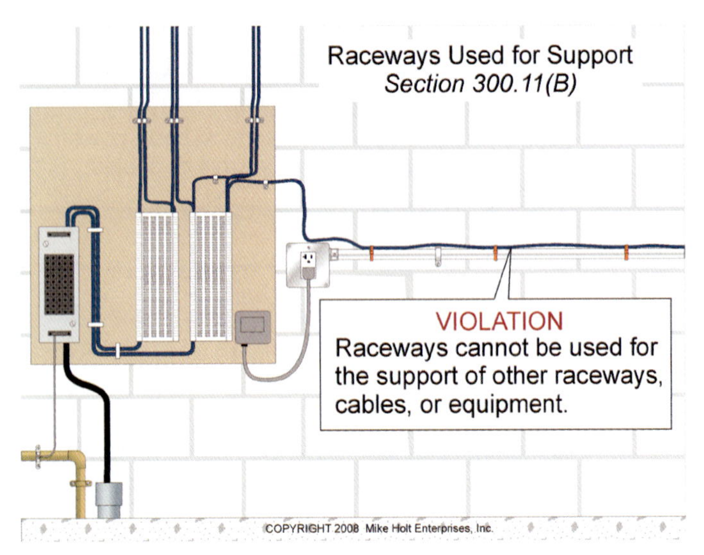

Figure 300–34

(1) Identified. Where the raceway or means of support is identified for the purpose.

(2) Class 2 and 3 Circuits. Class 2 and 3 cable can be supported by the raceway that supplies power to the equipment controlled by the Class 2 or 3 circuit. Figure 300–35

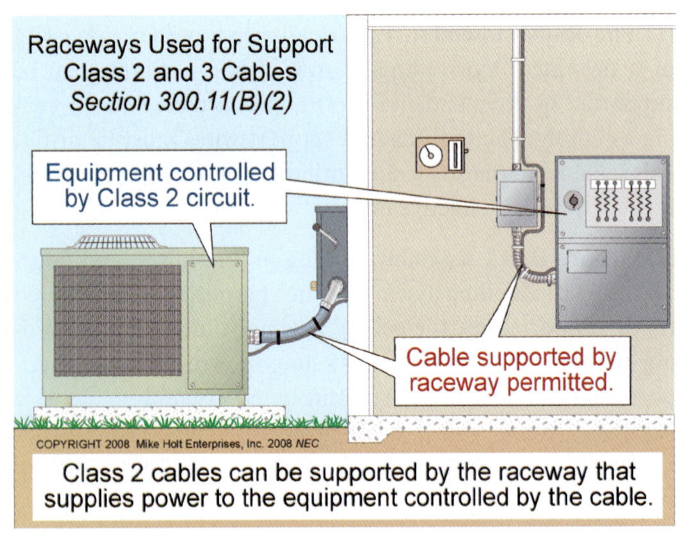

Figure 300–35

(3) Boxes Supported by Raceways. Raceways are permitted as a means of support for threaded boxes and conduit bodies in accordance with 314.23(E) and (F), or to support luminaires in accordance with 410.36(E).

(C) Cables Not Used as Means of Support. Cables must not be used to support other cables, raceways, or nonelectrical equipment. Figure 300–36

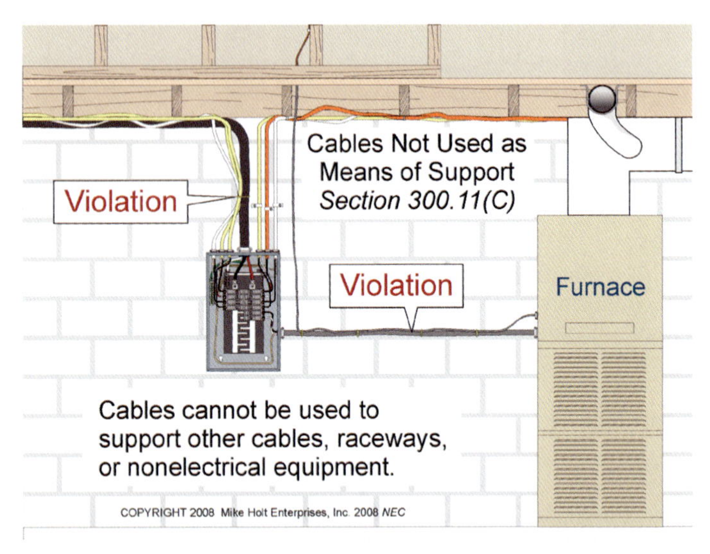

Figure 300–36

300.12 Mechanical Continuity.

Raceways and cable sheaths must be mechanically continuous between boxes, cabinets, and fittings. Figure 300–37

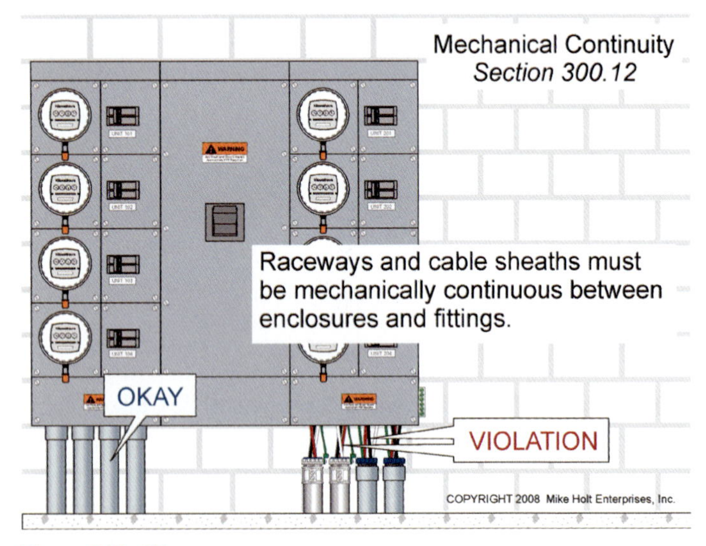

Figure 300–37

Exception No. 1: Short sections of raceways used to provide support or protection of cable from physical damage aren't required to be mechanically continuous [250.86 Ex 2 and 300.10 Ex 1]. **Figure 300–38**

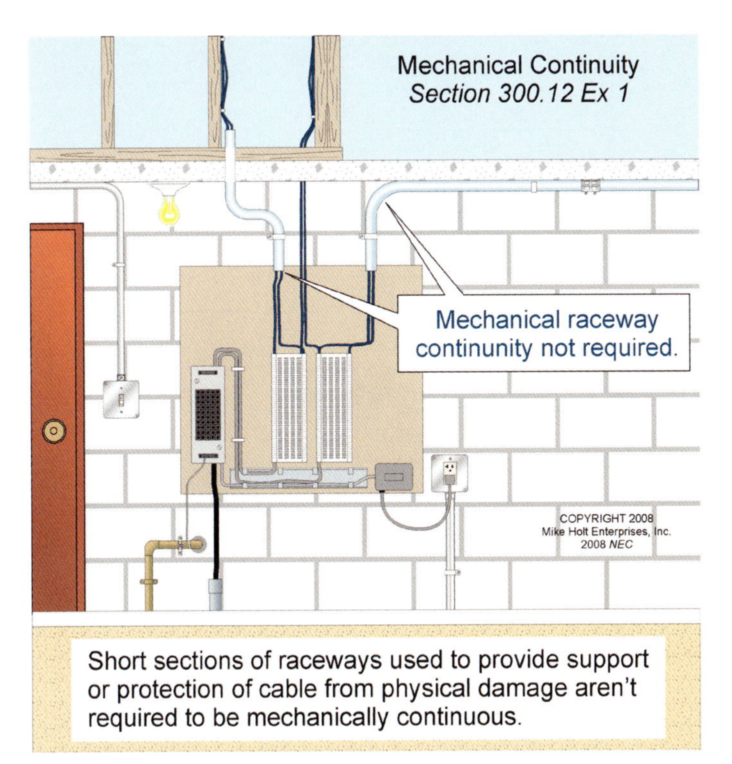

Figure 300–38

Exception No. 2: Raceways at the bottom of open-bottom equipment, such as switchboards, motor control centers, and transformers, are not required to be mechanically secured to the equipment. **Figure 300–39**

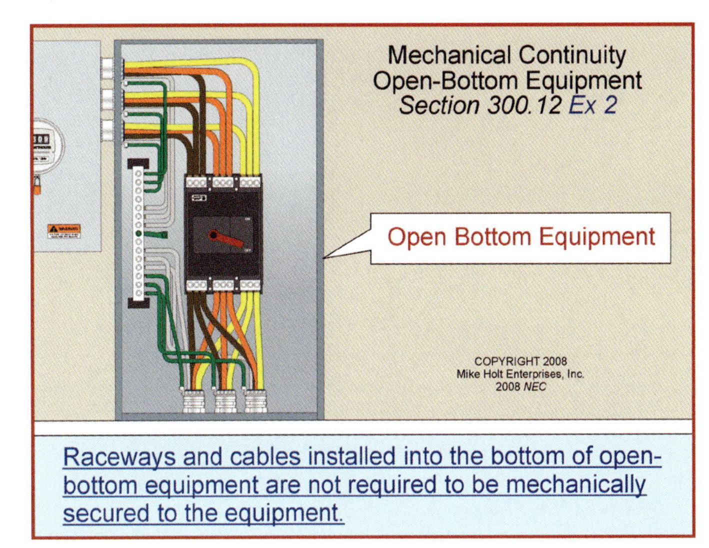

Figure 300–39

Author's Comment: When raceways are stubbed into an open-bottom switchboard or other apparatus, the raceway, including the end fitting, can't rise more than 3 in. above the bottom of the switchboard enclosure [408.5].

300.13 Splices and Pigtails.

(A) Conductor Splices. Conductors in raceways must be continuous between all points of the system, which means splices must not be made in raceways, except as permitted by 376.56, 378.56, 384.56, 386.56, or 388.56. See 300.15. **Figure 300–40**

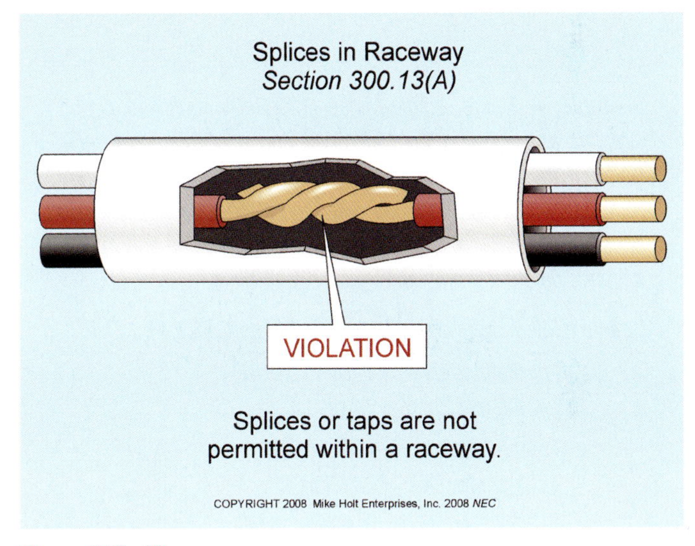

Figure 300–40

(B) Conductor Continuity. Continuity of the neutral conductor of a multiwire branch circuit must not be interrupted by the removal of a wiring device. In these applications the neutral conductors must be spliced together, and a "pigtail" must be provided for the wiring device. **Figure 300–41**

Author's Comment: The opening of the ungrounded conductors, or the neutral conductor of a 2-wire circuit during the replacement of a device, doesn't cause a safety hazard, so pigtailing of these conductors isn't required [110.14(B)].

CAUTION: *If the continuity of the neutral conductor of a multiwire circuit is interrupted (open), the resultant over- or undervoltage can cause a fire and/or destruction of electrical equipment.*

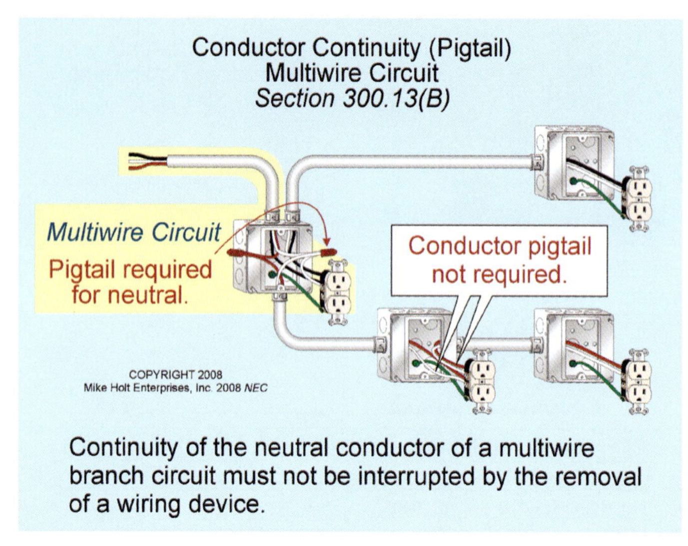

Continuity of the neutral conductor of a multiwire branch circuit must not be interrupted by the removal of a wiring device.

Figure 300–41

Example: *A 3-wire, single-phase, 120/240V multiwire circuit supplies a 1,200W, 120V hair dryer and a 600W, 120V television. If the neutral conductor of the multiwire circuit is interrupted, it will cause the 120V television to operate at 160V and consume 1,067W of power (instead of 600W) for only a few seconds before it burns up.* **Figure 300–42**

Step 1: Determine the resistance of each appliance, $R = E^2/P$.

R of the hair dryer = $120V^2/1{,}200W$
R of the hair dryer = 12 ohms

R of the television = $120V^2/600W$
R of the television = 24 ohms

Step 2: Determine the current of the circuit, I = E/R.

E = 240V

R = 36 ohms (12 ohms + 24 ohms)
I = 240V/36 ohms
I = 6.70A

Step 3: Determine the operating voltage for each appliance, E = I x R.

I – 6.70A

R = 12 ohms for dryer and 24 ohms for TV

Voltage of hair dryer = 6.70A x 12 ohms
Voltage of hair dryer = 80V

Voltage of television = 6.70A x 24 ohms
Voltage of Television = 160V

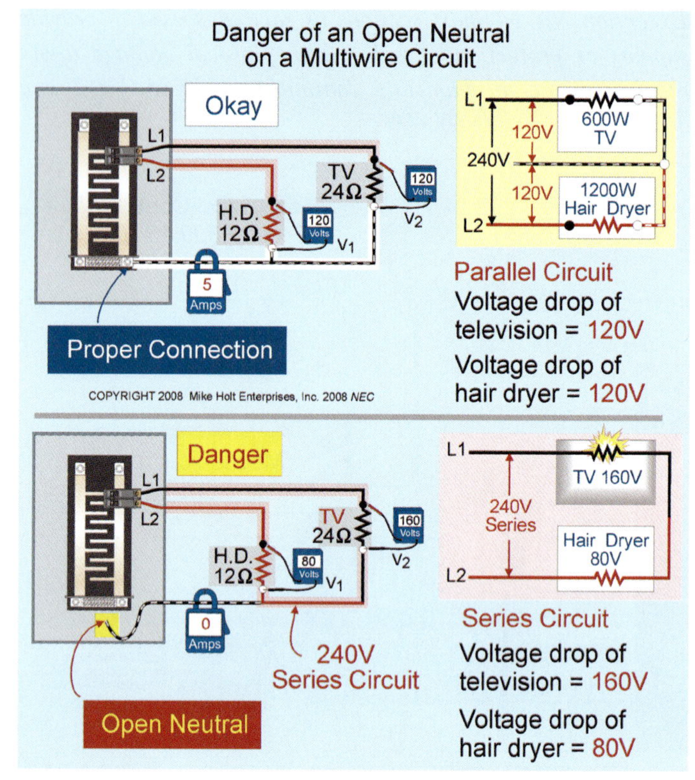

Figure 300–42

WARNING: *Failure to terminate the ungrounded conductors to separate phases can cause the neutral conductor to become overloaded, and the insulation can be damaged or destroyed by excessive heat. Conductor overheating is known to decrease the service life of insulating materials, which creates the potential for arcing faults in hidden locations, and could ultimately lead to fires. It isn't known just how long conductor insulation lasts, but heat does decrease its life span.* **Figure 300–43**

300.14 Length of Free Conductors. At least 6 in. of free conductor, measured from the point in the box where the conductors enter the enclosure, must be left at each outlet, junction, and switch point for splices or terminations of luminaires or devices. **Figure 300–44**

Boxes that have openings less than 8 in. in any dimension, must have at least 6 in. of free conductor, measured from the point where the conductors enter the box, and at least 3 in. of free conductor outside the box opening. **Figure 300–45**

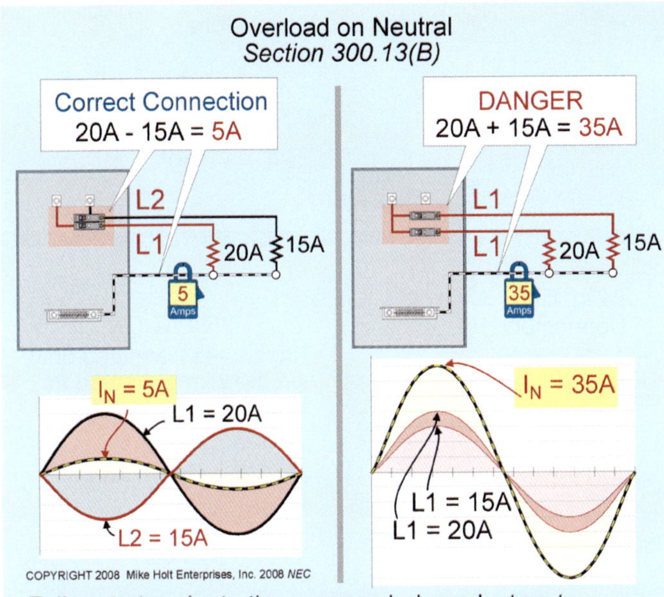

Failure to terminate the ungrounded conductors to different phases or lines can cause the neutral conductor to be overloaded, which can cause a fire.

Figure 300–43

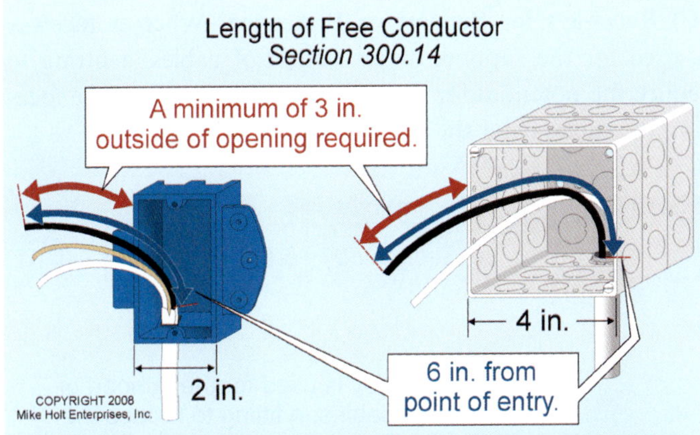

Boxes with openings less than 8 in. must have at least 6 in. of free conductor and at least 3 in. of free conductor outside the box opening.

Figure 300–44

Author's Comment: The *NEC* doesn't limit the number of extension rings permitted on an outlet box, and the UL White Book (product category QCIT) specifically states that multiple extensions can be used. When multiple extensions are used, however, there must be not less than 3 in. of free conductors outside the opening of the final extension ring.

Exception: This rule doesn't apply to conductors that pass through a box without a splice or termination.

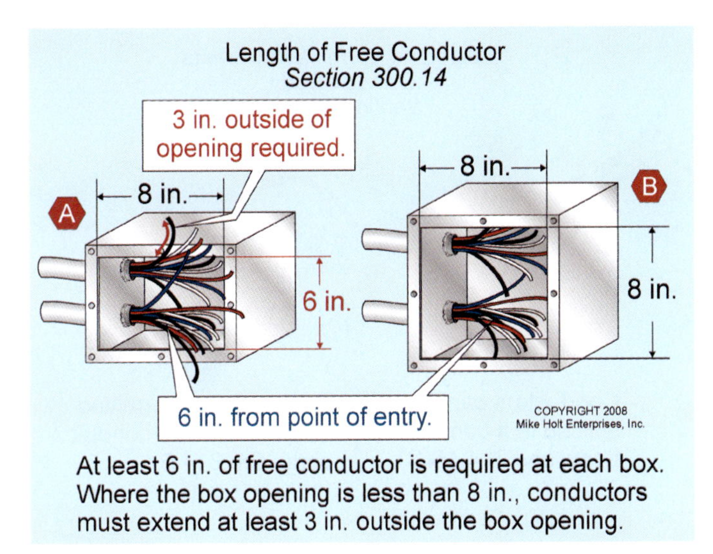

At least 6 in. of free conductor is required at each box. Where the box opening is less than 8 in., conductors must extend at least 3 in. outside the box opening.

Figure 300–45

300.15 Boxes or Conduit Bodies. A box must be installed at each splice or termination point, except as permitted for: Figure 300–46

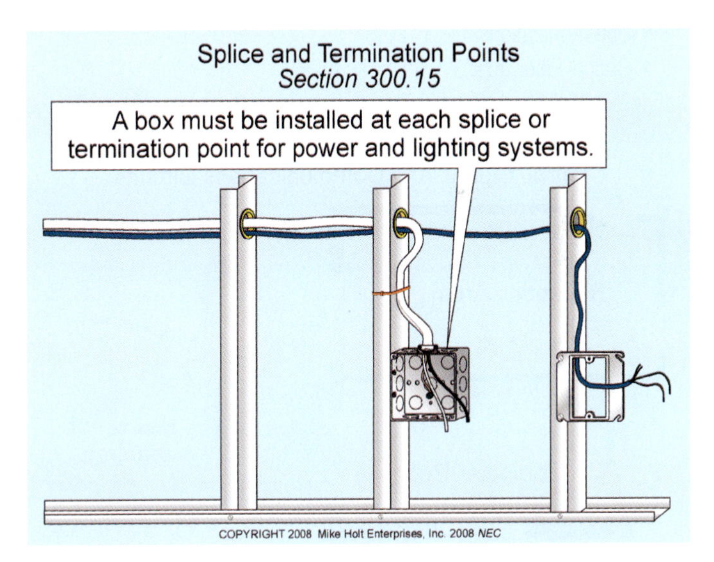

Figure 300–46

- Cabinet or Cutout Boxes, 312.8
- Conduit Bodies, 314.16(C) Figure 300–47
- Luminaires, 410.64
- Surface Raceways, 386.56 and 388.56
- Wireways, 376.56

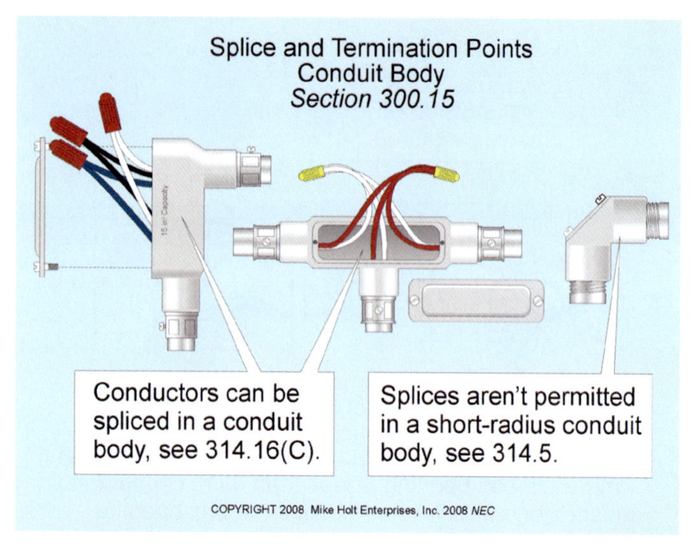

Figure 300–47

Author's Comment: Boxes aren't required for the following signaling and communications cables or raceways: Figure 300–48

- CATV, 90.3
- Class 2 and 3 Control and Signaling, 725.3
- Communications, 90.3
- Optical Fiber, 770.3

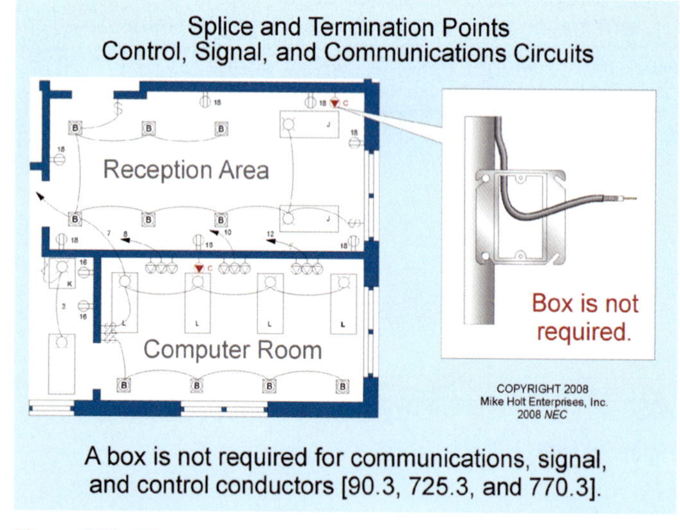

Figure 300–48

Fittings and Connectors. Fittings can only be used with the specific wiring methods for which they are listed and designed. For example, Type NM cable connectors must not be used with Type AC cable, and electrical metallic tubing fittings must not be used with rigid metal conduit or intermediate metal conduit, unless listed for the purpose. Figure 300–49

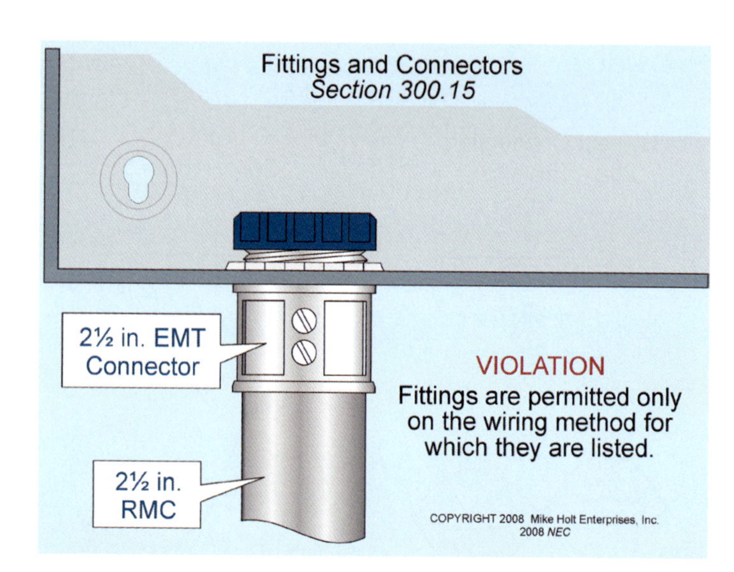

Figure 300–49

Author's Comment: PVC conduit couplings and connectors are permitted with electrical nonmetallic tubing if the proper glue is used in accordance with manufacturer's instructions [110.3(B)]. See 362.48.

(C) Raceways for Support or Protection. When a raceway is used for the support or protection of cables, a fitting to reduce the potential for abrasion must be placed at the location the cables enter the raceway. Figure 300–50

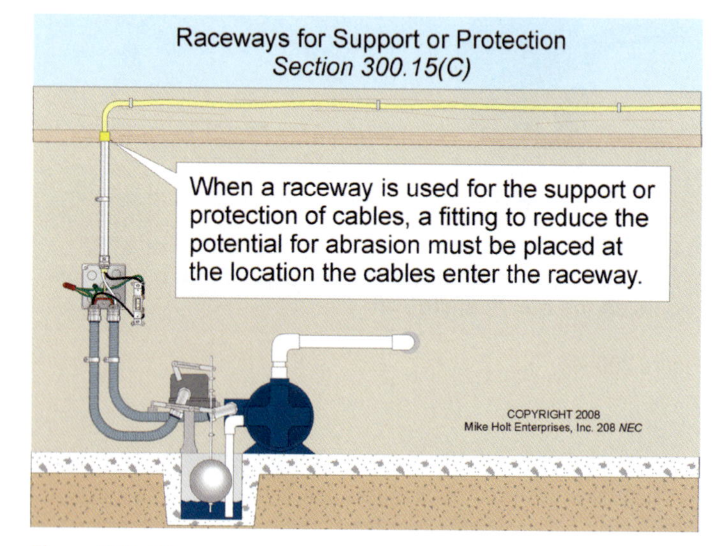

Figure 300–50

(F) Fitting. A fitting is permitted in lieu of a box or conduit body where conductors are not spliced or terminated within the fitting if it's accessible after installation. Figure 300–51

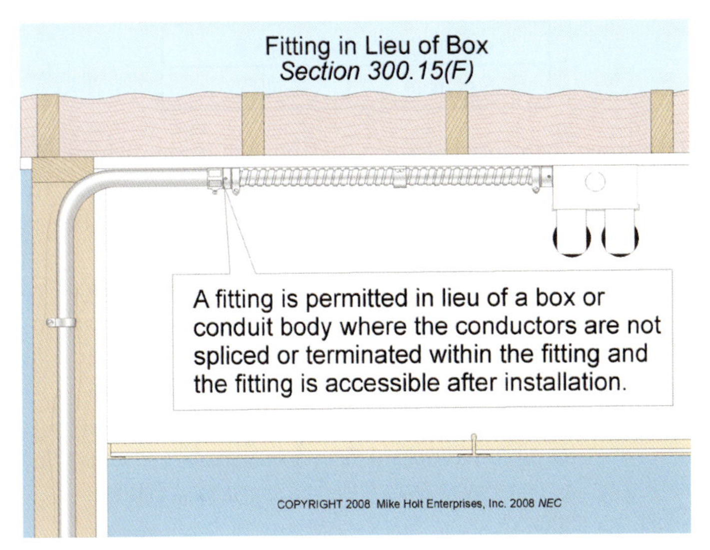

Figure 300–51

(G) Underground Splices. A box or conduit body isn't required where a splice is made underground if the conductors are spliced with a splicing device listed for direct burial. See 110.14(B) and 300.5(E).

Author's Comment: See the definition of "Conduit Body" in Article 100.

(I) Enclosures. A box or conduit body isn't required where a splice is made in a cabinet or in cutout boxes containing switches or overcurrent devices if the splices or taps don't fill the wiring space at any cross section to more than 75 percent, and the wiring at any cross section doesn't exceed 40 percent. See 312.8 and 404.3(B). Figure 300–52

Author's Comment: See the definitions of "Cabinet" and "Cutout Box" in Article 100.

(J) Luminaires. A box or conduit body isn't required where a luminaire is used as a raceway as permitted in 410.64 and 410.65.

(L) Handhole Enclosures. A box or conduit body isn't required for conductors installed in a handhole enclosure. Splices must be made in accordance with 314.30. Figure 300–53

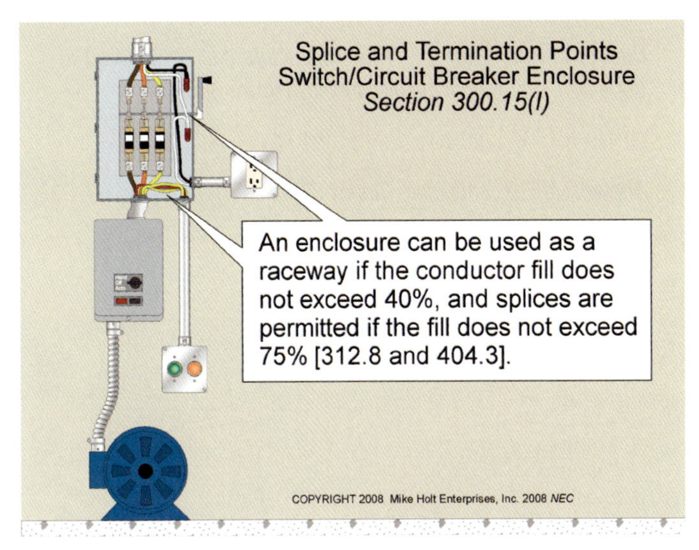

Figure 300–52

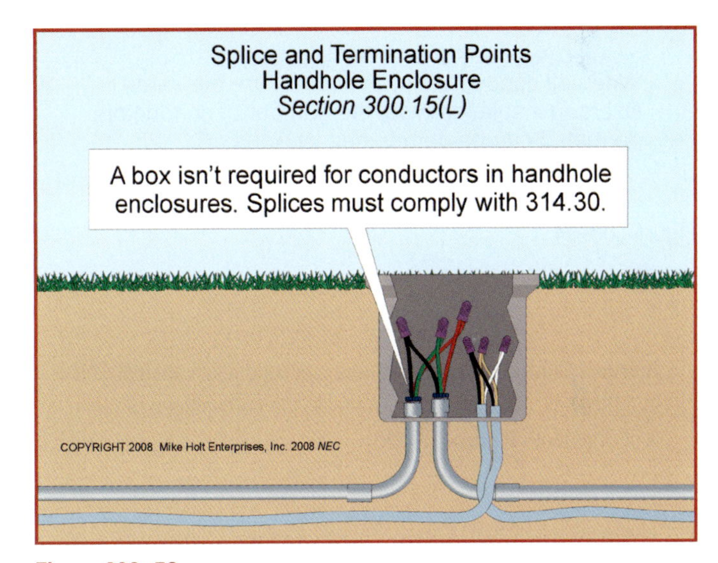

Figure 300–53

Author's Comment: Splices or terminations within a handhole must be accomplished by the use of fittings listed as suitable for wet locations [110.14(B) and 314.30(C)].

300.17 Raceway Sizing.

Raceways must be large enough to permit the installation and removal of conductors without damaging the conductor's insulation.

Author's Comment: When all conductors in a raceway are the same size and of the same insulation type, the number of conductors permitted can be determined by Annex C.

Question: *How many 12 THHN conductors can be installed in trade size ¾ electrical metallic tubing?* **Figure 300–54**

(a) 12 *(b) 13* *(c) 14* *(d) 16*

Answer: *(d) 16 conductors [Annex C, Table C1]*

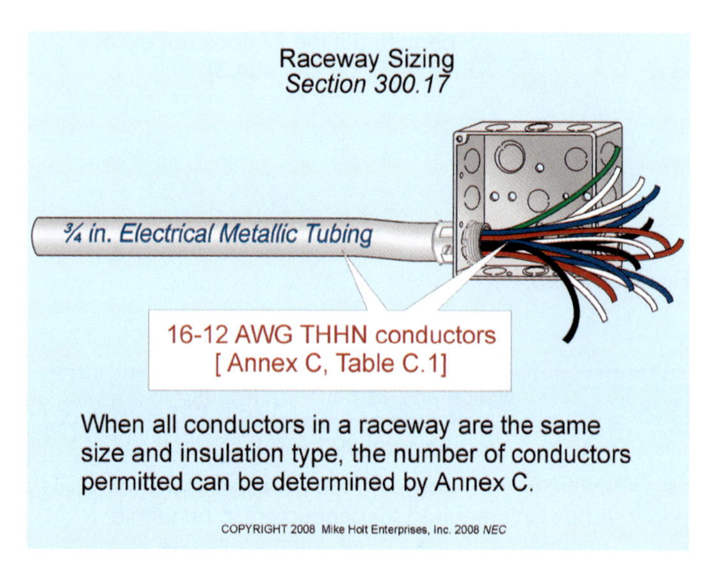

Figure 300–54

Author's Comment: When different size conductors are installed in a raceway, conductor fill is limited to the percentages of Table 1 of Chapter 9. **Figure 300–55**

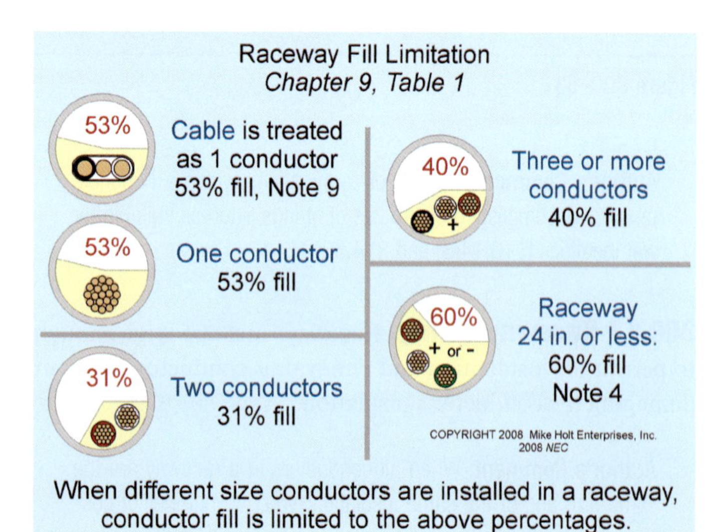

Figure 300–55

Table 1, Chapter 9	
Number	**Percent Fill**
1 Conductor	53%
2 Conductors	31%
3 or more	40%

The above percentages are based on conditions where the length of the conductor and number of raceway bends are within reasonable limits [Chapter 9, Table 1, FPN No. 1].

Author's Comment: Where a raceway has a maximum length of 24 in., it can be filled to 60 percent of its total cross-sectional area [Chapter 9, Table 1, Note 4]. **Figure 300–56**

Step 1: When sizing a raceway, first determine the total area of conductors (Chapter 9, Table 5 for insulated conductors and Chapter 9, Table 8 for bare conductors). **Figure 300–57**

Step 2: Select the raceway from Chapter 9, Table 4, in accordance with the percent fill listed in Chapter 9, Table 1. **Figure 300–58**

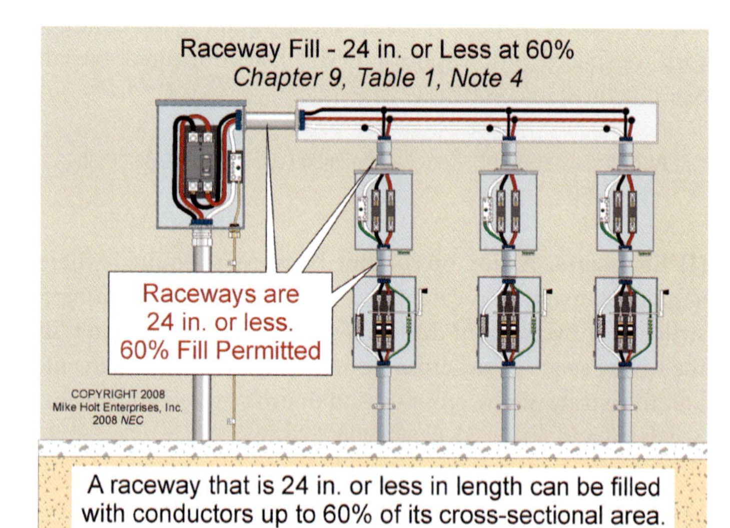

Figure 300–56

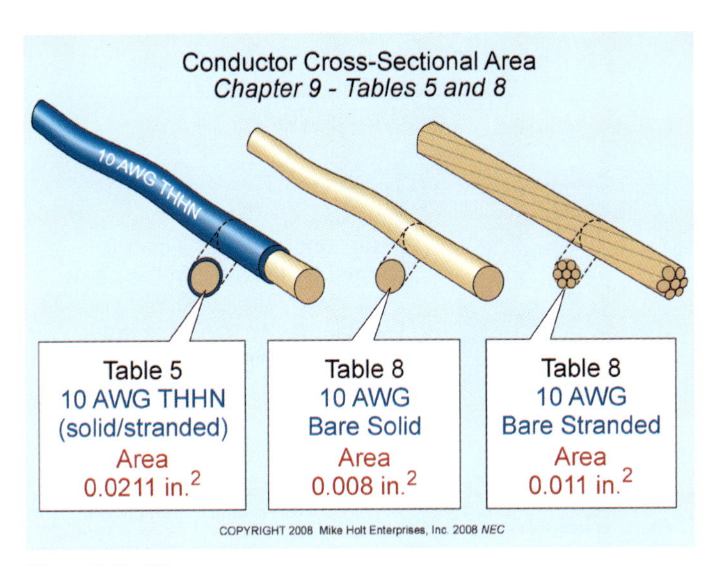

Figure 300–57

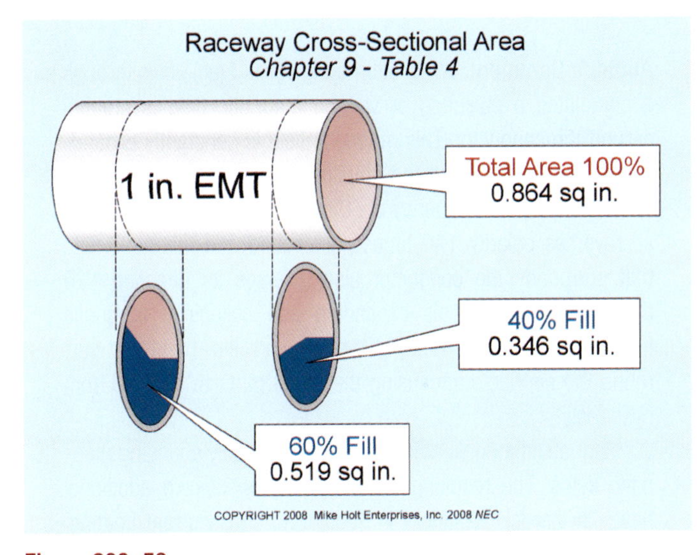

Figure 300–58

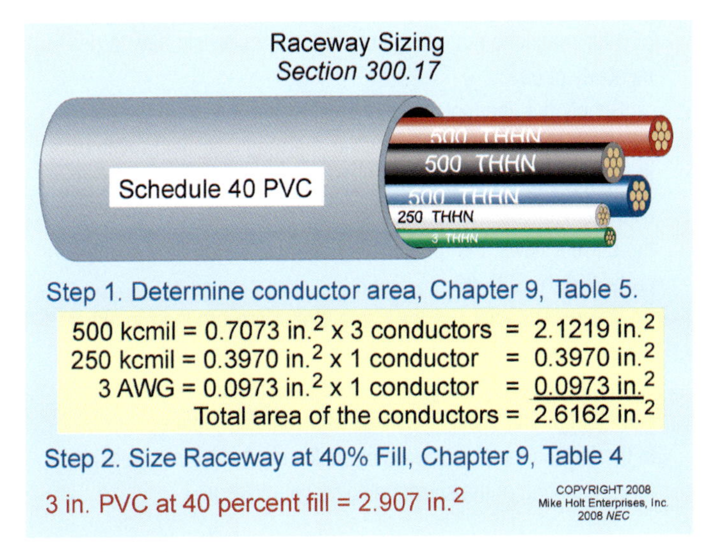

Figure 300–59

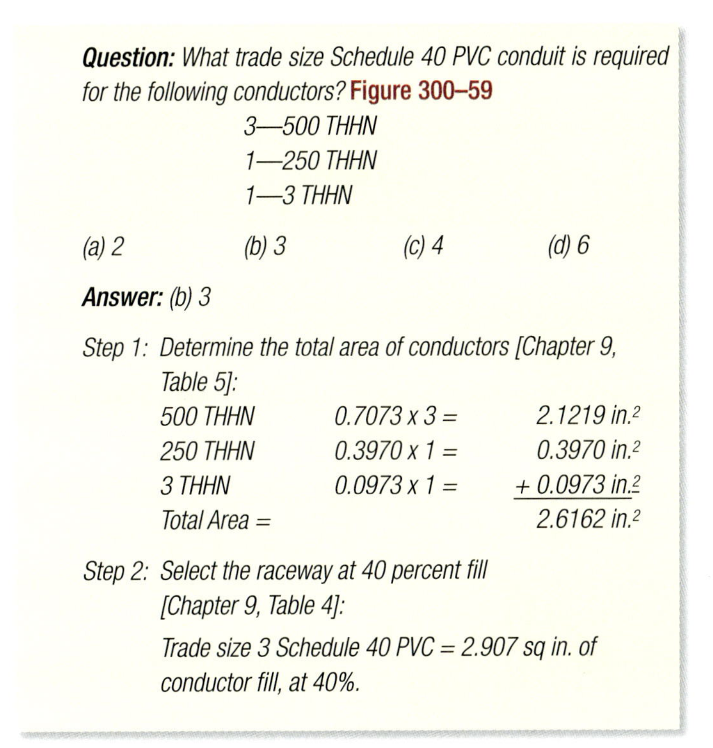

Question: *What trade size Schedule 40 PVC conduit is required for the following conductors?* Figure 300–59

3—500 THHN
1—250 THHN
1—3 THHN

(a) 2 *(b) 3* *(c) 4* *(d) 6*

Answer: *(b) 3*

Step 1: Determine the total area of conductors [Chapter 9, Table 5]:

500 THHN	*0.7073 x 3 =*	*2.1219 in.²*
250 THHN	*0.3970 x 1 =*	*0.3970 in.²*
3 THHN	*0.0973 x 1 =*	*+ 0.0973 in.²*
Total Area =		*2.6162 in.²*

Step 2: Select the raceway at 40 percent fill [Chapter 9, Table 4]:

Trade size 3 Schedule 40 PVC = 2.907 sq in. of conductor fill, at 40%.

300.18 Inserting Conductors in Raceways.

(A) Complete Runs. To protect conductor insulation from abrasion during installation, raceways must be mechanically completed between the pulling points before conductors are installed. See 300.10 and 300.12. Figure 300–60

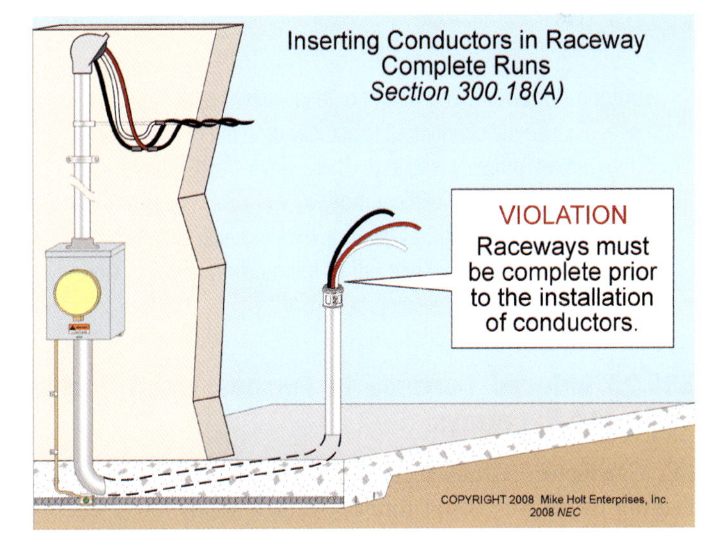

Figure 300–60

Exception: Short sections of raceways used for protection of cables from physical damage aren't required to be installed complete between outlet, junction, or splicing points.

300.19 Supporting Conductors in Vertical Raceways.

(A) Spacing Intervals. If the vertical rise of a raceway exceeds the values of Table 300.19(A), the conductors must be supported at the top, or as close to the top as practical. Intermediate support must also be provided in increments that do not exceed the values of Table 300.19(A). Figure 300–61

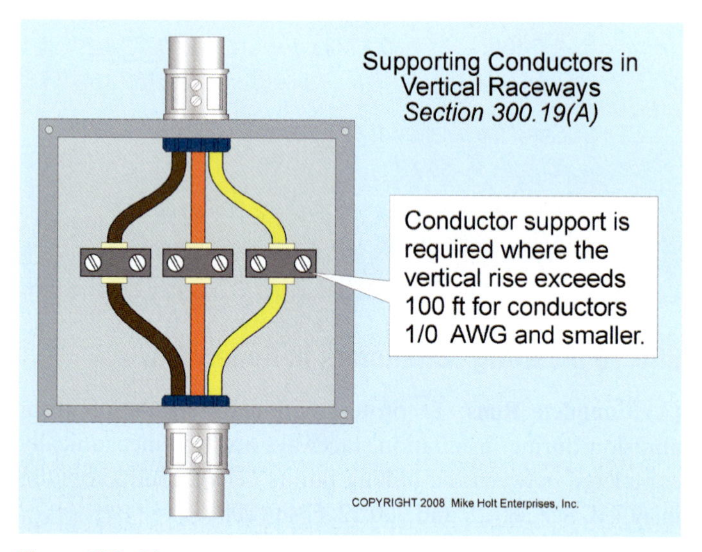

Figure 300–61

Author's Comment: The weight of long vertical runs of conductors can cause the conductors to actually drop out of the raceway if they aren't properly secured. There have been many cases where conductors in a vertical raceway were released from the pulling "basket" or "grip" (at the top) without being secured, and the conductors fell down and out of the raceway, injuring those at the bottom of the installation.

300.20 Induced Currents in Ferrous Metal Enclosures and Raceways.

(A) Conductors Grouped Together. To minimize induction heating of ferrous metal raceways and ferrous metal enclosures for alternating-current circuits, and to maintain an effective ground-fault current path, all conductors of a circuit must be installed in the same raceway, cable, trench, cord, or cable tray. See 250.102(E), 300.3(B), 300.5(I), and 392.8(D). Figure 300–62

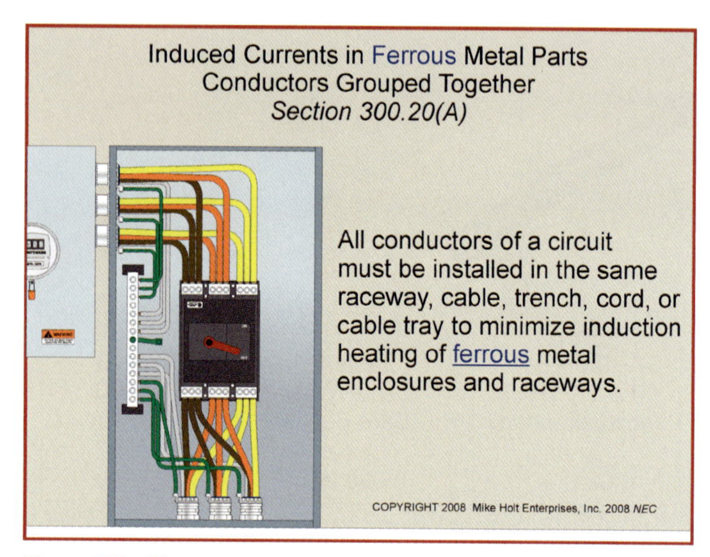

Figure 300–62

Author's Comment: When alternating current (ac) flows through a conductor, a pulsating or varying magnetic field is created around the conductor. This magnetic field is constantly expanding and contracting with the amplitude of the ac current. In the United States, the frequency is 60 cycles per second (Hz). Since ac reverses polarity 120 times per second, the magnetic field that surrounds the conductor also reverses its direction 120 times per second. This expanding and collapsing magnetic field induces eddy currents in the ferrous metal parts that surround the conductors, causing the metal parts to heat up from hysteresis.

Magnetic materials naturally resist the rapidly changing magnetic fields. The resulting friction produces its own additional heat - hysteresis heating - in addition to eddy current heating. A metal which offers high resistance is said to have high magnetic "permeability." Permeability can vary on a scale of 100 to 500 for magnetic materials; nonmagnetic materials have a permeability of one.

Simply put, the molecules of steel and iron align to the polarity of the magnetic field and when the magnetic field reverses, the molecules reverse their polarity as well. This back-and-forth alignment of the molecules heats up the metal, and the more the current flows, the greater the heat rises in the ferrous metal parts. Figure 300–63

When conductors of the same circuit are grouped together, the magnetic fields of the different conductors tend to cancel each other out, resulting in a reduced magnetic field around the conductors. The lower magnetic field reduces induced currents in the ferrous metal raceways or enclosures, which reduces hysteresis heating of the surrounding metal enclosure.

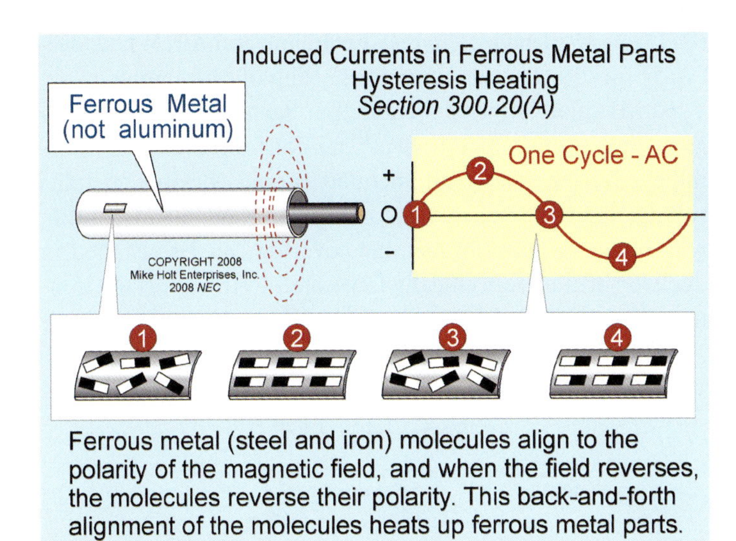

Figure 300–63

WARNING: *There has been much discussion in the press on the effects of electromagnetic fields on humans. According to the Institute of Electrical and Electronics Engineers (IEEE), there's insufficient information at this time to define an unsafe electromagnetic field level.*

(B) Single Conductors. When single conductors are installed in nonmetallic raceways as permitted in 300.5(I) Ex 2, the inductive heating of the metal enclosure must be minimized by the use of aluminum locknuts and by cutting a slot between the individual holes through which the conductors pass. Figure 300–64

FPN: Because aluminum is a nonmagnetic metal, aluminum parts don't heat up due to hysteresis.

Author's Comment: Aluminum conduit, locknuts, and enclosures carry eddy currents, but because aluminum is nonferrous, it doesn't heat up [300.20(B) FPN].

300.21 Spread of Fire or Products of Combustion.

Electrical circuits and equipment must be installed in such a way that the spread of fire or products of combustion will not be substantially increased. Openings in fire-rated walls, floors, and ceilings for electrical equipment must be firestopped using methods approved by the authority having jurisdiction to maintain the fire-resistance rating of the fire-rated assembly.

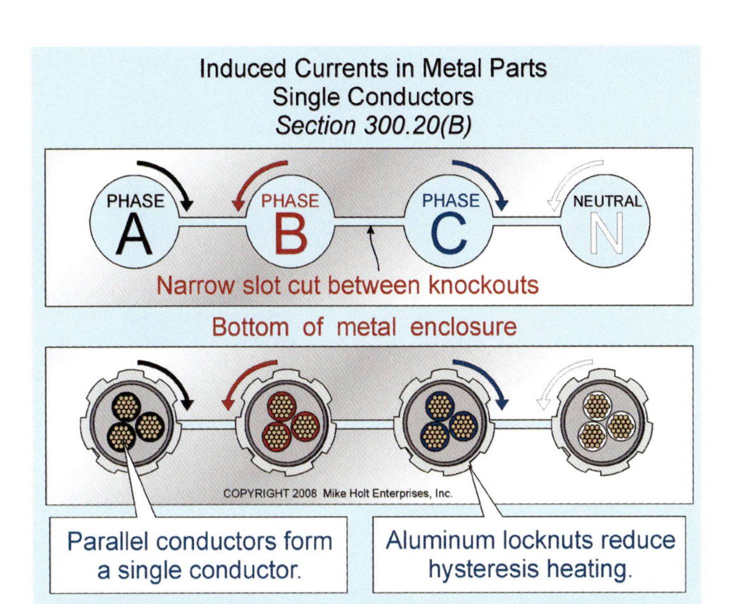

Figure 300–64

Author's Comment: Fire-stopping materials are listed for the specific types of wiring methods and the construction of the assembly that they penetrate.

FPN: Directories of electrical construction materials published by qualified testing laboratories contain listing and installation restrictions necessary to maintain the fire-resistive rating of assemblies. Outlet boxes must have a horizontal separation not less than 24 in. when installed in a fire-rated assembly, unless an outlet box is listed for closer spacing or protected by fire-resistant "putty pads" in accordance with manufacturer's instructions.

Author's Comments:

- Boxes installed in fire-resistance rated assemblies must be listed for the purpose. Where steel boxes are used, they must be secured to the framing member, so cut-in type boxes are not allowed (UL White Book, product category QCIT).
- This rule also applies to control, signal, and communications cables or raceways. Figure 300–65
  - CATV, 820.26
  - Communications, 800.26
  - Control and Signaling, 725.25
  - Fire Alarm, 760.3(A)
  - Optical Fiber, 770.26
  - Sound Systems, 640.3(A)

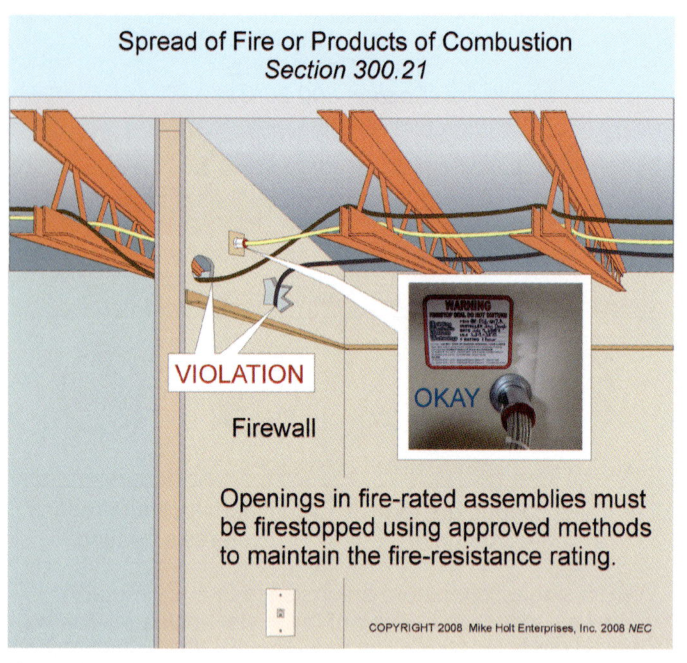

Figure 300–65

300.22 Ducts, Plenums, and Other Air-Handling Spaces.

(A) Ducts Used for Dust, Loose Stock, or Vapor. Ducts that transport dust, loose stock, or vapors must not have any wiring method installed within them. Figure 300–66

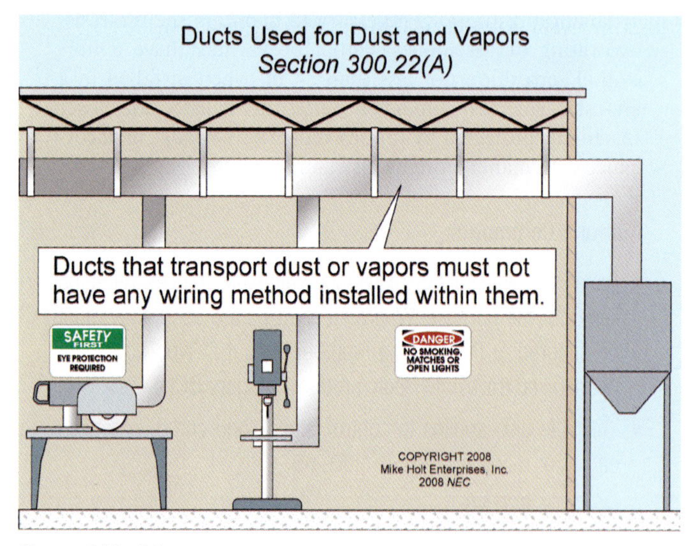

Figure 300–66

(B) Ducts or Plenums Used for Environmental Air. Where necessary for direct action upon, or sensing of, the contained air, Type MI cable, Type MC cable that has a smooth or corrugated impervious metal sheath without an overall nonmetallic covering, electrical metallic tubing, flexible metallic tubing, intermediate metal conduit, or rigid metal conduit without an overall nonmetallic covering can be installed in ducts or plenums specifically fabricated to transport environmental air. Figure 300–67

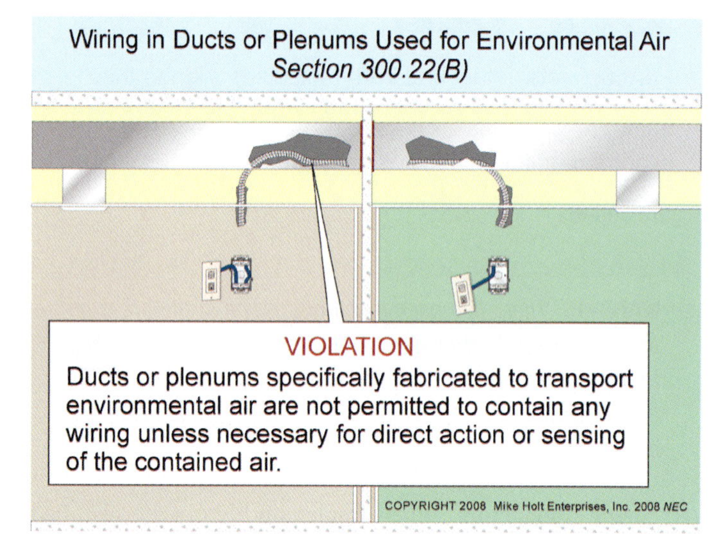

Figure 300–67

Author's Comment: See the definition of "Plenum" in Article 100.

Flexible metal conduit in lengths not exceeding 4 ft can be used to connect physically adjustable equipment and devices, provided any openings are effectively closed.

Where equipment or devices are installed and illumination is necessary to facilitate maintenance and repair, enclosed gasketed-type luminaires are permitted.

(C) Other Space Used for Environmental Air Space. Wiring and equipment in spaces used for environmental air-handling purposes must comply with (1) and (2). This requirement doesn't apply to habitable rooms or areas of buildings, the prime purpose of which isn't air handling.

FPN: The spaces above a suspended ceiling or below a raised floor used for environmental air are examples of the type of space to which this section applies. Figure 300–68

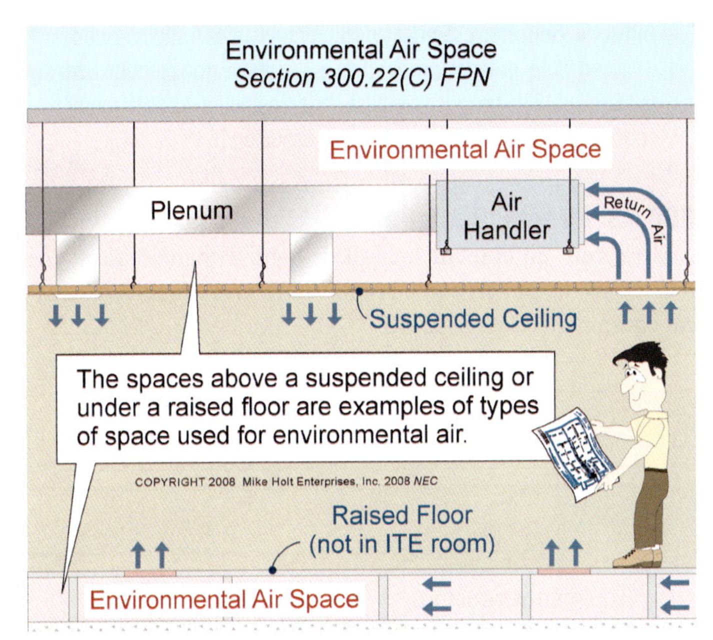

Figure 300–68

(1) Wiring Methods Permitted. Electrical metallic tubing, rigid metal conduit, intermediate metal conduit, armored cable, metal-clad cable without a nonmetallic cover, and flexible metal conduit can be installed in environmental air spaces.

Where accessible, surface metal raceways, metal wireways with metal covers, or solid bottom metal cable tray with solid metal covers can be installed in environmental air spaces.

Author's Comments:

- PVC conduit [Article 352], electrical nonmetallic tubing [Article 362], liquidtight flexible conduit, and nonmetallic cables are not permitted to be installed in spaces used for environmental air because they give off deadly toxic fumes when burned or superheated.
- Control, signaling, and communications cables installed in spaces used for environmental air must be suitable for plenum use. Figure 300–69
  - CATV, 820.179(A)
  - Communications, 800.21
  - Control and Signaling, 725.154(A)
  - Fire Alarm, 760.7
  - Optical Fiber Cables and Raceways, 770.154(A)
  - Sound Systems, 640.9(C) and 725.154(A)
- Any wiring method suitable for the condition can be used in a space not used for environmental air-handling purposes. Figure 300–70

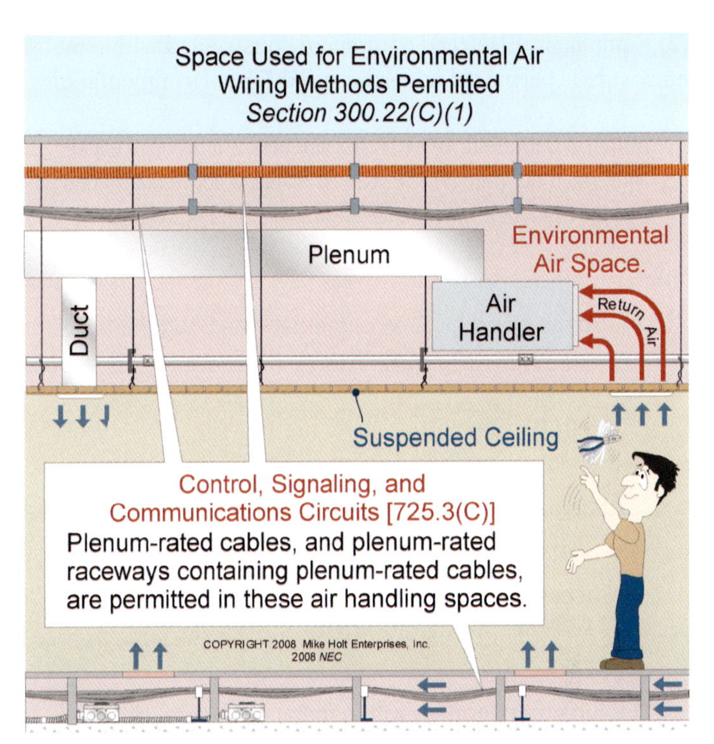

Figure 300–69

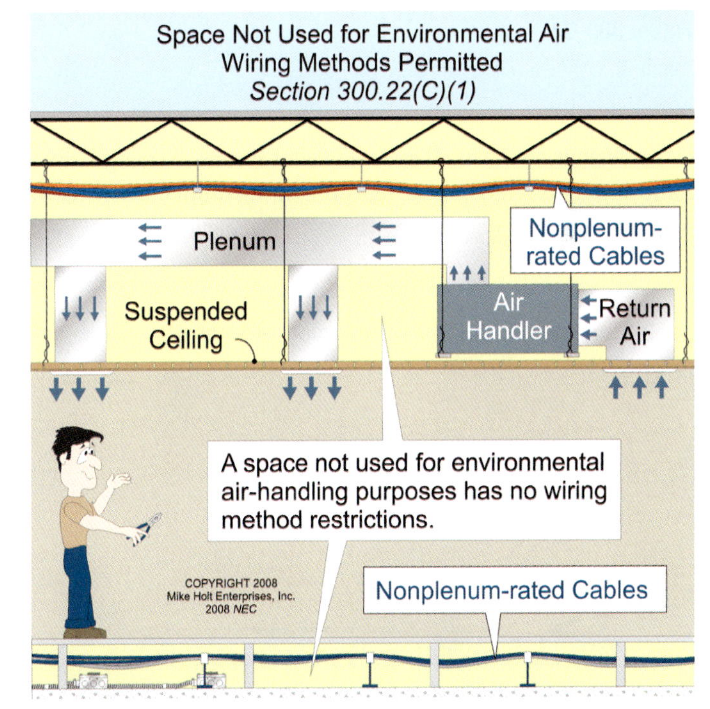

Figure 300–70

(2) Equipment. Electrical equipment constructed with a metal enclosure is permitted in a space used for environmental air.

Author's Comment: Dry-type transformers with a metal enclosure, rated not over 50 kVA, can be installed above suspended ceilings used for environmental air [450.13(B)]. Figure 300–71

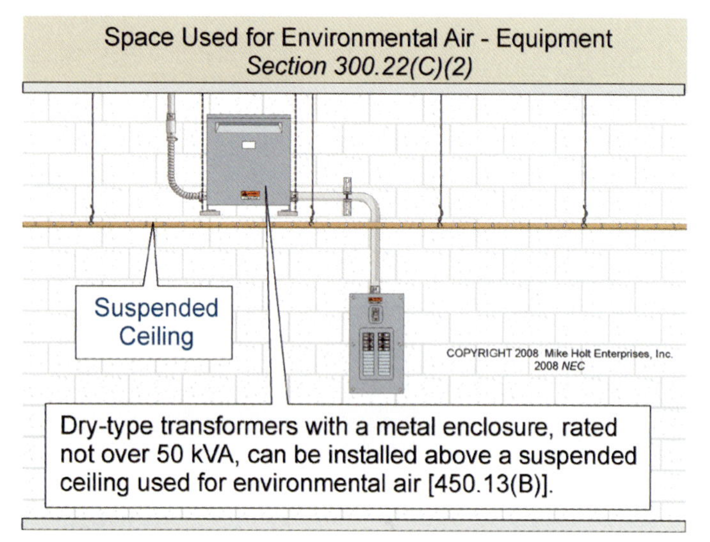

Figure 300–71

(D) Information Technology Equipment Rooms. Wiring in air-handling areas under a raised floor in an information technology room must comply with 645.5(D). Figure 300–72

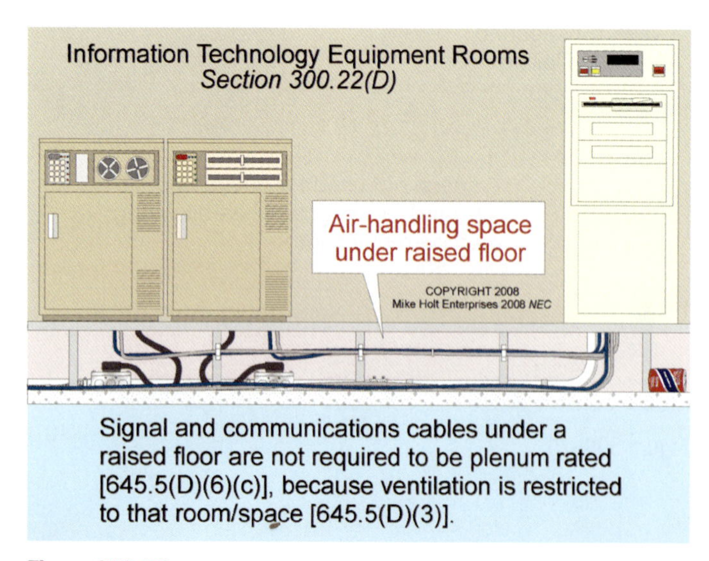

Figure 300–72

Author's Comment: Signal and communications cables under a raised floor are not required to be suitable for plenum use [645.5(D)(6)(c)], because ventilation is restricted to that room/space, and the space is not normally occupied [645.5(D)(4)].

300.23 Panels Designed to Allow Access. Wiring, cables, and equipment installed behind panels must be located so the panels can be removed to give access to electrical equipment. Figure 300–73

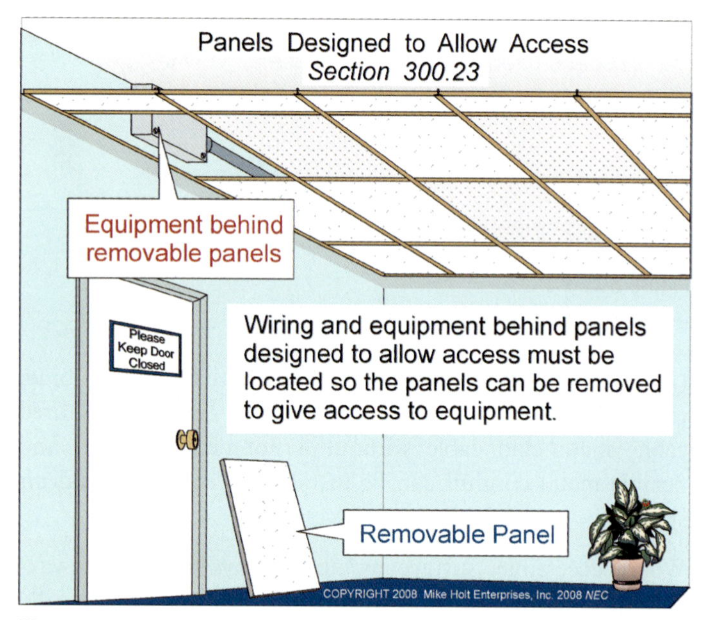

Figure 300–73

Author's Comment: Access to equipment must not be hindered by an accumulation of cables that prevent the removal of suspended-ceiling panels. Control, signaling, and communications cables must be located and supported so the suspended-ceiling panels can be moved to provide access to electrical equipment.

- CATV Coaxial Cable, 820.21
- Communications Cable, 800.21
- Control & Signaling Cable, 725.21
- Fire Alarm Cable, 760.7
- Optical Fiber Cable, 770.21
- Audio Cable, 640.5

ARTICLE 300 Practice Questions

ARTICLE 300. WIRING METHODS—PRACTICE QUESTIONS

1. Conductors of ac and dc circuits, rated 600V or less, can occupy the same _____ provided that all conductors have an insulation rating equal to the maximum voltage applied to any conductor.

   (a) enclosure
   (b) cable
   (c) raceway
   (d) all of these

2. Where Type NM cable passes through factory or field openings in metal members, it shall be protected by _____ bushings or _____ grommets that cover metal edges.

   (a) approved
   (b) identified
   (c) listed
   (d) none of these

3. Wiring methods installed behind panels that allow access shall be _____ according to their applicable articles.

   (a) supported
   (b) painted
   (c) in a metal raceway
   (d) all of these

4. Cables or raceways installed under metal-corrugated sheet roof decking shall be supported so the nearest outside surface of the cable or raceway is not less than _____ from the nearest surface of the roof decking.

   (a) ½ in.
   (b) 1 in.
   (c) 1½ in.
   (d) 2 in.

5. Rigid metal conduit that is directly buried outdoors shall have at least _____ of cover.

   (a) 6 in.
   (b) 12 in.
   (c) 18 in.
   (d) 24 in.

6. Type UF cable used with a 24V landscape lighting system can have a minimum cover of _____.

   (a) 6 in.
   (b) 12 in.
   (c) 18 in.
   (d) 24 in.

7. Where direct-buried conductors and cables emerge from grade, they shall be protected by enclosures or raceways to a point at least _____ above finished grade.

   (a) 3 ft
   (b) 6 ft
   (c) 8 ft
   (d) 10 ft

8. Backfill used for underground wiring shall not _____.

   (a) damage the wiring method
   (b) prevent compaction of the fill
   (c) contribute to the corrosion of the raceway
   (d) all of these

9. All conductors of the same circuit shall be _____, unless otherwise specifically permitted in the *Code*.

   (a) in the same raceway or cable
   (b) in close proximity in the same trench
   (c) the same size
   (d) a or b

10. Raceways, cable trays, cablebus, auxiliary gutters, cable armor, boxes, cable sheathing, cabinets, elbows, couplings, fittings, supports, and support hardware shall be of materials suitable for _____.

    (a) corrosive locations
    (b) wet locations
    (c) the environment in which they are to be installed
    (d) none of these

11. Aluminum raceways, cable trays, cablebus, auxiliary gutters, cable armor, boxes, cable sheathing, cabinets, elbows, couplings, nipples, fittings, supports, and support hardware _____ shall be provided with supplementary corrosion protection.

    (a) embedded or encased in concrete
    (b) in direct contact with the earth
    (c) likely to become energized
    (d) a or b

12. An exposed wiring system for indoor wet locations where walls are frequently washed shall be mounted so that there is at least a _____ between the mounting surface and the electrical equipment.

    (a) ¼ in. airspace
    (b) separation by insulated bushings
    (c) separation by noncombustible tubing
    (d) none of these

13. Raceways shall be provided with expansion fittings where necessary to compensate for thermal expansion and contraction.

    (a) True
    (b) False

14. Metal raceways, cable armors, and other metal enclosures shall be _____ joined together into a continuous electric conductor so as to provide effective electrical continuity.

    (a) electrically
    (b) permanently
    (c) metallically
    (d) none of these

15. Electrical wiring within the cavity of a fire-rated floor-ceiling or roof-ceiling assembly shall not be supported by the ceiling assembly or ceiling support wires.

    (a) True
    (b) False

16. Raceways can be used as a means of support of Class 2 circuit conductors or cables that connect to the same equipment.

    (a) True
    (b) False

17. Raceways and cables installed into the _____ of open-bottom equipment shall not be required to be mechanically secured to the equipment.

    (a) bottom
    (b) sides
    (c) top
    (d) any of these

18. When the opening to an outlet, junction, or switch point is less than 8 in. in any dimension, each conductor shall be long enough to extend at least _____ outside the opening of the enclosure.

    (a) 0 in.
    (b) 3 in.
    (c) 6 in.
    (d) 12 in.

19. A box or conduit body shall not be required for splices and taps in direct-buried conductors and cables as long as the splice is made with a splicing device that is identified for the purpose.

    (a) True
    (b) False

20. Raceways shall be _____ between outlet, junction, or splicing points prior to the installation of conductors.

    (a) installed complete
    (b) tested for ground faults
    (c) a minimum of 80 percent complete
    (d) none of these

21. Metal raceways shall not be _____ by welding to the raceway.

    (a) supported
    (b) terminated
    (c) connected
    (d) all of these

22. _____ is a nonferrous, nonmagnetic metal that has no heating due to hysteresis heating.

    (a) Steel
    (b) Iron
    (c) Aluminum
    (d) all of these

23. Openings around electrical penetrations through fire-resistant-rated walls, partitions, floors, or ceilings shall _____ to maintain the fire-resistance rating.

    (a) be documented
    (b) not be permitted
    (c) be firestopped using approved methods
    (d) be enlarged

24. The space above a hung ceiling used for environmental air-handling purposes is an example of _____, and the wiring limitations of _____ apply.

    (a) a plenum used for environmental air, 300.22(B)
    (b) other space used for environmental air, 300.22(C)
    (c) a duct used for environmental air, 300.22(B)
    (d) none of these

310 Conductors for General Wiring

INTRODUCTION TO ARTICLE 310—CONDUCTORS FOR GENERAL WIRING

This article contains the general requirements for conductors, such as insulation markings, ampacity ratings, and conditions of use. Article 310 doesn't apply to conductors that are part of flexible cords, fixture wires, or to conductors that are an integral part of equipment [90.7 and 300.1(B)].

People often make errors in applying the ampacity tables contained in Article 310. If you study the explanations carefully, you'll avoid common errors such as applying Table 310.17 when you should be applying Table 310.16.

Why so many tables? Why does Table 310.17 list the ampacity of 6 THHN as 105 amperes, yet Table 310.16 lists the same conductor as having an ampacity of only 75 amperes? To answer that, go back to Article 100 and review the definition of ampacity. Notice the phrase "conditions of use." What these tables do is set a maximum current value at which you can ensure the installation won't undergo premature failure of the conductor insulation in normal use, in the conditions described in the tables.

The designations THHN, THHW, RHH, and so on, are insulation types. Every type of insulation has a heat withstand limit. When current flows through a conductor, it creates heat. How well the insulation around a conductor can dissipate that heat depends on factors such as whether that conductor is in free air or not. Think what happens to you if you put on a sweater, a jacket, and then a coat—all at the same time. You heat up. Your skin can't dissipate heat with all that clothing on nearly as well as it dissipates heat in free air. The same principal applies to conductors.

Conductor insulation also fails with age. That's why we conduct cable testing and take other measures to predict failure and replace certain conductors (for example, feeders or critical equipment conductors) while they're still within design specifications. But conductor insulation failure takes decades under normal use—and it's a maintenance issue. However, if a conductor is forced to exceed the ampacity listed in the appropriate table, and as a result its design temperature is exceeded, insulation failure happens much more rapidly—often catastrophically. Exceeding the allowable ampacity is a serious safety issue.

310.1 Scope. Article 310 contains the general requirements for conductors, such as insulation markings, ampacity ratings, and their use. This article doesn't apply to conductors that are an integral part of equipment [90.7 and 300.1(B)].

310.2 Conductors.

(A) Insulated. Conductors must be insulated and installed in a recognized wiring method [110.8]. Figure 310–1

Exception: Where covered or bare conductors are specifically permitted elsewhere in this Code. Figure 310–2

310.3 Stranded Conductors. Conductors 8 AWG and larger must be stranded when installed in a raceway. Figure 310–3

Author's Comment: Solid conductors are often used for the grounding electrode conductor [250.62] and for the bonding of pools, spas, and outdoor hot tubs [680.26(C)]. Technically, the practice of installing 8 AWG and larger solid conductors in a raceway for protection of grounding and bonding conductors is a violation of this rule.

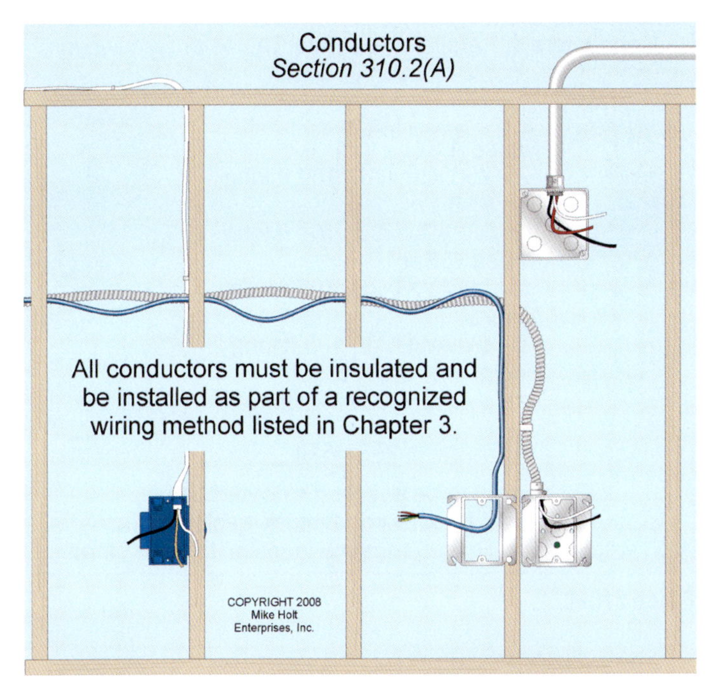

Figure 310–1

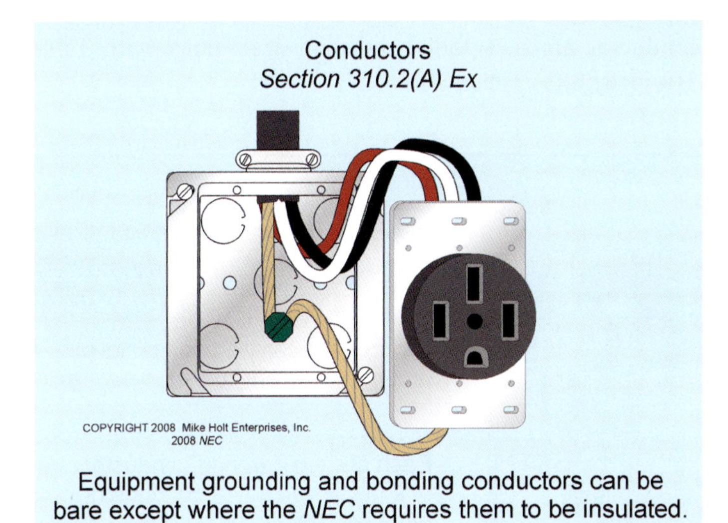

Figure 310–2

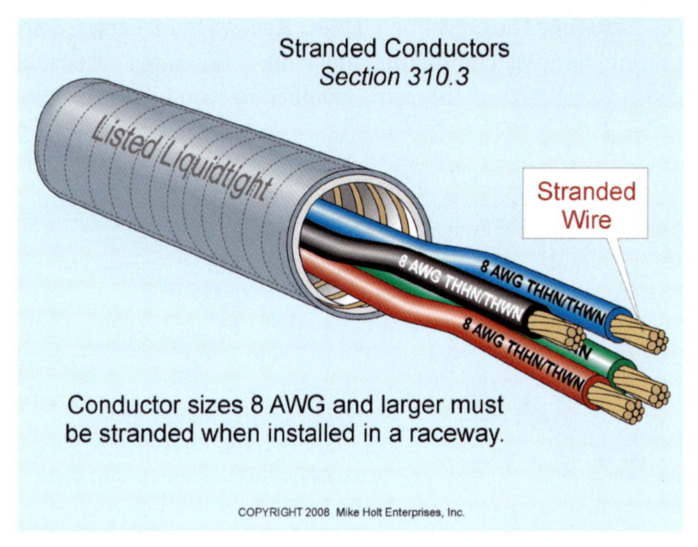

Figure 310–3

310.4 Conductors in Parallel.

(A) General. Ungrounded and neutral conductors sized 1/0 AWG and larger can be connected in parallel.

(B) Conductor Characteristics. When circuit conductors are run in parallel, the current must be evenly distributed between the individual parallel conductors by requiring all circuit conductors within a parallel set to: Figure 310–4

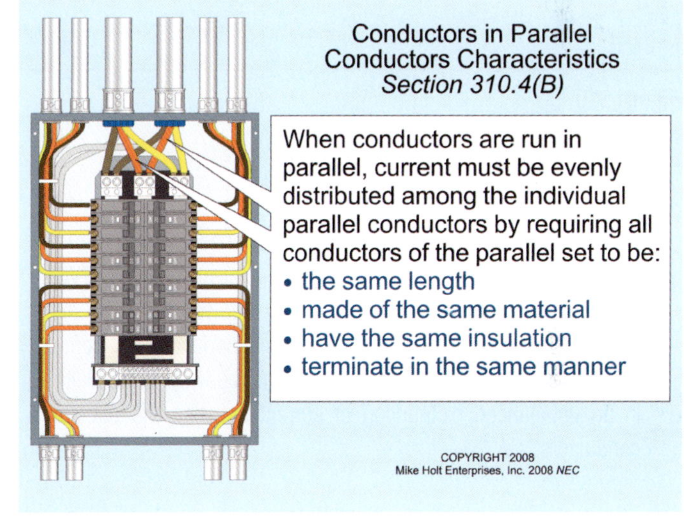

Figure 310–4

(1) Be the same length.

(2) Be made of the same conductor material (copper/aluminum).

(3) Be the same size in circular mil area (minimum 1/0 AWG).

(4) Have the same insulation (like THHN).

(5) Terminate in the same method (set screw versus compression).

Author's Comment: Conductors aren't required to have the same physical characteristics as those of another phase or neutral conductor to achieve balance [310.14(C)].

(C) Separate Raceways or Cables. Raceways or cables containing parallel conductors must have the same electrical characteristics and the same number of conductors. Figure 310–5

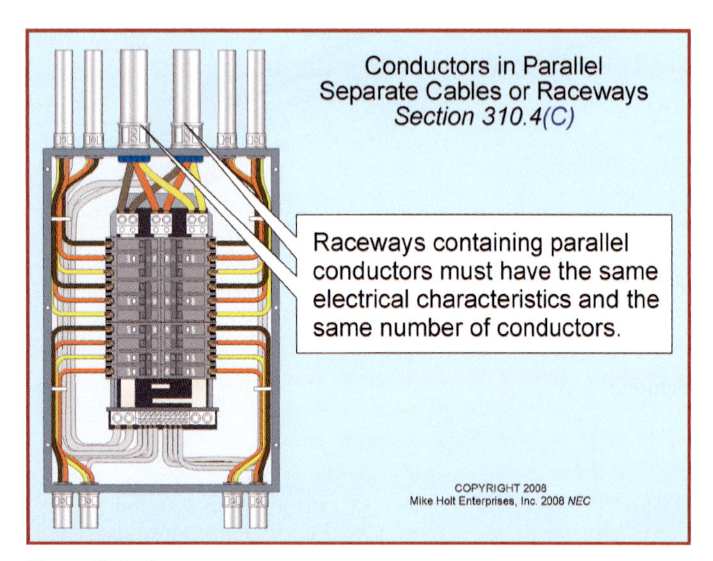

Figure 310–5

Author's Comment: If one set of parallel conductors is run in a metallic raceway and the other conductors are run in PVC conduit, the conductors in the metallic raceway will have an increased opposition to current flow (impedance) as compared to the conductors in the nonmetallic raceway. This results in an unbalanced distribution of current between the parallel conductors.

Paralleling is done in sets. Parallel sets of conductors aren't required to have the same physical characteristics as those of another set to achieve balance.

Author's Comment: For example, a 400A feeder with a neutral load of 240A can be paralleled as follows. Figure 310–6.

- Phase A, Two—250 kcmil THHN aluminum, 100 ft
- Phase B, Two—3/0 THHN copper, 104 ft
- Phase C, Two—3/0 THHN copper, 102 ft
- Neutral, Two—1/0 THHN aluminum, 103 ft
- Equipment Grounding Conductor, Two—3 AWG copper, 101 ft*

**The minimum 1/0 AWG size requirement doesn't apply to equipment grounding conductors [310.4(E)].*

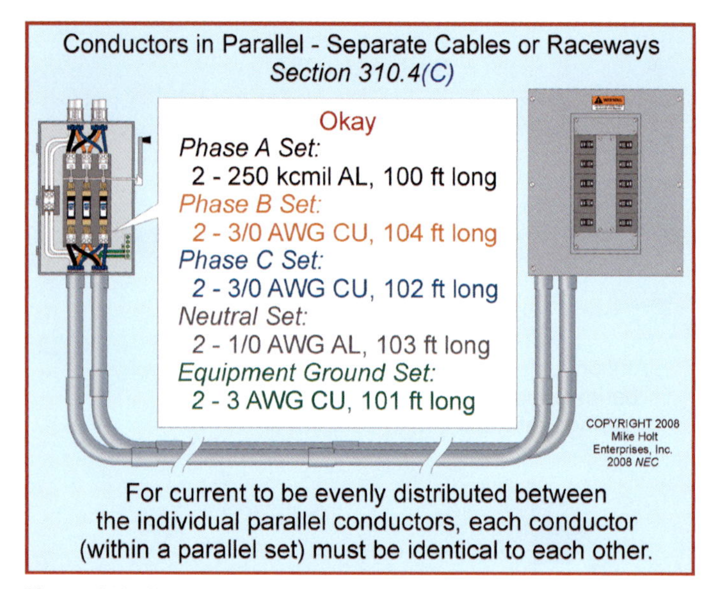

Figure 310–6

(D) Conductor Ampacity Adjustment. Each current-carrying conductor of a paralleled set of conductors must be counted as a current-carrying conductor for the purpose of conductor ampacity adjustment, in accordance with Table 310.15(B)(2)(a). Figure 310–7

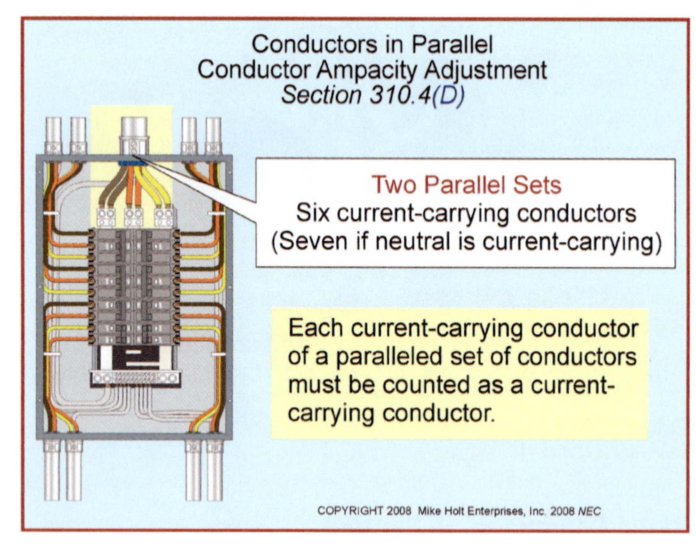

Figure 310–7

(E) Equipment Grounding Conductors. The equipment grounding conductors for circuits in parallel must be identical to each other in length, material, size, insulation, and termination. In addition, each raceway, where required, must have an equipment grounding conductor sized in accordance with 250.122. The minimum 1/0 AWG rule of 310.4 doesn't apply to equipment grounding conductors. See 250.122(F)(1) for more information on equipment grounding conductors installed in parallel. Figure 310–8

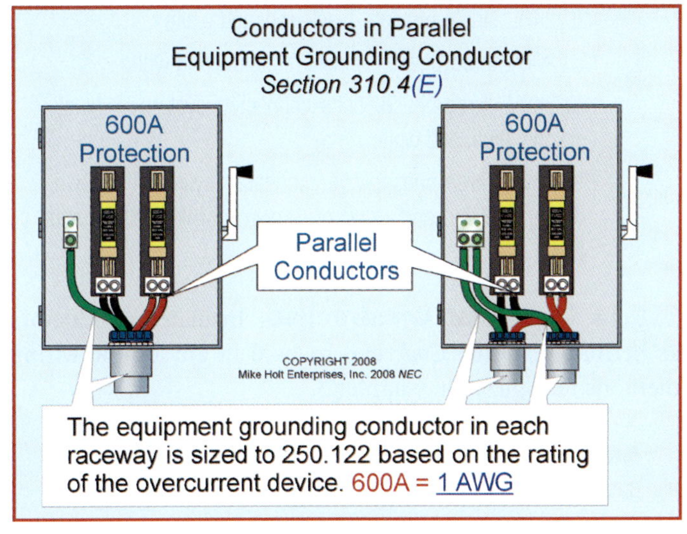

Figure 310–8

310.5 Minimum Size Conductors.

The smallest conductor permitted for branch circuits for residential, commercial, and industrial locations is 14 AWG copper, except as permitted elsewhere in this *Code*.

Author's Comment: There's a misconception that 12 AWG copper is the smallest conductor permitted for commercial or industrial facilities. Although this isn't true based on *NEC* rules, it may be a local code requirement.

310.8 Location.

(B) Dry and Damp Locations. Insulated conductors used in dry and damp locations must be Types THHN, THHW, THWN, or THWN-2.

Author's Comment: These conductor types are only the most common. Refer to the *Code* for a complete list of conductors that may be installed in a dry or damp location.

(C) Wet Locations. Insulated conductors used in wet locations must be:

(2) Types THHW, THWN, THWN-2, XHHW, or XHHW-2

Author's Comment: These conductor types are only the most common. Refer to the *Code* for a complete list of conductors that may be installed in a wet location.

(D) Locations Exposed to Direct Sunlight. Insulated conductors and cables exposed to the direct rays of the sun must comply with (1) or (2):

(1) Listed as sunlight resistant or marked as being sunlight resistant. Figure 310–9

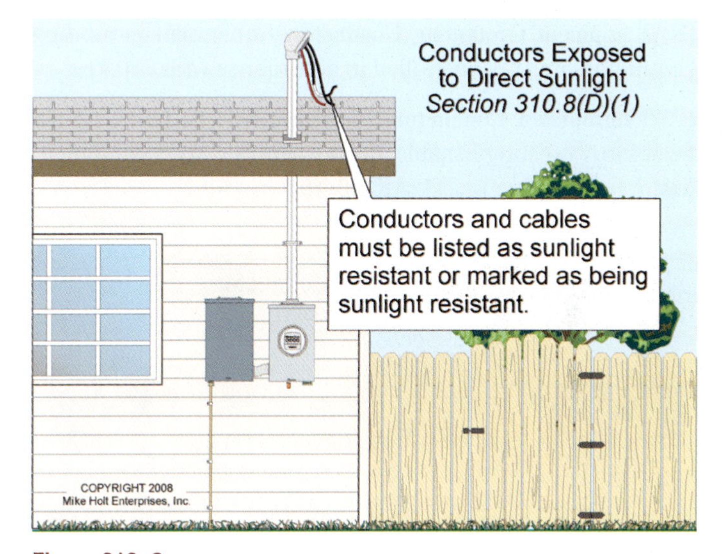

Figure 310–9

Author's Comment: SE cable and the conductors contained in the cable are listed as sunlight resistant. However, according to the UL listing standard, the conductors contained in SE cable aren't required to be marked as sunlight resistant.

(2) Covered with insulating material, such as tape or sleeving materials that are listed as being sunlight resistant or marked as being sunlight resistant.

310.9 Corrosive Conditions.

Conductor insulation must be suitable for any substance to which it may be exposed that may have a detrimental effect on the conductor's insulation, such as oil, grease, vapor, gases, fumes, liquids, or other substances. See 110.11.

310.10 Insulation Temperature Limitation.

Conductors must not be used where the operating temperature exceeds that designated for the type of insulated conductor involved.

FPN No. 1: The insulation temperature rating of a conductor is the maximum temperature a conductor can withstand over a prolonged time without serious degradation. The main factors to consider for conductor operating temperature include ambient temperature, heat generated internally from current flow through the conductor, the rate at which heat can dissipate, and adjacent load-carrying conductors.

310.12 Conductor Identification.

(A) Neutral Conductor. Neutral conductors must be identified in accordance with 200.6.

(B) Equipment Grounding Conductor. Equipment grounding conductors must be identified in accordance with 250.119.

(C) Ungrounded Conductors. Ungrounded conductors must be clearly distinguishable from neutral and equipment grounding conductors. Figure 310–10

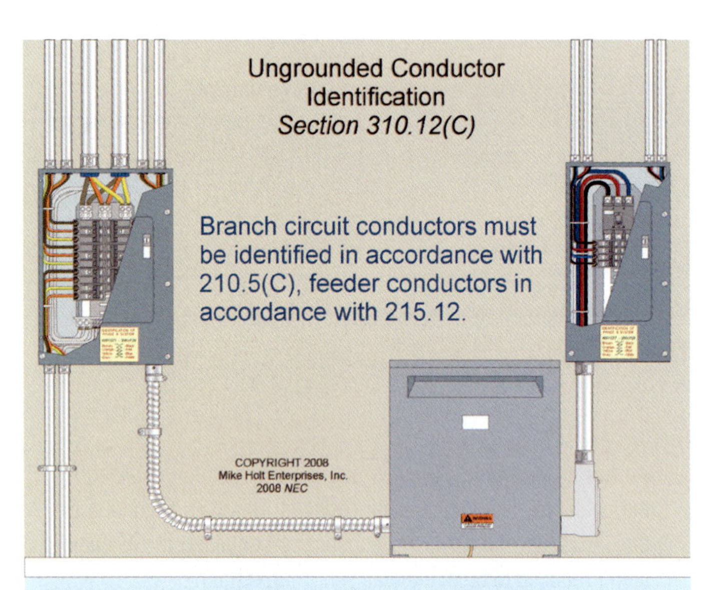

Ungrounded conductors must have a finish to be clearly distinguishable from the neutral and grounding conductor.

Figure 310–10

Author's Comments:

- Where the premises wiring system has branch circuits or feeders supplied from more than one nominal voltage system, each ungrounded conductor of the branch circuit or feeder, where accessible, must be identified by system. The means of identification can be by separate color coding, marking tape, tagging, or other means approved by the authority having jurisdiction. Such identification must be permanently posted at each panelboard. See 210.5(C) and 215.12 for specific requirements.
- The *NEC* doesn't require color coding of ungrounded conductors, except for the high-leg conductor when a neutral conductor is present [110.15 and 230.56]. Although not required, electricians often use the following color system for power and lighting conductor identification:
  - 120/240V, single-phase—black, red, and white
  - 120/208V, three-phase—black, red, blue, and white
  - 120/240V, three-phase, delta-connected system —black, orange, blue, and white
  - 277/480V, three-phase, wye-connected system —brown, orange, yellow, and gray; or, brown, purple, yellow, and gray

310.13 Conductor Construction. Insulated conductors as permitted by the *Code* can be used in any of the wiring methods recognized in Chapter 3.

Author's Comment: The following explains the lettering on conductor insulation: Figure 310–11

- F — Fixture wires (solid or 7 strands) [Table 402.3]
- FF — Flexible fixture wire (19 strands) [Table 402.3]
- No H — 60°C insulation rating [Table 310.13(A)]
- H — 75°C insulation rating [Table 310.13(A)]
- HH — 90°C insulation rating [Table 310.13(A)]
- N — Nylon outer cover [Table 310.13(A)]
- R — Thermoset insulation [Table 310.13(A)]
- T — Thermoplastic insulation [Table 310.13(A)]
- U — Underground [Table 310.13(A)]
- W — Wet or damp locations [Table 310.13(A)]
- X — Cross-linked polyethylene insulation [Table 310.13(A)]

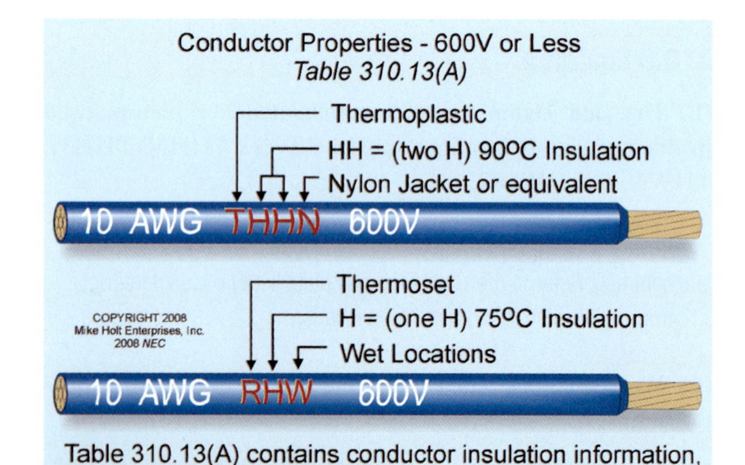

Table 310.13(A) contains conductor insulation information, such as operating temperature and applications. These conductors can be used in any Chapter 3 wiring method.

Figure 310–11

Equipment grounding conductors can be sectioned within a listed multiconductor cable, such as SE cable, provided the combined circular mil area complies with 250.122.

310.15 Conductor Ampacity.

(A) General Requirements.

(1) Tables for Engineering Supervision. The ampacity of a conductor can be determined either by using the tables in accordance with 310.15(B), or under engineering supervision as provided in 310.15(C).

FPN No. 1: Ampacities provided by this section don't take voltage drop into consideration. See 210.19(A) FPN No. 4, for branch circuits and 215.2(D) FPN No. 2, for feeders.

(2) Conductor Ampacity—Lower Rating. If a single length of conductor is routed in a manner that two or more ampacity ratings apply to a single conductor length, the lower ampacity must be used for the entire circuit. See 310.15(B). Figure 310–12

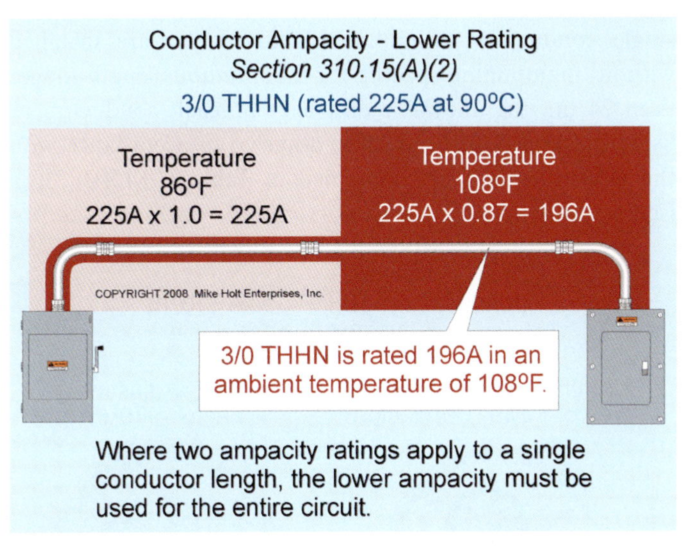

Figure 310–12

Exception: When different ampacities apply to a length of conductor, the higher ampacity is permitted for the entire circuit if the reduced ampacity length doesn't exceed 10 ft and its length doesn't exceed 10 percent of the length of the higher ampacity. Figure 310–13

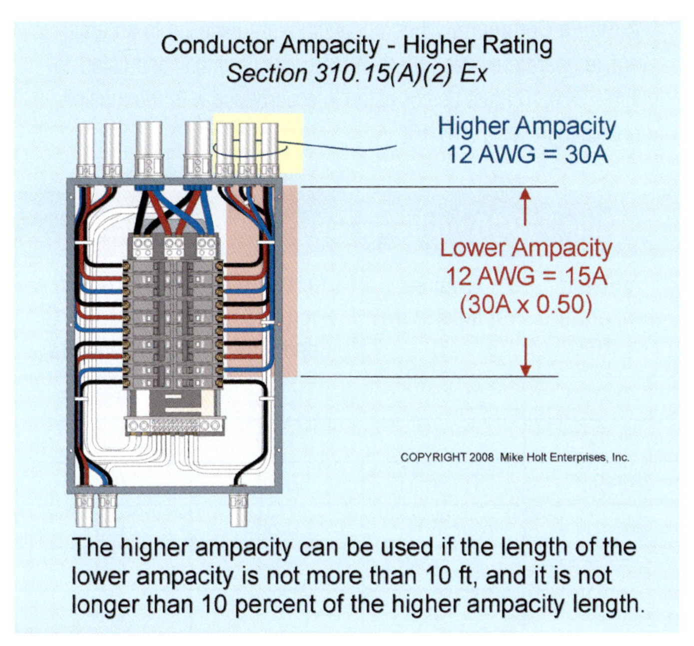

Figure 310–13

(B) Ampacity Table. The allowable conductor ampacities listed in Table 310.16 are based on conditions where the ambient temperature isn't over 86°F, and no more than three current-carrying conductors are bundled together. Figure 310–14

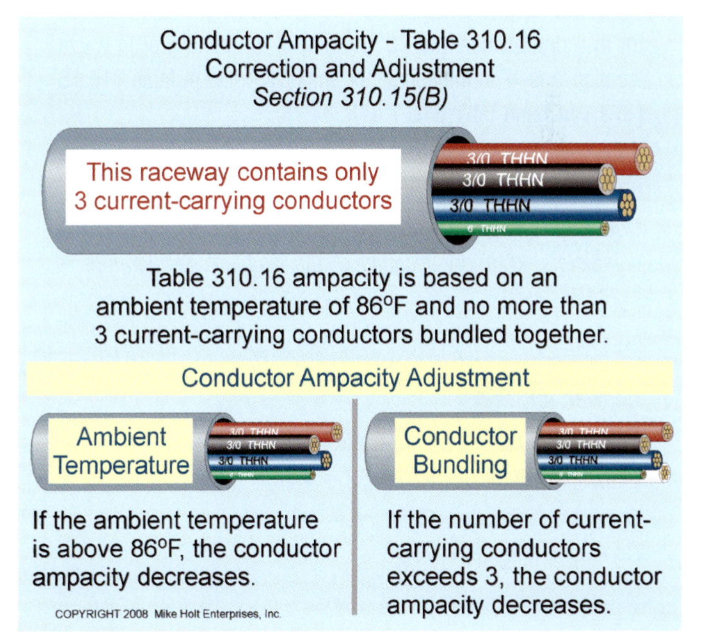

Figure 310–14

Author's Comment: When conductors are installed in an ambient temperature other than 78°F to 86°F, ampacities listed in Table 310.16 must be corrected in accordance with the multipliers listed in Table 310.16.

Table 310.16 Ambient Temperature Correction

Ambient Temperature °F	Ambient Temperature °C	Correction Factor 75°C Conductors	Correction Factor 90°C Conductors
70–77°F	21–25°C	1.05	1.04
78–86°F	26–30°C	1.00	1.00
87–95°F	31–35°C	0.94	0.96
96–104°F	36–40°C	0.88	0.91
105–113°F	41–45°C	0.82	0.87
114–122°F	46–-50°C	0.75	0.82
123–131°F	51–-55°C	0.67	0.76
132–140°F	56–60°C	0.58	0.71
141–158°F	61–70°C	0.33	0.58
159–176°F	71–80°C	0.00	0.41

Author's Comment: When correcting conductor ampacity for elevated ambient temperature, the correction factor used for THHN conductors is based on the 90°C rating of the conductor in a dry location and 75°C rating of the conductor in a wet location, based on the conductor ampacity listed in Table 310.16 [110.14(C) and Table 310.13(A)].

Question: *What is the corrected ampacity of 3/0 THHN/THWN conductors in a dry location if the ambient temperature is 108°F?*

(a) 173A *(b) 196A* *(c) 213A* *(d) 241A*

Answer: *(b) 196A*

Conductor Ampacity [90°C] = 225A
Correction Factor [Table 310.16] = 0.87

Corrected Ampacity = 225A x 0.87
Corrected Ampacity = 196A

Question: *What is the corrected ampacity of 3/0 THHN/THWN conductors in a wet location if the ambient temperature is 108°F?*

(a) 164A *(b) 196A* *(c) 213A* *(d) 241A*

Answer: *(a) 164A*

Conductor Ampacity [75°C] = 200A
Correction Factor [Table 310.16] = 0.82

Corrected Ampacity = 200A x 0.82
Corrected Ampacity = 164A

Author's Comment: When adjusting conductor ampacity, the ampacity is based on the temperature insulation rating of the conductor as listed in Table 310.16, not the temperature rating of the terminal [110.14(C)].

(2) Adjustment Factors.

(a) Conductor Bundle. Where the number of current-carrying conductors in a raceway or cable exceeds three, or where single conductors or multiconductor cables are installed without maintaining spacing for a continuous length longer than 24 in., the allowable ampacity of each conductor, as listed in Table 310.16, must be adjusted in accordance with the adjustment factors contained in Table 310.15(B)(2)(a). **Figure 310–15**

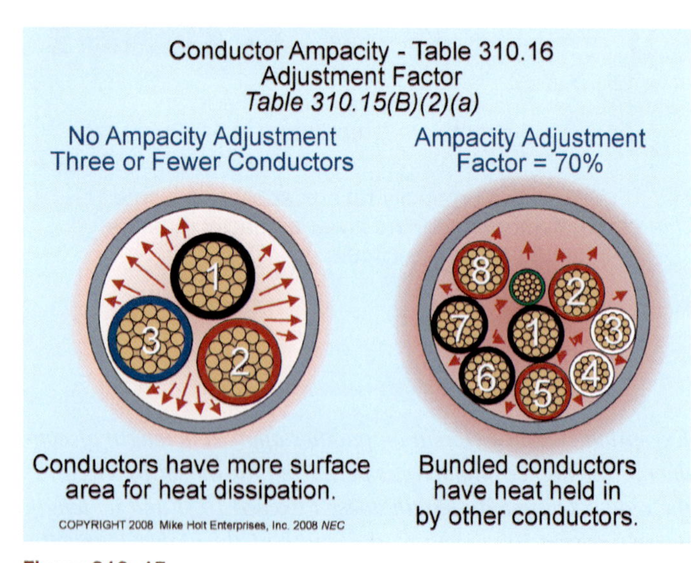

Figure 310–15

Each current-carrying conductor of a paralleled set of conductors must be counted as a current-carrying conductor. Figure 310–16

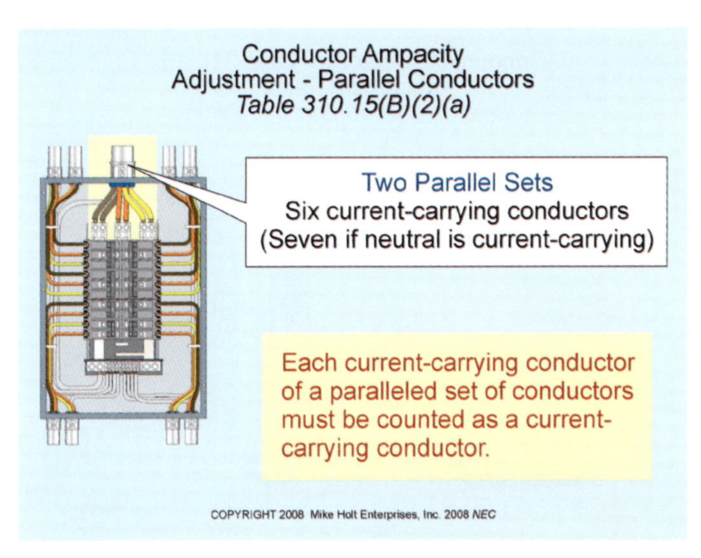

Figure 310–16

Table 310.15(B)(2)(a)	
Number of Current–Carrying	Adjustment Factor
1–3 Conductors	1.00
4–6 Conductors	0.80
7–9 Conductors	0.70*
10–20 Conductors	0.50

Author's Comment: The neutral conductor might be a current-carrying conductor, but only under the conditions specified in 310.15(B)(4). Equipment grounding conductors are never considered current carrying [310.15(B)(5)].

Question: *What is the adjusted ampacity of 3/0 THHN/THWN conductors in a dry location if the raceway contains a total of four current-carrying conductors?* Figure 310–17

(a) 180A (b) 196A (c) 213A (d) 241A

Answer: *(a) 180A*

Table 310.16 ampacity if 3/0 THHN/THWN in a dry location is 225A

Adjustment Factor [Table 310.15(B)(2)(a)] = 0.80

Adjusted Ampacity = 225A x 0.80
Adjusted Ampacity = 180A

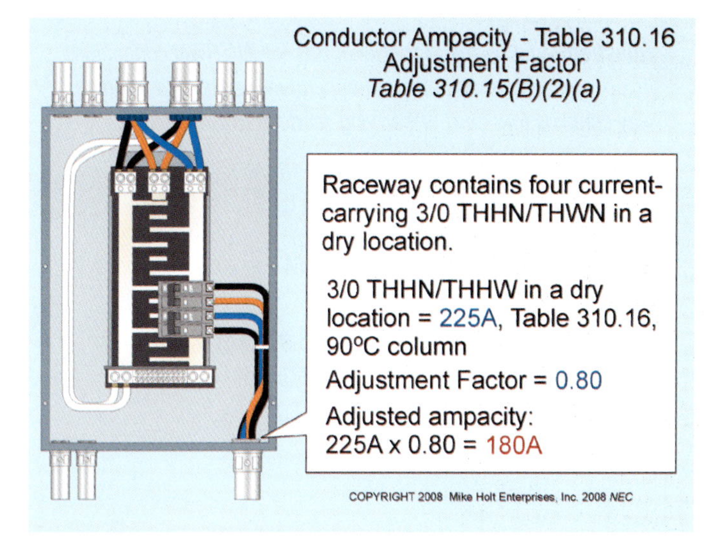

Figure 310–17

Question: *What is the adjusted ampacity of 3/0 THHN/THWN conductors in a wet location if the raceway contains a total of four current-carrying conductors?*

(a) 160A (b) 196A (c) 213A (d) 241A

Answer: *(a) 160A*

Table 310.16 ampacity if 3/0 THHN/THWN in a wet location is 200A

Adjustment Factor [Table 310.15(B)(2)(a)] = 0.80

Adjusted Ampacity = 200A x 0.80
Adjusted Ampacity = 160A

Author's Comments:

- When correcting or adjusting conductor ampacity, the ampacity is based on the temperature insulation rating of the conductor as listed in Table 310.16, not the temperature rating of the terminal [110.14(C)].
- Where more than three current-carrying conductors are present and the ambient temperature isn't between 78 and 86°F, the ampacity listed in Table 310.16 must be corrected and adjusted for both conditions.

Question: *What is the ampacity of 3/0 THHN/THWN conductors in a dry location at an ambient temperature of 108°F if the raceway contains four current-carrying conductors?*

(a) 157A (b) 176A (c) 199A (d) 214A

Answer: *(a) 157A*

Table 310.16 ampacity if 3/0 THHN/THWN in a dry location is 225A

Ambient Temperature Correction [Table 310.16] = 0.87
Conductor Bundle Adjustment [310.15(B)(2)(a)] = 0.80
Adjusted Ampacity = 225A x 0.87 x 0.80
Adjusted Ampacity = 157A

Author's Comment: When adjusting or correcting conductor ampacity, the ampacity of THHN/THWN conductors in a dry location is based on the 90°C rating of the conductor [110.14(C) and Table 310.13(A)].

Question: *What is the ampacity of 3/0 THHN/THWN conductors in a wet location at an ambient temperature of 108°F if the raceway contains four current-carrying conductors?*

(a) 131A (b) 176A (c) 199A (d) 214A

Answer: *(a) 131A*

Table 310.16 ampacity if 3/0 THHN/THWN in a wet location is 200A

Ambient Temperature Correction [Table 310.16] = 0.82

Conductor Bundle Adjustment [310.15(B)(2)(a)] = 0.80

Adjusted Ampacity = 200A x 0.82 x 0.80
Adjusted Ampacity = 131.20A

Author's Comment: When adjusting or correcting conductor ampacity, the ampacity of THHN/THWN conductors in a wet location is based on the 75°C rating of the conductor [110.14(C) and Table 310.13(A)].

FPN No. 2: See 376.22(B) for conductor ampacity adjustment factors for conductors in metal wireways.

Author's Comment: Conductor ampacity adjustment only applies when more than 30 current-carrying conductors are installed in any cross-sectional area of a metal wireway.

Exception No. 3: The conductor ampacity adjustment factors of Table 310.15(B)(2)(a) don't apply to conductors installed in raceways not exceeding 24 in. in length. **Figure 310–18**

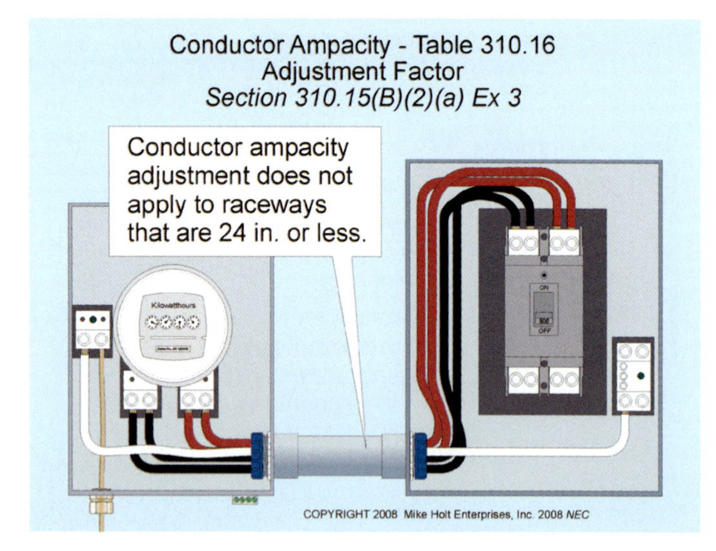

Figure 310–18

Exception No. 5: The conductor ampacity adjustment factors of Table 310.15(B)(2)(a) don't apply to Type AC or Type MC cable when: **Figure 310–19**

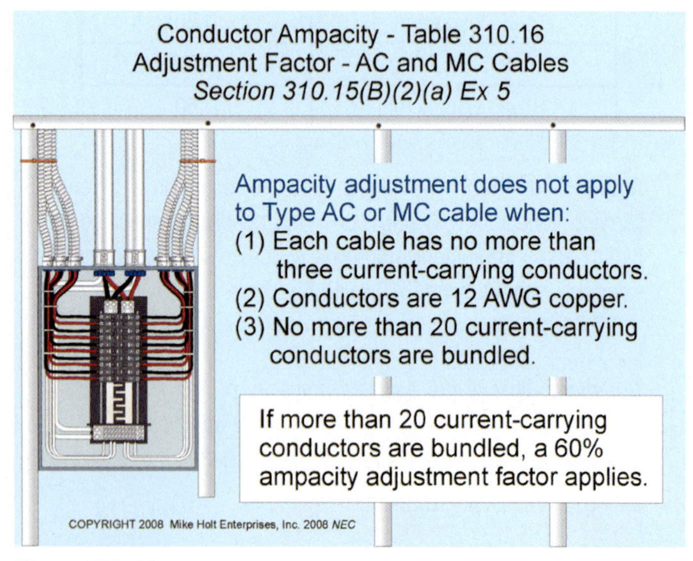

Figure 310–19

(1) Each cable has not more than three current-carrying conductors,

(2) The conductors are 12 AWG copper, and

(3) No more than 20 current-carrying conductors (ten 2-wire cables or six 3-wire cables) are bundled.

Author's Comment: When eleven or more 2-wire cables or seven or more 3-wire cables (more than 20 current-carrying conductors) are bundled or stacked for more than 24 in., an ampacity adjustment factor of 60 percent must be applied.

(c) Raceways Exposed to Sunlight on Rooftops. The ambient temperature adjustment contained in Table 310.15(B)(2)(c) is added to the outdoor ambient temperature for conductors or cables that are installed in raceways exposed to direct sunlight on or above rooftops when applying ampacity adjustment correction factors contained in Table 310.16.

FPN No. 1: See ASHRAE Handbook—*Fundamentals* (www.ashrae.org) as a source for the average ambient temperatures in various locations.

FPN No. 2: The temperature adders in Table 310.15(B)(2)(c) are based on the results of averaging the ambient temperatures.

Table 310.15(B)(2)(c) Ambient Temperature Adder for Raceways On or Above Rooftops

Distance of Raceway Above Roof	C°	F°
0 to ½ in.	33	60
Above ½ in. to 3 ½ in.	22	40
Above 3 ½ in. to 12 in.	17	30
Above 12 in. to 36 in.	14	25

Author's Comment: This section requires the ambient temperature used for ampacity correction to be adjusted where conductors or cables are installed in conduit on or above a rooftop and the conduit is exposed to direct sunlight. The reasoning is that the air inside conduits in direct sunlight is significantly hotter than the surrounding air, and appropriate ampacity corrections must be made in order to comply with 310.10.

For example, a conduit with three 6 THWN-2 conductors with direct sunlight exposure that is ¾ in. above the roof will require 40°F to be added to the correction factors at the bottom of Table 310.16. Assuming an ambient temperature of 90°F, the temperature to use for conductor correction will be 130°F (90°F + 40°F), the 6 THWN-2 conductor ampacity after correction will be 57A (75A x 0.76). Figure 310–20

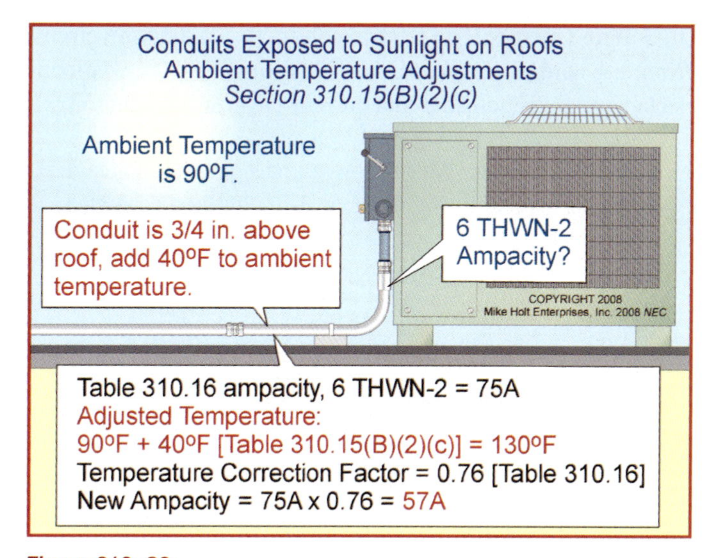

Figure 310–20

When adjusting conductor ampacity, use the conductor ampacity as listed in Table 310.16 based on the conductors' insulation rating; in this case, it's 75A at 90°F. Conductor ampacity adjustment is not based on the temperature terminal rating as per 110.14(C).

(4) Neutral Conductors.

(a) Balanced Circuits. The neutral conductor of a 3-wire, single-phase, 120/240V system, or 4-wire, three-phase, 120/208V or 277/480V wye-connected system, isn't considered a current-carrying conductor. Figure 310–21

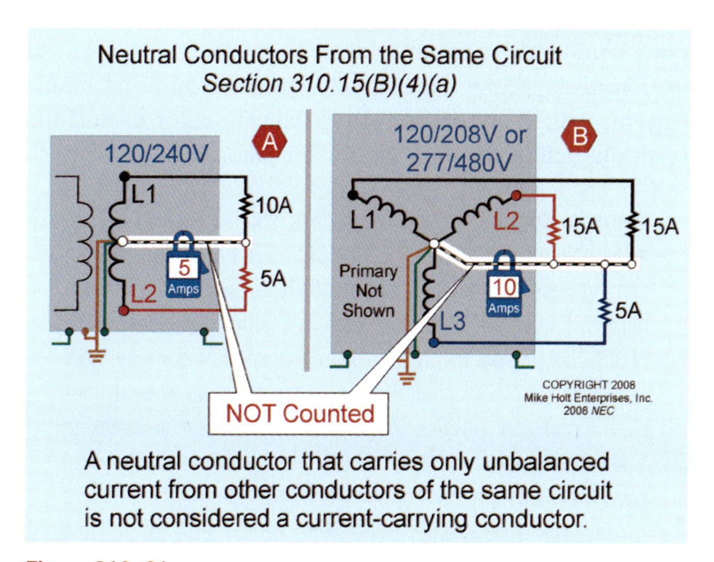

Figure 310–21

(b) 3-Wire Circuits. The neutral conductor of a 3-wire circuit from a 4-wire, three-phase, 120/208V or 277/480V wye-connected system is considered a current-carrying conductor.

Author's Comment: When a 3-wire circuit is supplied from a 4-wire, three-phase, 120/208V or 277/480V wye-connected system, the neutral conductor carries approximately the same current as the ungrounded conductors. Figure 310–22

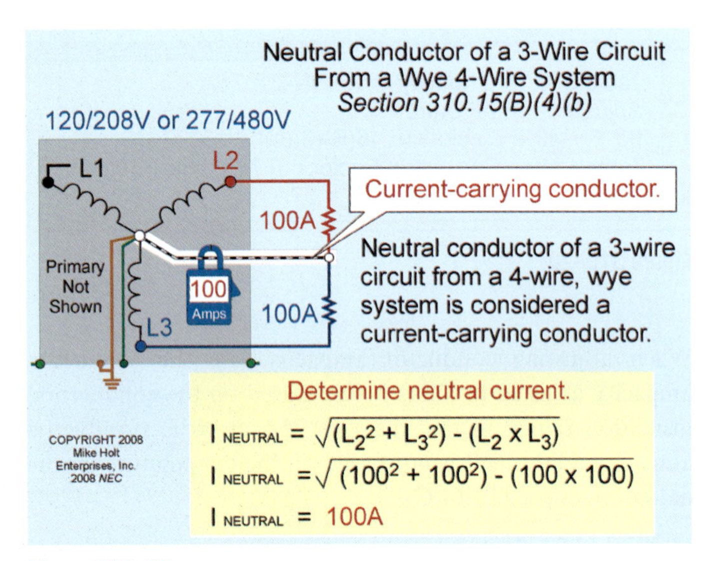

Figure 310–22

(c) Wye 4-Wire Circuits That Supply Nonlinear Loads. The neutral conductor of a 4-wire, three-phase, 120/208V or 277/480V wye-connected system is considered a current-carrying conductor where more than 50 percent of the neutral load consists of nonlinear loads. This is because harmonic currents will be present in the neutral conductor, even if the loads on each of the three phases are balanced. Figure 310–23

Author's Comment: Nonlinear loads supplied by a 4-wire, three-phase, 120/208V or 277/480V wye-connected system can produce unwanted and potentially hazardous odd triplen harmonic currents (3rd, 9th, 15th, etc.) that can add on the neutral conductor. To prevent fire or equipment damage from excessive harmonic neutral current, the designer should consider increasing the size of the neutral conductor or installing a separate neutral for each phase. For more information, visit www.MikeHolt.com, click on the 'Technical' link, then the 'Power Quality' link. Also see 210.4(A) FPN, 220.61 FPN No. 2, and 450.3 FPN No. 2.

(5) Grounding Conductors. Grounding and bonding conductors aren't considered current carrying.

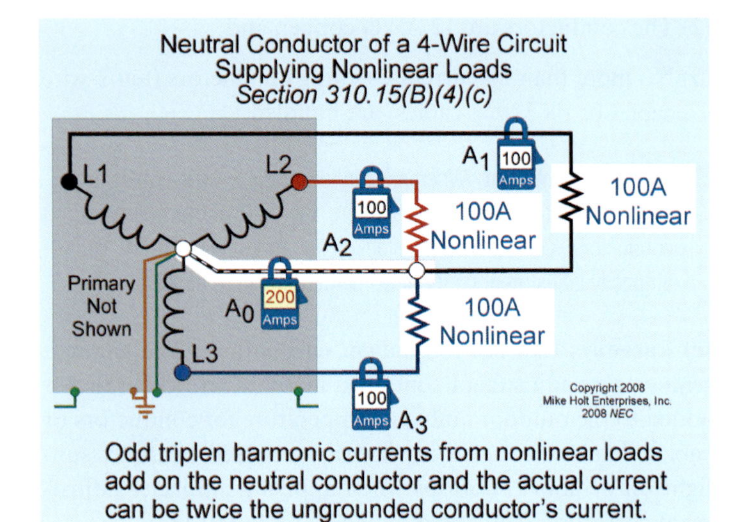

Figure 310–23

(6) Dwelling Unit Feeder/Service Conductors. For individual dwelling units of one-family, two-family, and multifamily dwellings, Table 310.15(B)(6) can be used to size 3-wire, single-phase, 120/240V service or feeder conductors that supply all loads that are part of, or associated with, the dwelling unit. Figure 310–24

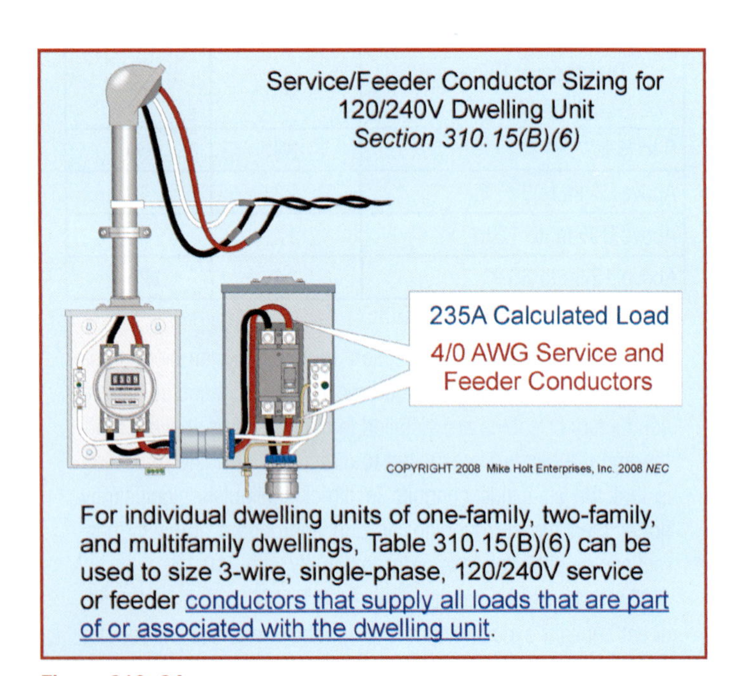

Figure 310–24

Author's Comment: Table 310.15(B)(6) cannot be used for service conductors for two-family or multifamily buildings. Figure 310–25

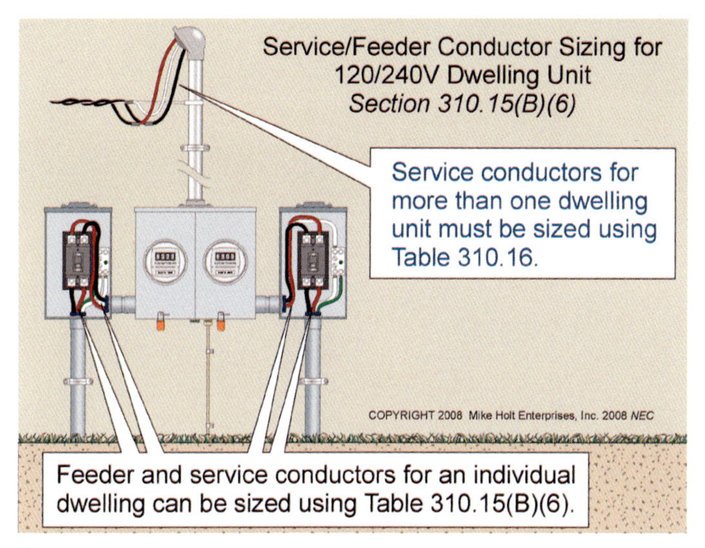

Figure 310–25

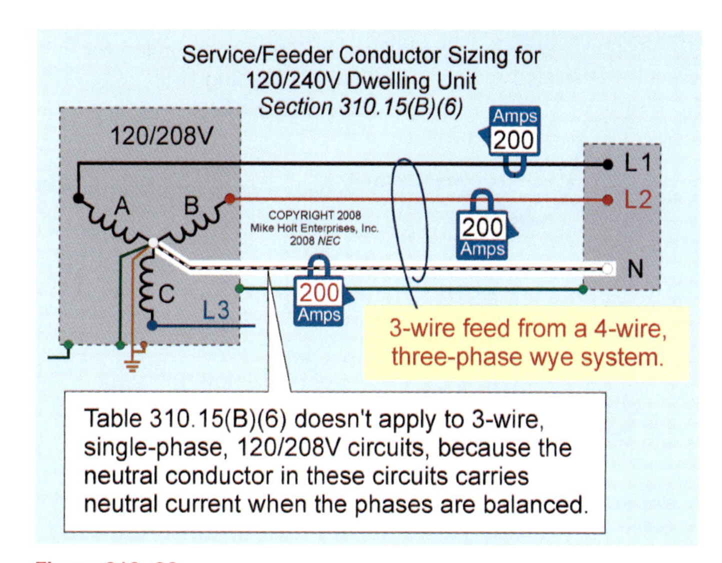

Figure 310–26

Feeder conductors are not required to have an ampacity rating more than the service conductors [215.2(A)(3)].

Table 310.15(B)(6) Conductor Sizes for 120/240V, 3-Wire, Single-Phase Dwelling Services and Feeders		
Amperes	Copper	Aluminum
100	4 AWG	2 AWG
110	3 AWG	1 AWG
125	2 AWG	1/0 AWG
150	1 AWG	2/0 AWG
175	1/0 AWG	3/0 AWG
200	2/0 AWG	4/0 AWG
225	3/0 AWG	250 kcmil
250	4/0 AWG	300 kcmil
300	250 kcmil	350 kcmil
350	350 kcmil	500 kcmil
400	400 kcmil	600 kcmil

WARNING: *Table 310.15(B)(6) doesn't apply to 3-wire feeder/service conductors connected to a three-phase, 120/208V system, because the neutral conductor in these systems always carries neutral current, even when the load on the phases is balanced [310.15(B)(4)(b)]. For more information on this topic, see 220.61(C)(1).* **Figure 310–26**

Neutral Conductor Sizing. Table 310.15(B)(6) can be used to size the neutral conductor of a 3-wire, single-phase, 120/240V service or feeder that carries all loads associated with the dwelling unit, based on the calculated load in accordance with 220.61.

CAUTION: *Because the service neutral conductor is required to serve as the effective ground-fault current path, it must be sized so it can safely carry the maximum fault current likely to be imposed on it [110.10 and 250.4(A)(5)]. This is accomplished by sizing the neutral conductor in accordance with Table 250.66, based on the area of the largest ungrounded service conductor [250.24(C)(1)].*

Question: *What size service conductors are required if the calculated load for a dwelling unit equals 195A, and the maximum unbalanced neutral load is 100A?* **Figure 310–27**

(a) 1/0 AWG and 6 AWG *(b) 2/0 AWG and 4 AWG*
(c) 3/0 AWG and 2 AWG *(d) 4/0 AWG and 1 AWG*

Answer: *(b) 2/0 AWG and 4 AWG*

Service Conductor: 2/0 AWG rated 200A [Table 310.15(B)(6)]

Neutral Conductor: 4 AWG is rated 100A in accordance with Table 310.15(B)(6). In addition, 250.24(C) requires the neutral conductor to be sized no smaller than 4 AWG based on 2/0 AWG service conductors in accordance with Table 250.66.

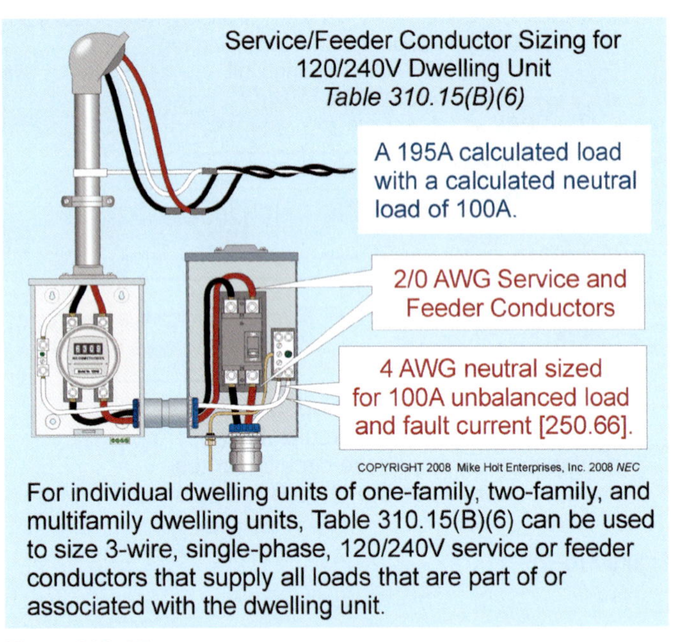

For individual dwelling units of one-family, two-family, and multifamily dwelling units, Table 310.15(B)(6) can be used to size 3-wire, single-phase, 120/240V service or feeder conductors that supply all loads that are part of or associated with the dwelling unit.

Figure 310–27

ARTICLE 310 Practice Questions

ARTICLE 310. CONDUCTORS FOR GENERAL WIRING—PRACTICE QUESTIONS

1. In general, the minimum size conductor permitted for use in parallel installations is _____ AWG.

   (a) 10
   (b) 4
   (c) 1
   (d) 1/0

2. Parallel conductors shall _____.

   (a) be the same length and conductor material
   (b) have the same circular mil area and insulation type
   (c) be terminated in the same manner
   (d) all of these

3. No conductor shall be used where its operating temperature exceeds that designated for the type of insulated conductor involved.

   (a) True
   (b) False

4. THWN insulated conductors are rated _____.

   (a) 75°C
   (b) for wet locations
   (c) a and b
   (d) not enough information

5. The ampacity of a conductor can be different along the length of the conductor. The higher ampacity can be used beyond the point of transition for a distance of no more than _____ ft, or no more than _____ percent of the circuit length figured at the higher ampacity, whichever is less.

   (a) 10, 10
   (b) 10, 20
   (c) 15, 15
   (d) 20, 10

6. Conductor derating factors shall not apply to conductors in nipples having a length not exceeding _____

   (a) 12 in.
   (b) 24 in.
   (c) 36 in.
   (d) 48 in.

7. A _____ conductor that carries only the unbalanced current from other conductors of the same circuit shall not be required to be counted when applying the provisions of 310.15(B)(2)(a).

   (a) neutral
   (b) grounded
   (c) grounding
   (d) none of these

8. For individual dwelling units of _____ dwellings, Table 310.15(B)(6) can be used to size 3-wire, single-phase, 120/240V service or feeder conductors that serve as the main power feeder.

   (a) one-family
   (b) two-family
   (c) multifamily
   (d) any of these

ARTICLE

312 Cabinets, Cutout Boxes, and Meter Socket Enclosures

INTRODUCTION TO ARTICLE 312—CABINETS, CUTOUT BOXES, AND METER SOCKET ENCLOSURES

This article addresses the installation and construction specifications for the items mentioned in its title. In Article 310, we observed that you need different ampacities for the same conductor, depending on conditions of use. The same thing applies to these items—just in a different way. For example, you can't use just any enclosure in a wet location or in a hazardous (classified) location. The conditions of use impose special requirements for these situations.

For all such enclosures, certain requirements apply—regardless of the use. For example, you must cover any openings, protect conductors from abrasion, and allow sufficient bending room for conductors.

Part I is where you'll find the requirements most useful to the electrician in the field. Part II applies to manufacturers. If you use name brand components that are listed or labeled, you don't need to be concerned with Part II. However, if you are specifying custom enclosures, you need to be familiar with these requirements to help ensure that the authority having jurisdiction approves the enclosures.

312.1 Scope. Article 312 covers the installation and construction specifications for cabinets, cutout boxes, and meter socket enclosures. Figure 312–1

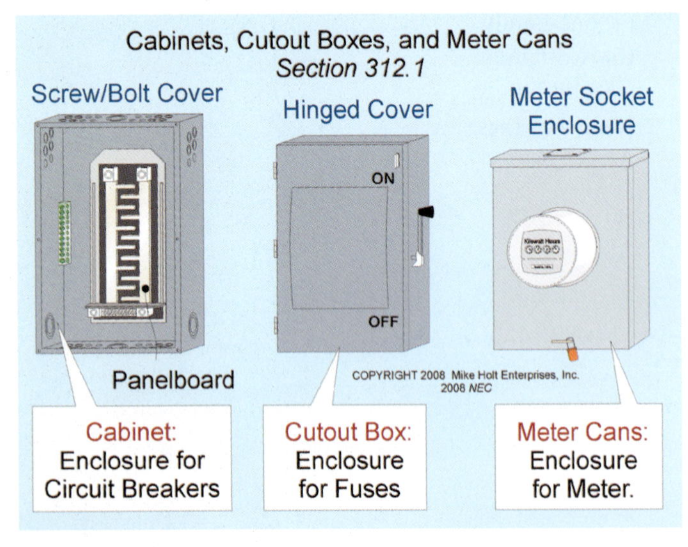

Figure 312–1

> **Author's Comment:** A cabinet is an enclosure for either surface mounting or flush mounting provided with a frame in which a door may be hung. A cutout box is designed for surface mounting with a swinging door [Article 100]. The industry name for a meter socket enclosure is "meter can."

PART I. INSTALLATION

312.2 Damp or Wet Locations.

Enclosures in damp or wet locations must prevent moisture or water from entering or accumulating within the enclosure, and must be weatherproof. When the enclosure is surface mounted in a wet location, the enclosure must be mounted with not less than a ¼ in. air space between it and the mounting surface. See 300.6(D).

Where raceways or cables enter above the level of uninsulated live parts of an enclosure in a wet location, a fitting listed for wet locations must be used for termination.

> **Author's Comment:** A fitting listed for use in a wet location with a sealing locknut is suitable for this application.

Exception: The ¼ in. air space isn't required for nonmetallic equipment, raceways, or cables.

312.3 Installed in Walls. Cabinets or cutout boxes installed in walls of concrete, tile, or other noncombustible material must be installed so that the front edge of the enclosure is set back no more than ¼ in. from the finished surface. In walls constructed of wood or other combustible material, cabinets or cutout boxes must be flush with the finished surface or project outward.

312.4 Repairing Gaps. Gaps around cabinets and cutout boxes that are recessed in noncombustible surfaces (plaster, drywall, or plasterboard) having a flush-type cover, must be repaired so that there will be no gap more than ⅛ in. at the edge of the cabinet or cutout box. Figure 312–2

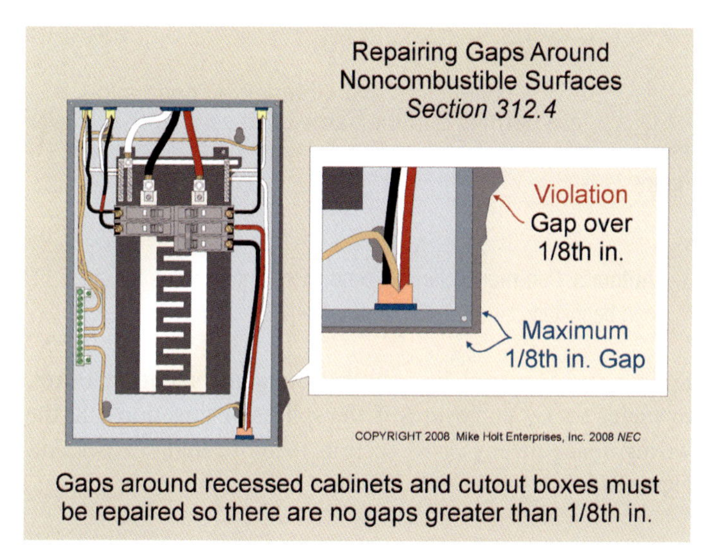

Figure 312–2

312.5 Enclosures.

(A) Unused Openings. Openings intended to provide entry for conductors must be adequately closed. Figure 312–3

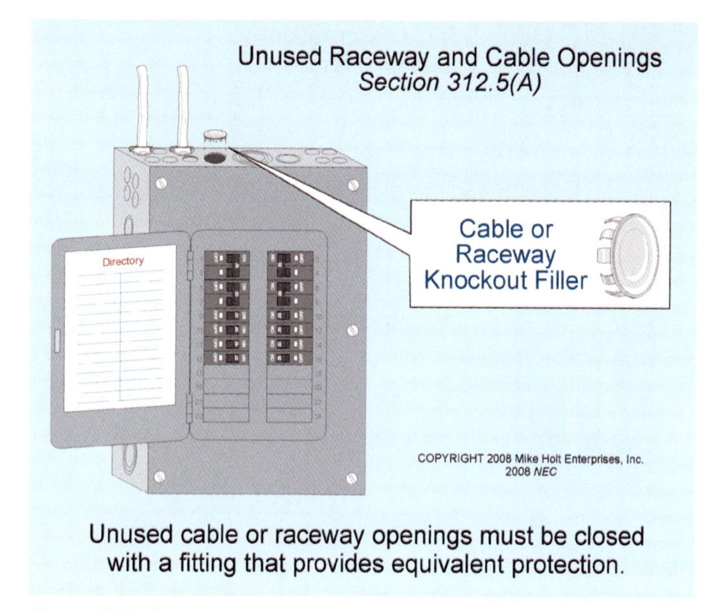

Figure 312–3

Author's Comment: Unused openings for circuit breakers must be closed by means that provide protection substantially equivalent to the wall of the enclosure [408.7]. Figure 312–4

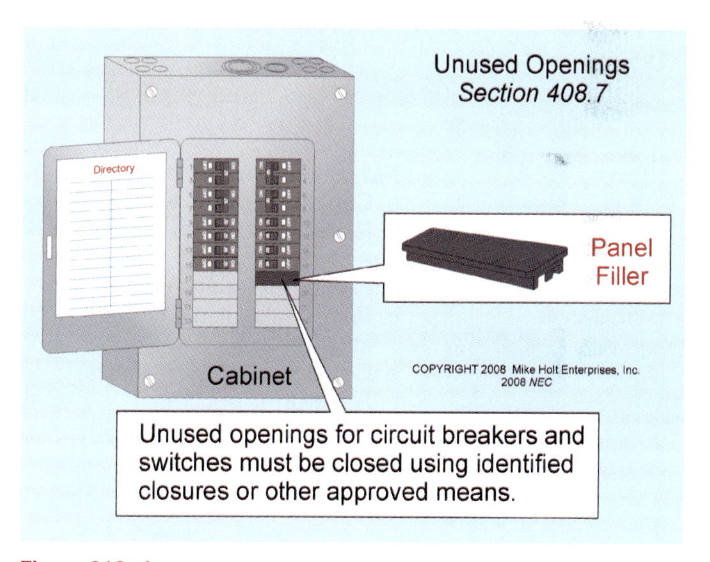

Figure 312–4

(C) Cable Termination. Cables must be secured to the enclosure with fittings designed and listed for the cable. See 300.12 and 300.15. Figure 312–5

Author's Comment: Cable clamps or cable connectors must be used with only one cable, unless that clamp or fitting is identified for more than one cable. Some Type NM cable clamps are listed for two Type NM cables within a single fitting (UL White Book, product category PXJV).

Exception: Cables with nonmetallic sheaths aren't required to be secured to the enclosure if the cables enter the top of a surface-mounted enclosure through a nonflexible raceway not less than 18 in. or more than 10 ft long, if all of the following conditions are met: Figure 312–6

(a) Each cable is fastened within 1 ft from the raceway.

(b) The raceway doesn't penetrate a structural ceiling.

(c) Fittings are provided on the raceway to protect the cables from abrasion.

(d) The raceway is sealed.

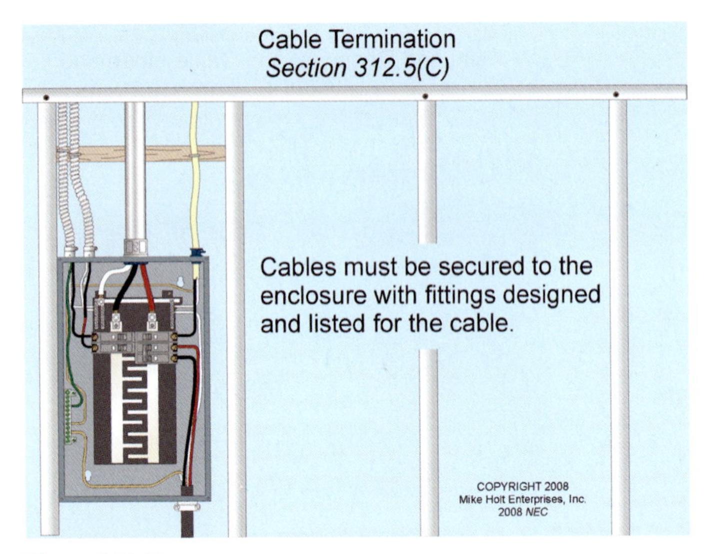

Figure 312–5

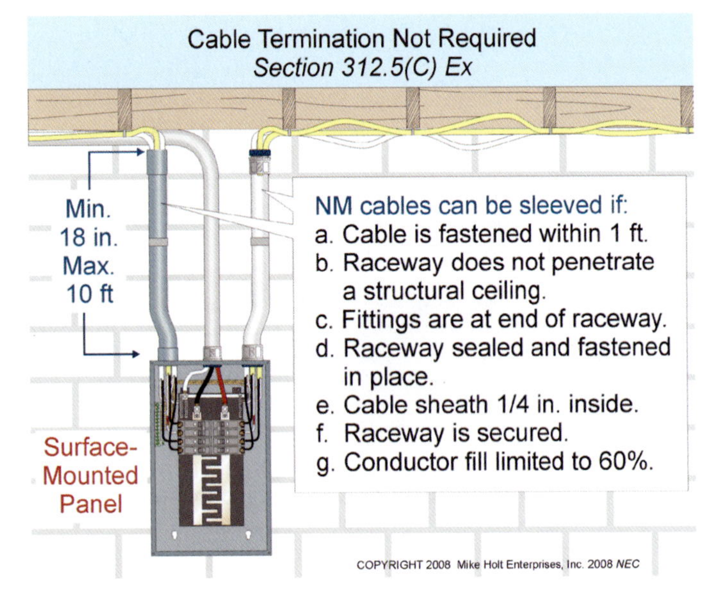

Figure 312–6

(e) Each cable sheath extends not less than ¼ in. into the panelboard.

(f) The raceway is properly secured.

(g) Conductor fill is limited to Chapter 9, Table 1 percentages.

312.8 Used for Raceway and Splices. Cabinets, cutout boxes, and meter socket enclosures can be used as a raceway for conductors that feed through if the conductors don't fill the wiring space at any cross section to more than 40 percent. **Figure 312–7**

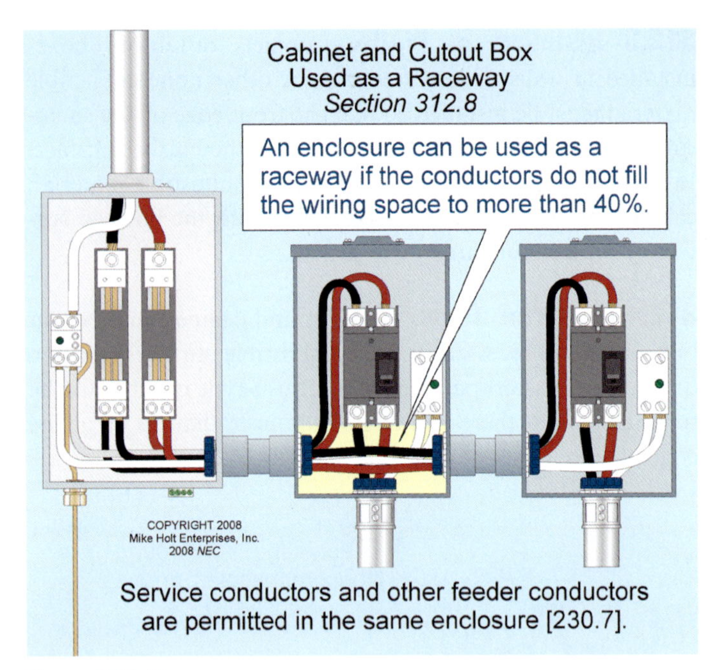

Figure 312–7

Author's Comment: Service conductors and other conductors can be installed in the same enclosure [230.7].

Splices and taps can be installed in cabinets, cutout boxes, or meter socket enclosures if the splices or taps don't fill the wiring space at any cross section to more than 75 percent. **Figure 312–8**

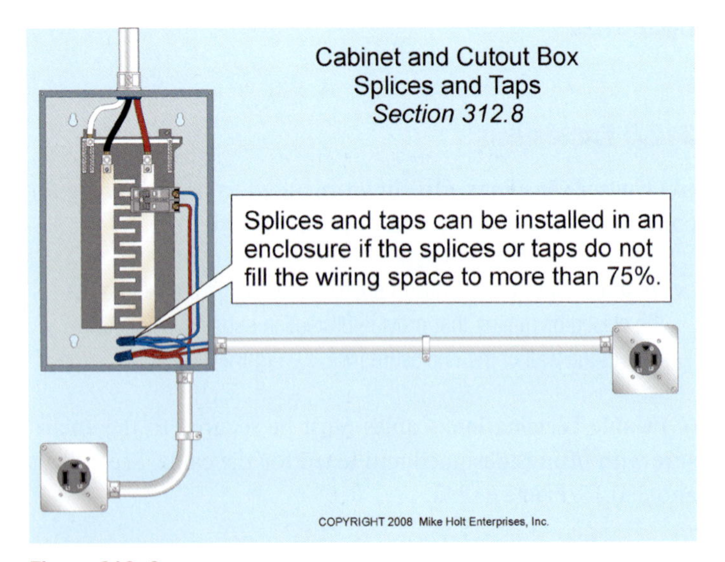

Figure 312–8

ARTICLE 312 Practice Questions

ARTICLE 312. CABINETS, CUTOUT BOXES, AND METER SOCKET ENCLOSURES—PRACTICE QUESTIONS

1. Cabinets, cutout boxes, and meter socket enclosures installed in wet locations shall be _____.

   (a) waterproof
   (b) raintight
   (c) weatherproof
   (d) watertight

2. Noncombustible surfaces that are broken or incomplete shall be repaired so there will be no gaps or open spaces greater than _____ at the edge of a cabinet or cutout box employing a flush-type cover.

   (a) 1/32 in.
   (b) 1/16 in.
   (c) 1/8 in.
   (d) 1/4 in.

3. Nonmetallic cables can enter the top of surface-mounted cabinets, cutout boxes, and meter socket enclosures through nonflexible raceways not less than 18 in. or more than _____ ft in length if all of the required conditions are met.

   (a) 3
   (b) 10
   (c) 25

Outlet, Device, Pull, and Junction Boxes; Conduit Bodies; and Handhole Enclosures

INTRODUCTION TO ARTICLE 314—OUTLET, DEVICE, PULL, AND JUNCTION BOXES; CONDUIT BODIES; AND HANDHOLE ENCLOSURES

Article 314 contains installation requirements for outlet boxes, pull and junction boxes, conduit bodies, and handhole enclosures.

As with Article 312, conditions of use apply. If you're running a raceway in a hazardous (classified) location, for example, you must use the correct fittings and the proper installation methods. But consider something as simple as a splice. It makes sense you wouldn't put a splice in the middle of a raceway—doing so means you can't get to it. But if you put a splice in a conduit body, you're fine, right? Not necessarily. Suppose the conduit body is a "short radius" version (think of it as an elbow with the bend chopped off). You don't have much room inside such an enclosure, and for that reason you can't put a splice inside a short-radius conduit body.

Properly applying Article 314 means you'll need to account for the internal volume of all boxes and fittings, and then determine the maximum conductor fill. You'll also need to understand many other requirements, which we'll cover. If you start to get confused, take a break. Look carefully at the illustrations, and you'll learn more quickly and with more retention.

PART I. SCOPE AND GENERAL

314.1 Scope. Article 314 contains the installation requirements for outlet boxes, conduit bodies, pull and junction boxes, and handhole enclosures. **Figure 314–1**

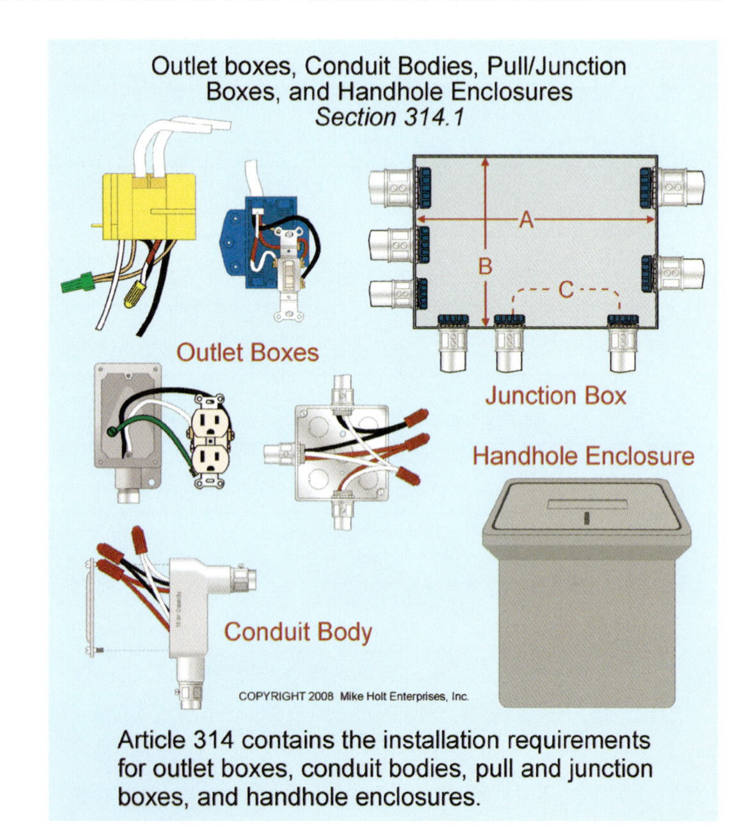

Figure 314–1

314.3 Nonmetallic Boxes. Nonmetallic boxes can only be used with nonmetallic cables and raceways because there is no way to maintain the electrical continuity of the effective ground-fault current path [250.2 and 250.4(A)(3)].

Exception No. 1: Metal raceways and metal cables can be used with nonmetallic boxes, but only if an internal bonding means is provided in the box between all metal entries.

314.4 Metal Boxes. Metal boxes containing circuits that operate at 50V or more must be connected to an equipment grounding conductor of a type listed in 250.118 [250.112(I)]. **Figure 314–2**

314.5 Short-Radius Conduit Bodies. Short-radius conduit bodies, such as capped elbows, handy ells, and service-entrance elbows must not contain any splices or taps. **Figure 314–3**

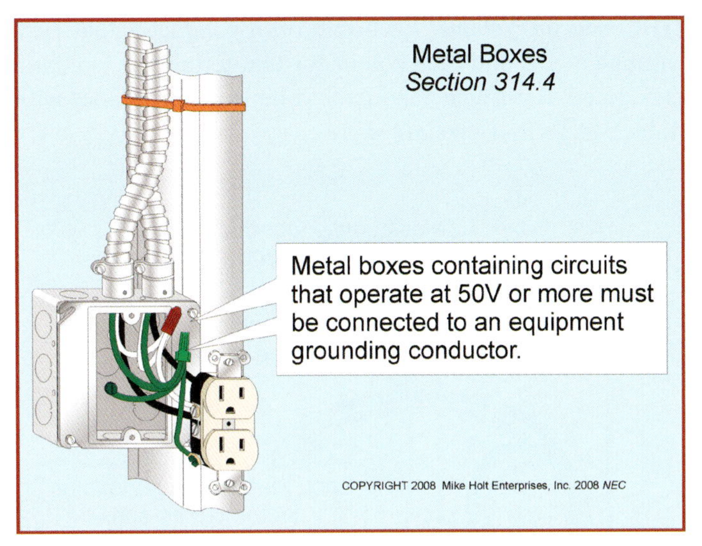

Figure 314–2

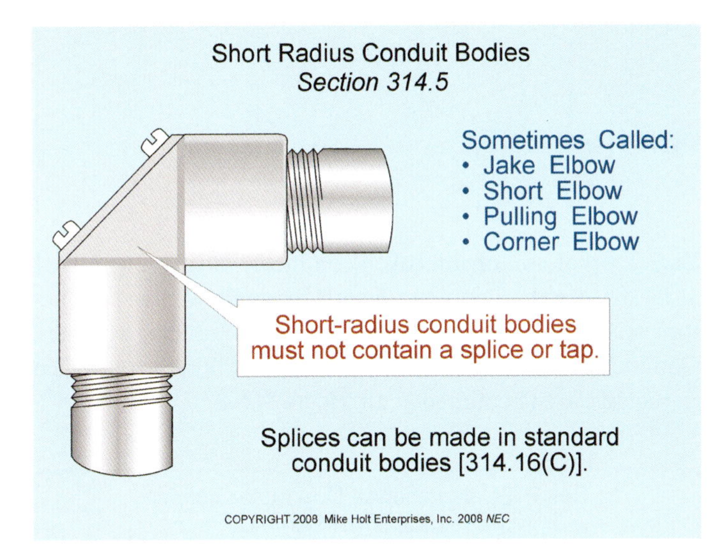

Figure 314–3

Author's Comment: Splices and taps can be made in standard conduit bodies. See 314.16(C) for specific requirements.

PART II. INSTALLATION

314.15 Damp or Wet Locations. Boxes and conduit bodies in damp or wet locations must prevent moisture or water from entering or accumulating within the enclosure. Boxes, conduit bodies, and fittings installed in wet locations must be listed for use in wet locations. Figure 314–4

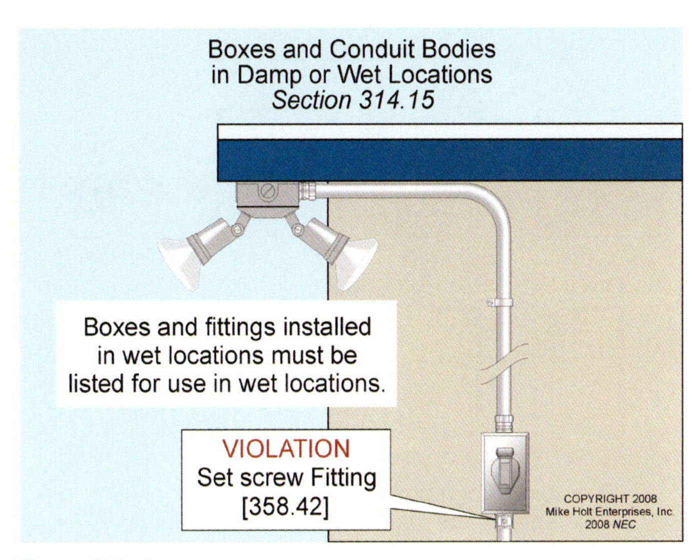

Figure 314–4

Author's Comment: Where handhole enclosures without bottoms are installed, all enclosed conductors and any splices or terminations must be listed as suitable for wet locations [314.30(C)].

314.16 Number of 6 AWG and Smaller Conductors in Boxes and Conduit Bodies. Boxes containing 6 AWG and smaller conductors must be sized to provide sufficient free space for all conductors, devices, and fittings. In no case can the volume of the box, as calculated in 314.16(A), be less than the volume requirement as calculated in 314.16(B).

Conduit bodies must be sized according to 314.16(C).

Author's Comment: The requirements for sizing boxes and conduit bodies containing conductors 4 AWG and larger are contained in 314.28. The requirements for sizing handhole enclosures are contained in 314.30(A).

(A) Box Volume Calculations. The volume of a box includes the total volume of its assembled parts, including plaster rings, extension rings, and domed covers that are either marked with their volume in cubic inches (cu in.), or are made from boxes listed in Table 314.16(A). Figure 314–5

(B) Box Fill Calculations. The calculated conductor volume determined by (1) through (5) and Table 314.16(B) are added together to determine the total volume of the conductors, devices, and fittings. Raceway and cable fittings, including locknuts and bushings, are not counted for box fill calculations. Figure 314–6

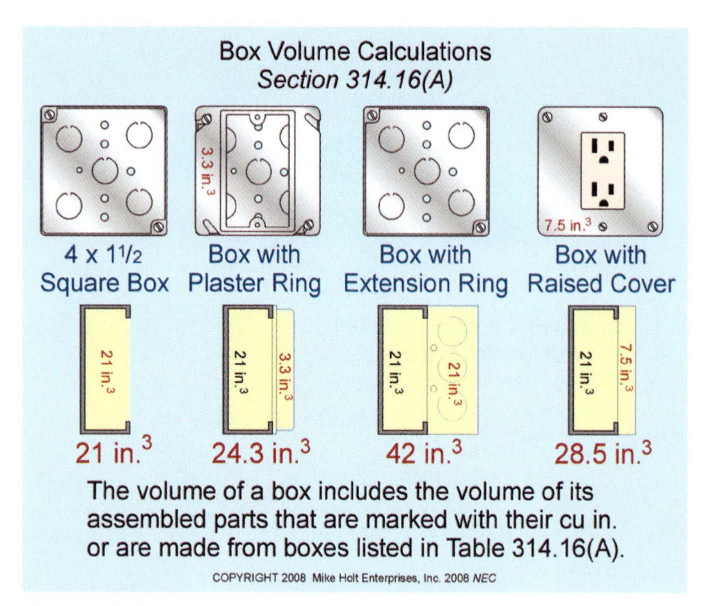

Figure 314–5

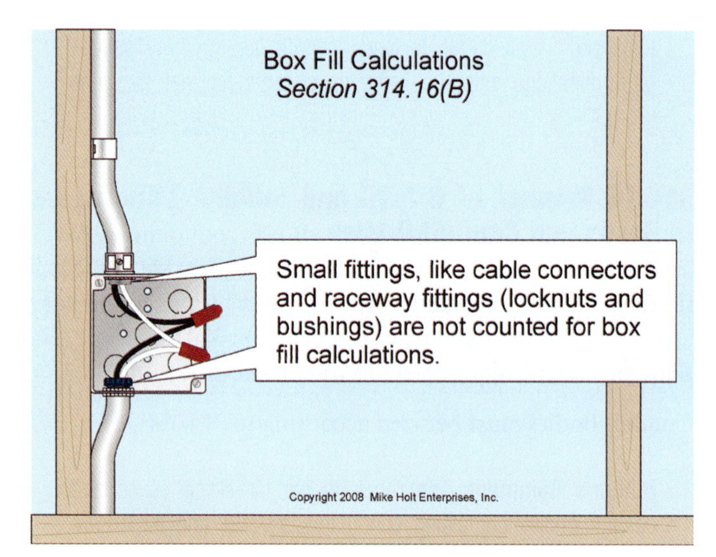

Figure 314–6

Table 314.16(B) Volume Allowance Required per Conductor	
Conductor AWG	Volume cu in.
18	1.50
16	1.75
14	2.00
12	2.25
10	2.50
8	3.00
6	5.00

(1) Conductor Volume. Each unbroken conductor that runs through a box, and each conductor that terminates in a box, is counted as a single conductor volume in accordance with Table 314.16(B). Figure 314–7

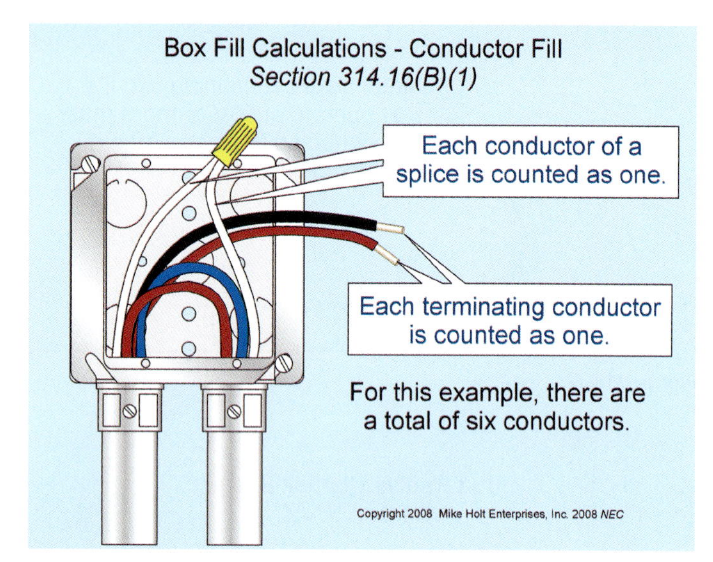

Figure 314–7

Each loop or coil of unbroken conductor having a length of at least twice the minimum length required for free conductors in 300.14 must be counted as two conductor volumes. Conductors that originate and terminate within the box, such as pigtails, aren't counted at all. Figure 314–8

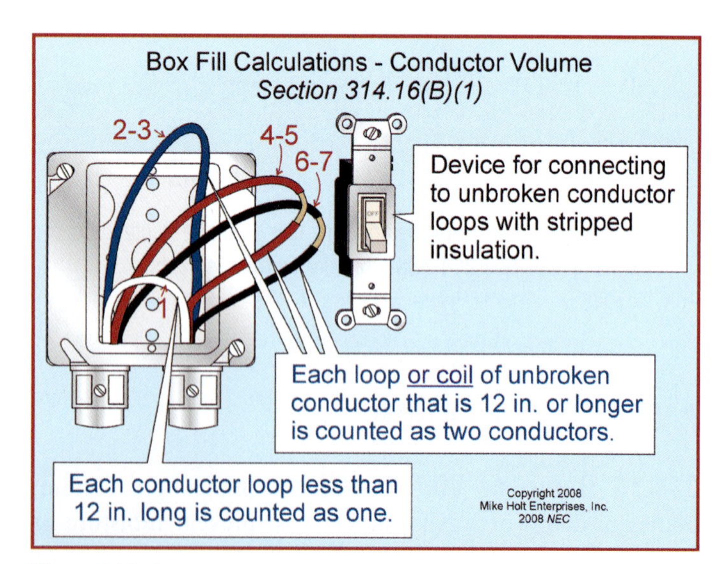

Figure 314–8

Author's Comment: According to 300.14, at least 6 in. of free conductor, measured from the point in the box where the conductors enter the enclosure, must be left at each outlet, junction, and switch point for splices or terminations of luminaires or devices.

Exception: Equipment grounding conductors, and up to four 16 AWG and smaller fixture wires, can be omitted from box fill calculations if they enter the box from a domed luminaire or similar canopy, such as a ceiling paddle fan canopy. Figure 314–9

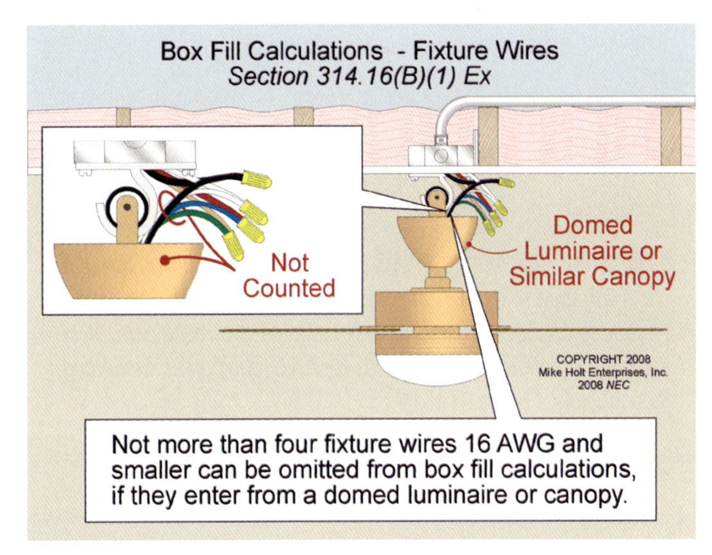

Figure 314–9

(2) Cable Clamp Volume. One or more internal cable clamps count as a single conductor volume in accordance with Table 314.16(B), based on the largest conductor that enters the box. Cable connectors that have their clamping mechanism outside of the box aren't counted. Figure 314–10

(3) Support Fitting Volume. Each luminaire stud or luminaire hickey counts as a single conductor volume in accordance with Table 314.16(B), based on the largest conductor that enters the box. Figure 314–11

Author's Comment: Luminaire stems do not need to be counted as a conductor volume.

(4) Device Yoke Volume. Each single gang device yoke (regardless of the ampere rating of the device) counts as two conductor volumes, based on the largest conductor that terminates on the device in accordance with Table 314.16(B). Figure 314–12

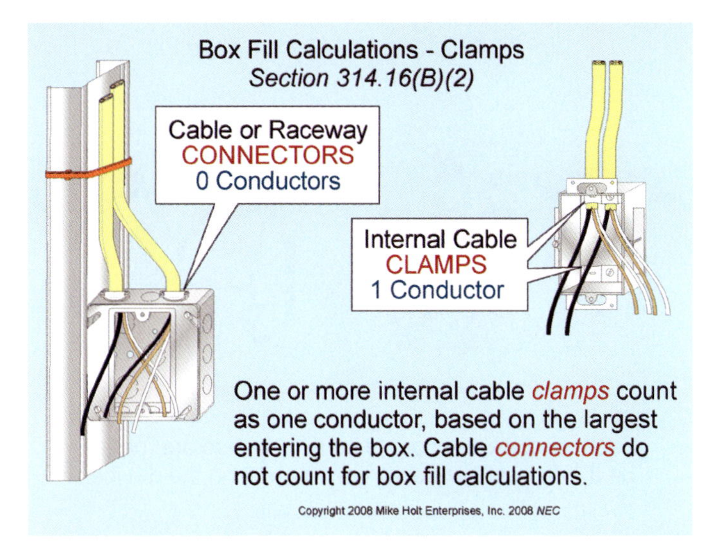

Figure 314–10

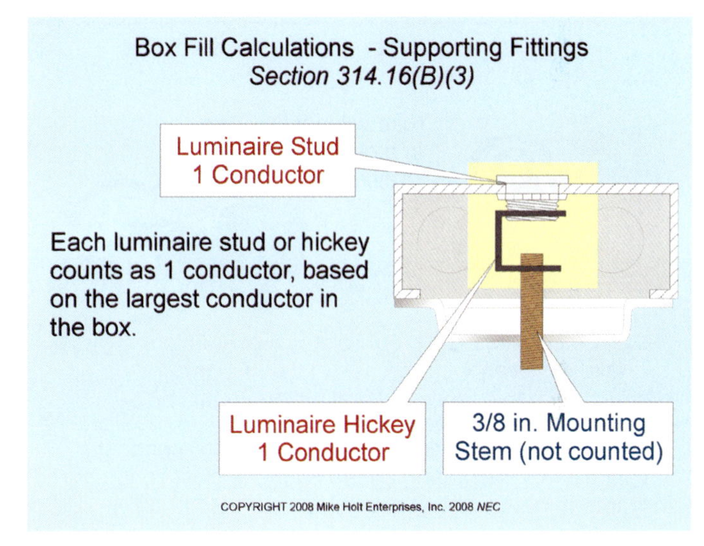

Figure 314–11

Each multigang device yoke counts as two conductor volumes for each gang, based on the largest conductor that terminates on the device in accordance with Table 314.16(B). Figure 314–13

Author's Comment: A device that is too wide for mounting in a single gang box, as described in Table 314.16(A), is counted based on the number of gangs required for the device.

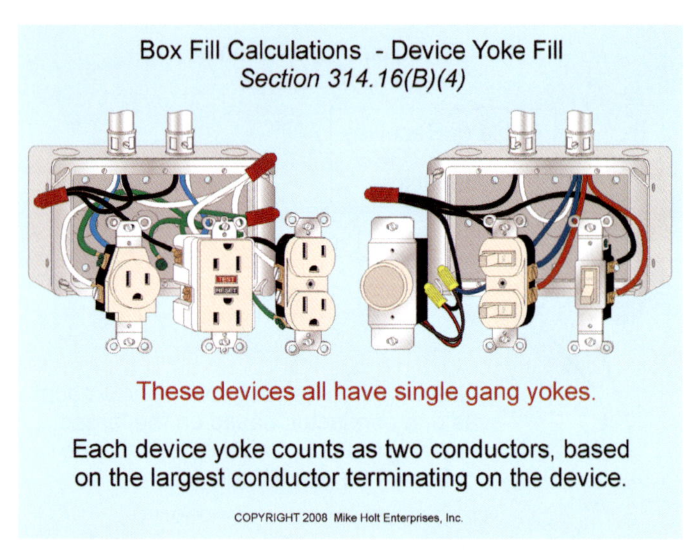

Figure 314–12

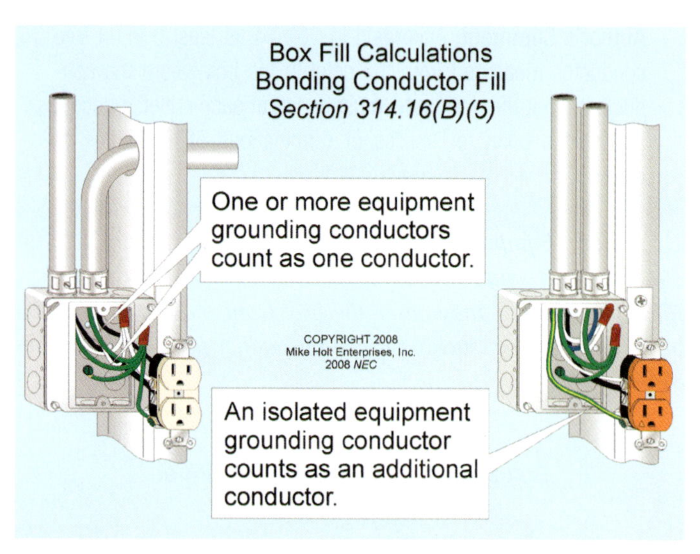

Figure 314–14

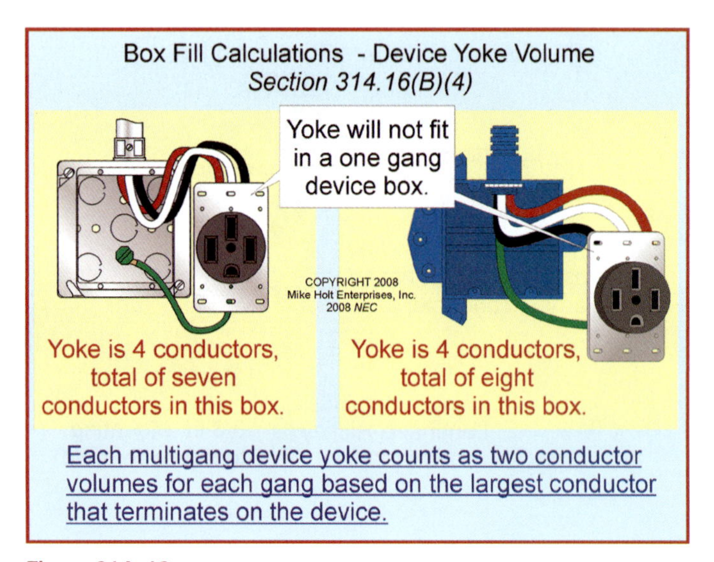

Figure 314–13

(5) Equipment Grounding Conductor Volume. All equipment grounding conductors in a box count as a single conductor volume in accordance with Table 314.16(B), based on the largest equipment grounding conductor that enters the box. Insulated equipment grounding conductors for receptacles having insulated grounding terminals (isolated ground receptacles) [250.146(D)] count as a single conductor volume in accordance with Table 314.16(B). Figure 314–14

Author's Comment: Conductor insulation is not a factor considered when determining box volume calculations.

Question: *How many 14 AWG conductors can be pulled through a 4 in. square x 2 1/8 in. deep box with a plaster ring with a marking of 3.60 cu in.? The box contains two receptacles, five 12 AWG conductors, and two 12 AWG equipment grounding conductors.* Figure 314–15

(a) 3 *(b) 5* *(c) 7* *(d) 9*

Answer: *(b) 5*

Step 1: Determine the volume of the box assembly [314.16(A)]:

Box 30.30 cu in. + 3.60 cu in. plaster ring = 33.90 cu in.

A 4 x 4 x 2 1/8 in. box will have a gross volume of 34 cu in., but the interior volume is 30.30 cu in., as listed in Table 314.16(A).

Step 2: Determine the volume of the devices and conductors in the box:

Two—receptacles	*4—12 AWG*
Five—12 AWG	*5—12 AWG*
Two—12 AWG Grounds	*1—12 AWG*

Total 10—12 AWG x 2.25 cu in. = 22.50 cu in.

Step 3: Determine the remaining volume permitted for the 14 AWG conductors:

33.90 cu in.—22.50 cu in. = 11.40 cu in.

Step 4: Determine the number of 14 AWG conductors permitted in the remaining volume:

14 AWG = 2.00 cu in. each [Table 314.16(B)]
11.40 cu in./2.00 cu in. = 5 conductors

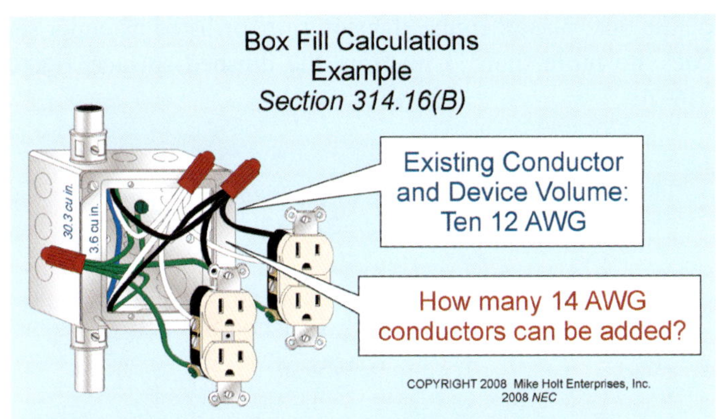

Figure 314–15

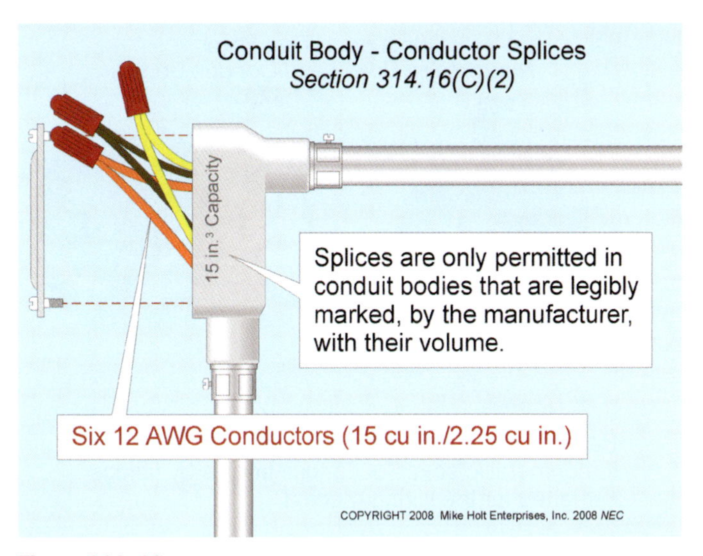

Figure 314–16

(C) Conduit Bodies.

(2) Splices. Splices are only permitted in conduit bodies that are legibly marked, by the manufacturer, with their volume. The maximum number of conductors permitted in a conduit body is limited in accordance with 314.16(B).

Question: *How many 12 AWG conductors can be spliced in a 15 cu in. conduit body?* Figure 314–16

(a) 4 *(b) 6* *(c) 8* *(d) 10*

Answer: *(b) 6 conductors (15 cu in./2.25 cu in.)*

12 AWG = 2.25 cu in. [Table 314.16(B)]
15 cu in./2.25 cu in. = 6

314.17 Conductors That Enter Boxes or Conduit Bodies.

(A) Openings to Be Closed. Openings through which cables or raceways enter must be adequately closed.

Author's Comment: Unused cable or raceway openings in electrical equipment must be effectively closed by fittings that provide protection substantially equivalent to the wall of the equipment [110.12(A)]. Figure 314–17

(B) Metal Boxes and Conduit Bodies. Raceways and cables must be mechanically fastened to metal boxes or conduit bodies by fittings designed for the wiring method. See 300.12 and 300.15.

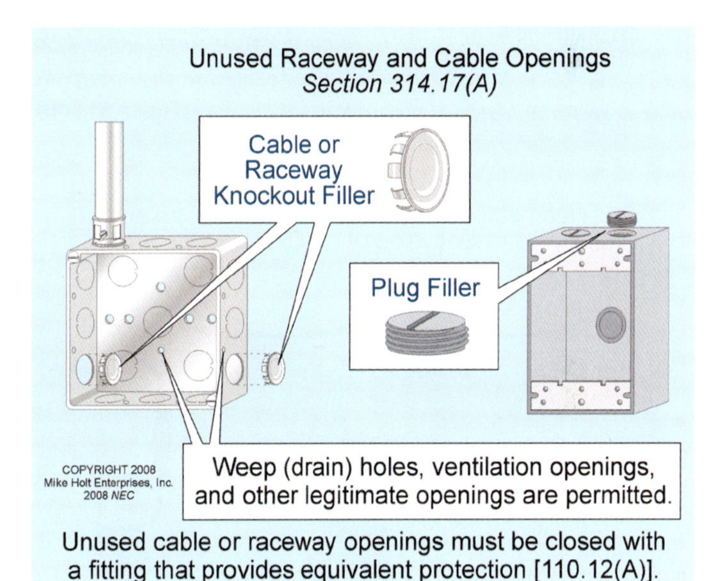

Figure 314–17

(C) Nonmetallic Boxes and Conduit Bodies. Raceways and cables must be securely fastened to nonmetallic boxes or conduit bodies by fittings designed for the wiring method [300.12 and 300.15]. Figure 314–18

The sheath of type NM cable must extend not less than ¼ in. into the nonmetallic box.

Author's Comment: Two Type NM cables can terminate in a single cable clamp, if the clamp is listed for this purpose.

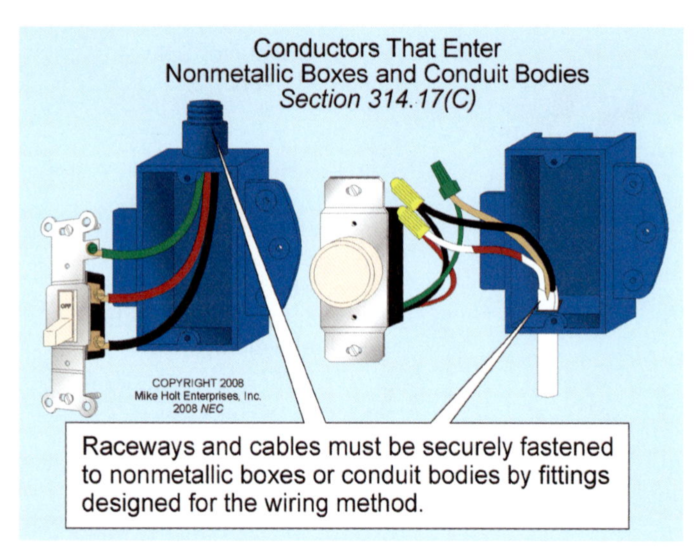

Figure 314–18

Exception: Type NM cable terminating to a single gang (2¼ x 4 in.) device box isn't required to be secured to the box if the cable is securely fastened within 8 in. of the box. Figure 314–19

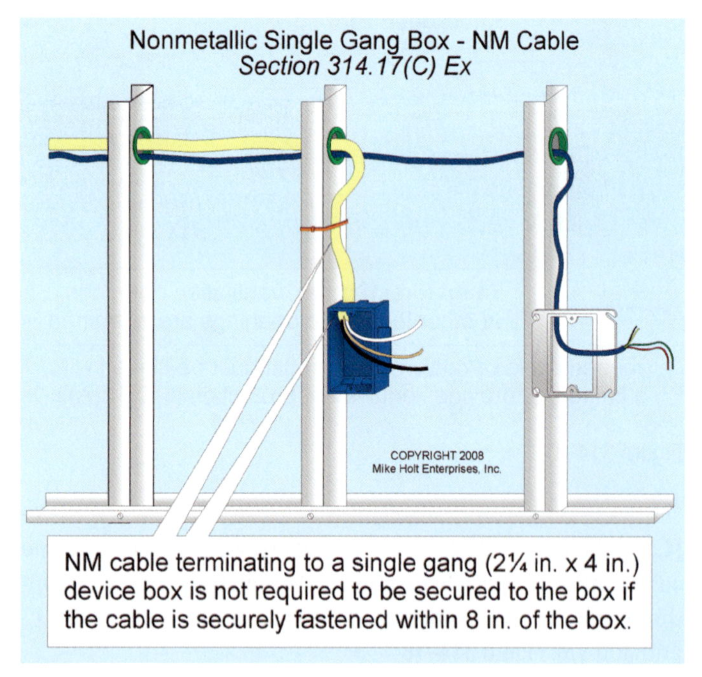

Figure 314–19

314.20 Boxes Recessed in Walls or Ceilings.

Boxes having flush-type covers that are recessed in walls or ceilings of noncombustible material must have the front edge of the box, plaster ring, extension ring, or listed extender set back no more than ¼ in. from the finished surface. Figure 314–20

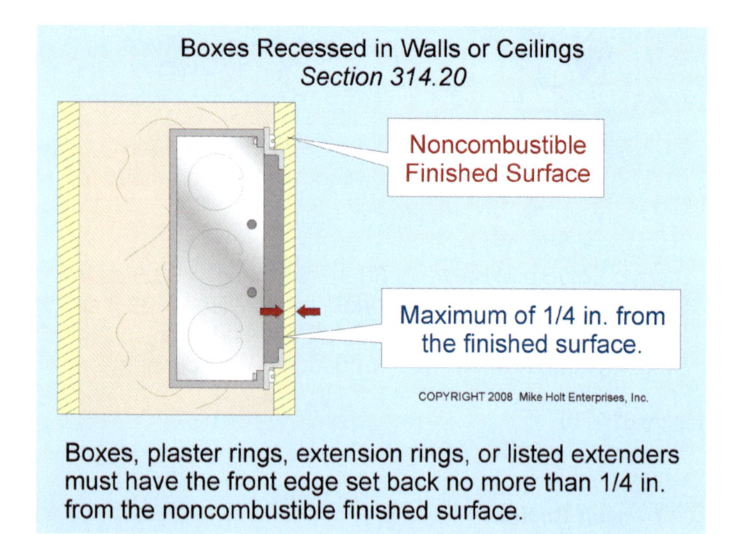

Figure 314–20

In walls or ceilings that are constructed of wood or other combustible material, boxes must be installed so the front edge of the enclosure, plaster ring, extension ring, or listed extender is flush with, or projects out from, the finished surface. Figure 314–21

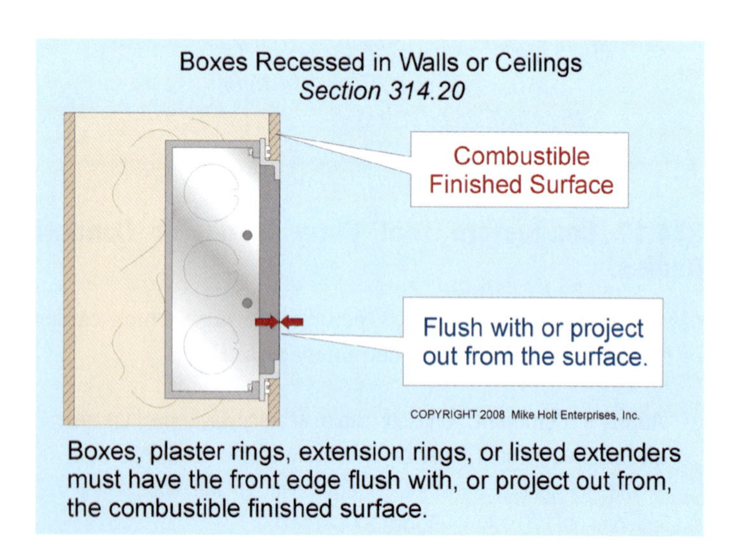

Figure 314–21

Author's Comment: Plaster rings and extension rings are available in a variety of depths to meet the above requirements.

314.21 Repairing Gaps Around Boxes. Gaps around boxes that are recessed in plaster, drywall, or plasterboard having a flush-type cover must be repaired so there will be no gap more than 1/8 in. at the edge of the box. Figure 314–22

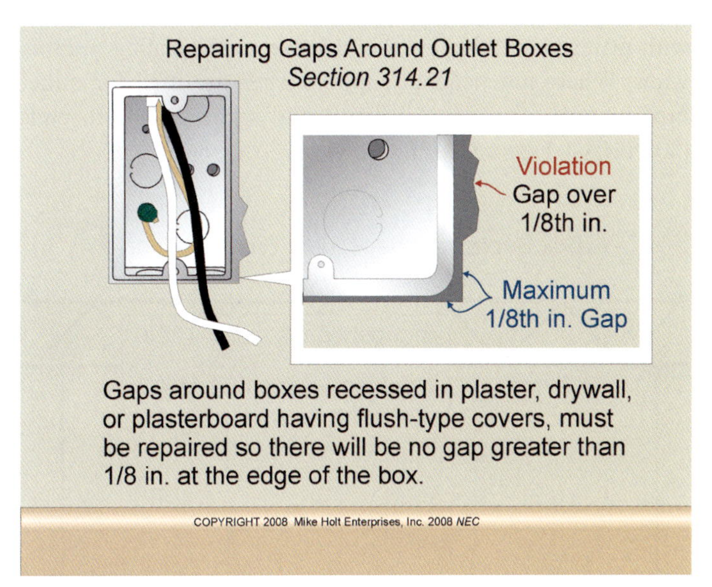

Figure 314–22

314.22 Surface Extensions. Surface extensions can only be made from an extension ring mounted over a flush-mounted box. Figure 314–23

Exception: A surface extension can be made from the cover of a flush-mounted box if the cover is designed so it's unlikely to fall off if the mounting screws become loose. The surface extension wiring method must be flexible to permit the removal of the cover and provide access to the box interior, and equipment grounding continuity must be independent of the connection between the box and the cover. Figure 314–24

314.23 Support of Boxes and Conduit Bodies. Boxes must be securely supported by one of the following methods:

(A) Surface. Boxes can be fastened to any surface that provides adequate support.

(B) Structural Mounting. Boxes can be supported from a structural member of a building or supported from grade by a metal, plastic, or wood brace.

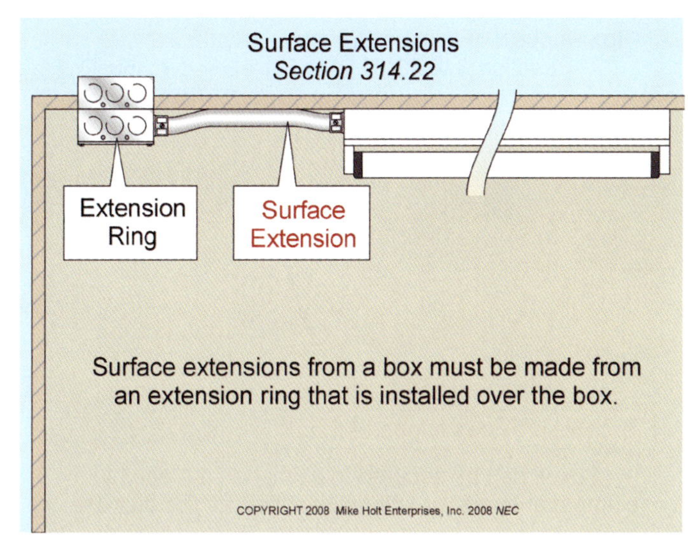

Figure 314–23

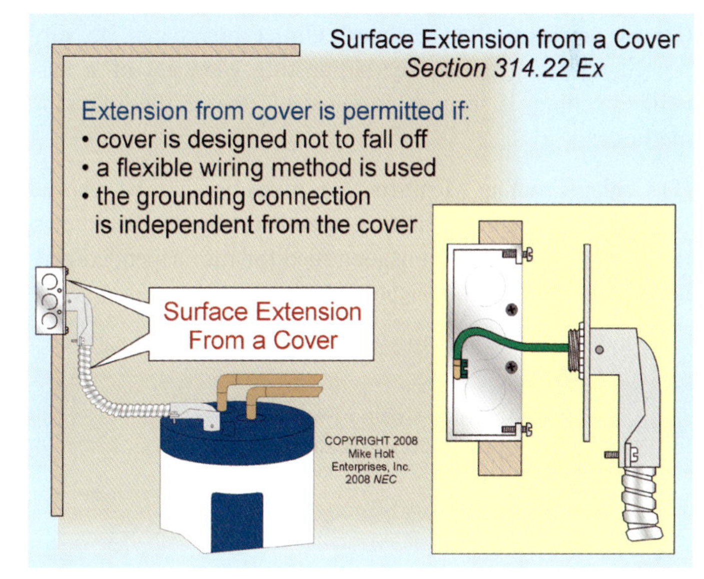

Figure 314–24

(1) Nails and Screws. Nails or screws can be used to fasten boxes, provided the exposed threads of screws are protected to prevent abrasion of conductor insulation.

(2) Braces. Metal braces no less than 0.020 in. thick and wood braces not less than a nominal 1 x 2 in. can support a box.

(C) Finished Surface Support. Boxes can be secured to a finished surface (drywall or plaster walls or ceilings) by clamps, anchors, or fittings identified for the purpose. Figure 314–25

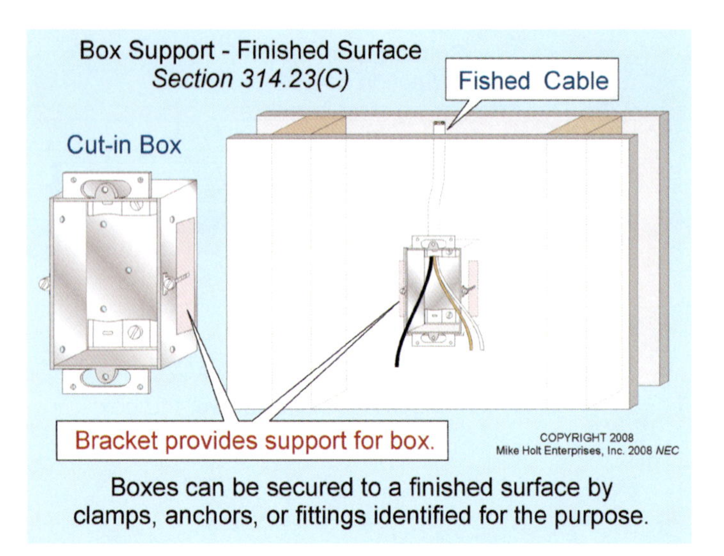

Figure 314–25

(D) Suspended-Ceiling Support. Outlet boxes can be supported to the structural or supporting elements of a suspended ceiling, if securely fastened by one of the following methods:

(1) Ceiling-Framing Members. An outlet box can be secured to suspended-ceiling framing members by bolts, screws, rivets, clips, or other means identified for the suspended-ceiling framing member(s). Figure 314–26

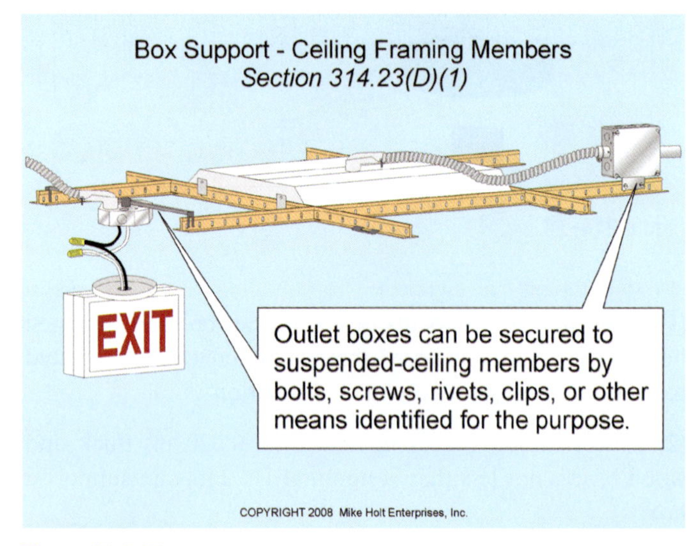

Figure 314–26

Author's Comment: Where framing members of suspended ceiling systems are used to support luminaires, they must be securely fastened to each other and must be securely attached to the building structure at appropriate intervals. In addition, luminaires must be attached to the suspended-ceiling framing members with screws, bolts, rivets, or clips listed and identified for such use [410.36(B)].

(2) Independent Support Wires. Outlet boxes can be secured, with fittings identified for the purpose, to the ceiling support wires. Where independent support wires are used for outlet box support, they must be taut and secured at both ends [300.11(A)]. Figure 314–27

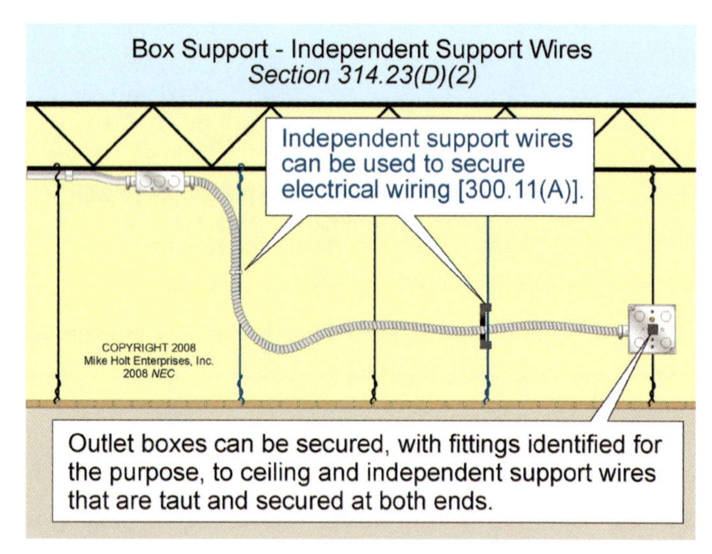

Figure 314–27

Author's Comment: See 300.11(A) on the use of independent support wires to support raceways and cables.

(E) Raceway—Boxes and Conduit Bodies Without Devices or Luminaires. Two intermediate metal or rigid metal conduits, threaded wrenchtight into the enclosure, can be used to support an outlet box that doesn't contain a device or luminaire, if each raceway is supported within 36 in. of the box or within 18 in. if all conduit entries are on the same side. Figure 314–28

(F) Raceway—Boxes and Conduit Bodies with Devices or Luminaires. Two intermediate metal or rigid metal conduits, threaded wrenchtight into the enclosure, can be used to support an outlet box containing devices or luminaires, if each raceway is supported within 18 in. of the box. Figure 314–29

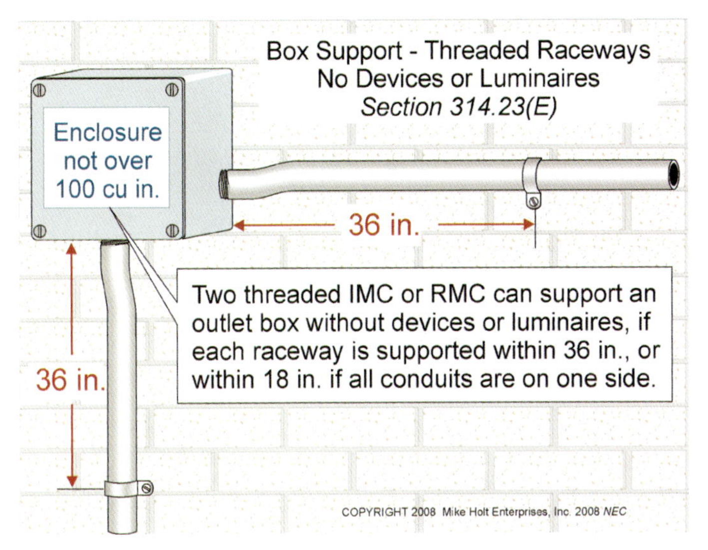

Figure 314–28

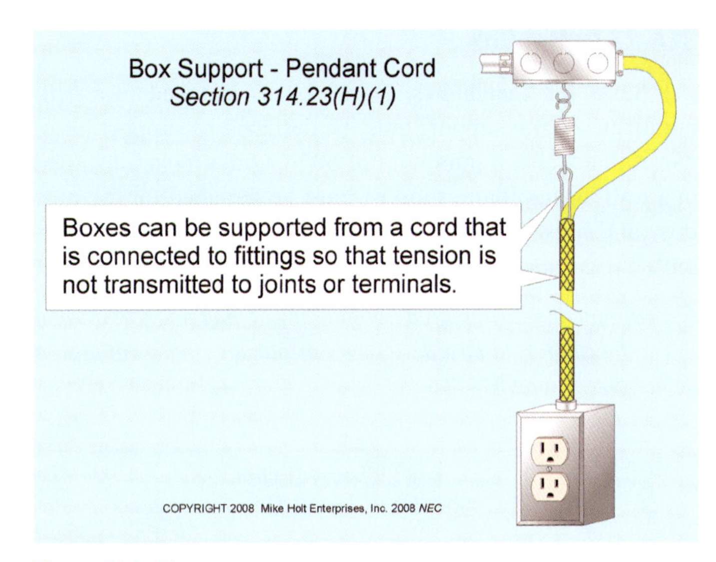

Figure 314–30

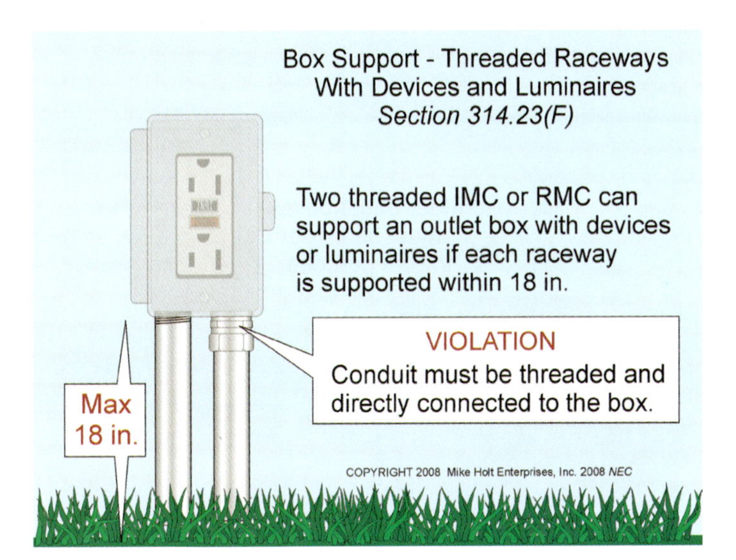

Figure 314–29

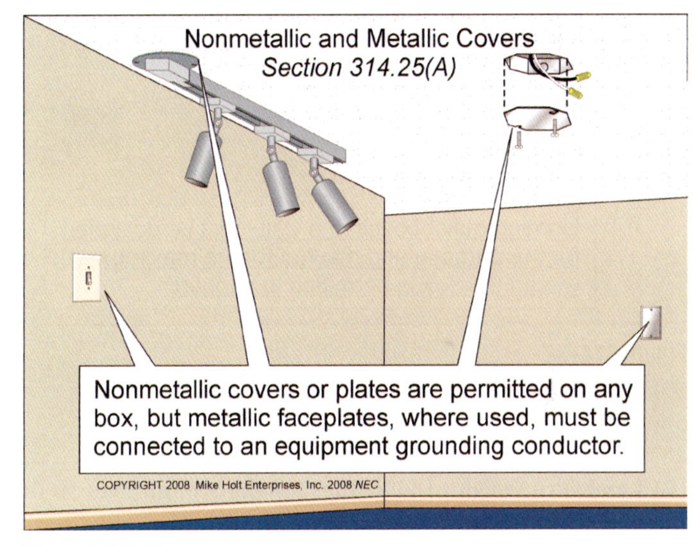

Figure 314–31

(H) Pendant Boxes.

(1) Flexible Cord. Boxes containing a hub can be supported from a cord connected to fittings that prevent tension from being transmitted to joints or terminals [400.10]. Figure 314–30

314.25 Covers and Canopies.

When the installation is complete, each outlet box must be provided with a cover or faceplate, unless covered by a fixture canopy, lampholder, or similar device. Figure 314–31

(A) Nonmetallic or Metallic. Nonmetallic covers are permitted on any box, but metal covers are only permitted where they can be connected to an equipment grounding conductor of a type recognized in 250.118, in accordance with 250.110 [250.4(A)(3)].

Author's Comment: Metal switch faceplates [404.9(B)] and metal receptacle faceplates [406.5(A)] must be connected to an equipment grounding conductor.

314.27 Outlet Box.

(A) Boxes for Luminaires. Outlet boxes installed in a ceiling for luminaire support must be designed to support luminaires that weigh a minimum of 50 lb. Outlet boxes installed in a wall for luminaire support must be designed for the purpose and must indicate the maximum luminaire weight permitted to be supported by the box.

Exception: A wall-mounted luminaire weighing no more than 6 lb can be supported to a device box or plaster ring secured to a device box. **Figure 314–32**

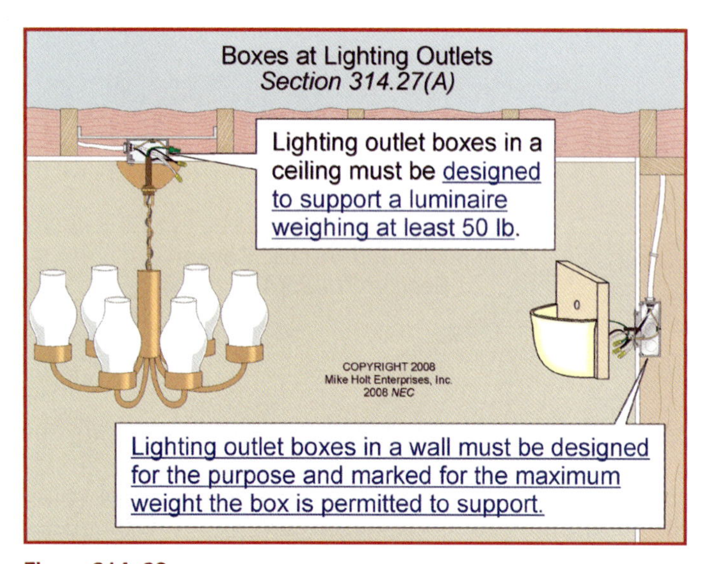

Figure 314–32

(B) Luminaire Weight. Luminaires weighing more than 50 lb must be supported independently of the lighting outlet box, unless the lighting outlet box is listed and marked for the maximum weight of the luminaire. **Figure 314–33**

(C) Floor Box. Floor boxes must be specifically listed for the purpose. **Figure 314–34**

(D) Ceiling Paddle Fan Box. Outlet boxes for a ceiling paddle fan must be listed and marked as suitable for the purpose, and must not support a fan weighing more than 70 lb. Outlet boxes for a ceiling paddle fan that weighs more than 35 lb must include the maximum weight to be supported in the required marking. **Figure 314–35**

> **Author's Comment:** Where the maximum weight isn't marked on the box, and the fan weighs over 35 lb, the fan must be supported independently of the outlet box. Ceiling paddle fans over 70 lb must be supported independently of the outlet box.

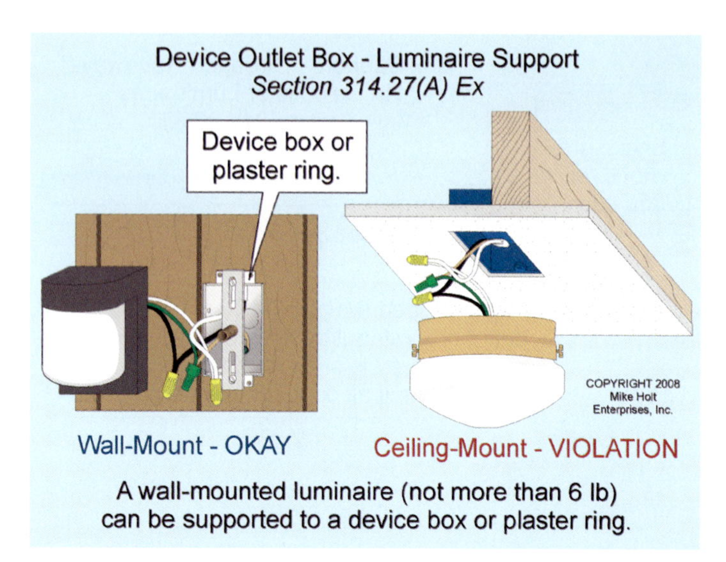

Figure 314–33

Figure 314–34

(E) Utilization Equipment. Boxes used for the support of utilization equipment must be designed to support equipment that weighs a minimum of 50 lb [314.27(A)].

Exception: Utilization equipment weighing 6 lb or less is permitted to be supported by any box or plaster ring secured to a box, provided the equipment is secured with no fewer than two No. 6 or larger screws. **Figure 314–36**

314.28 Boxes and Conduit Bodies for Conductors 4 AWG and Larger. Boxes and conduit bodies containing conductors 4 AWG and larger that are required to be insulated must be sized so the conductor insulation will not be damaged.

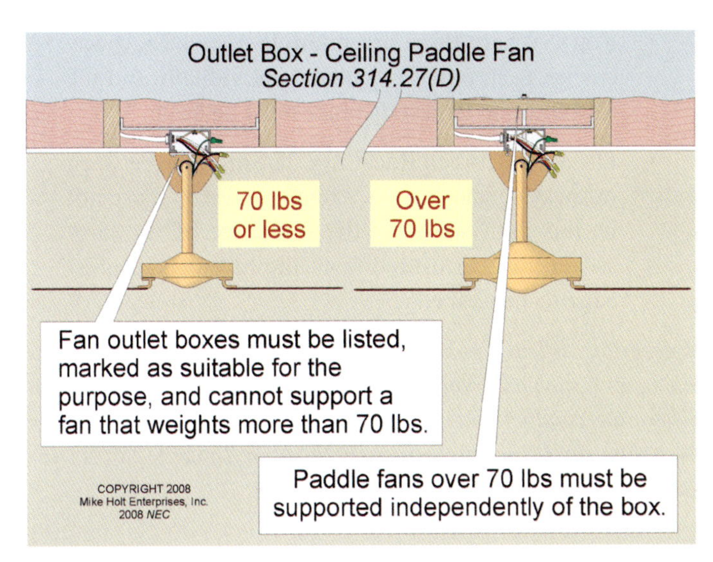

Figure 314–35

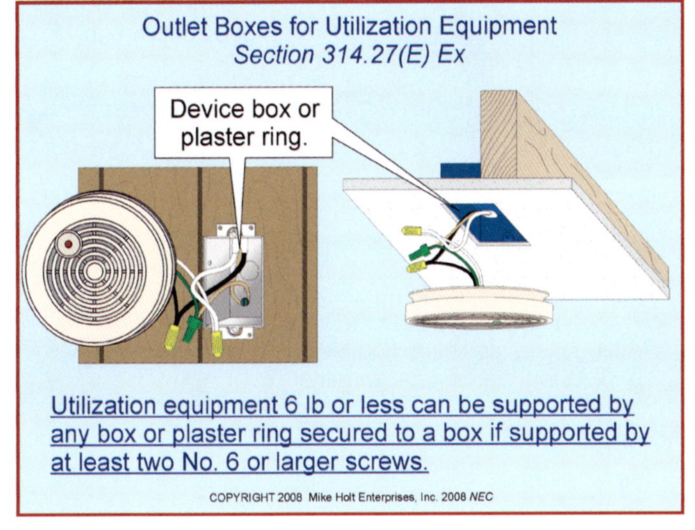

Figure 314–36

Author's Comments:

- The requirements for sizing boxes and conduit bodies containing conductors 6 AWG and smaller are contained in 314.16.
- Where conductors 4 AWG or larger enter a box or other enclosure, a fitting that provides a smooth, rounded, insulating surface, such as a bushing or adapter, is required to protect the conductors from abrasion during and after installation [300.4(G)].

(A) Minimum Size. For raceways containing conductors 4 AWG or larger, the minimum dimensions of boxes and conduit bodies must comply with the following:

(1) Straight Pulls. The minimum distance from where the conductors enter to the opposite wall must not be less than eight times the trade size of the largest raceway. Figure 314–37

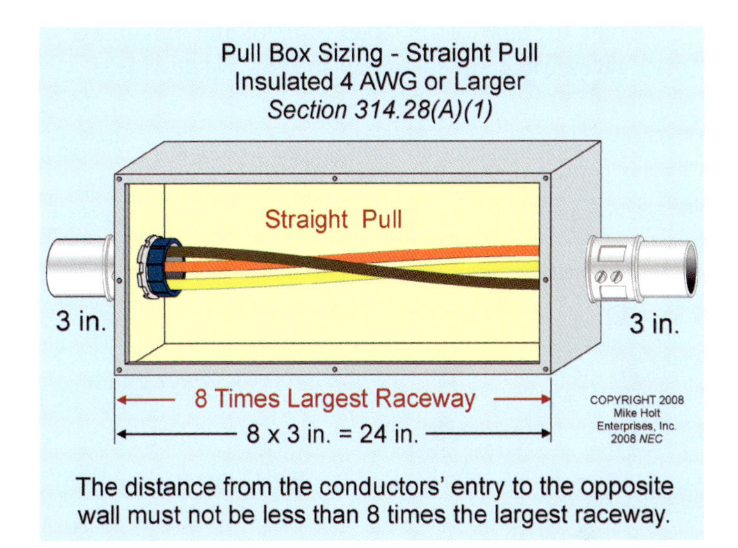

Figure 314–37

(2) Angle Pulls, U Pulls, or Splices.

- Angle Pulls. The distance from the raceway entry to the opposite wall must not be less than six times the trade size of the largest raceway, plus the sum of the trade sizes of the remaining raceways on the same wall and row. Figure 314–38

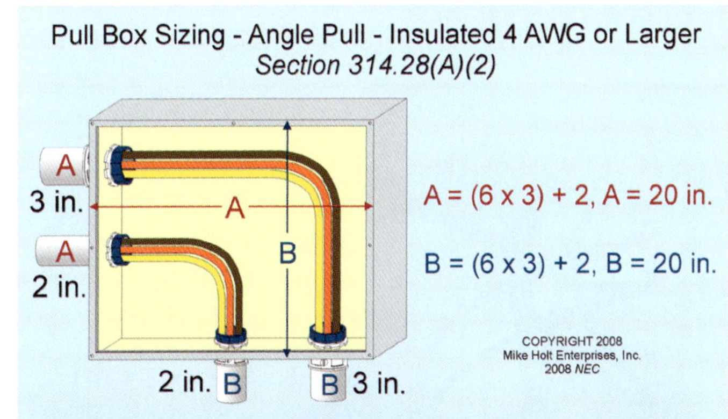

Figure 314–38

- U Pulls. When a conductor enters and leaves from the same wall, the distance from where the raceways enter to the opposite wall must not be less than six times the trade size of the largest raceway, plus the sum of the trade sizes of the remaining raceways on the same wall and row. Figure 314–39

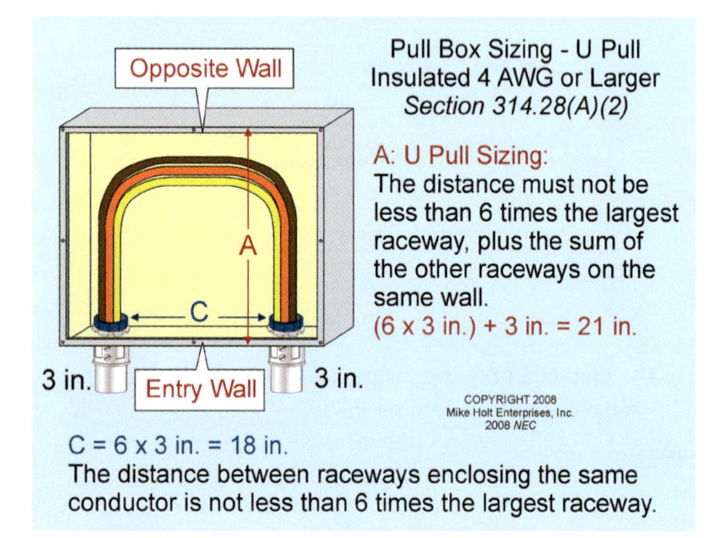

Figure 314–39

- Splices. When conductors are spliced, the distance from where the raceways enter to the opposite wall must not be less than six times the trade size of the largest raceway, plus the sum of the trade sizes of the remaining raceways on the same wall and row. Figure 314–40

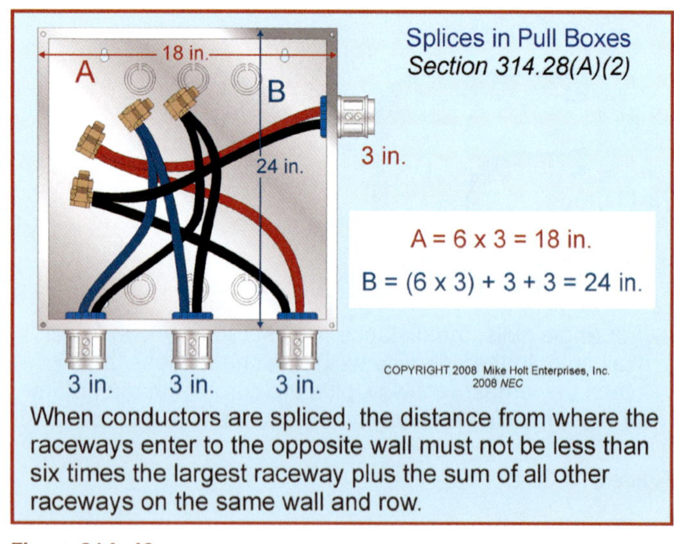

Figure 314–40

- Rows. Where there are multiple rows of raceway entries, each row is calculated individually and the row with the largest distance must be used.

- Distance Between Raceways. The distance between raceways enclosing the same conductor must not be less than six times the trade size of the largest raceway, measured from the raceways' nearest edge-to-nearest edge.

Exception: When conductors enter an enclosure with a removable cover, such as a conduit body or wireway, the distance from where the conductors enter to the removable cover must not be less than the bending distance as listed in Table 312.6(A) for one conductor per terminal. Figure 314–41

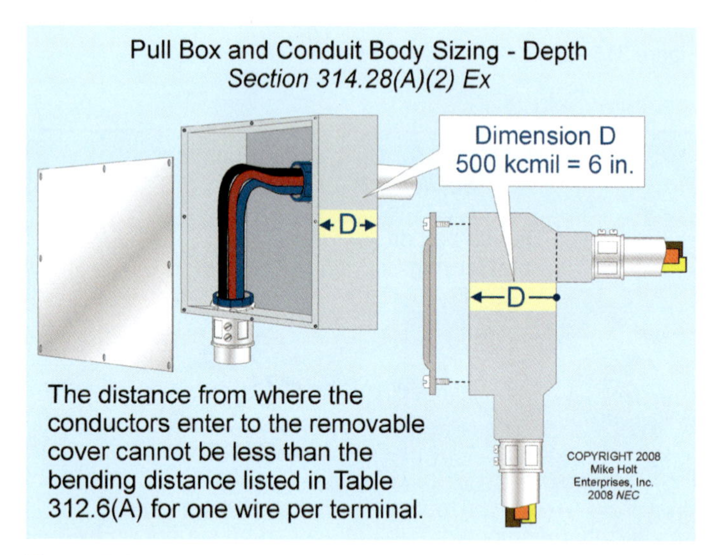

Figure 314–41

(3) Smaller Dimensions. Boxes or conduit bodies smaller than those required in 314.28(A)(1) and 314.28(A)(2) are permitted, if the enclosure is permanently marked with the maximum number and maximum size of conductors.

(C) Covers. Pull boxes, junction boxes, and conduit bodies must have a cover suitable for the conditions. Nonmetallic covers are permitted on any box, but metal covers are only permitted where they can be connected to an equipment grounding conductor of a type recognized in 250.118, in accordance with 250.110 [250.4(A)(3)]. Figure 314–42

314.29 Wiring to be Accessible. Boxes, conduit bodies, and handhole enclosures must be installed so that the wiring is accessible without removing any part of the building, sidewalks, paving, or earth. Figure 314–43

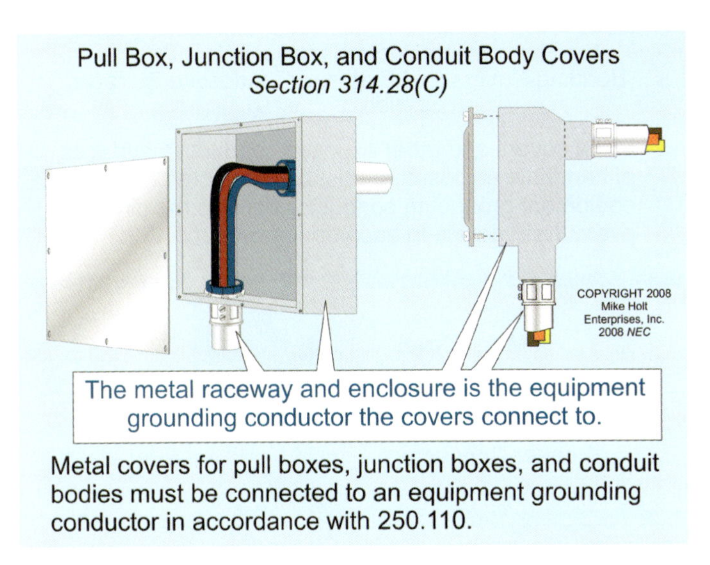

Figure 314–42

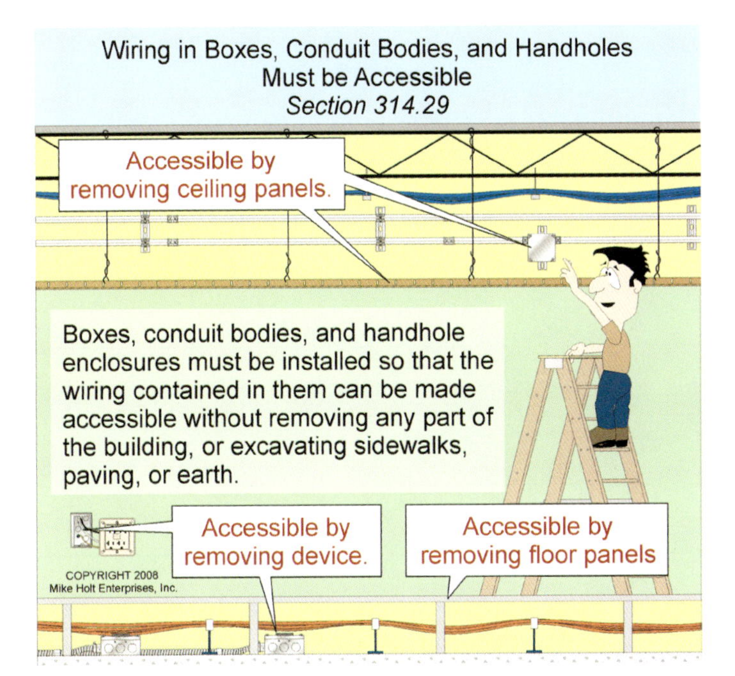

Figure 314–43

Exception: Listed boxes and handhole enclosures can be buried if covered by gravel, light aggregate, or noncohesive granulated soil, and their location is effectively identified and accessible for excavation.

314.30 Handhole Enclosures. Handhole enclosures must be identified for underground use, and be designed and installed to withstand all loads likely to be imposed on them. Figure 314–44

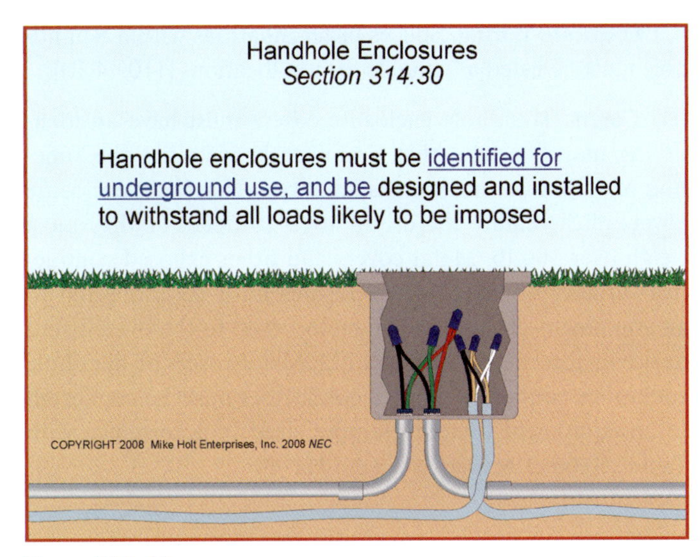

Figure 314–44

(A) Size. Handhole enclosures must be sized in accordance with 314.28(A). For handhole enclosures without bottoms, the measurement to the removable cover is taken from the end of the raceway or cable assembly. When the measurement is taken from the end of the raceway or cable assembly, the values in Table 312.6(A) for one wire to terminal can be used [314.28(A)(2) Ex].

(B) Mechanical Raceway and Cable Connection. Underground raceways and cables entering a handhole enclosure aren't required to be mechanically connected to the handhole enclosure. Figure 314–45

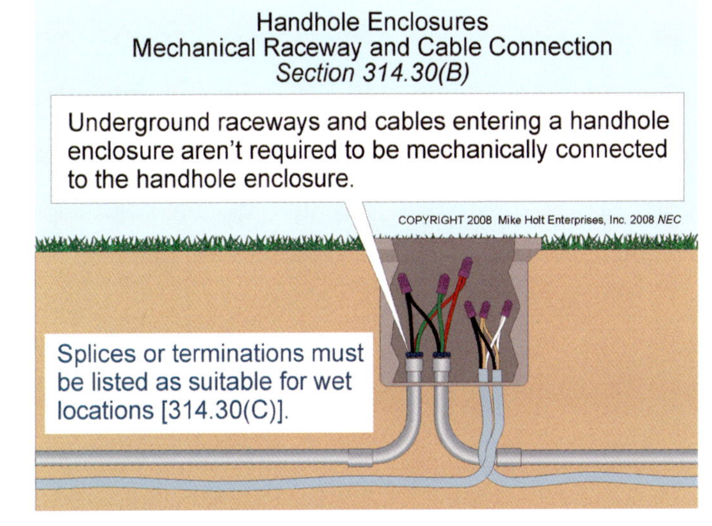

Figure 314–45

(C) Enclosure Wiring. Splices or terminations within a handhole must be listed as suitable for wet locations [110.14(B)].

(D) Covers. Handhole enclosure covers must have an identifying mark or logo that prominently identifies the function of the enclosure, such as "electric." Handhole enclosure covers must require the use of tools to open, or they must weigh over 100 lb. Metal covers and other exposed conductive surfaces of handhole enclosures must be connected to an equipment grounding conductor sized to the overcurrent device in accordance with 250.122. Metal covers of handhole enclosures containing service conductors must be connected to an equipment bonding jumper sized in accordance with Table 250.66 [250.102(C)]. Figure 314–46

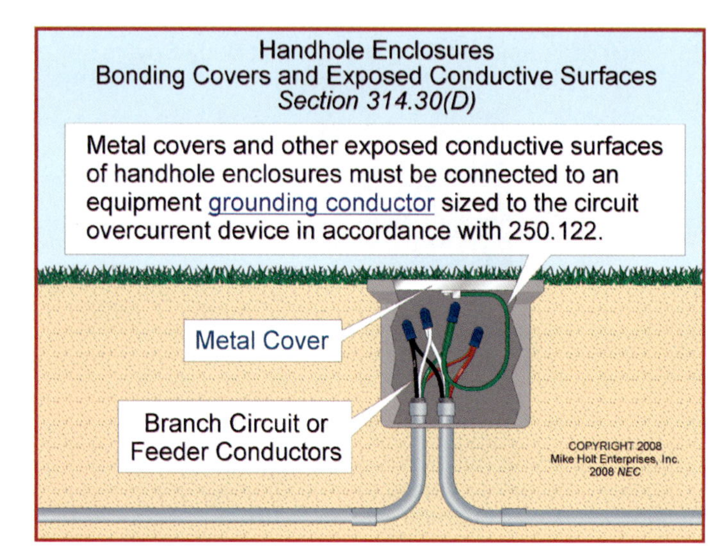

Figure 314–46

ARTICLE 314 Practice Questions

ARTICLE 314. OUTLET, DEVICE, PULL, AND JUNCTION BOXES; CONDUIT BODIES; AND HANDHOLE ENCLOSURES—PRACTICE QUESTIONS

1. Nonmetallic boxes can be used with _____.

   (a) nonmetallic cables
   (b) nonmetallic raceways
   (c) flexible cords
   (d) all of these

2. Boxes, conduit bodies, and fittings installed in wet locations shall be required to be listed for use in wet locations.

   (a) True
   (b) False

3. When counting the number of conductors in a box, a conductor running through the box with an unbroken loop or coil not less than twice the minimum length required for free conductors shall be counted as _____ conductor(s).

   (a) one
   (b) two
   (c) three
   (d) four

4. Where a luminaire stud or hickey is present in the box, a _____ volume allowance in accordance with Table 314.16(b) shall be made for each type of fitting, based on the largest conductor present in the box.

   (a) single
   (b) double
   (c) single allowance for each gang
   (d) none of these

5. Where one or more equipment grounding conductors enter a box, a _____ volume allowance in accordance with Table 314.16(b) shall be made, based on the largest equipment grounding conductor.

   (a) single
   (b) double
   (c) triple
   (d) none of these

6. In noncombustible walls or ceilings, the front edge of a box, plaster ring, extension ring, or listed extender employing a flush-type cover, shall be set back not more than _____ from the finished surface.

   (a) 1/8 in.
   (b) 1/4 in.
   (c) 3/8 in.
   (d) 1/2 in.

7. Surface extensions shall be made by mounting and mechanically securing an extension ring over the box, unless otherwise permitted.

   (a) True
   (b) False

8. Surface-mounted enclosures shall be _____.

   (a) rigidly and securely fastened in place
   (b) supported by cables that protrude from the box
   (c) supported by cable entries from the top and permitted to rest against the supporting surface
   (d) none of these

9. When mounting an enclosure in a finished surface, the enclosure shall be _____ secured to the surface by clamps, anchors, or fittings identified for the application.

   (a) temporarily
   (b) partially
   (c) never
   (d) rigidly

10. Enclosures not over 100 cu in. having threaded entries and not containing a device shall be considered to be adequately supported where _____ or more conduits are threaded wrenchtight into the enclosure and each conduit is secured within 3 ft of the enclosure.

    (a) one
    (b) two
    (c) three
    (d) none of these

11. Boxes used at luminaire or lampholder outlets in a ceiling shall be designed for the purpose and shall be required to support a luminaire weighing a minimum of _____.

    (a) 20 lb
    (b) 30 lb
    (c) 40 lb
    (d) 50 lb

12. Floor boxes shall be _____ specifically for this application.

    (a) identified
    (b) listed
    (c) marked
    (d) none of these

13. In straight pulls, the length of the box shall not be less than _____ times the trade size of the largest raceway.

    (a) six
    (b) eight
    (c) twelve
    (d) none of these

14. Listed boxes and handhole enclosures designed for underground installation can be directly buried when covered by _____, if their location is effectively identified and accessible.

    (a) concrete
    (b) gravel
    (c) noncohesive granulated soil
    (d) b or c

15. Conductors, splices or terminations in a handhole enclosure shall be listed as _____.

    (a) suitable for wet locations
    (b) suitable for damp locations
    (c) suitable for direct burial in the earth
    (d) none of these

16. Handhole enclosure covers shall require the use of tools to open, or they shall weigh over _____.

    (a) 45 lb
    (b) 70 lb
    (c) 100 lb
    (d) 200 lb

ARTICLE

320 Armored Cable (Type AC)

INTRODUCTION TO ARTICLE 320—ARMORED CABLE (TYPE AC)

Armored cable is an assembly of insulated conductors, 14 AWG through 1 AWG, individually wrapped within waxed paper and contained within a flexible spiral metal sheath. The outside appearance of armored cable looks like flexible metal conduit as well as metal-clad cable to the casual observer.

PART I. GENERAL

320.1 Scope. This article covers the use, installation, and construction specifications of armored cable, Type AC.

320.2 Definition.

Armored Cable (Type AC). A fabricated assembly of conductors in a flexible metal sheath with an internal bonding strip in intimate contact with the armor for its entire length. See 320.100. Figure 320–1

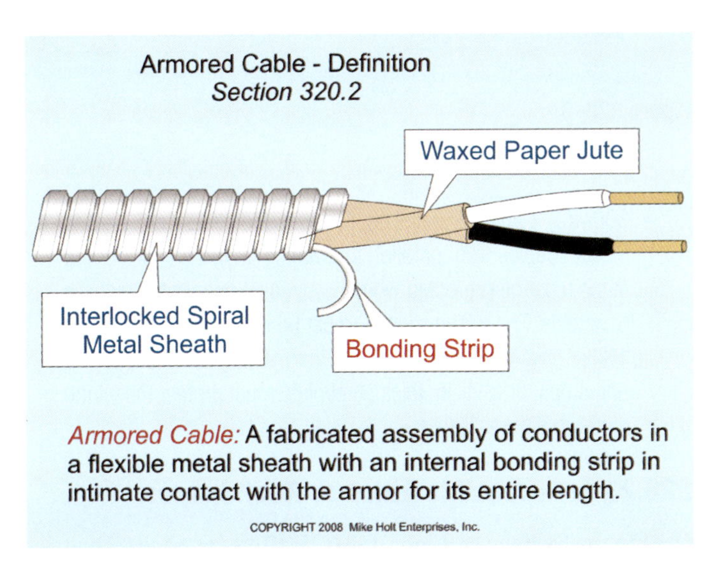

Figure 320–1

Author's Comment: The conductors are contained within a flexible metal sheath that interlocks at the edges, with an internal aluminum bonding strip, giving the cable an outside appearance similar to that of flexible metal conduit. Many electricians call this metal cable BX®. The advantages of any flexible cables, as compared to raceway wiring methods, are that there is no limit to the number of bends between terminations and the cable can be quickly installed.

PART II. INSTALLATION

320.10 Uses Permitted. Type AC cable is permitted only where not subject to physical damage in the following locations, and in other locations and conditions not prohibited by 320.12 or elsewhere in the *Code*:

(1) For feeders and branch circuits in both exposed and concealed work.

(2) Cable trays.

(3) Dry locations.

(4) Embedded in plaster or brick, except in damp or wet locations.

(5) Air voids where not exposed to excessive moisture or dampness.

FPN: The "Uses Permitted" is not an all-inclusive list, which indicates that other suitable uses are permitted if approved by the authority having jurisdiction.

Author's Comment: Type AC cable is also permitted to be installed above a suspended ceiling used for environmental air [300.22(C)(1)].

320.12 Uses Not Permitted. Type AC cable must not be installed in any of the following locations:

(1) Where subject to physical damage.

(2) In damp or wet locations.

(3) In air voids of masonry block or tile walls where such walls are exposed or subject to excessive moisture or dampness.

320.15 Exposed Work. Exposed Type AC cable must closely follow the surface of the building finish or running boards. Type AC cable run on the bottom of floor or ceiling joists must be secured at every joist, and must not be subject to physical damage. Figure 320–2

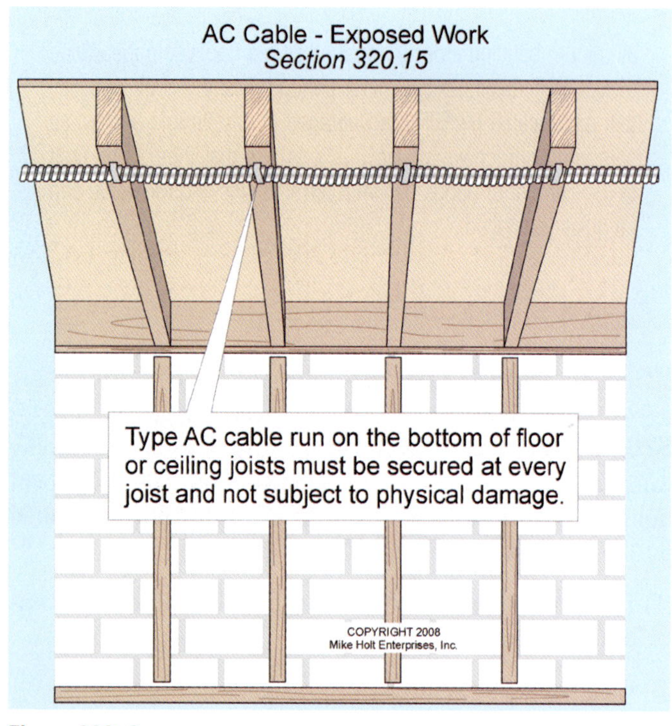

Figure 320–2

320.17 Through or Parallel to Framing Members. Type AC cable installed through, or parallel to, framing members or furring strips must be protected against physical damage from penetration by screws or nails by maintaining 1¼ in. of separation, or by installing a suitable metal plate in accordance with 300.4(A), (C), and (D):

Author's Comments:

- 300.4(A)(1) Drilling Holes in Wood Members. When drilling holes through wood framing members for cables, the edge of the holes must be not less than 1¼ in. from the edge of the wood member. Figure 320–3A

  If the edge of the hole is less than 1¼ in. from the edge, a 1⁄16 in. thick steel plate of sufficient length and width must be installed to protect the wiring method from screws and nails. Figure 320–3B

- 300.4(A)(2) Notching Wood Members. Where notching of wood framing members for cables is permitted by the building code, a 1⁄16 in. thick steel plate of sufficient length and width must be installed to protect the cables and raceways from screws and nails. Figure 320–3C

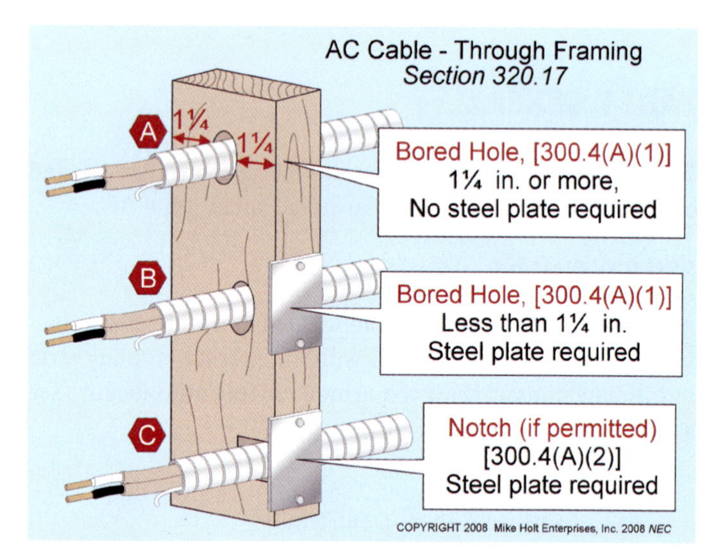

Figure 320–3

- 300.4(D) Cables Parallel to Framing Members and Furring Strips. Cables run parallel to framing members or furring strips must be protected where likely to be penetrated by nails or screws. The wiring method must be installed so it's at least 1¼ in. from the nearest edge of the framing members or furring strips, or a 1⁄16 in. thick steel plate must protect the wiring method. Figure 320–4

320.23 In Accessible Attics or Roof Spaces.

(A) On the Surface of Floor Joists, Rafters, or Studs. In attics and roof spaces that are accessible, substantial guards must protect the cables run across the top of floor joists, or across the face of rafters or studding within 7 ft of floor or floor joists. Where this space isn't accessible by permanent stairs or ladders, protection is required only within 6 ft of the nearest edge of the scuttle hole or attic entrance.

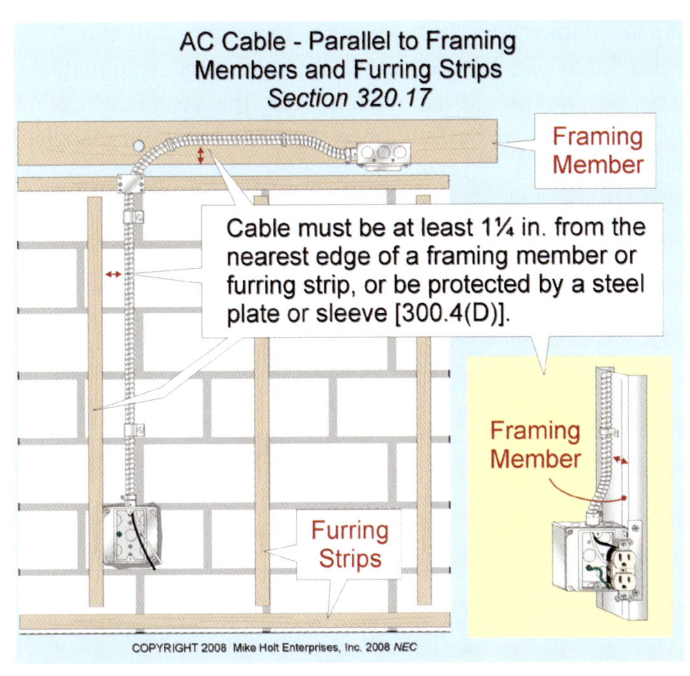

Figure 320–4

(B) Along the Side of Framing Members. When Type AC cable is run on the side of rafters, studs, or floor joists, no protection is required if the cable is installed and supported so the nearest outside surface of the cable or raceway is at least 1¼ in. from the nearest edge of the framing member where nails or screws are likely to penetrate [300.4(D)].

320.24 Bends. Type AC cable must not be bent in a manner that will damage the cable. This is accomplished by limiting bending of the inner edge of the cable to a radius of not less than five times the diameter of the cable.

320.30 Securing and Supporting.

(A) General. Type AC cable must be supported and secured by staples, cable ties, straps, hangers, or similar fittings, designed and installed so as not to damage the cable.

(B) Securing. Type AC cable must be secured within 12 in. of every outlet box, junction box, cabinet, or fitting, and at intervals not exceeding 4½ ft. Figure 320–5

> **Author's Comment:** Type AC cable is considered secured when installed horizontally through openings in wooden or metal framing members [320.30(C)].

(C) Supporting. Type AC cable must be supported at intervals not exceeding 4½ ft.

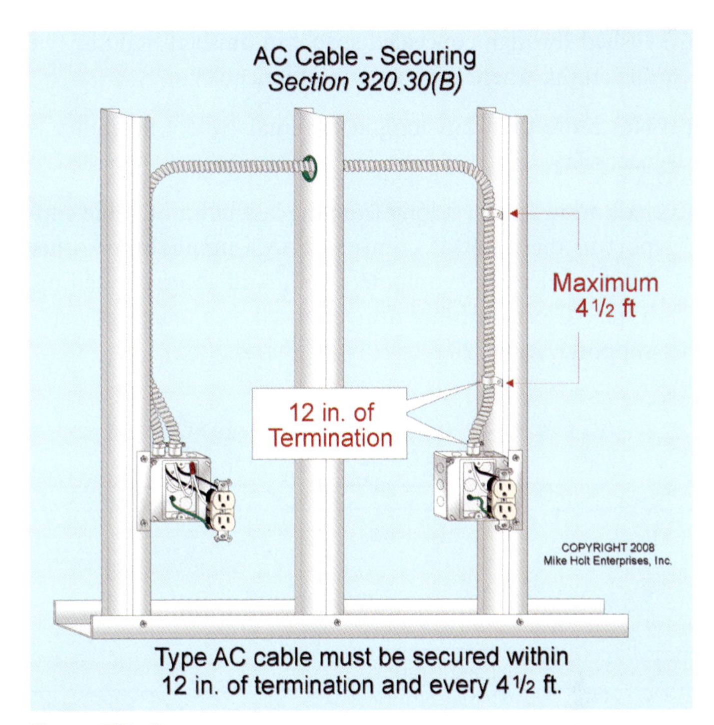

Figure 320–5

Cables installed horizontally through wooden or metal framing members are considered supported where support doesn't exceed 4½ ft. Figure 320–6

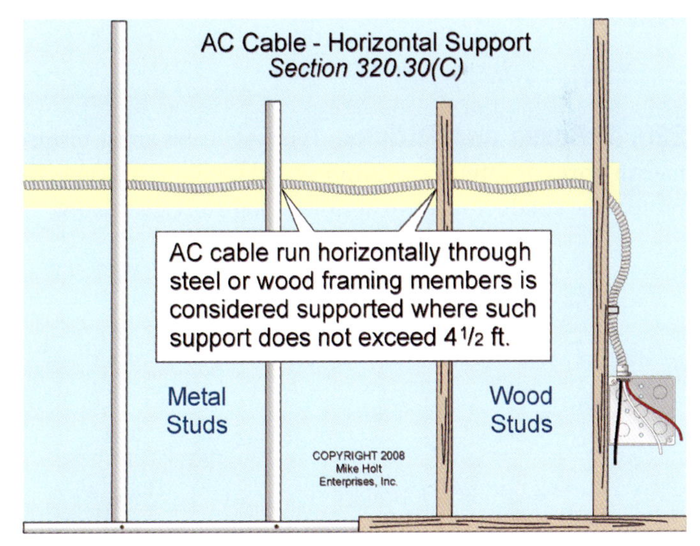

Figure 320–6

(D) Unsupported Cables. Type AC cable can be unsupported where the cable is:

(1) Fished through concealed spaces in finished buildings or structures, where support is impracticable; or

(2) Not more than 2 ft long at terminals where flexibility is necessary; or

(3) Not more than 6 ft long from the last point of cable support to the point of connection to a luminaire or other piece of electrical equipment within an accessible ceiling. Type AC cable fittings are permitted as a means of cable support. Figure 320–7

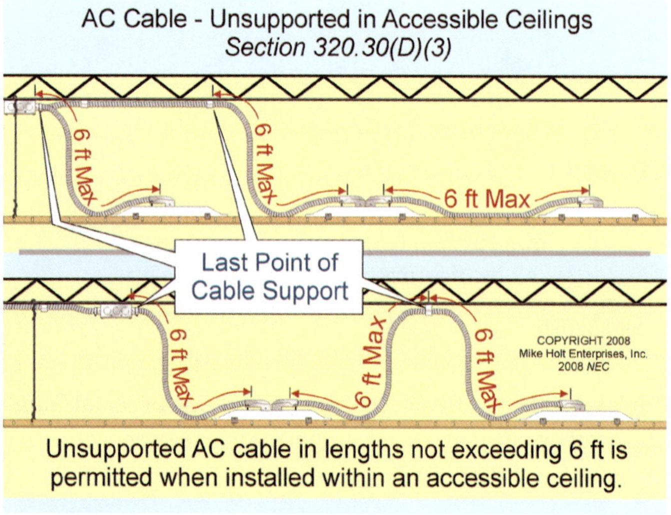

Figure 320–7

320.40 Boxes and Fittings. Type AC cable must terminate in boxes or fittings specifically listed for Type AC cable to protect the conductors from abrasion [300.15]. Figure 320–8

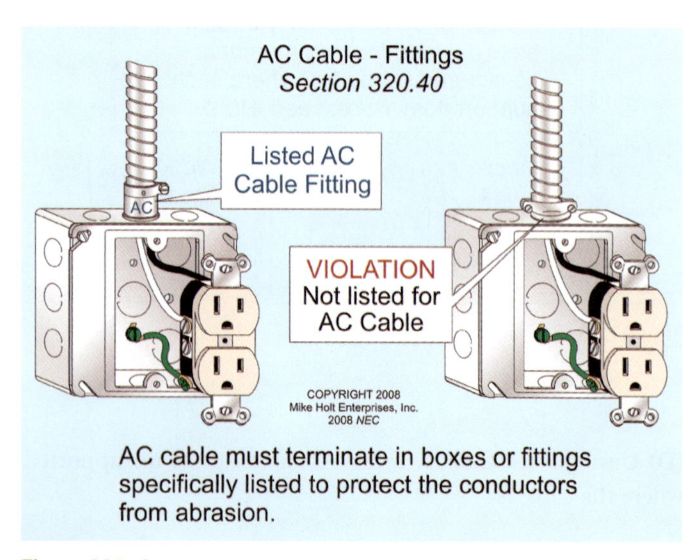

Figure 320–8

An insulating anti-short bushing, sometimes called a "red-head," must be installed at all Type AC cable terminations. The termination fitting must permit the visual inspection of the anti-short bushing once the cable has been installed. Figure 320–9

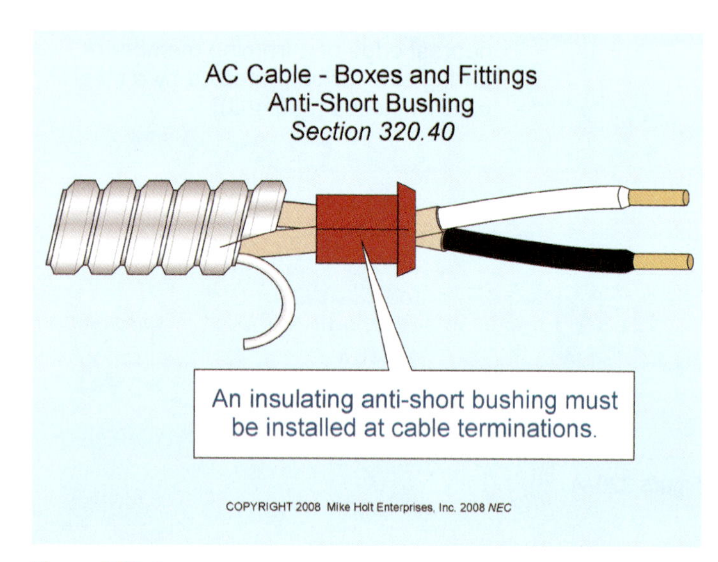

Figure 320–9

Author's Comments:

- The internal aluminum bonding strip within the cable serves no electrical purpose once outside the cable, and can be cut off, but many electricians use it to secure the anti-short bushing to the cable. See 320.108.
- Conductors 4 AWG and larger that enter an enclosure must be protected from abrasion during and after installation by a fitting that provides a smooth, rounded, insulating surface, such as an insulating bushing unless the design of the box, fitting, or enclosure provides equivalent protection in accordance with 300.4(G).

320.80 Conductor Ampacities. Conductor ampacity is calculated in accordance with 310.15, based on the insulation rating of the conductors, provided the adjusted or corrected ampacity doesn't exceed the temperature ratings of terminations and equipment [110.14(C)].

Question: What is the ampacity of four 12 THHN current-carrying conductors installed in Type AC cable?

(a) 18A (b) 24A (c) 27A (d) 30A

***Answer:** (b) 24A*

Table 310.16 ampacity if 12 THHN is 30A

Conductor Adjusted Ampacity = 0.80 [Table 310.15(B)(2)(a)]

Conductor Adjusted Ampacity = 30A x 0.80

Conductor Adjusted Ampacity = 24A

(A) Thermal Insulation. Type AC cable installed in thermal insulation must have conductors rated 90°C. The ampacity rating at 90°C can be used for conductor ampacity adjustment and/or correction, but the adjusted ampacity must not exceed the 60°C ampacity listed in Table 310.16. Figure 320–10

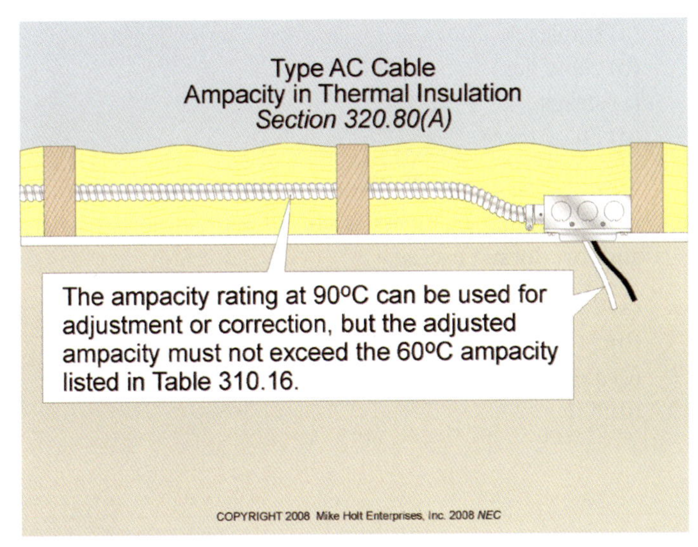

Figure 320–10

PART III. CONSTRUCTION SPECIFICATIONS

320.100 Construction. Type AC cable has an armor of flexible metal tape with an internal aluminum bonding strip in intimate contact with the armor for its entire length.

Author's Comments:

- The best method of cutting Type AC cable is to use a tool specially designed for the purpose, such as a rotary armor cutter.
- When cutting Type AC cable with a hacksaw, be sure to cut only one spiral of the cable and be careful not to nick the conductors; this is done by cutting the cable at an angle. Breaking the cable spiral (bending the cable very sharply), then cutting the cable with a pair of dikes isn't a good practice.

320.108 Equipment Grounding Conductor. Type AC cable must provide an adequate path for fault current as required by 250.4(A)(5) or 250.4(B)(4) to act as an equipment grounding conductor. Figure 320–11

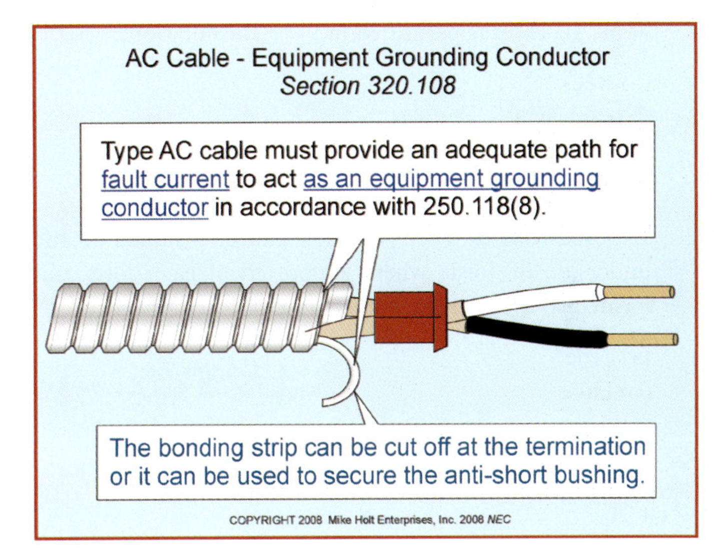

Figure 320–11

Author's Comments: The internal aluminum bonding strip isn't an equipment grounding conductor, but it allows the interlocked armor to serve as an equipment grounding conductor because it reduces the impedance of the armored spirals to ensure that a ground fault will be cleared. It's the combination of the aluminum bonding strip and the cable armor that creates the equipment grounding conductor. Once the bonding strip exits the cable, it can be cut off because it no longer serves any purpose. The effective ground-fault current path must be maintained by the use of fittings specifically listed for Type AC cable [320.40]. See 300.12, 300.15, and 300.100.

ARTICLE 320 Practice Questions

ARTICLE 320. ARMORED CABLE (TYPE AC)—PRACTICE QUESTIONS

1. Type AC cable is permitted in _____ installations.

   (a) wet
   (b) cable tray
   (c) exposed
   (d) b and c

2. Exposed runs of Type AC cable can be installed on the underside of joists where supported at each joist and located so they are not subject to physical damage.

   (a) True
   (b) False

3. When Type AC cable is run across the top of a floor joist in an attic without permanent ladders or stairs, substantial guard strips within _____ of the scuttle hole, or attic entrance, shall protect the cable.

   (a) 3 ft
   (b) 4 ft
   (c) 5 ft
   (d) 6 ft

4. Type AC cable can be supported and secured by _____.

   (a) staples
   (b) cable ties
   (c) straps
   (d) all of these

5. Type AC cable installed horizontally through wooden or metal framing members is considered supported where support doesn't exceed _____.

   (a) 2 ft
   (b) 3 ft
   (c) 4½ ft
   (d) 6 ft

Metal-Clad Cable (Type MC)

INTRODUCTION TO ARTICLE 330—METAL-CLAD CABLE (TYPE MC)

Metal-clad cable encloses insulated conductors in a metal sheath of either corrugated or smooth copper or aluminum tubing, or spiral interlocked steel or aluminum. The physical characteristics of Type MC cable make it a versatile wiring method that you can use in almost any location, and for almost any application. The most common Type MC cable is the interlocking type, which looks similar to armored cable or flexible metal conduit.

PART I. GENERAL

330.1 Scope. Article 330 covers the use, installation, and construction specifications of metal-clad cable.

330.2 Definition.

Metal-Clad Cable (Type MC). A factory assembly of insulated circuit conductors, with or without optical fiber members, enclosed in an armor of interlocking metal tape or a smooth or corrugated metallic sheath. Figure 330–1

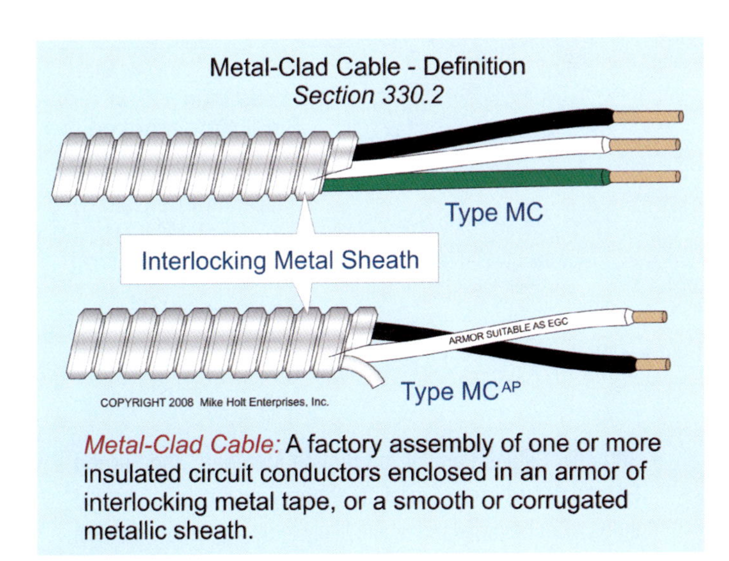

Figure 330–1

Author's Comment: Because the outer sheath of interlocked Type MC cable isn't listed as an equipment grounding conductor, it contains an equipment grounding conductor [330.108].

PART II. INSTALLATION

330.10 Uses Permitted.

(A) General Uses. Type MC cable is permitted only where not subject to physical damage, and in other locations and conditions not prohibited by 330.12, or elsewhere in the *Code*:

(1) In branch circuits, feeders and services.

(2) In power, lighting, control, and signal circuits.

(3) Indoors or outdoors.

(4) Exposed or concealed.

(5) Directly buried (if identified for the purpose).

(6) In a cable tray.

(7) In a raceway.

(8) As aerial cable on a messenger.

(9) In hazardous (classified) locations as permitted in 501.10(B), 502.10(B), and 503.10.

(10) Embedded in plaster or brick.

(11) In wet locations, if any the following are met:

a. The metallic covering is impervious to moisture.

b. A lead sheath or moisture-impervious jacket is provided under the metal covering.

c. The insulated conductors under the metallic covering are listed for use in wet locations and a corrosion-resistant jacket is provided over the metallic sheath.

(12) Where single-conductor cables are used, all ungrounded conductors and, where used, the neutral conductor must be grouped together to minimize induced voltage on the sheath [300.3(B)].

(B) Specific Uses. Type MC cable can be installed in compliance with Parts II and III of Article 725 and 770.133 as applicable, and in accordance with (1) through (4).

(1) Cable Tray. Type MC cable installed in a cable tray must comply with 392.3, 392.4, 392.6, and 392.8 through 392.80.

(2) Direct Buried. Direct-buried cables must be protected in accordance with 300.5.

(3) Installed as Service-Entrance Cable. Type MC cable is permitted for service entrances, when installed in accordance with 230.43.

(4) Installed Outside of Buildings or Structures. Type MC cable installed outside of buildings or structures must comply with 225.10, 396.10, and 396.12.

FPN: The "Uses Permitted" is not an all-inclusive list, which indicates that other suitable uses are permitted if approved by the authority having jurisdiction.

Author's Comment: Type MC cable is also permitted to be installed above a suspended ceiling used for environmental air [300.22(C)(1)].

330.12 Uses Not Permitted.

Type MC cable must not be used where:

(1) Subject to physical damage.

(2) Exposed to the destructive corrosive conditions in (a) or (b), unless the metallic sheath or armor is resistant or protected by material resistant to the conditions:

(a) Direct burial in the earth or embedded in concrete unless identified for the application.

(b) Exposed to cinder fills, strong chlorides, caustic alkalis, or vapors of chlorine or of hydrochloric acids.

330.17 Through or Parallel to Framing Members.

Type MC cable installed through or parallel to framing members or furring strips must be protected against physical damage from penetration of screws or nails by maintaining a 1¼ in. separation, or by installing a suitable metal plate in accordance with 300.4(A) and (D).

Author's Comments:

- 300.4(A)(1) Drilling Holes in Wood Members. When drilling holes through wood framing members for cables, the edge of the holes must be not less than 1¼ in. from the edge of the wood member. **Figure 330–2A**
- If the edge of the hole is less than 1¼ in. from the edge, a 1⁄16 in. thick steel plate of sufficient length and width must be installed to protect the wiring method from screws and nails. **Figure 330–2B**
- 300.4(A)(2) Notching Wood Members. Where notching of wood framing members for cables is permitted by the building code, a 1⁄16 in. thick steel plate of sufficient length and width must be installed to protect the cables and raceways from screws and nails. **Figure 330–2C**

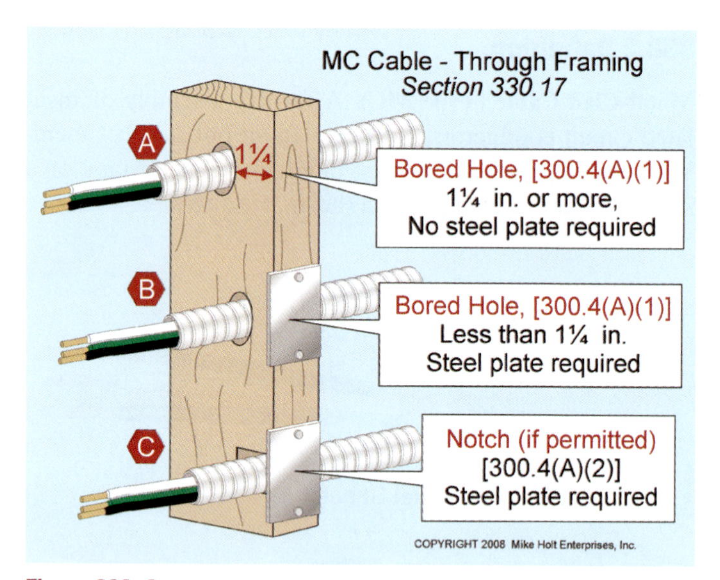

Figure 330–2

- **300.4(D) Cables Parallel to Framing Members and Furring Strips.** Cables run parallel to framing members or furring strips must be protected where likely to be penetrated by nails or screws. The wiring method must be installed so it's at least 1¼ in. from the nearest edge of the framing member or furring strips, or a 1⁄16 in. thick steel plate must protect it. **Figure 330–3**

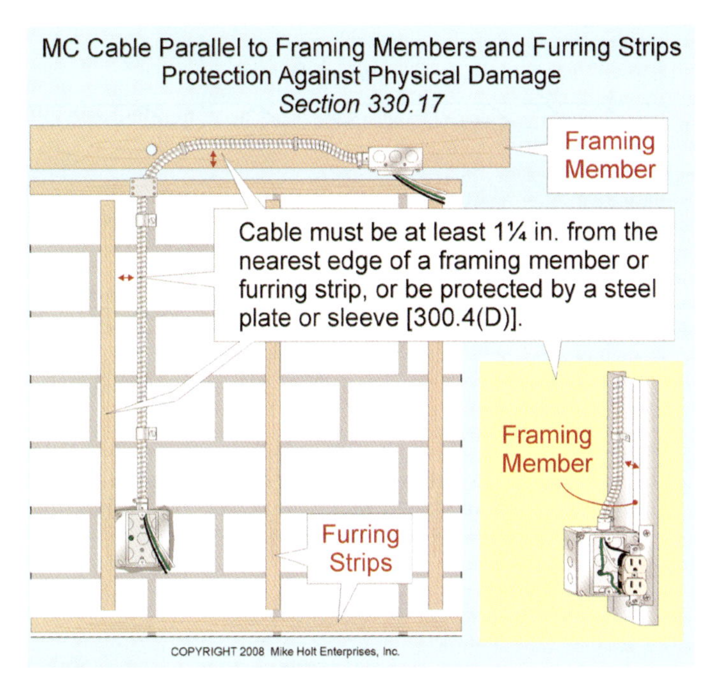

Figure 330–3

330.23 In Accessible Attics or Roof Spaces.

Type MC cable installed in accessible attics or roof spaces must comply with 320.23.

Author's Comments:

- On the Surface of Floor Joists, Rafters, or Studs. In attics and roof spaces that are accessible, substantial guards must protect cables run across the top of floor joists, or across the face of rafters or studding within 7 ft of floor or floor joists. Where this space isn't accessible by permanent stairs or ladders, protection is required only within 6 ft of the nearest edge of the scuttle hole or attic entrance [320.23(A)].
- Along the Side of Framing Members [320.23(B)]. When Type MC cable is run on the side of rafters, studs, or floor joists, no protection is required if the cable is installed and supported so the nearest outside surface of the cable or raceway is at least 1¼ in. from the nearest edge of the framing member where nails or screws are likely to penetrate [300.4(D)].

330.24 Bends.

Bends must be made so that the cable will not be damaged, and the radius of the curve of any bend at the inner edge of the cable must not be less than what is dictated in each of the following instances:

(A) Smooth-Sheath Cables. Smooth-sheath Type MC cables must not be bent so the bending radius of the inner edge of the cable is less than ten times the external diameter of the metallic sheath for cable up to ¾ in. in external diameter.

(B) Interlocked or Corrugated Sheath. Interlocked- or corrugated-sheath Type MC cable must not be bent so the bending radius of the inner edge of the cable is less than seven times the external diameter of the cable. Figure 330–4

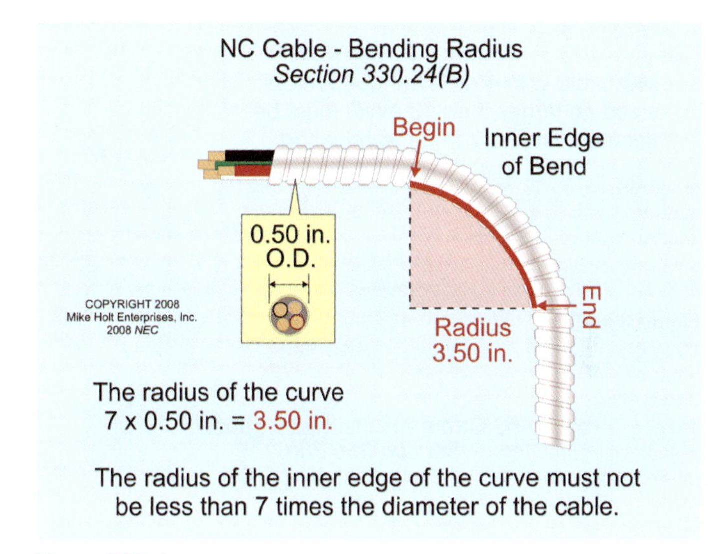

Figure 330–4

330.30 Securing and Supporting.

(A) General. Type MC cable must be supported and secured by staples, cable ties, straps, hangers, or similar fittings, designed and installed so as not to damage the cable.

(B) Securing. Type MC cable with four or less conductors sized no larger than 10 AWG, must be secured within 12 in. of every outlet box, junction box, cabinet, or fitting and at intervals not exceeding 6 ft. Figure 330–5

(C) Supporting. Type MC cable must be supported at intervals not exceeding 6 ft. Cables installed horizontally through wooden or metal framing members are considered secured and supported where such support doesn't exceed 6 ft intervals. Figure 330–6

(D) Unsupported Cables. Type MC cable can be unsupported where the cable is:

(1) Fished through concealed spaces in finished buildings or structures, where support is impracticable, or

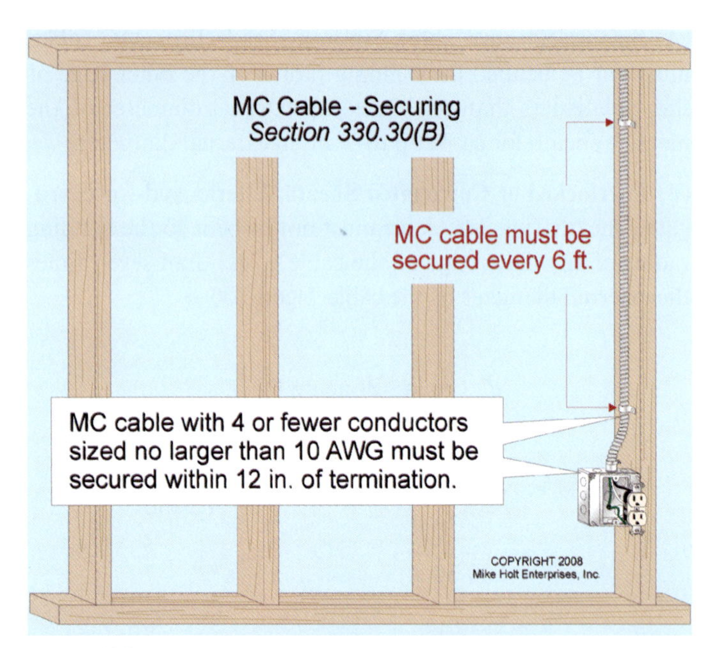

Figure 330–5

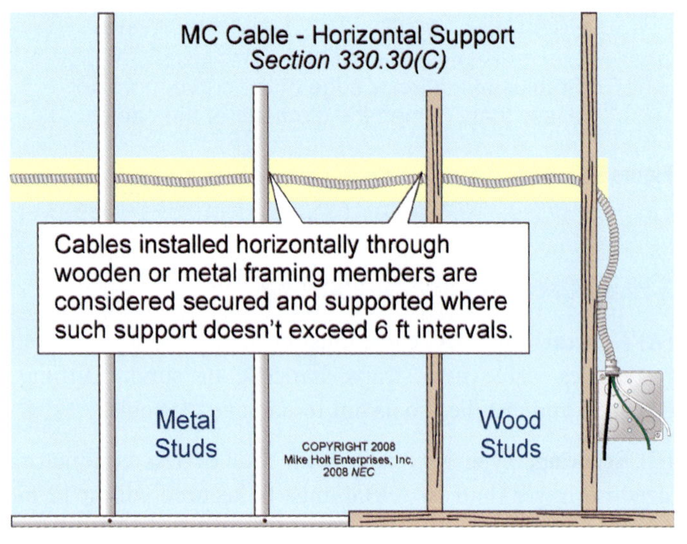

Figure 330–6

(2) Not more than 6 ft long from the last point of cable support to the point of connection to luminaires or other electrical equipment within an accessible ceiling. Type MC cable fittings are permitted as a means of cable support. Figure 330–7

330.40 Fittings. Fittings used to secure Type MC cable to boxes or other enclosures must be listed and identified for such use [300.15]. Figure 330–8

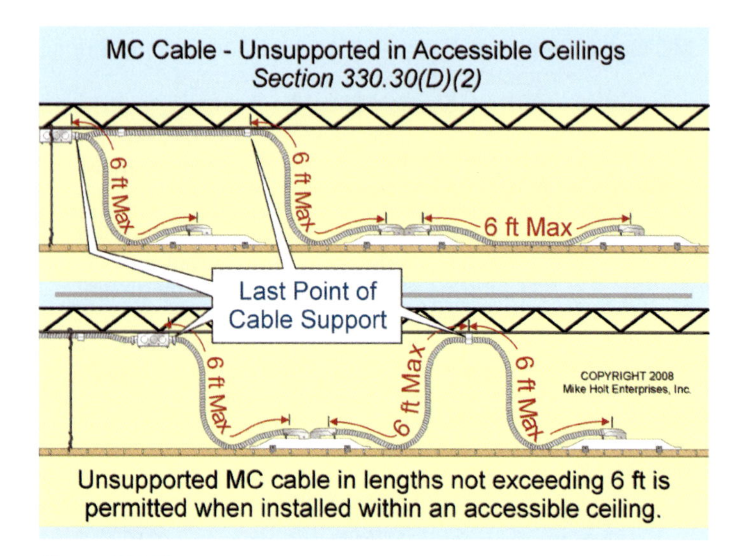

Figure 330–7

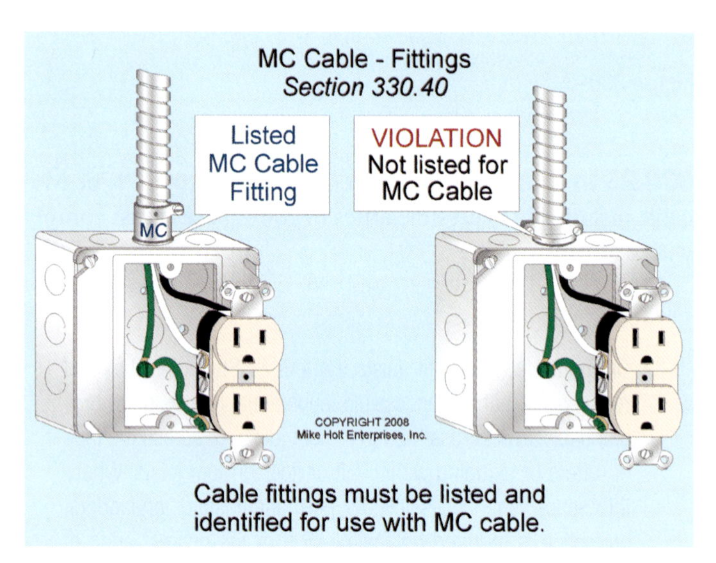

Figure 330–8

Author's Comments:

- The *NEC* doesn't require anti-short bushings (red heads) at the termination of Type MC cable, but if they are supplied it's considered by many to be a good practice to use them.
- Conductors 4 AWG and larger that enter an enclosure must be protected from abrasion during and after installation by a fitting that provides a smooth, rounded, insulating surface, such as an insulating bushing unless the design of the box, fitting, or enclosure provides equivalent protection in accordance with 300.4(G).

330.80 Conductor Ampacities. Conductor ampacity is calculated in accordance with 310.15, based on the insulation rating of the conductors, provided the adjusted or corrected ampacity doesn't exceed the temperature ratings of terminations and equipment [110.14(C)].

PART III. CONSTRUCTION SPECIFICATIONS

330.108 Equipment Grounding Conductor. Where Type MC cable is to serve as an equipment grounding conductor, it must comply with 250.118 and 250.122.

Author's Comment: The outer sheath of:

- Traditional interlocked Type MC cable is not permitted to serve as an equipment grounding conductor, therefore this cable must contain an insulated equipment grounding conductor in accordance with 250.118(1). **Figure 330–9**

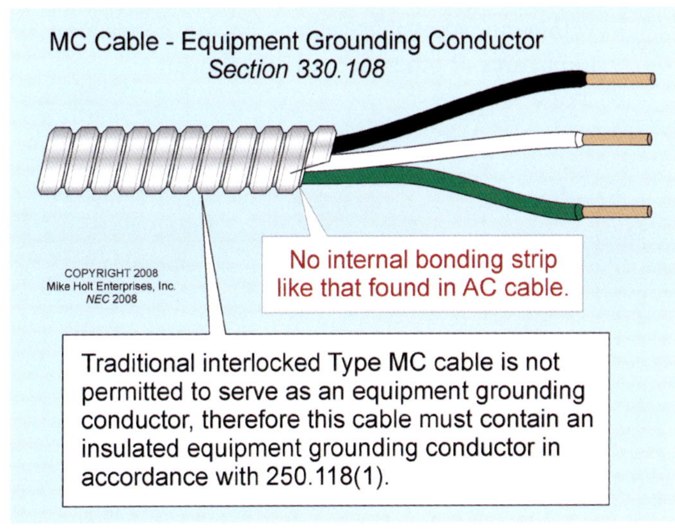

Figure 330–9

- Interlocked Type MC[AP] cable containing an aluminum grounding/bonding conductor running just below the metal armor is listed to serve an equipment grounding conductor [250.118(10)(a)]. **Figure 330–10**
- Smooth or corrugated-tube Type MC cable with or without an equipment grounding conductor [250.118(10)(b)].

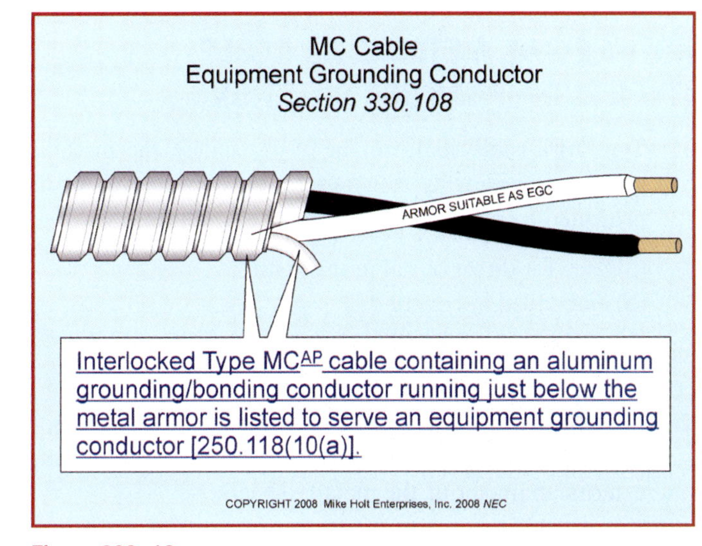

Figure 330–10

ARTICLE 330 Practice Questions

ARTICLE 330. METAL-CLAD CABLE (TYPE MC)—PRACTICE QUESTIONS

1. Type MC cable shall not be _____ unless the metallic sheath or armor is resistant to the conditions, or is protected by material resistant to the conditions.

   (a) used for direct burial in the earth
   (b) embedded in concrete
   (c) exposed to cinder fill
   (d) all of these

2. Bends made in interlocked or corrugated sheath Type MC cable shall have a radius of at least _____ times the external diameter of the metallic sheath.

   (a) 5
   (b) 7
   (c) 10
   (d) 12

3. Type MC cable shall be secured at intervals not exceeding _____.

   (a) 3 ft
   (b) 4 ft
   (c) 6 ft
   (d) 8 ft

4. Type MC cable can be unsupported where _____.

   (a) fished between concealed access points in finished buildings or structures and support is impracticable
   (b) not more than 2 ft in length at terminals where flexibility is necessary
   (c) not more than 6 ft from the last point of support within an accessible ceiling for the connection of luminaires or other electrical equipment
   (d) a or c

ARTICLE

334 Nonmetallic-Sheathed Cable (Types NM and NMC)

INTRODUCTION TO ARTICLE 334—NONMETALLIC-SHEATHED CABLE (TYPES NM AND NMC)

Nonmetallic-sheathed cable is flexible, inexpensive, and easily installed. It provides very limited physical protection of the conductors, so the installation restrictions are strict. Its low cost and relative ease of installation make it a common wiring method for residential and commercial branch circuits.

PART I. GENERAL

334.1 Scope. Article 334 covers the use, installation, and construction specifications of nonmetallic-sheathed cable.

334.2 Definition.

Nonmetallic-Sheathed Cable (Types NM and NMC). A wiring method that encloses two or more insulated conductors, 14 AWG through 2 AWG, within a nonmetallic jacket.

- NM has insulated conductors enclosed within an overall nonmetallic jacket.
- NMC has insulated conductors enclosed within an overall, corrosion resistant, nonmetallic jacket. Figure 334–1

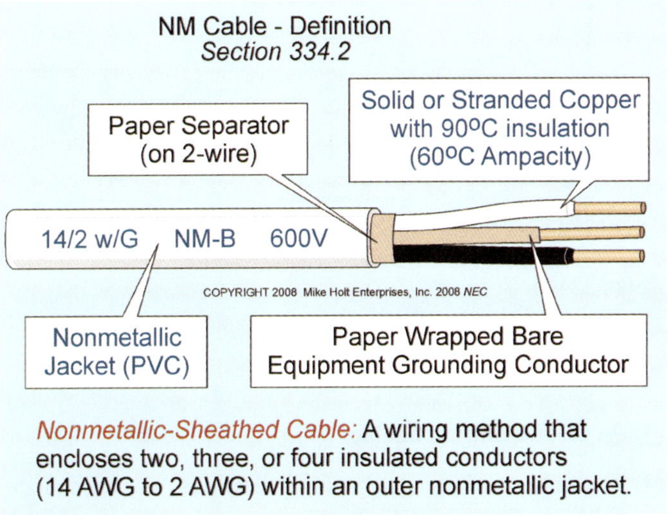

Figure 334–1

Author's Comment: It's the generally accepted practice in the electrical industry to call Type NM cable "Romex®," a registered trademark of the Southwire Company.

334.6 Listed. Types NM and NMC cables must be listed.

PART II. INSTALLATION

334.10 Uses Permitted.

Type NM and Type NMC cables can be used in the following:

(1) One- and two-family dwellings. Figure 334–2

(2) Multifamily dwellings permitted to be of Types III, IV, and V construction. Figure 334–3

(3) Other structures permitted to be of Types III, IV, and V construction, except as prohibited in 334.12. Cables must be concealed within walls, floors, or ceilings that provide a thermal barrier of material with at least a 15-minute finish rating, as identified in listings of fire-rated assemblies. Figure 334–4

Author's Comment: See the definition of "Concealed" in Article 100.

FPN No. 1: Building constructions are defined in NFPA 220, Standard on Types of Building Construction, the applicable building code, or both.

FPN No. 2: See Annex E for the determination of building types [NFPA 220, Table 3-1].

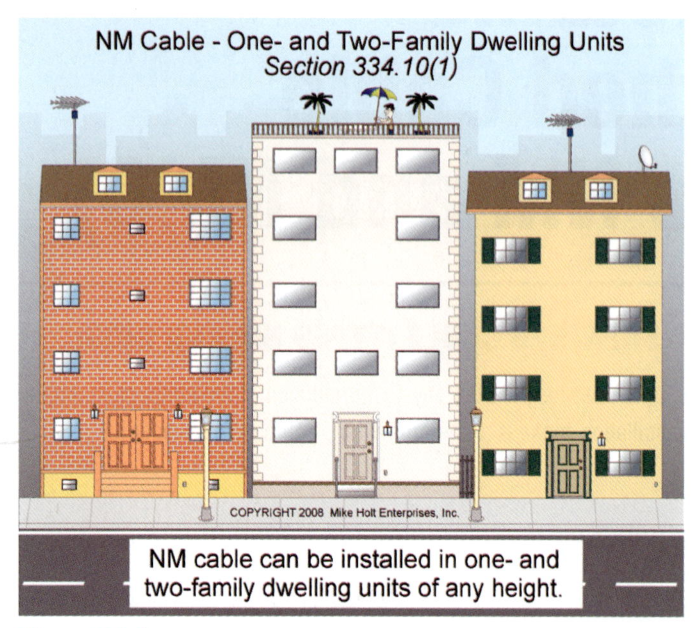

Figure 334–2

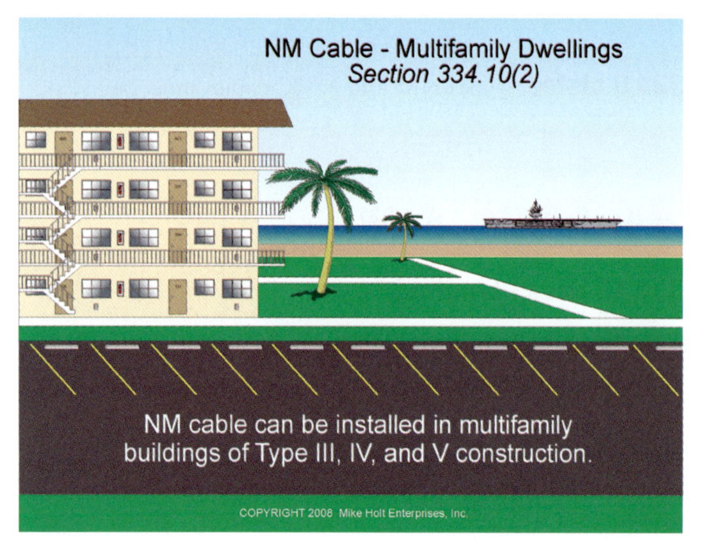

Figure 334–3

(4) Cable trays, where the cables are identified for this use.

FPN: See 310.10 for temperature limitation of conductors.

334.12 Uses Not Permitted.

(A) Types NM, NMC, and NMS.

(1) In any dwelling or structure not specifically permitted in 334.10(1), (2), and (3).

Exception: NM, NMC, and NMS cable is permitted in Type I and II construction when installed within a raceway.

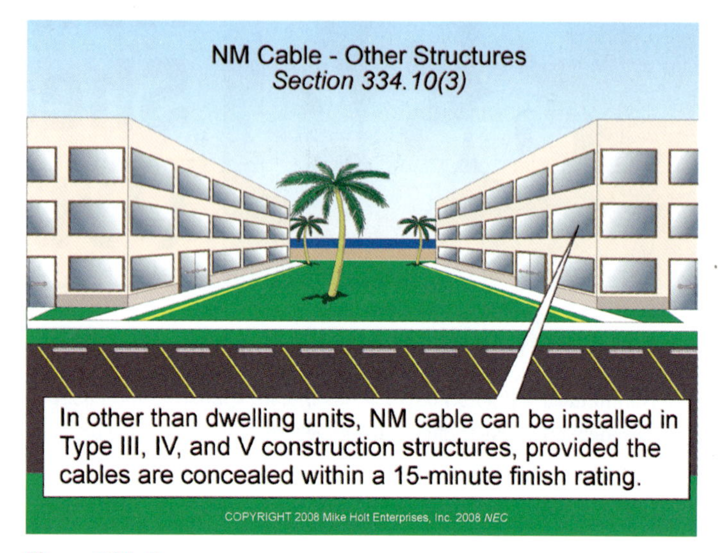

Figure 334–4

(2) Exposed in dropped or suspended ceilings in other than one- and two-family and multifamily dwellings. Figure 334–5

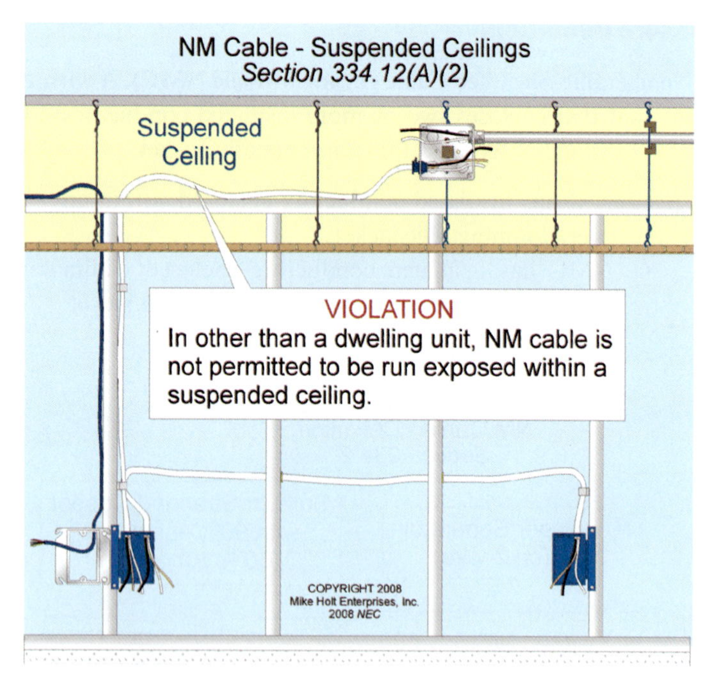

Figure 334–5

(3) As service-entrance cable.

(4) In commercial garages having hazardous (classified) locations, as defined in 511.3.

(5) In theaters and similar locations, except where permitted in 518.4(B).

(6) In motion picture studios.

(7) In storage battery rooms.

(8) In hoistways, or on elevators or escalators.

(9) Embedded in poured cement, concrete, or aggregate.

(10) In any hazardous (classified) location, except where permitted by the following:

a. 501.10(B)(3)
b. 502.10(B)(3)
c. 504.20

(B) Type NM. Type NM cables must not be used under the following conditions, or in the following locations:

(1) Where exposed to corrosive fumes or vapors.

(2) Where embedded in masonry, concrete, adobe, fill, or plaster.

(3) In a shallow chase in masonry, concrete, or adobe and covered with plaster, adobe, or similar finish.

(4) In wet or damp locations.

Author's Comment: Type NM cable isn't permitted in ducts, plenums, or other environmental air spaces [300.22], or for wiring in patient care areas [517.13].

334.15 Exposed.

(A) Surface of the Building. Exposed Type NM cable must closely follow the surface of the building.

(B) Protected from Physical Damage. Nonmetallic-sheathed cable must be protected from physical damage by rigid metal conduit, intermediate metal conduit, Schedule 80 PVC conduit [352.10(F)], electrical metallic tubing, guard strips, or other means approved by the authority having jurisdiction.

Author's Comment: When installed in a raceway, the cable must be protected from abrasion by a fitting installed on the end of the raceway [300.15(C)].

Type NMC cable installed in shallow chases in masonry, concrete, or adobe, must be protected against nails or screws by a steel plate not less than 1/16 in. thick [300.4(F)] and covered with plaster, adobe, or similar finish.

Author's Comment: Where Type NM cable is installed in a metal raceway, the raceway isn't required to be connected to an equipment grounding conductor [250.86 Ex 2 and 300.12 Ex].

(C) In Unfinished Basements and Crawl Spaces. Where Type NM cable is run at angles with joists in unfinished basements and crawl spaces, it's permissible to secure cables containing conductors not smaller than two 6 AWG or three 8 AWG conductors directly to the lower edges of the joists. Smaller cables must be run through bored holes in joists or on running boards. Figure 334–6

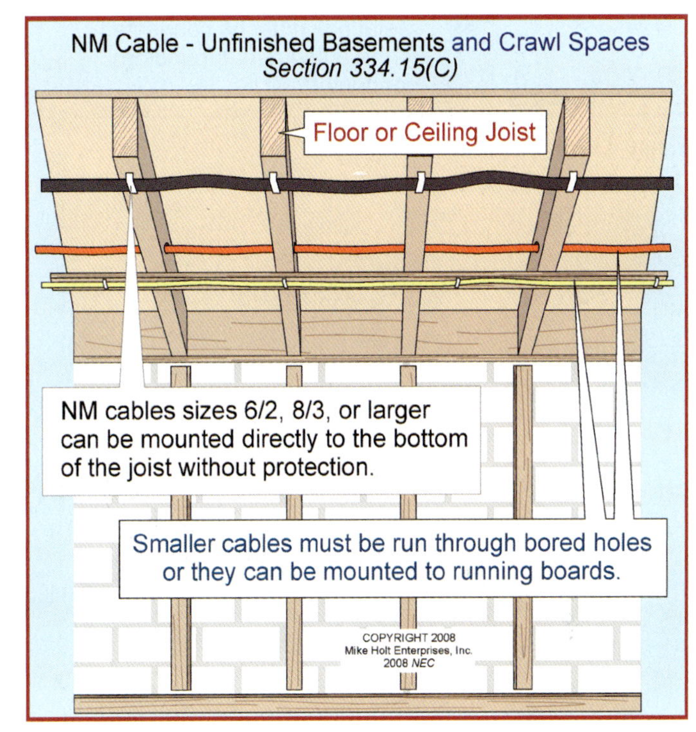

Figure 334–6

Type NM cable installed on a wall of an unfinished basement or crawl space subject to physical damage must be protected in accordance with 300.4, or be installed in a raceway with a nonmetallic bushing or adapter at the point where the cable enters the raceway, and the cable must be secured within 12 in. of the point where the cable enters the raceway. Figure 334–7

334.17 Through or Parallel to Framing Members.

Type NM cable installed through or parallel to framing members or furring strips must be protected against physical damage from penetration by screws or nails by 1¼ in. of separation or by a suitable metal plate [300.4(A) and (D)]. Figures 334–8 and 334–9

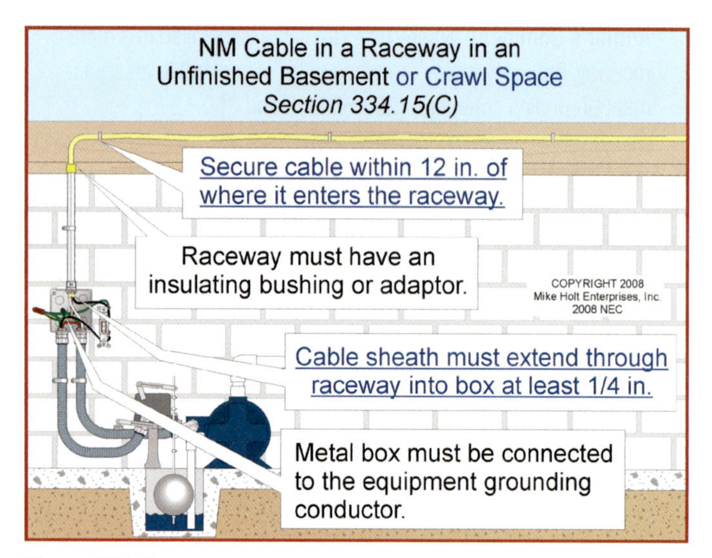

Figure 334–7

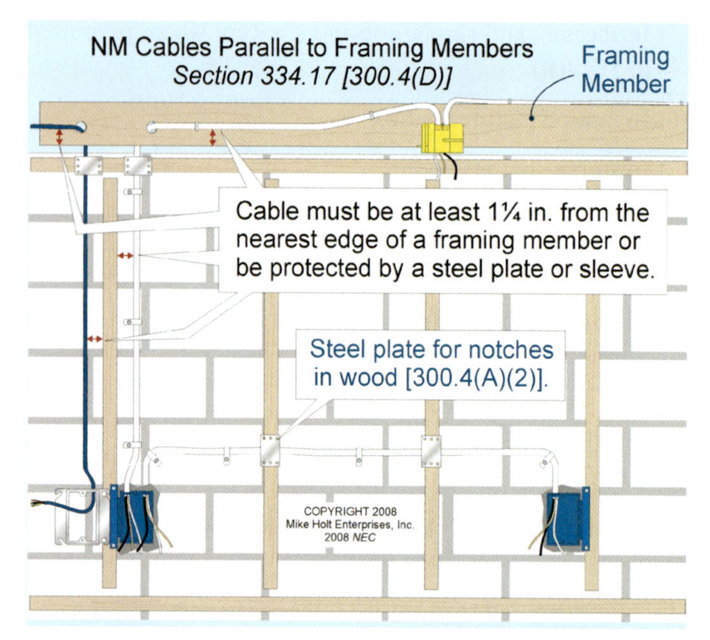

Figure 334–9

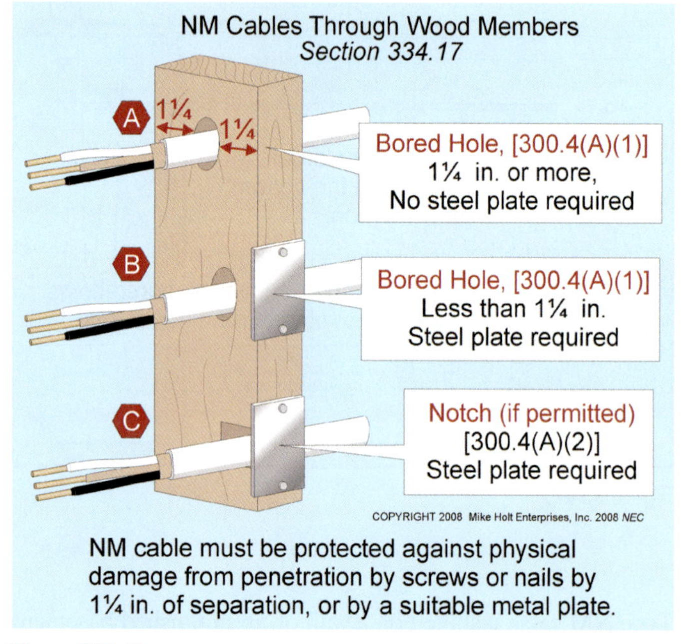

Figure 334–8

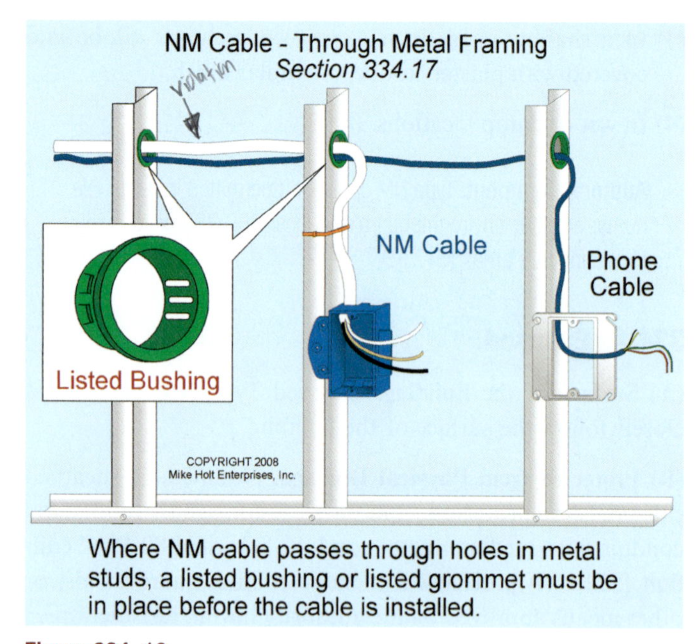

Figure 334–10

Where Type NM cable passes through holes in metal studs, a listed bushing or listed grommet is required [300.4(B)(1)] to be in place before the cable is installed. Figure 334–10

334.23 Attics and Roof Spaces. Type NM cable installed in accessible attics or roof spaces must comply with 320.23.

Author's Comments:

- On the Surface of Floor Joists , Rafters, or Studs. In attics and roof spaces that are accessible, substantial guards must protect cables run across the top of floor joists, or across the face of rafters or studding within 7 ft of floor or floor joists. Where this space isn't accessible by permanent stairs or ladders, protection is required only within 6 ft of the nearest edge of the scuttle hole or attic entrance [320.23(A)].

- Along the Side of Framing Members [320.23(B)]. When Type NM cable is run on the side of rafters, studs, or floor joists, no protection is required if the cable is installed and supported so the nearest outside surface of the cable or raceway is at least 1¼ in. from the nearest edge of the framing member where nails or screws are likely to penetrate [300.4(D)].

334.24 Bends. When the cable is bent, it must not be damaged. The radius of the curve of the inner edge of any bend must not be less than five times the diameter of the cable. Figure 334–11

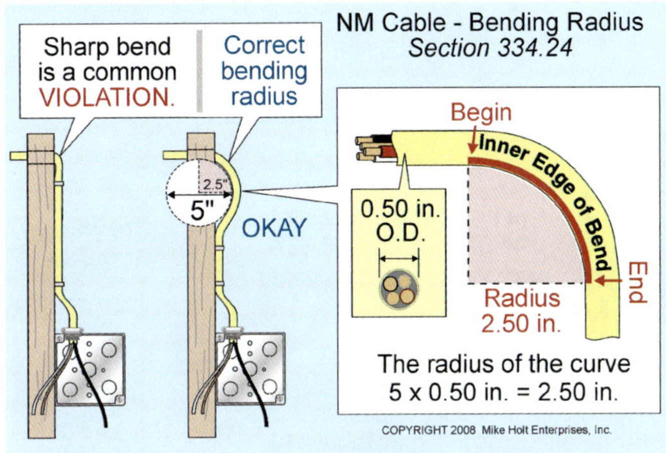

Figure 334–11

334.30 Securing and Supporting. Staples, straps, cable ties, hangers, or similar fittings must secure Type NM cable in a manner that will not damage the cable. Type NM cable must be secured within 12 in. of every box, cabinet, enclosure, or termination fitting, except as permitted by 314.17(C) Ex or 312.5(C) Ex, and at intervals not exceeding 4½ ft. Two-wire (flat) Type NM cable is not permitted to be stapled on edge. Figure 334–12

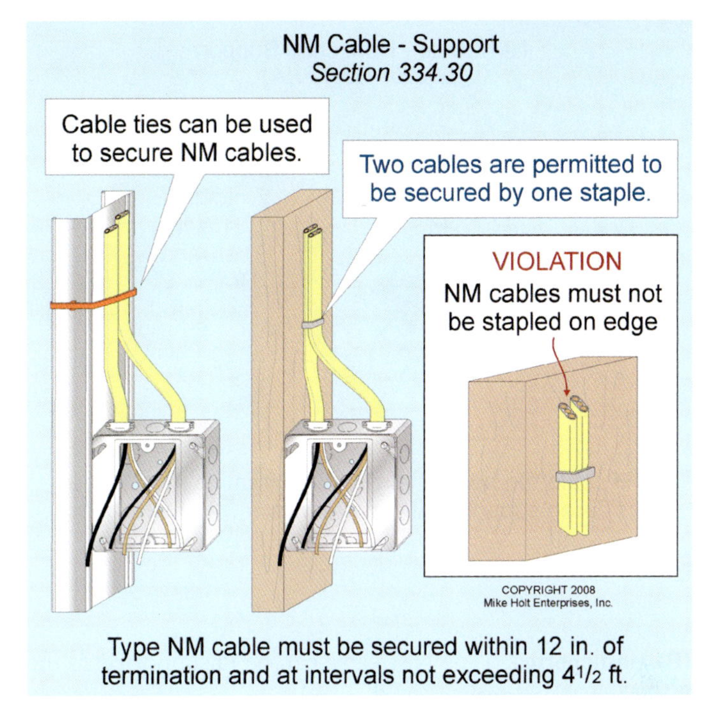

Figure 334–12

Type NM cable installed in a raceway isn't required to be secured within the raceway. Figure 334–13

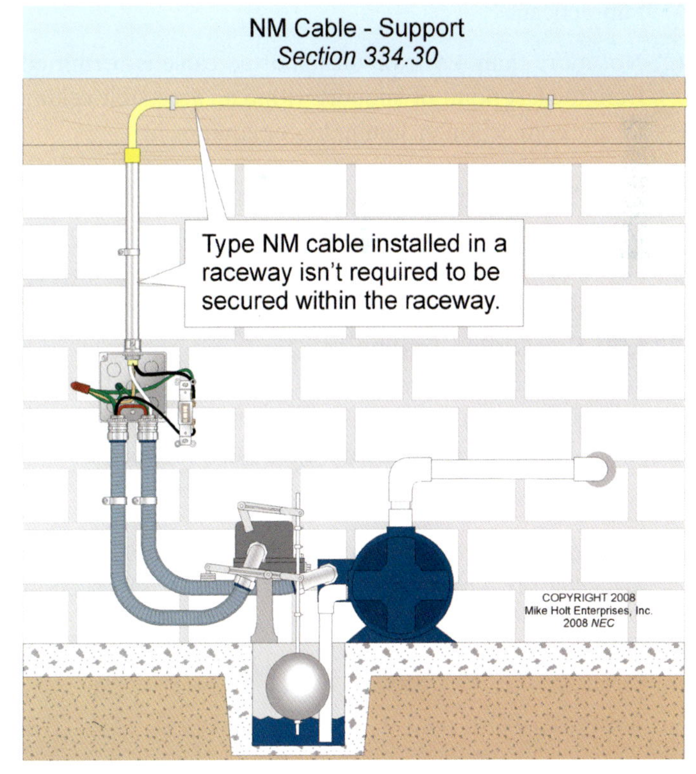

Figure 334–13

(A) Horizontal Runs. Type NM cable installed horizontally in bored or punched holes in wood or metal framing members, or notches in wooden members is considered secured and supported, but the cable must be secured within 1 ft of termination. Figure 334–14

> **FPN:** See 314.17(C) for support where nonmetallic boxes are used.

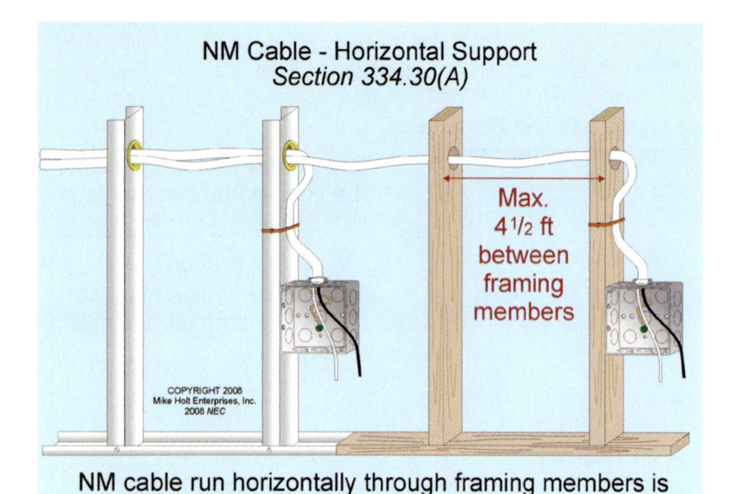

Figure 334–14

(B) Unsupported. Type NM cable can be unsupported in the following situations:

(1) Where Type NM cable is fished between concealed access points in finished buildings or structures, and support is impracticable.

(2) Not more than 4½ ft of unsupported cable is permitted from the last point of support within an accessible ceiling for the connection of luminaires or equipment.

Author's Comment: Type NM cable isn't permitted as a wiring method above accessible ceilings, except in dwellings [334.12(A)(2)].

334.80 Conductor Ampacity. Conductor ampacity used for adjustment or correction as per 310.15 is based on the 90ºC conductor insulation rating, provided the adjusted or corrected ampacity doesn't exceed that for a 60°C rated conductor.

Question: *What size Type NM cable is required to supply a 10 kW, 240V, single-phase fixed space heater with a 3A blower motor? The terminals are rated 75°C.* Figure 334–15

(a) 2 AWG *(b) 4 AWG* *(c) 6 AWG* *(d) 8 AWG*

Answer: *(b) 4 AWG*

Step 1: Determine the total load in amperes:

I = VA/E
I = 10,000W/240V + 3A
I = 44.67A

Step 2: Conductor and Protection Size [424.3(B)]. Size the ungrounded conductors and overcurrent device at no less than 125 percent of the total heating load.

Conductor/Protection Size = Load x 1.25
Conductor/Protection Size = 44.67A x 1.25
Conductor/Protection Size = 56A

According to Table 310.16, a 4 AWG conductor rated 70A at 60°C would be required to be protected with a 60A overcurrent device [240.6(A)].

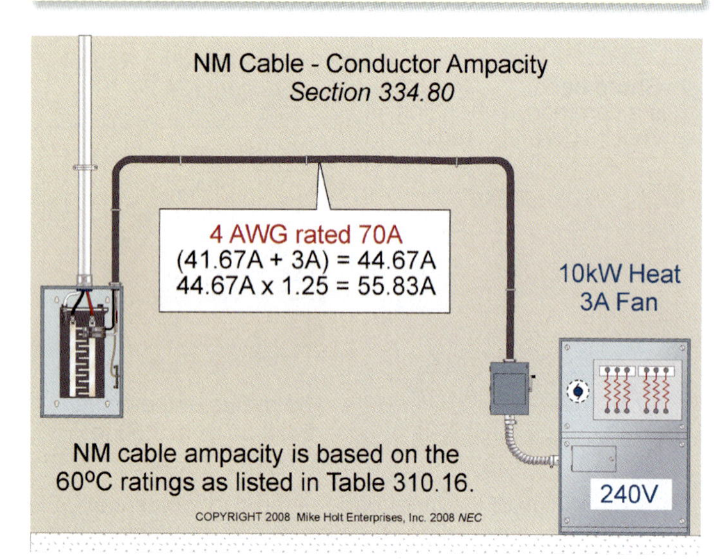

Figure 334–15

Where multiple Type NM cables pass through the same wood framing opening to be fire- or draft-stopped using thermal insulation, caulking, or sealing foam, the allowable ampacity of each conductor must be adjusted in accordance with Table 310.15(B)(2)(a). Figure 334–16

Author's Comment: This requirement has no effect on conductor sizing if you bundle no more than nine current-carrying 14 or 12 AWG conductors together. For example, three 14/2 cables and one 14/3 cable (nine current-carrying 14 THHN conductors) are bundled together in a dry location, the ampacity for each conductor (25A at 90°C, Table 310.16) is adjusted by a 70 percent adjustment factor [Table 310.15(B)(2)(a)].

Adjusted Conductor Ampacity = 25A x 0.70
Adjusted Conductor Ampacity = 17.50A

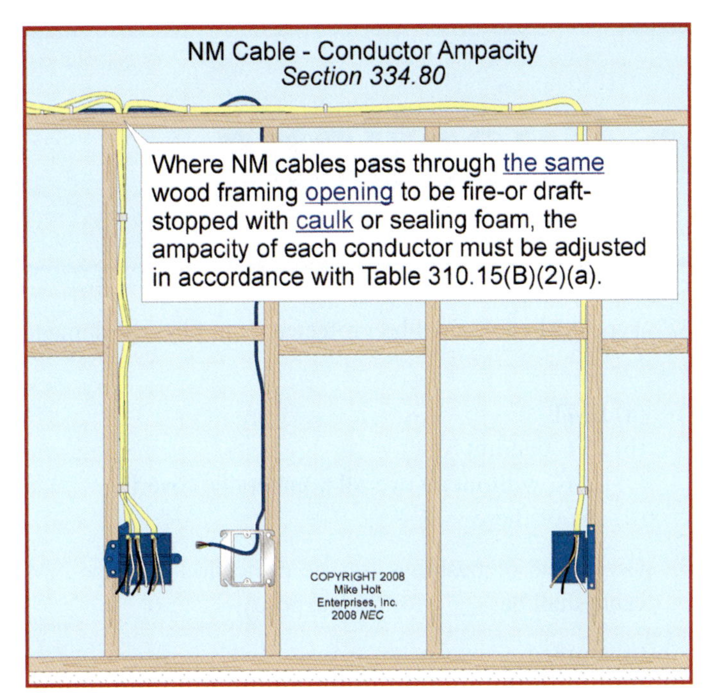

Figure 334–16

Where multiple Type NM cables are installed in contact with thermal insulation without maintaining spacing between cables, the allowable ampacity of each conductor must be adjusted in accordance with Table 310.15(B)(2)(a). Figure 334–17

Author's Comment: This requirement has no effect on conductor sizing if you bundle fewer than nine current-carrying conductors together. For example, if three 12/2 cables and one 12/3 cable (nine current-carrying 12 THHN conductors) are bundled together, the ampacity for each conductor (30A at 90°C, Table 310.16) is adjusted by a 70 percent adjustment factor [Table 310.15(B)(2)(a)].

Adjusted Conductor Ampacity = 30A x 0.70
Adjusted Conductor Ampacity = 21A

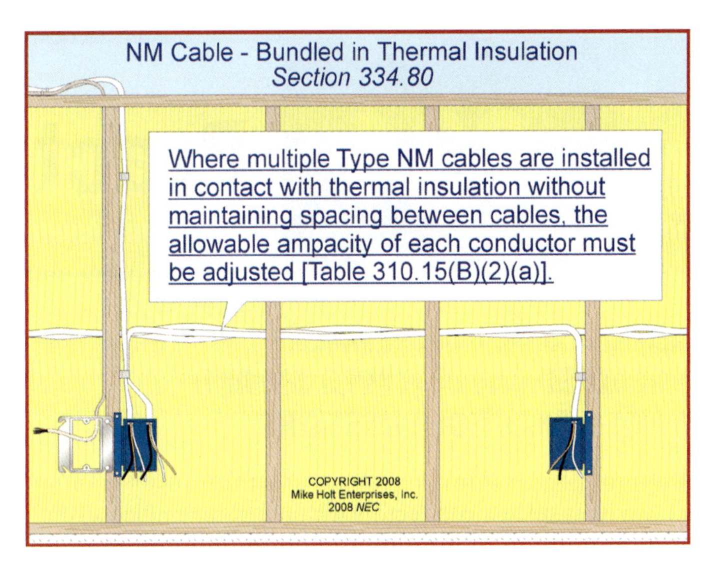

Figure 334–17

PART III. CONSTRUCTION SPECIFICATIONS

334.100 Construction. The outer cable sheath of Type NM cable must be constructed with nonmetallic material.

334.104 Conductors. The conductors must be 14 AWG through 2 AWG copper, or 12 AWG through 2 AWG aluminum or copper-clad aluminum.

334.108 Equipment Grounding Conductor. Type NM cable must have an insulated, covered or bare equipment grounding conductor.

334.112 Insulation. NM conductor insulation must be rated 90°C (194°F).

ARTICLE 334 Practice Questions

ARTICLE 334. NONMETALLIC-SHEATHED CABLE (TYPES NM AND NMC)—PRACTICE QUESTIONS

1. Type _____ cable is a wiring method that encloses two or more insulated conductors within a nonmetallic jacket.

   (a) AC
   (b) MC
   (c) NM
   (d) b and c

2. Type NM cable can be installed as open runs in dropped or suspended ceilings in other than one- and two-family and multifamily dwellings.

   (a) True
   (b) False

3. Type NM cable shall be protected from physical damage by _____.

   (a) EMT
   (b) PVC conduit
   (c) RMC without an overall nonmetallic covering
   (d) any of these

4. Grommets or bushings for the protection of Type NM cable shall be _____ for the purpose.

   (a) marked
   (b) approved
   (c) identified
   (d) listed

5. Flat Type NM cables shall not be stapled on edge.

   (a) True
   (b) False

Service-Entrance Cable (Types SE and USE)

INTRODUCTION TO ARTICLE 338—SERVICE-ENTRANCE CABLE (TYPES SE AND USE)

Service-entrance cable is a single conductor or multiconductor assembly with or without an overall moisture-resistant covering. This cable is used primarily for services not over 600V, but can also be used for feeders and branch circuits as limited by this article.

PART I. GENERAL

338.1 Scope. Article 338 covers the use, installation, and construction specifications of service-entrance cable, Types SE and USE.

338.2 Definitions.

Service-Entrance Cable. Service-entrance cable is a single or multiconductor assembly, with or without an overall covering, used primarily for services not over 600V. Figure 338–1

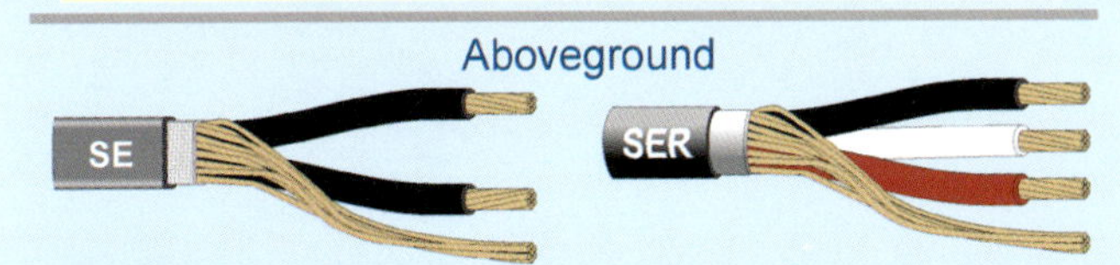

Figure 338–1

Type SE. SE and SER cables have a flame-retardant, moisture-resistant covering and are permitted only in aboveground installations. These cables are permitted for branch circuits or feeders when installed in accordance with 338.10(B).

> **Author's Comment:** SER cable is SE cable with an insulated neutral, resulting in three insulated conductors with an uninsulated equipment grounding conductor. SER cable is round, while 2-wire SE cable is flat.

Type USE. USE cable is identified as a wiring method permitted for underground use. Its covering is moisture resistant, but not required to be flame retardant.

> **Author's Comment:** USE cable is not permitted to be run indoors [338.10(B)], except single-conductor USE dual rated as RHH/RHW.

PART II. INSTALLATION

338.10 Uses Permitted.

(A) Service-Entrance Conductors. Service-entrance cable used as service-entrance conductors must be installed in accordance with Article 230.

(B) Branch Circuits or Feeders.

(2) Uninsulated Conductor. SE cable is permitted for use where the insulated conductors are used for circuit wiring, and the uninsulated conductor is only used for equipment grounding. Figure 338–2

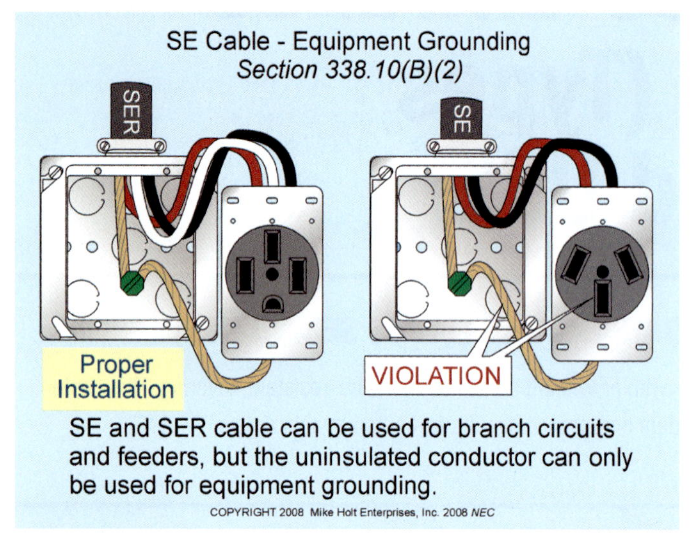

Figure 338–2

(3) Temperature Limitations. SE cable must not be subjected to conductor temperatures exceeding its insulation rating.

(4) Installation Methods for Branch Circuits and Feeders. SE cable used for branch circuits or feeders must comply with (a) and (b).

(a) Interior Installations. SE cable used for interior branch circuit or feeder wiring must be installed in accordance with the same requirements as Type NM Cable—Article 334. Figure 338–3

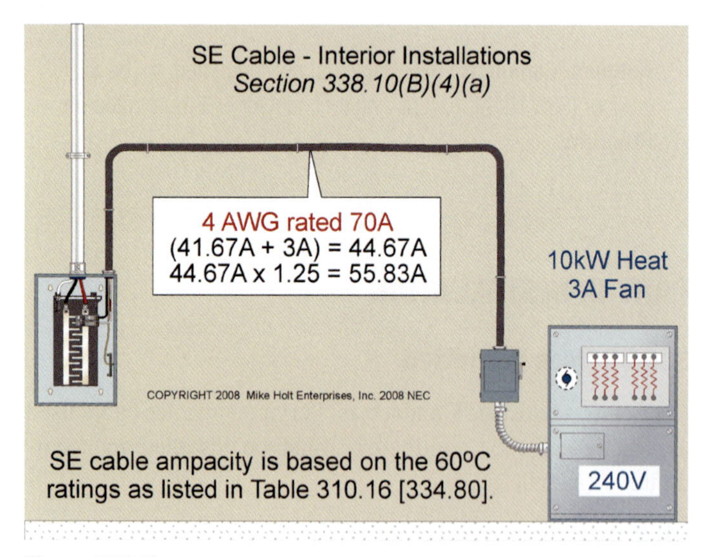

Figure 338–3

Author's Comment: Where Type SE cable is used for interior wiring, its ampacity is limited to the 60°C insulation rating listed in Table 310.16 [334.80].

CAUTION: *Underground service-entrance cable (USE) must not be used for interior wiring because it doesn't have flame-retardant insulation.*

(b) Exterior Installations. Service-entrance cable used for exterior branch circuits or feeders must be installed in accordance with Part I of Article 225.

338.12 Uses Not Permitted.

(A) Service-Entrance Cable. SE cable is not permitted under the following conditions or locations.

(1) Where subject to physical damage unless protected in accordance with 230.50(A).

(2) Underground with or without a raceway.

(B) Underground Service-Entrance Cable. USE cable is not permitted under the following conditions or locations:

(1) Interior wiring

(2) Above ground, except where protected against physical damage in accordance with 300.5(D).

338.24 Bends. Bends in cable must be made so the protective coverings of the cable are not damaged, and the radius of the curve of the inner edge is at least five times the diameter of the cable.

ARTICLE 338 Practice Questions

ARTICLE 338. SERVICE-ENTRANCE CABLE (TYPES SE AND USE)—PRACTICE QUESTIONS

1. Type _____ cable is an assembly primarily for services not over 600V.

   (a) NM
   (b) TC
   (c) SE
   (d) none of these

2. Type SE cables can be used for branch circuits or feeders where the insulated conductors are used for circuit wiring and the uninsulated conductor is used only for _____ purposes.

   (a) grounded connection
   (b) equipment grounding
   (c) remote control and signaling
   (d) none of these

3. Type USE cable is not permitted for _____ wiring.

   (a) underground
   (b) interior
   (c) a or b
   (d) a and b

ARTICLE

340 Underground Feeder and Branch-Circuit Cable (Type UF)

INTRODUCTION TO ARTICLE 340—UNDERGROUND FEEDER AND BRANCH-CIRCUIT CABLE (TYPE UF)

UF cable is a moisture-, fungus-, and corrosion-resistant cable suitable for direct burial in the earth.

PART I. GENERAL

340.1 Scope. Article 340 covers the use, installation, and construction specifications of underground feeder and branch-circuit cable, Type UF.

340.2 Definition.

Underground Feeder and Branch-Circuit Cable (Type UF). A factory assembly of insulated conductors with an integral or an overall covering of nonmetallic material suitable for direct burial in the earth.

Author's Comments:

- UF cable is a moisture-, fungus-, and corrosion-resistant cable suitable for direct burial in the earth. It comes in sizes 14 AWG through 4/0 AWG [340.104]. The covering of multiconductor Type UF cable is molded plastic that encases the insulated conductors.
- Because the covering of Type UF cable encapsulates the insulated conductors, it's difficult to strip off the outer jacket to gain access to the conductors, but this covering provides excellent corrosion protection. Be careful not to damage the conductor insulation or cut yourself when you remove the outer cover.

340.6 Listing Requirements. Type UF cable must be listed.

PART II. INSTALLATION

340.10 Uses Permitted.

(1) Underground, in accordance with 300.5.

(2) As a single conductor in the same trench or raceway with circuit conductors.

(3) As interior or exterior wiring in wet, dry, or corrosive locations.

(4) As Type NM cable, when installed in accordance with Article 334.

(5) For solar photovoltaic systems, in accordance with 690.31.

(6) As single-conductor cables for nonheating leads for heating cables, as provided in 424.43.

(7) Supported by cable trays.

340.12 Uses Not Permitted.

(1) Services [230.43].

(2) Commercial garages [511.3].

(3) Theaters [520.5].

(4) Motion picture studios [530.11].

(5) Storage battery rooms [Article 480].

(6) Hoistways [Article 620].

(7) Hazardous (classified) locations, except as otherwise permitted.

(8) Embedded in concrete.

(9) Exposed to direct sunlight unless identified.

(10) Where subject to physical damage. Figure 340–1

(11) Overhead messenger-supported wiring.

> **Author's Comment:** UF cable isn't permitted in ducts, plenums, other environmental air spaces [300.22], or in patient care areas [517.13].

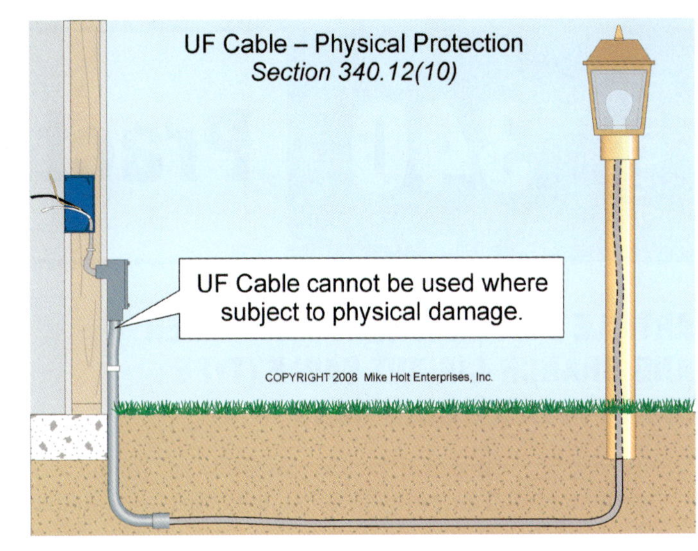

Figure 340–1

340.24 Bends. Bends in cables must be made so that the protective coverings of the cable are not damaged, and the radius of the curve of the inner edge must not be less than five times the diameter of the cable.

340.80 Ampacity. The ampacity of conductors contained in UF cable is based on 60°C insulation rating listed in Table 310.16.

340.112 Insulation. The conductors of UF cable must be one of the moisture-resistant types listed in Table 310.13(A) suitable for branch-circuit wiring. Where installed as a substitute wiring method for Type NM cable, the conductor insulation must be rated 90°C (194°F).

ARTICLE 340 Practice Questions

ARTICLE 340. UNDERGROUND FEEDER AND BRANCH-CIRCUIT CABLE (TYPE UF)—PRACTICE QUESTIONS

1. Type _____ cable is a factory assembly of conductors with an overall covering of nonmetallic material suitable for direct burial in the earth.

    (a) NM
    (b) UF
    (c) SE
    (d) TC

2. Type UF cable can be used for service conductors.

    (a) True
    (b) False

3. Type UF cable shall not be used in _____.

    (a) motion picture studios
    (b) storage battery rooms
    (c) hoistways
    (d) all of these

4. Type UF cable shall not be used where subject to physical damage.

    (a) True
    (b) False

ARTICLE

342 Intermediate Metal Conduit (Type IMC)

INTRODUCTION TO ARTICLE 342—INTERMEDIATE METAL CONDUIT (TYPE IMC)

Intermediate metal conduit (IMC) is a circular metal raceway with an outside diameter equal to that of rigid metal conduit. The wall thickness of intermediate metal conduit is less than that of rigid metal conduit (RMC), so it has a greater interior cross-sectional area for containing conductors. Intermediate metal conduit is lighter and less expensive than rigid metal conduit, but it can be used in all of the same locations as rigid metal conduit. Intermediate metal conduit also uses a different steel alloy that makes it stronger than rigid metal conduit, even though the walls are thinner. Intermediate metal conduit is manufactured in both galvanized steel and aluminum; the steel type is much more common.

PART I. GENERAL

342.1 Scope. Article 342 covers the use, installation, and construction specifications of intermediate metal conduit and associated fittings.

342.2 Definition.

Intermediate Metal Conduit (Type IMC). A listed steel raceway of circular cross section that can be threaded with integral or associated couplings. It's listed for the installation of electrical conductors, and is used with listed fittings to provide electrical continuity.

Author's Comment: The type of steel from which intermediate metal conduit is manufactured, the process by which it's made, and the corrosion protection applied are all equal, or superior, to that of rigid metal conduit.

342.6 Listing Requirements. Intermediate metal conduit and its associated fittings, such as elbows and couplings, must be listed.

PART II. INSTALLATION

342.10 Uses Permitted.

(A) All Atmospheric Conditions and Occupancies. Intermediate metal conduit is permitted in all atmospheric conditions and occupancies.

(B) Corrosion Environments. Intermediate metal conduit, elbows, couplings, and fittings can be installed in concrete, in direct contact with the earth, or in areas subject to severe corrosive influences where provided with corrosion protection and judged suitable for the condition in accordance with 300.6.

(D) Wet Locations. Support fittings, such as screws, straps, and so forth, installed in a wet location must be made of corrosion-resistant material, or be protected by corrosion-resistant coatings in accordance with 300.6.

CAUTION: *Supplementary coatings for corrosion protection have not been investigated by a product testing and listing agency, and these coatings are known to cause cancer in laboratory animals. There is a documented case where an electrician was taken to the hospital for lead poisoning after using a supplemental coating product (asphalted paint) in a poorly ventilated area. As with all products, be sure to read and follow all product instructions, including material data safety sheets, particularly when petroleum-based chemicals (volatile organic compounds) may be in the material.*

342.14 Dissimilar Metals. Where practical, contact with dissimilar metals should be avoided to prevent the deterioration of the metal because of galvanic action. Aluminum fittings and enclosures, however, are permitted with steel intermediate metal conduit.

342.20 Trade Size.

(A) Minimum. Intermediate metal conduit smaller than trade size ½ must not be used.

(B) Maximum. Intermediate metal conduit larger than trade size 4 must not be used.

342.22 Number of Conductors. The number of conductors in IMC is not permitted to exceed the percentage fill specified in Table 1, Chapter 9. Raceways must be large enough to permit the installation and removal of conductors without damaging the conductor insulation. When all conductors in a raceway are the same size and insulation, the number of conductors permitted can be found in Annex C for the raceway type.

Question: *How many 10 THHN conductors can be installed in trade size 1 IMC?*

(a) 12 *(b) 14* *(c) 16* *(d) 18*

Answer: *(d) 18 conductors [Annex C, Table C4]*

Author's Comment: See 300.17 for additional examples on how to size raceways when conductors aren't all the same size.

Cables can be installed in intermediate metal conduit, as long as the number of cables does not exceed the allowable percentage fill specified in Table 1, Chapter 9.

342.24 Bends. Raceway bends must not be made in any manner that would damage the raceway, or significantly change its internal diameter (no kinks). The radius of the curve of the inner edge of any field bend must not be less than shown in Table 2, Chapter 9.

Author's Comment: This is usually not a problem, because benders are made to comply with this table. However, when using a hickey bender (short-radius bender), be careful not to over-bend the raceway.

342.26 Number of Bends (360°). To reduce the stress and friction on conductor insulation, the maximum number of bends (including offsets) between pull points must not exceed 360°. Figure 342–1

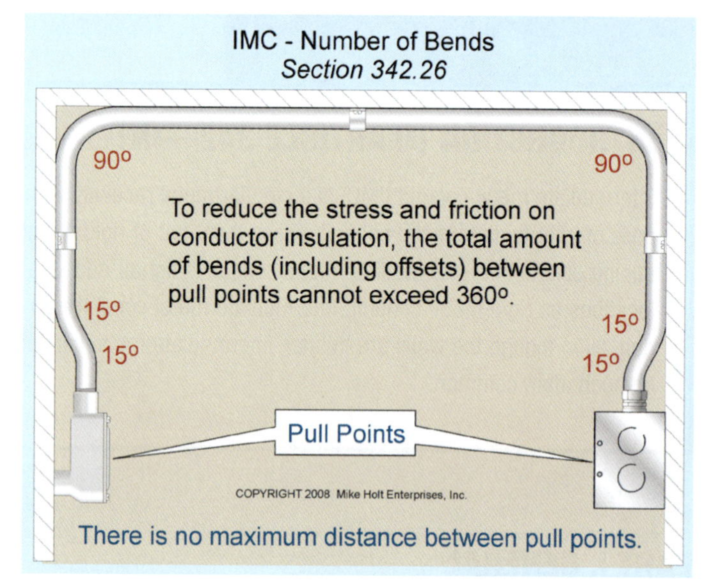

Figure 342–1

Author's Comment: There is no maximum distance between pull boxes because this is a design issue, not a safety issue.

342.28 Reaming. When the raceway is cut in the field, reaming is required to remove the burrs and rough edges.

Author's Comment: It's a commonly accepted practice to ream small raceways with a screwdriver or the backside of pliers. However, when the raceway is cut with a three-wheel pipe cutter, a reaming tool is required to remove the sharp edge of the indented raceway. When conduits are threaded in the field, the threads must be coated with an electrically conductive, corrosion-resistant compound approved by the authority having jurisdiction, in accordance with 300.6(A).

342.30 Securing and Supporting. Intermediate metal conduit must be installed as a complete system in accordance with 300.18 [300.10 and 300.12], and it must be securely fastened in place and supported in accordance with (A) and (B), or unsupported as permitted in (C).

(A) Securely Fastened. Intermediate metal conduit must generally be securely fastened within 3 ft of every box, cabinet, or termination fitting. Figure 342–2

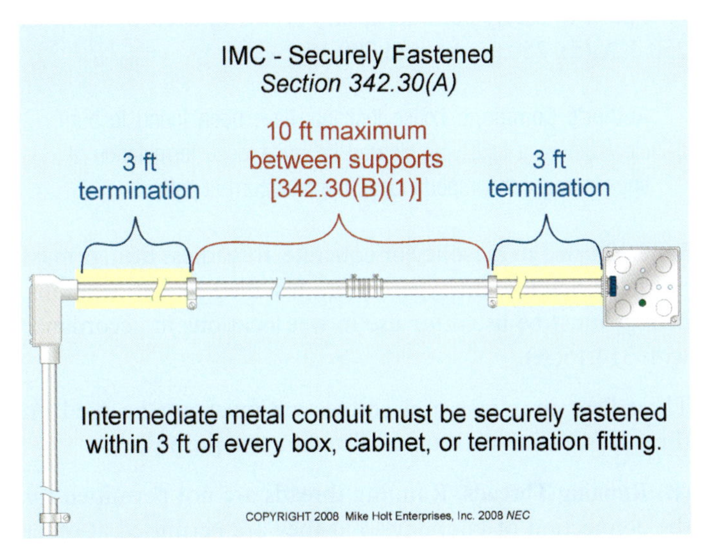

Figure 342–2

Author's Comment: Fastening is required within 3 ft of terminations, not within 3 ft of each coupling.

When structural members don't permit the raceway to be secured within 3 ft of a box or termination fitting, the raceway must be secured within 5 ft of the termination. Figure 342–3

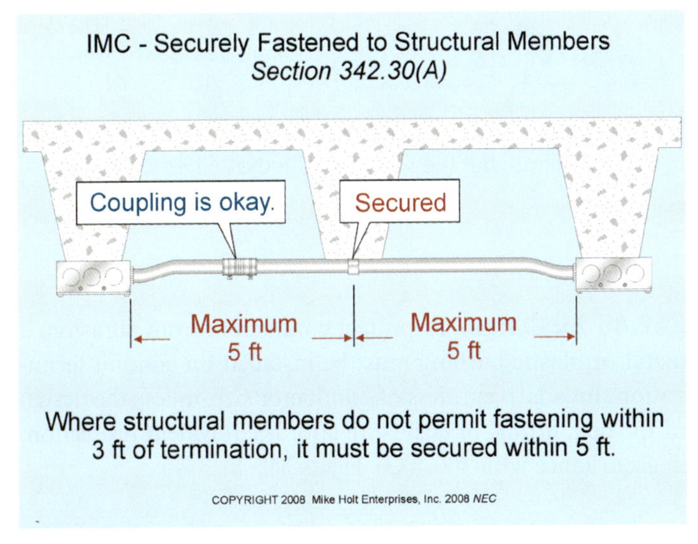

Figure 342–3

(B) Supports.

(1) General. Intermediate metal conduit must generally be supported at intervals not exceeding 10 ft.

(2) Straight Horizontal Runs. Straight horizontal runs made with threaded couplings can be supported in accordance with the distances listed in Table 344.30(B)(2). Figure 342–4

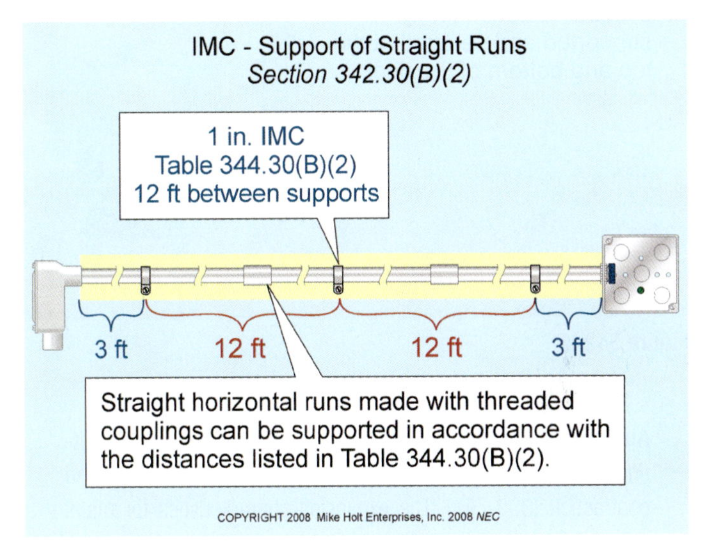

Figure 342–4

Table 344.30(B)(2)

Trade Size	Support Spacing
1/2–3/4	10 ft
1	12 ft*
1 1/4–1 1/2	14 ft
2–2 1/2	16 ft
3 and larger	20 ft

(3) Vertical Risers. Exposed vertical risers for fixed equipment can be supported at intervals not exceeding 20 ft, if the conduit is made up with threaded couplings, firmly supported, securely fastened at the top and bottom of the riser, and if no other means of support is available. Figure 342–5

(4) Horizontal Runs. Conduits installed horizontally in bored or punched holes in wood or metal framing members, or notches in wooden members are considered supported, but the raceway must be secured within 3 ft of termination.

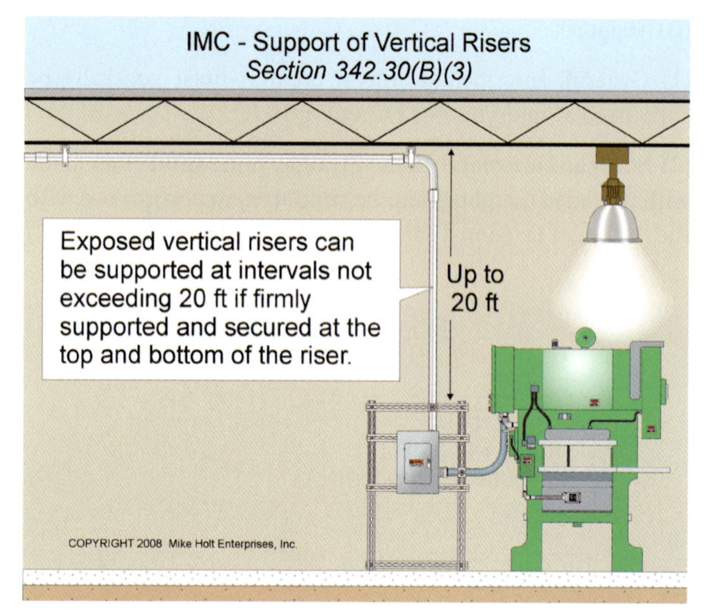

Figure 342–5

Author's Comment: IMC must be provided with expansion fittings where necessary to compensate for thermal expansion and contraction [300.7(B)]. The expansion characteristics for metal raceways are determined by multiplying the values from Table 352.44 by 0.20, and the expansion characteristics for aluminum raceways is determined by multiplying the values from Table 352.44 by 0.40 [300.7 FPN].

(C) Unsupported Raceways. Where oversized, concentric or eccentric knockouts are not encountered, intermediate metal conduit between enclosures is permitted to be unsupported where the raceway is unbroken and not more than 18 in. in length. Figure 342–6

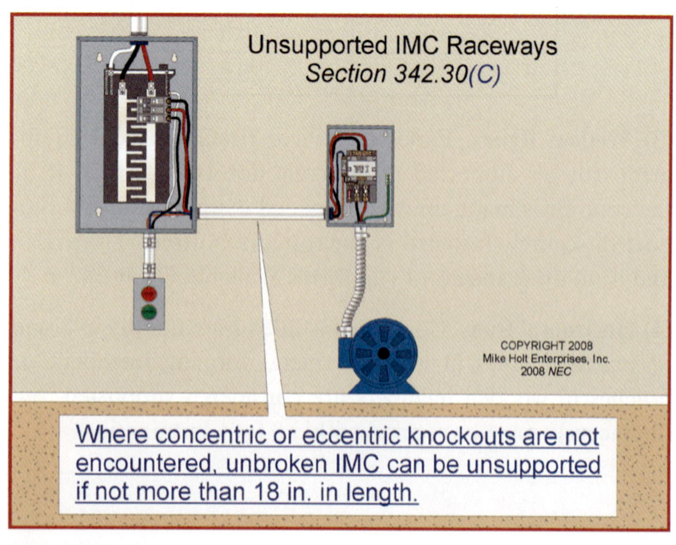

Figure 342–6

342.42 Couplings and Connectors.

(A) Installation. Threadless couplings and connectors must be made up tight to maintain an effective ground-fault current path to safely conduct fault current in accordance with 250.4(A)(5), 250.96(A), and 300.10.

Author's Comment: Loose locknuts have been found to burn clear before a fault was cleared because loose termination fittings increase the impedance of the fault current path.

Where buried in masonry or concrete, threadless fittings must be the concrete-tight type. Where installed in wet locations, fittings must be listed for use in wet locations in accordance with 314.15(A).

Threadless couplings and connectors must not be used on threaded conduit ends unless listed for the purpose.

(B) Running Threads. Running threads are not permitted for the connection of couplings, but they are permitted at other locations. Figure 342–7

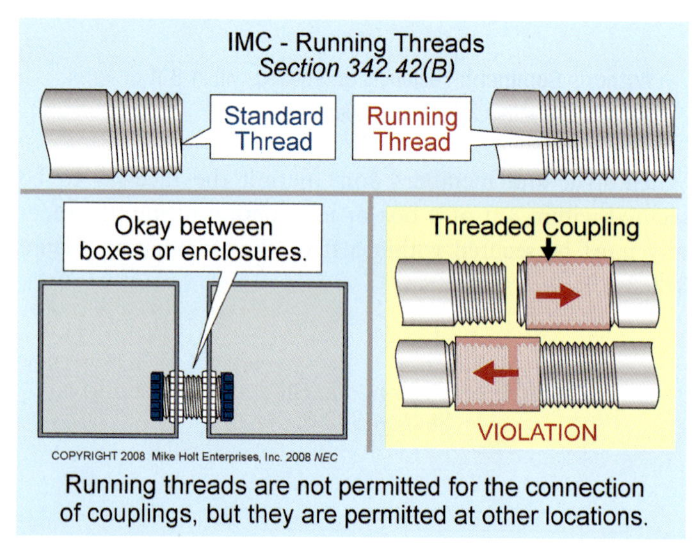

Figure 342–7

342.46 Bushings. To protect conductors from abrasion, a metal or plastic bushing must be installed on conduit termination threads, regardless of conductor size, unless the design of the box, fitting, or enclosure affords equivalent protection, in accordance with 300.4(G). Figure 342–8

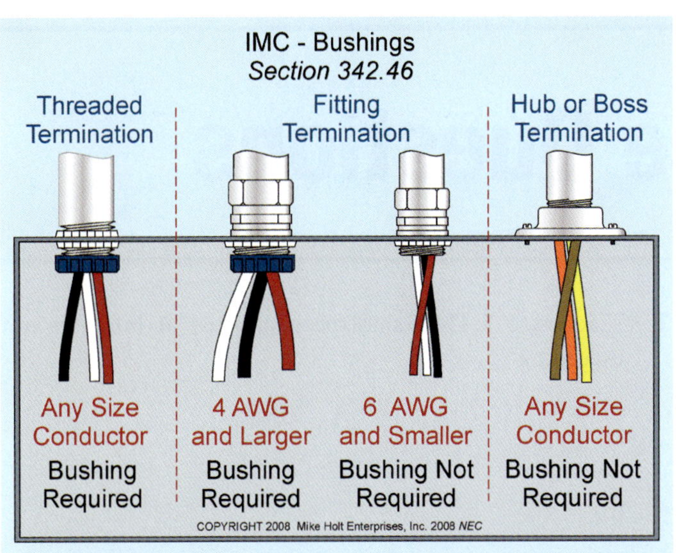

Conductors 4 AWG and larger must be protected by a fitting that provides a smooth, rounded, insulating surface, such as an insulating bushing [300.4(G)].

Figure 342–8

Author's Comment: Insulated circuit conductors 4 AWG and larger that enter an enclosure must be protected from abrasion during and after installation by a fitting that provides a smooth, rounded insulating surface, such as an insulating bushing, unless the design of the box, fitting, or enclosure provides equivalent protection in accordance with 300.4(G).

ARTICLE 342 Practice Questions

ARTICLE 342. INTERMEDIATE METAL CONDUIT (TYPE IMC)—PRACTICE QUESTIONS

1. Materials such as straps, bolts, screws, etc., that are associated with the installation of IMC in wet locations shall be _____.

   (a) weatherproof
   (b) weathertight
   (c) corrosion resistant
   (d) none of these

2. A run of IMC shall not contain more than the equivalent of _____ quarter bends between pull points such as conduit bodies and boxes.

   (a) one
   (b) two
   (c) three
   (d) four

3. Trade size 1 IMC shall be supported at intervals not exceeding _____.

   (a) 8 ft
   (b) 10 ft
   (c) 12 ft
   (d) 14 ft

4. Threadless couplings and connectors used on threaded IMC ends shall be listed for the purpose.

   (a) True
   (b) False

5. Running threads shall not be used on IMC for connection at couplings.

   (a) True
   (b) False

Rigid Metal Conduit (Type RMC)

INTRODUCTION TO ARTICLE 344—RIGID METAL CONDUIT (TYPE RMC)

Rigid metal conduit, commonly called "rigid," has long been the standard raceway for providing protection from physical impact and from difficult environments. The outside diameter of rigid metal conduit is the same as intermediate metal conduit. However, the wall thickness of rigid metal conduit is greater than intermediate metal conduit; therefore it has a smaller interior cross-sectional area. Rigid metal conduit is heavier and more expensive than intermediate metal conduit, and it can be used in any location. Rigid metal conduit is manufactured in both galvanized steel and aluminum; the steel type is much more common.

PART I. GENERAL

344.1 Scope. Article 344 covers the use, installation, and construction specifications of rigid metal conduit and associated fittings.

344.2 Definition.

Rigid Metal Conduit (Type RMC). A listed metal raceway of circular cross section with integral or associated couplings, listed for the installation of electrical conductors, and used with listed fittings to provide electrical continuity.

Author's Comment: When the mechanical and physical characteristics of rigid metal conduit are desired and a corrosive environment is anticipated, a PVC-coated raceway system is commonly used. This type of raceway is frequently used in the petrochemical industry. The common trade name of this coated raceway is Plasti-bond®, and it's commonly referred to as "Rob Roy conduit." The benefits of the improved corrosion protection can be achieved only when the system is properly installed. Joints must be sealed in accordance with the manufacturer's instructions, and the coating must not be damaged with tools such as benders, pliers, and pipe wrenches. Couplings are available with an extended skirt that can be properly sealed after installation.

344.6 Listing Requirements. Rigid metal conduit, elbows, couplings, and associated fittings must be listed.

PART II. INSTALLATION

344.10 Uses Permitted.

(A) Atmospheric Conditions and Occupancies.

(1) Galvanized Steel and Stainless Steel. Galvanized steel and stainless steel rigid metal conduit is permitted in all atmospheric conditions and occupancies.

(2) Red Brass. Red brass rigid metal conduit is permitted for direct burial and swimming pool applications.

(3) Aluminum. Rigid aluminum conduit is permitted where judged suitable for the environment.

(B) Corrosion Environments.

(1) Galvanized Steel and Stainless Steel. Rigid metal conduit fittings, elbows, and couplings can be installed in concrete, in direct contact with the earth, or in areas subject to severe corrosive influences judged suitable for the condition.

(2) Aluminum. Rigid aluminum conduit must be provided with supplementary corrosion protection approved by the authority having jurisdiction where encased in concrete or in direct contact with the earth.

(D) Wet Locations. Support fittings, such as screws, straps, and so forth, installed in a wet location must be made of corrosion-resistant material or protected by corrosion-resistant coatings in accordance with 300.6.

CAUTION: *Supplementary coatings (asphalted paint) for corrosion protection have not been investigated by a product testing and listing agency, and these coatings are known to cause cancer in laboratory animals.*

344.14 Dissimilar Metals. Where practical, contact with dissimilar metals should be avoided to prevent the deterioration of the metal because of galvanic action. Aluminum fittings and enclosures are permitted, however, with rigid metal conduit.

344.20 Trade Size.

(A) Minimum. Rigid metal conduit smaller than trade size ½ must not be used.

(B) Maximum. Rigid metal conduit larger than trade size 6 must not be used.

344.22 Number of Conductors. Raceways must be large enough to permit the installation and removal of conductors without damaging the conductors' insulation. When all conductors in a raceway are the same size and insulation, the number of conductors permitted can be found in Annex C for the raceway type.

Question: *How many 8 THHN conductors can be installed in trade size 1 1/2 RMC?*

(a) 16 *(b) 18* *(c) 20* *(d) 22*

Answer: *22 conductors [Annex C, Table C8]*

Author's Comment: See 300.17 for additional examples on how to size raceways when conductors aren't all the same size.

Cables can be installed in rigid metal conduit, as long as the number of cables does not exceed the allowable percentage fill specified in Table 1, Chapter 9.

344.24 Bends. Raceway bends must not be made in any manner that would damage the raceway, or significantly change its internal diameter (no kinks). The radius of the curve of the inner edge of any field bend must not be less than shown in Table 2, Chapter 9.

Author's Comment: This is usually not a problem because benders are made to comply with this table. However, when using a hickey bender (short-radius bender), be careful not to over-bend the raceway.

344.26 Number of Bends (360°). To reduce the stress and friction on conductor insulation, the maximum number of bends (including offsets) between pull points must not exceed 360°. Figure 344–1

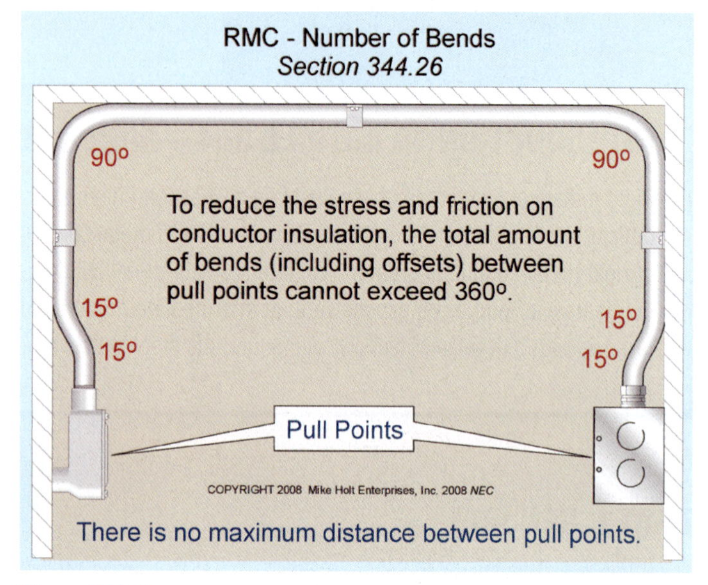

Figure 344–1

Author's Comment: There is no maximum distance between pull boxes because this is a design issue, not a safety issue.

344.28 Reaming. When the raceway is cut in the field, reaming is required to remove the burrs and rough edges.

Author's Comment: It's a commonly accepted practice to ream small raceways with a screwdriver or the backside of pliers. However, when the raceway is cut with a three-wheel pipe cutter, a reaming tool is required to remove the sharp edge of the indented raceway. When conduit is threaded in the field, the threads must be coated with an electrically conductive, corrosion-resistant compound approved by the authority having jurisdiction, in accordance with 300.6(A).

344.30 Securing and Supporting. Rigid metal conduit must be installed as a complete system in accordance with 300.18 [300.10 and 300.12], and it must be securely fastened in place and supported in accordance with (A) and (B), or unsupported as permitted in (C).

(A) Securely Fastened. Rigid metal conduit must generally be securely fastened within 3 ft of every box, cabinet, or termination fitting. Figure 344–2

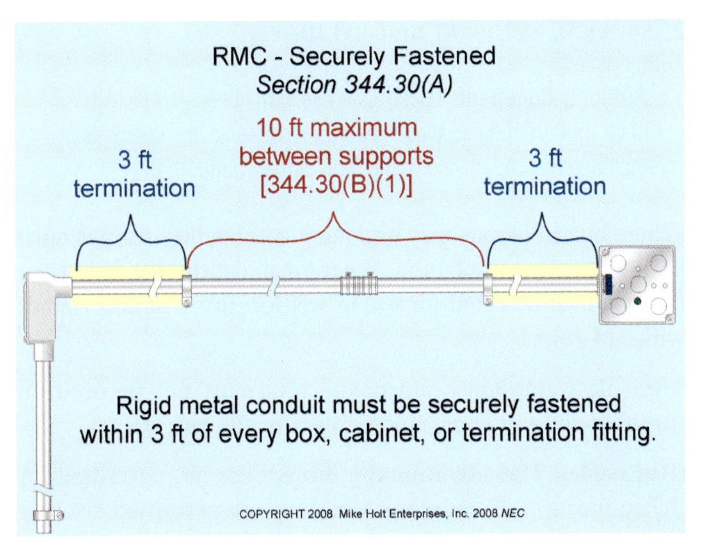

Figure 344–2

When structural members don't permit the raceway to be secured within 3 ft of a box or termination fitting, the raceway must be secured within 5 ft of termination. Figure 344–3

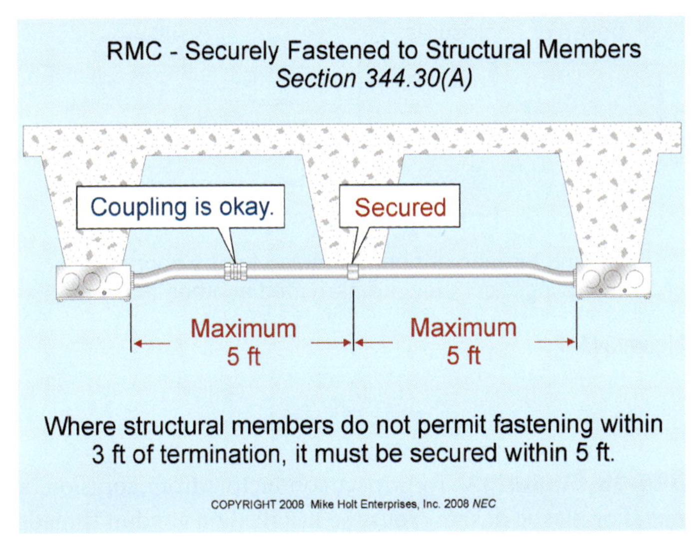

Figure 344–3

Author's Comment: Fastening is required within 3 ft of terminations, not within 3 ft of each coupling.

(B) Supports.

(1) General. Rigid metal conduit must be supported at intervals not exceeding 10 ft.

(2) Straight Horizontal Runs. Straight horizontal runs made with threaded couplings can be supported in accordance with the distances listed in Table 344.30(B)(2). Figure 344–4

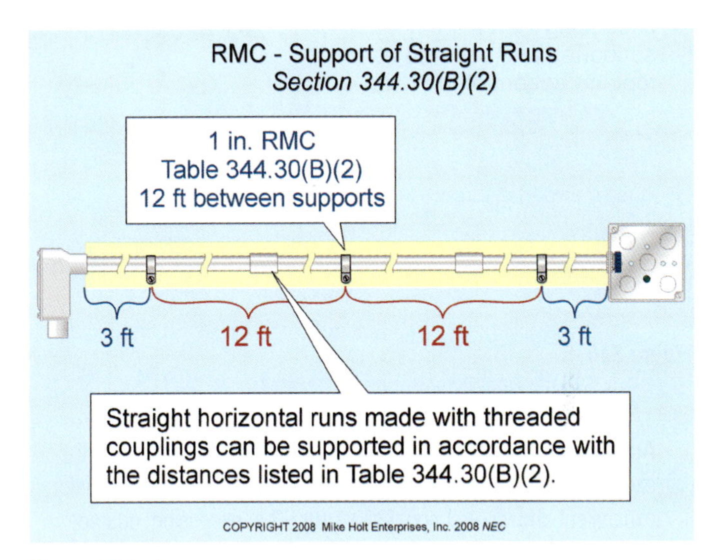

Figure 344–4

Table 344.30(B)(2)

Trade Size	Support Spacing
½–¾	10 ft
1	12 ft*
1¼–1½	14 ft
2–2½	16 ft
3 and larger	20 ft

(3) Vertical Risers. Exposed vertical risers for fixed equipment can be supported at intervals not exceeding 20 ft, if the conduit is made up with threaded couplings, firmly supported, securely fastened at the top and bottom of the riser, and if no other means of support is available. Figure 344–5

(4) Horizontal Runs. Conduits installed horizontally in bored or punched holes in wood or metal framing members, or notches in wooden members, are considered supported, but the raceway must be secured within 3 ft of termination.

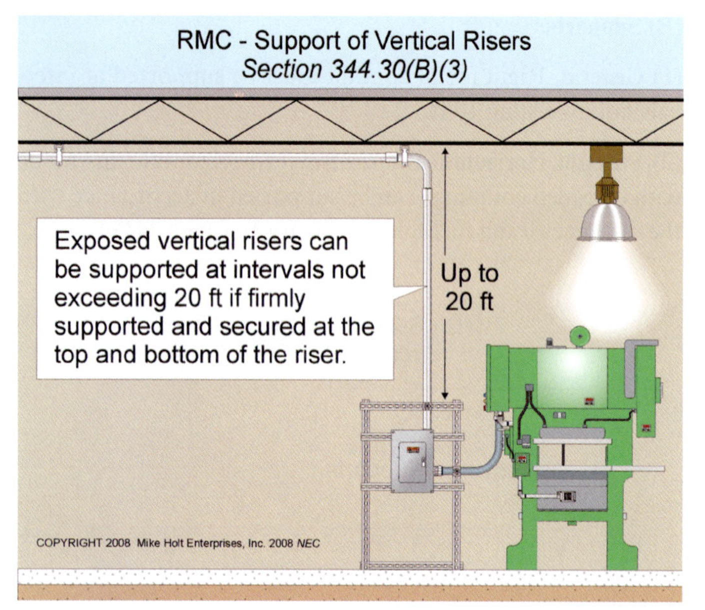

Figure 344–5

Author's Comment: Rigid metal conduit must be provided with expansion fittings where necessary to compensate for thermal expansion and contraction [300.7(B)]. The expansion characteristics for metal raceways are determined by multiplying the values from Table 352.44 by 0.20, and the expansion characteristics for aluminum raceways is determined by multiplying the values from Table 352.44 by 0.40 [300.7 FPN].

(C) Unsupported Raceways. Where oversized concentric or eccentric knockouts are not encountered, rigid metal conduit between enclosures is permitted to be unsupported where the raceway is unbroken and not more than 18 in. in length. Figure 344–6

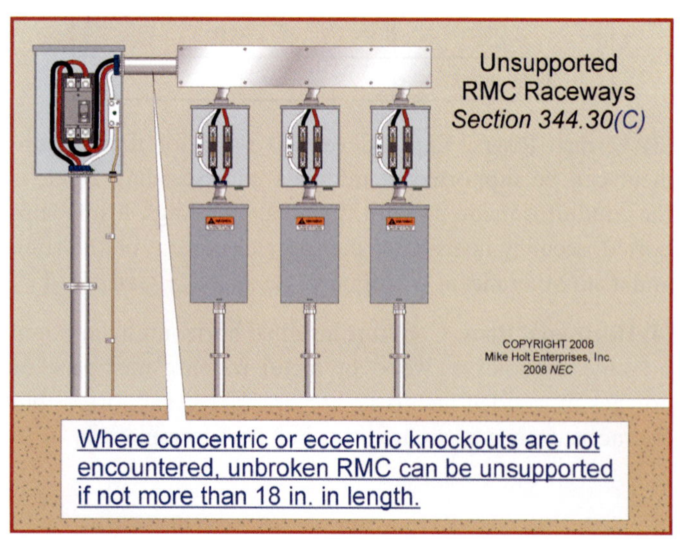

Figure 344–6

344.42 Couplings and Connectors.

(A) Installation. Threadless couplings and connectors must be made up tight to maintain an effective ground-fault current path to safely conduct fault current in accordance with 250.4(A)(5), 250.96(A), and 300.10.

Author's Comment: Loose locknuts have been found to burn clear before a fault was cleared because loose connections increase the impedance of the fault current path.

Where buried in masonry or concrete, threadless fittings must be the concrete-tight type. Where installed in wet locations, fittings must be listed for use in wet locations, in accordance with 314.15(A).

Threadless couplings and connectors must not be used on threaded conduit ends, unless listed for the purpose.

(B) Running Threads. Running threads are not permitted for the connection of couplings, but they are permitted at other locations. Figure 344–7

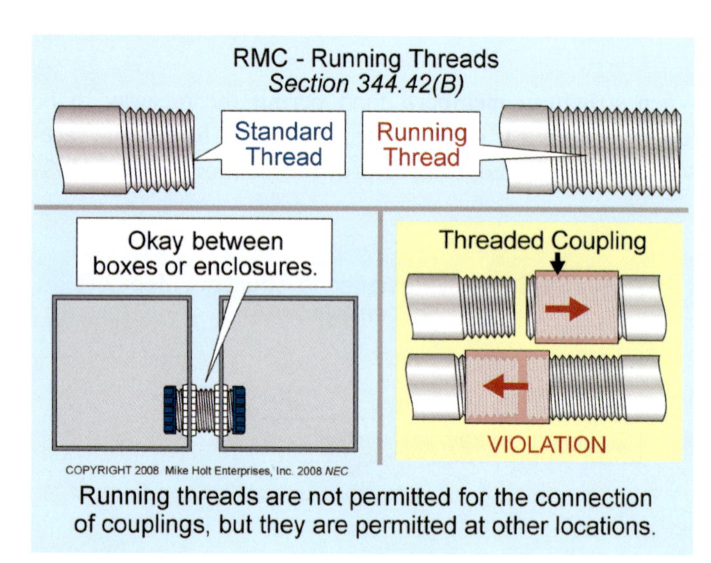

Figure 344–7

344.46 Bushings.

To protect conductors from abrasion, a metal or plastic bushing must be installed on conduit threads at terminations, regardless of conductor size, in accordance with 300.4(G), unless the design of the box, fitting, or enclosure affords equivalent protection. Figure 344–8

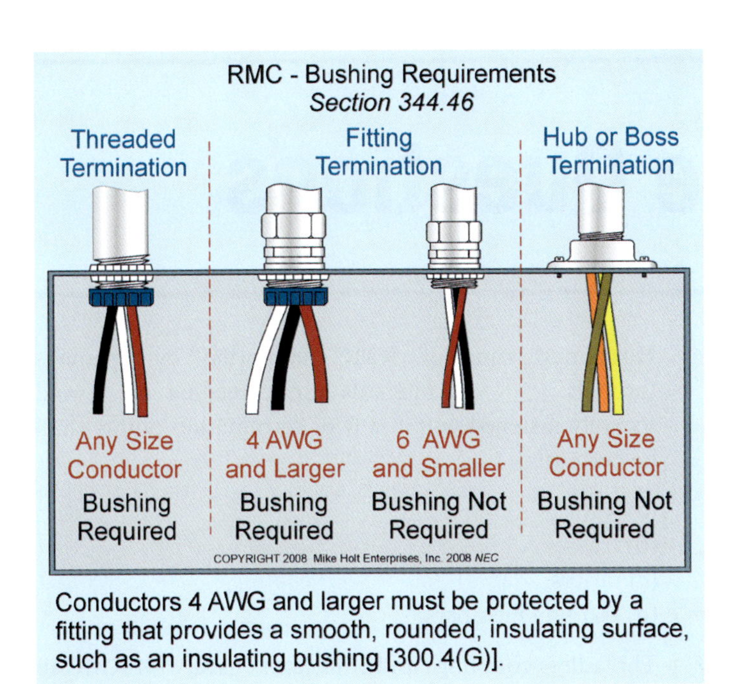

Conductors 4 AWG and larger must be protected by a fitting that provides a smooth, rounded, insulating surface, such as an insulating bushing [300.4(G)].

Figure 344–8

Author's Comment: Conductors 4 AWG and larger that enter an enclosure must be protected from abrasion, during and after installation, by a fitting that provides a smooth, rounded, insulating surface, such as an insulating bushing, unless the design of the box, fitting, or enclosure provides equivalent protection, in accordance with 300.4(G).

ARTICLE 344 Practice Questions

ARTICLE 344. RIGID METAL CONDUIT (TYPE RMC)—PRACTICE QUESTIONS

1. Galvanized steel, stainless steel and red brass RMC can be installed in concrete, in direct contact with the earth, or in areas subject to severe corrosive influences when protected by _____ and judged suitable for the condition.

   (a) ceramic
   (b) corrosion protection
   (c) backfill
   (d) a natural barrier

2. Aluminum fittings and enclosures can be used with _____ conduit where not subject to severe corrosive influences.

   (a) steel rigid metal
   (b) aluminum rigid metal
   (c) PVC-coated rigid conduit only
   (d) a and b

3. A run of RMC shall not contain more than the equivalent of _____ quarter bends between pull points such as conduit bodies and boxes.

   (a) one
   (b) two
   (c) three
   (d) four

4. Horizontal runs of RMC supported by openings through _____ at intervals not exceeding 10 ft and securely fastened within 3 ft of termination points shall be permitted.

   (a) walls
   (b) trusses
   (c) rafters
   (d) framing members

5. Threadless couplings and connectors used with RMC in wet locations shall be _____.

   (a) listed for wet locations
   (b) listed for damp locations
   (c) nonabsorbent
   (d) weatherproof

6. Running threads shall not be used on RMC for connection at _____.

   (a) boxes
   (b) cabinets
   (c) couplings
   (d) meter sockets

ARTICLE

348 Flexible Metal Conduit (Type FMC)

INTRODUCTION TO ARTICLE 348—FLEXIBLE METAL CONDUIT (TYPE FMC)

Flexible metal conduit (FMC), commonly called Greenfield or "flex," is a raceway of an interlocked metal strip of either steel or aluminum. It's primarily used for the final 6 ft or less of raceways between a more rigid raceway system and equipment that moves, shakes, or vibrates. Examples of such equipment include pump motors and industrial machinery.

PART I. GENERAL

348.1 Scope. Article 348 covers the use, installation, and construction specifications for flexible metal conduit and associated fittings.

348.2 Definition.

Flexible Metal Conduit (Type FMC). A raceway of circular cross section made of a helically wound, formed, interlocked metal strip of either steel or aluminum.

348.6 Listing Requirements. Flexible metal conduit and associated fittings must be listed.

PART II. INSTALLATION

348.10 Uses Permitted. Flexible metal conduit is permitted exposed or concealed.

348.12 Uses Not Permitted.

(1) In wet locations.

(2) In hoistways, other than as permitted in 620.21(A)(1).

(3) In storage battery rooms.

(4) In any hazardous (classified) location, except as permitted by 501.10(B).

(5) Exposed to material having a deteriorating effect on the installed conductors.

(6) Underground or embedded in poured concrete.

(7) Where subject to physical damage.

348.20 Trade Size.

(A) Minimum. Flexible metal conduit smaller than trade size ½ must not be used, except trade size ⅜ can be used for the following applications:

(1) For enclosing the leads of motors,

(2) Not exceeding 6 ft in length: Figure 348–1

a. For utilization equipment,

b. As part of a listed assembly, or

c. For luminaire tap connections, in accordance with 410.117(C).

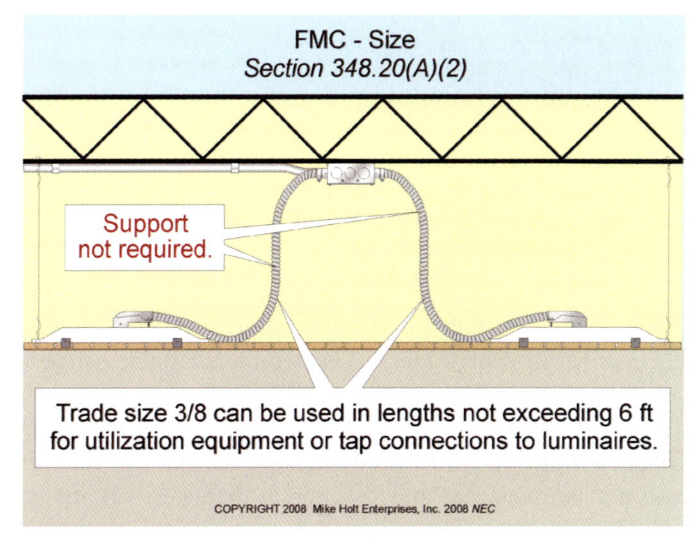

Figure 348–1

(3) In manufactured wiring systems, 604.6(A).

(4) In hoistways, 620.21(A)(1).

(5) As part of a listed luminaire assembly in accordance with 410.137(C).

(B) Maximum. Flexible metal conduit larger than trade size 4 must not be used.

348.22 Number of Conductors.

Trade Size ½ and Larger. Flexible metal conduit must be large enough to permit the installation and removal of conductors without damaging the conductors' insulation. When all conductors in a raceway are the same size and insulation, the number of conductors permitted can be found in Annex C for the raceway type.

> ***Question:*** *How many 6 THHN conductors can be installed in trade size 1 flexible metal conduit?*
>
> *(a) 2 (b) 4 (c) 6 (d) 8*
>
> ***Answer:*** *(c) 6 conductors [Annex C, Table C3]*

Author's Comment: See 300.17 for additional examples on how to size raceways when conductors aren't all the same size.

Trade Size ⅜. The number and size of conductors in trade size ⅜ flexible metal conduit must comply with Table 348.22.

> ***Question:*** *How many 12 THHN conductors can be installed in trade size ⅜ flexible metal conduit that uses outside fittings?*
>
> *(a) 1 (b) 3 (c) 5 (d) 7*
>
> ***Answer:*** *(b) 3 conductors [Table 348.22]*

One insulated, covered, or bare equipment grounding conductor of the same size is permitted with the circuit conductors. See the "*" note at the bottom of Table 348.22.

Cables can be installed in flexible metal conduit as long as the number of cables does not exceed the allowable percentage fill specified in Table 1, Chapter 9.

348.24 Bends. Bends must be made so that the conduit will not be damaged, and its internal diameter will not be effectively reduced. The radius of the curve of the inner edge of any field bend must not be less than shown in Table 2, Chapter 9 using the column "Other Bends."

348.26 Number of Bends (360°). To reduce the stress and friction on conductor insulation, the maximum number of bends (including offsets) between pull points must not exceed 360°.

Author's Comment: There is no maximum distance between pull boxes because this is a design issue, not a safety issue.

348.28 Trimming. The cut ends of flexible metal conduit must be trimmed to remove the rough edges, but this isn't necessary where fittings are threaded into the convolutions.

348.30 Securing and Supporting. Flexible metal conduit must be installed as a complete system [300.10, 300.12, and 300.18(A)], and it must be securely fastened in place and supported in accordance with (A) and (B).

(A) Securely Fastened. Flexible metal conduit must be securely fastened by a means approved by the authority having jurisdiction within 1 ft of termination, and it must be secured and supported at intervals not exceeding 4½ ft. Figure 348–2

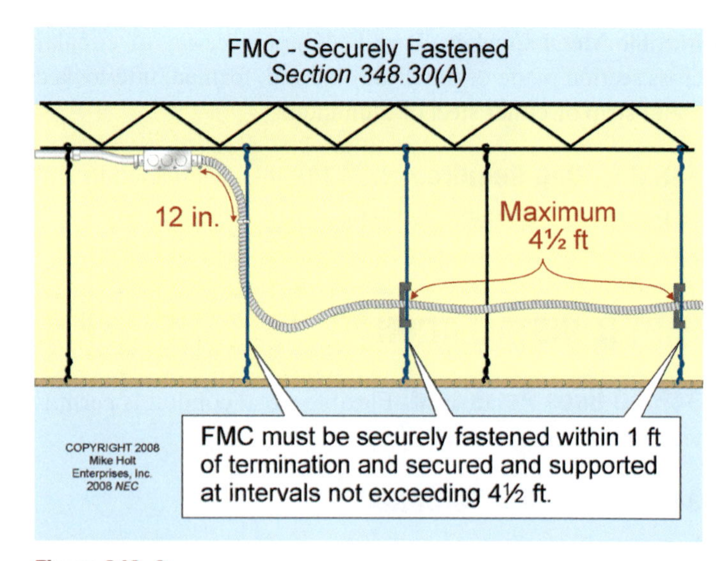

Figure 348–2

Exception No. 1: Flexible metal conduit is not required to be securely fastened or supported where fished between access points through concealed spaces and supporting is impractical.

Exception No. 2: Where flexibility is necessary after installation, unsecured lengths must not exceed: Figure 348–3

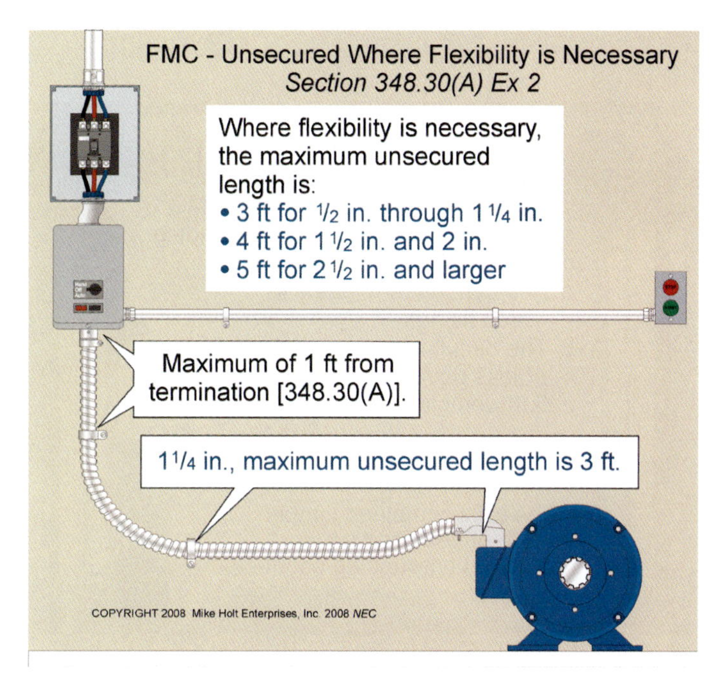

Figure 348–3

(1) 3 ft for trade sizes ½ through 1¼

(2) 4 ft for trade sizes 1½ through 2

(3) 5 ft for trade size 2½ and larger

Exception No. 4: FMC to a luminaire or electrical equipment within an accessible ceiling is permitted to be unsupported for not more than 6 ft from the last point where the raceway is securely fastened. Figure 348–4

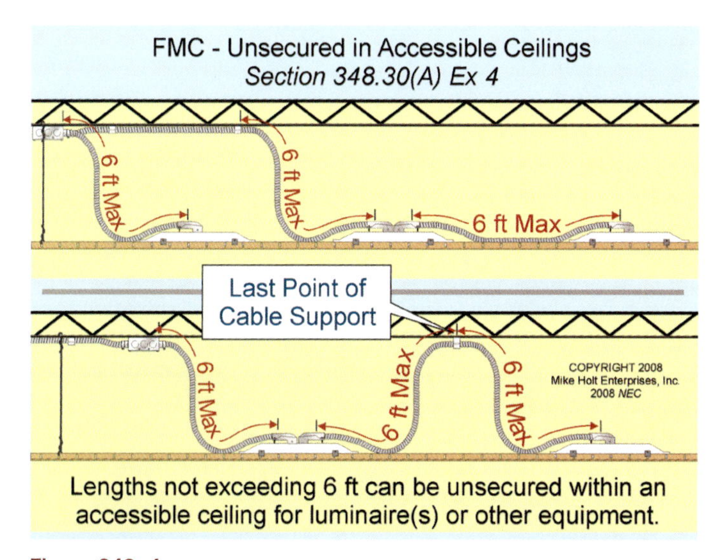

Figure 348–4

(B) Horizontal Runs. Flexible metal conduit installed horizontally in bored or punched holes in wood or metal framing members, or notches in wooden members, is considered supported, but the raceway must be secured within 1 ft of terminations. Figure 348–5

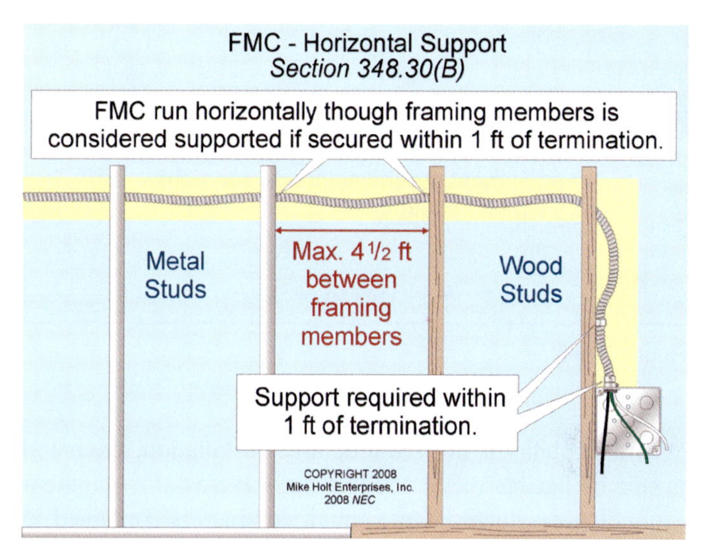

Figure 348–5

348.42 Fittings.

Angle connectors must not be installed in concealed locations.

Author's Comment: Conductors 4 AWG and larger that enter an enclosure must be protected from abrasion during and after installation by a fitting that provides a smooth, rounded insulating surface, such as an insulating bushing, unless the design of the box, fitting, or enclosure provides equivalent protection in accordance with 300.4(G).

348.60 Grounding and Bonding.

Where flexibility is required after installation, an equipment grounding conductor of the wire type must be installed with the circuit conductors in accordance with 250.118(5), based on the rating of the circuit overcurrent device in accordance with 250.122. Figure 348–6

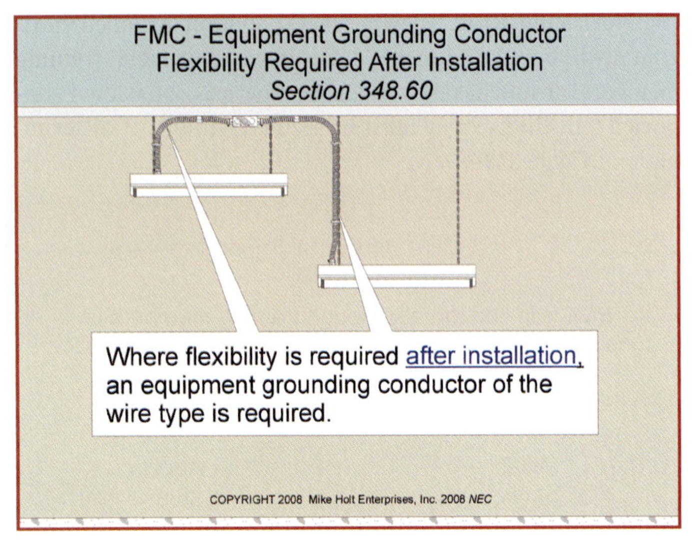

Figure 348–6

Where flexibility is not required after installation, the metal armor of flexible metal conduit can serve as an equipment grounding conductor if the circuit conductors contained in the raceway are protected by an overcurrent device rated 20A or less, and the combined length of the flexible metal raceway in the same ground-fault return path doesn't exceed 6 ft [250.118(5)]. Figure 348–7

Where an equipment bonding jumper is installed outside of a raceway, the length of the equipment bonding jumper must not exceed 6 ft, and it must be routed with the raceway or enclosure in accordance with 250.102(E).

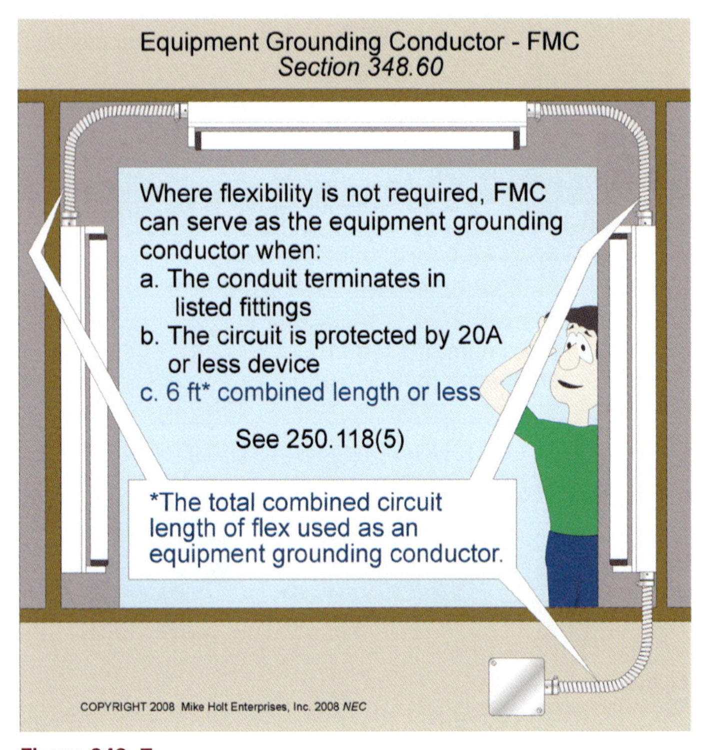

Figure 348–7

ARTICLE 348 Practice Questions

ARTICLE 348. FLEXIBLE METAL CONDUIT (TYPE FMC)—PRACTICE QUESTIONS

1. _____ is a raceway of circular cross section made of a helically wound, formed, interlocked metal strip of either steel or aluminum.

   (a) Type MC cable
   (b) Type AC Cable
   (c) LFMC
   (d) FMC

2. FMC can be installed exposed or concealed where not subject to physical damage.

   (a) True
   (b) False

3. Bends in FMC _____ between pull points.

   (a) shall not be made
   (b) need not be limited (in degrees)
   (c) shall not exceed 360 degrees
   (d) shall not exceed 180 degrees

4. FMC shall be supported and secured _____.

   (a) at intervals not exceeding 4½ ft
   (b) within 8 in. on each side of a box where fished
   (c) where fished
   (d) at intervals not exceeding 6 ft

5. In a concealed FMC installation, _____ connectors shall not be used.

   (a) straight
   (b) angle
   (c) grounding-type
   (d) none of these

ARTICLE

350 Liquidtight Flexible Metal Conduit (Type LFMC)

INTRODUCTION TO ARTICLE 350—LIQUIDTIGHT FLEXIBLE METAL CONDUIT (TYPE LFMC)

Liquidtight flexible metal conduit (LFMC), with its associated connectors and fittings, is a flexible raceway commonly used for connections to equipment that vibrate or are required to move occasionally. Liquidtight flexible metal conduit is commonly called Sealtight® or "liquidtight." Liquidtight flexible metal conduit is of similar construction to flexible metal conduit, but it also has an outer liquidtight thermoplastic covering. It has the same primary purpose as flexible metal conduit, but it also provides protection from moisture and some corrosive effects.

PART I. GENERAL

350.1 Scope. Article 350 covers the use, installation, and construction specifications of liquidtight flexible metal conduit and associated fittings.

350.2 Definition.

Liquidtight Flexible Metal Conduit (Type LFMC). A raceway of circular cross section, having an outer liquidtight, nonmetallic, sunlight-resistant jacket over an inner flexible metal core, with associated connectors and fittings for the installation of electric conductors.

350.6 Listing Requirements. Liquidtight flexible metal conduit and its associated fittings must be listed. Figure 350–1

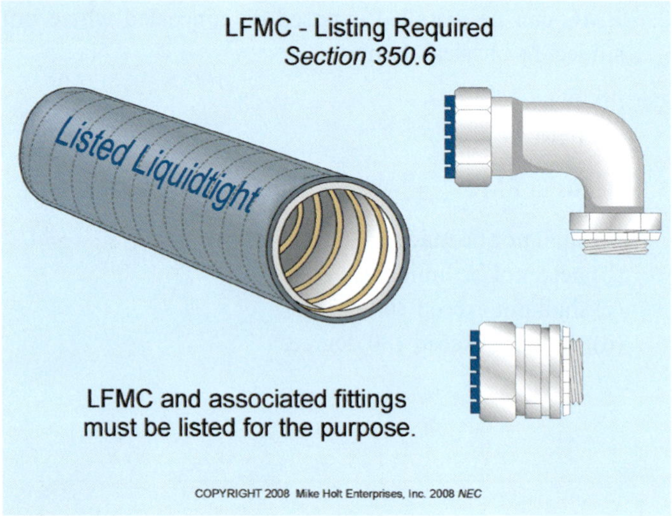

Figure 350–1

PART II. INSTALLATION

350.10 Uses Permitted.

(A) Permitted Use. Listed liquidtight flexible metal conduit is permitted, either exposed or concealed, at any of the following locations:

(1) Where flexibility or protection from liquids, vapors, or solids is required.

(2) In hazardous (classified) locations, as permitted in 501.10(B), 502.10(A)(2), 502.10(B)(2), or 503.10(A)(2).

(3) For direct burial, if listed and marked for this purpose.

350.12 Uses Not Permitted.

(1) Where subject to physical damage.

(2) Where the combination of the ambient and conductor operating temperatures exceeds the rating of the raceway.

350.20 Trade Size.

(A) Minimum. Liquidtight flexible metal conduit smaller than trade size ½ must not be used.

Exception: Liquidtight flexible metal conduit can be smaller than trade size ½ if installed in accordance with 348.20(A).

Author's Comment: According to 348.20(A), LFMC smaller than trade size ½ is permitted for the following:

(1) For enclosing the leads of motors.

(2) Not exceeding 6 ft in length:

a. For utilization equipment,
b. As part of a listed assembly, or
c. For tap connections to luminaires as permitted by 410.117(C).

(3) In manufactured wiring systems, 604.6(A).

(4) In hoistways, 620.21(A)(1).

(5) As part of a listed assembly to connect wired luminaire sections, 410.137(C).

(B) Maximum. Liquidtight flexible metal conduit larger than trade size 4 must not be used.

350.22 Number of Conductors.

(A) Raceway Trade Size ½ and Larger. Raceways must be large enough to permit the installation and removal of conductors without damaging the insulation. When all conductors in a raceway are the same size and insulation, the number of conductors permitted can be found in Annex C for the raceway type.

Question: *How many 6 THHN conductors can be installed in trade size 1 LFMC?* Figure 350–2

(a) 3 *(b) 5* *(c) 7* *(d) 9*

Answer: *(c) 7 conductors [Annex C, Table C.7]*

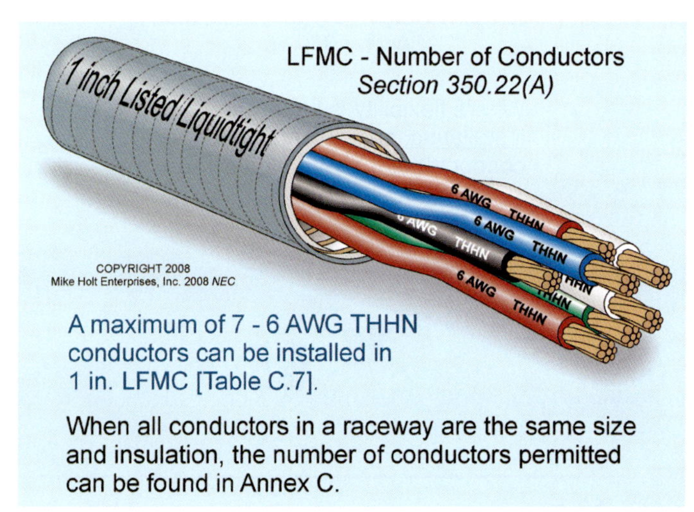

Figure 350–2

Author's Comment: See 300.17 for additional examples on how to size raceways when conductors aren't all the same size.

Cables can be installed in liquidtight flexible metal conduit as long as the number of cables does not exceed the allowable percentage fill specified in Table 1, Chapter 9.

(B) Raceway Trade Size ⅜. The number and size of conductors in a trade size ⅜ liquidtight flexible metal conduit must comply with Table 348.22.

Question: *How many 12 THHN conductors can be installed in trade size ⅜ LFMC that uses outside fittings?*

(a) 1 *(b) 3* *(c) 5* *(d) 7*

Answer: *(b) 3 conductors [Table 348.22]*

One insulated, covered, or bare equipment grounding conductor of the same size is permitted with the circuit conductors. See the "*" note at the bottom of Table 348.22.

350.24 Bends. Bends must be made so that the conduit will not be damaged and the internal diameter of the conduit will not be effectively reduced. The radius of the curve of the inner edge of any field bend must not be less than shown in Table 2, Chapter 9 using the column "Other Bends."

350.26 Number of Bends (360°). To reduce the stress and friction on conductor insulation, the maximum number of bends (including offsets) between pull points must not exceed 360°.

Author's Comment: There is no maximum distance between pull boxes because this is a design issue, not a safety issue.

350.30 Securing and Supporting. Liquidtight flexible metal conduit must be securely fastened in place and supported in accordance with (A) and (B).

(A) Securely Fastened. Liquidtight flexible metal conduit must be securely fastened by a means approved by the authority having jurisdiction within 1 ft of termination, and must be secured and supported at intervals not exceeding 4½ ft. Figure 350–3

Exception No. 1: Liquidtight flexible metal conduit is not required to be securely fastened or supported where fished between access points through concealed spaces and supporting is impractical.

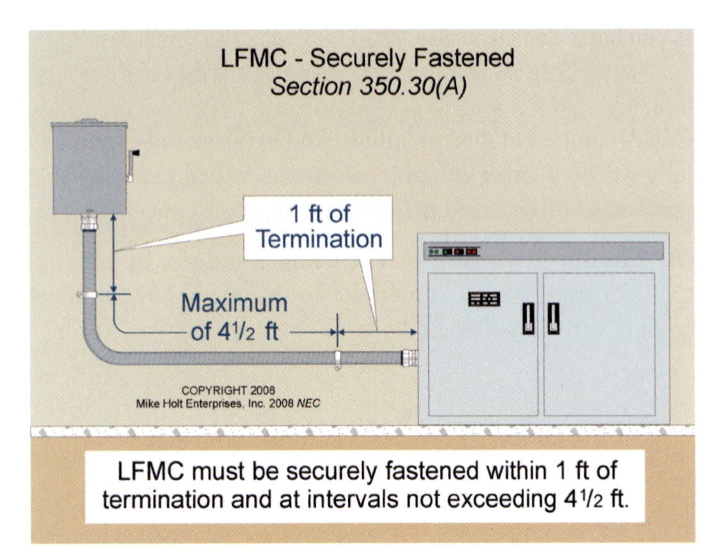

Figure 350–3

Exception No. 2: Where flexibility is necessary after installation, unsecured lengths of liquidtight flexible metal conduit can run unsupported as follows: Figure 350–4

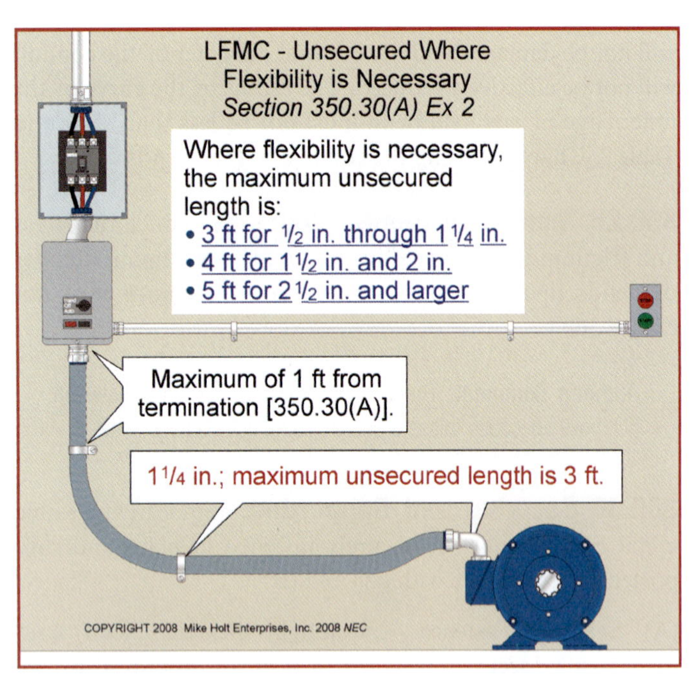

Figure 350–4

(1) 3 ft for trade sizes ½ through 1¼

(2) 4 ft for trade sizes 1½ through 2

(3) 5 ft for trade size 2½ and larger

Exception No. 4: Lengths not exceeding 6 ft can be unsecured within an accessible ceiling for luminaire(s) or other equipment.

(B) Horizontal Runs. Liquidtight flexible metal conduit installed horizontally in bored or punched holes in wood or metal framing members, or notches in wooden members, is considered supported, but the raceway must be secured within 1 ft of termination.

350.42 Fittings. Angle connector fittings must not be installed in concealed locations.

Author's Comment: Conductors 4 AWG and larger that enter an enclosure must be protected from abrasion, during and after installation, by a fitting that provides a smooth, rounded, insulating surface, such as an insulating bushing, unless the design of the box, fitting, or enclosure provides equivalent protection, in accordance with 300.4(G).

350.60 Grounding and Bonding.

Where flexibility is required after installation, an equipment grounding conductor of the wire type must be installed with the circuit conductors in accordance with 250.118(6), based on the rating of the circuit overcurrent device in accordance with 250.122. Figure 350–5

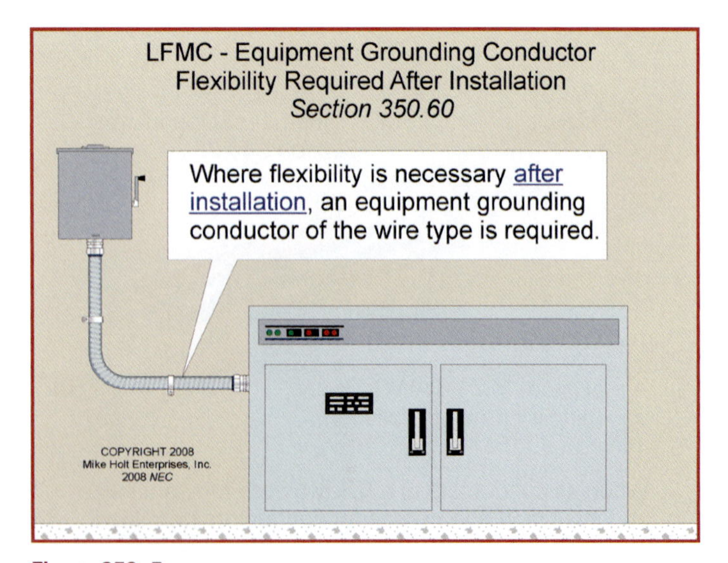

Figure 350–5

Where flexibility is not required after installation, the metal armor of liquidtight flexible metal conduit can serve as an equipment grounding conductor, if the circuit conductors are protected by an overcurrent device rated 20A or less for trade size ½ or smaller, or 60A or less for trade size ¾ through 1¼, and the combined length of flexible raceway in the same ground-fault return path doesn't exceed 6 ft [250.118(6)].

Where an equipment bonding jumper is installed outside a raceway, the length of the equipment bonding jumper can't exceed 6 ft and it must be routed with the raceway or enclosure in accordance with 250.102(E).

ARTICLE 350 Practice Questions

ARTICLE 350. LIQUIDTIGHT FLEXIBLE METAL CONDUIT (TYPE LFMC)—PRACTICE QUESTIONS

1. _____ is a raceway of circular cross section having an outer liquidtight, nonmetallic, sunlight-resistant jacket over an inner flexible metal core.

   (a) FMC
   (b) LFNMC
   (c) LFMC
   (d) none of these

2. LFMC shall not be required to be secured or supported where fished between access points through _____ spaces in finished buildings or structures and supporting is impractical.

   (a) concealed
   (b) exposed
   (c) hazardous
   (d) completed

3. _____ connectors shall not be used for concealed installations of LFMC.

   (a) Straight
   (b) Angle
   (c) Grounding-type
   (d) none of these

ARTICLE

352 Rigid Polyvinyl Chloride Conduit Type (PVC)

INTRODUCTION TO ARTICLE 352—RIGID POLYVINYL CHLORIDE CONDUIT TYPE (PVC)

Rigid polyvinyl chloride conduit (PVC) is a rigid nonmetallic conduit that provides many of the advantages of rigid metal conduit, while allowing installation in areas that are wet or corrosive. It's an inexpensive raceway, and easily installed. It's lightweight, easily cut, glued together, and relatively strong. However, conduits manufactured from polyvinyl chloride (PVC) are brittle when cold, and they sag when hot. This type of conduit is commonly used as an underground raceway because of its low cost, ease of installation, and resistance to corrosion and decay.

PART I. GENERAL

352.1 Scope. Article 352 covers the use, installation, and construction specifications of PVC conduit and associated fittings.

352.2 Definition.

Rigid Polyvinyl Chloride Conduit (PVC). A rigid nonmetallic raceway of circular cross section with integral or associated couplings, listed for the installation of electrical conductors and cables. Figure 352–1

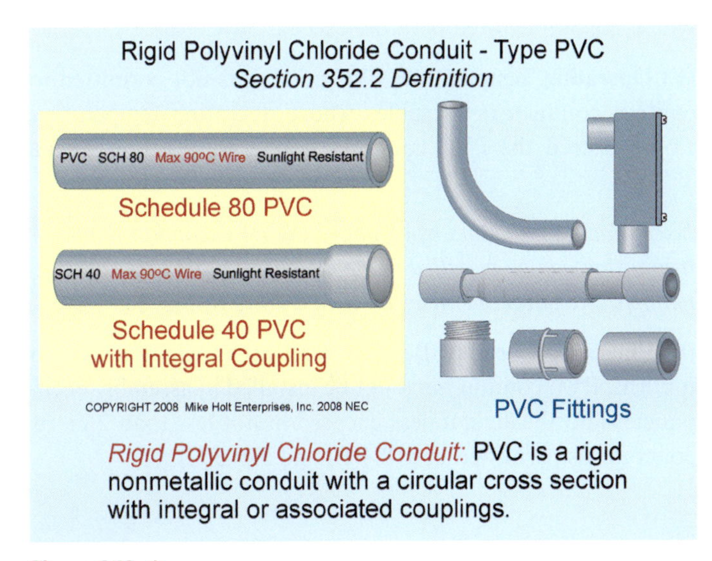

Figure 352–1

PART II. INSTALLATION

352.10 Uses Permitted.

> **FPN:** In extreme cold, PVC conduit can become brittle, and is more susceptible to physical damage.

(A) Concealed. PVC conduit can be concealed within walls, floors, or ceilings, directly buried or embedded in concrete in buildings of any height.

(B) Corrosive Influences. PVC conduit is permitted in areas subject to severe corrosion for which the material is specifically approved by the authority having jurisdiction.

> **Author's Comment:** Where subject to exposure to chemical solvents, vapors, splashing, or immersion, materials or coatings must either be inherently resistant to chemicals based upon their listing, or be identified for the specific chemical reagent [300.6(C)(2)].

(D) Wet Locations. PVC conduit is permitted in wet locations such as dairies, laundries, canneries, car washes, and other areas frequently washed or in outdoor locations. Support fittings such as straps, screws, and bolts must be made of corrosion-resistant materials, or must be protected with a corrosion-resistant coating, in accordance with 300.6(A).

(E) Dry and Damp Locations. PVC conduit is permitted in dry and damp locations, except where limited in 352.12.

(F) Exposed. Schedule 40 PVC conduit is permitted for exposed locations where not subject to physical damage. Figure 352–2

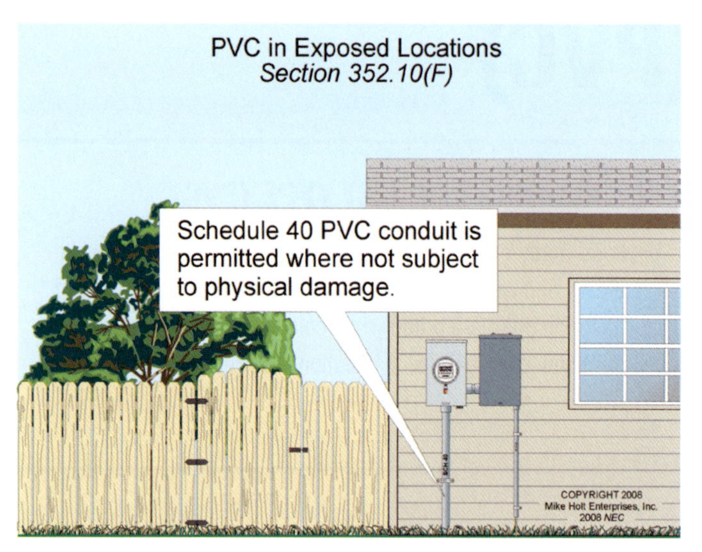

Figure 352–2

Where PVC conduit is exposed to physical damage, the raceway must be identified for the application.

FPN: PVC Schedule 80 conduit is identified for use in areas subject to physical damage. Figure 352–3

Figure 352–3

(G) Underground. PVC conduit installed underground must comply with the burial requirements of 300.5.

(H) Support of Conduit Bodies. PVC conduit is permitted to support nonmetallic conduit bodies that are not larger than the largest trade size of an entering raceway. These conduit bodies can't support luminaires or other equipment, and are not permitted to contain devices, other than splicing devices permitted by 110.14(B) and 314.16(C)(2).

352.12 Uses Not Permitted.

(A) Hazardous (Classified) Locations. PVC conduit is not allowed to be used in hazardous (classified) locations except as permitted by 501.10(A)(1)(a) Ex, 503.10(A), 504.20, 514.8 Ex 2, and 515.8.

(2) In Class I, Division 2 locations, except as permitted in 501.10(B)(7).

(B) Support of Luminaires. PVC conduit must not be used for the support of luminaires or other equipment not described in 352.10(H).

Author's Comment: PVC conduit is permitted to support conduit bodies in accordance with 314.23(E) Ex.

(C) Physical Damage. Schedule 40 PVC conduit must not be installed where subject to physical damage, unless identified for the application.

Author's Comment: PVC Schedule 80 conduit is identified for use in areas subject to physical damage [352.10(F) FPN].

(D) Ambient Temperature. PVC conduit must not be installed where the ambient temperature exceeds 50°C (122°F).

(E) Operating Temperature. PVC conduit is not permitted to contain conductors or cables whose operating temperature would exceed the PVC conduit's temperature rating. Figure 352–4

Exception: Conductors rated above the PVC conduit temperature rating can be installed if the conductors don't operate at a temperature above the raceway temperature rating.

(F) Places of Assembly. Because it releases toxic fumes when it burns, PVC conduit must not be installed in assembly occupancies and theaters, unless encased in not less than 2 in. of concrete, as permitted in 518.4(B) and 520.5(C).

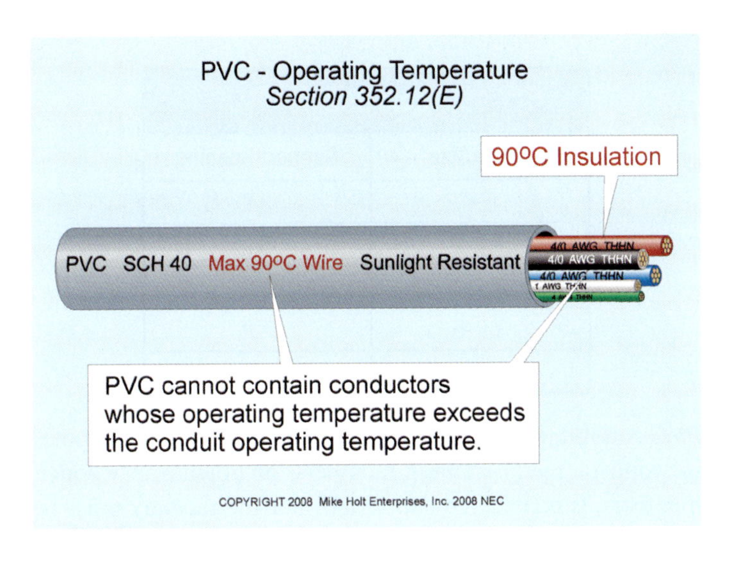

Figure 352–4

Author's Comment: PVC conduit is prohibited as a wiring method for patient care areas in health care facilities [517.13(A)] or in ducts, plenums, and other environmental air spaces [300.22].

352.20 Trade Size.

(A) Minimum. PVC conduit smaller than trade size ½ must not be used.

(B) Maximum. PVC conduit larger than trade size 6 must not be used.

352.22 Number of Conductors. Raceways must be large enough to permit the installation and removal of conductors without damaging the conductors' insulation, and the number of conductors must not exceed that permitted by the percentage fill specified in Table 1, Chapter 9.

When all conductors in a raceway are the same size and insulation, the number of conductors permitted can be found in Annex C for the raceway type.

Question: *How many 4/0 THHN conductors can be installed in trade size 2 Schedule 40 PVC?*

(a) 2 *(b) 4* *(c) 6* *(d) 8*

Answer: *(b) 4 conductors [Annex C, Table C10]*

Author's Comment: Schedule 80 PVC conduit has the same outside diameter as Schedule 40 PVC conduit, but the wall thickness of Schedule 80 PVC conduit is greater, which results in a reduced interior area for conductor fill.

Question: *How many 4/0 THHN conductors can be installed in trade size 2 Schedule 80 PVC conduit?*

(a) 3 *(b) 5* *(c) 7* *(d) 9*

Answer: *(a) 3 conductors [Annex C, Table C9]*

Author's Comment: See 300.17 for additional examples on how to size raceways when conductors aren't all the same size.

Cables can be installed in PVC conduit, as long as the number of cables does not exceed the allowable percentage fill specified in Table 1, Chapter 9.

352.24 Bends. Raceway bends must not be made in any manner that would damage the raceway, or significantly change its internal diameter (no kinks). The radius of the curve of the inner edge of any field bend must not be less than shown in Table 2, Chapter 9.

Author's Comment: Be sure to use equipment designed for heating the nonmetallic raceway so it's pliable for bending (for example, a "hot box"). Don't use open-flame torches.

352.26 Number of Bends (360°). To reduce the stress and friction on conductor insulation, the maximum number of bends (including offsets) between pull points must not exceed 360°. Figure 352–5

352.28 Trimming. The cut ends of PVC conduit must be trimmed (inside and out) to remove the burrs and rough edges. Trimming PVC conduit is very easy; most of the burrs rub off with fingers, and a knife will smooth the rough edges.

352.30 Securing and Supporting. PVC conduit must be securely fastened and supported in accordance with (A) and (B), or unsupported in accordance with (C).

(A) Secured. PVC conduit must be secured within 3 ft of every box, cabinet, or termination fitting, such as a conduit body. Figure 352–6

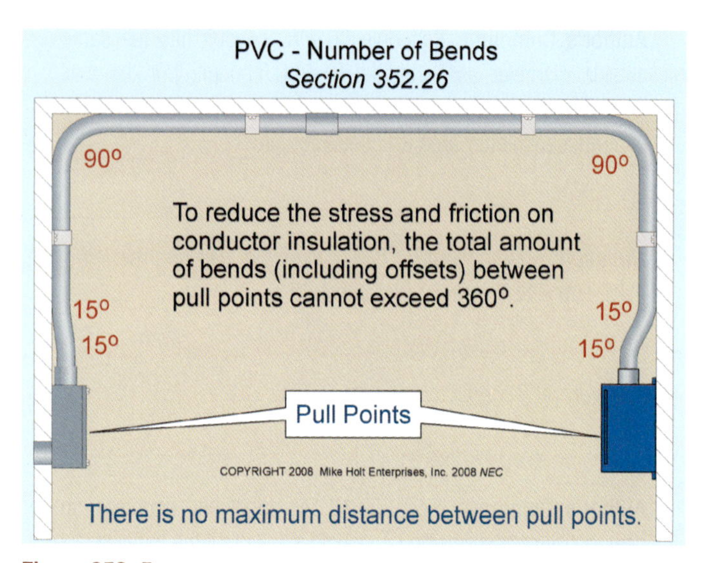

Figure 352–5

(B) Supports. PVC conduit must be supported at intervals not exceeding the values in Table 352.30, and the raceway must be fastened in a manner that permits movement from thermal expansion or contraction. See Figure 352–6.

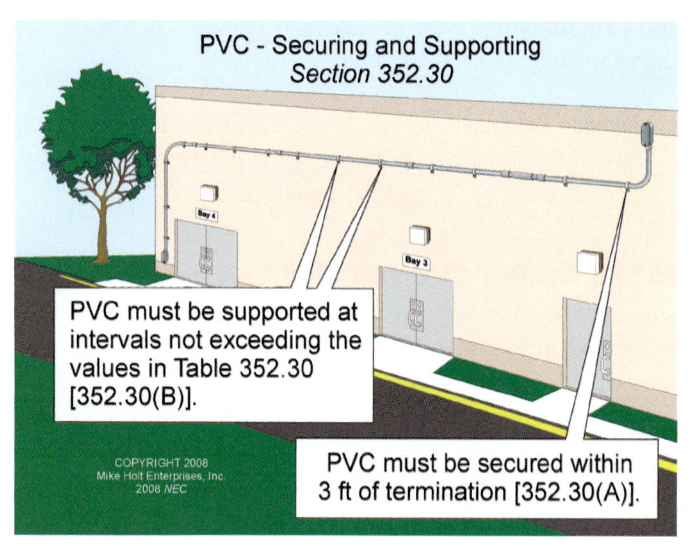

Figure 352–6

Table 352.30

Trade Size	Support Spacing
½–1	3 ft
1¼–2	5 ft
2½–3	6 ft
3½–5	7 ft
6	8 ft

PVC conduit installed horizontally in bored or punched holes in wood or metal framing members, or notches in wooden members, is considered supported, but the raceway must be secured within 3 ft of termination.

(C) Unsupported Raceways. Where oversized concentric or eccentric knockouts are not encountered, PVC conduit is permitted to be unsupported between enclosures where the raceway is unbroken and not more than 18 in. in length. Figure 352–7

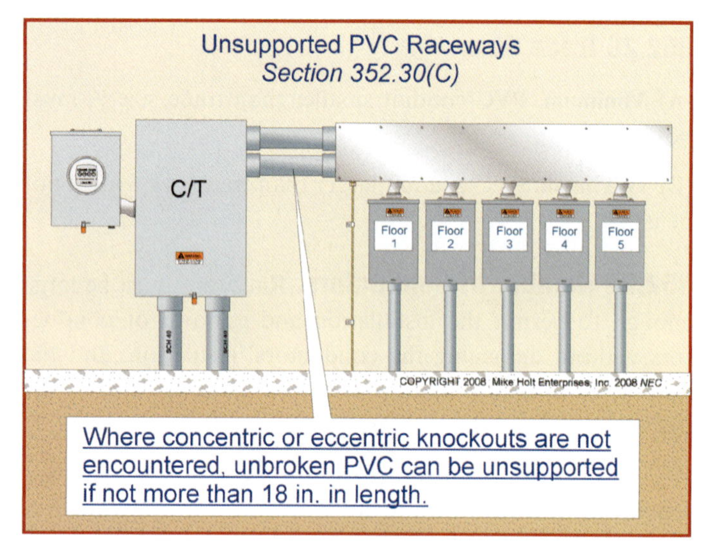

Figure 352–7

352.44 Expansion Fittings. Where PVC conduit is installed in a straight run between securely mounted items, such as boxes, cabinets, elbows, or other conduit terminations, expansion fittings must be provided to compensate for thermal expansion and contraction of the raceway in accordance with Table 352.44, if the length change is determined to be ¼ in. or greater. Figure 352–8

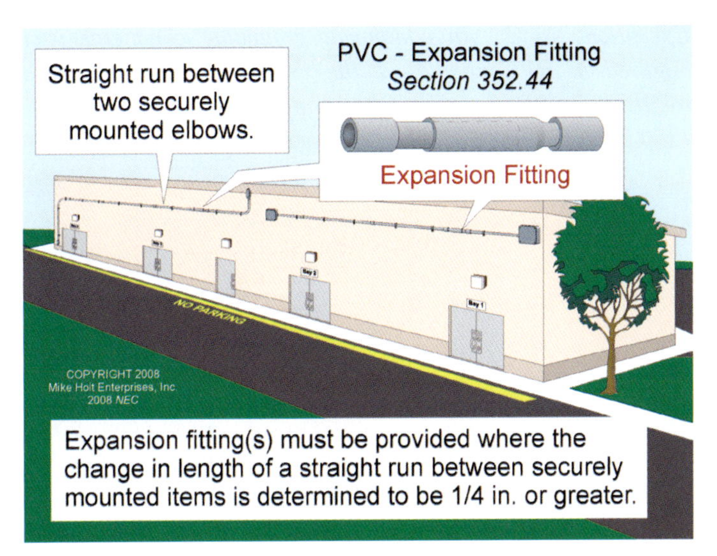

Figure 352–8

Author's Comment: Table 352.44 in the *NEC* was created based on the following formula: Figure 352–9

Expansion/Contraction Inches =
Raceway Length/100 x [(Temp Change/100) x 4.00]

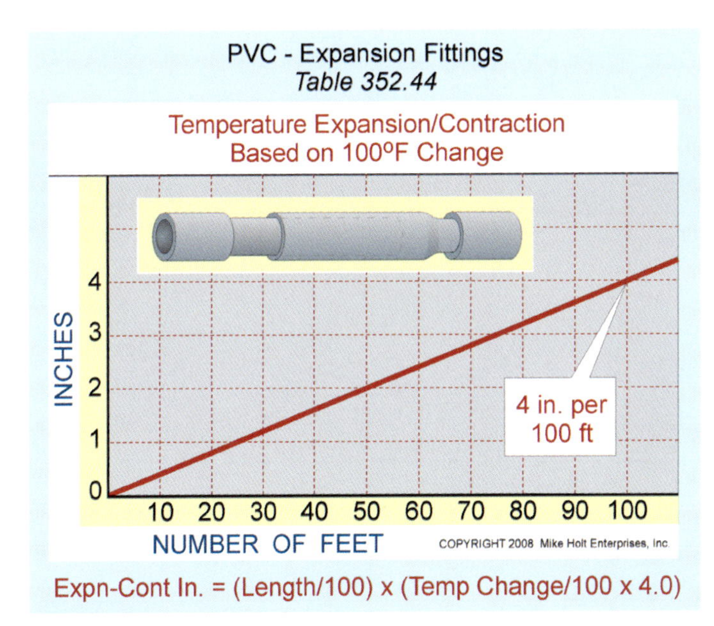

Figure 352–9

Example: *What is the contraction for a 25 ft run of PVC conduit located in an ambient temperature change of 25°F?*

(a) 1 in. *(b) 2 in.* *(c) 3 in.* *(d) 4 in.*

Answer: *(a) 1 in.*

Expansion/Contraction Inches =
Raceway Length/100 x ((Temp °F Change/100) x 4.00)
Expansion/Contraction Inches = (25/100) x ((25/100) x 4.00)
Expansion/Contraction Inches = 0.25 in.

352.46 Bushings. Conductors 4 AWG and larger that enter an enclosure must be protected from abrasion, during and after installation, by a fitting that provides a smooth, rounded insulating surface, such as an insulating bushing, unless the design of the box, fitting, or enclosure provides equivalent protection, in accordance with 300.4(G). PVC bell-ends provide the conductor protection required in this section. Figure 352–10

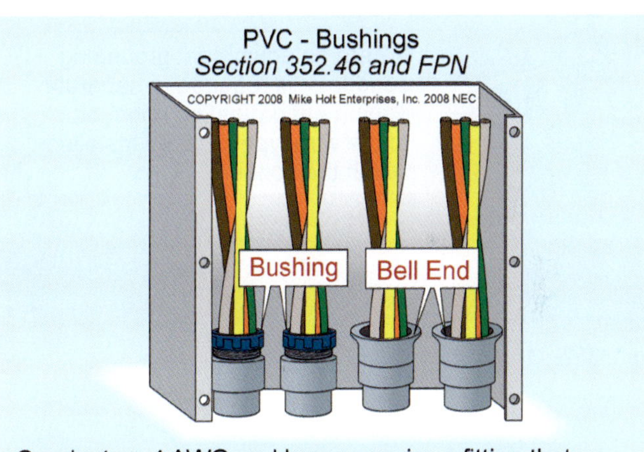

Figure 352–10

Author's Comment: When PVC conduit is stubbed into an open-bottom switchboard or other apparatus, the raceway, including the end fitting (bell-end), must not rise more than 3 in. above the bottom of the switchboard enclosure [300.16(B) and 408.5].

352.48 Joints. Joints, such as couplings and connectors, must be made in a manner approved by the authority having jurisdiction.

Author's Comment: Follow the manufacturer's instructions for the raceway, fittings, and glue. Some glue requires the raceway surface to be cleaned with a solvent before the application of the glue. After applying glue to both surfaces, a quarter turn of the fitting is required.

352.60 Equipment Grounding Conductor. Where equipment grounding is required, a separate equipment grounding conductor of the wire type must be installed within the conduit [300.2(B)]. Figure 352–11

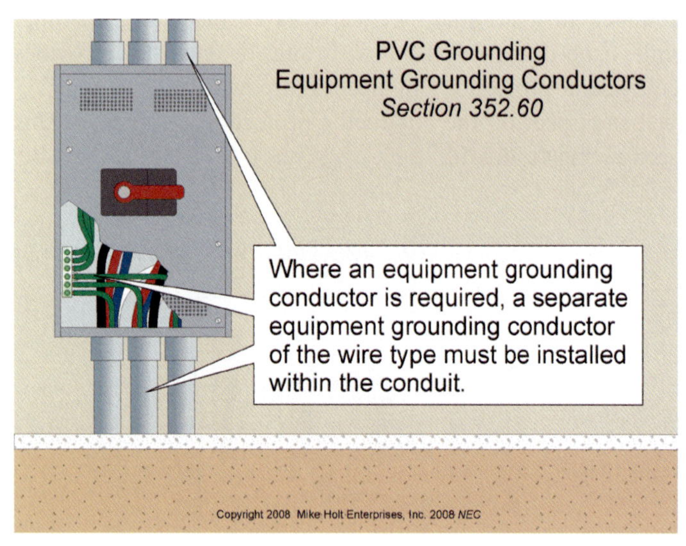

Figure 352–11

Exception No. 2: An equipment grounding conductor isn't required in PVC conduit if the neutral conductor is used to ground service equipment, as permitted in 250.142(A) [250.24(C)]. Figure 352–12

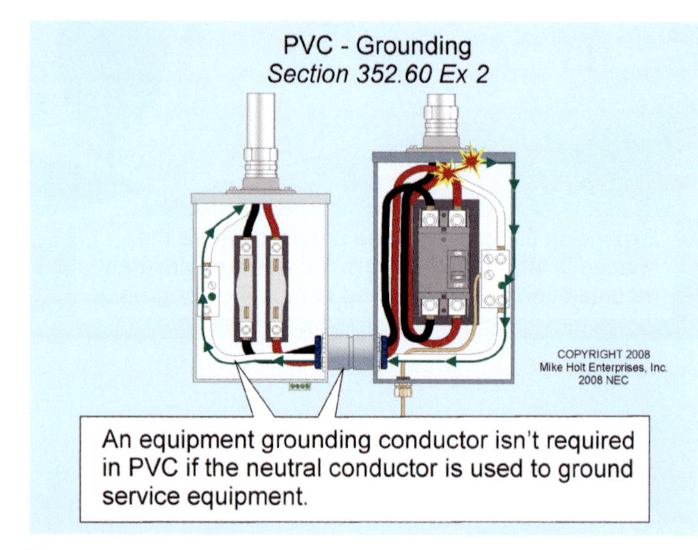

Figure 352–12

ARTICLE 352 Practice Questions

ARTICLE 352. RIGID POLYVINYL CHLORIDE CONDUIT (TYPE PVC)—PRACTICE QUESTIONS

1. Extreme _____ may cause PVC conduit to become brittle, and therefore more susceptible to damage from physical contact.

   (a) sunlight
   (b) corrosive conditions
   (c) heat
   (d) cold

2. PVC conduit shall be permitted for exposed work where subject to physical damage if identified for such use.

   (a) True
   (b) False

3. The number of conductors permitted in PVC conduit shall not exceed the percentage fill specified in _____.

   (a) Chapter 9, Table 1
   (b) Table 250.66
   (c) Table 310.16
   (d) 240.6

4. Bends in PVC conduit shall _____ between pull points.

   (a) not be made
   (b) not be limited in degrees
   (c) be limited to 360 degrees
   (d) be limited to 180 degrees

5. PVC conduit shall be securely fastened within _____ of each box.

   (a) 6 in.
   (b) 12 in.
   (c) 24 in.
   (d) 36 in.

ARTICLE 356 Liquidtight Flexible Nonmetallic Conduit (Type LFNC)

INTRODUCTION TO ARTICLE 356—LIQUIDTIGHT FLEXIBLE NONMETALLIC CONDUIT (TYPE LFNC)

Liquidtight flexible nonmetallic conduit (LFNC) is a listed raceway of circular cross section having an outer liquidtight, nonmetallic, sunlight-resistant jacket over an inner flexible core with associated couplings, connectors, and fittings.

PART I. GENERAL

356.1 Scope. Article 356 covers the use, installation, and construction specifications of liquidtight flexible nonmetallic conduit and associated fittings.

356.2 Definition.

Liquidtight Flexible Nonmetallic Conduit (Type LFNC). A listed raceway of circular cross section, having an outer liquidtight, nonmetallic, sunlight-resistant jacket over a flexible inner core, with associated couplings, connectors, and fittings, listed for the installation of electrical conductors.

(1) Type LFNC-A (orange color). A smooth seamless inner core and cover having reinforcement layers between the core and cover.

(2) Type LFNC-B (gray color). A smooth inner surface with integral reinforcement within the conduit wall.

(3) Type LFNC-C (black color). A corrugated internal and external surface without integral reinforcement.

356.6 Listing Requirement. Liquidtight flexible nonmetallic conduit, and its associated fittings, must be listed. Figure 356–1

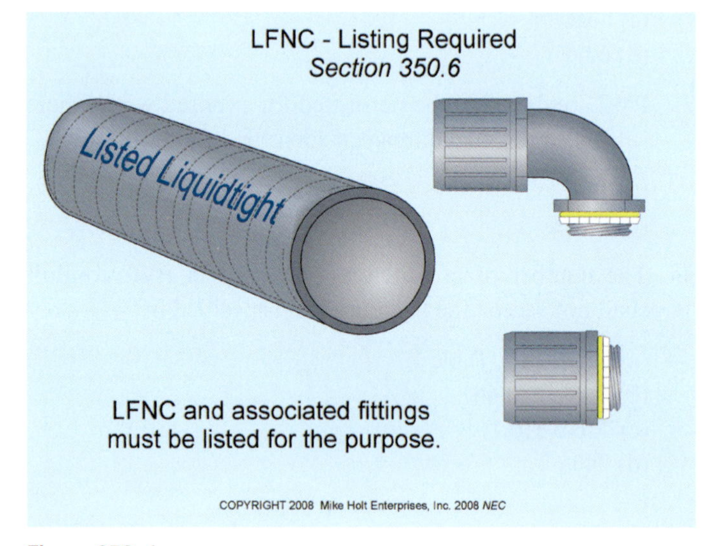

Figure 356–1

PART II. INSTALLATION

356.10 Uses Permitted. Listed liquidtight flexible nonmetallic conduit is permitted, either exposed or concealed, at any of the following locations:

(1) Where flexibility is required.

(2) Where protection from liquids, vapors, or solids is required.

(3) Outdoors, if listed and marked for this purpose.

(4) Directly buried in the earth, if listed and marked for this purpose.

(5) LFNC-B (gray color) is permitted in lengths over 6 ft where secured according to 356.30.

(6) LFNC-B (black color) as a listed manufactured prewired assembly.

(7) Encasement in concrete where listed for direct burial.

356.12 Uses Not Permitted.

(1) Where subject to physical damage.

(2) Where the combination of ambient and conductor temperature will produce an operating temperature above the rating of the raceway.

(3) Longer than 6 ft, except if approved by the authority having jurisdiction as essential for a required degree of flexibility.

(4) Where the operating voltage of the contained conductors exceeds 600 volts, nominal.

(5) In any hazardous (classified) location, except as permitted by 501.10(B), 502.10(A) and (B), and 504.20.

356.20 Trade Size.

(A) Minimum. Liquidtight flexible nonmetallic conduit smaller than trade size ½ is not permitted, except as permitted in the following:

(1) Enclosing the leads of motors, 430.245(B).

(2) Tap connections to lighting fixtures as permitted by 410.117(C).

(B) Maximum. Liquidtight flexible nonmetallic conduit larger than trade size 4 is not permitted.

356.22 Number of Conductors.

Raceways must be large enough to permit the installation and removal of conductors without damaging the insulation. When all conductors in a raceway are the same size and insulation, the number of conductors permitted can be found in Annex C for the raceway type. Figure 356–2

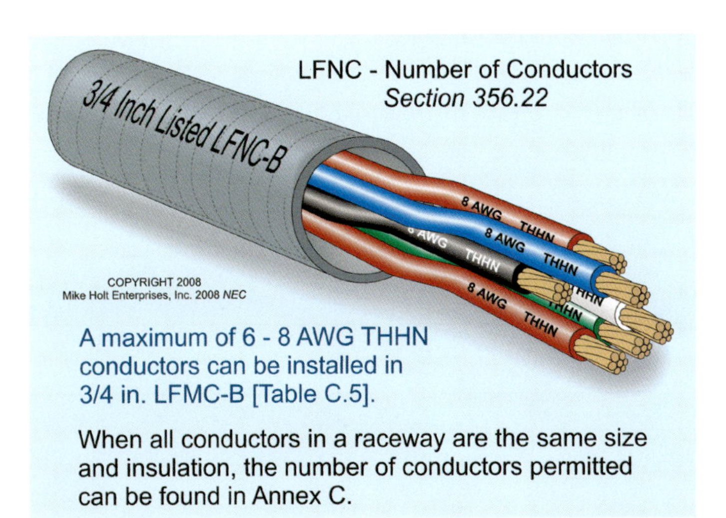

Figure 356–2

Question: How many 8 THHN conductors can be installed in trade size ¾ LFNC-B?

Answer: Six conductors [Annex C, Table C5]

Cables can be installed in liquidtight flexible nonmetallic conduit, as long as the number of cables does not exceed the allowable percentage fill specified in Table 1, Chapter 9.

356.24 Bends. Raceway bends must not be made in any manner that would damage the raceway or significantly change its internal diameter (no kinks). The radius of the curve of the inner edge of any field bend must not be less than shown in Table 2, Chapter 9 using the column "Other Bends."

356.26 Number of Bends (360°). To reduce the stress and friction on conductor insulation, the maximum number of bends (including offsets) between pull points must not exceed 360°.

Author's Comment: There is no maximum distance between pull boxes because this is a design issue, not a safety issue.

356.30 Securing and Supporting. LFNC-B (gray color) must be securely fastened and supported in accordance with one of the following: Figure 356–3

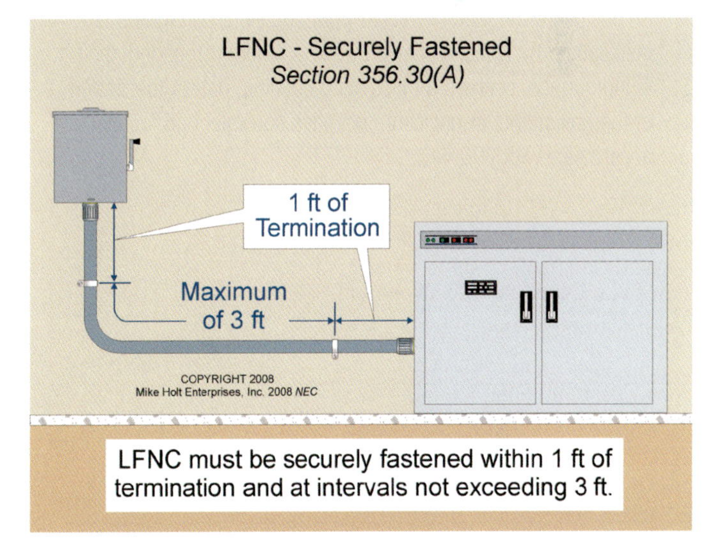

Figure 356–3

(1) The conduit must be securely fastened at intervals not exceeding 3 ft, and within 1 ft of termination when installed longer than 6 ft.

(2) Securing or supporting isn't required where it's fished, installed in lengths not exceeding 3 ft at terminals where flexibility is required, or installed in lengths not exceeding 6 ft for tap conductors to luminaires, as permitted in 410.117(C).

(3) Horizontal runs of liquidtight flexible nonmetallic conduit installed horizontally in bored or punched holes in wood or metal framing members, or notches in wooden members, are considered supported, but the raceway must be secured within 1 ft of termination.

(4) Securing or supporting of LFNC-B (gray color) isn't required where installed in lengths not exceeding 6 ft from the last point where the raceway is securely fastened for connections within an accessible ceiling to luminaire(s) or other equipment.

356.42 Fittings. Only fittings listed for use with liquidtight flexible nonmetallic conduit can be used [300.15]. Angle connector fittings must not be used in concealed raceway installations. Straight liquidtight flexible nonmetallic conduit fittings are permitted for direct burial or encasement in concrete.

Author's Comment: Conductors 4 AWG and larger that enter an enclosure must be protected from abrasion, during and after installation, by a fitting that provides a smooth, rounded, insulating surface, such as an insulating bushing, unless the design of the box, fitting, or enclosure provides equivalent protection, in accordance with 300.4(G).

356.60 Equipment Grounding Conductor. Where equipment grounding is required, a separate equipment grounding conductor of the wire type must be installed within the conduit [300.2(B)]. Figure 356–4

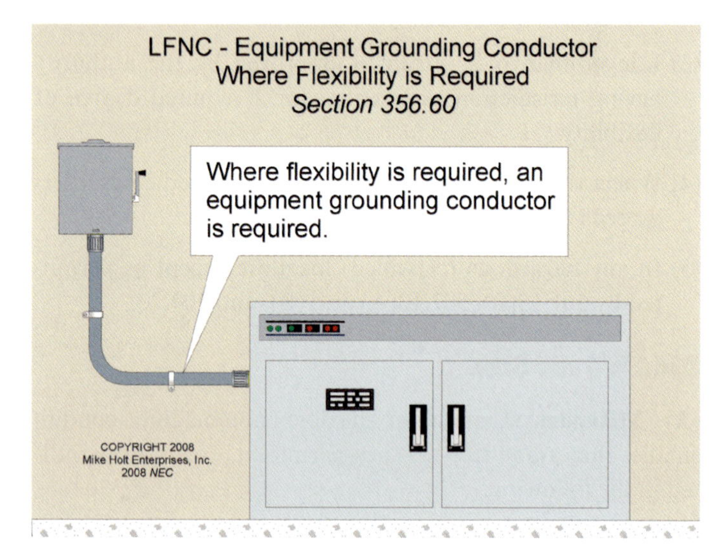

Figure 356–4

Where an equipment bonding jumper is installed outside of a raceway, the length of the equipment bonding jumper must not exceed 6 ft, and it must be routed with the raceway or enclosure in accordance with 250.102(E).

ARTICLE 356 Practice Questions

ARTICLE 356. LIQUIDTIGHT FLEXIBLE NONMETALLIC CONDUIT (TYPE LFNC)—PRACTICE QUESTIONS

1. _____ is a listed raceway of circular cross section having an outer liquidtight, nonmetallic, sunlight-resistant jacket over a flexible inner core.

   (a) LFNC
   (b) ENT
   (c) NUCC
   (d) RTRC

2. LFNC shall be permitted for _____.

   (a) direct burial where listed and marked for the purpose
   (b) exposed work
   (c) outdoors where listed and marked for this purpose
   (d) all of these

3. Bends in LFNC shall be made so that the conduit will not be damaged and the internal diameter of the conduit will not be effectively reduced. Bends can be made _____.

   (a) manually without auxiliary equipment
   (b) with bending equipment identified for the purpose
   (c) with any kind of conduit bending tool that will work
   (d) by the use of an open flame torch

ARTICLE

358 Electrical Metallic Tubing (Type EMT)

INTRODUCTION TO ARTICLE 358—ELECTRICAL METALLIC TUBING (TYPE EMT)

Electrical metallic tubing (EMT) is a lightweight raceway that is relatively easy to bend, cut, and ream. Because it isn't threaded, all connectors and couplings are of the threadless type and provide quick, easy, and inexpensive installation when compared to other metallic conduit systems, which makes it very popular. Electrical metallic tubing is manufactured in both galvanized steel and aluminum; the steel type is the most common type used.

PART I. GENERAL

358.1 Scope. Article 358 covers the use, installation, and construction specifications of electrical metallic tubing.

358.2 Definition.

Electrical Metallic Tubing (Type EMT). A metallic tubing of circular cross section used for the installation and physical protection of electrical conductors when joined together with fittings.

358.6 Listing Requirement. Electrical metallic tubing, elbows, and associated fittings must be listed.

PART II. INSTALLATION

358.10 Uses Permitted.

(A) Exposed and Concealed. Electrical metallic tubing is permitted exposed or concealed.

(B) Corrosion Protection. Electrical metallic tubing, elbows, couplings, and fittings can be installed in concrete, in direct contact with the earth, or in areas subject to severe corrosive influences where protected by corrosion protection and judged suitable for the condition. Figure 358–1

CAUTION: Supplementary coatings for corrosion protection (asphalted paint) have not been investigated by a product testing and listing agency, and these coatings are known to cause cancer in laboratory animals.

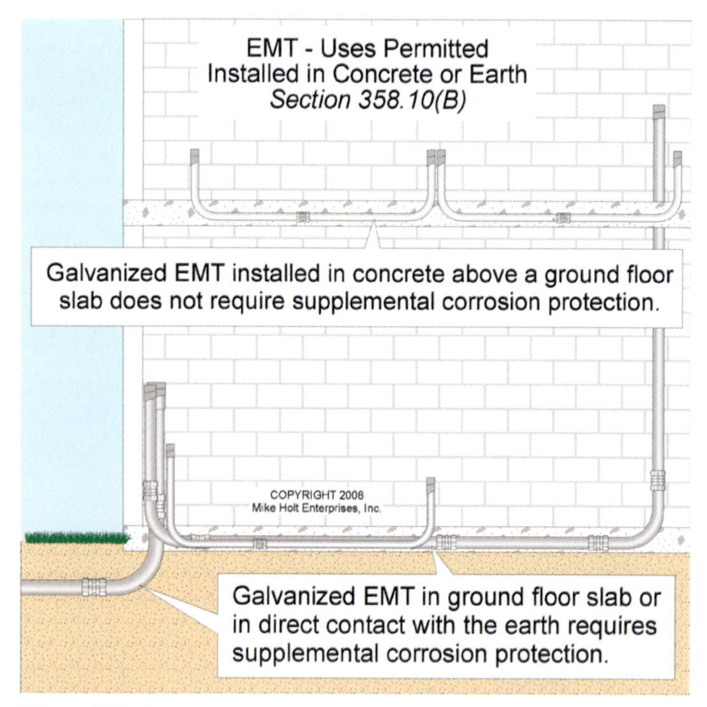

Figure 358–1

(C) Wet Locations. Support fittings, such as screws, straps, etc., installed in a wet location must be made of corrosion-resistant material, or a corrosion-resistant coating must protect them in accordance with 300.6.

Author's Comment: Fittings used in wet locations must be listed for the application (wet location). For more information, visit http://www.etpfittings.com/.

358.12 Uses Not Permitted.

EMT must not be used under the following conditions:

(1) Where, during installation or afterward, it will be subject to severe physical damage.

(2) Where protected from corrosion solely by enamel.

(3) In cinder concrete or cinder fill where subject to permanent moisture, unless encased in not less than 2 in. of concrete.

(4) In any hazardous (classified) location, except as permitted by 502.10, 503.10, and 504.20.

(5) For the support of luminaires or other equipment (like boxes), except conduit bodies no larger than the largest trade size of the tubing can be supported by the raceway. Figure 358–2

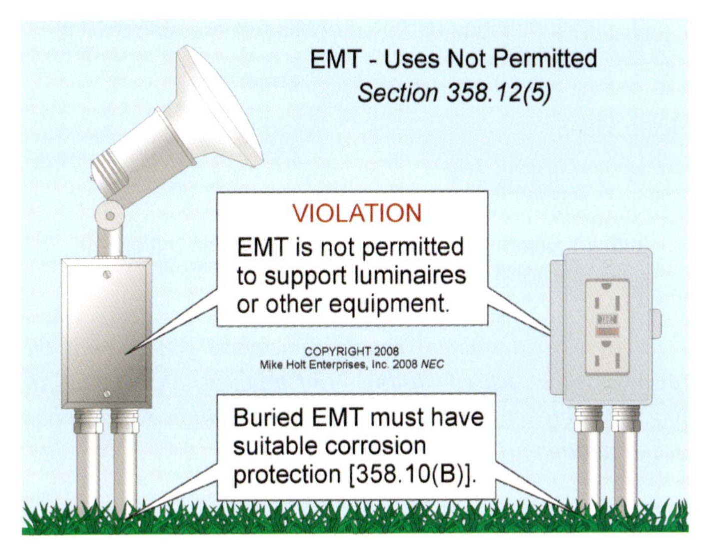

Figure 358–2

(6) Where practical, contact with dissimilar metals must be avoided to prevent the deterioration of the metal because of galvanic action.

Exception: Aluminum fittings are permitted on steel electrical metallic tubing, and steel fittings are permitted on aluminum EMT.

358.20 Trade Size.

(A) Minimum. Electrical metallic tubing smaller than trade size ½ is not permitted.

(B) Maximum. Electrical metallic tubing larger than trade size 4 is not permitted.

358.22 Number of Conductors.

Raceways must be large enough to permit the installation and removal of conductors without damaging the conductor insulation. When all conductors in a raceway are the same size and insulation, the number of conductors permitted can be found in Annex C for the raceway type.

Question: *How many 12 THHN conductors can be installed in trade size 1 EMT?* Figure 358–3

(a) 26 *(b) 28* *(c) 30* *(d) 32*

Answer: *(a) 26 [Annex C, Table C.1]*

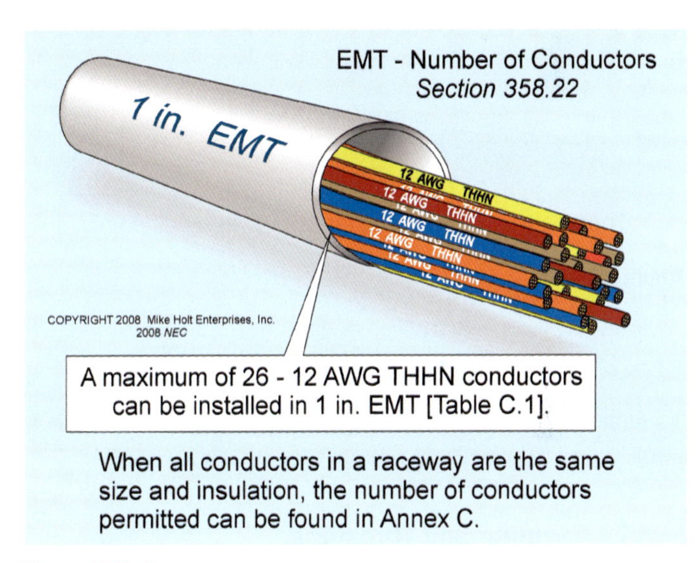

Figure 358–3

Author's Comment: See 300.17 for additional examples on how to size raceways when conductors aren't all the same size.

Cables can be installed in electrical metallic tubing, as long as the number of cables does not exceed the allowable percentage fill specified in Table 1, Chapter 9.

358.24 Bends.

Raceway bends must not be made in any manner that would damage the raceway, or significantly change its internal diameter (no kinks). The radius of the curve of the inner edge of any field bend must not be less than shown in Chapter 9, Table 2 for one-shot and full shoe benders.

Author's Comment: This typically isn't a problem, because most benders are made to comply with this table.

358.26 Number of Bends (360°). To reduce the stress and friction on conductor insulation, the maximum number of bends (including offsets) between pull points can't exceed 360°. Figure 358–4

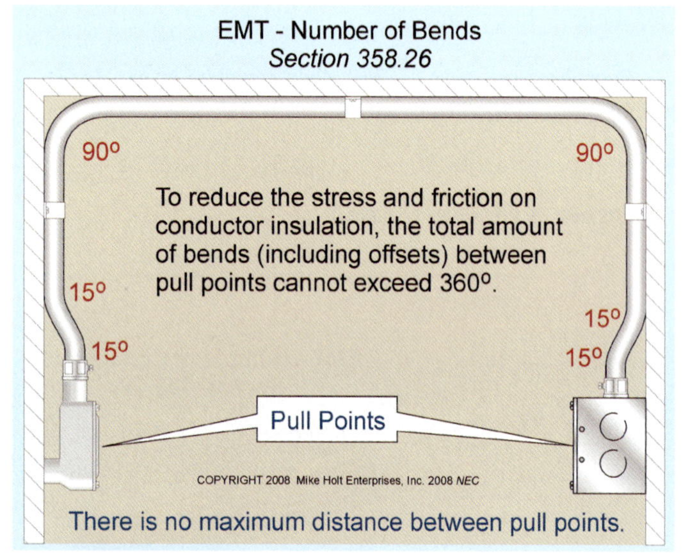

Figure 358–4

Author's Comment: There is no maximum distance between pull boxes because this is a design issue, not a safety issue.

358.28 Reaming and Threading.

(A) Reaming. Reaming to remove the burrs and rough edges is required when the raceway is cut.

Author's Comment: It's considered an accepted practice to ream small raceways with a screwdriver or the backside of pliers.

(B) Threading. Electrical metallic tubing must not be threaded.

358.30 Securing and Supporting. Electrical metallic tubing must be installed as a complete system in accordance with 300.18 [300.10 and 300.12], and it must be securely fastened in place and supported in accordance with (A) and (B), or unsupported as permitted in (C).

(A) Securely Fastened. Electrical metallic tubing must generally be securely fastened within 3 ft of every box, cabinet, or termination fitting, and at intervals not exceeding 10 ft. Figure 358–5

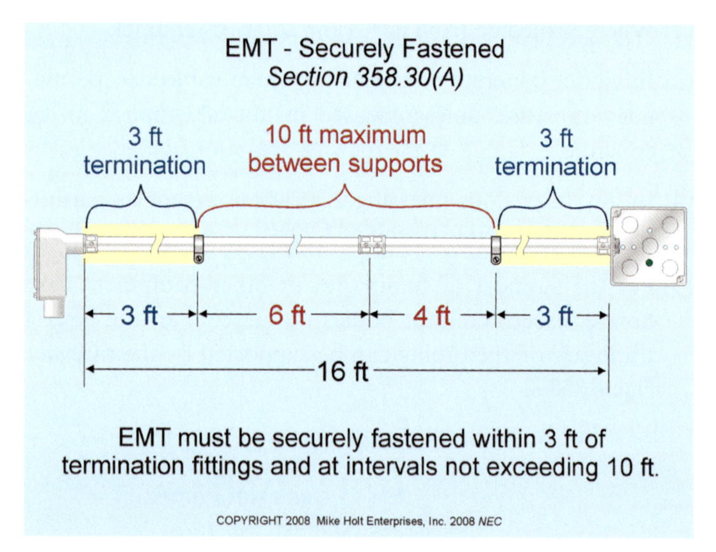

Figure 358–5

Author's Comment: Fastening is required within 3 ft of termination, not within 3 ft of a coupling.

Exception No. 1: When structural members don't permit the raceway to be secured within 3 ft of a box or termination fitting, an unbroken raceway can be secured within 5 ft of a box or termination fitting. Figure 358–6

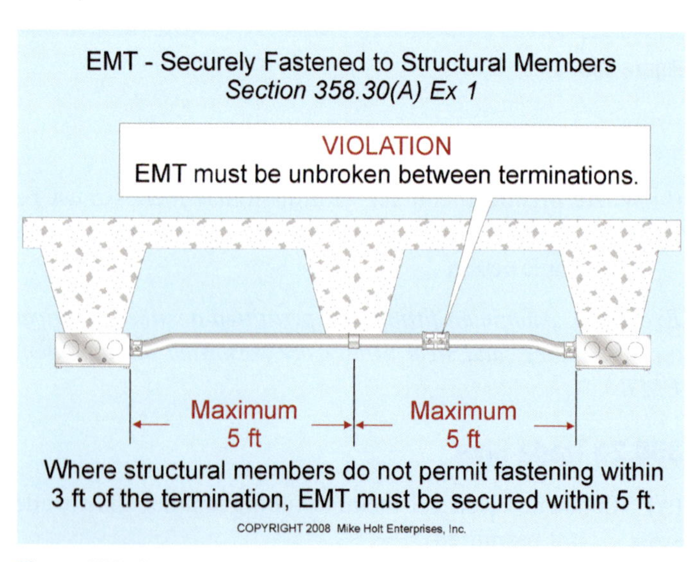

Figure 358–6

(B) Horizontal Runs. Electrical metallic tubing installed horizontally in bored or punched holes in wood or metal framing members, or notches in wooden members, is considered supported, but the raceway must be secured within 3 ft of termination.

(C) Unsupported Raceways. Where oversized concentric or eccentric knockouts are not encountered, electrical metallic tubing is permitted to be unsupported between enclosures where the raceway is unbroken and not more than 18 in. in length. Figure 358–7

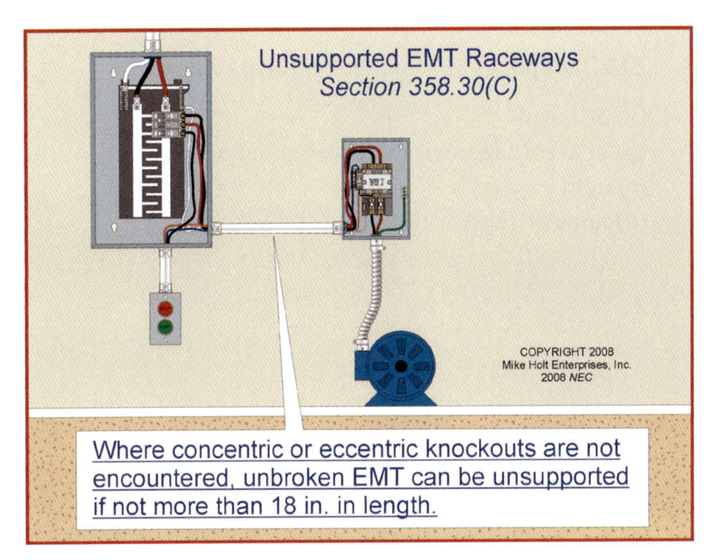

Figure 358–7

358.42 Couplings and Connectors. Couplings and connectors must be made up tight to maintain an effective ground-fault current path to safely conduct fault current in accordance with 250.4(A)(5), 250.96(A), and 300.10.

Where buried in masonry or concrete, threadless electrical metallic tubing fittings must be of the concrete-tight type. Where installed in wet locations, fittings must be listed for use in wet locations in accordance with 314.15(A).

> **Author's Comment:** Conductors 4 AWG and larger that enter an enclosure must be protected from abrasion, during and after installation, by a fitting that provides a smooth, rounded, insulating surface, such as an insulating bushing, unless the design of the box, fitting, or enclosure provides equivalent protection, in accordance with 300.4(G). Figure 358–8

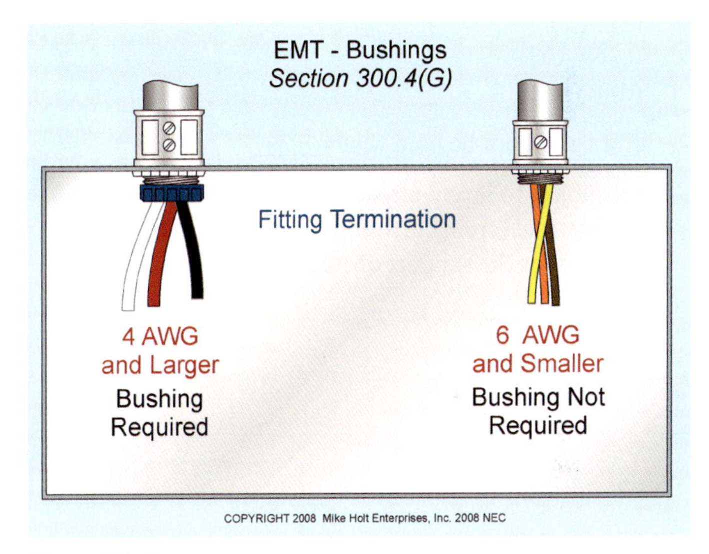

Figure 358–8

ARTICLE 358 Practice Questions

ARTICLE 358. ELECTRICAL METALLIC TUBING (TYPE EMT)—PRACTICE QUESTIONS

1. _____ is a listed thinwall, metallic tubing of circular cross section used for the installation and physical protection of electrical conductors when joined together with listed fittings.

   (a) LFNC
   (b) EMT
   (c) NUCC
   (d) RTRC

2. EMT shall not be used where _____.

   (a) subject to severe physical damage
   (b) protected from corrosion only by enamel
   (c) used for the support of luminaires
   (d) any of these

3. EMT shall not be threaded.

   (a) True
   (b) False

4. EMT couplings and connectors shall be made up _____.

   (a) of metal
   (b) in accordance with industry standards
   (c) tight
   (d) none of these

ARTICLE

362 Electrical Nonmetallic Tubing (Type ENT)

INTRODUCTION TO ARTICLE 362—ELECTRICAL NONMETALLIC TUBING (TYPE ENT)

Electrical nonmetallic tubing is a pliable, corrugated, circular raceway made of polyvinyl chloride. In some parts of the country, the field name for electrical nonmetallic tubing is "Smurf Pipe" or "Smurf Tube," because it was only available in blue when it originally came out at the time the children's cartoon characters "The Smurfs" were most popular. Today, the raceway is available in a rainbow of colors such as white, yellow, red, green, and orange, and is sold in both fixed lengths and on reels.

PART I. GENERAL

362.1 Scope. Article 362 covers the use, installation, and construction specifications of electrical nonmetallic tubing and associated fittings.

362.2 Definition.

Electrical Nonmetallic Tubing (Type ENT). A pliable corrugated raceway of circular cross section, with integral or associated couplings, connectors, and fittings listed for the installation of electrical conductors.

Electrical nonmetallic tubing can be bent by hand with a reasonable force, but without other assistance.

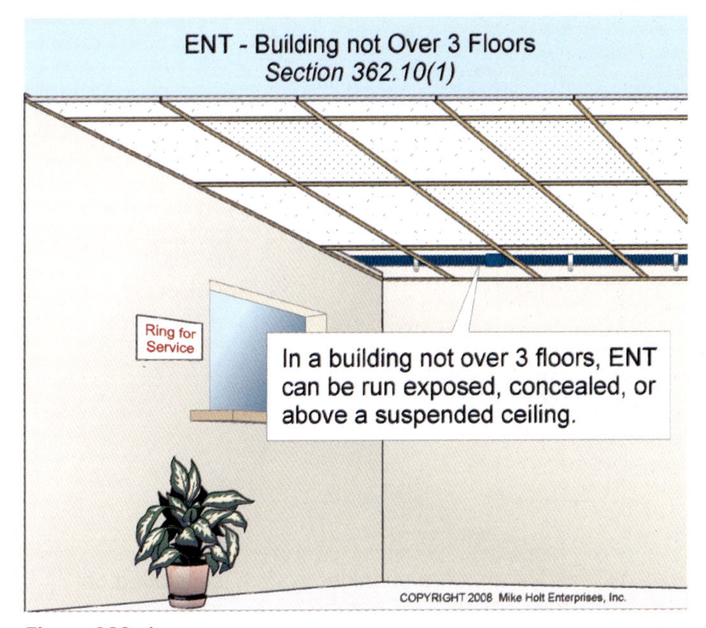

Figure 362–1

PART II. INSTALLATION

362.10 Uses Permitted.

Definition of First Floor: The first floor of a building is the floor with 50 percent or more of the exterior wall surface area level with or above finished grade. If one additional level not designed for human habitation and used only for vehicle parking, storage, or similar use is at ground level, then the first of the three permissible floors can be the next higher floor.

(1) In buildings not exceeding three floors. Figure 362–1

a. Exposed, where not prohibited by 362.12.

b. Concealed within walls, floors, and ceilings.

(2) In buildings exceeding three floors, electrical nonmetallic tubing can be installed concealed in walls, floors, or ceilings that provide a thermal barrier having a 15-minute finish rating, as identified in listings of fire-rated assemblies. Figure 362–2

Exception to (2): Where a fire sprinkler system is installed on all floors, in accordance with NFPA 13, Standard for the Installation of Sprinkler Systems, electrical nonmetallic tubing is permitted exposed or concealed in buildings of any height. Figure 362–3

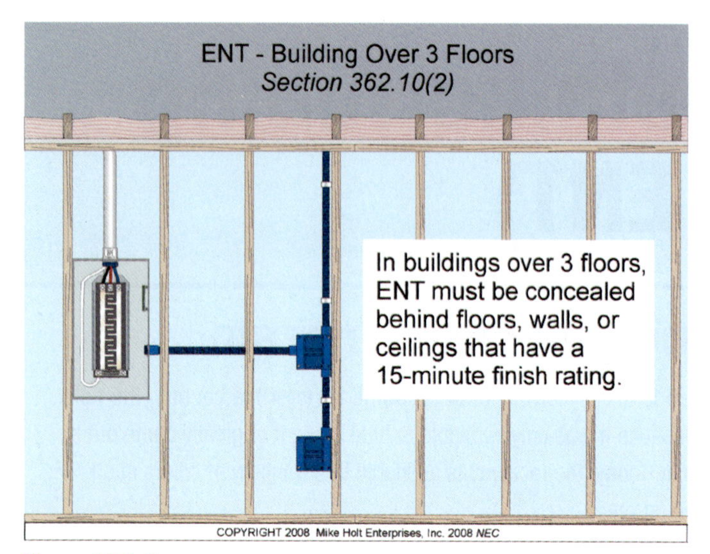

Figure 362–2

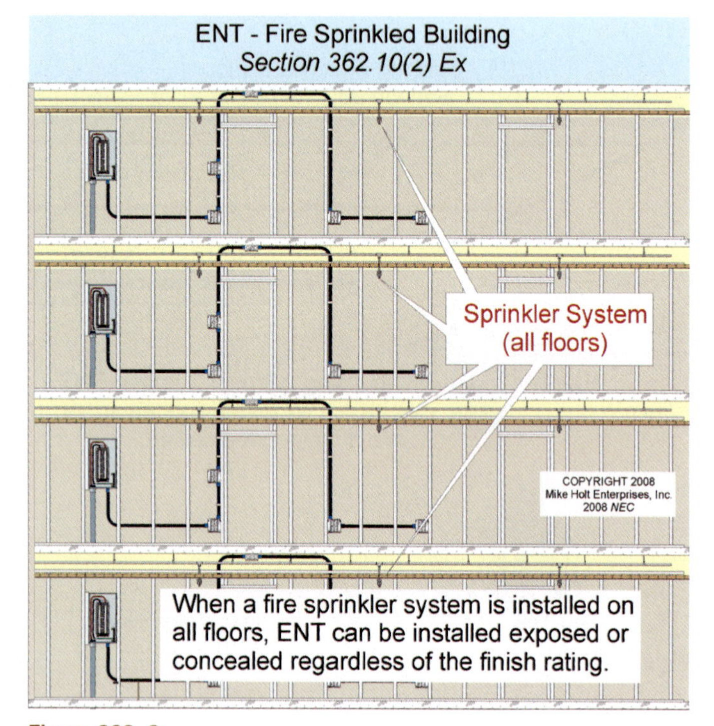

Figure 362–3

(3) Electrical nonmetallic tubing is permitted in severe corrosive and chemical locations, when identified for this use.

(4) Electrical nonmetallic tubing is permitted in dry and damp concealed locations, where not prohibited by 362.12.

(5) Electrical nonmetallic tubing is permitted above a suspended ceiling, if the suspended ceiling provides a thermal barrier having a 15-minute finish rating, as identified in listings of fire-rated assemblies. Figure 362–4

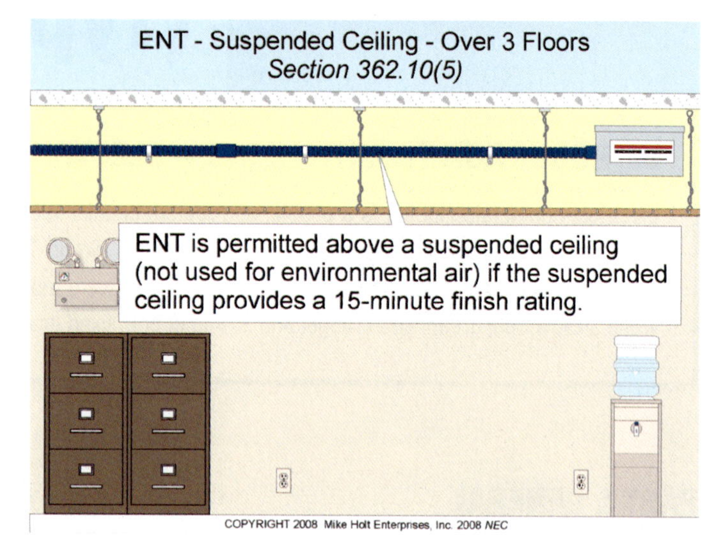

Figure 362–4

Exception: Where a fire sprinkler system is installed on all floors, in accordance with NFPA 13, Standard for the Installation of Sprinkler Systems, electrical nonmetallic tubing is permitted above a suspended ceiling that doesn't have a 15-minute finish rated thermal barrier material. Figure 362–5

(6) Electrical nonmetallic tubing can be encased or embedded in a concrete slab provided fittings identified for the purpose are used.

Author's Comment: Electrical nonmetallic tubing is not permitted in the earth [362.12(5)].

(7) Electrical nonmetallic tubing is permitted in wet locations indoors, or in a concrete slab on or below grade, with fittings listed for the purpose.

(8) Listed prewired electrical nonmetallic tubing with conductors is permitted in trade sizes ½, ¾, and 1.

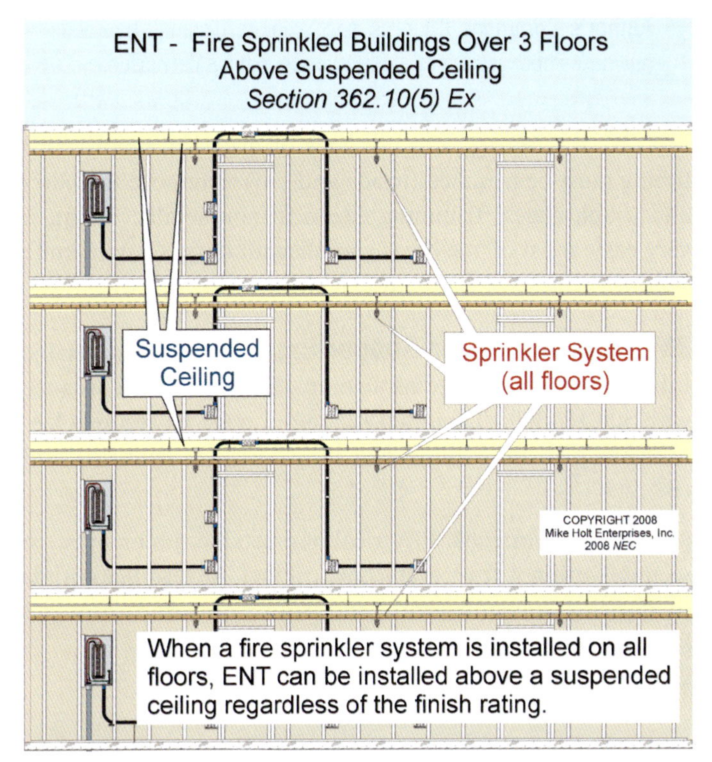

Figure 362–5

362.12 Uses Not Permitted.

(1) In any hazardous (classified) location, except as permitted by 504.20 and 505.15(A)(1).

(2) For the support of luminaires or equipment. See 314.2.

(3) Where the ambient temperature exceeds 50°C (122°F).

(4) Where conductors operate at a temperature above the temperature rating of the raceway. Figure 362–6

Exception: Conductors rated at a temperature above the electrical nonmetallic tubing temperature rating are permitted, provided the conductors don't operate at a temperature above the electrical nonmetallic tubing's listed temperature rating.

(5) For direct earth burial.

Author's Comment: Electrical nonmetallic tubing can be encased in concrete [362.10(6)].

(6) As a wiring method for systems over 600V.

(7) Exposed in buildings over three floors, except as permitted by 362.10(2) and (5) Ex.

(8) In assembly occupancies or theaters, except as permitted by 518.4 and 520.5.

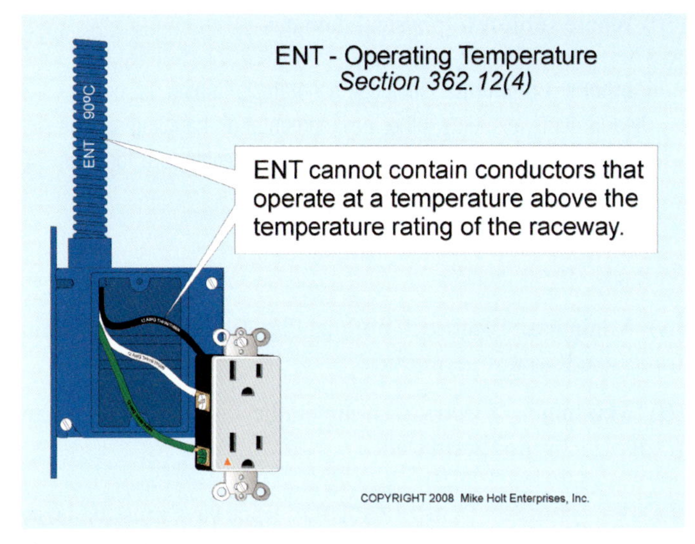

Figure 362–6

(9) Exposed to the direct rays of the sun for an extended period, unless listed as sunlight resistant.

Author's Comment: Exposing electrical nonmetallic tubing exposed to the direct rays of the sun for an extended time may result in the product becoming brittle, unless it is listed to resist the effects of ultraviolet (UV) radiation. Figure 362–7

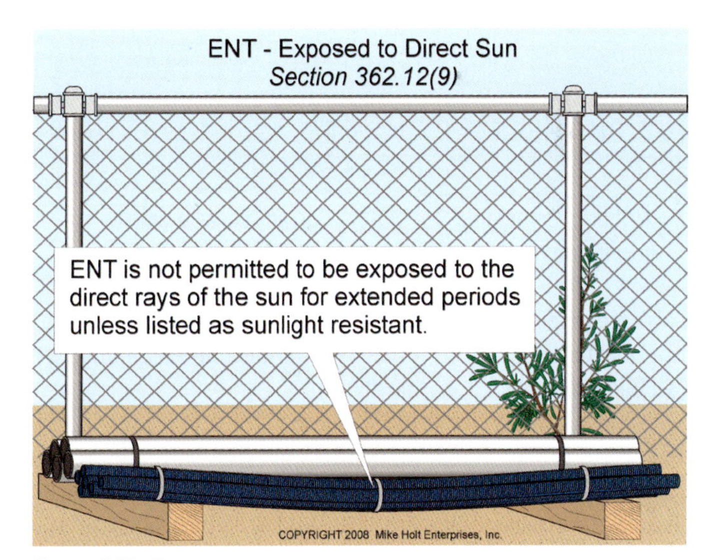

Figure 362–7

(10) Where subject to physical damage.

Author's Comment: Electrical nonmetallic tubing is prohibited in ducts, plenums, other environmental air-handling spaces [300.22], and patient care area circuits in health care facilities [517.13(A)].

362.20 Trade Sizes.

(A) Minimum. Electrical nonmetallic tubing smaller than trade size ½ is not permitted.

(B) Maximum. Electrical nonmetallic tubing larger than trade size 2 is not permitted.

362.22 Number of Conductors. Raceways must be large enough to permit the installation and removal of conductors without damaging the conductors' insulation, and the number of conductors must not exceed that permitted by the percentage fill specified in Table 1, Chapter 9.

When all conductors in a raceway are the same size and insulation, the number of conductors permitted can be found in Annex C for the raceway type.

Question: *How many 12 THHN conductors can be installed in trade size 1/2 ENT?*

(a) 5 *(b) 7* *(c) 9* *(d) 11*

Answer: *(b) 7 conductors [Annex C, Table C2]*

Author's Comment: See 300.17 for additional examples on how to size raceways when conductors aren't all the same size.

Cables can be installed in electrical nonmetallic tubing, as long as the cables don't exceed the allowable percentage fill specified in Table 1, Chapter 9.

362.24 Bends. Raceway bends must not be made in any manner that would damage the raceway, or significantly change its internal diameter (no kinks). The radius of the curve to the centerline of any field bend must not be less than shown in Chapter 9, Table 2, using the column "Other Bends."

362.26 Number of Bends (360°). To reduce the stress and friction on conductor insulation, the maximum number of bends (including offsets) between pull points can't exceed 360°.

Author's Comment: There is no maximum distance between pull boxes because this is a design issue, not a safety issue.

362.28 Trimming. The cut ends of electrical nonmetallic tubing must be trimmed (inside and out) to remove the burrs and rough edges. Trimming electrical nonmetallic tubing is very easy; most of the burrs rub off with fingers, and a knife can be used to smooth the rough edges.

362.30 Securing and Supporting. Electrical nonmetallic tubing must be installed as a complete system in accordance with 300.18 [300.10 and 300.12], and it must be securely fastened in place and supported in accordance with (A) and (B).

(A) Securely Fastened. Electrical nonmetallic tubing must be secured within 3 ft of every box, cabinet, or termination fitting, such as a conduit body, and at intervals not exceeding 3 ft. Figure 362–8

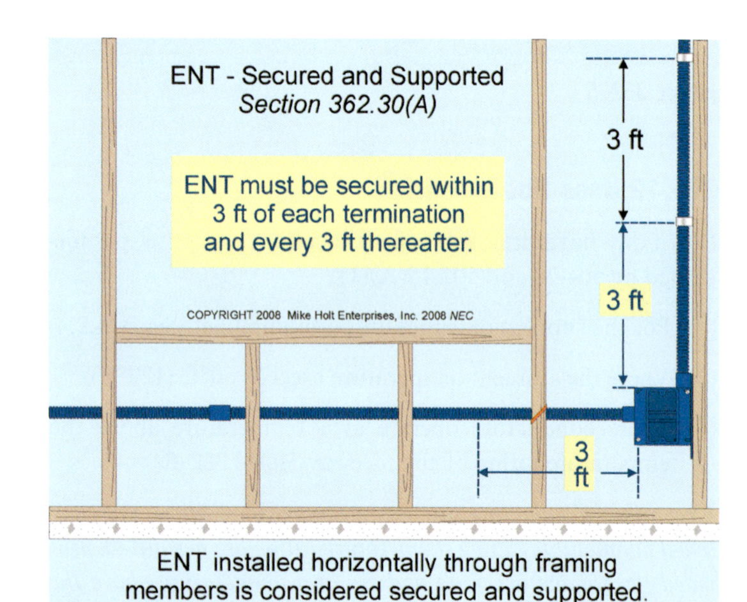

Figure 362–8

Exception No. 2: Lengths not exceeding 6 ft from the last point where the raceway is securely fastened within an accessible ceiling to luminaire(s) or other equipment.

Exception No. 3: Where fished between access points through concealed spaces and supporting is impractical.

(B) Horizontal Runs. Electrical nonmetallic tubing installed horizontally in bored or punched holes in wood or metal framing members, or notches in wooden members, is considered supported, but the raceway must be secured within 3 ft of terminations.

362.46 Bushings. Conductors 4 AWG and larger that enter an enclosure from a fitting must be protected from abrasion, during and after installation, by a fitting that provides a smooth, rounded, insulating surface, such as an insulating bushing, unless the design of the box, fitting, or enclosure provides equivalent protection, in accordance with 300.4(G).

362.48 Joints. Joints, such as couplings and connectors, must be made in a manner approved by the authority having jurisdiction.

Author's Comment: Follow the manufacturer's instructions for the raceway, fittings, and glue. According to product listings, PVC conduit fittings are permitted with electrical nonmetallic tubing.

CAUTION: *Glue used with electrical nonmetallic tubing must be listed for ENT. Glue for PVC conduit must not be used with electrical nonmetallic tubing because it damages the plastic from which electrical nonmetallic tubing is manufactured.*

362.60 Equipment Grounding Conductor. Where equipment grounding is required, a separate equipment grounding conductor of the wire type must be installed within the raceway. Figure 362–9

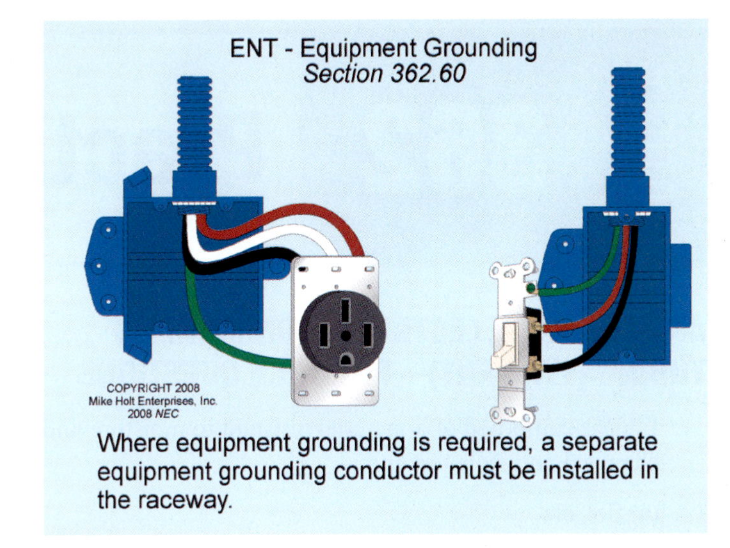

Figure 362–9

ARTICLE 362 Practice Questions

ARTICLE 362. ELECTRICAL NONMETALLIC TUBING (TYPE ENT)—PRACTICE QUESTIONS

1. ENT is composed of a material resistant to moisture and chemical atmospheres, and is _____.

   (a) flexible
   (b) flame retardant
   (c) fireproof
   (d) flammable

2. ENT is not permitted in hazardous (classified) locations, unless permitted in other articles of the *Code*.

   (a) True
   (b) False

3. ENT shall not be used where exposed to the direct rays of the sun, unless identified as _____.

   (a) high-temperature rated
   (b) sunlight resistant
   (c) Schedule 80
   (d) never can be

4. Cut ends of ENT shall be trimmed inside and _____ to remove rough edges.

   (a) outside
   (b) tapered
   (c) filed
   (d) beveled

5. Bushings or adapters shall be provided at ENT terminations to protect the conductors from abrasion, unless the box, fitting, or enclosure design provides equivalent protection.

   (a) True
   (b) False

Metal Wireways

INTRODUCTION TO ARTICLE 376—METAL WIREWAYS

Metal wireways are commonly used where access to the conductors within the raceway is required to make terminations, splices, or taps to several devices at a single location. High cost precludes their use for other than short distances, except in some commercial or industrial occupancies where the wiring is frequently revised.

Author's Comment: Both metal wireways and nonmetallic wireways are often called "troughs" or "gutters" in the field.

PART I. GENERAL

376.1 Scope. Article 376 covers the use, installation, and construction specifications of metal wireways and associated fittings.

376.2 Definition.

Metal Wireway. A sheet metal raceway with hinged or removable covers for housing and protecting electric conductors and cable, and in which conductors are placed after the wireway has been installed. Figure 376–1

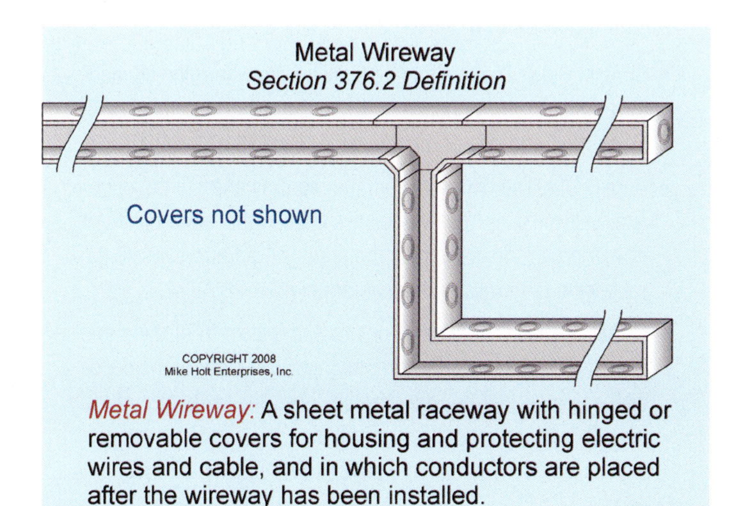

Figure 376–1

PART II. INSTALLATION

376.10 Uses Permitted.

(1) Exposed.

(2) Concealed, as permitted by 376.10(4).

(3) In any hazardous (classified) locations, as permitted by 501.10(B), 502.10(B), or 504.20.

(4) Unbroken through walls, partitions, and floors.

376.12 Uses Not Permitted.

(1) Subject to severe physical damage.

(2) Subject to corrosive environments.

376.21 Conductors—Maximum Size. The maximum size conductor permitted in a wireway must not be larger than that for which the wireway is designed.

376.22 Number of Conductors and Ampacity. The number of conductors and their ampacity must comply with 376.22(A) and (B).

(A) Number of Conductors. The maximum number of conductors permitted in a wireway is limited to 20 percent of the cross-sectional area of the wireway. Figure 376–2

Author's Comment: Splices and taps must not fill more than 75 percent of the wiring space at any cross section [376.56].

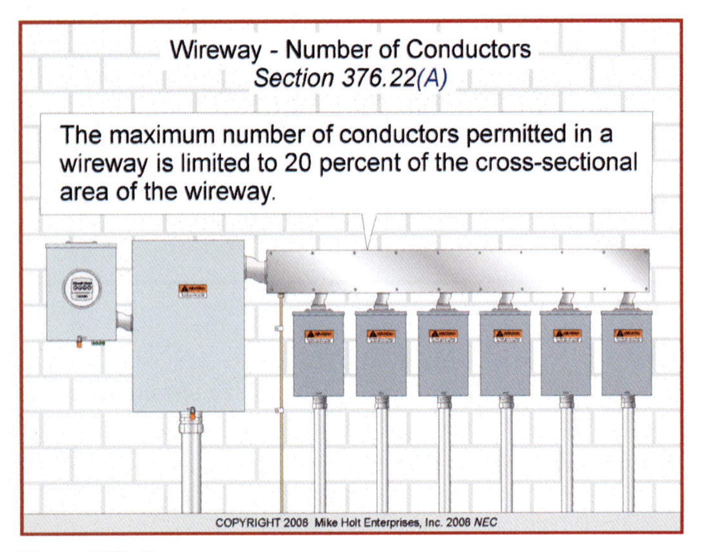

Figure 376–2

(B) Conductor Ampacity Adjustment Factors. When more than 30 current-carrying conductors are installed in any cross-sectional area of the wireway, the conductor ampacity, as listed in Table 310.16, must be adjusted in accordance with Table 310.15(B)(2)(a). Figure 376–3

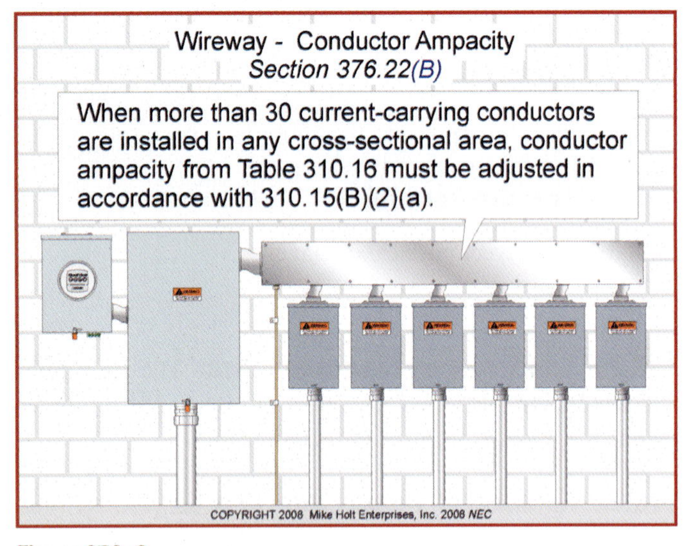

Figure 376–3

Signaling and motor control conductors between a motor and its starter used only for starting duty aren't considered current carrying for conductor ampacity adjustment.

376.23 Wireway Sizing.

(A) Sizing for Conductor Bending Radius. Where conductors are bent within a metal wireway, the wireway must be sized to meet the bending radius requirements contained in Table 312.6(A), based on one wire per terminal. Figure 376–4

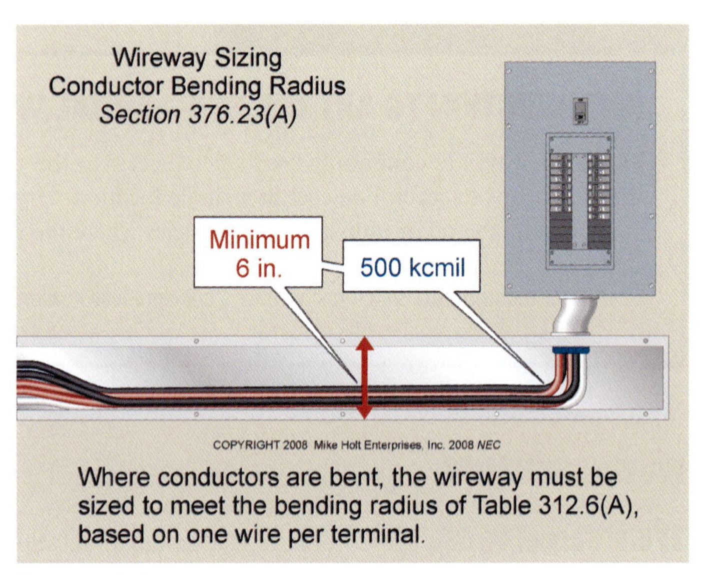

Figure 376–4

(B) Wireway Used as Pull Box. Where insulated conductors 4 AWG or larger are pulled through a metal wireway, the distance between raceway and cable entries enclosing the same conductor must not be less than required by 314.28(A)(1) and 314.28(A)(2). Figure 376–5

Author's Comments:

- Straight Pulls. The minimum distance from where the conductors enter to the opposite wall must not be less than eight times the trade size of the largest raceway [314.28(A)(1)].
- Angle Pulls. The distance from the raceway entry to the opposite wall must not be less than six times the trade diameter of the largest raceway, plus the sum of the trade sizes of the remaining raceways on the same wall [314.28(A)(2)].
- U Pulls. When a conductor enters and leaves from the same wall, the distance from where the raceways enter to the opposite wall must not be less than six times the trade size of the largest raceway, plus the sum of the trade sizes of the remaining raceways on the same wall [314.28(A)(2)].
- The distance between raceways enclosing the same conductor must not be less than six times the trade size of the largest raceway [314.28(A)(2)].

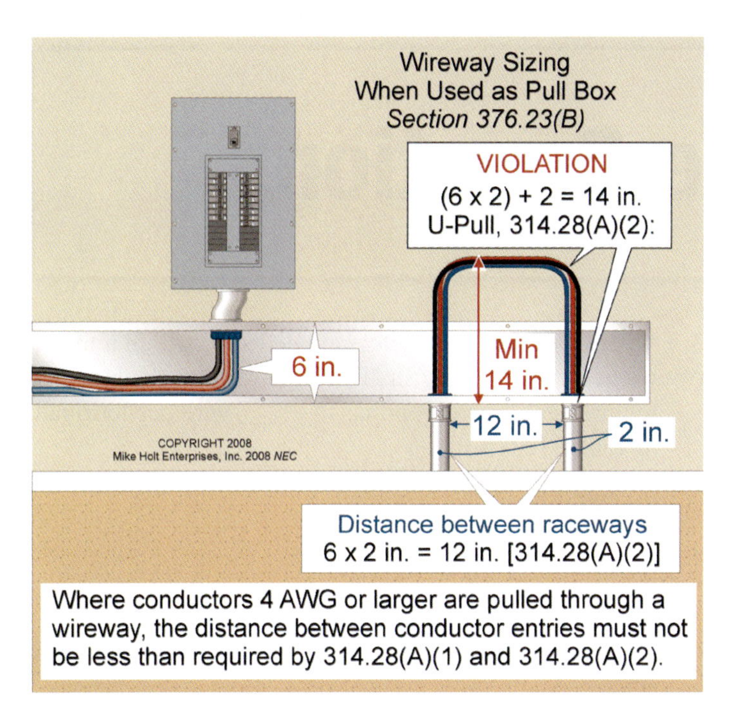

Figure 376–5

376.30 Supports.

Wireways must be supported in accordance with (A) and (B).

(A) Horizontal Support. Where run horizontally, metal wireways must be supported at each end and at intervals not exceeding 5 ft.

(B) Vertical Support. Where run vertically, metal wireways must be securely supported at intervals not exceeding 15 ft, with no more than one joint between supports.

376.56 Splices, Taps, and Power Distribution Blocks.

(A) Splices and Taps. Splices and taps in metal wireways must be accessible, and they must not fill the wireway to more than 75 percent of its cross-sectional area. Figure 376–6

Author's Comment: The maximum number of conductors permitted in a metal wireway is limited to 20 percent of its cross-sectional area at any point [376.22(A)].

(B) Power Distribution Blocks.

(1) Installation. Power distribution blocks installed in wireways must be listed.

(2) Size of Enclosure. In addition to the wiring space requirements [376.56(A)], the power distribution block must be installed in a metal wireway not smaller than specified in the installation instructions of the power distribution block.

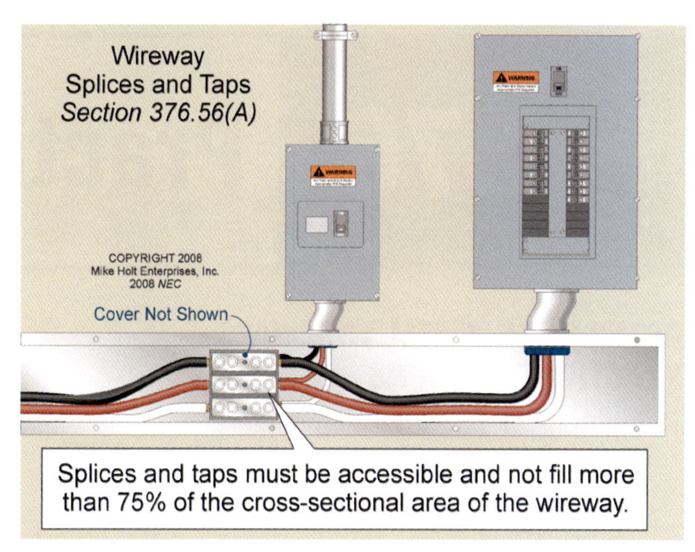

Figure 376–6

(3) Wire Bending Space. Wire bending space at the terminals of power distribution blocks must comply with 312.6(B).

(4) Live Parts. Power distribution blocks must not have uninsulated exposed live parts in the metal wireway after installation, whether or not the wireway cover is installed. Figure 376–7

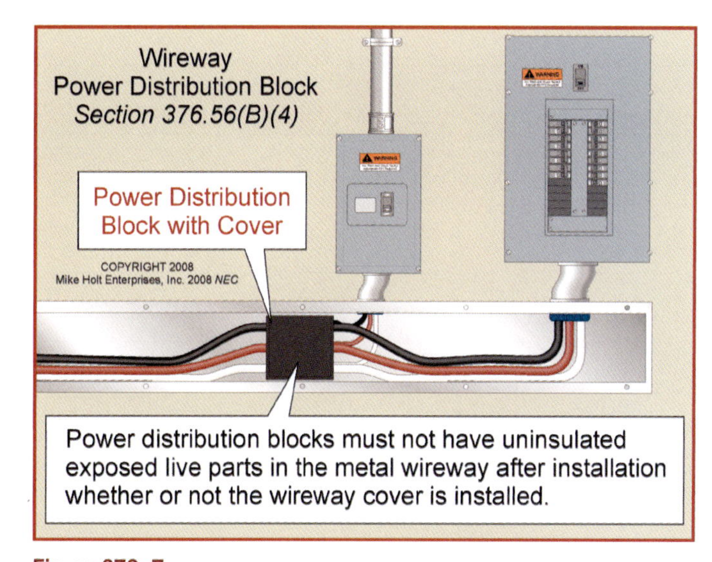

Figure 376–7

ARTICLE 376 Practice Questions

ARTICLE 376. METAL WIREWAYS— PRACTICE QUESTIONS

1. Metal wireways are sheet metal troughs with _____ for housing and protecting electric conductors and cable.

   (a) removable covers
   (b) hinged covers
   (c) a or b
   (d) none of these

2. Wireways can pass transversely through a wall _____.

   (a) if the length passing through the wall is unbroken
   (b) if the wall is of fire-rated construction
   (c) in hazardous (classified) locations
   (d) if the wall is not of fire-rated construction

3. Where insulated conductors are deflected within a metal wireway, the wireway shall be sized to meet the bending requirements corresponding to _____ wire per terminal in Table 312.6(A).

   (a) one
   (b) two
   (c) three
   (d) none of these

4. Power distribution blocks installed in metal wireways shall _____.

   (a) allow for sufficient wire-bending space at terminals
   (b) not have uninsulated exposed live parts
   (c) a or b
   (d) a and b

ARTICLE 380

Multioutlet Assemblies

INTRODUCTION TO ARTICLE 380—MULTIOUTLET ASSEMBLIES

A multioutlet assembly is a surface, flush, or freestanding raceway designed to hold conductors and receptacles, and is assembled in the field or at the factory [Article 100]. It's not limited to systems commonly referred to by trade names of "Plugtrak®" or "Plugmold®."

380.1 Scope. Article 380 covers the use, installation, and construction specifications of multioutlet assemblies. Figure 380–1

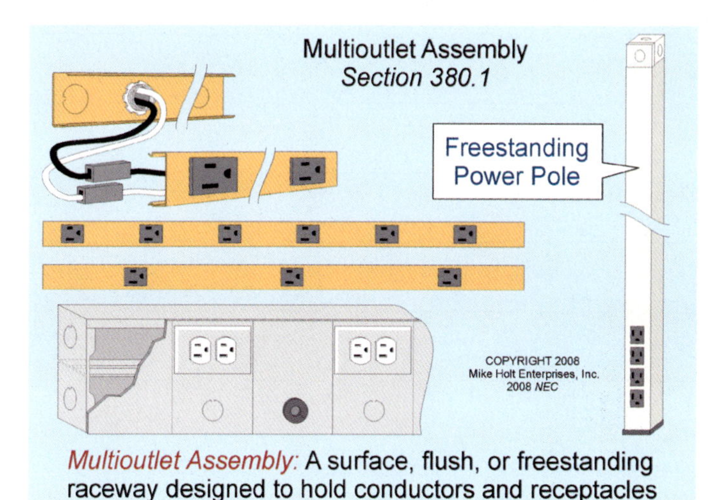

Figure 380–1

380.2 Uses.

(A) Permitted. Dry locations only.

(B) Not Permitted.

(1) Concealed.

(2) Where subject to severe physical damage.

(3) Where the voltage is 300V or more between conductors, unless the metal has a thickness not less than 0.040 in.

(4) Where subject to corrosive vapors.

(5) In hoistways.

(6) In any hazardous (classified) location, except as permitted by 501.10(B).

380.3 Through Partitions. Metal multioutlet assemblies can pass through a dry partition, provided no receptacle is concealed in the wall, and the cover of the exposed portion of the system can be removed.

ARTICLE 380 Practice Questions

ARTICLE 380. MULTIOUTLET ASSEMBLIES—PRACTICE QUESTIONS

1. A multioutlet assembly can be installed in _____.

   (a) dry locations
   (b) wet locations
   (c) a and b
   (d) damp locations

2. Metal multioutlet assemblies can pass through a dry partition, provided no receptacle is concealed in the partition and the cover of the exposed portion of the system can be removed.

   (a) True
   (b) False

Surface Metal Raceways

INTRODUCTION TO ARTICLE 386—SURFACE METAL RACEWAYS

A surface metal raceway is a common method of adding a raceway when exposed traditional raceway systems are not acceptable, and concealing the raceway is not economically feasible. It comes in several colors, and is now available with colored or real wood inserts designed to make it look like molding rather than a raceway. Surface metal raceway is commonly known as "Wiremold®" in the field.

PART I. GENERAL

386.1 Scope. This article covers the use, installation, and construction specifications of surface metal raceways and associated fittings.

386.2 Definition.

Surface Metal Raceway. A metallic raceway intended to be mounted to the surface, with associated accessories, in which conductors are placed after the raceway has been installed as a complete system [300.18(A)]. Figure 386–1

Surface Metal Raceways
Section 386.2 Definition

Surface Raceway: A raceway intended to be mounted to the surface, in which conductors are placed after the raceway has been installed as a complete system.

COPYRIGHT 2008 Mike Holt Enterprises, Inc. 2008 *NEC*

Figure 386–1

Author's Comment: Surface raceways are available in different shapes and sizes and can be mounted on walls, ceilings, or floors. Some surface raceways have two or more separate compartments, which permit the separation of power and lighting conductors from low-voltage or limited-energy conductors or cables (control, signal, and communications cables and conductors) [386.70].

386.6 Listing Requirements. Surface metal raceways and associated fittings must be listed.

Author's Comment: Enclosures for switches, receptacles, luminaires, and other devices are identified by the markings on their packaging, which identify the type of surface metal raceway that can be used with the enclosure.

PART II. INSTALLATION

386.10 Uses Permitted.

(1) In dry locations. Figure 386–2

(2) In Class I, Division 2 locations, as permitted in 501.10(B)(3).

(3) Under raised floors, as permitted in 645.5(D)(2).

(4) Run through walls and floors, if access to the conductors is maintained on both sides of the wall, partition, or floor.

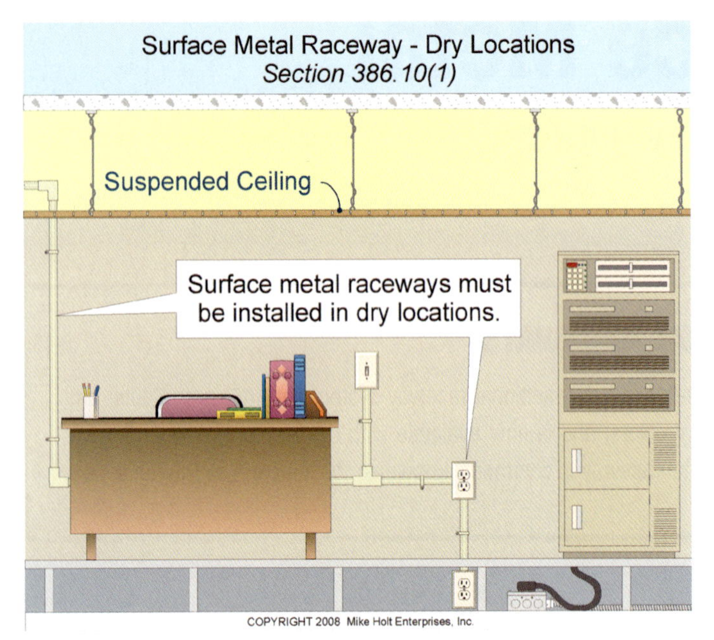

Figure 386–2

386.12 Uses Not Permitted.

(1) Where subject to severe physical damage, unless otherwise approved by the authority having jurisdiction.

(2) Where the voltage is 300V or more between conductors, unless the metal has a thickness not less than 0.040 in.

(3) Where subject to corrosive vapors.

(4) In hoistways.

(5) Where concealed, except as permitted in 386.10.

386.21 Size of Conductors. The maximum size conductor permitted in a surface metal wireway must not be larger than that for which the wireway is designed.

Author's Comment: Because partial packages are often purchased, you may not always receive this information.

386.22 Number of Conductors. The number of conductors or cables installed in a surface metal raceway must not be more than the number for which the raceway is designed. Cables can be installed in surface metal raceways as long as the number of cables does not exceed the allowable percentage fill specified in Table 1, Chapter 9.

The ampacity adjustment factors of 310.15(B)(2)(a) don't apply to conductors installed in surface metal raceways where all of the following conditions are met: Figure 386–3

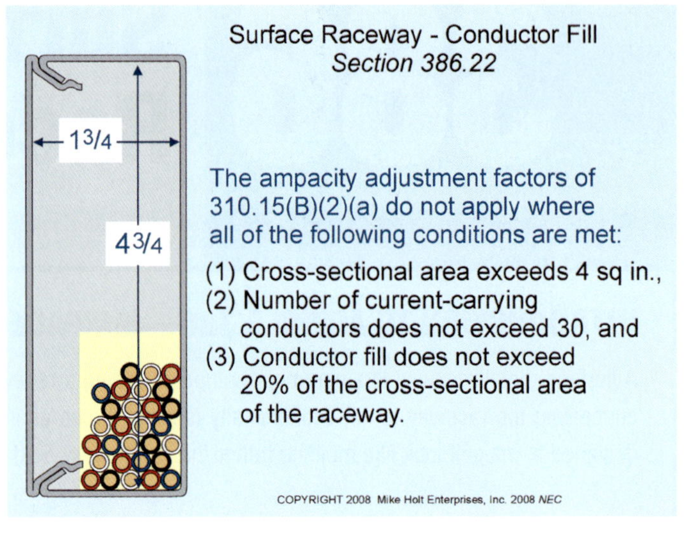

Figure 386–3

(1) The cross-sectional area of the raceway exceeds 4 square inches,

(2) The number of current-carrying conductors doesn't exceed 30, and

(3) The sum of the cross-sectional areas of all contained conductors doesn't exceed 20 percent of the interior cross-sectional area of the raceways.

386.30 Securing and Supporting. Surface metal raceways must be secured and supported at intervals in accordance with the manufacturer's installation instructions.

386.56 Splices and Taps. Splices and taps must be accessible, and must not fill the raceway to more than 75 percent of its cross-sectional area.

386.60 Equipment Grounding Conductor. Surface metal raceway fittings must be mechanically and electrically joined together in a manner that doesn't subject the conductors to abrasion. Surface metal raceways that allow a transition to another wiring method, such as knockouts for connecting raceways, must have a means for the termination of an equipment grounding conductor. A surface metal raceway is considered suitable as an equipment grounding conductor, in accordance with 250.118(14).

386.70 Separate Compartments. Where surface metal raceways have separate compartments within a single raceway, power and lighting conductors can occupy one compartment, and the other compartment may contain control, signaling, or communications wiring. Stamping, imprinting, or color coding of the interior finish must identify the separate compartments, and the same relative position of compartments must be maintained throughout the premises.

Author's Comments:

- Separation from power conductors is required by the *NEC* for the following low-voltage and limited-energy systems:
  - CATV, 820.44(F)(1)
  - Communications, 800.133(A)(1)
  - Control and Signaling, 725.136(B)
  - Fire Alarm, 760.136(B)
  - Intrinsically Safe Systems, 504.30(A)(2)
  - Instrumentation Tray Cable, 727.5
  - Radio and Television, 810.18(C)
  - Sound Systems, 640.9(C)
- Nonconductive optical fiber cables can occupy the same cable tray or raceway as conductors for electric light, power, Class 1, or nonpower-limited fire alarm circuits [770.133(A)].

ARTICLE 386 Practice Questions

ARTICLE 386. SURFACE METAL RACEWAYS—PRACTICE QUESTIONS

1. Surface metal raceway is a metallic raceway that is intended to be mounted to the surface of a structure, with associated couplings, connectors, boxes, and fittings for the installation of electrical conductors.

   (a) True
   (b) False

2. Surface metal raceways shall not be used _____.

   (a) where subject to severe physical damage
   (b) where subject to corrosive vapors
   (c) in hoistways
   (d) all of these

3. The maximum size conductors permitted in a metal surface raceway shall not be larger than that for which the wireway is designed.

   (a) True
   (b) False

4. Surface metal raceways shall be secured and supported at intervals _____.

   (a) in accordance with the manufacturer's installation instructions
   (b) appropriate for the building design
   (c) not exceeding 4 ft
   (d) not exceeding 8 ft

5. Where combination surface metal raceways are used for both signaling conductors and lighting and power circuits, the different systems shall be run in separate compartments identified by _____ of the interior finish.

   (a) stamping
   (b) imprinting
   (c) color coding
   (d) any of these

ARTICLE

392 Cable Trays

INTRODUCTION TO ARTICLE 392—CABLE TRAYS

A cable tray system is a unit or an assembly of units or sections with associated fittings that forms a structural system used to securely fasten or support cables and raceways. Cable tray systems include ladder, ventilated trough, ventilated channel, solid bottom, and other similar structures. Cable trays are manufactured in many forms, from a simple hanger or wire mesh to a substantial, rigid, steel support system. Cable trays are designed and manufactured to support specific wiring methods, as identified in 392.3(A).

PART I. GENERAL

392.1 Scope. Article 392 covers cable tray systems, including ladder, ventilated trough, ventilated channel, solid bottom, and other similar structures.

392.2 Definition.

Cable Tray System. A unit or assembly of units or sections with associated fittings forming a rigid structural system used to securely fasten or support cables, raceways, and boxes.

> **Author's Comment:** Cable tray isn't a type of raceway. It's a support system for cables and raceways.

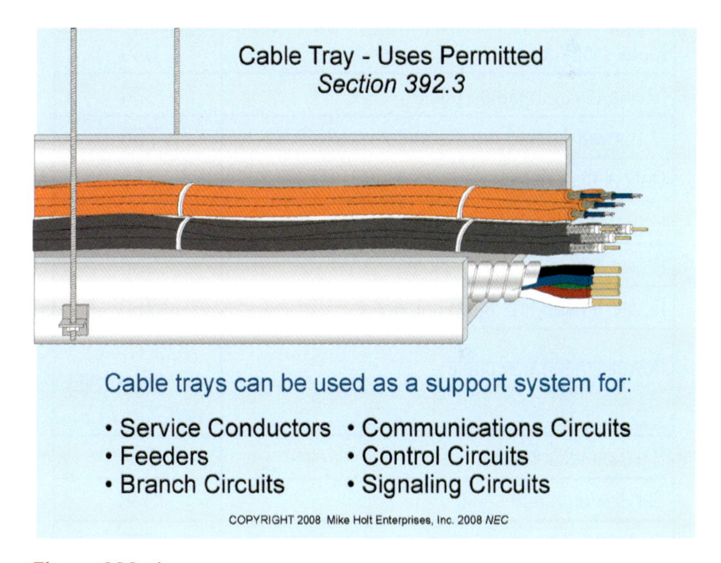

Figure 392–1

PART II. INSTALLATION

392.3 Uses Permitted. Cable trays can be used as a support system for service, feeder, or branch-circuit conductors, as well as communications circuits, control circuits, and signaling circuits. Figure 392–1

Author's Comments:

- Cable trays used to support service-entrance conductors must contain only service-entrance conductors unless a solid fixed barrier separates the service-entrance conductors from other conductors [230.44].
- Cable tray installations aren't limited to industrial establishments.
- Where exposed to the direct rays of the sun, insulated conductors and jacketed cables must be identified as being sunlight resistant. The manufacturer must identify cable trays and associated fittings for their intended use.

(A) Wiring Methods. Any of the following wiring methods can be installed in a cable tray:

Armored Cable	320
CATV cables	820
CATV raceways	820
Class 2 & 3 cables	725
Communications cables	800
Communications raceways	800
Electrical metallic tubing	358
Electrical nonmetallic tubing	362
Fire alarm cables	760
Flexible metal conduit	348
High-density polyethylene conduit	353
Instrumentation tray cable	727
Intermediate metal conduit	342
Liquidtight flexible metal conduit	350
Liquidtight flexible nonmetallic conduit	356
Metal-clad cable	330
Nonmetallic-sheathed cable	334
Nonpower-limited fire alarm cable	760
Optical fiber cables and raceways	770
Polyvinyl chloride PVC conduit	352
Power and control tray cable	336
Power-limited fire alarm cable	760
Power-limited tray cable	725.154(C) and 725.179(E) and 725.71(F)
Rigid metal conduit	344
Service-entrance cable	338
Signaling raceway	725
Underground feeder and branch-circuit cable	340

Author's Comment: Control, signal, and communications cables must be separated from the power conductors by a barrier or maintain a 2 in. separation.

- Coaxial Cables, 820.133(A)(1)(b) Ex 1
- Class 2 and 3 Cables, 725.136(B) and 725.136(I)
- Communications Cables, 800.133(A)(2) Ex 1
- Fire Alarm Cables, 760.136(G)
- Optical Fiber Cables, 770.133(B)
- Intrinsically Safe Systems Cables, 504.30(A)(2) Ex 1
- Radio and Television Cables, 810.18(B) Ex 1

(B) In Industrial Establishments. Where conditions of maintenance and supervision ensure that only qualified persons service the installed cable tray system:

(1) Single Conductors. Single-conductor cables:

(a) 1/0 AWG or larger listed and marked for use in cable trays.

(c) Single conductors used as equipment grounding conductors must be insulated, covered, or bare, and they must be 4 AWG or larger.

(C) Equipment Grounding Conductor. Metal cable trays can serve as an equipment grounding conductor where maintenance and supervision ensure that only qualified persons service the cable tray system, and the cable tray is bonded in accordance with 392.7. Figure 392–2

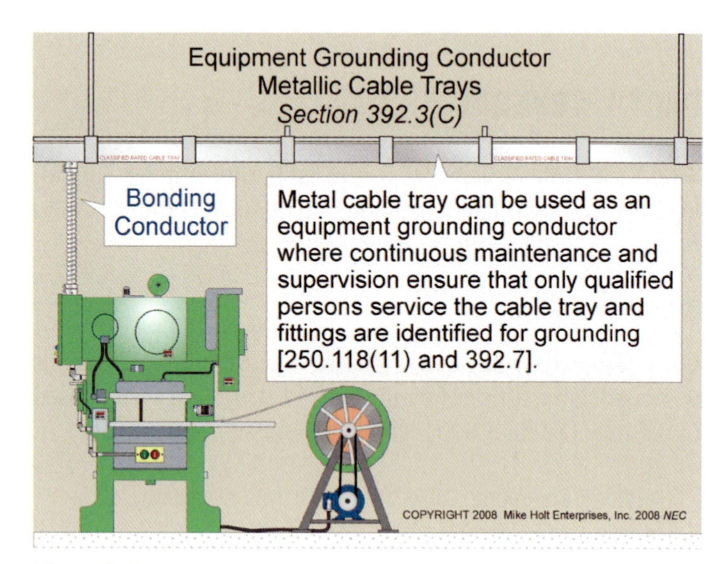

Figure 392–2

(D) Hazardous (Classified) Locations. Cable trays in hazardous (classified) locations must contain only the cable types permitted in 501.10, 502.10, 503.10, 504.20, and 505.15.

(E) Nonmetallic Cable Trays. In addition to the uses permitted elsewhere in Article 392, nonmetallic cable trays can be installed in corrosive areas, and in areas requiring voltage isolation.

392.4 Uses Not Permitted. Cable tray systems are not permitted in hoistways, or where subject to severe physical damage.

392.6 Installation.

(A) Complete System. Cable trays must be installed as a complete system, except mechanically discontinuous segments between cable tray runs, or between cable tray runs and equipment are permitted. A bonding jumper, sized in accordance with 250.102 and installed in accordance with 250.96, must bond the sections of cable tray, or the cable tray and the raceway or equipment.

(B) Completed Before Installation. Each run of cable tray must be completed before the installation of cables or conductors.

(C) Support. Supports for cable trays must be provided to prevent stress on cables where they enter raceways or other enclosures from cable tray systems. Cable trays must be supported in accordance with the manufacturer's installation instructions.

(G) Through Partitions and Walls. Cable trays can extend through partitions and walls, or vertically through platforms and floors where the installation is made in accordance with the firestopping requirements of 300.21.

(H) Exposed and Accessible. Cable trays must be exposed and accessible, except as permitted by 392.6(G).

(I) Adequate Access. Sufficient space must be provided and maintained about cable trays to permit adequate access for installing and maintaining the cables.

(J) Raceways, Cables, and Boxes Supported from Cable Trays. In industrial facilities where conditions of maintenance and supervision ensure only qualified persons will service the installation, and where the cable tray system is designed and installed to support the load, cable tray systems can support raceways, cables, boxes, and conduit bodies. Figure 392–3

For raceways terminating at the tray, a listed cable tray clamp or adapter must be used to securely fasten the raceway to the cable tray system. The raceway must be supported in accordance with the appropriate raceway article.

Raceways or cables running parallel to the cable tray system can be attached to the bottom or side of a cable tray system. The raceway or cable must be fastened and supported in accordance with the appropriate raceway or cable's *Code* article.

Boxes and conduit bodies attached to the bottom or side of a cable tray system must be fastened and supported in accordance with 314.23.

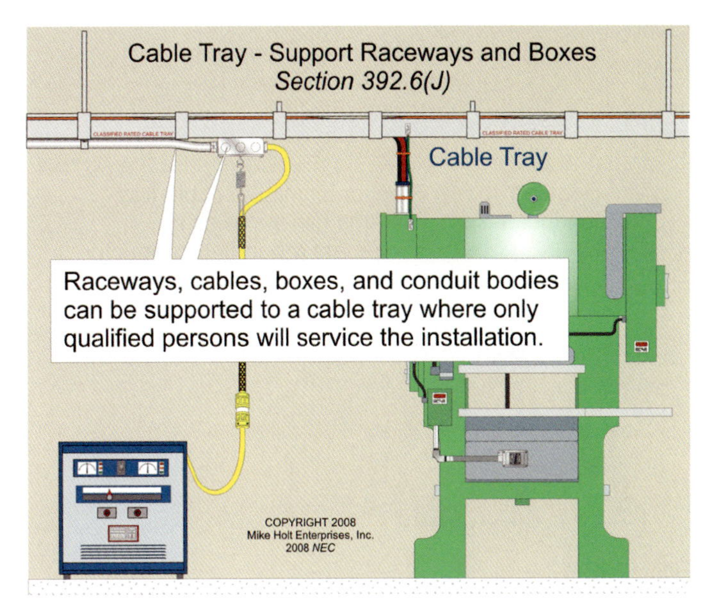

Figure 392–3

392.7 Equipment Grounding Conductor.

(A) Metallic Cable Trays. Metallic cable trays must be bonded together to ensure that they have the capacity to conduct safely any fault current likely to be imposed on them, in accordance with 250.96(A) and Part IV of Article 250.

Author's Comment: Nonconductive coatings such as paint, lacquer, and enamel on equipment must be removed to ensure an effective ground-fault current path, or the termination fittings must be designed so as to make such removal unnecessary [250.12].

(B) Serve as Equipment Grounding Conductor. Metal cable trays can serve as equipment grounding conductors where maintenance and supervision ensure that qualified persons service the installed cable tray system, and the following requirements have been met [392.3(C)]: Figure 392–4

(1) Cable tray sections and fittings are identified for grounding.

Author's Comment: Identification will be marked on each cable tray section.

(4) Cable tray sections, fittings, and connected raceways are effectively bonded to each other to ensure electrical continuity and the capacity to conduct safely any fault current likely to be imposed on them [250.96(A)]. This is accomplished by using bolted mechanical connectors or bonding jumpers sized in accordance with 250.102.

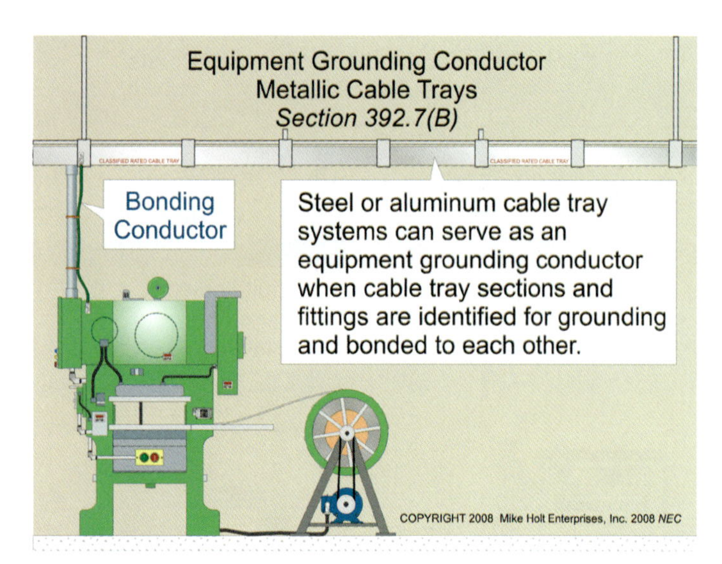

Figure 392–4

392.8 Cable Installation.

(A) Cable Splices. Splices are permitted in a cable tray if the splice is accessible and insulated by a method approved by the authority having jurisdiction. Splices can project above the side rails of the cable tray where not subject to physical damage. Figure 392–5

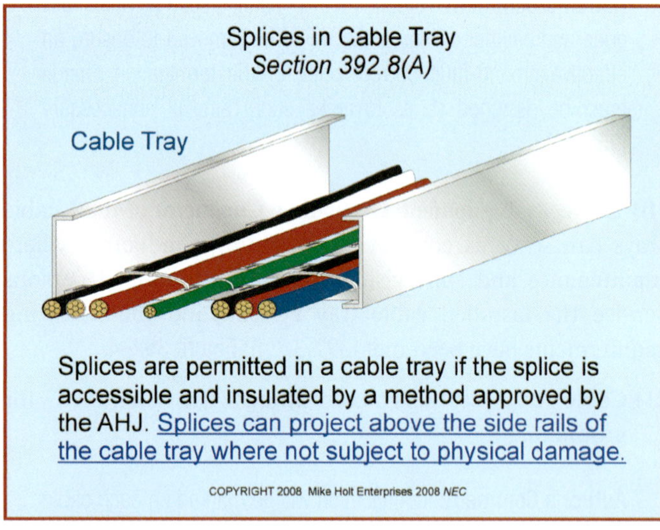

Figure 392–5

(B) Fastened Securely. Cables run vertically must be securely fastened to the cable tray.

(C) Bushed Raceway. A box isn't required where cables or conductors exit a bushed raceway used for the support or protection of the conductors.

(D) Connected in Parallel. To prevent unbalanced current in the parallel conductors due to inductive reactance, all circuit conductors of a parallel set (A, B, C, N) [310.4] must be bundled together and secured to prevent excessive movement due to fault current magnetic forces.

(E) Single Conductors. Single conductors of a circuit not connected in parallel must be installed in a single layer, unless the conductors are bound together.

392.9 Number of Multiconductor Cables in Cable Trays.

(A) Any Mixture of Cables. Where ladder or ventilated trough cable trays contain multiconductor power or lighting cables, the maximum number of cables must conform to the following:

(1) Where all of the cables are 4/0 AWG or larger, the sum of the diameters of all cables must not exceed the cable tray width, and the cables must be installed in a single layer.

392.11 Conductor Ampacity of Multiconductor Cables in Cable Trays.

(A) Multiconductor Cables. The allowable ampacity of multiconductor cables installed in a single layer [392.9(A)(1)] must be as given in Table 310.16 and Table 310.18.

(1) The conductor ampacity adjustment factors of 310.15(B)(2)(a) apply to a given cable if it contains more than three current-carrying conductors. The conductor adjustment factors only apply to the number of current-carrying conductors in the cable and not to the number of conductors in the cable tray.

ARTICLE 392 Practice Questions

ARTICLE 392. CABLE TRAYS—PRACTICE QUESTIONS

1. A cable tray is a unit or assembly of units or sections and associated fittings forming a _____ system used to securely fasten or support cables and raceways.

   (a) structural
   (b) flexible
   (c) movable
   (d) secure

2. Where exposed to the direct rays of the sun, insulated conductors and jacketed cables installed in cable trays shall be _____ as being sunlight resistant.

   (a) listed
   (b) approved
   (c) identified
   (d) none of these

3. Cable tray systems shall not be used _____.

   (a) in hoistways
   (b) where subject to severe physical damage
   (c) in hazardous (classified) locations
   (d) a and b

4. In industrial facilities where conditions of maintenance and supervision ensure that only qualified persons will service the installation, cable tray systems can be used to support _____.

   (a) raceways
   (b) cables
   (c) boxes and conduit bodies
   (d) all of these

5. Cable _____ made and insulated by approved methods can be located within a cable tray provided they are accessible, and do not project above the side rails where the splices are subject to physical damage.

   (a) connections
   (b) jumpers
   (c) splices
   (d) conductors

6. The conductor adjustment factors only apply to the number of current-carrying conductors in the cable and not to the number of conductors in the cable tray.

   (a) True
   (b) False

CHAPTER 3

Notes

EQUIPMENT FOR GENERAL USE

INTRODUCTION TO CHAPTER 4—EQUIPMENT FOR GENERAL USE

With the first three chapters behind you, the final chapter in the *NEC* for building a solid foundation in general work is Chapter 4. This chapter helps you apply the first three chapters to installations involving general equipment. These first four chapters follow a natural sequential progression. Each of the next four *NEC* Chapters—5, 6, 7, and 8—builds upon the first four, but in no particular order. You don't need to understand any of the other "next chapters" to work with any of the others, but you do need to understand all of the first four chapters to properly apply any of the next four.

Chapter 4 has some logical arrangements of its own. Here are the groupings:

- Flexible cords and cables, fixture wires, switches, and receptacles
- Switchboards and panelboards
- Lamps, luminaires, appliances, and space heaters
- Motors, refrigeration equipment, generators, and transformers
- Capacitors and other components

These groupings make sense. For example, motors, refrigeration equipment, generators, and transformers are all inductive equipment.

This logical arrangement of the *NEC* is something to keep in mind when you're searching for a particular item. You know, for example, that transformers are general equipment. So you'll find the *Code* requirements for them in Chapter 4. You know they are wound devices. So you'll find transformer requirements located somewhere near motor requirements.

- **Article 400.** Flexible Cords and Flexible Cables. Article 400 covers the general requirements, applications, and construction specifications for flexible cords and flexible cables.

- **Article 402.** Fixture Wires. This article covers the general requirements and construction specifications for fixture wires.

- **Article 404.** Switches. The requirements of Article 404 apply to switches of all types. These include snap (toggle) switches, dimmer switches, fan switches, knife switches, circuit breakers used as switches, and automatic switches such as time clocks, timers, and switches and circuit breakers used for disconnecting means.

- **Article 406.** Receptacles, Cord Connectors, and Attachment Plugs (Caps). This article covers the rating, type, and installation of receptacles, cord connectors, and attachment plugs (cord caps). It also covers flanged surface inlets.

- **Article 408.** Switchboards and Panelboards. Article 408 covers specific requirements for switchboards, panelboards, and distribution boards that supply lighting and power circuits.

- **Article 410.** Luminaires, Lampholders, and Lamps. This article contains the requirements for luminaires, lampholders, and lamps. Because of the many types and applications of luminaires, manufacturer's instructions are very important and helpful for proper installation. Underwriters Laboratories produces a pamphlet called the Luminaire Marking Guide, which provides information for properly installing common types of incandescent, fluorescent, and high-intensity discharge (HID) luminaires.

- **Article 411.** Lighting Systems Operating at 30V or Less. Article 411 covers lighting systems, and their associated components, that operate at 30V or less.

- **Article 422.** Appliances. This article covers electric appliances used in any occupancy.

- **Article 424.** Fixed Electric Space-Heating Equipment. Article 422 covers fixed electric equipment used for space-heating. For the purpose of this article, heating equipment includes heating cable, unit heaters, boilers, central systems, and other fixed electric space-heating equipment. This article does not apply to process heating and room air-conditioning.

- **Article 430.** Motors, Motor Circuits, and Controllers. Article 430 contains the specific requirements for conductor sizing, overcurrent protection, control circuit conductors, motor controllers, and disconnecting means. The installation requirements for motor control centers are covered in Article 430, Part VIII.

- **Article 440.** Air-Conditioning and Refrigeration Equipment. This article applies to electrically driven air-conditioning and refrigeration equipment with a motorized hermetic refrigerant compressor. The requirements in this article are in addition to, or amend, the requirements in Article 430 and other articles.

- **Article 445.** Generators. Article 445 contains the electrical installation requirements for generators, such as where they can be installed, nameplate markings, conductor ampacity, and disconnecting means.

- **Article 450.** Transformers. This article covers the installation of transformers.

- **Article 460.** Capacitors. Article 460 covers the installation of capacitors, including those in hazardous (classified) locations as modified by Articles 501 through 503.

Chapter 4 doesn't end with Article 460. The remaining articles are also important, but they don't address topics the typical electrician deals with. These are:

- Article 480. Batteries.
- Article 490. Equipment, Over 600 Volts, Nominal.

You may want to read through these, just to see what's there. After you finish Article 460 in this book, you'll have completed your study of the first four chapters of the *NEC*, and will have a solid foundation for properly applying the Code to your work.

ARTICLE

400 Flexible Cords and Flexible Cables

INTRODUCTION TO ARTICLE 400—FLEXIBLE CORDS AND FLEXIBLE CABLES

This article covers the general requirements, applications, and construction specifications for flexible cords and flexible cables. The *NEC* doesn't consider flexible cords to be wiring methods like those defined in Chapter 3.

Always use a cord (and fittings) identified for the application. For example, use cords listed for a wet location if you're using them outdoors. The jacket material of any cord is tested to maintain its insulation properties and other characteristics in the environments for which it has been listed.

400.1 Scope. Article 400 covers the general requirements, applications, and construction specifications for flexible cords and flexible cables as contained in Table 400.4.

Author's Comment: Extension cords must not be used as a substitute for fixed wiring [400.8(1)], but they can be used for temporary wiring if approved by the authority having jurisdiction in accordance with 590.2(B).

400.3 Suitability. Flexible cords and flexible cables, as well as their fittings must be suitable for the use and location. Figure 400–1

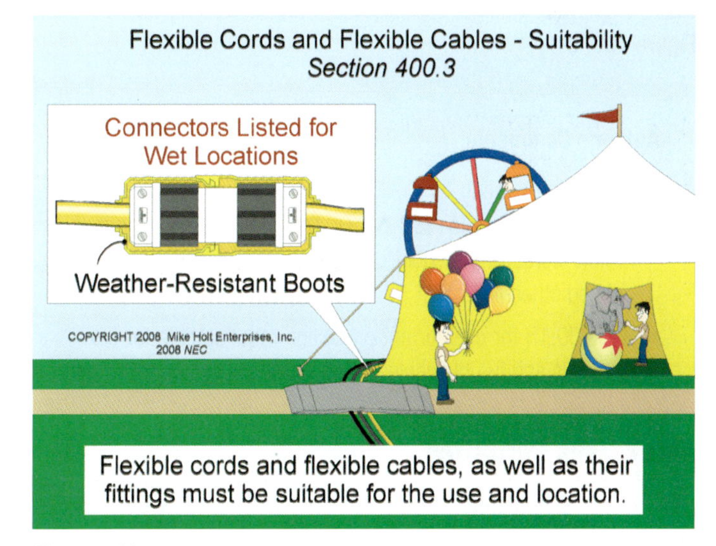

Figure 400–1

400.4 Types of Flexible Cords and Flexible Cables. The use of flexible cords and flexible cables must conform to the descriptions contained in Table 400.4.

400.5 Ampacity of Flexible Cords and Flexible Cables.

(A) Ampacity Tables. Tables 400.5(A) and 400.5(B) list the allowable ampacity for copper conductors in flexible cords and flexible cables with not more than three current-carrying conductors at an ambient temperature of 86°F.

Where the number of current-carrying conductors in a cable or raceway exceeds three, the allowable ampacity of each conductor must be adjusted in accordance with the following multipliers:

Table 400.5 Adjustment Factor

Current Carrying	Ampacity Multiplier
4–6 Conductors	0.80
7–9 Conductors	0.70
10–20 Conductors	0.50

Where the ambient temperature exceeds 86°F, the flexible cord or flexible cable ampacity, as listed in Table 400.5(A) or 400.5(B), must be adjusted by using the temperature correction factors listed in Table 310.16. Figure 400–2

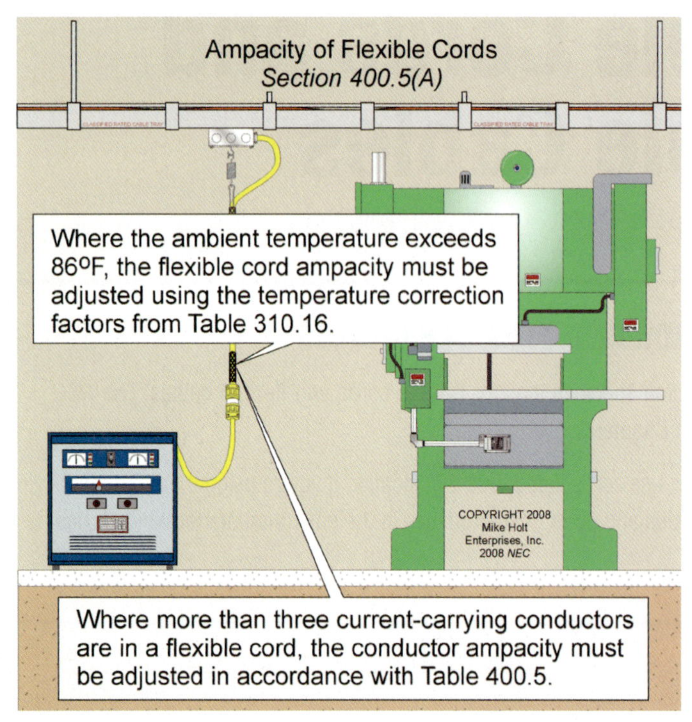

Figure 400–2

Author's Comments:

- Temperature rating for flexible cords and flexible cables are not contained in the *NEC*, but UL listing standards state that flexible cords and flexible cables are rated for 60°C unless marked otherwise.
- See 400.13 for overcurrent protection requirements for flexible cords and flexible cables.

400.7 Uses Permitted.

(A) Uses Permitted. Flexible cords and flexible cables within the scope of this article can be used for the following applications:

(1) Pendants [210.50(A) and 314.23(H)].

Author's Comment: Only cords identified for use as pendants in Table 400.4 may be used for pendants.

(2) Wiring of luminaires [410.24(A) and 410.62(B)].

(3) Connection of portable luminaires, portable and mobile signs, or appliances [422.16].

(4) Elevator cables.

(5) Wiring of cranes and hoists.

(6) Connection of utilization equipment to facilitate frequent interchange [422.16]. Figure 400–3

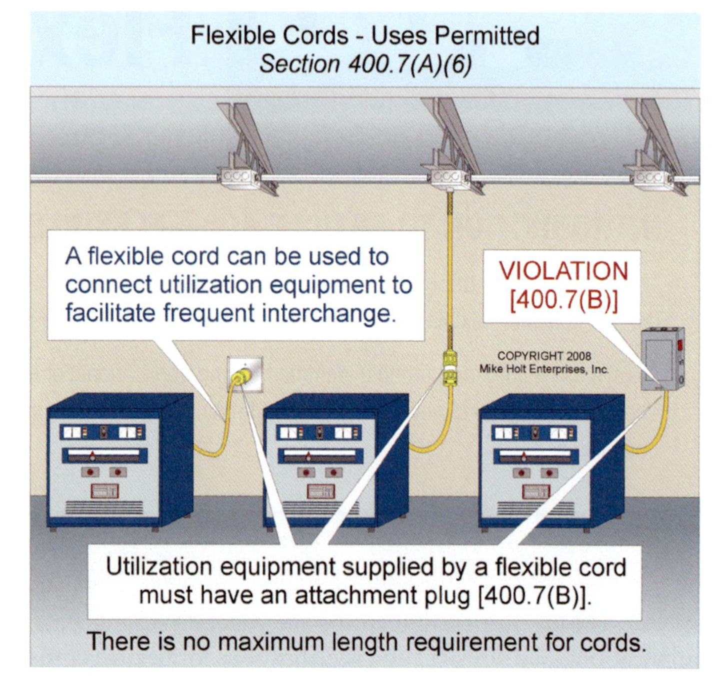

Figure 400–3

(7) Prevention of the transmission of noise or vibration [422.16].

(8) Appliances where the fastening means and mechanical connections are specifically designed to permit ready removal for maintenance and repair, and the appliance is intended or identified for flexible cord connections [422.16].

(9) Connection of moving parts.

(10) Where specifically permitted elsewhere in this *Code*.

Author's Comment: Flexible cords and flexible cables are permitted for fixed permanent wiring by 501.10(A)(2) and (B)(2), 501.140, 502.4(A)(1)(e), 502.4(B)(2), 503.3(A)(2), 550.10(B), 553.7(B), and 555.13(A)(2).

(B) Attachment Plugs. Attachment plugs are required for flexible cords used in any of the following applications: Figure 400–4

- Portable luminaires, portable and mobile signs, or appliances [400.7(A)(3)].

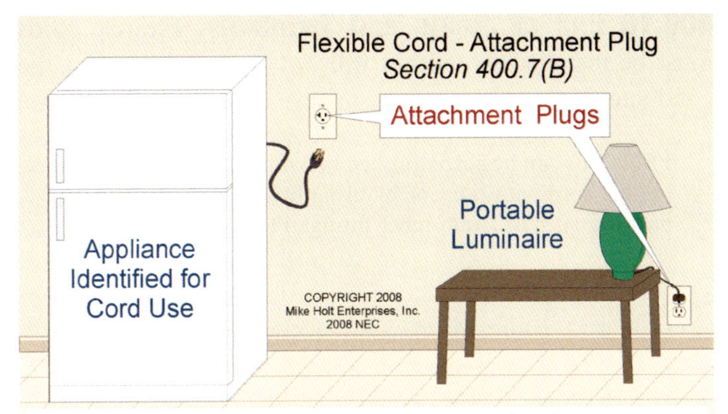

Figure 400–4

- Stationary equipment to facilitate its frequent interchange [400.7(A)(6)]. See 422.16.
- Appliances specifically designed to permit ready removal for maintenance and repair, and identified for flexible cord connection [400.7(A)(8)].

Author's Comment: An attachment plug can serve as the disconnecting means for stationary appliances [422.33] and room air conditioners [440.63].

400.8 Uses Not Permitted. Unless specifically permitted in 400.7, flexible cords must not be:

(1) Used as a substitute for the fixed wiring of a structure.

(2) Run through holes in walls, structural ceilings, suspended/dropped ceilings, or floors. Figure 400–5

Author's Comment: According to an article in the International Association of Electrical Inspectors magazine (IAEI News), a flexible cord run through a cabinet for an appliance isn't considered as being run through a wall. Figure 400–6

(3) Run through doorways, windows, or similar openings.

(4) Attached to building surfaces.

(5) Concealed by walls, floors, or ceilings, or located above suspended or dropped ceilings. Figure 400–7

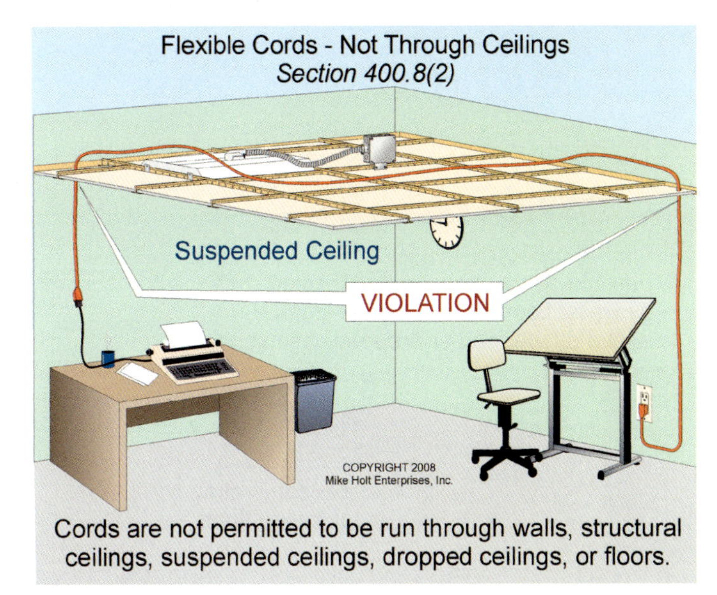

Figure 400–5

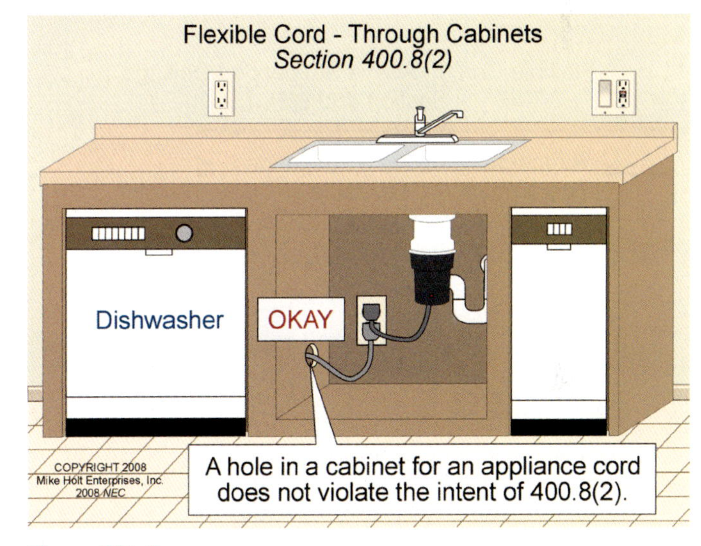

Figure 400–6

Author's Comments:

- Flexible cords are permitted under a raised floor (with removable panels) used for environmental air, because this area isn't considered a concealed space. See the definition of "Exposed" in Article 100.
- Receptacles are permitted above a suspended ceiling, but a flexible cord is not. Why install a receptacle above a ceiling if the flexible cord is not permitted in this space? Because the receptacle can be used for portable tools; it just can't be used for cord-and-plug-connected equipment fastened in place, such as a projector. Figure 400–8

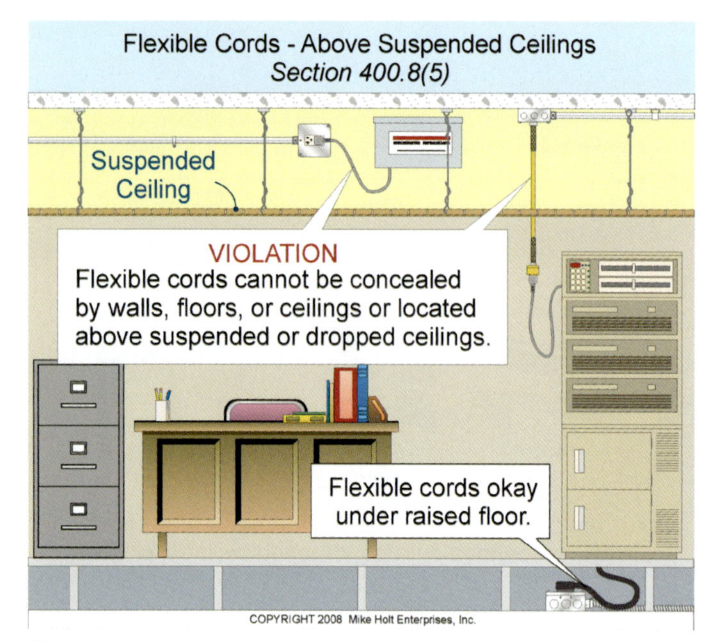

Figure 400–7

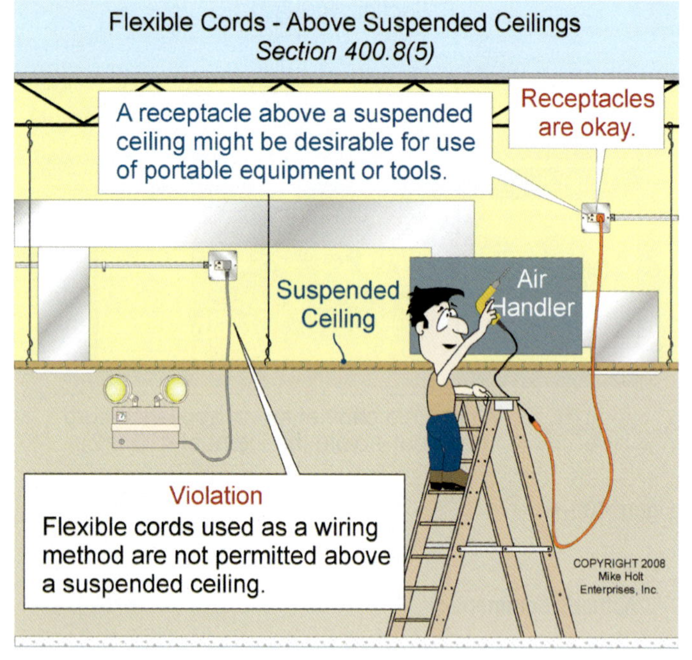

Figure 400–8

(6) Installed in raceways, except as permitted elsewhere in the *Code*.

(7) Where subject to physical damage.

Author's Comment: Even cords listed as "extra-hard usage" must not be used where subject to physical damage.

400.10 Pull at Joints and Terminals.

Flexible cords must be installed so tension will not be transmitted to the conductor terminals.

FPN: This can be accomplished by knotting the cord, winding the cord with tape, or by using fittings designed for the purpose, such as strain-relief fittings. Figure 400–9

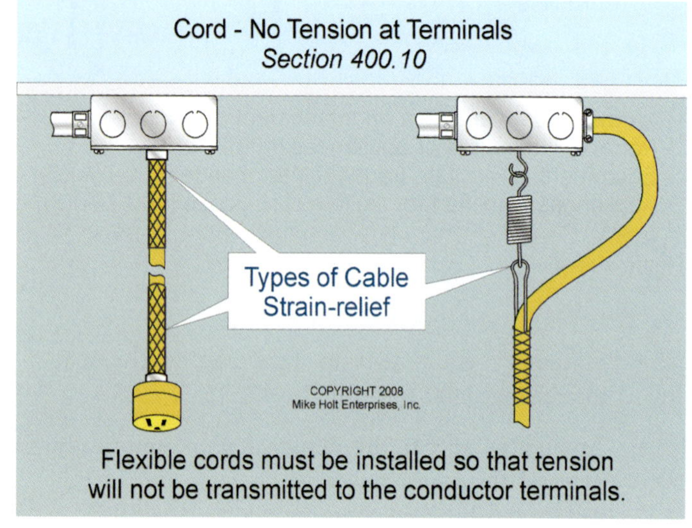

Figure 400–9

Author's Comment: When critical health and economic activities are dependent on flexible cord supplied equipment, the best method is a factory-made, stress-relieving, listed device, not an old-timer's knot.

400.13 Overcurrent Protection.

Flexible cords and flexible cables must be protected against overcurrent in accordance with 240.5.

Author's Comment: Section 240.5 contains the following requirements:

- Overcurrent devices must not be rated higher than the cord's ampacity as specified in Table 400.5(A) and Table 400.5(B) [240.5(A)].
- Flexible cord for listed utilization equipment is considered protected when used in accordance with the equipment listing requirements [240.5(B)(1)].
- Extension cord sets are considered protected when used in accordance with the extension cord listing requirements [240.5(B)(3)].

- Flexible cord used in field-installed extension cords, made with separately listed and installed components, can be supplied by a 20A branch circuit for 16 AWG and larger conductors [240.5(B)(4)].

400.14 Protection from Damage. Flexible cords must be protected by bushings or fittings where passing through holes in covers, outlet boxes, or similar enclosures.

In industrial establishments where the conditions of maintenance and supervision ensure that only qualified persons will service the installation, flexible cords or flexible cables not exceeding 50 ft can be installed in aboveground raceways.

400.22 Neutral Conductor Identification. The neutral conductor of a flexible cord must be identified in one of the following methods:

(A) Colored Braid. A white or gray colored braid.

(B) Tracer in Braid. A colored tracer located in the braid.

(C) Colored Insulation. A white, gray, or light blue conductor.

(F) Surface Marking. Ridges, grooves or white stripes on the exterior of the cord.

400.23 Equipment Grounding Conductor Identification. A conductor intended to be used as an equipment grounding conductor must have a continuous green color or a continuous identifying marker distinguishing it from the other conductor(s). Conductors with green insulation, or green with one or more yellow stripes must not be used for an ungrounded or neutral conductor [250.119].

ARTICLE 400 Practice Questions

ARTICLE 400. FLEXIBLE CORDS AND CABLES—PRACTICE QUESTIONS

1. HPD cord shall be permitted for _____.

   (a) not hard usage
   (b) hard usage
   (c) extra-hard usage
   (d) all of these

2. A 3-conductor SJE cable (one conductor is used for grounding) has a maximum ampacity of _____ for each 16 AWG conductor.

   (a) 9A
   (b) 11A
   (c) 13A
   (d) 15A

3. Flexible cords and cables shall not be used where _____.

   (a) run through holes in walls, ceilings, or floors
   (b) run through doorways, windows, or similar openings
   (c) attached to building surfaces, unless permitted by 368.56(B)
   (d) all of these

4. Flexible cords and cables shall be protected by _____ where passing through holes in covers, outlet boxes, or similar enclosures.

   (a) bushings
   (b) fittings
   (c) a or b
   (d) none of these

5. A flexible cord conductor intended to be used as a(n) _____ conductor shall have a continuous identifying marker readily distinguishing it from the other conductor or conductors.

   (a) ungrounded
   (b) equipment grounding
   (c) service
   (d) high-leg

6. The number of fixture wires in a single conduit or tubing shall not exceed the percentage fill specified in _____.

   (a) Chapter 9, Table 1
   (b) Table 250.66
   (c) Table 310.16
   (d) 240.6

ARTICLE

402 Fixture Wires

INTRODUCTION TO ARTICLE 402—FIXTURE WIRES

This article covers the general requirements and construction specifications for fixture wires. One such requirement is that no fixture wire can be smaller than 18 AWG. Another requirement is that fixture wires must be of a type listed in Table 402.3. That table makes up the bulk of Article 402.

402.1 Scope. Article 402 covers the general requirements and construction specifications for fixture wires.

402.3 Types. Fixture wires must be of a type contained in Table 402.3.

402.5 Allowable Ampacity of Fixture Wires. The allowable ampacity of fixture wires is as follows:

Table 402.5 Allowable Ampacity for Fixture Wires

Wire AWG	Wire Ampacity
18	6A
16	8A
14	17A
12	23A
10	28A

402.6 Minimum Size. Fixture wires must not be smaller than 18 AWG.

402.7 Raceway Size. Raceways must be large enough to permit the installation and removal of conductors without damaging conductor insulation. The number of fixture wires permitted in a single raceway must not exceed the percentage fill specified in Table 1, Chapter 9.

Author's Comment: When all conductors in a raceway are the same size and insulation, the number of conductors permitted can be found in Annex C for the raceway type.

Question: *How many 18 TFFN conductors can be installed in trade size ½ electrical metallic tubing?* **Figure 402–1**

(a) 12 (b) 14 (c) 19 (d) 22

Answer: *(d) 22 conductors [Annex C, Table C.1]*

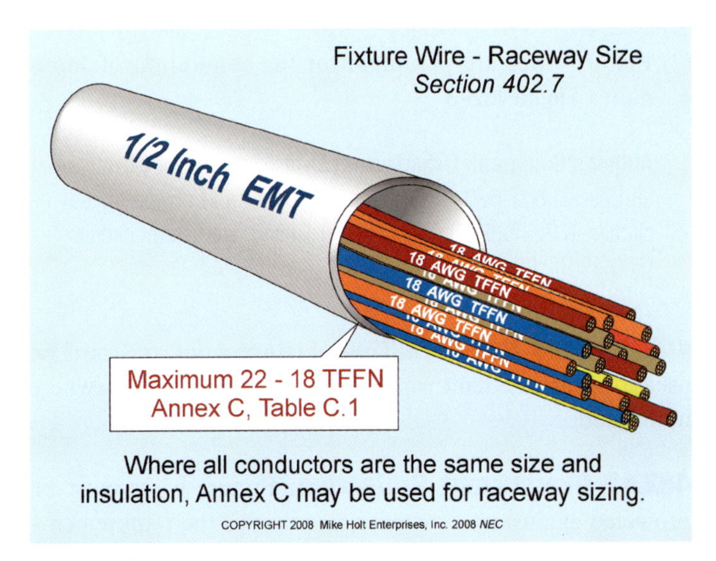

Figure 402–1

Author's Comment: See 300.17 for additional examples on how to size raceways when conductors aren't all the same size.

402.8 Neutral Conductor. Fixture wire used as a neutral conductor must be identified by continuous white stripes.

Author's Comment: To prevent electric shock, the screw shell of a luminaire or lampholder must be connected to the neutral conductor [200.10(C) and 410.50]. Figure 402–2

Figure 402–2

402.10 Uses Permitted.

(2) Fixture wires are permitted for the connection of luminaires. Figure 402–3

Author's Comment: Fixture wires can also be used for elevators and escalators [620.11(C)], Class 1 control and power-limited circuits [725.49(B)], and nonpower-limited fire alarm circuits [760.49(B)]. Figure 402–4

402.11 Uses Not Permitted. Fixture wires must not be used for branch-circuit wiring, except as permitted elsewhere in the Code.

402.12 Overcurrent Protection. Fixture wires must be protected against overcurrent according to the requirements contained in 240.5.

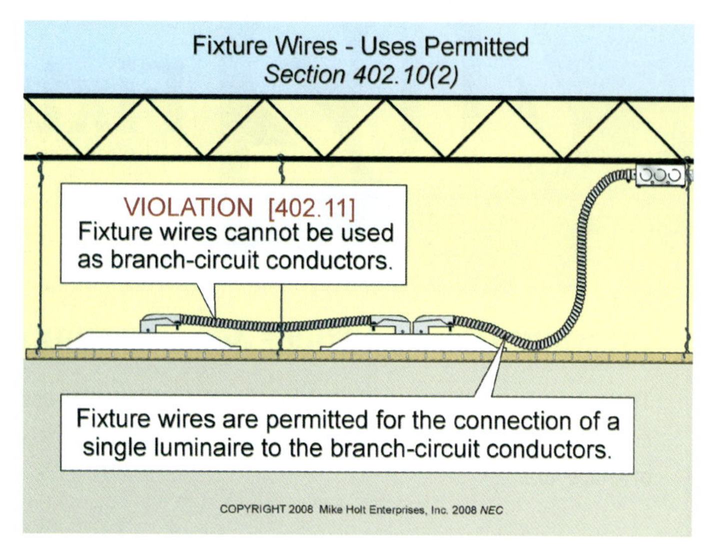

Figure 402–3

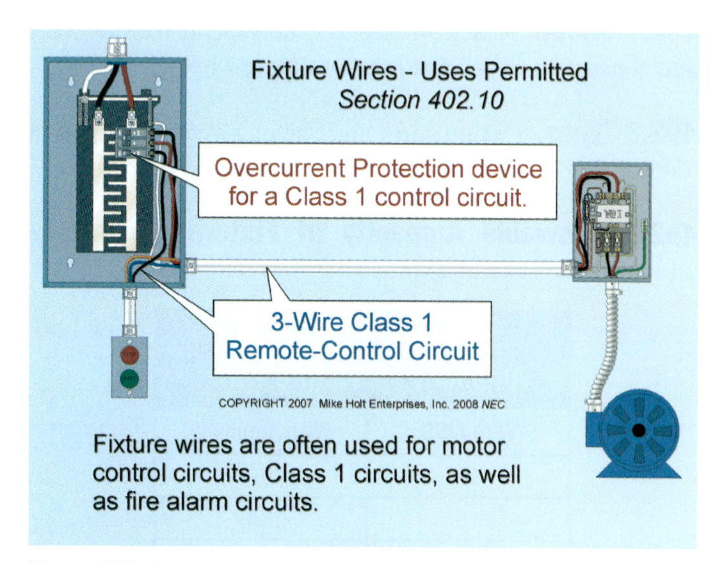

Figure 402–4

Author's Comment: Fixture wires used for motor control circuit taps must have overcurrent protection in accordance with 430.72(A), and Class 1 remote-control circuits must have overcurrent protection in accordance with 725.43.

ARTICLE 402 Practice Questions

ARTICLE 402. FIXTURE WIRES—PRACTICE QUESTIONS

1. The ampacity of 18 TFFN is _____.

   (a) 6A
   (b) 8A
   (c) 10A
   (d) 14A

2. Fixture wires are used to connect luminaires to the _____ conductors supplying the luminaires.

   (a) service
   (b) branch-circuit
   (c) feeder
   (d) none of these

3. The number of fixture wires in a single conduit or tubing shall not exceed the percentage fill specified in _____.

   (a) Chapter 9, Table 1
   (b) Table 250.66
   (c) Table 310.16
   (d) 240.6

ARTICLE

404 Switches

INTRODUCTION TO ARTICLE 404—SWITCHES

The requirements of Article 404 apply to switches of all types, including snap (toggle) switches, dimmer switches, fan switches, knife switches, circuit breakers used as switches, and automatic switches, such as time clocks and timers.

404.1 Scope. The requirements of Article 404 apply to all types of switches, switching devices, and circuit breakers where used as switches. Figure 404–1

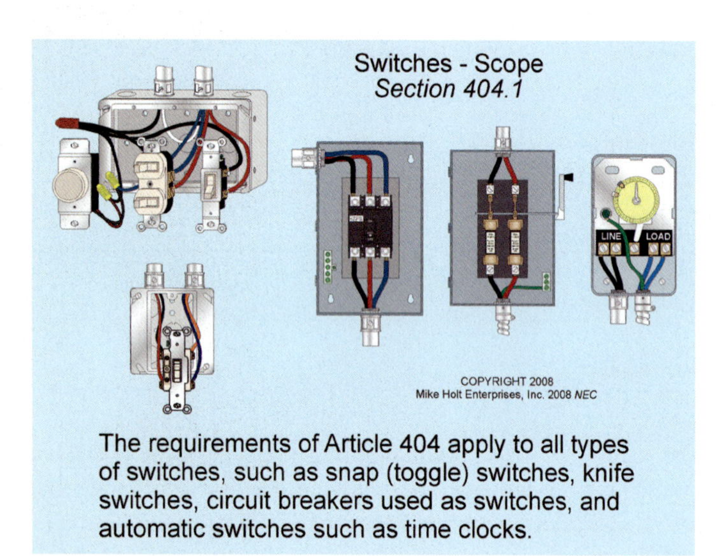

Figure 404–1

404.2 Switch Connections.

(A) Three-Way and Four-Way Switches. Wiring for 3-way and 4-way switching must be done so that only the ungrounded conductors are switched. Figure 404–2

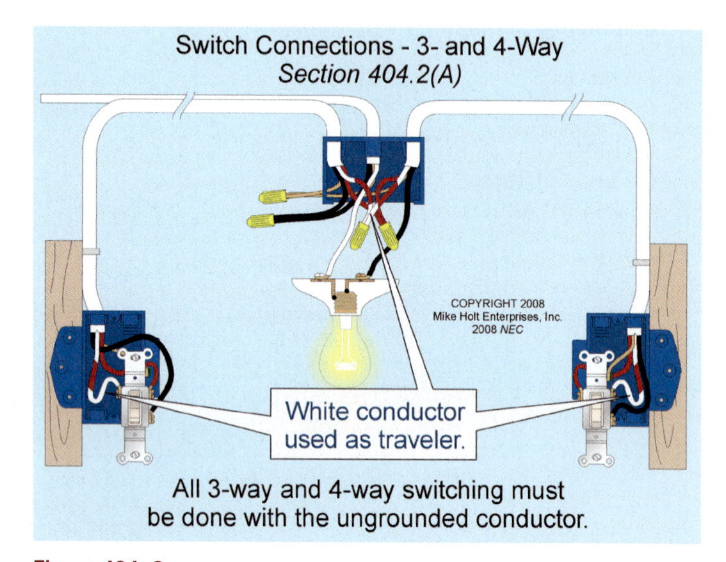

Figure 404–2

Author's Comment: In other words, the neutral conductor must not be switched. The white insulated conductor within a cable assembly can be used for single-pole, 3-way, or 4-way switch loops if it's permanently reidentified to indicate its use as an ungrounded conductor at each location where the conductor is visible and accessible [200.7(C)(2)].

Where a metal raceway or metal-clad cable contains the ungrounded conductors for switches, the wiring must be arranged to avoid heating the surrounding metal by induction. This is accomplished by installing all circuit conductors in the same raceway in accordance with 300.3(B) and 300.20(A), or ensuring that they're all within the same cable.

Exception: A neutral conductor isn't required in the same raceway or cable with travelers and switch leg (switch loops) conductors. Figure 404–3

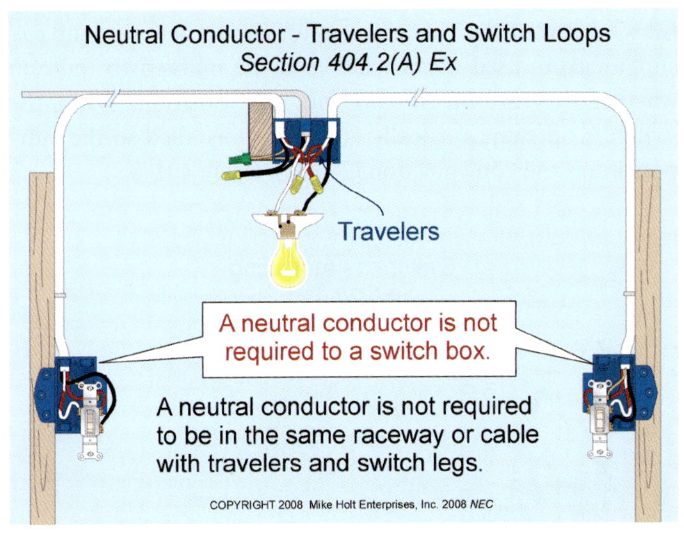

Figure 404–3

(B) Switching Neutral Conductors. Only the ungrounded conductor is permitted to be used for switching. Figure 404–4

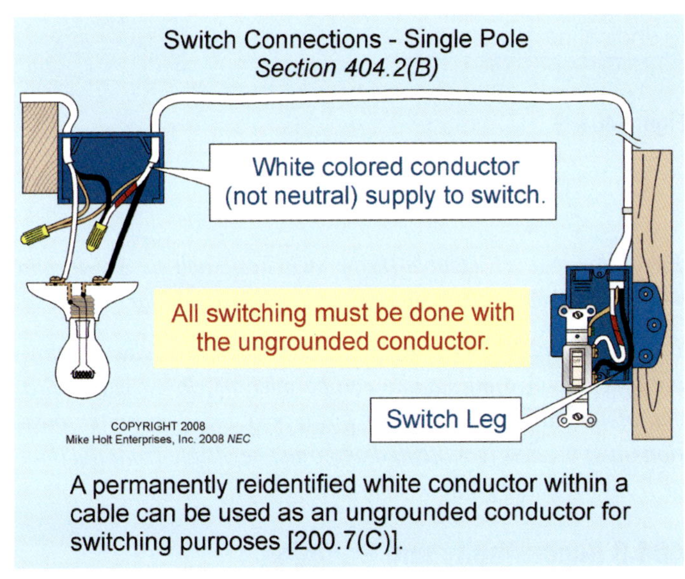

Figure 404–4

404.3 Switch Enclosures.

(A) General. Switches and circuit breakers used as switches must be of the externally operable type mounted in an enclosure listed for the intended use.

(B) Used for Raceways or Splices. Switch or circuit-breaker enclosures can contain splices and taps if the splices and/or taps don't fill the wiring space at any cross section to more than 75 percent. Figure 404–5

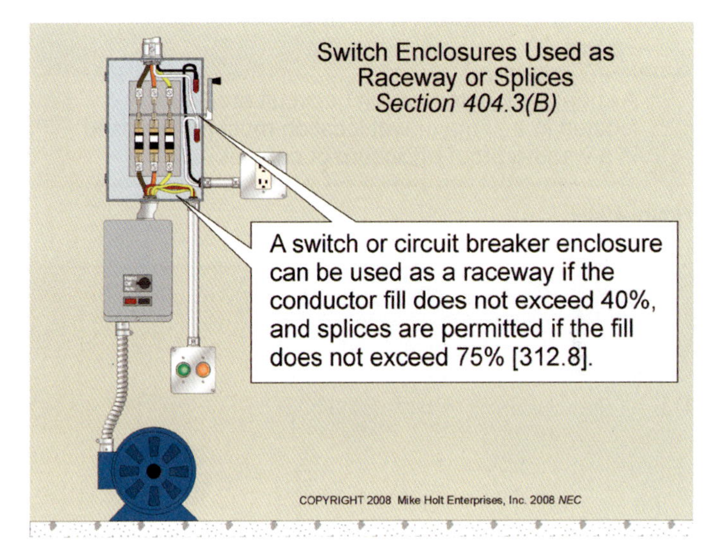

Figure 404–5

Switch or circuit-breaker enclosures can have conductors feed through them if the wiring does not fill the wiring space at any cross section to more than 40 percent [312.8].

404.4 Damp or Wet Locations.

Surface-mounted switches and circuit breakers in a damp or wet location must be installed in a weatherproof enclosure. The enclosure must be installed so not less than ¼ in. airspace is provided between the enclosure and the wall or other supporting surface [312.2]. Figure 404–6

A flush-mounted switch or circuit breaker in a damp or wet location must have a weatherproof cover. Figure 404–7

Switches can be located next to, but not within, a bathtub, hydromassage bathtub, or shower space [404.4, 680.70, and 680.72]. Figure 404–8

Author's Comment: Switches must be located not less than 5 ft from pools [680.22(D)], outdoor spas or hot tubs [680.40], and indoor spas or hot tubs [680.43(C)].

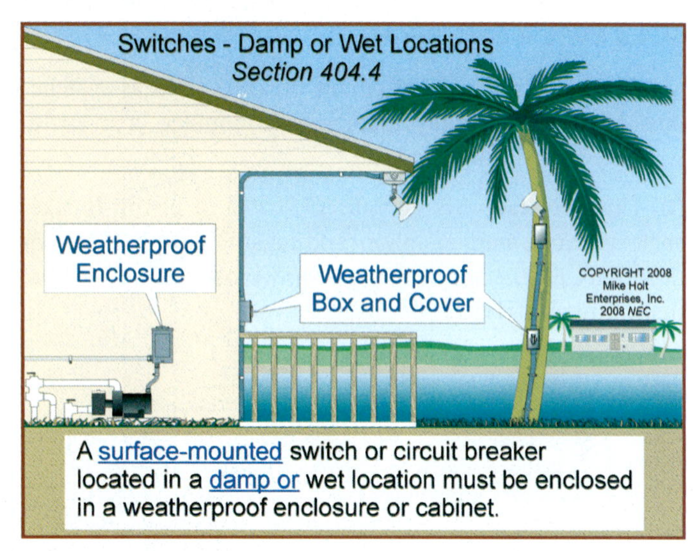

Figure 404–6

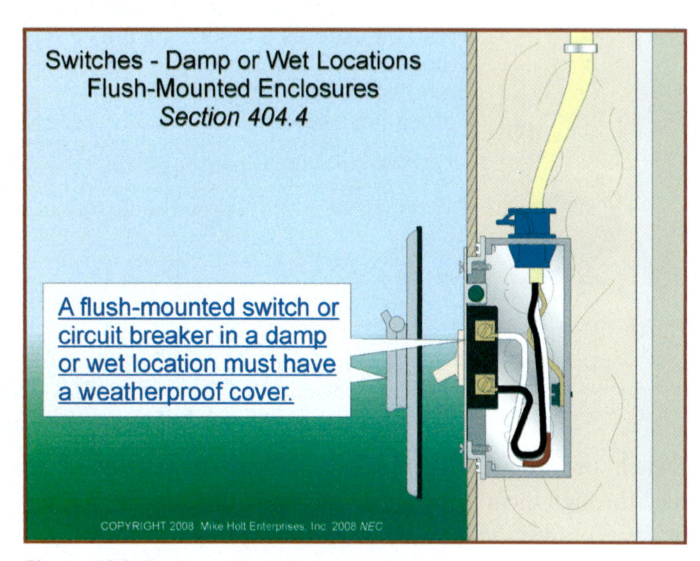

Figure 404–7

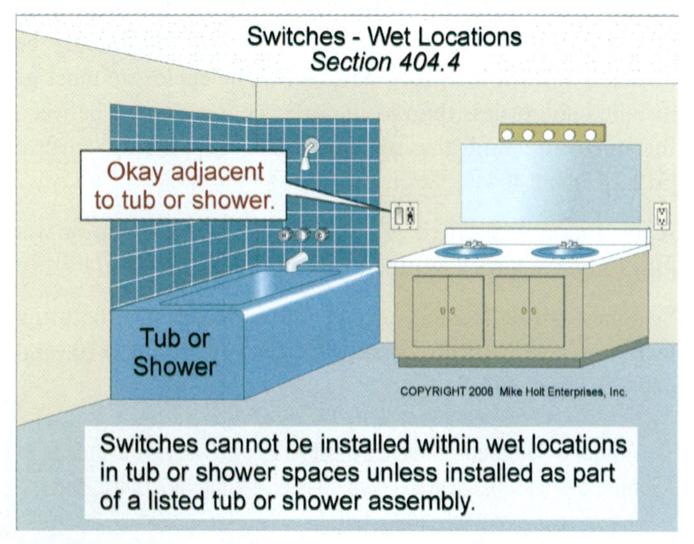

Figure 404–8

404.6 Position of Knife Switches.

(A) Single-Throw Knife Switch. Single-throw knife switches must be installed so gravity will not tend to close them.

404.7 Indicating. Switches, motor circuit switches, and circuit breakers used as switches must be marked to indicate whether they are in the "on" or "off" position. When the switch is operated vertically, it must be installed so the "up" position is the "on" position [240.81]. Figure 404–9

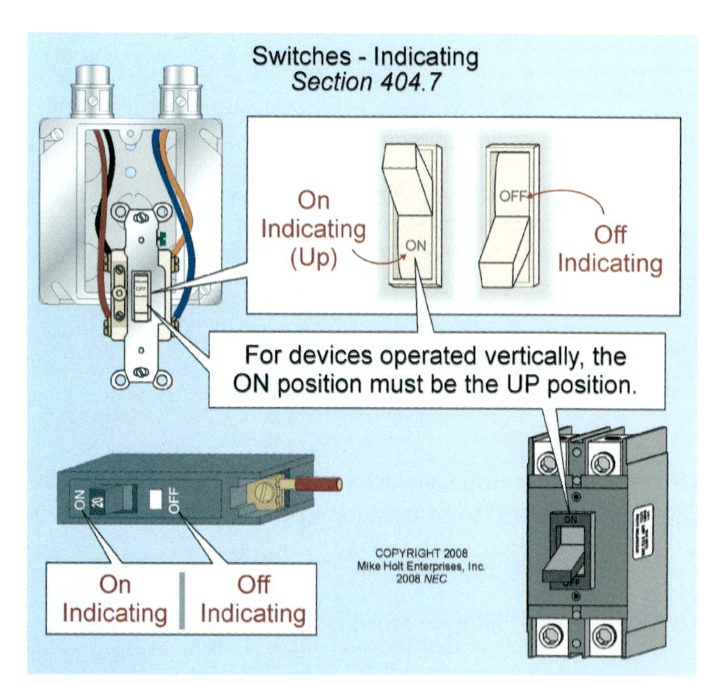

Figure 404–9

Exception No. 1: Double-throw switches, such as 3-way and 4-way switches, aren't required to be marked "on" or "off."

Exception No. 2: On busway installations, tap switches employing a center-pivoting handle can be open or closed with either end of the handle in the up or down position. The switch position must be clearly indicated and must be visible from the floor or from the usual point of operation.

404.8 Accessibility and Grouping.

(A) Location. Switches and circuit breakers used as switches must be capable of being operated from a readily accessible location. They must also be installed so the center of the grip of the operating handle of the switch or circuit breaker, when in its highest position, isn't more than 6 ft 7 in. above the floor or working platform [240.24(A)]. Figure 404–10

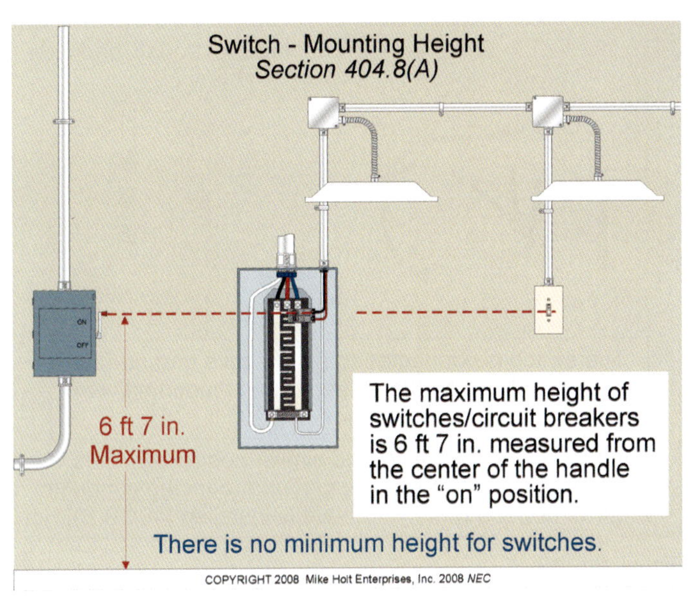

Figure 404–10

Author's Comment: The disconnecting means for a mobile home must be installed so the bottom of the enclosure isn't less than 2 ft above the finished grade or working platform [550.32(F)].

Exception No. 1: On busways, fusible switches, and circuit breakers where suitable means is provided to operate the handle of the device from the floor.

Exception No. 2: Switches and circuit breakers used as switches can be mounted above 6 ft 7 in. if they are next to the equipment they supply, and are accessible by portable means [240.24(A)(4)]. Figure 404–11

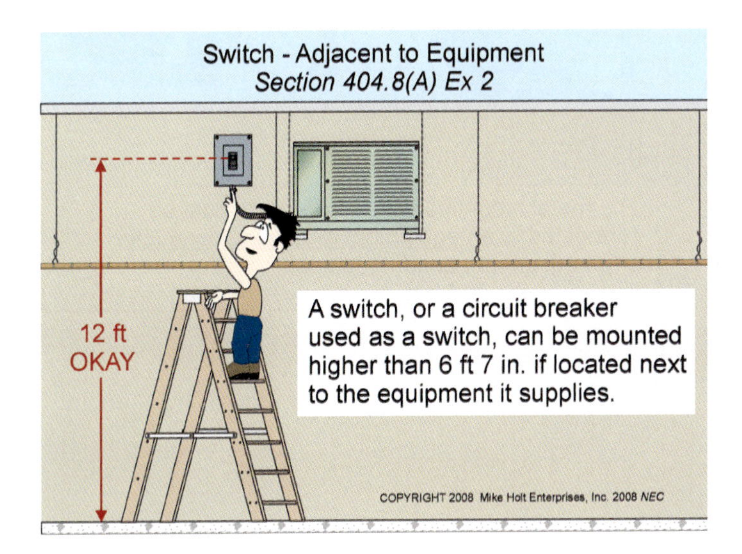

Figure 404–11

(B) Voltage Between Devices. Snap switches must not be grouped or ganged in enclosures with other snap switches, receptacles, or similar devices if the voltage between devices exceeds 300V, unless the devices are separated by barriers. Figures 404–12 and 404–13

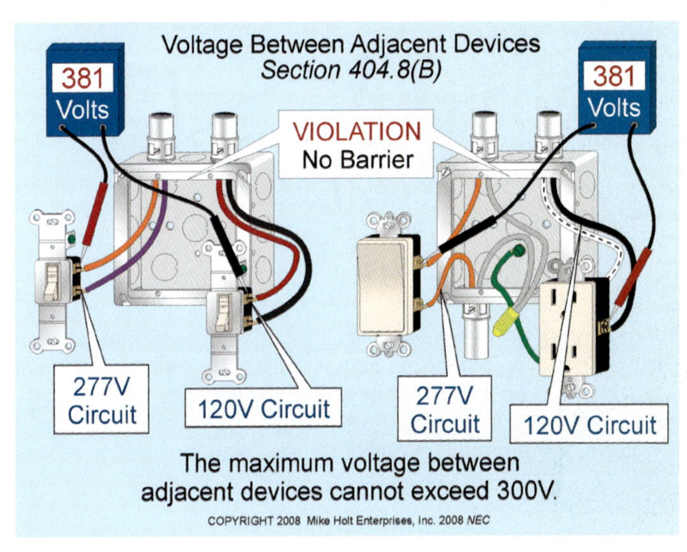

Figure 404–12

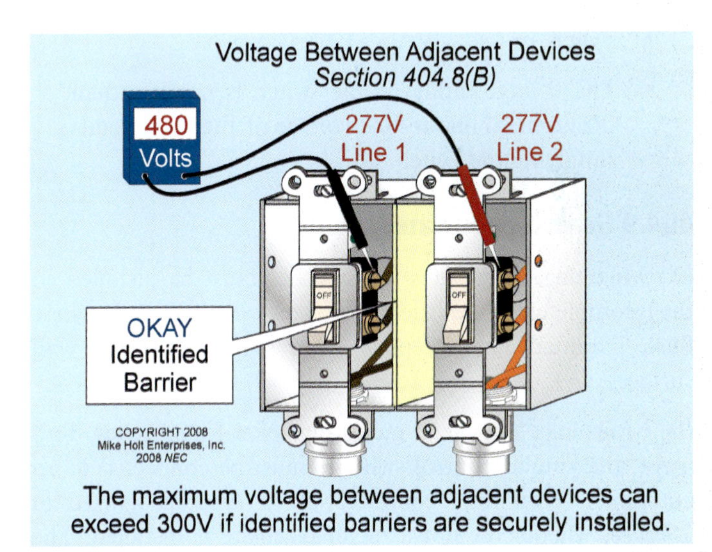

Figure 404–13

(C) Multipole Snap Switch. A multipole, general-use snap switch is not permitted to be fed by more than a single circuit unless the switch: Figure 404–14

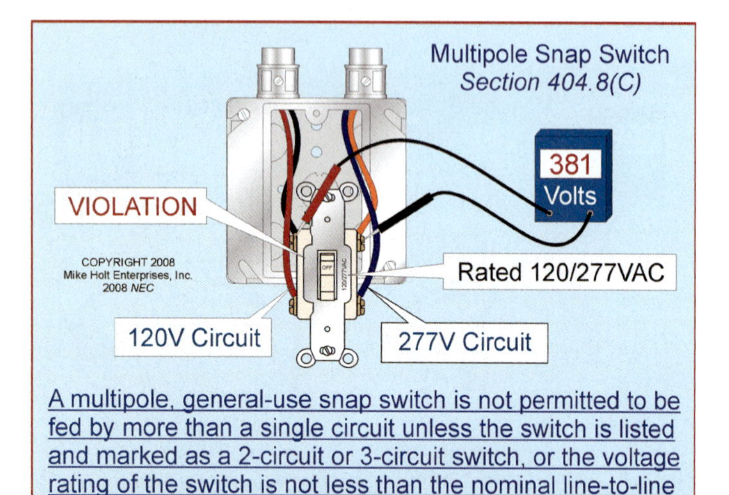

Figure 404–14

- Is listed and marked as a 2-circuit or 3-circuit switch, or
- The voltage rating of the switch is not less than the nominal line-to-line voltage of the two circuits supplying the switch.

404.9 Switch Faceplates.

(A) Mounting. Faceplates for switches must be installed so they completely cover the outlet box opening, and, where flush mounted, the faceplate must seat against the wall surface.

(B) Grounding. The metal mounting yokes for switches, dimmers, and similar control switches must be connected to an equipment grounding conductor of a type recognized in 250.118, whether or not a metal faceplate is installed. The metal mounting yoke is considered part of the effective ground-fault current path [250.2] by one of the following means:

(1) Mounting Screw. The switch is mounted with metal screws to a metal box or a metal cover that is connected to an equipment grounding conductor of a type recognized in 250.118. Figure 404–15

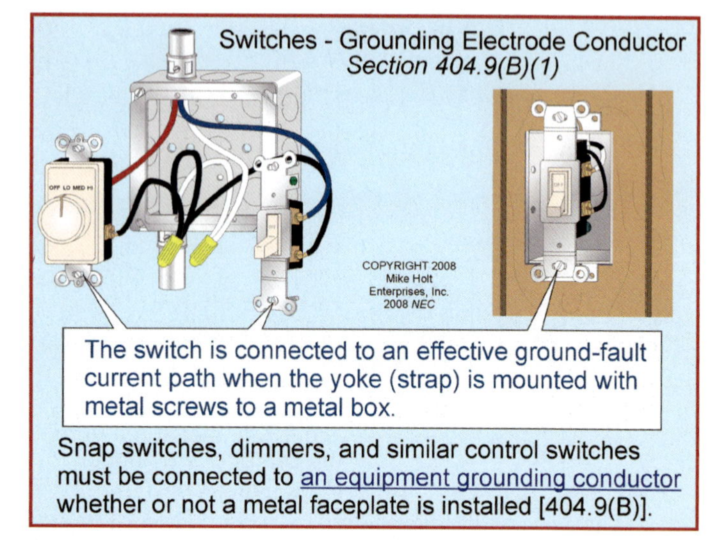

Figure 404–15

Author's Comment: Direct metal-to-metal contact between the device yoke of a switch and the box isn't required.

(2) Equipment Grounding Conductor. An equipment grounding conductor or equipment bonding jumper is connected to the grounding terminal of the metal mounting yoke. Figure 404–16

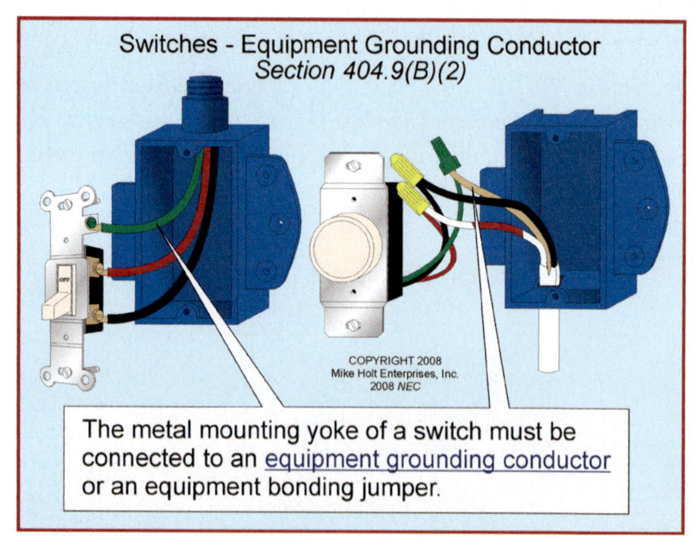

Figure 404–16

Exception: The metal mounting yoke of a replacement switch isn't required to be connected to an equipment grounding conductor of a type recognized in 250.118 if the wiring to the existing switch doesn't contain an equipment grounding conductor, and the switch faceplate is nonmetallic or the replacement switch is GFCI protected.

404.10 Mounting Snap Switches.

(A) Mounting of Snap Switches. Snap switches installed in recessed boxes must have the ears of the switch yoke seated firmly against the finished wall surface.

> **Author's Comment:** In walls or ceilings of noncombustible material, such as drywall, boxes must not be set back more than ¼ in. from the finished surface. In combustible walls or ceilings, boxes must be flush with, or project slightly from, the finished surface [314.20]. There must not be any gaps more than ⅛ in. at the edge of the box [314.21].

404.11 Circuit Breakers Used as Switches. A manually operable circuit breaker used as a switch must show when it's in the "on" (closed) or "off" (open) position [404.7].

> **Author's Comment:** Circuit breakers used to switch fluorescent lighting must be listed and marked "SWD" or "HID." Circuit breakers used to switch high-intensity discharge lighting must be listed and must be marked "HID" [240.83(D)]. Figure 404–17

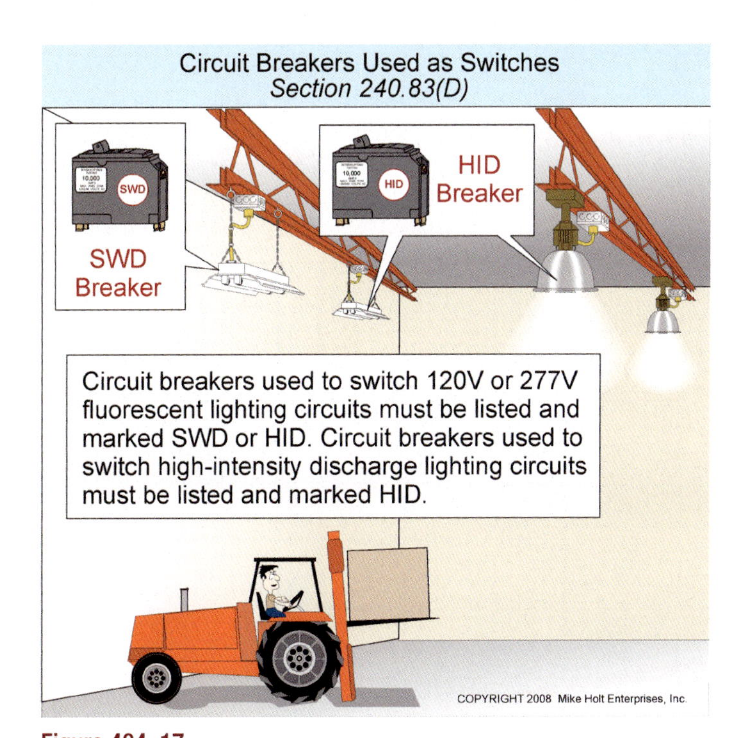

Figure 404–17

404.12 Grounding of Enclosures. Metal enclosures for switches and circuit breakers used as switches must be connected to an equipment grounding conductor of a type recognized in 250.118 [250.4(A)(3)].

404.14 Rating and Use of Snap Switches.

(A) AC General-Use Snap Switches. Alternating-current general-use snap switches can control:

(1) Resistive and inductive loads, including electric-discharge lamps that do not exceed the ampere rating of the switch, at the voltage involved.

(2) Tungsten-filament lamp loads not exceeding the ampere rating of the switch at 120V.

(3) Motor loads rated 2 hp or less that don't exceed 80 percent of the ampere rating of the switch. See 430.109(C).

(C) CO/ALR Snap Switches. Snap switches rated 15A or 20A connected to aluminum wire must be marked CO/ALR. See 406.2(C).

> **Author's Comment:** According to UL listing requirements, aluminum conductors must not terminate in screwless (push-in) terminals of a snap switch (UL White Book, product category AALZ).

(E) Dimmers. General-use dimmer switches are only permitted to control permanently installed incandescent luminaires. Figure 404–18

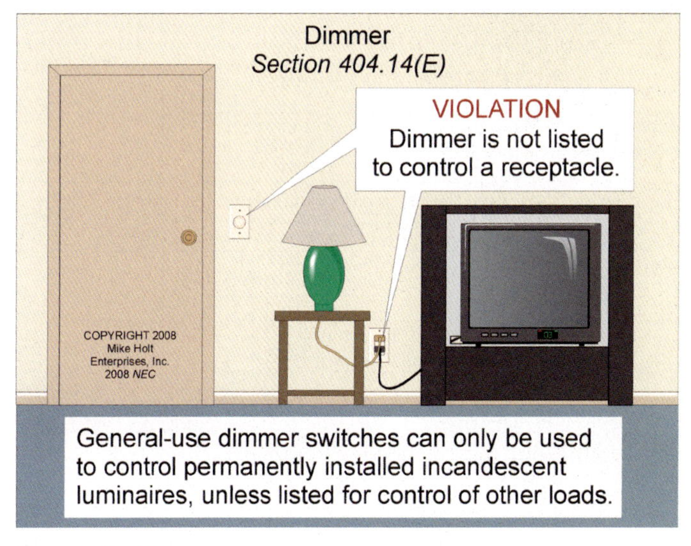

Figure 404–18

404.15 Switch Marking.

(A) Markings. Switches must be marked with the current, voltage, and, if horsepower rated, the maximum rating for which they are designed.

(B) Off Indication. Where in the off position, a switching device with a marked "off" position must completely disconnect all ungrounded conductors of the load it controls.

> **Author's Comment:** Where an electronic occupancy sensor is used for switching, voltage will be present and a small current of 0.05 mA can flow through the circuit when the switch is in the "off" position. This small amount of current can startle a person, perhaps causing a fall. To solve this problem, manufacturers have simply removed the word "off" from the switch. Figure 404–19

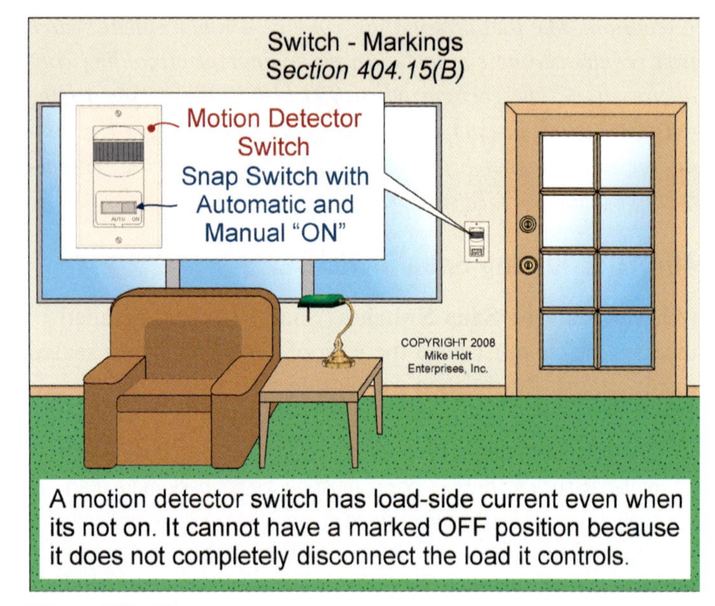

Figure 404–19

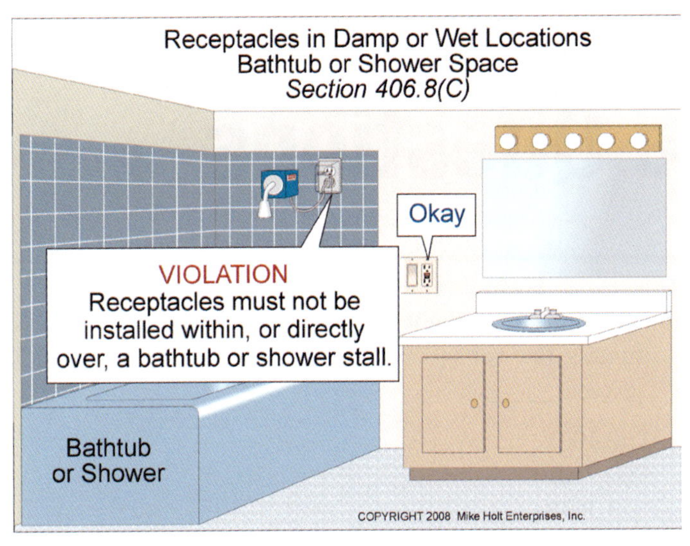

Figure 406–20

406.10 Connecting Receptacle Grounding Terminal to Equipment Grounding Conductor. The grounding terminal of receptacles must be connected to an equipment grounding conductor in accordance with 250.146.

406.11 Tamper-Resistant Receptacles in Dwelling Units. All 15A and 20A, 125V receptacles required in the following areas of a dwelling unit [210.52] must be listed as tamper resistant.

- Wall Space—210.52(A)
- Small-Appliance Circuit—210.52(B)
- Countertop Space—210.52(C)
- Bathroom Area—210.52(D)
- Outdoors—210.52(E)
- Laundry Area—210.52(F)
- Garage and Outbuildings—210.52(G)
- Hallways—210.52(H)

Author's Comments:

- From 1991 through 2001 over 24,000 children less than 10 years of age were admitted to the emergency rooms of study participating hospitals for injuries involving electrical receptacles. Most of the victims were male and used a hairpin or other metal object, in the home, and sustained first or second degree burns.
- This rule applies to 15A and 20A, 125V receptacles installed behind appliances, above countertops and other locations out of the reach of children.

ARTICLE 406 Practice Questions

ARTICLE 406. RECEPTACLES, CORD CONNECTORS, AND ATTACHMENT PLUGS (CAPS)—PRACTICE QUESTIONS

1. Receptacles and cord connectors shall be rated not less than _____ at 125V, or at 250V, and shall be of a type not suitable for use as lampholders.

   (a) 10A
   (b) 15A
   (c) 20A
   (d) 30A

2. Isolated ground receptacles installed in nonmetallic boxes shall be covered with a nonmetallic faceplate, unless the box contains a feature or accessory that permits the effective grounding of the faceplate.

   (a) True
   (b) False

3. When replacing receptacles in locations that would require GFCI protection under the current NEC, _____ receptacles shall be installed.

   (a) dedicated
   (b) isolated ground
   (c) GFCI-protected
   (d) grounding

4. Receptacles mounted in boxes flush with the finished surface or projecting beyond it shall be installed so that the mounting yoke or strap of the receptacle is _____.

   (a) held rigidly against the box or box cover
   (b) mounted behind the wall surface
   (c) held rigidly at the finished surface
   (d) none of these

5. Receptacles shall not be grouped or ganged in enclosures unless the voltage between adjacent devices does not exceed _____.

   (a) 100V
   (b) 200V
   (c) 300V
   (d) 400V

6. Receptacles installed outdoors, in a location protected from the weather or other damp locations, shall be in an enclosure that is _____ when the receptacle is covered.

   (a) raintight
   (b) weatherproof
   (c) rainproof
   (d) weathertight

7. _____, 125V and 250V receptacles installed in a wet location shall have an enclosure that is weatherproof whether or not the attachment plug cap is inserted.

   (a) 15A
   (b) 20A
   (c) a and b
   (d) none of these

8. A 30A, 208V receptacle installed in a wet location, where the product intended to be plugged into it is not attended while in use, shall have an enclosure that is weatherproof with the attachment plug cap inserted or removed.

   (a) True
   (b) False

9. Grounding-type attachment plugs shall be used only with a cord having a(n) _____ conductor.

   (a) equipment grounding
   (b) isolated
   (c) computer circuit
   (d) insulated

ARTICLE

408 Switchboards and Panelboards

INTRODUCTION TO ARTICLE 408—SWITCHBOARDS AND PANELBOARDS

Article 408 covers the specific requirements for switchboards, panelboards, and distribution boards that control power and lighting circuits. Some key points to remember:

- One objective of Article 408 is that the installation prevents contact between current-carrying conductors and people or equipment.
- The circuit directory of a panelboard must clearly identify the purpose or use of each circuit that originates in the panelboard.
- You must understand the detailed grounding and overcurrent protection requirements for panelboards.

PART I. GENERAL

408.1 Scope.

(1) Article 408 covers the specific requirements for switchboards, panelboards, and distribution boards that control power and lighting circuits. Figure 408–1

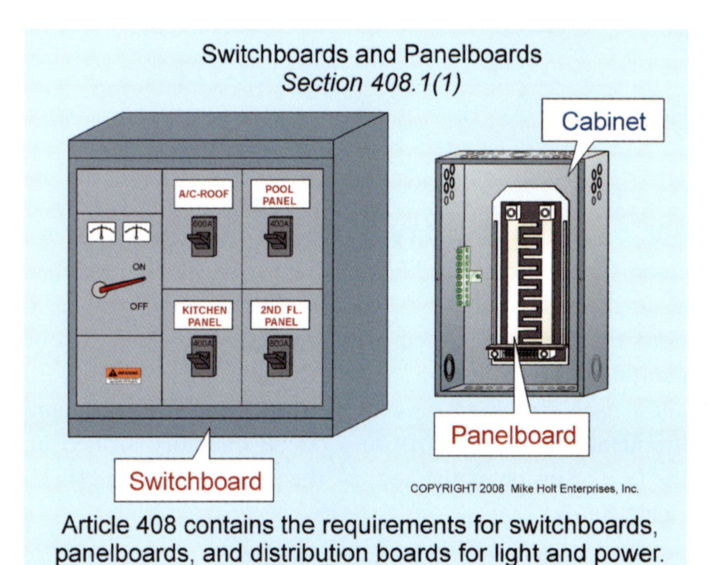

Article 408 contains the requirements for switchboards, panelboards, and distribution boards for light and power.

Figure 408–1

(2) Article 408 covers battery-charging panels supplied from light or power circuits.

Author's Comment: For the purposes of this textbook, we'll only cover the requirements for panelboards.

408.3 Arrangement of Busbars and Conductors.

(C) Used as Service Equipment. Panelboards suitable for use as service equipment must be provided with a main bonding jumper to connect the service neutral conductor to the panelboard's metal frame.

(D) Terminals. In switchboards and panelboards, terminals for neutral and equipment grounding conductors must be located so it's not necessary to reach beyond live parts in order to make connections.

(E) Panelboard Phase Arrangement. Panelboards supplied by a 4-wire, delta-connected, three-phase (high-leg) system must have the high-leg conductor (which operates at 208V to ground) terminate to the "B" phase of the panelboard. Figure 408–2

Exception: The high-leg conductor can terminate to the "C" phase when the meter is located in the same section of a switchboard or panelboard.

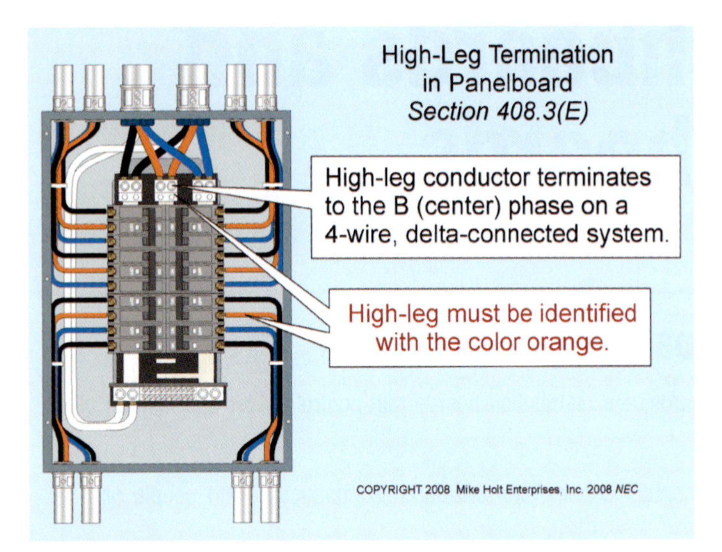

Figure 408–2

FPN: Orange identification, or some other effective means, is required for the high-leg conductor [110.15 and 230.56].

WARNING: *The ANSI standard for meter equipment requires the high-leg conductor (208V to neutral) to terminate on the "C" (right) phase of the meter socket enclosure. This is because the demand meter needs 120V, and it gets it from the "B" phase.* Figure 408–3

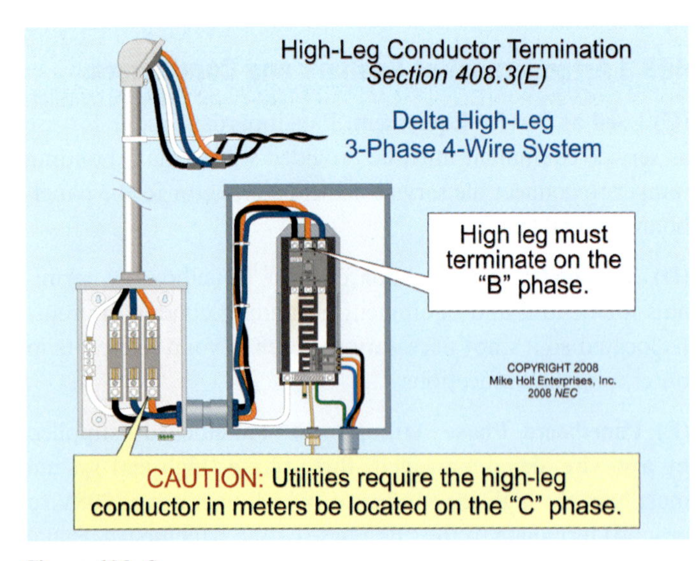

Figure 408–3

WARNING: *When replacing equipment in existing facilities that contain a high-leg conductor, use care to ensure that the high-leg conductor is replaced in the original location. Prior to 1975, the high-leg conductor was required to terminate on the "C" phase of panelboards and switchboards. Failure to re-terminate the high leg in accordance with the existing installation can result in 120V circuits inadvertently connected to the 208V high leg, with disastrous results.*

(F) High-Leg Identification. Switchboards and panelboards containing a 4-wire, delta-connected system where the midpoint of one phase winding is grounded (high-leg system), must be legibly and permanently field-marked with the following: Figure 408–4

"CAUTION B PHASE HAS 208V TO GROUND"

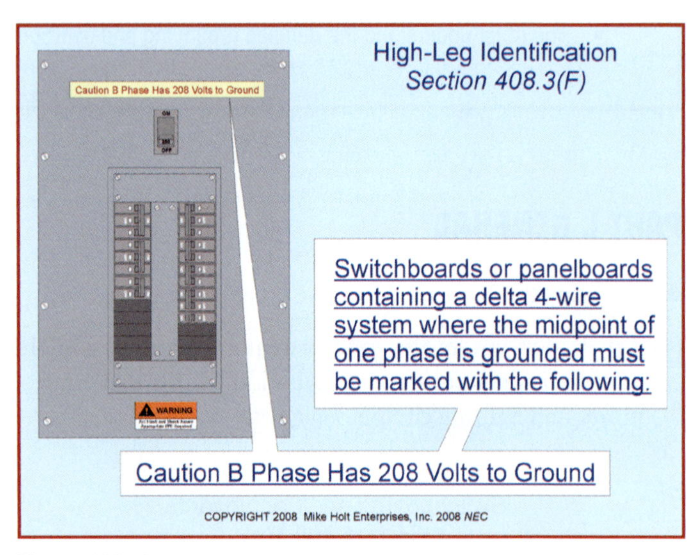

Figure 408–4

408.4 Circuit Directory or Circuit Identification.

All circuits, and circuit modifications, must be legibly identified as to their clear, evident, and specific purpose. Spare positions that contain unused overcurrent devices must also be identified. Identification must include sufficient detail to allow each circuit to be distinguished from all others, and the identification must be on a circuit directory located on the face or inside of the door of the panelboard. See 110.22. Figure 408–5

Circuit identification must not be based on transient conditions of occupancy, such as Steven's, or Brittney's bedroom. Figure 408–6

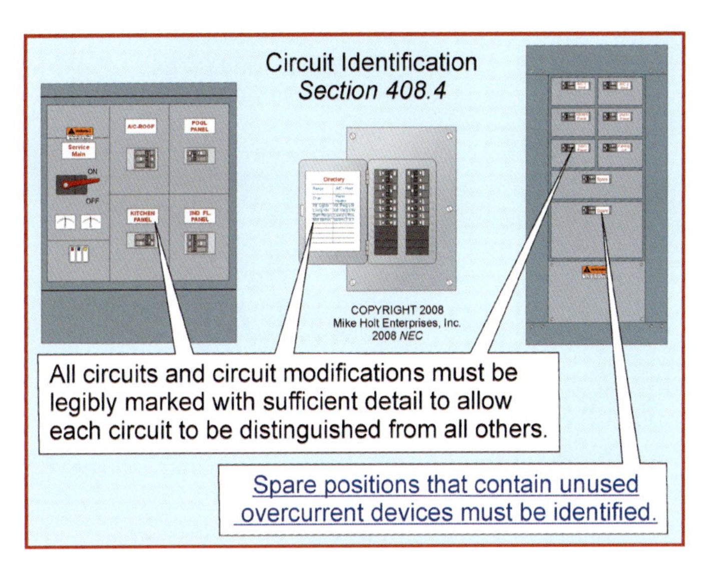

Figure 408–5

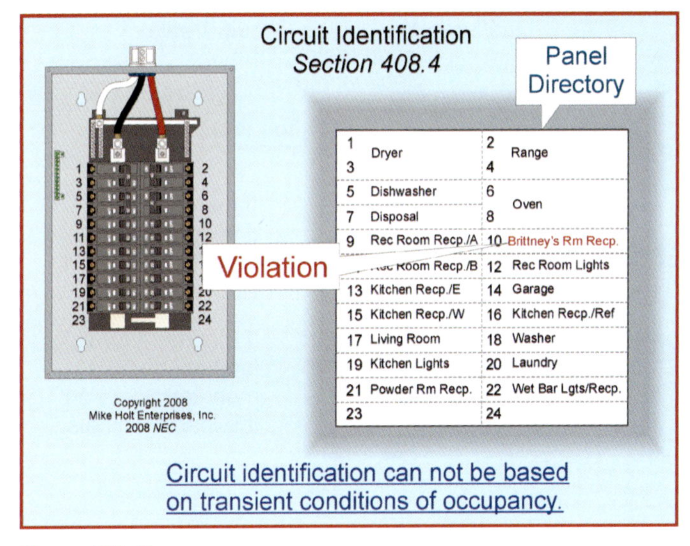

Figure 408–6

408.5 Clearance for Conductors Entering Bus Enclosures. Where raceways enter a switchboard, floor-standing panelboard, or similar enclosure, the raceways, including end fittings, must not rise more than 3 in. above the bottom of the enclosure.

408.7 Unused Openings. Unused openings for circuit breakers and switches must be closed using identified closures, or other means approved by the authority having jurisdiction, that provide protection substantially equivalent to the wall of the enclosure. Figure 408–7

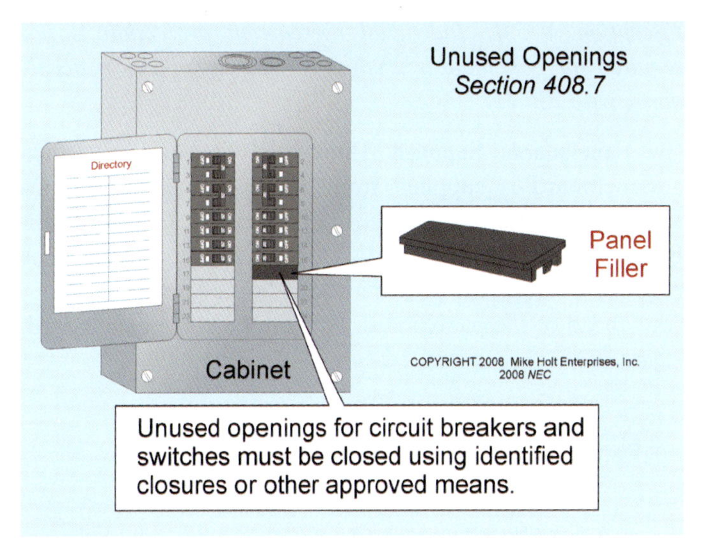

Figure 408–7

PART III. PANELBOARDS

408.36 Overcurrent Protection of Panelboards.

Each panelboard must be provided with overcurrent protection located within, or at any point on the supply side of, the panelboard. The overcurrent device must have a rating not greater than that of the panelboard, and it can be located within, or on the supply side of, the panelboard. Figure 408–8

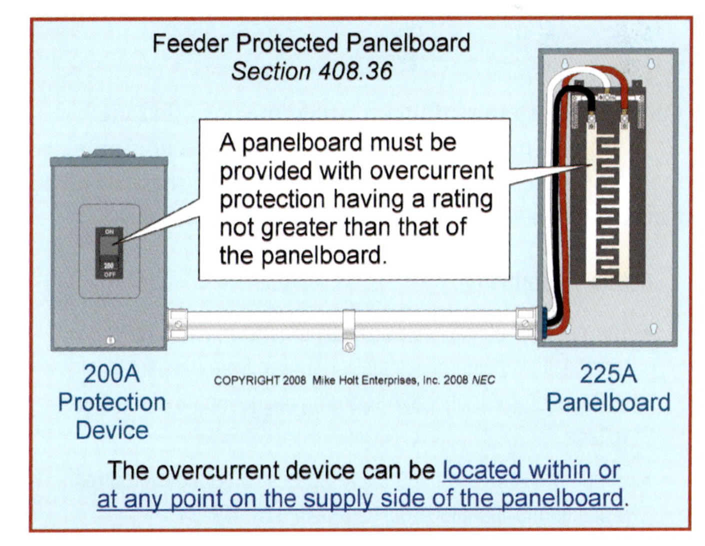

Figure 408–8

Exception No. 1: Individual overcurrent protection is not required for panelboards used as service equipment in accordance with 230.71.

(B) Panelboards Supplied Through a Transformer. When a panelboard is supplied from a transformer, as permitted in 240.21(C), the overcurrent protection for the panelboard must be on the secondary side of the transformer. The required overcurrent protection can be in a separate enclosure ahead of the panelboard, or it can be in the panelboard. Figure 408–9

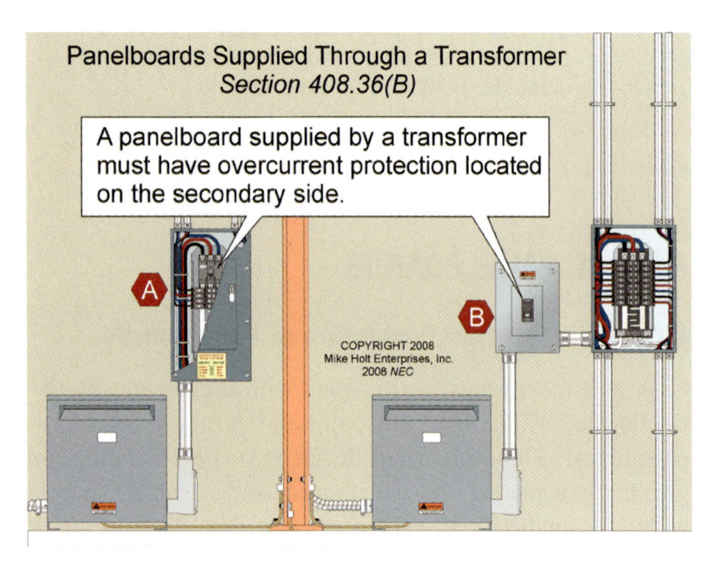

Figure 408–9

(D) Back-Fed Devices. Plug-in circuit breakers that are back-fed from field installed conductors must be secured in place by an additional fastener that requires other than a pull to release the breaker from the panelboard. Figure 408–10

Author's Comments:

- The purpose of the breaker fastener is to prevent the circuit breaker from being accidentally removed from the panelboard while energized, thereby exposing someone to dangerous voltage.
- Circuit breakers are often back-fed to provide the overcurrent protection for panelboards, as required by 408.36.

CAUTION: *Circuit breakers marked "Line" and "Load" must be installed in accordance with listing or labeling instructions [110.3(B)]; therefore, these types of devices must not be back-fed.* Figure 408–11

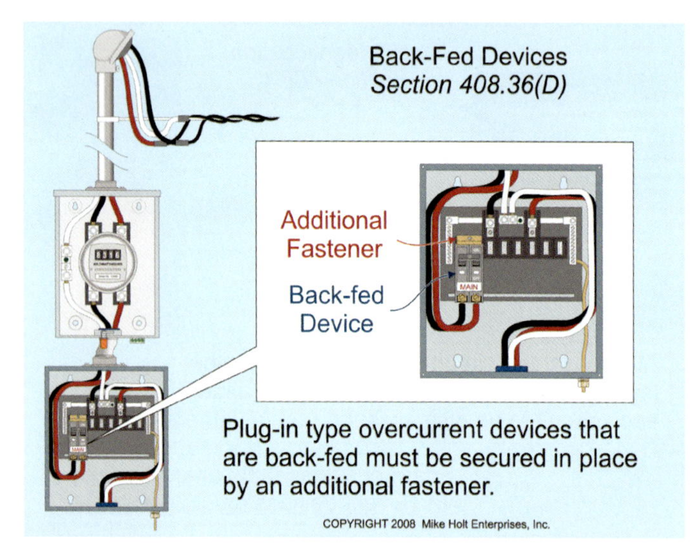

Figure 408–10

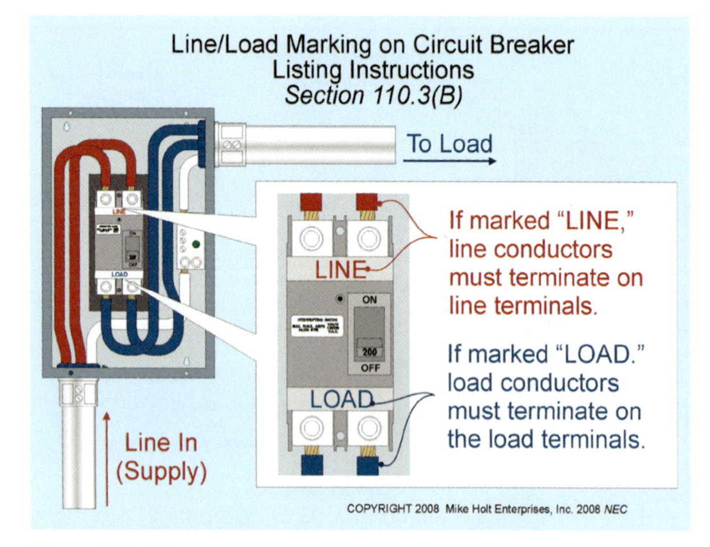

Figure 408–11

408.37 Panelboards in Damp or Wet Locations.

The enclosures (cabinets) for panelboards must prevent moisture or water from entering or accumulating within the enclosure, and they must be weatherproof when located in a wet location. When the enclosure is surface mounted in a wet location, the enclosure must be mounted with not less than ¼ in. air space between it and the mounting surface [312.2].

408.40 Equipment Grounding Conductor.

Metal panelboard cabinets and frames must be connected to an equipment grounding conductor of a type recognized in 250.118 [215.6 and 250.4(A)(3)].

Where the panelboard cabinet is used with nonmetallic raceways or cables, or where separate equipment grounding conductors are provided, a terminal bar for the circuit equipment grounding conductors must be bonded to the metal cabinet. Figure 408–12

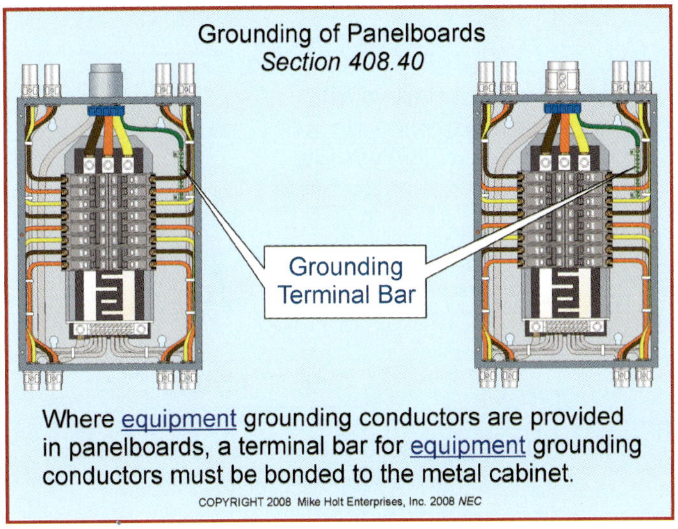

Grounding of Panelboards
Section 408.40

Grounding Terminal Bar

Where equipment grounding conductors are provided in panelboards, a terminal bar for equipment grounding conductors must be bonded to the metal cabinet.

COPYRIGHT 2008 Mike Holt Enterprises, Inc. 2008 *NEC*

Figure 408–12

Exception: Insulated equipment grounding conductors for receptacles having insulated grounding terminals (isolated ground receptacles) [250.146(D)] can pass through the panelboard without terminating onto the equipment grounding terminal of the panelboard cabinet. Figure 408–13

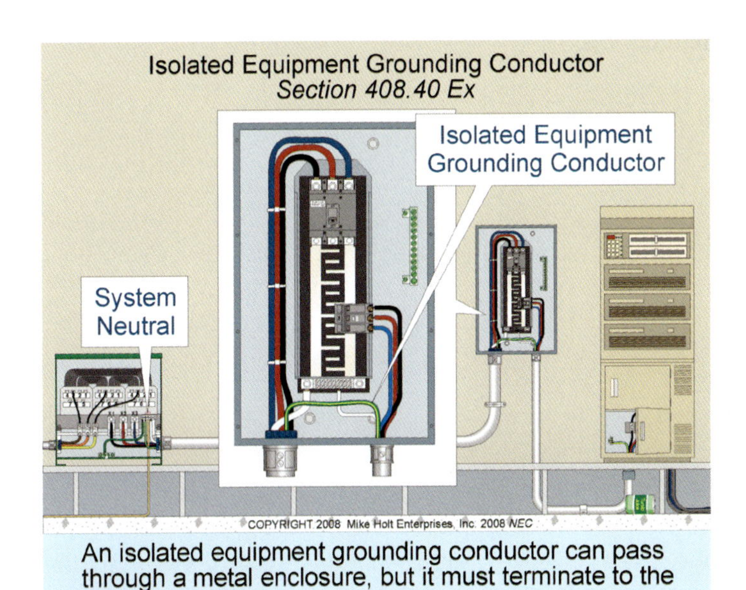

Figure 408–13

Equipment grounding conductors must not terminate on the neutral terminal bar, and neutral conductors must not terminate on the equipment grounding terminal bar, except as permitted by 250.142 for services and separately derived systems. Figure 408–14

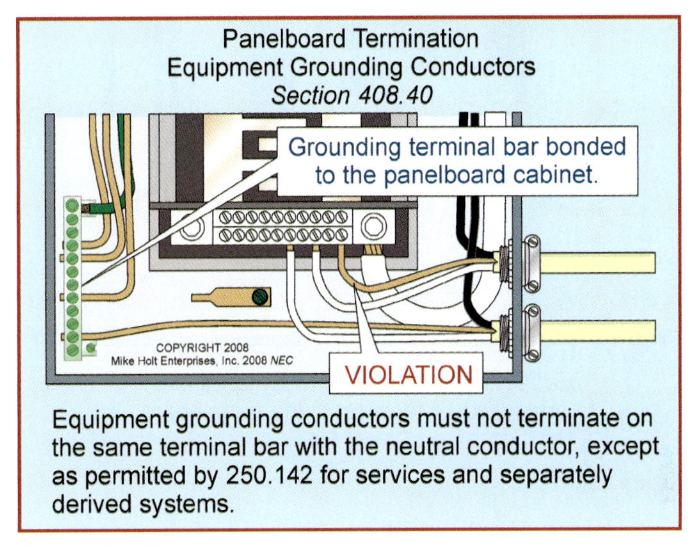

Figure 408–14

Author's Comment: See the definition of "Separately Derived System" in Article 100.

CAUTION: *Most panelboards are rated as suitable for use as service equipment, which means they are supplied with a main bonding jumper [250.28]. This screw or strap must not be installed except when the panelboard is used for service equipment [250.24(A)(5)] or separately derived systems [250.30(A)(1)]. In addition, a panelboard marked "suitable only for use as service equipment" means the neutral bar or terminal of the panelboard has been bonded to the case at the factory, and this panelboard is restricted to being used only for service equipment or on separately derived systems according to 250.142(B).*

408.41 Neutral Conductor Terminations. Each neutral conductor within a panelboard must terminate to an individual terminal. Figure 408–15

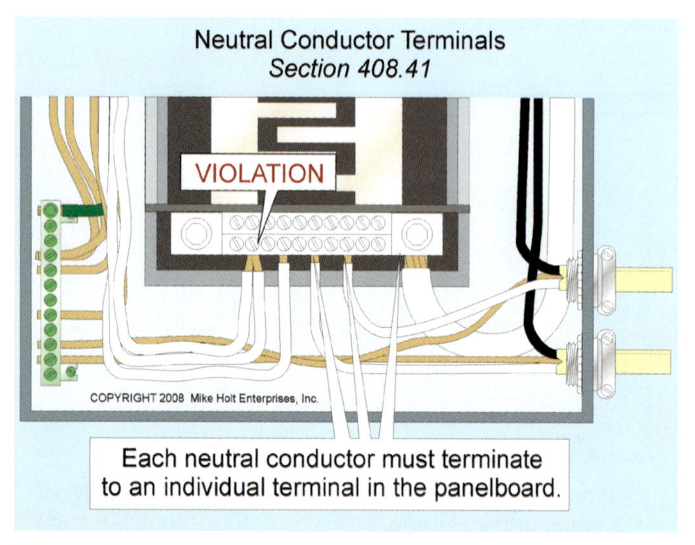

Figure 408–15

Author's Comment: If two neutral conductors are connected to the same terminal, and someone removes one of them, the other neutral conductor might unintentionally be removed as well. If that happens to the neutral conductor of a multiwire circuit, it can result in excessive line-to-neutral voltage for one of the circuits, as well as undervoltage for the other circuit. See 300.13(B) of this textbook for details. Figure 408–16

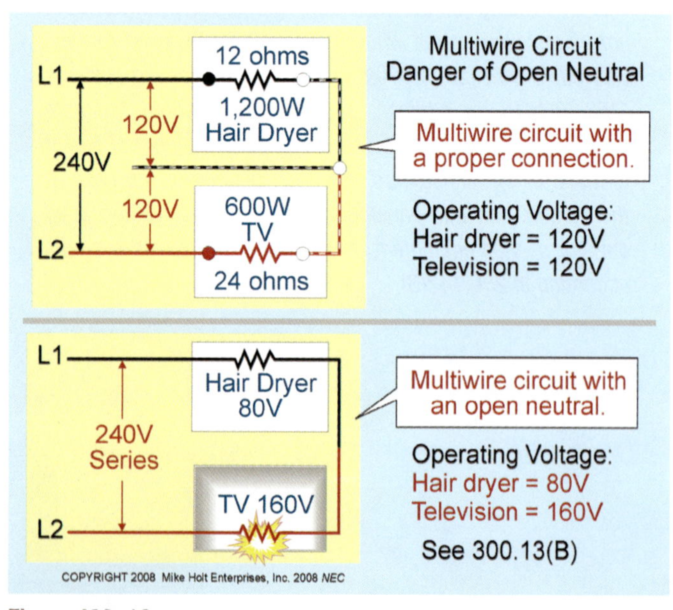

Figure 408–16

This requirement doesn't apply to equipment grounding conductors, because the voltage of a circuit is not affected if an equipment grounding conductor is accidentally removed. Figure 408–17

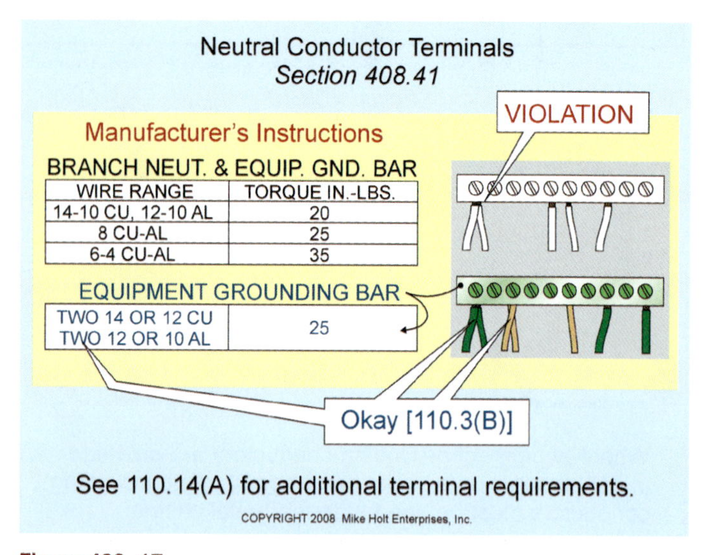

Figure 408–17

408.54 Maximum Number of Overcurrent Devices. A panelboard must prevent the installation of more overcurrent devices than the number for which the panelboard was designed, rated, and listed. When applying this rule, a 2-pole circuit breaker is considered to be two overcurrent devices, and a 3-pole circuit breaker is considered to be three overcurrent devices.

ARTICLE 408 Practice Questions

ARTICLE 408. SWITCHBOARDS AND PANELBOARDS—PRACTICE QUESTIONS

1. Each switchboard or panelboard used as service equipment shall be provided with a main bonding jumper within the panelboard, or within one of the sections of the switchboard, for connecting the grounded service conductor on its _____ side to the switchboard or panelboard frame.

   (a) load
   (b) supply
   (c) phase
   (d) high leg

2. A switchboard or panelboard containing a 4-wire, _____ system where the midpoint of one phase winding is grounded, shall be legibly and permanently field-marked to caution that one phase has a higher voltage-to-ground.

   (a) wye-connected
   (b) delta-connected
   (c) solidly grounded
   (d) ungrounded

3. Unused openings for circuit breakers and switches in switchboards and panelboards shall be closed using _____ or other approved means that provide protection substantially equivalent to the wall of the enclosure.

   (a) duct seal and tape
   (b) identified closures
   (c) exothermic welding
   (d) sheet metal

4. Plug-in-type circuit breakers that are back-fed shall be _____ by an additional fastener that requires more than a pull to release.

   (a) grounded
   (b) secured in place
   (c) shunt tripped
   (d) none of these

5. A panelboard shall be provided with physical means to prevent the installation of more _____ devices than that number for which the panelboard was designed, rated, and listed.

   (a) overcurrent
   (b) equipment
   (c) circuit breaker
   (d) all of these

ARTICLE

410 Luminaires, Lampholders, and Lamps

INTRODUCTION TO ARTICLE 410—LUMINAIRES, LAMPHOLDERS, AND LAMPS

This article covers luminaires, lampholders, lamps, decorative lighting products, lighting accessories for temporary seasonal and holiday use, including portable flexible lighting products, and the wiring and equipment of such products and lighting installations.

Even though Article 410 is highly detailed, it's broken down into sixteen parts. The first five are sequential, and apply to all luminaires, lampholders, and lamps:

- GENERAL, PART I
- LOCATION, PART II
- BOXES AND COVERS, PART III
- SUPPORTS, PART IV
- EQUIPMENT GROUNDING CONDUCTOR, PART V

This is mostly mechanical information, and it's not hard to follow or absorb. Part VI, Wiring, ends the sequence. The seventh, ninth, and tenth parts provide requirements for manufacturers to follow—use only equipment that conforms to these requirements. Part VIII provides requirements for installing Lampholders. The rest of Article 410 addresses specific types of lighting.

Author's Comment: Article 411 addresses "Lighting Systems Operating at 30 Volts or Less."

PART I. GENERAL

410.1 Scope. This article covers luminaires, lampholders, lamps, decorative lighting products, lighting accessories for temporary seasonal and holiday use, portable flexible lighting products, and the wiring and equipment of such products and lighting installations. Figure 410–1

Author's Comment: Because of the many types and applications of luminaires, manufacturers' instructions are very important and helpful for proper installation. UL produces a pamphlet called the Luminaire Marking Guide, which provides information for properly installing common types of incandescent, fluorescent, and high-intensity discharge (HID) luminaires.

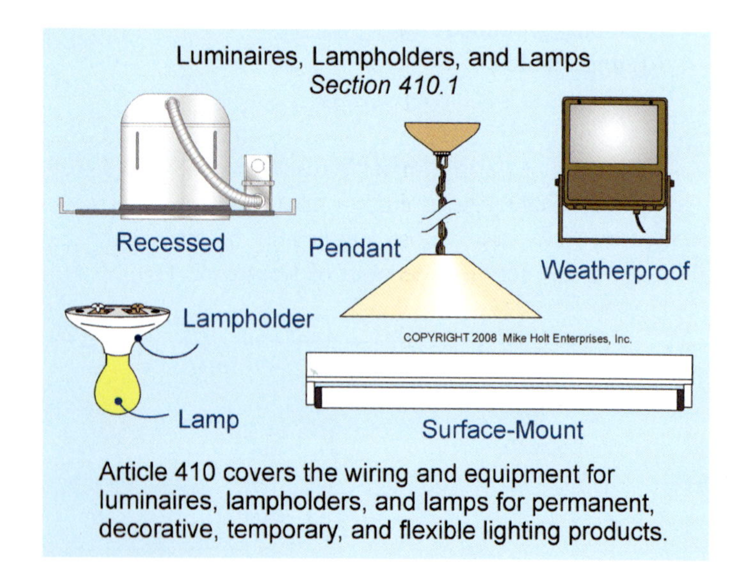

Figure 410–1

410.2 Definitions.

Closet Storage Space. Storage space is defined as a volume bounded by the sides and back closet walls, extending from the closet floor vertically to a height of 6 ft or the highest clothes-hanging rod at a horizontal distance of 2 ft from the sides and back of the closet walls. Storage space continues vertically to the closet ceiling for a distance of 1 ft or the width of the shelf, whichever is greater. Figure 410–2

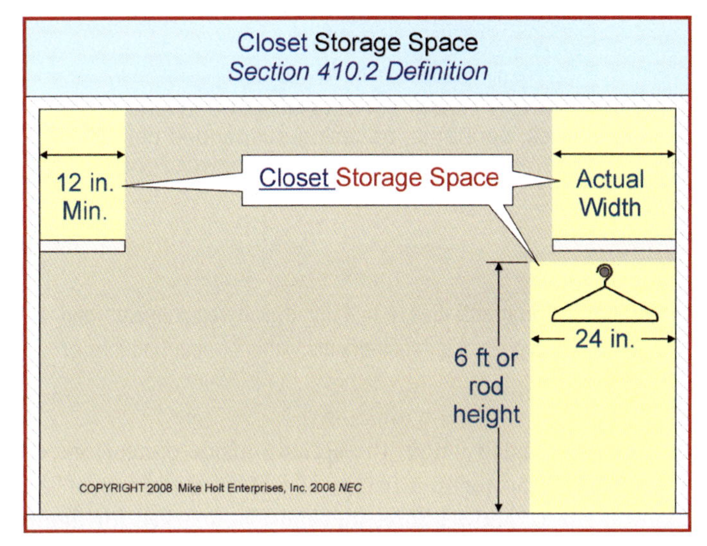

Figure 410–2

Author's Comment: The definition consists of approximately 125 words in one sentence, perhaps the longest sentence in the *Code*. Take a breath and don't get lost when reading it!

Lighting Track. This is a manufactured assembly, designed to support and energize luminaires, that can be readily repositioned on track, and whose length may be altered by the addition or subtraction of sections of track. Figure 410–3

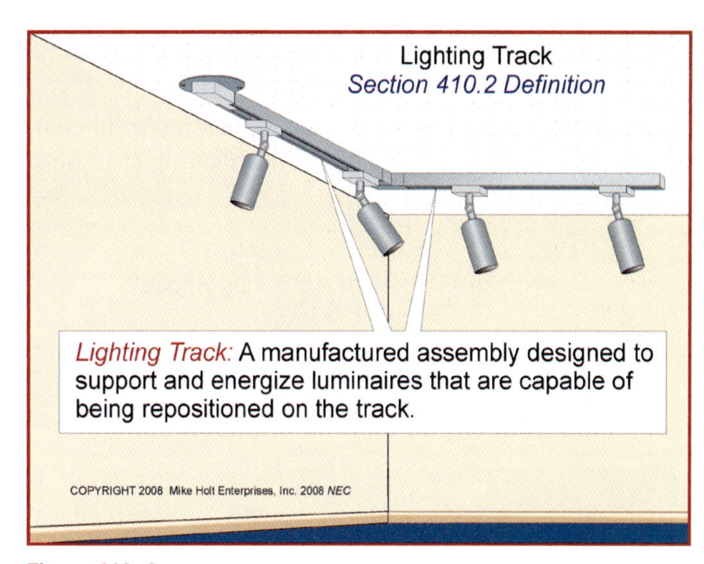

Figure 410–3

410.6 Listing Required. Luminaires and lampholders must be listed.

PART II. LUMINAIRE LOCATIONS

410.10 Luminaires in Specific Locations.

(A) Wet or Damp Locations. Luminaires in wet or damp locations must be installed in a manner that prevents water from accumulating in any part of the luminaire. Luminaires marked "Suitable for Dry Locations Only" must be installed only in a dry location; luminaires marked "Suitable for Damp Locations" can be installed in either a damp or dry location; and luminaires marked "Suitable for Wet Locations" can be installed in a dry, damp, or wet location. Figure 410–4

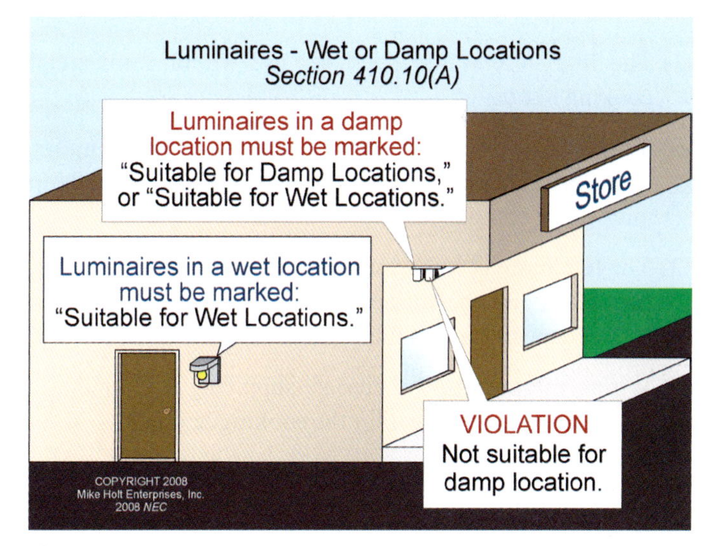

Figure 410–4

Author's Comment: A dry location can be subjected to occasional dampness or wetness. See the definition of "Location, Dry" in Article 100.

(B) Corrosive Locations. Luminaires installed in corrosive locations must be suitable for the location.

(C) In Ducts or Hoods. Luminaires can be installed in commercial cooking hoods where all of the following conditions are met: Figure 410–5

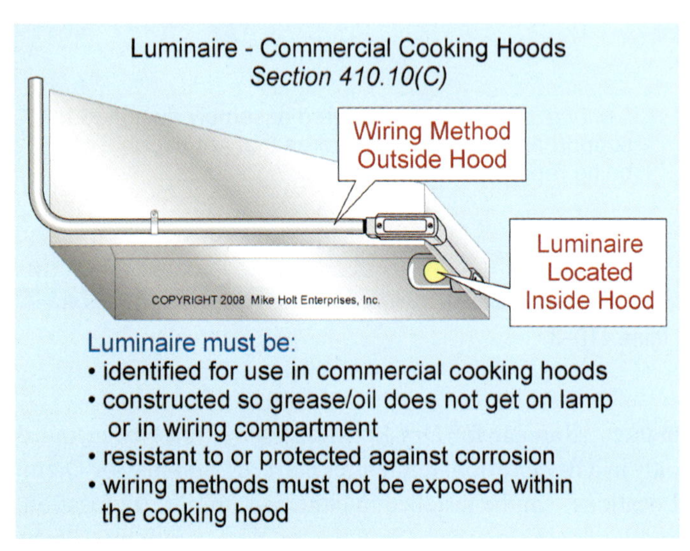

Figure 410–5

(1) The luminaire is identified for use within commercial cooking hoods.

(2) The luminaire is constructed so that all exhaust vapors, grease, oil, or cooking vapors are excluded from the lamp and wiring compartment.

(3) The luminaire is corrosion resistant, or protected against corrosion, and the surface must be smooth so as not to collect deposits and to facilitate cleaning.

(4) Wiring methods and materials supplying the luminaire must not be exposed within the cooking hood.

Author's Comment: Standard gasketed luminaires must not be installed in a commercial cooking hood because accumulations of grease and oil can result in a fire caused by high temperatures on the glass globe.

(D) Bathtub and Shower Areas. No part of chain-, cable-, or cord-suspended luminaires, track luminaires, or ceiling paddle fans can be located within 3 ft horizontally and 8 ft vertically from the top of the bathtub rim or shower stall threshold. Figure 410–6

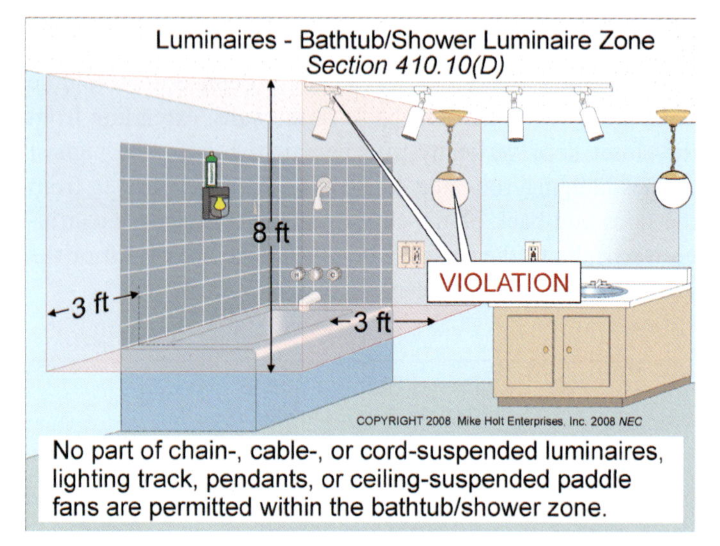

Figure 410–6

Author's Comment: See 404.4 for switch requirements and 406.8(C) for receptacle requirements within or near bathtubs or shower stalls.

Luminaires located within the actual outside dimensions of a bathtub or shower to a height of 8 ft from the top of the bathtub rim or shower threshold must be marked for damp locations. Where subject to shower spray, the luminaires must be marked for wet locations. Figure 410–7

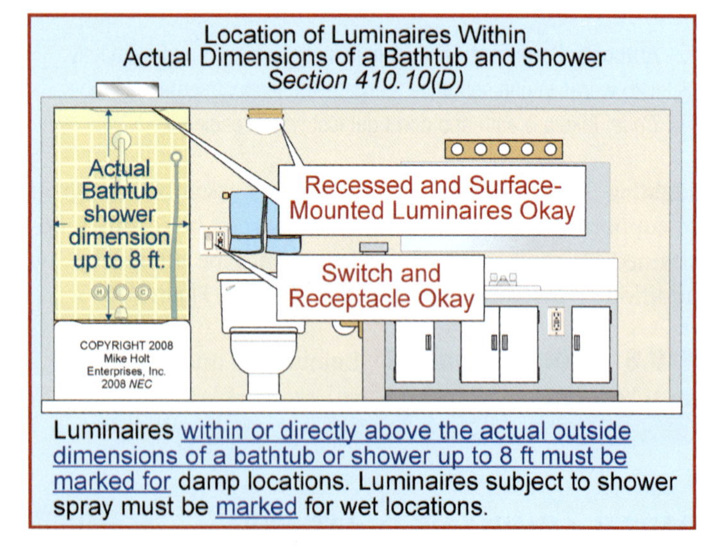

Figure 410–7

(E) Luminaires in Indoor Sports, Mixed-Use, and All-Purpose Facilities. Luminaires using a mercury vapor or metal halide lamp that are subject to physical damage and are installed in playing and spectator seating areas of indoor sports, mixed-use, or all-purpose facilities must be of the type that has a glass or plastic lamp shield. Such luminaires can have an additional guard. Figure 410–8

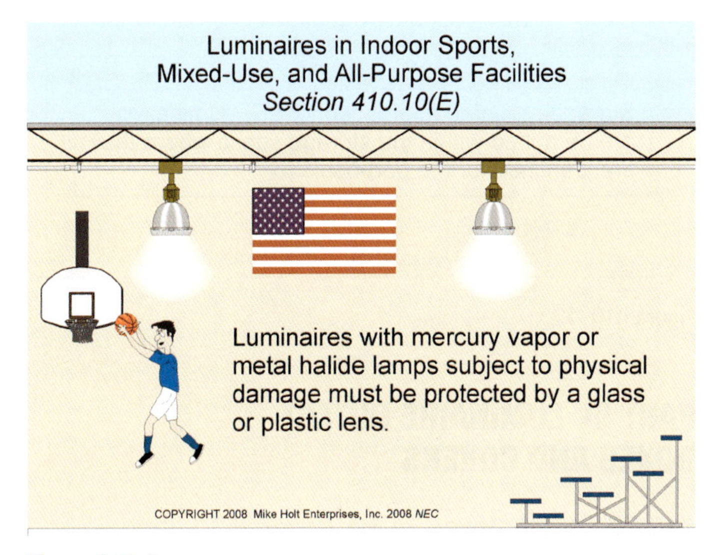

Figure 410–8

WARNING: *Metal halide lamps can cause serious skin burns and eye inflammation from shortwave ultraviolet radiation if the outer envelope of the lamp is broken or punctured. They should not be used where people will remain more than a few minutes unless adequate shielding or other safety precautions are used. Lamps that will automatically extinguish when the outer envelope is broken are commercially available.*

If a metal halide or mercury vapor lamp is broken during use:

- *Turn off the light immediately,*
- *Move people out of the area as quickly a possible,*
- *Advise people exposed to the damaged lamp to see a doctor if symptoms of skin burns or eye irritation occur.*

410.11 Luminaires Near Combustible Material.

Luminaires must be installed or be equipped with shades or guards so that combustible material is not subjected to temperatures in excess of 90°C (194°F).

410.16 Clothes Closets.

(A) Luminaires Permitted in Clothes Closets. The following types of luminaires can be installed in a clothes closet:

(1) A surface or recessed incandescent luminaire with an enclosed lamp.

(2) A surface or recessed fluorescent luminaire.

(3) A surface-mounted or recessed LED luminaire with a completely enclosed light source that is identified for use in clothes closets.

(B) Luminaires Not Permitted in Clothes Closets. Incandescent luminaires with open or partially open lamps and pendant-type luminaires must not be installed in a clothes closet. Figure 410–9

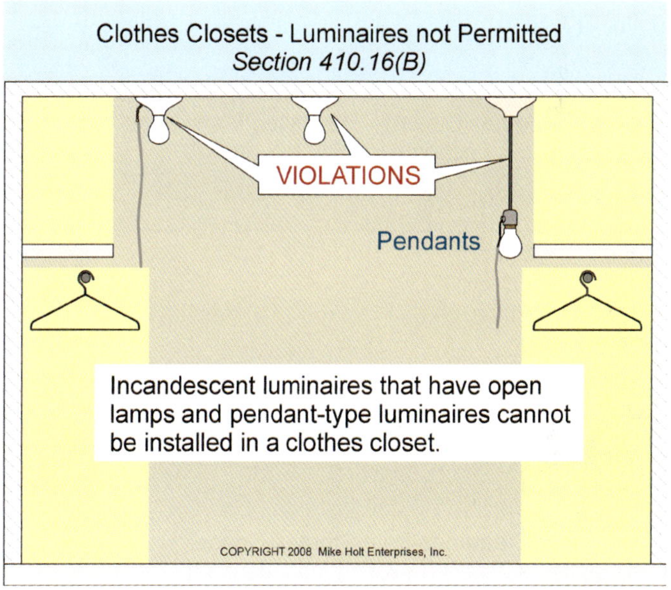

Figure 410–9

(C) Installation of Luminaires in Clothes Closets. Luminaires must maintain a minimum clearance between luminaires and the storage space as follows:

(1) 12 in. for surface-mounted incandescent or LED luminaires within an enclosed light source. Figure 410–10

(2) 6 in. for surface-mounted fluorescent luminaires.

(3) 6 in. for recessed incandescent or LED luminaires within an enclosed light source. Figure 410–11

(4) 6 in. for recessed fluorescent luminaires.

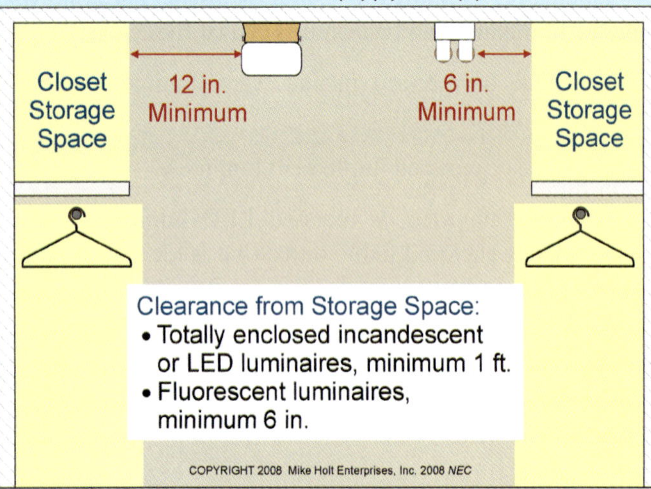

Figure 410–10

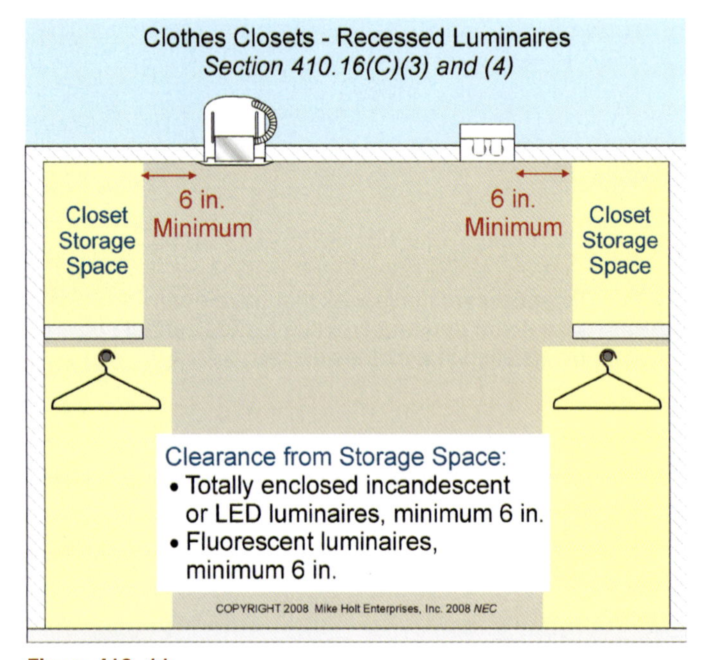

Figure 410–11

(5) Surface-mounted fluorescent or LED luminaires are permitted within the storage space where identified for this use. Figure 410–12

410.18 Space for Cove Lighting. Coves must have adequate space so that lamps and equipment can be properly installed and maintained.

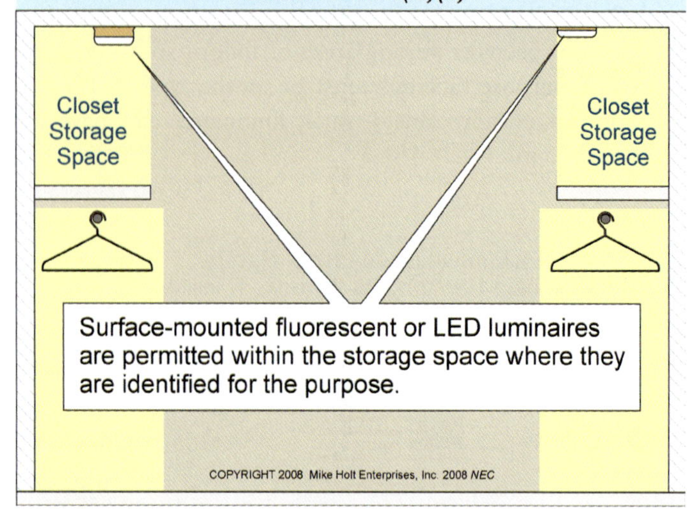

Figure 410–12

PART III. LUMINAIRE OUTLET BOXES AND COVERS

410.22 Outlet Boxes to be Covered. Outlet boxes for luminaires must be covered with a luminaire, lampholder, or blank faceplate. See 314.25. Figure 410–13

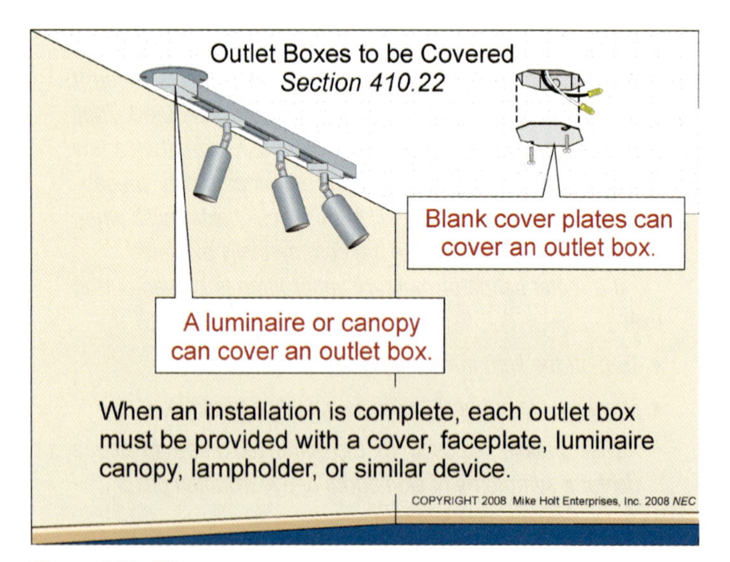

Figure 410–13

410.24 Connection of Electric-Discharge Luminaires.

(A) Luminaires Supported Independently of the Outlet Box. Electric-discharge luminaires supported independently of the outlet box must be connected to the branch circuit with a raceway, or with Types MC, AC, or NM cable. Figure 410–14

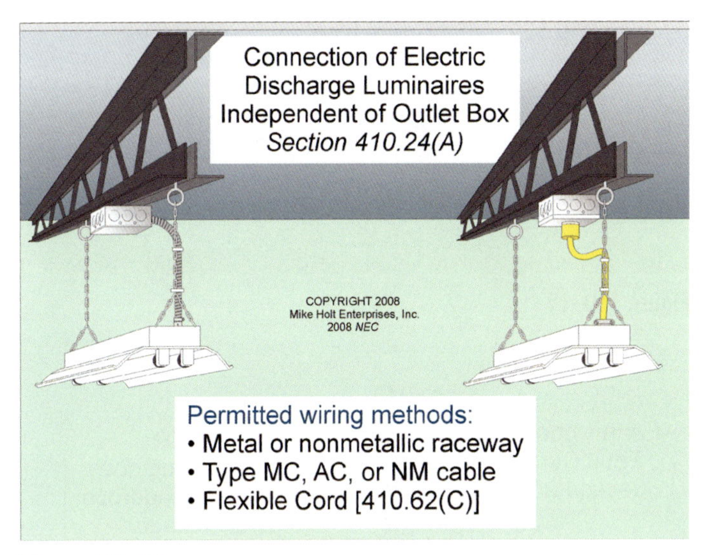

Figure 410–14

Electric-discharge luminaires can be cord-connected if the luminaires are provided with internal adjustments to position the lamp [410.62(B)].

Electric-discharge luminaires can be cord-connected if the cord is visible for its entire length and is plugged into a receptacle, and the installation complies with 410.62(C).

(B) Access to Outlet Box. When an electric-discharge luminaire is surface mounted over a concealed outlet box, and not supported by the outlet box, the luminaire must be provided with suitable openings that permit access to the branch-circuit wiring within the outlet box. Figure 410–15

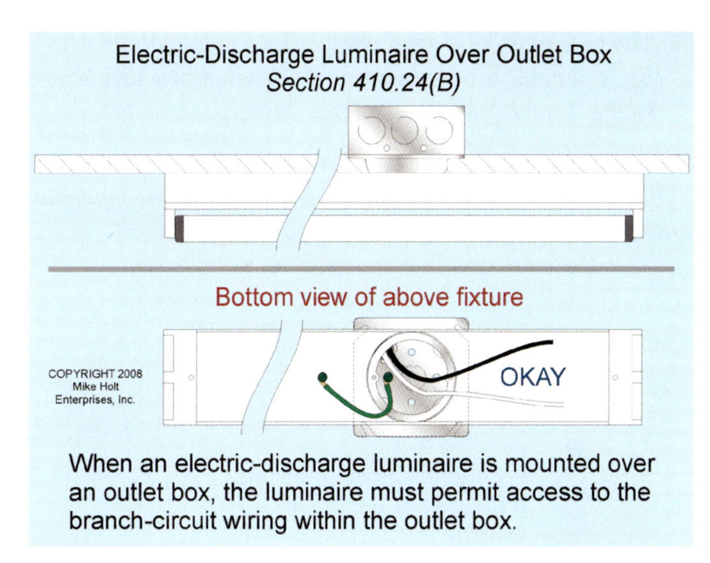

Figure 410–15

Author's Comment: With security being a high priority, many owners want to install security cameras on existing parking lot poles. However, 820.133(A)(1)(b) prohibits the mixing of power and communications conductors in the same raceway. Figure 410–16

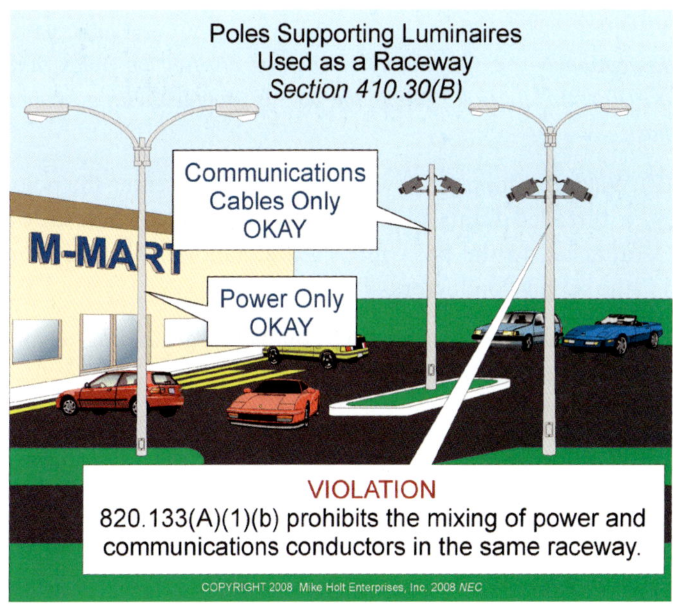

Figure 410–16

PART IV. LUMINAIRE SUPPORTS

410.30 Supports.

(A) General Support Requirements. Luminaires and lampholders must be securely supported.

(B) Metallic or Nonmetallic Poles. Metallic or nonmetallic poles can be used to support luminaires, and they can be used as a raceway.

In addition, they must comply with the requirements of (1) through (6).

(1) The pole must have an accessible 2 x 4 in. handhole with a cover suitable for use in wet locations that provides access to the supply conductors within the pole.

Exception No. 1: The handhole isn't required for a pole that is 8 ft or less in height, if the supply conductors for the luminaire are accessible by removing the luminaire. **Figure 410–17**

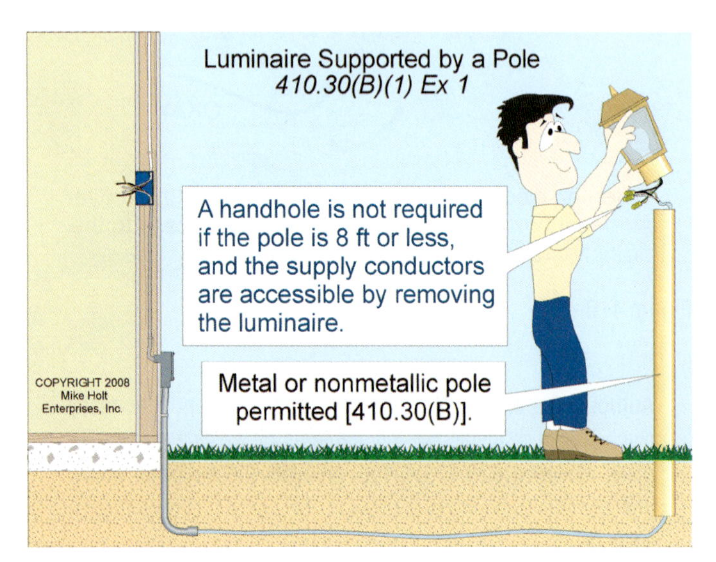

Figure 410–17

Exception No. 2: The handhole can be omitted on poles that are 20 ft or less in height, if the pole is provided with a hinged base.

(2) When the supply raceway or cable doesn't enter the pole, a threaded fitting or nipple must be welded, brazed, or attached to the pole opposite the handhole opening for the supply conductors.

(3) A metal pole must have an equipment grounding terminal accessible from the handhole.

Exception: A grounding terminal is not required in a pole that is 8 ft or less in height above grade where the splices are accessible by removing the luminaire.

(5) Metal poles used for the support of luminaires must be connected to an equipment grounding conductor of a type recognized in 250.118 [250.4(A)(5)]. **Figure 410–18**

DANGER: *Because the contact resistance of an electrode to the earth is so high, very little fault current returns to the power supply if the earth is the only fault current return path. Result—the circuit overcurrent device will not open and clear the ground fault, and the metal pole will become and remain energized by the circuit voltage.* **Figure 410–19**

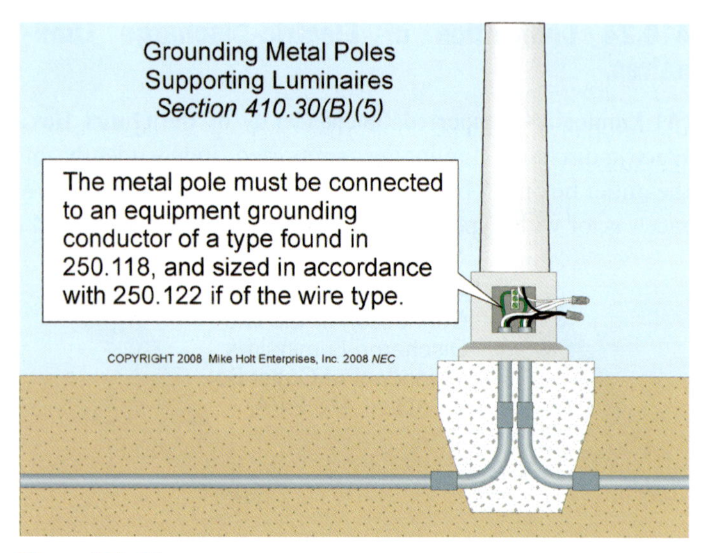

Figure 410–18

Figure 410–19

(6) Conductors in vertical metal poles must be supported when the vertical rise exceeds 100 ft [Table 300.19(A)].

Author's Comment: When provided by the manufacturer of roadway lighting poles, so-called J-hooks must be used to support conductors, as they are part of the listing instructions [110.3(B)].

410.36 Means of Support.

(A) Outlet Boxes. Outlet boxes designed for the support of luminaires must be supported by one of the following methods:

- Fastened to any surface that provides adequate support [314.23(A)].
- Supported from a structural member of a building or from grade by a metal, plastic, or wood brace [314.23(B)].
- Secured to a finished surface (drywall or plaster walls or ceilings) by clamps, anchors, or fittings identified for the application [314.23(C)].
- Secured to the structural or supporting elements of a suspended ceiling [314.23(D)].
- Supported by two intermediate metal conduits or rigid metal conduits threaded wrenchtight [314.23(E) and (F)].
- Embedded in concrete or masonry [314.23(G)].

Maximum Luminaire Weight. Outlet boxes for luminaires can support a luminaire that weighs up to 50 lb, unless the box is listed for the luminaire's actual weight [314.27(B)].

(B) Suspended-Ceiling Framing Members. Where framing members of suspended ceiling systems are used to support luminaires, they must be securely fastened to each other and they must be securely attached to the building structure at appropriate intervals. Luminaires must be attached to the suspended-ceiling framing members with screws, bolts, rivets, or clips that are listed and identified for such use. Figure 410–20

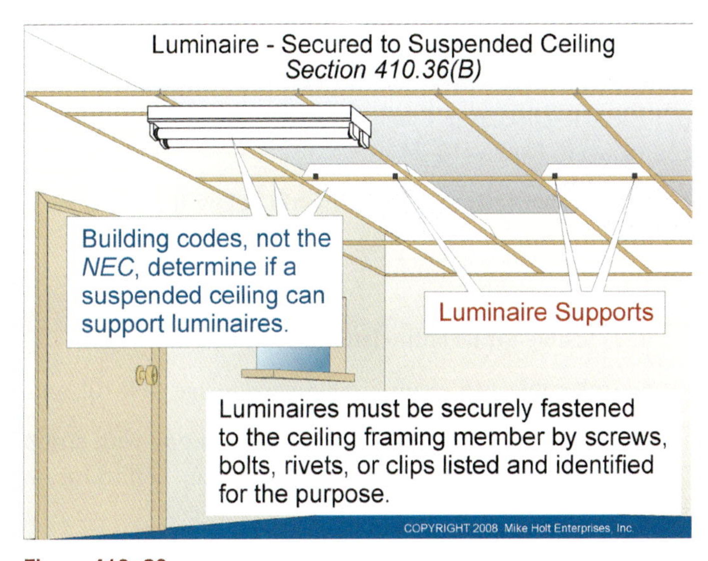

Figure 410–20

Author's Comment:

- The *NEC* doesn't require independent support wires for suspended ceiling luminaires that aren't installed in a fire-rated ceiling; however, building codes often do. Figure 410–21

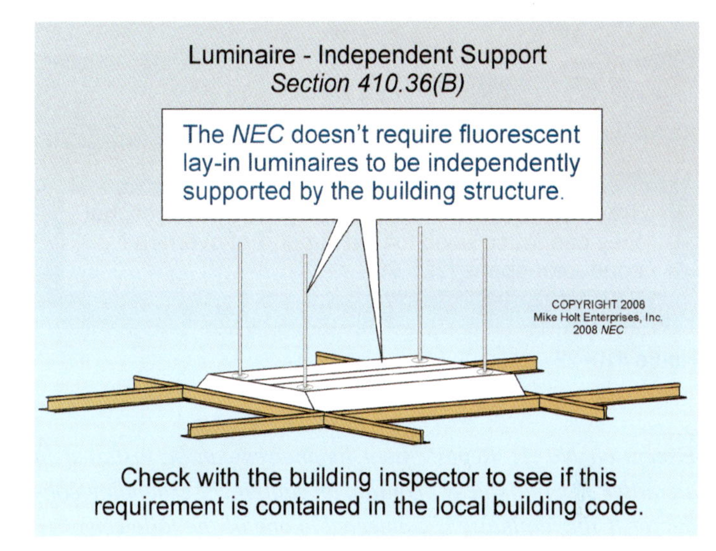

Figure 410–21

- Raceways and cables within a suspended ceiling must be supported in accordance with 300.11(A). Outlet boxes can be secured to the ceiling-framing members by bolts, screws, rivets, clips, or independent support wires that are taut and secured at both ends [314.23(D)].

(G) Luminaires Supported by Trees. Trees can be used to support luminaires, but they must not be used to support overhead conductor spans [225.26]. Figure 410–22

PART V. GROUNDING (BONDING)

410.42 Exposed Luminaire Parts.

(A) Exposed Conductive Parts. Exposed metal parts of luminaires must be connected to an equipment grounding conductor of a type recognized in 250.118.

(B) Made of Insulating Material. Where an equipment grounding conductor isn't present in the outlet box for a luminaire, the luminaire must be made of insulating material and must not have any exposed conductive parts.

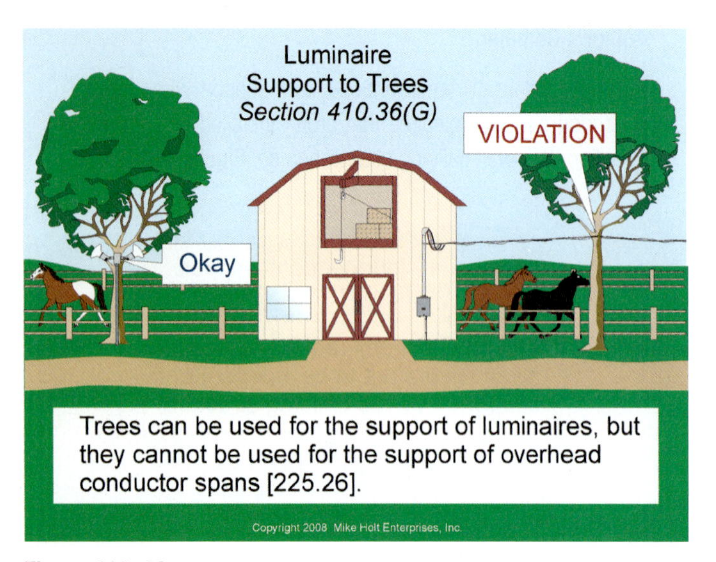

Figure 410–22

Exception No. 1: Replacement luminaires can be installed in an outlet box that doesn't contain an equipment grounding conductor if the luminaire is connected to one of the following:

(1) Grounding electrode system [250.50].

(2) Grounding electrode conductor.

(3) Panelboard equipment grounding terminal.

(4) Service neutral conductor within the service equipment enclosure.

Exception No. 2: GFCI-protected replacement luminaires aren't required to be connected to an equipment grounding conductor of a type recognized in 250.118 if no equipment grounding conductor exists at the outlet box.

Author's Comment: This is similar to the rule for receptacle replacements in locations where an equipment grounding conductor isn't present in the outlet box [406.3(D)(3)].

410.46 Methods of Grounding. Luminaires must be connected to an equipment grounding conductor of a type recognized in 250.118. Where of the wire type, the circuit equipment grounding conductor must be sized in accordance with Table 250.122, based on the rating of the overcurrent device.

PART VI. WIRING OF LUMINAIRES

410.50 Polarization of Luminaires. Luminaires must have the neutral conductor connected to the screw shell of the lampholder [200.10(C)], and the neutral conductor must be properly identified in accordance with 200.6.

410.62 Cord-Connected Luminaires.

(B) Adjustable Luminaires. Luminaires that require adjusting or aiming after installation can be cord connected, with or without an attachment plug, provided the exposed cord is of the hard usage or extra-hard usage type. The cord must not be longer than necessary for luminaire adjustment, and it must not be subject to strain or physical damage [400.10]. Figure 410–23

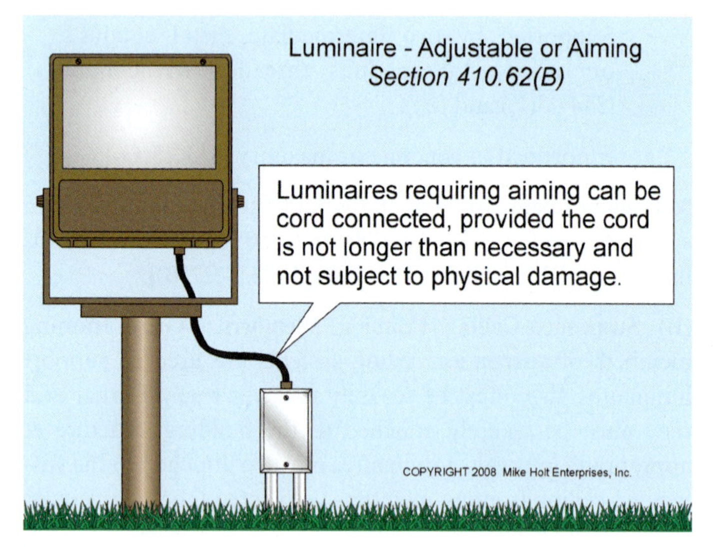

Figure 410–23

(C) Electric-Discharge Luminaires. A luminaire can be cord connected if: Figure 410–24

(1) The luminaire is mounted directly below the outlet box, and

(2) The flexible cord:

a. Is visible for its entire length,

b. Isn't subject to strain or physical damage [400.10], and

c. Terminates in an attachment plug, canopy with strain relief, or manufactured wiring system connector in accordance with 604.6(C).

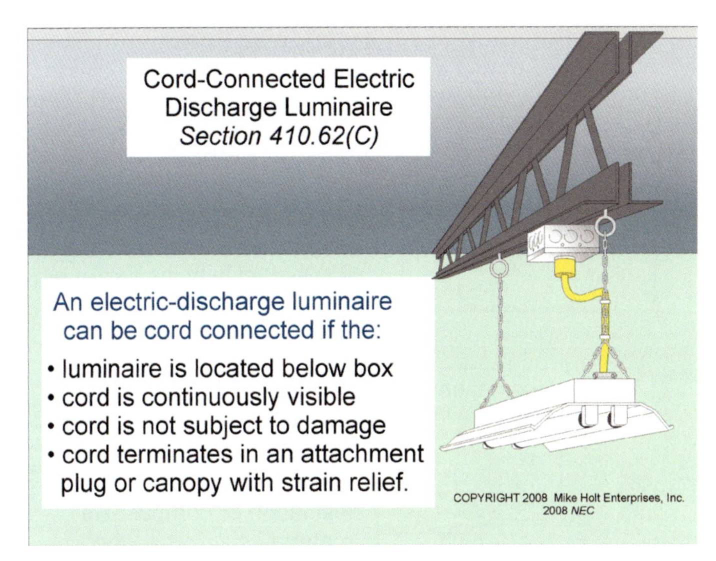

Figure 410–24

Author's Comment: The *Code* doesn't require twist-lock receptacles for this application.

410.64 Luminaires Used as a Raceway. Luminaires must not be used as a raceway for circuit conductors, unless the luminaire is listed and marked for use as a raceway. Figure 410–25

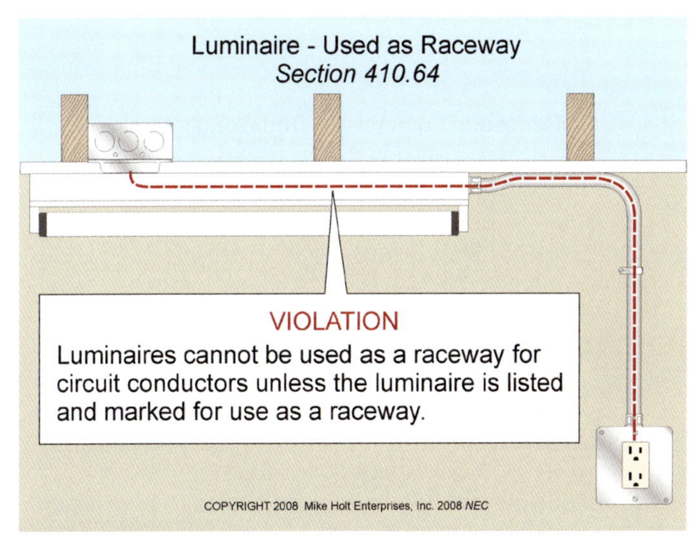

Figure 410–25

410.65 Luminaires Connected Together. Luminaires designed for end-to-end assembly, or luminaires connected together by recognized wiring methods, can contain a 2-wire branch circuit, or one multiwire branch circuit, supplying the connected luminaires. One additional 2-wire branch circuit supplying a night light is permitted. Figure 410–26

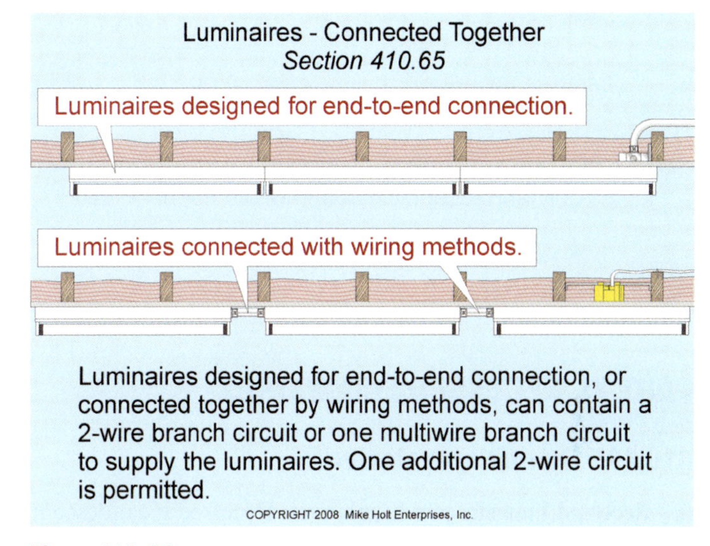

Figure 410–26

410.68 Branch-Circuit Conductors and Ballasts. Conductors within 3 in. of luminaire ballasts must have an insulation rating not less than 90°C.

PART VIII. LAMPHOLDERS

410.90 Screw-Shell Lampholders. Lampholders of the screw-shell type must be installed for use as lampholders only.

Author's Comment: A receptacle adapter that screws into a lampholder is a violation of this section. Figure 410–27

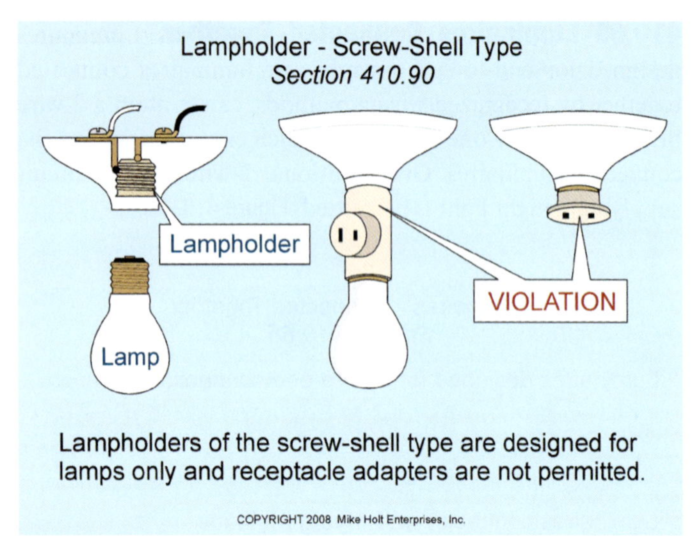

Figure 410–27

PART XI. RECESSED LUMINAIRES

410.115 Thermally Protected.

(C) Recessed Incandescent Luminaires. Recessed incandescent luminaires must be identified as thermally protected.

> **Author's Comment:** When higher-wattage lamps or improper trims are installed, the lampholder contained in a recessed luminaire can overheat, activating the thermal overcurrent device and causing the luminaire to cycle on and off.

Exception No. 2: Thermal protection isn't required for recessed Type IC luminaires whose design, construction, and thermal performance characteristics are equivalent to a thermally protected luminaire.

410.116 Recessed Luminaire Clearances.

(A) Clearances From Combustible Materials.

(1) Non-Type IC Luminaires. A recessed luminaire that isn't identified for contact with insulation must have all recessed parts, except the points of supports, spaced not less than ¼ in. from combustible materials. Figure 410–28A

(2) Type IC Luminaires. A Type IC luminaire (identified for contact with insulation) can be in contact with combustible materials. Figure 410–28B

(B) Installation. Thermal insulation must not be installed above a recessed luminaire or within 3 in. of the recessed luminaire's enclosure, wiring compartment, or ballast unless identified for contact with insulation; Type IC.

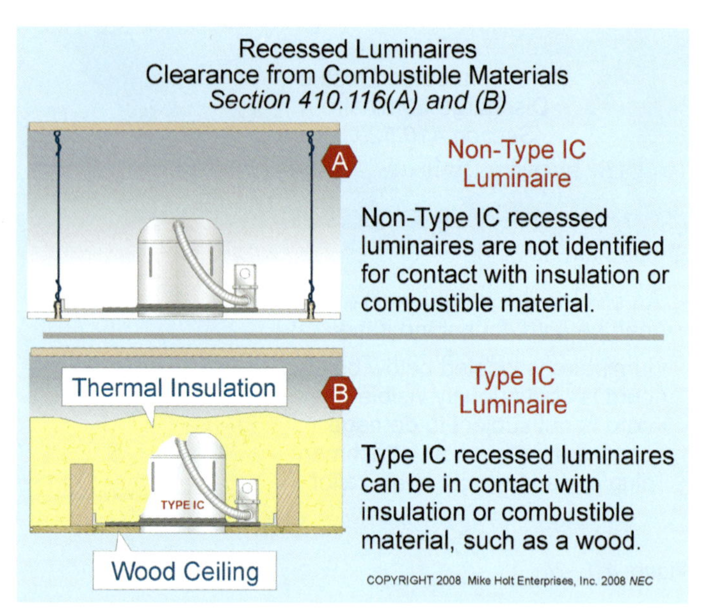

Figure 410–28

410.117 Wiring.

(C) Tap Conductors. Fixture wires installed in accordance with Article 402 and protected against overcurrent in accordance with 240.5(B)(2), are allowed to run from the luminaire to an outlet box located at least 1 ft away from the luminaire, as long as the conductors are not over 6 ft long. Figure 410–29

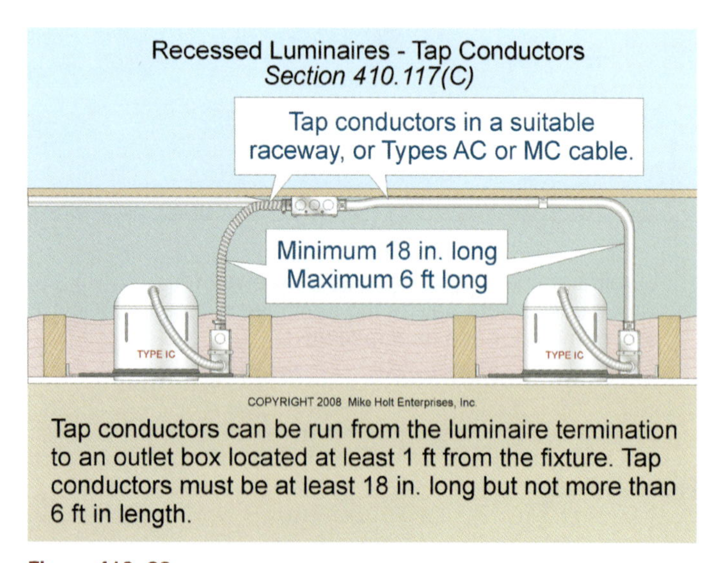

Figure 410–29

PART XIII. ELECTRIC-DISCHARGE LIGHTING

410.130 General.

(F) High-Intensity Discharge Luminaires.

(5) Metal Halide Lamp Containment. Luminaires containing a metal halide lamp, other than a thick-glass parabolic reflector lamp (PAR), must be provided with a containment barrier that encloses the lamp, or the luminaire must only allow the use of a Type "O" lamp that has an internal arc-tube shield. Figure 410–30

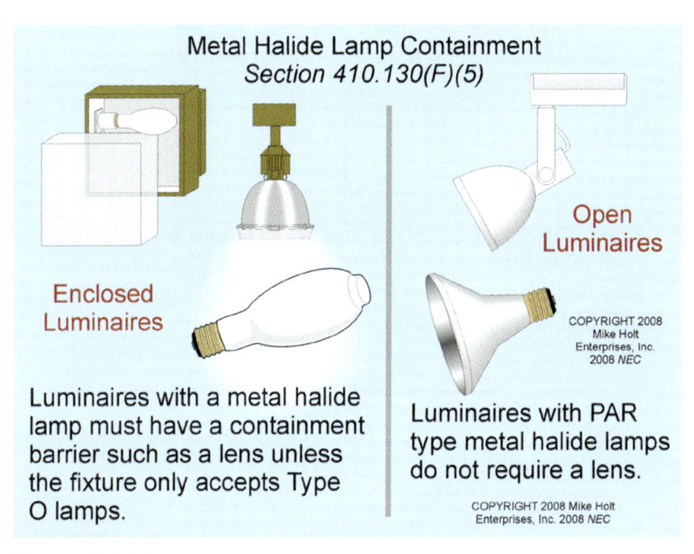

Figure 410–30

Author's Comments:

- Fires have resulted from arc lamps exploding on startup, shattering the lamp globe and showering glass and hot quartz fragments from metal halide lamp failures. The possibility of failure increases significantly as the lamp approaches and exceeds its rated life. It's projected that one violent rupture occurs in every 100,000 failures.
- "O-rated" lamps, which have an internal arc-tube shield, have been designed to meet ANSI containment standards for the installation of a metal halide lamp in an open fixture.

(G) Disconnecting Means.

(1) General. In indoor locations, other than dwellings and associated accessory structures, fluorescent luminaires that utilize double-ended lamps (typical fluorescent lamps) and contain ballasts that can be serviced in place must have a disconnecting means.

Author's Comment: Changing the ballast while the circuit feeding the luminaire is energized has become a regular practice because a local disconnect isn't available.

Exception No. 2: A disconnecting means isn't required for the emergency illumination required in 700.16.

Exception No. 3: For cord-and-plug-connected luminaires, an accessible separable connector, or an accessible plug and receptacle, is permitted to serve as the disconnecting means.

Exception No. 4: A disconnecting means isn't required in industrial establishments with restricted public access where written procedures and conditions of maintenance and supervision ensure that only qualified persons will service the installation.

Exception No. 5: Where more than one luminaire is installed and is supplied by a branch circuit that is not of the multiwire type, a disconnecting means is not required for every luminaire; but only when the light switch for the space ensures that some of the luminaires in the space will still provide illumination.

(2) Multiwire Branch Circuits. When connected to multiwire branch circuits, the fluorescent luminaire disconnect must simultaneously break all circuit conductors of the ballast, including the neutral conductor.

Author's Comment: This new rule requires the disconnecting means to open "all circuit conductors of a multiwire branch circuit," including the neutral conductor. If the neutral conductor in a multiwire circuit is not disconnected at the same time as the ungrounded conductors, a false sense of security can result in an unexpected shock from the neutral conductor.

(3) Location. The fluorescent luminaire disconnecting means must be accessible to qualified persons, and where the disconnecting means is external to the luminaire, it must be a single device and must be located in sight from the luminaire.

410.136 Luminaire Mounting.

(B) Surface-Mounted Luminaires with Ballasts. Surface-mounted luminaires containing ballasts must have a minimum of 1½ in. clearance from combustible low-density fiberboard. Figure 410–31

Author's Comment: This rule doesn't apply to the mounting of surface-mounted fluorescent luminaires onto wood, plaster, concrete, or drywall. Figure 410–32

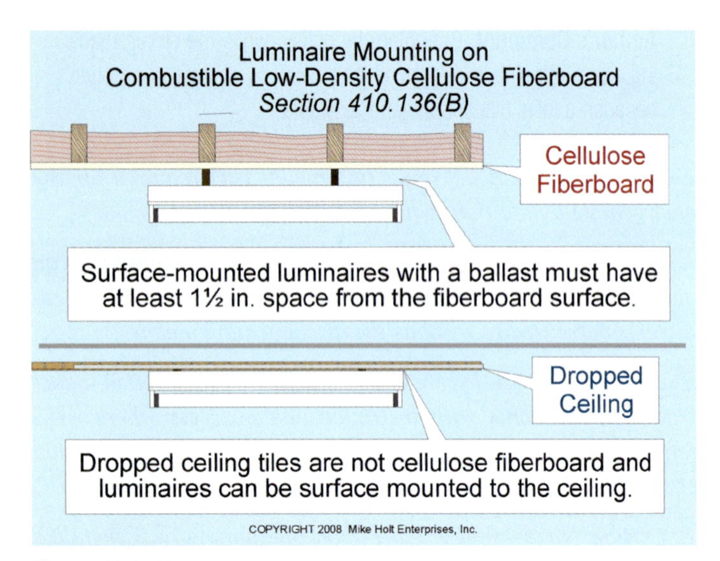

Figure 410–31

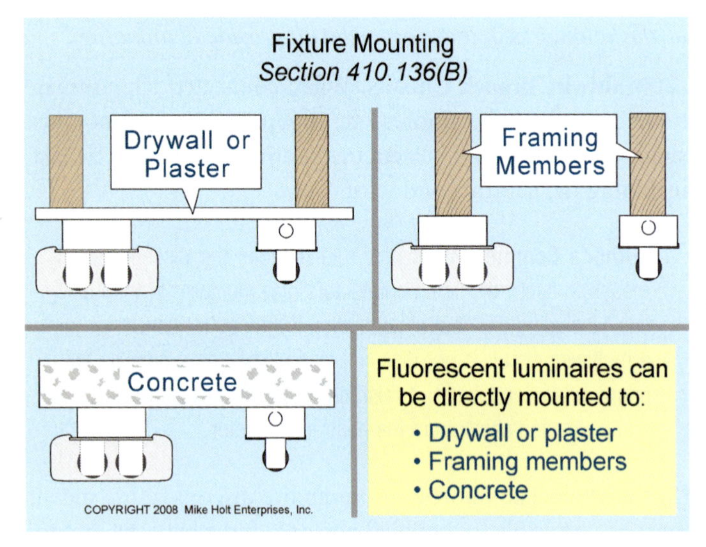

Figure 410–32

PART XV. TRACK LIGHTING

410.151 Installation.

(A) Track Lighting. Track lighting must be permanently installed and permanently connected to the branch-circuit wiring. Lampholders for track lighting are designed for lamps only, so a receptacle adapter isn't permitted [410.90].

(B) Circuit Rating. The connected load on a lighting track must not exceed the rating of the track, and an overcurrent device whose rating exceeds the rating of the track must not supply the track. Figure 410–33

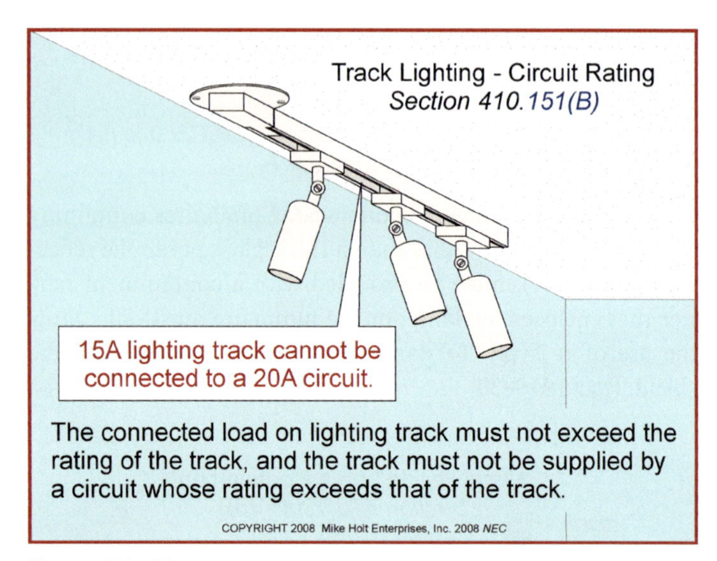

Figure 410–33

FPN: The feeder or service load calculations of 220.43(B) do not limit the number of feet of track on a circuit, nor do they limit the number of luminaires mounted on an individual track. Figure 410–34

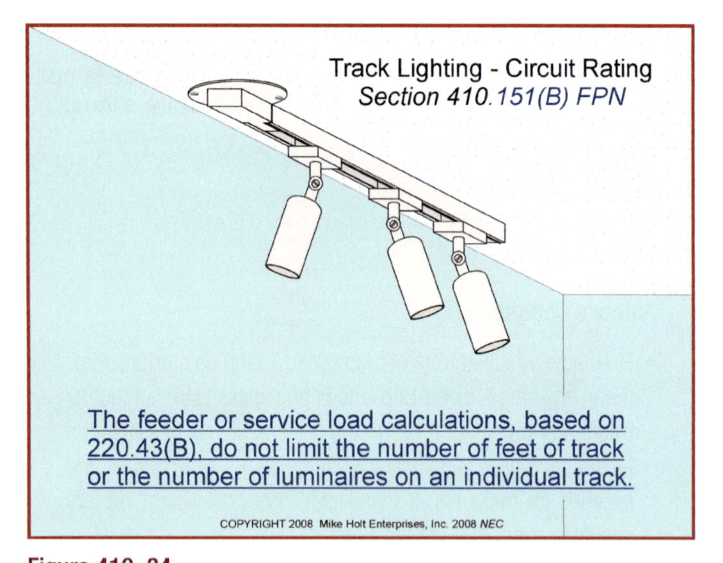

Figure 410–34

(C) Locations Not Permitted. Track lighting must not be installed:

(1) Where it's likely to be subjected to physical damage.

(2) In wet or damp locations.

(3) Where subject to corrosive vapors.

(4) In storage battery rooms.

(5) In any hazardous (classified) location.

(6) Where concealed.

(7) Where extended through walls, partitions, or floors.

(8) Less than 5 ft above the finished floor, except where protected from physical damage or where the track operates below 30V RMS open-circuit voltage.

(9) Within 3 ft horizontally and 8 ft vertically from the top of a bathtub rim or shower space [410.4(D)].

410.154 Fastening. Track lighting must be securely mounted to support the weight of the luminaires. A single track section 4 ft or shorter in length must have two supports, and, where installed in a continuous row, each individual track section of not more than 4 ft in length must have one additional support.

ARTICLE 410 Practice Questions

ARTICLE 410 LUMINAIRES, LAMPHOLDERS, AND LAMPS—PRACTICE QUESTIONS

1. Article 410 covers luminaires, portable luminaires, lampholders, pendants, incandescent filament lamps, arc lamps, electric discharge lamps, and _____, and the wiring and equipment forming part of such products and lighting installations.

   (a) decorative lighting products
   (b) lighting accessories for temporary seasonal and holiday use
   (c) portable flexible lighting products
   (d) all of these

2. Lighting track is a manufactured assembly designed to support and _____ luminaires that are capable of being readily repositioned on the track.

   (a) connect
   (b) protect
   (c) energize
   (d) all of these

3. No part of cord-connected luminaires, chain-, cable-, or cord-suspended luminaires, lighting track, pendants, or paddle fans shall be located within a zone measured 3 ft horizontally and _____ vertically from the top of the bathtub rim or shower stall threshold.

   (a) 4 ft
   (b) 6 ft
   (c) 8 ft
   (d) 10 ft

4. The NEC requires a lighting outlet on the wall in clothes closets.

   (a) True
   (b) False

5. Surface-mounted fluorescent luminaires in clothes closets shall be permitted on the wall above the door, or on the ceiling, provided there is a minimum clearance of _____ between the luminaire and the nearest point of a storage space.

   (a) 3 in.
   (b) 6 in.
   (c) 9 in.
   (d) 12 in.

6. Electric-discharge luminaires supported independently of the outlet box shall be connected to the branch circuit through _____.

   (a) raceways
   (b) Types MC, AC, MI, or NM cable
   (c) flexible cords
   (d) any of these

7. Handholes in poles supporting luminaires shall not be required for poles _____ or less in height above finished grade, if the pole is provided with a hinged base.

   (a) 5 ft
   (b) 10 ft
   (c) 15 ft
   (d) 20 ft

8. Luminaires attached to the framing of a suspended ceiling shall be secured to the framing member(s) by mechanical means such as _____.

   (a) bolts
   (b) screws
   (c) rivets
   (d) any of these

9. Exposed conductive parts of luminaires shall be _____.

   (a) connected to an equipment grounding conductor
   (b) painted
   (c) removed
   (d) a and b

10. Luminaires that require adjustment or aiming after installation can be cord-connected without an attachment plug, provided the exposed cord is of the hard-usage type and is not longer than that required for maximum adjustment.

    (a) True
    (b) False

11. Branch-circuit conductors within _____ of a ballast shall have an insulation temperature rating not lower than 90°C (194°F).

    (a) 1 in.
    (b) 3 in.
    (c) 6 in.
    (d) 8 in.

12. A recessed luminaire not identified for contact with insulation shall have all recessed parts spaced not less than _____ from combustible materials, except for points of support.

    (a) ¼ in.
    (b) ½ in.
    (c) 1¼ in.
    (d) 6 in.

13. The minimum distance that an outlet box containing tap supply conductors is permitted to be placed from a recessed luminaire is _____.

    (a) 1 ft
    (b) 2 ft
    (c) 3 ft
    (d) 4 ft

14. In indoor locations, other than dwellings, fluorescent luminaires that utilize double-ended lamps and contain ballast(s) shall have a disconnecting means either internal or external to each luminaire.

    (a) True
    (b) False

15. The connected load on lighting track is permitted to exceed the rating of the track under some conditions.

    (a) True
    (b) False

16. Lighting track shall not be installed within the zone measured 3 ft horizontally and _____ vertically from the top of the bathtub rim or shower stall threshold.

    (a) 2 ft
    (b) 3 ft
    (c) 4 ft
    (d) 8 ft

ARTICLE

411 Lighting Systems Operating at 30V or Less

INTRODUCTION TO ARTICLE 411—LIGHTING SYSTEMS OPERATING AT 30V OR LESS

Article 411 provides the requirements for lighting systems operating at 30V or less, which are often found in such applications as landscaping, jewelry stores, and museums. Don't let the half-page size of Article 411 give you the impression that 30V lighting isn't something you need to be concerned about. These systems are limited in their voltage, but the current rating can be as high as 25A, which means they are still a potential source of fire. Installation of these systems is widespread and becoming more so.

Many of these systems now use LEDs, and 30V halogen lamps are also fairly common. All 30V lighting systems have an ungrounded secondary circuit supplied by an isolating transformer. These systems have restrictions that affect where they can be located, and they can have a maximum supply breaker size of 25A.

411.1 Scope. Article 411 covers the installation of lighting systems that operate at 30V or less, as well as their associated components.

411.2 Definition.

Lighting Systems Operating at 30 Volts or Less. A lighting system consisting of an isolating power supply, luminaires, and associated equipment identified for the use. The lighting system power supply must be rated not more than 25A and not more than 30V. Figure 411–1

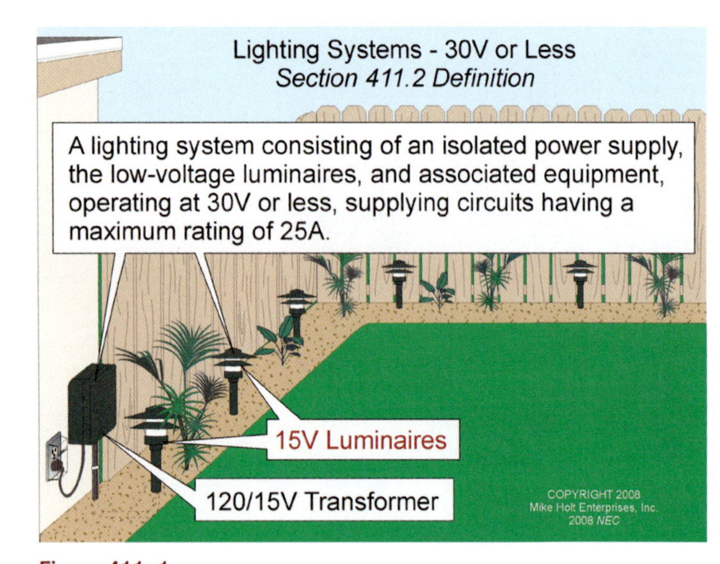

Figure 411–1

411.3 Listing Required. Low-voltage lighting systems operating at 30V or less must comply with (A) or (B).

(A) Listed System. Be listed as a complete system, including the power supply and luminaires.

(B) Assembly of Listed Parts. A lighting system assembled from the following listed parts is permitted:

(1) Low-voltage luminaires.

(2) Low-voltage luminaire power supply.

(3) Class 2 power supply.

(4) Low-voltage luminaire fitting.

(5) Cords that the luminaires and power supply are listed for use with.

(6) Cable, conductors in raceway, or other fixed wiring method for the secondary circuit.

The luminaires, power supply, and luminaire fittings of an exposed bare conductor lighting system must be listed for use as part of the same identified lighting system.

411.4 Specific Location Requirements.

(A) Walls, Floors, and Ceilings. Conductors concealed or run through a wall, floor, or ceiling must comply with (1) or (2):

(1) Lighting system conductors must be installed within a Chapter 3 wiring method.

(2) Lighting system conductors supplied by a listed Class 2 power supply can use Class 2 cables, installed in accordance with 725.130.

(B) Pools, Spas, Fountains, and Similar Locations. Low-voltage lighting systems must not be installed less than 10 ft from the edge of the water. Figure 411–2

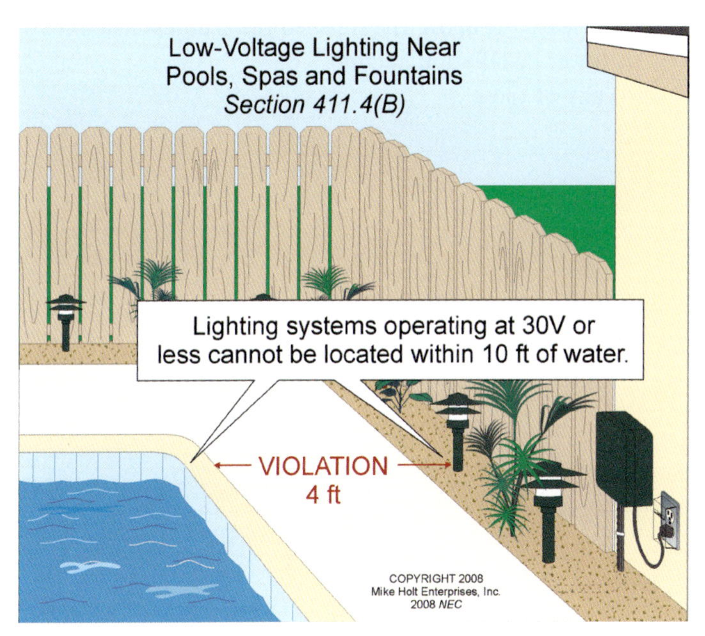

Figure 411–2

411.5 Secondary Circuits.

(A) Grounding. Secondary circuits are not permitted to be grounded.

(B) Isolation. The secondary circuit must be insulated from the branch circuit by an isolating transformer.

(C) Bare Conductors. Exposed bare conductors and current-carrying parts must not be installed less than 7 ft above the finished floor, unless listed for a lower height.

(D) Insulated Conductors. Exposed insulated secondary circuit conductors must be:

(1) Supplied by a Class 2 power source with Class 2 cable in accordance with Article 725.

(2) Installed at least 7 ft above the finished floor unless listed for a lower installation height.

(3) Installed in a Chapter 3 wiring method.

ARTICLE 411 Practice Questions

ARTICLE 411. LIGHTING SYSTEMS OPERATING AT 30 VOLTS OR LESS—PRACTICE QUESTIONS

1. A lighting system covered by Article 411 shall have a power supply rated not more than _____ and not more than 30V.

   (a) 15A
   (b) 20A
   (c) 25A
   (d) 30A

2. Lighting systems operating at 30V or less can be concealed or extended through a building wall, floor or ceiling without regard to the wiring method used.

   (a) True
   (b) False

3. Exposed secondary circuits of lighting systems operating at 30V or less shall be _____.

   (a) installed in a Chapter 3 wiring method
   (b) Class 2 cable supplied by a Class 2 power source in accordance with Article 725
   (c) at least 7 ft above the finished floor unless listed for a lower installation height
   (d) any of these

ARTICLE

422 Appliances

INTRODUCTION TO ARTICLE 422—APPLIANCES

Article 422 covers electric appliances used in any occupancy. The meat of this article is contained in Parts II and III. Parts IV and V are primarily for manufacturers, but you should examine appliances for compliance before installing them. If the appliance has a label from a recognized labeling authority (for example, UL), it complies [90.7].

PART I. GENERAL

422.1 Scope. The scope of Article 422 includes appliances in any occupancy that are fastened in place, permanently connected, or cord-and-plug-connected. Figure 422–1

Author's Comment: Appliances are electrical equipment, other than industrial equipment, built in standardized sizes, such as ranges, ovens, cooktops, refrigerators, drinking water coolers, or beverage dispensers [Article 100].

422.3 Other Articles.

Motor-operated appliances must comply with Article 430, and appliances containing hermetic refrigerant motor compressors must comply with Article 440.

Author's Comment: Room air-conditioning equipment must be installed in accordance with Part VII of Article 440.

Figure 422–1

PART II. BRANCH-CIRCUIT REQUIREMENTS

422.10 Branch-Circuit Rating.

(A) Individual Circuits. The branch-circuit ampere rating for an individual appliance must not be less than the branch-circuit rating marked on the appliance [110.3(B)].

The branch-circuit rating for motor-operated appliances must be in accordance with 430.6(A) and 430.22(A).

A branch-circuit for an appliance that is a continuous load must be rated not less than 125 percent of the marked ampere rating of the appliance [210.19(A)(1)].

Branch circuits for household ranges and cooking appliances can be sized in accordance with Table 220.55, and 210.19(A)(3).

(B) Circuits Supplying Two or More Loads. Branch circuits supplying appliances and other loads must be sized in accordance with the following:

- Cord-and-plug-connected equipment must not be rated more than 80 percent of the branch-circuit ampere rating [210.23(A)(1)]. Figure 422–2

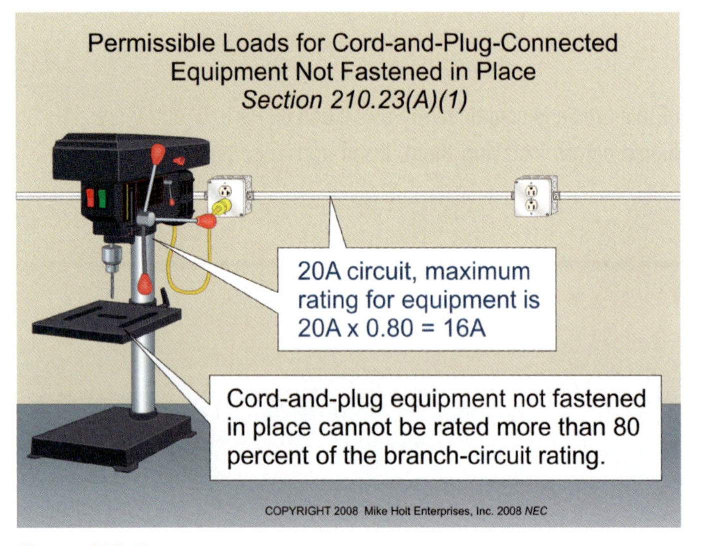

Figure 422–2

- Equipment fastened in place must not be rated more than 50 percent of the branch-circuit ampere rating, if the circuit supplies both luminaires and receptacles [210.23(A)(2)]. Figure 422–3

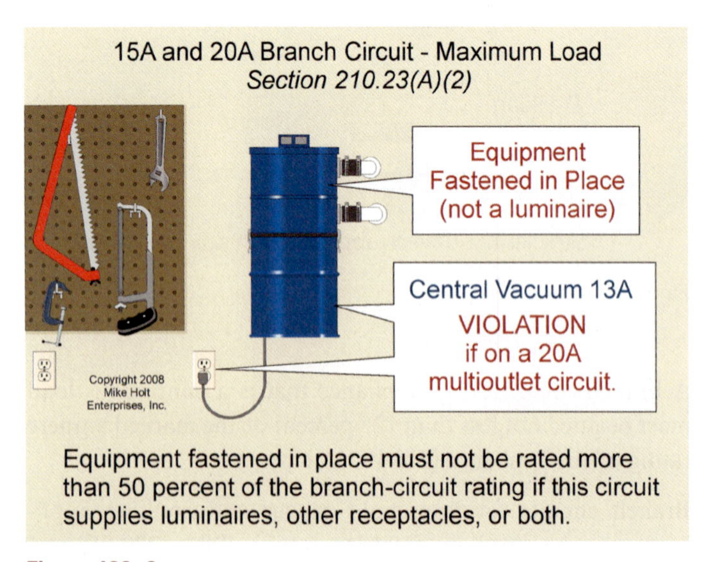

Figure 422–3

422.11 Overcurrent Protection.

(A) Branch-Circuits. Branch-circuit conductors must have overcurrent protection in accordance with 240.4, and the overcurrent device rating must not exceed the rating marked on the appliance.

(E) Nonmotor Appliances. The appliance overcurrent device must:

(1) Not exceed the rating marked on the appliance.

(2) Not exceed 20A if the overcurrent device rating isn't marked, and the appliance is rated 13.30A or less, or

(3) Not exceed 150 percent of the appliance rated current if the overcurrent device rating isn't marked, and the appliance is rated over 13.30A. Where 150 percent of the appliance rating doesn't correspond to a standard overcurrent device ampere rating listed in 240.6(A), the next higher standard rating is permitted.

Question: *What is the maximum size overcurrent protection for a 4,500W, 240V water heater?* Figure 422–4

(a) 20A *(b) 30A* *(c) 40A* *(d) 50A*

Answer: *(b) 30A*

Conductor/Protection Size = 4,500W/240V
Conductor/Protection Size = 18.75A x 1.50
Conductor/Protection Size = 28A, next size up, 30A [240.6(A)]

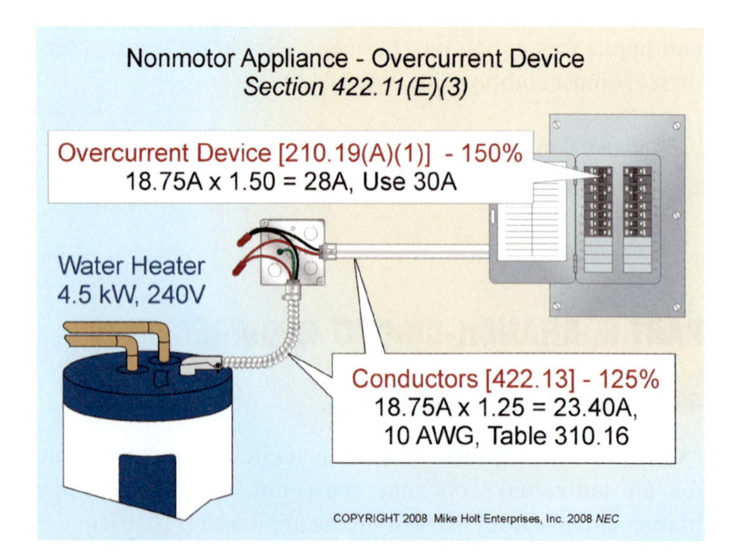

Figure 422–4

422.12 Central Heating Equipment (Furnaces).

An individual branch circuit must supply central heating equipment, such as gas, oil, or coal furnaces.

Author's Comment: This rule isn't intended to apply to a listed wood-burning fireplace with a fan, since the fireplace isn't central heating equipment.

Exception No. 1: Auxiliary equipment to the central heating equipment, such as pumps, valves, humidifiers, and electrostatic air cleaners, can be connected to the central heater circuit.

Author's Comment: Electric space-heating equipment must be installed in accordance with Article 424 Electric Space-Heating Equipment.

Exception No. 2: Permanently connected air-conditioning equipment can be connected to the individual branch circuit that supplies central heating equipment.

422.13 Storage Water Heaters.

An electric water heater having a capacity of 120 gallons or less is considered a continuous load, for the purpose of sizing branch circuits.

Author's Comment: Branch-circuit conductors and overcurrent devices must have an ampacity of at least 125 percent of the ampere rating of a continuous load [210.19(A)(1) and 210.20(A). Figure 422–5

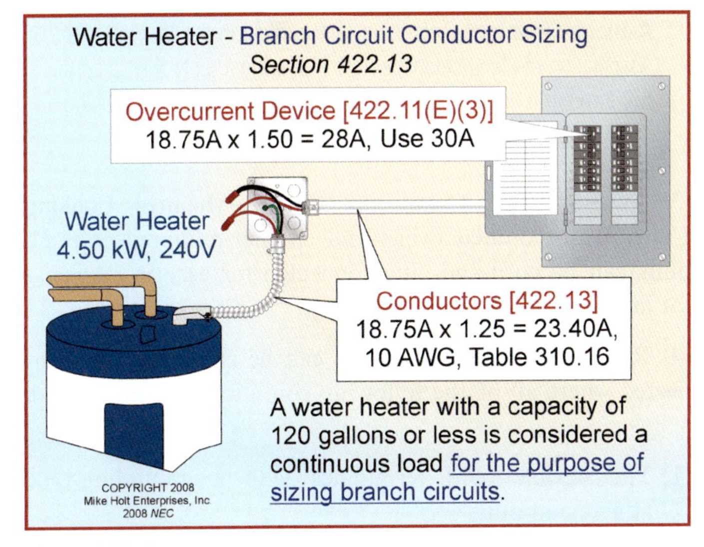

Figure 422–5

Question: *What is the calculated load for conductor sizing and overcurrent protection for a 4,500W, 230V water heater?*

(a) 15A (b) 20A (c) 25A (d) 30A

Answer: *(c) 25A*

I = P/E
P = 4,500W
E = 230V

I = 4,500W/230V
I = 20A

Calculated Continuous Load for Conductor Sizing and Protection = 20A x 1.25
Calculated Continuous Load for Conductor Sizing and Protection = 25A

422.15 Central Vacuums.

(A) Circuit Loading. A separate circuit is not required for a central vacuum, if the rating of the equipment doesn't exceed 50 percent of the ampere rating of the circuit.

Author's Comment: 210.23(A)(2) specifies that equipment fastened in place, other than luminaires, must not be rated more than 50 percent of the branch-circuit ampere rating if this circuit supplies both luminaires and receptacles. Due to this requirement, a separate 15A circuit is required for a central vacuum receptacle outlet if the rating of the central vacuum exceeds 7.50A. A separate 20A circuit is required for a central vacuum receptacle outlet if the rating of the central vacuum exceeds 10A, but not 16A [210.23(A)(2)]. Figure 422–6

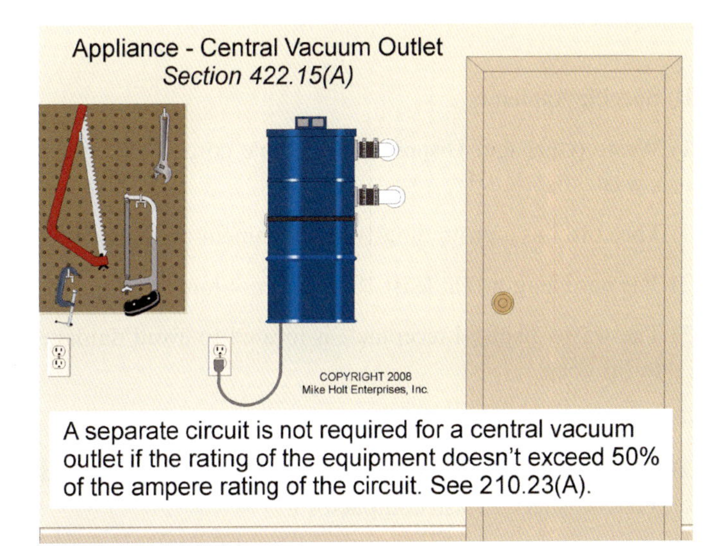

Figure 422–6

422.16 Flexible Cords.

(A) General. Flexible cords are permitted to:

(1) Facilitate frequent interchange, or to prevent the transmission of noise and vibration [400.7(A)(6) and 400.7(A)(7)].

(2) Facilitate the removal of appliances fastened in place, where the fastening means and mechanical connections are specifically designed to permit ready removal [400.8(A)(8)].

Author's Comment: Flexible cords must not be used for the connection of water heaters, furnaces, and other appliances fastened in place, unless the appliances are specifically identified to be used with a flexible cord. Figure 422–7

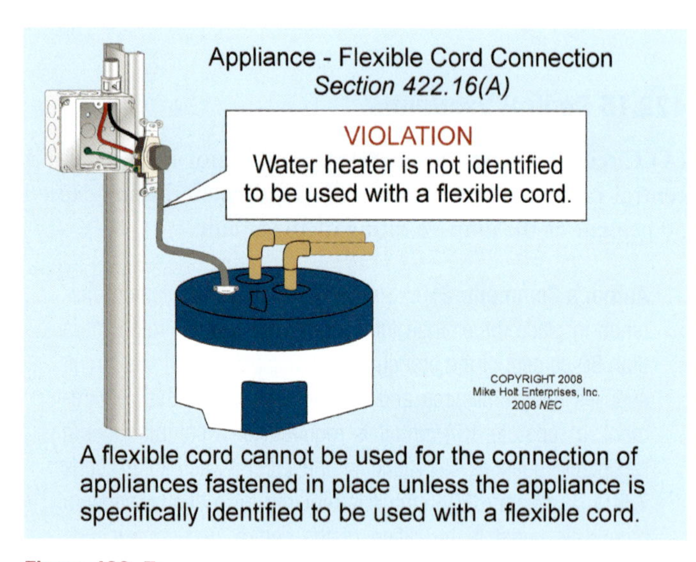

Figure 422–7

(B) Specific Appliances.

(1) Waste (Garbage) Disposals. A flexible cord is permitted for a waste disposal if:

(1) The cord has a grounding-type attachment plug.

(2) The cord length is at least 18 in. and not longer than 3 ft.

(3) The waste disposal receptacle is located to avoid damage to the cord.

(4) The waste disposal receptacle is accessible.

(2) Dishwashers and Trash Compactors. A cord is permitted for a dishwasher or trash compactor if:

(1) The cord has a grounding-type attachment plug.

(2) The cord length is at least 3 ft and not longer than 4 ft, measured from the rear plane of the appliance. Figure 422–8

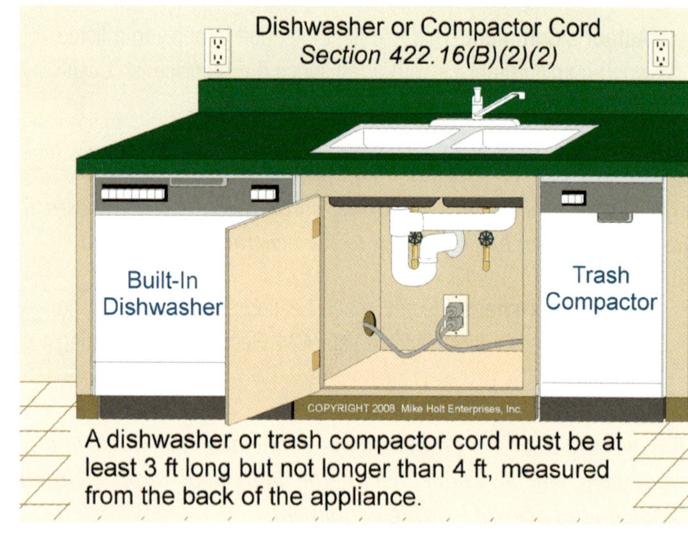

Figure 422–8

(3) The appliance receptacle is located to avoid damage to the cord.

(4) The receptacle is located in the space occupied by the appliance or in the space adjacent to the appliance.

(5) The receptacle is accessible.

Author's Comment: According to an article in the International Association of Electrical Inspectors magazine (IAEI News), a cord run through a cabinet for an appliance isn't considered as being run through a wall.

(3) Wall-Mounted Ovens and Counter-Mounted Cooking Units. Wall-mounted ovens and counter-mounted cooking units can be cord-and-plug-connected for ease in servicing for installation.

(4) Range Hoods. Range hoods can be cord-and-plug-connected where all of the following conditions are met: Figure 422–9

(1) The flexible cord terminates with a grounding-type attachment plug.

(2) The length of the cord must not be less than 18 in. or longer than 36 in.

(3) The range hood receptacle must be located to avoid physical damage to the flexible cord.

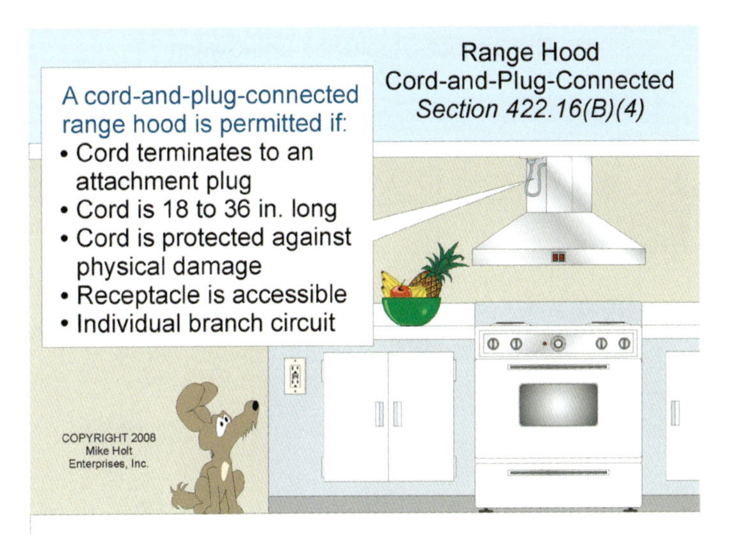

Figure 422–9

(4) The range hood receptacle must be accessible.

(5) The range hood receptacle must be supplied by an individual branch circuit.

Author's Comment: An above the range microwave that contains a fan listed as a range hood must comply with this section, if it is cord-and-plug-connected.

422.18 Support of Ceiling Paddle Fans.

Ceiling paddle fans must be supported by a listed fan outlet box, or outlet box system, in accordance with 314.27(D).

Author's Comment: An outlet box identified for use with a ceiling paddle fan is permitted for a ceiling paddle fan that doesn't weigh more than 70 lb. An outlet box for ceiling paddle fans that weigh between 35 lb and 70 lb must have the maximum weight marked on the box. Where the maximum isn't marked on the box, and the fan weighs over 35 lb, the fan must be supported independently of the outlet box. Ceiling paddle fans over 70 lb must be supported independently of the outlet box [314.27(D)].

PART III. DISCONNECT

422.31 Permanently Connected Appliance Disconnects.

(A) Appliances Rated at Not Over 300 VA or ⅛ Horsepower. The branch-circuit overcurrent device, such as a plug fuse or circuit breaker, can serve as the appliance disconnect.

(B) Appliances Rated Over 300 VA or ⅛ Horsepower. A switch or circuit breaker located within sight from the appliance can serve as the appliance disconnecting means. If the switch or circuit breaker is capable of being locked in the open position, it doesn't need to be within sight. The provision for locking or adding a lock to the disconnecting means must be on the switch or circuit breaker and remain in place with or without the lock installed. A portable locking means doesn't meet the "locked in the open position" requirement. Figure 422–10

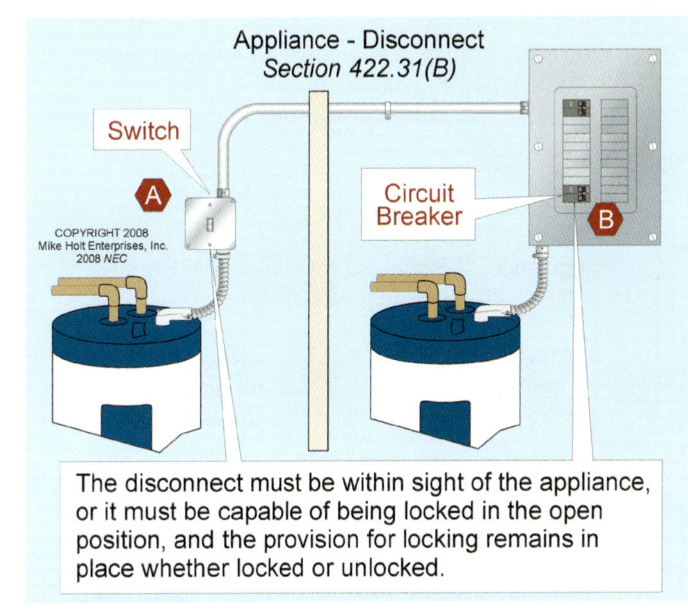

Figure 422–10

Author's Comment: "Within Sight" is visible and not more than 50 ft from one to the other [Article 100].

422.33 Cord-and-Plug-Connected Appliance Disconnects.

(A) Attachment Plugs and Receptacles. A plug and receptacle can serve as the disconnecting means for a cord-and-plug-connected appliance. Figure 422–11

(B) Cord-and-Plug-Connected Ranges. The plug and receptacle of a cord-and-plug-connected household electric range can serve as the range disconnecting means, if the plug is accessible from the front of the range by the removal of a drawer. Figure 422–12

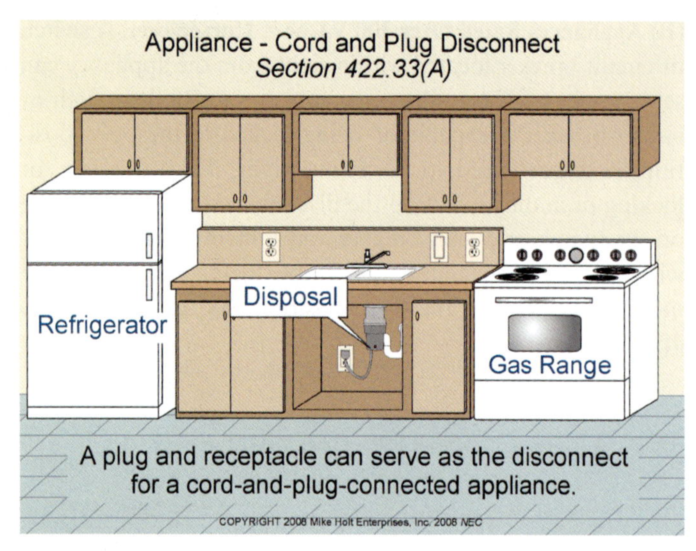

Figure 422–11

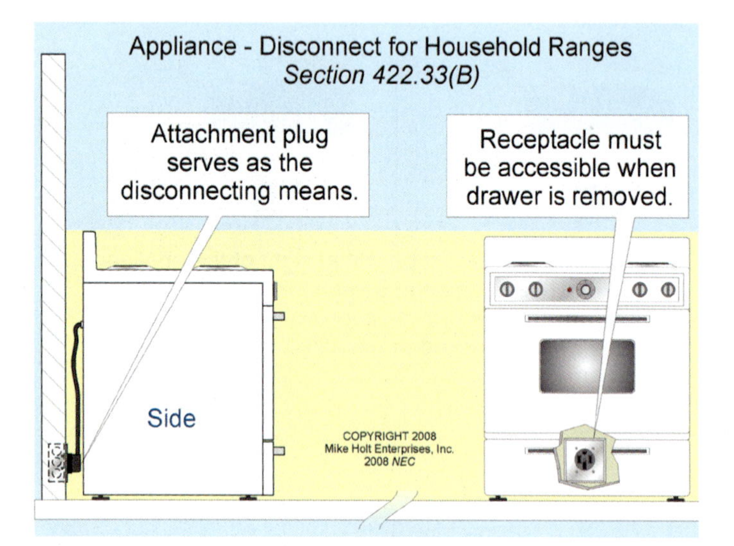

Figure 422–12

422.34 Unit Switches as Disconnects.

A unit switch with a marked "off" position that is a part of the appliance can serve as the appliance disconnect, if it disconnects all ungrounded conductors. Figure 422–13

422.51 Cord-and-Plug-Connected Vending Machines.

Cord-and-plug-connected vending machines must include a GFCI as an integral part of the attachment plug, or within 12 in. of the attachment plug. Older machines that are not so equipped must be connected to a GFCI-protected outlet. Figure 422–14

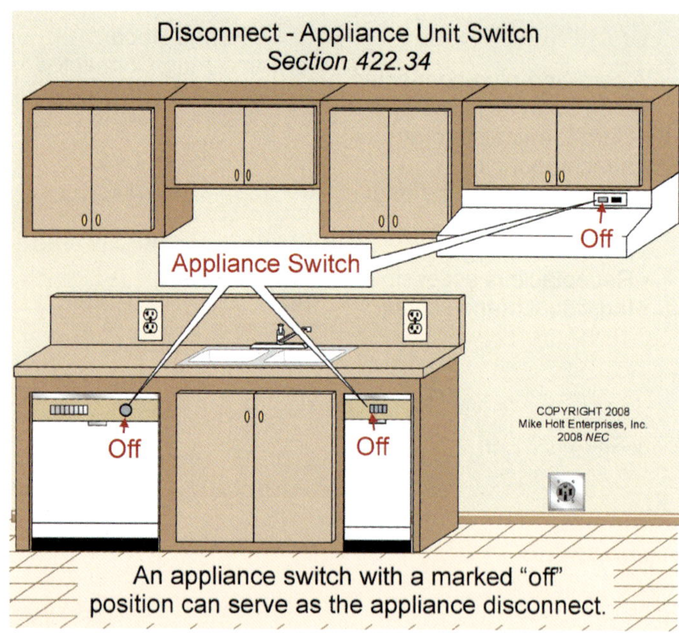

Figure 422–13

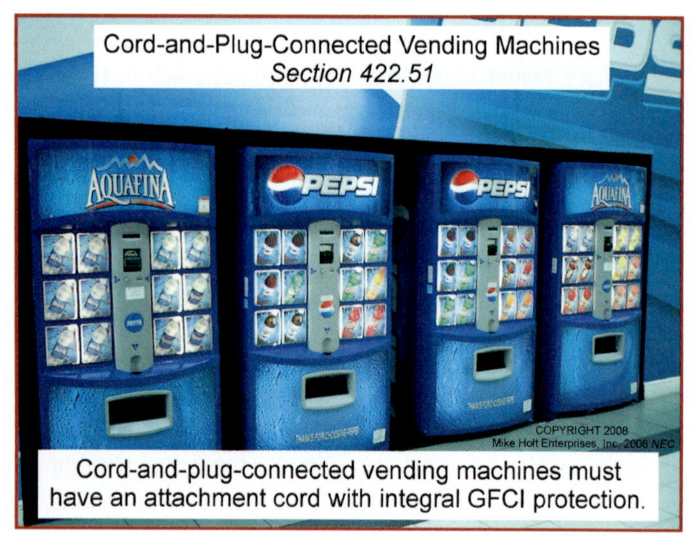

Figure 422–14

The term "vending machine" means a self-service device that dispenses products or merchandise and requires coin, paper currency, token, card, key, or receipt of payment by other means.

> **Author's Comment:** Because electric vending machines are often located in damp or wet locations in public places, and are used by people standing on the ground, reliance on an equipment grounding conductor for protection against electrocution is insufficient.

422.52 Electric Drinking Fountains. Electric drinking fountains must be GFCI protected. Figure 422–15

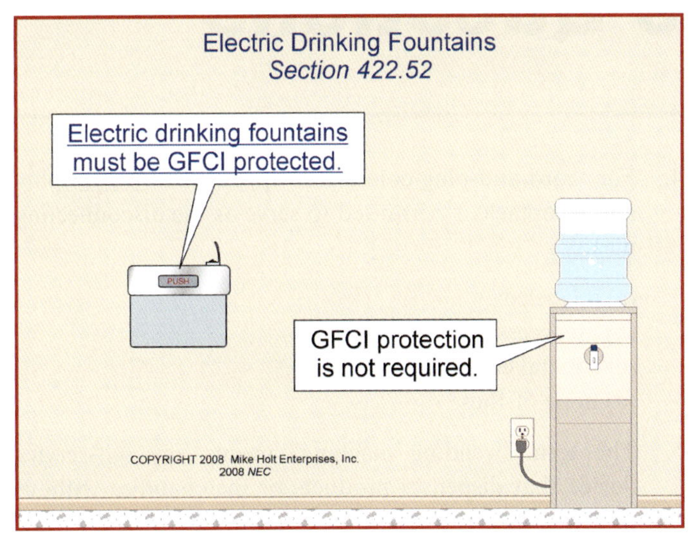

Figure 422–15

ARTICLE 422 Practice Questions

ARTICLE 422. APPLIANCES—PRACTICE QUESTIONS

1. If a protective device rating is marked on an appliance, the branch-circuit overcurrent device rating shall not exceed _____ percent of the protective device rating marked on the appliance.

   (a) 50
   (b) 80
   (c) 100
   (d) 115

2. Central heating equipment, other than fixed electric space-heating equipment, shall be supplied by a(n) _____ branch circuit.

   (a) multiwire
   (b) individual
   (c) multipurpose
   (d) small-appliance

3. The length of the cord for a dishwasher or trash compactor shall not be longer than _____, measured from the rear of the appliance.

   (a) 2 ft
   (b) 4 ft
   (c) 6 ft
   (d) 8 ft

4. For cord-and-plug-connected appliances, _____ plug and receptacle is permitted to serve as the disconnecting means.

   (a) a labeled
   (b) an accessible
   (c) a metal enclosed
   (d) none of these

5. The term "vending machine" means any self-service device that dispenses products or merchandise without the necessity of replenishing the device between each vending operation and is designed to require insertion of a _____.

   (a) coin or paper currency
   (b) token or card
   (c) key or payment by other means
   (d) all of these

Fixed Electric Space-Heating Equipment

INTRODUCTION TO ARTICLE 424—FIXED ELECTRIC SPACE-HEATING EQUIPMENT

Many people are surprised to see how many pages Article 424 has. This is a nine-part article on fixed electric space heaters. Why so much text for what seems to be a simple application? The answer is that Article 424 covers a variety of applications—heaters come in various configurations for various uses. Not all of these parts are for the electrician in the field—the requirements in Part IV are for manufacturers.

Most electricians should focus on Part III, Part V, and Part VI. Fixed space heaters (wall-mounted, ceiling-mounted, or free-standing) are common in many utility buildings and other small structures, as well as in some larger structures. When used to heat floors, space-heating cables address the thermal layering problem typical of forced-air systems—so it's likely you'll encounter them. Duct heaters are very common in large office and educational buildings. These provide a distributed heating scheme. Locating the heater in the ductwork, but close to the occupied space, eliminates the waste of transporting heated air through sheet metal routed in unheated spaces, so it's likely you'll encounter those as well.

PART I. GENERAL

424.1 Scope. Article 424 contains the installation requirements for fixed electrical equipment used for space heating, such as heating cables, unit heaters, boilers, or central systems.

> **Author's Comment:** Wiring for fossil-fuel heating equipment, such as gas, oil, or coal central furnaces, must be installed in accordance with Article 422, specifically 422.12.

424.3 Branch Circuits.

(B) Branch-Circuit Sizing. For the purpose of sizing branch-circuit conductors, fixed electric space-heating equipment is considered a continuous load.

> **Author's Comment:** The branch-circuit conductors and overcurrent devices for fixed electric space-heating equipment must have an ampacity not less than 125 percent of the total heating load [210.19(A)(1) and 210.20(A)].

Question: *What size conductor and overcurrent device (with 75°C terminals) are required for a 10 kW, 240V fixed electric space heater that has a 3A blower motor?* **Figure 424–1**

(a) 10 AWG, 30A *(b) 8 AWG, 40A*
(c) 6 AWG, 60A *(d) 4 AWG, 80A*

Answer: *(c) 6 AWG, 60A*

Step 1: Determine the total load:

I = VA/E
I = 10,000 VA/240V
I = 41.67A + 3A = 44.67A, round to 45A [220.5(B)]

Step 2: Size the conductors at 125 percent of the total current load [110.14(C), 210.19(A)(1), and Table 310.16]:

Conductor = 45A x 1.25
Conductor = 56A, 6 AWG, rated 65A at 75°C

Step 3: Size the overcurrent device at 125 percent of the total current load [210.20(A), 240.4(B) and 240.6(A)]:

Overcurrent Protection = 45A x 1.25
Overcurrent Protection = 56A, next size up is 60A

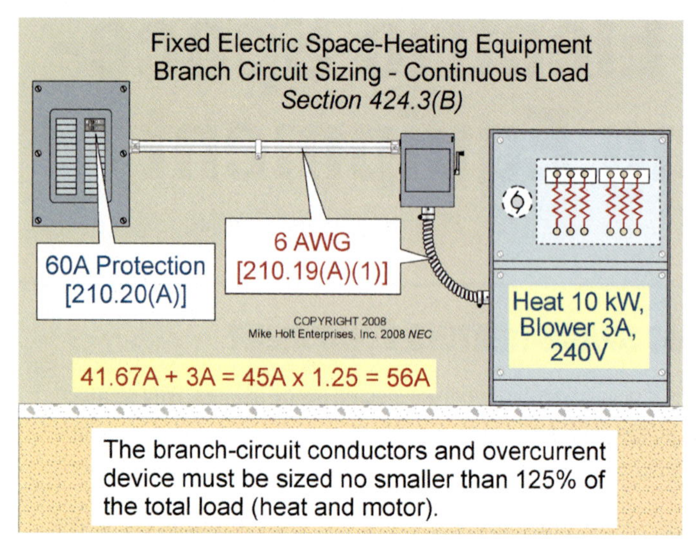

Figure 424–1

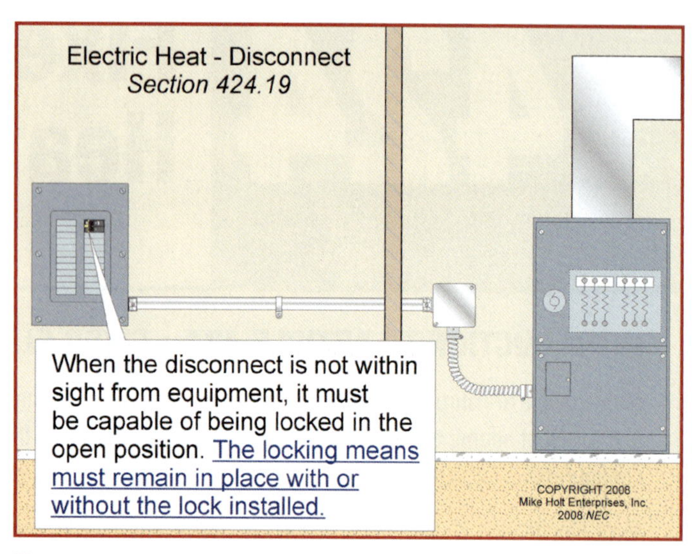

Figure 424–2

424.9 Permanently Installed Electric Baseboard Heaters with Receptacles. If a permanently installed electric baseboard heater has factory-installed receptacle outlets, the receptacles must not be connected to the heater circuits.

> **FPN:** Listed baseboard heaters include instructions that prohibit their installation below receptacle outlets.

PART III. ELECTRIC SPACE-HEATING EQUIPMENT

424.19 Disconnecting Means. Means must be provided to simultaneously disconnect the heater, motor controller, and supplementary overcurrent devices of all fixed electric space-heating equipment from all ungrounded conductors.

The disconnecting means must be capable of being locked in the open position. The provision for locking or adding a lock to the disconnecting means must be on the switch or circuit breaker, and it must remain in place with or without the lock installed. Figure 424–2

(A) Heating Equipment with Supplementary Overcurrent Protection. The disconnecting means for fixed electric space-heating equipment with supplementary overcurrent protection must be within sight from the supplementary overcurrent devices.

> **Author's Comment:** "Within Sight" means the specific equipment is visible and not more than 50 ft from one to the other [Article 100].

(B) Heating Equipment Without Supplementary Overcurrent Protection. For fixed electric space-heating equipment, the branch-circuit circuit breaker is permitted to serve as the disconnecting means where the circuit breaker is within sight from the heater or it's capable of being locked in the open position.

(C) Unit Switch as Disconnect. A unit switch with a marked "off" position that is an integral part of the equipment can serve as the heater disconnecting means, if it disconnects all ungrounded conductors of the circuit. Figure 424–3

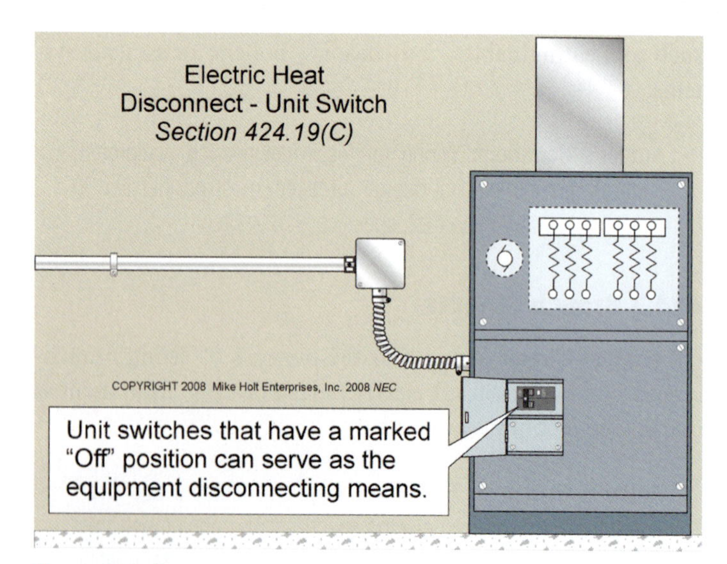

Figure 424–3

PART V. ELECTRIC SPACE-HEATING CABLES

424.44 Installation of Cables in Concrete or Poured Masonry Floors.

(G) GFCI Protection. GFCI protection is required for electric space-heating cables that are embedded in concrete or poured masonry floors of bathrooms and hydromassage bathtub locations. Figure 424–4

Author's Comment: See 680.28(C)(3) for restrictions on the installation of radiant heating cables for spas and hot tubs installed outdoors.

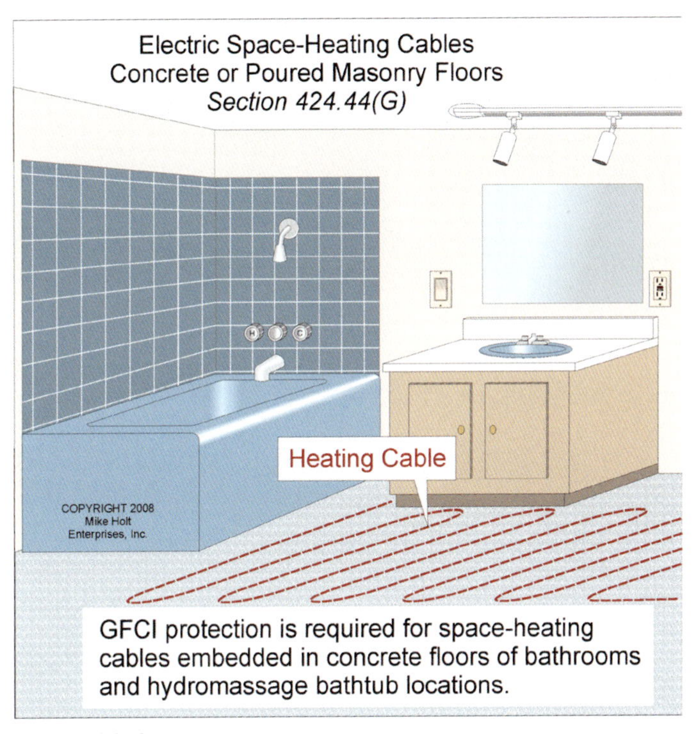

Figure 424–4

PART VI. DUCT HEATERS

424.65 Disconnect for Electric Duct Heater Controllers.

Means must be provided to disconnect the heater, motor controller, and supplementary overcurrent devices from all ungrounded conductors of the circuit. The disconnecting means must be within sight from the equipment, or it must be capable of being locked in the open position [424.19(A)]. The provision for locking or adding a lock to the disconnecting means must be on the switch or circuit breaker, and must remain in place with or without the lock installed. A portable locking means doesn't meet the "locked in the open position" requirement. Figure 424–5

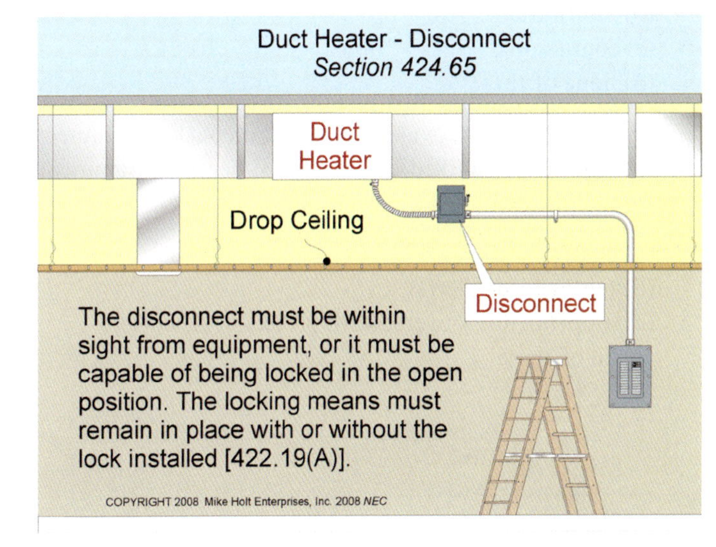

Figure 424–5

Author's Comment: The disconnecting means for a duct heater isn't required to be readily accessible. Therefore, it can be located within a suspended ceiling area adjacent to the duct heater as long as it's accessible by portable means [240.24(A)(4) and 404.8(A) Ex 2].

ARTICLE 424 Practice Questions

ARTICLE 424. FIXED ELECTRIC SPACE-HEATING EQUIPMENT—PRACTICE QUESTIONS

1. Fixed electric space-heating equipment shall be considered a(n) _____ load.

   (a) noncontinuous
   (b) intermittent
   (c) continuous
   (d) none of these

2. Means shall be provided to simultaneously disconnect the _____ of all fixed electric space-heating equipment from all ungrounded conductors.

   (a) heater
   (b) motor controller(s)
   (c) supplementary overcurrent protective device(s)
   (d) all of these

3. GFCI protection shall be provided for electrically heated floors in _____ locations.

   (a) bathroom
   (b) hydromassage bathtub
   (c) kitchen
   (d) a and b

ARTICLE

430 Motors, Motor Circuits, and Controllers

INTRODUCTION TO ARTICLE 430—MOTORS, MOTOR CIRCUITS, AND CONTROLLERS

Article 430 contains the specific rules for conductor sizing, overcurrent protection, control circuit conductors, controllers, and disconnecting means for electric motors. The installation requirements for motor control centers are covered in Part VIII, and air-conditioning and refrigeration equipment are covered in Article 440.

Article 430 is one of the longest articles in the *NEC.* It's also one of the most complex, but motors are also complex equipment. They are electrical and mechanical devices, but what makes motor applications complex is the fact that they are inductive loads with a high-current demand at startup that is typically six, or more, times the running current. This makes overcurrent protection and motor protection necessarily different from other pieces of equipment. So don't confuse general overcurrent protection with motor protection—you must calculate and apply them differently.

You might be uncomfortable with the allowances for overcurrent protection found in this article, such as protecting a 10 AWG conductor with a 60A device, but as you learn to understand how motor protection works, you'll understand why these allowances are not only safe, but necessary.

PART I. GENERAL

430.1 Scope. Article 430 covers motors, motor branch-circuit and feeder conductors and their protection, motor overload protection, motor control circuits, motor controllers, and motor control centers. This article is divided into many parts, the most important being: Figure 430–1

- General—Part I
- Circuit Conductors—Part II
- Overload Protection—Part III
- Branch Circuit Short-Circuit and Ground-Fault Protection—Part IV
- Feeder Circuit Short-Circuit and Ground-Fault Protection—Part V
- Control Circuits—Part VI
- Controllers—Part VII
- Motor Control Centers—Part VIII
- Disconnecting Means—Part IX

FPN No. 1: Article 440 contains the installation requirements for electrically driven air-conditioning and refrigeration equipment [440.1]. Also see 110.26(F) for dedicated space requirements for motor control centers.

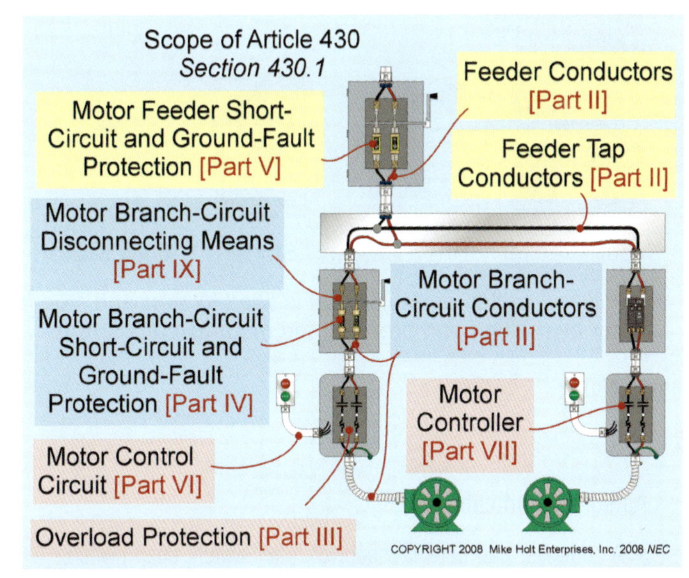

Figure 430–1

430.2 Definitions.

Adjustable-Speed Drive. A combination of the power converter, motor, and motor mounted auxiliary devices such as encoders, tachometers, thermal switches and detectors, air blowers, heaters, and vibration sensors.

Author's Comment: Adjustable-speed drives are often referred to as "variable-speed drives" or "variable-frequency drives."

Adjustable-Speed Drive System. An interconnected combination of equipment that provides a means of adjusting the speed of a mechanical load coupled to a motor. An adjustable-speed drive system typically consists of an adjustable-speed drive and auxiliary electrical apparatus.

Controller. A switch or device used to start and stop a motor by making and breaking the motor circuit current. Figure 430–2

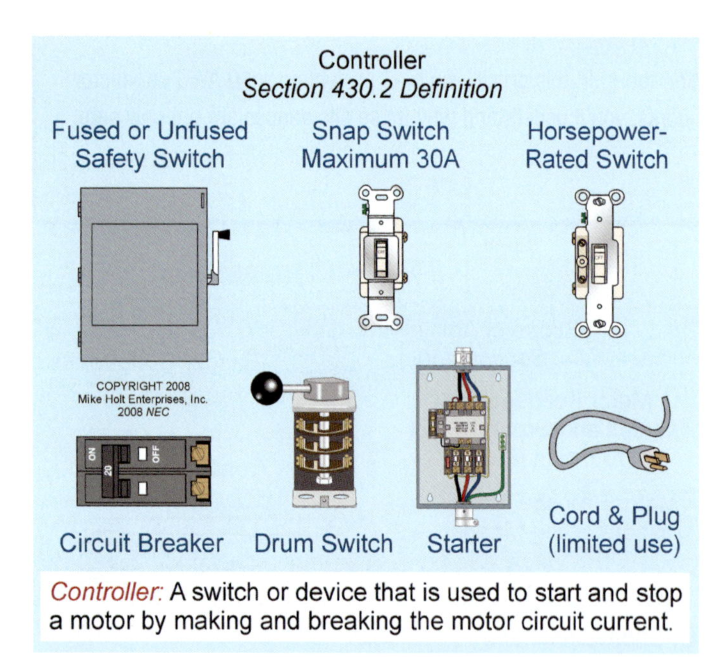

Figure 430–2

Author's Comments:

- A controller can be a horsepower-rated switch, snap switch, or circuit breaker. A pushbutton that operates an electromechanical relay isn't a controller because it doesn't meet the controller rating requirements of 430.83. Devices such as start-stop stations and pressure switches are control devices, not motor controllers. Figure 430–3

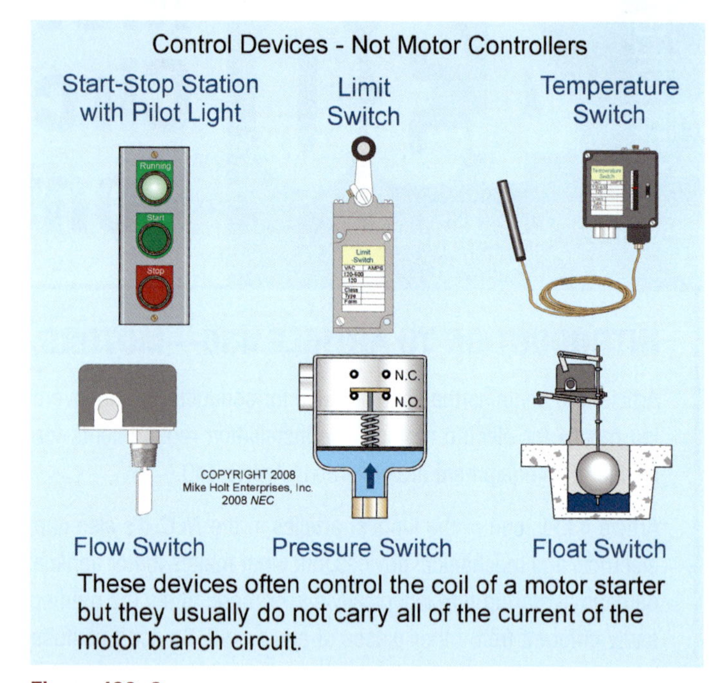

Figure 430–3

- Controllers discussed in Article 430 are those that meet this definition, not the definition of "Controller" in Article 100.

Motor Control Circuit. The circuit that carries the electric signals that direct the performance of the controller. Figure 430–4

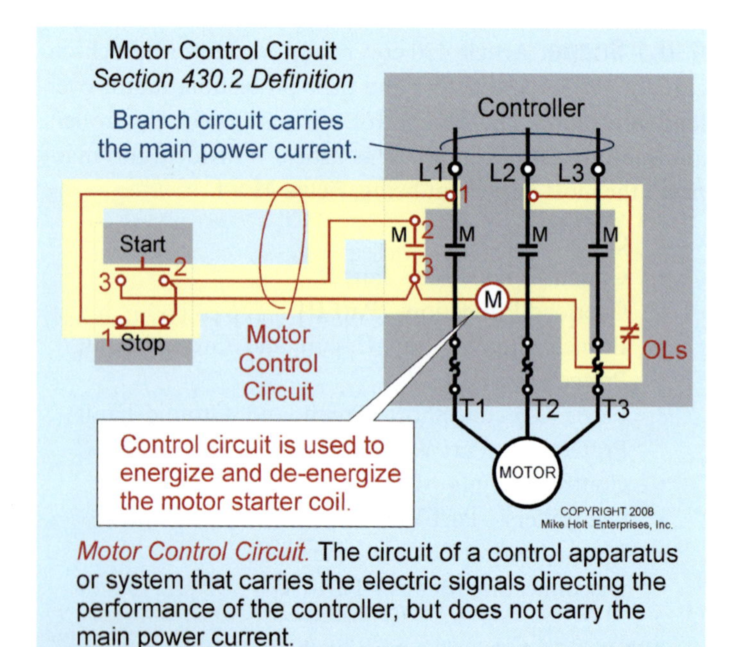

Figure 430–4

430.6 Table FLC Versus Motor Nameplate Current Rating.

(A) General Requirements. Figure 430–5

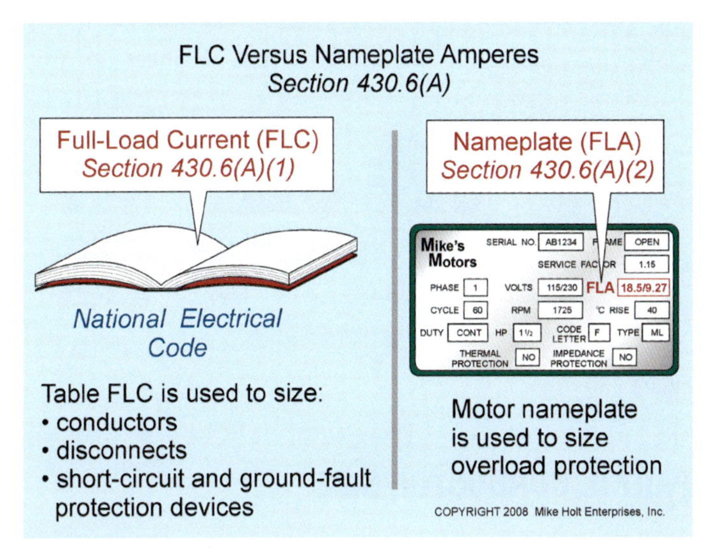

Figure 430–5

(1) Table Full-Load Current (FLC). The motor full-load current ratings listed in Tables 430.247, 430.248, and 430.250 are used to determine the conductor ampacity [430.22], the branch circuit short-circuit and ground-fault overcurrent device size [430.52 and 430.62], and the ampere rating of disconnecting switches [430.110].

Author's Comment: The actual current rating on the motor nameplate full-load amperes (FLA) [430.6(A)(2)] is not permitted to be used to determine the conductor ampacity, the branch-circuit short-circuit and ground-fault overcurrent device size, nor the ampere rating of disconnecting switches.

Motors built to operate at less than 1,200 RPM or that have high torques may have higher full-load currents, and multispeed motors have full-load current varying with speed, in which case the nameplate current ratings must be used.

Exception No. 3: For a listed motor-operated appliance, the motor full-load current marked on the nameplate of the appliance must be used instead of the horsepower rating on the appliance nameplate to determine the ampacity or rating of the disconnecting means, the branch-circuit conductors, the controller, and the branch-circuit short-circuit and ground-fault protection.

(2) Motor Nameplate Current Rating (FLA). Overload devices must be sized based on the motor nameplate current rating in accordance with 430.31.

Author's Comment: The motor nameplate full-load ampere rating is identified as full-load amperes (FLA). The FLA rating is the current in amperes the motor draws while producing its rated horsepower load at its rated voltage, based on its rated efficiency and power factor. Figure 430–6

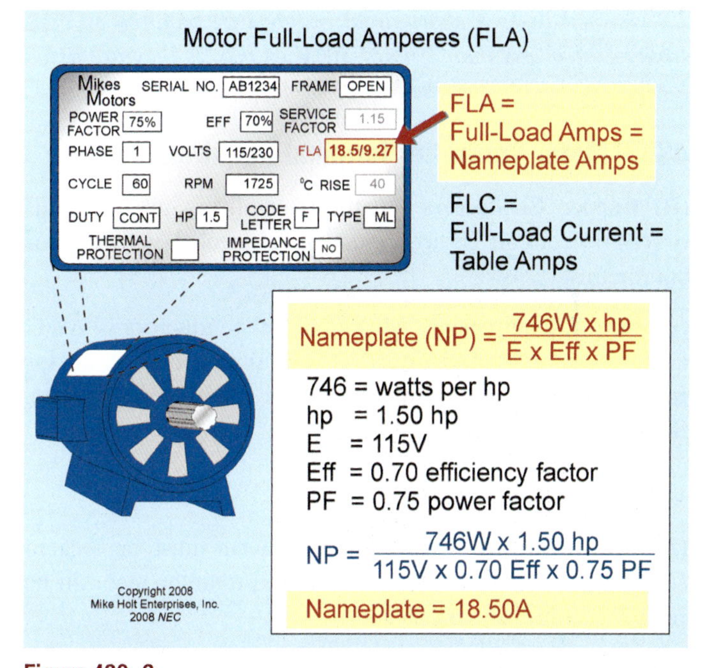

Figure 430–6

The actual current drawn by the motor depends upon the load on the motor and on the actual operating voltage at the motor terminals. That is, if the load increases, the current also increases, or if the motor operates at a voltage below its nameplate rating, the operating current will increase.

CAUTION: *To prevent damage to motor windings from excessive heat (caused by excessive current), never load a motor above its horsepower rating, and be sure the voltage source matches the motor's voltage rating.*

430.8 Marking on Controllers. A controller must be marked with the manufacturer's name or identification, the voltage, the current or horsepower rating, the short-circuit current rating, and other necessary data to properly indicate the applications for which it's suitable.

Exception No. 1: The short-circuit current rating isn't required for controllers applied in accordance with 430.81(A), 430.81(B), or 430.83(C).

Exception No. 2: The short-circuit rating isn't required on the controller when the short-circuit current rating of the controller is marked elsewhere on the assembly.

Exception No. 3: The short-circuit rating isn't required on the controller when the assembly into which it's installed has a marked short-circuit current rating.

Exception No. 4: A short-circuit rating isn't required on controllers rated less than 2 hp at 300V or less, if they are listed for use on general-purpose branch circuits.

430.9 Motor Controller Terminal Requirements.

(B) Copper Conductors. Motor controllers and terminals of control circuit devices must be connected with copper conductors.

(C) Torque Requirements. Motor control conductors 14 AWG and smaller must be torqued at a minimum of 7 lb-in. for screw-type pressure terminals, unless identified otherwise. See 110.3(B) and 110.14 FPN.

430.14 Location of Motors.

(A) Ventilation and Maintenance. Motors must be located so adequate ventilation is provided and maintenance can be readily accomplished.

430.17 The Highest Rated Motor. When sizing motor circuit conductors, the highest rated motor is the motor with the highest rated full-load current rating (FLC).

Question: *Which of the following motors has the highest FLC rating?* **Figure 430–7**

(a) 10 hp, three-phase, 208V *(b) 5 hp, single-phase, 208V*
(c) 3 hp, single-phase, 120V *(d) none of these*

Answer: *(c) 3 hp, single-phase, 120V*

10 hp = 30.80A [Table 430.250]
5 hp = 30.80A [Table 430.248]
3 hp = 34.00A [Table 430.248]

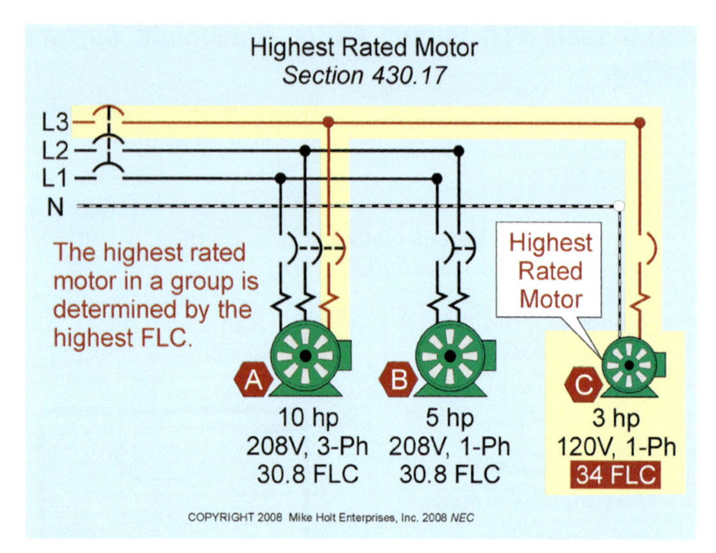

Figure 430–7

PART II. CONDUCTOR SIZE

430.22 Single Motor Conductor Size.

(A) Conductor Size. Conductors to a single motor must be sized not smaller than 125 percent of the motor FLC rating as listed in: Figure 430–8

- Table 430.247 Direct-Current Motors
- Table 430.248 Single-Phase Motors
- Table 430.250 Three-Phase motors

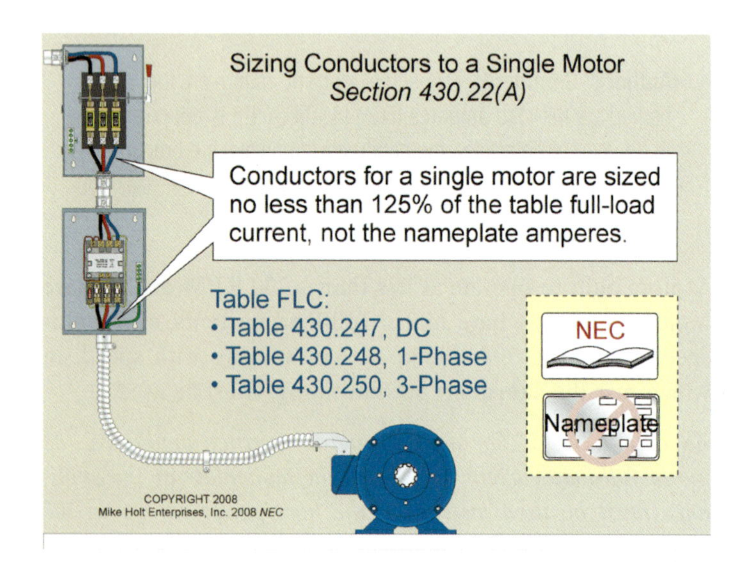

Figure 430–8

Question: *What size branch-circuit conductor is required for a 7½ hp, 230V, three-phase motor?* **Figure 430–9**

(a) 14 AWG *(b) 12 AWG*
(c) 10 AWG *(d) 8 AWG*

Answer: *(c) 10 AWG*

Motor FLC = 22A [Table 430.250]

Conductor's Size = 22A x 1.25
Conductor's Size = 27.50A, 10 AWG, rated 30A at 75°C [Table 310.16]

Note: The branch-circuit short-circuit and ground-fault protection device using an inverse time breaker is sized at 60A according to 430.52(C)(1) Ex 1:

Circuit Protection = 22A x 2.50
Circuit Protection = 55A, next size up 60A [240.6(A)]

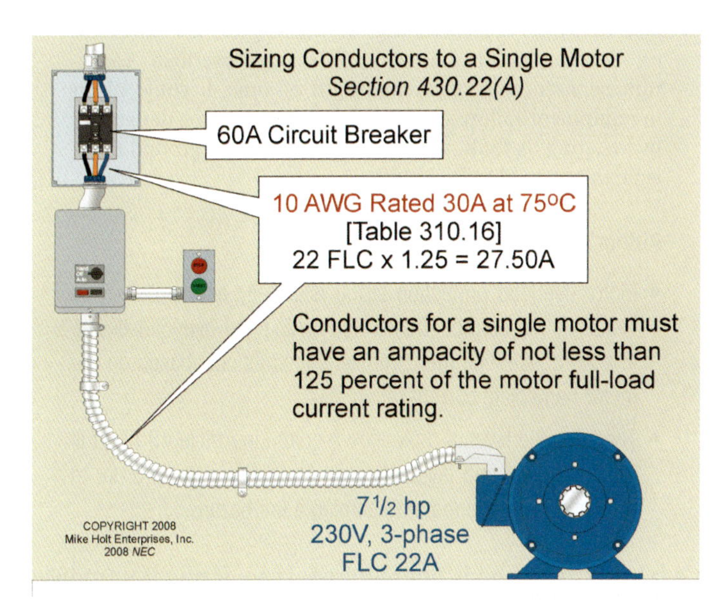

Figure 430–9

430.24 Several Motors—Conductor Size. Circuit conductors that supply several motors (typically feeders) must not be sized smaller than 125 percent of the largest motor FLC, plus the sum of the FLCs of the other motors.

Question: *What size feeder conductor is required for two 7½ hp, 230V, three-phase motors, if the terminals are rated for 75°C?* **Figure 430–10**

(a) 14 AWG *(b) 12 AWG* *(c) 10 AWG* *(d) 8 AWG*

Answer: *(d) 8 AWG*

Motor FLC = 22A [Table 430.250]

Motor Feeder Conductor = (22A x 1.25) + 22A
Motor Feeder Conductor = 49.50A, 8 AWG rated 50A at 75°C [Table 310.16]

The feeder overcurrent device (inverse time circuit breaker) must comply with 430.62 as follows:

Step 1: Determine the largest branch-circuit overcurrent device rating [240.6(A) and 430.52(C)(1) Ex 1]:

22A x 2.50 = 55A, next size up 60A

Step 2: Size the feeder overcurrent device in accordance with 240.6(A) and 430.62:

Feeder Inverse Time Breaker: 60A + 22A = 82A, next size down, 80A

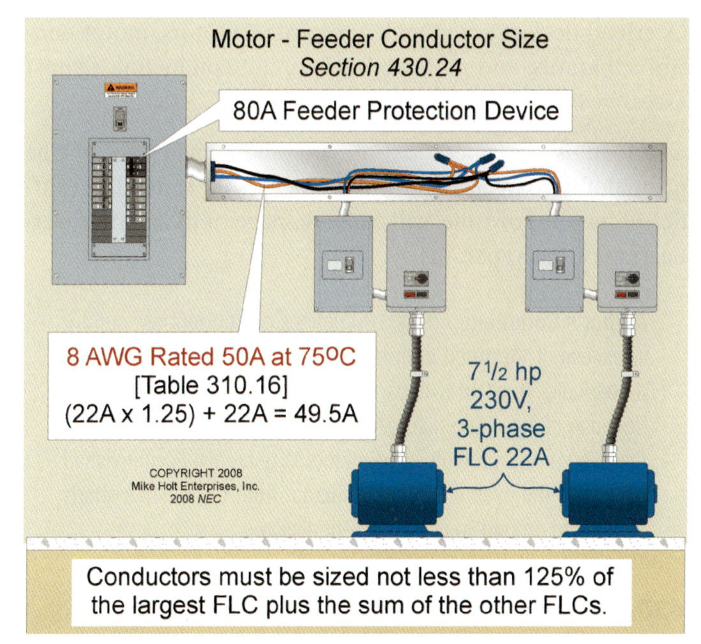

Figure 430–10

Author's Comment: The "next size up protection" rule for branch circuits [430.52(C)(1) Ex 1] doesn't apply to motor feeder short-circuit and ground-fault protection device sizing.

430.28 Motor Feeder Taps. Motor circuit conductors tapped from a feeder must have an ampacity in accordance with 430.22(A), and the tap conductors must terminate in a branch-circuit short-circuit and ground-fault protection device sized in accordance with 430.52. In addition, one of the following requirements must be met:

(1) 10 ft Tap. Tap conductors not over 10 ft long must have an ampacity not less than one-tenth the rating of the feeder protection device.

(2) 25 ft Tap. Tap conductors over 10 ft, but not over 25 ft, must have an ampacity not less than one-third the ampacity of the feeder conductor.

(3) Ampacity. Tap conductors must have an ampacity not less than the feeder conductors.

PART III. OVERLOAD PROTECTION

Part III contains the requirements for overload devices. Overload devices are intended to protect motors, motor control apparatus, and motor branch-circuit conductors against excessive heating due to motor overloads.

Overload is the operation of equipment in excess of the normal, full-load current rating, which, if it persists for a sufficient amount of time, will cause damage or dangerous overheating of the apparatus.

Author's Comment: Article 100 defines overcurrent as "current in excess of the rated current of equipment or the ampacity of a conductor from an overload, a short circuit, or a ground fault. Because of the difference between starting and running current, the overcurrent protection for motors is generally accomplished by having the overload device separate from the motor's short-circuit and ground-fault overcurrent device.

430.31 Overload. Overload devices (sometimes called "heaters") are intended to provide overload protection, and come in a variety of configurations; they can be conventional or electronic. In addition, a fuse sized in accordance with 430.32 can be used for circuit overload protection [430.55]. Figure 430–11

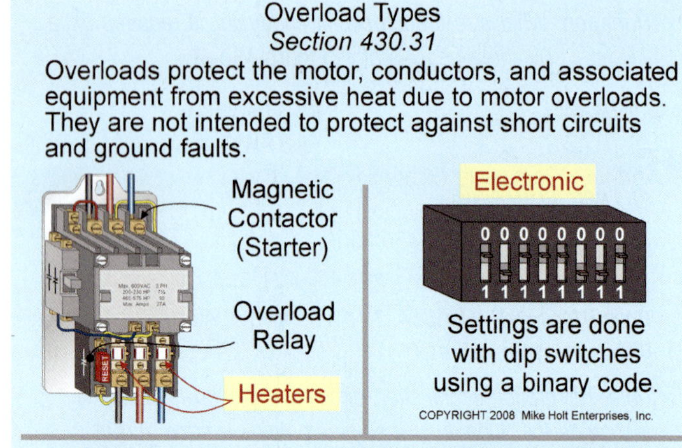

Figure 430–11

FPN: An overload is a condition where equipment is operated above its current rating, or where current is in excess of the conductor ampacity. When an overload condition persists for a sufficient length of time, it could result in equipment failure or a fire from damaging or dangerous overheating. A fault, such as a short circuit or ground fault, isn't an overload [Article 100].

Author's Comments:

- Motor overload protection sizing is usually accomplished by installing the correct "heater" or setting the overload device in accordance with the controller's instructions, based on the motor nameplate current rating.
- The intended level of protection required in Article 430 Part III is for overload and failure-to-start protection only, in order to protect against the motor becoming a fire hazard.

Overload protection is not required where it might introduce additional or increased hazards, as in the case of fire pumps.

FPN: See 695.7 for the protection requirements for fire pump supply conductors.

430.32 Overload Sizing for Continuous-Duty Motors.

(A) Motors Rated More Than One Horsepower. Motors rated more than 1 hp, used in a continuous-duty application without integral thermal protection, must have an overload device sized as follows:

(1) Separate Overload Device. A separate overload device must be selected to open at no more than the following percent of the motor nameplate full-load current rating:

Service Factor. Motors with a marked service factor (SF) of 1.15 or more on the nameplate must have the overload device sized no more than 125 percent of the motor nameplate current rating.

Author's Comment: A service factor of 1.15 means the motor is designed to operate continuously at 115 percent of its rated horsepower.

Temperature Rise. Motors with a nameplate temperature rise of 40°C or less must have the overload device sized no more than 125 percent of the motor nameplate current rating.

Author's Comment: A motor with a nameplate temperature rise of 40°C means the motor is designed to operate so that it will not heat up more than 40°C above its rated ambient temperature when operated at its rated load and voltage. Studies have shown that when the operating temperature of a motor is increased 10°C, the motor winding insulating material's anticipated life is reduced by 50 percent.

All Other Motors. No more than 115 percent of the motor "nameplate current rating."

430.36 Use of Fuses for Overload Protection.

Where fuses are used for overload protection, one must be provided for each ungrounded conductor of the circuit.

Author's Comment: If remote control isn't required for a motor, considerable savings can be achieved by using dual-element fuses (eliminate a motor controller) sized in accordance with 430.32 to protect the motor and the circuit conductors against overcurrent, which includes overload, short circuit, and ground faults. See 430.55 for more information.

430.37 Number of Overload Devices.

An overload device must be installed in each ungrounded conductor.

PART IV. BRANCH-CIRCUIT SHORT-CIRCUIT AND GROUND-FAULT PROTECTION

430.51 General.

A branch-circuit short-circuit and ground-fault protective device protects the motor, the motor control apparatus, and the conductors against short circuits or ground faults, but not against overload. Figure 430–12

Author's Comment: Overload protection must comply with the requirements contained in 430.32.

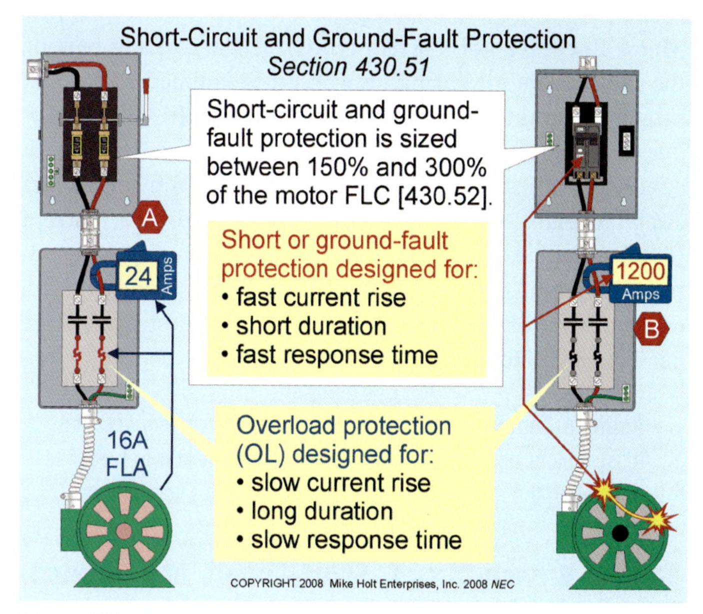

Figure 430–12

Motor-Starting Current. When voltage is first applied to the field winding of an induction motor, only the conductor resistance opposes the flow of current through the motor winding. Because the conductor resistance is so low, the motor will have a very large inrush current. Figure 430–13

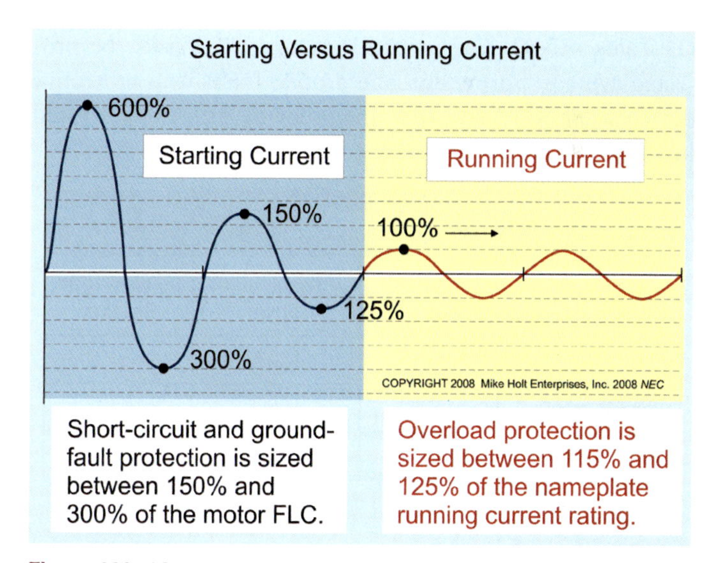

Figure 430–13

Motor-Running Current. Once the rotor begins turning, there is an increase in counter-electromotive force which reduces the starting current to running current.

Motor Locked-Rotor Current (LRC). If the rotating part of the motor winding (armature) becomes jammed so it can't rotate, no counter-electromotive force (CEMF) will be produced in the motor winding. This results in a decrease in conductor impedance to the point that it's effectively a short circuit. Result—the motor operates at locked-rotor current (LRC), often six times the full-load ampere rating, depending on the motor Code Letter rating [430.7(B)], and this will cause the motor winding to overheat and be destroyed if the current isn't quickly reduced or removed.

> **Author's Comment:** The *National Electrical Code* requires that most motors be provided with overcurrent protection to prevent damage to the motor winding because of locked-rotor current.

430.52 Branch-Circuit Short-Circuit and Ground-Fault Protection.

(A) General. The motor branch-circuit short-circuit and ground-fault protective device must comply with 430.52(B) and 430.52(C).

(B) All Motors. A motor branch-circuit short-circuit and ground-fault protective device must be capable of carrying the motor's starting current.

(C) Rating or Setting.

(1) Table 430.52. Each motor branch circuit must be protected against short circuit and ground faults by a protective device sized no greater than the following percentages listed in Table 430.52.

Table 430.52

Circuit Motor Type	Nontime Delay	Dual-Element Fuse	Inverse Time Breaker
Wound Rotor	150%	150%	150%
Direct Current	150%	150%	150%
All Other Motors	300%	175%	250%

Question: *What size conductor and inverse time circuit breaker are required for a 2 hp, 230V, single-phase motor?* **Figure 430–14**

(a) 14 AWG, 30A breaker *(b) 14 AWG, 35A breaker*
(c) 14 AWG, 40A breaker *(d) 14 AWG, 45A breaker*

Answer: *(a) 14 AWG, 30A breaker*

Step 1: Determine the branch-circuit conductor [Table 310.16, 430.22(A), and Table 430.248]:

12A x 1.25 = 15A, 14 AWG, rated 20A at 75°C [Table 310.16]

Step 2: Determine the branch-circuit protection [240.6(A), 430.52(C)(1), and Table 430.248]:

12A x 2.50 = 30A

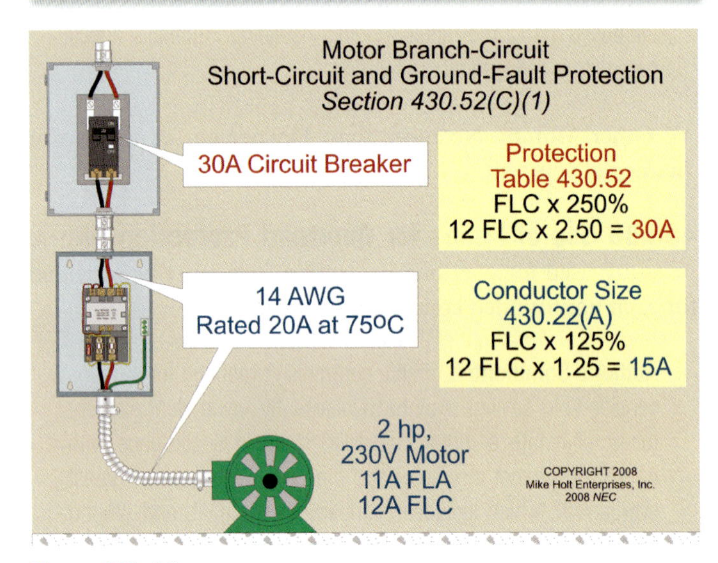

Figure 430–14

> **Author's Comment:** I know it bothers many in the electrical industry to see a 14 AWG conductor protected by a 30A circuit breaker, but branch-circuit conductors are protected against overloads by the overload device, which is sized between 115 and 125 percent of the motor nameplate current rating [430.32]. See 240.4(G) for details.

Exception No. 1: Where the motor short-circuit and ground-fault protective device values derived from Table 430.52 don't correspond with the standard overcurrent device ratings listed in 240.6(A), the next higher overcurrent device rating can be used.

Question: *What size conductor and inverse time circuit breaker are required for a 7½ hp, 230V, three-phase motor?* **Figure 430–15**

(a) 10 AWG, 50A breaker *(b) 10 AWG, 60A breaker*
(c) a or b *(d) none of these*

Answer: *(b) 10 AWG, 60A breaker*

Step 1: Determine the branch-circuit conductor [Table 310.16, 430.22(A), and Table 430.250]:

22A x 1.25 = 27.50A, 10 AWG, rated 30A at 75°C [Table 310.16]

Step 2: Determine the branch-circuit protection [240.6(A), 430.52(C)(1) Ex 1, and Table 430.250]:

22A x 2.50 = 55A, next size up = 60A

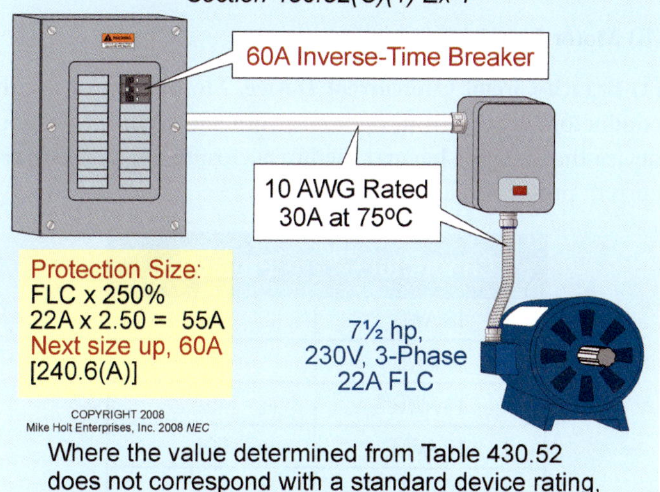

Figure 430–15

430.55 Single Overcurrent Device. A motor can be protected against overload, short circuit, and ground fault by a single overcurrent device sized to the overload requirements contained in 430.32.

Question: *What size dual-element fuse is permitted to protect a 5 hp, 230V, single-phase motor with a service factor of 1.20 and a nameplate current rating of 28A?* **Figure 430–16**

(a) 20A *(b) 25A* *(c) 30A* *(d) 35A*

Answer: *(d) 35A*

Overload Protection [430.32(A)(1)]
28A x 1.25 = 35A

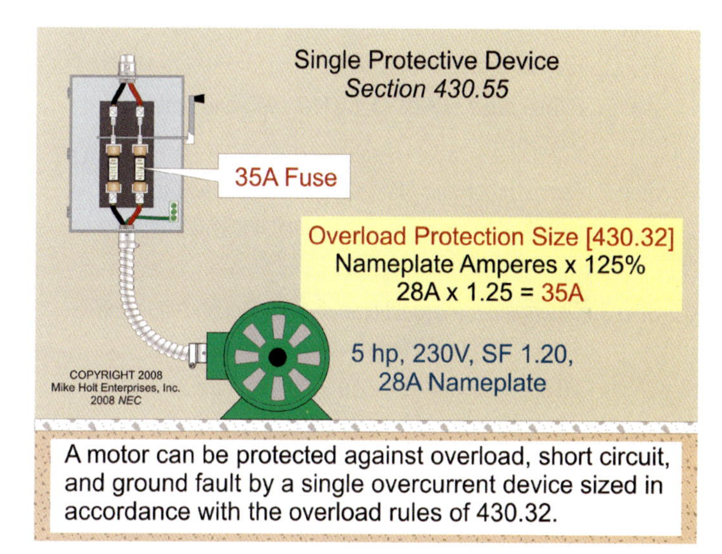

Figure 430–16

PART V. FEEDER SHORT-CIRCUIT AND GROUND-FAULT PROTECTION

430.62 Feeder Protection.

(A) Motors Only. Feeder conductors must be protected against short circuits and ground faults by a protective device sized not more than the largest rating of the branch-circuit short-circuit and ground-fault protective device for any motor, plus the sum of the full-load currents of the other motors in the group.

Question: *What size feeder protection (inverse time breakers with 75°C terminals) and conductors are required for the following two motors?* **Figure 430–17**

Motor 1—20 hp, 460V, three-phase = 27A
Motor 2—10 hp, 460V, three-phase = 14A

(a) 8 AWG, 70A breaker *(b) 8 AWG, 80A breaker*
(c) 8 AWG, 90A breaker *(d) 10 AWG, 90A breaker*

Answer: *(b) 8 AWG, 80A breaker*

Step 1: Determine the feeder conductor size [430.24]:

(27A x 1.25) + 14A = 48A

8 AWG rated 50A at 75°C [110.14(C) and Table 310.16]

Step 2: Feeder protection [430.62(A)] is not greater than the largest branch-circuit ground-fault and short-circuit protective device plus other motor FLC.

Step 3: Determine the largest branch-circuit ground-fault and short-circuit protective device [430.52(C)(1) Ex]:

20 hp Motor = 27A x 2.50 = 68, next size up = 70A
10 hp Motor = 14A x 2.50 = 35A

Step 4: Determine the size feeder protection:

Not more than 70A + 14A, = 84A, next size down = 80A [240.6(A)]

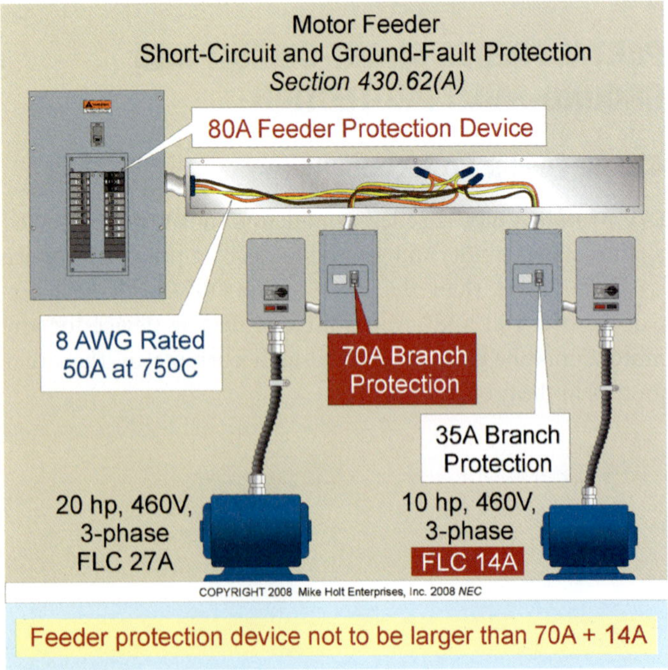

Figure 430–17

Author's Comment: The "next size up protection" rule for branch circuits [430.52(C)(1) Ex 1] doesn't apply to a motor feeder protection device rating.

PART VI. MOTOR CONTROL CIRCUITS

430.72 Overcurrent Protection for Control Circuits.

(A) Class 1 Control Conductors. Motor control conductors that are not tapped from the branch-circuit protective device are classified as a Class 1 remote-control circuit, and they must have overcurrent protection in accordance with 725.43.

Author's Comment: Section 725.43 states that overcurrent protection for conductors 14 AWG and larger must comply with the conductor ampacity from Table 310.16. Overcurrent protection for 18 AWG must not exceed 7A, and a 10A device must protect 16 AWG conductors.

(B) Motor Control Conductors.

(2) Branch-Circuit Overcurrent Device. Motor control circuit conductors tapped from the motor branch-circuit protection device that extends beyond the tap enclosure must have overcurrent protection as follows:

Conductor	Protection
18 AWG	7A
16 AWG	10A
14 AWG	45A
12 AWG	60A
10 AWG	90A

Author's Comment: The above limitations don't apply to the internal wiring of industrial control panels listed in UL *508 Standard for Practical Application Guidelines.*

(C) Control Circuit Transformer Protection. Transformers for motor control circuit conductors must have overcurrent protection on the primary side in accordance with 430.72(C)(1) through (5).

Author's Comment: Many control transformers have small iron cores, which result in very high inrush (excitation) current when the coil is energized. This high inrush current can cause standard fuses to blow, so you should only use the fuses recommended by the control transformer manufacturer.

430.73 Protection of Conductors from Physical Damage. Where physical damage would result in a hazard, the conductors of a remote motor control circuit run outside the control device must be protected by installing the conductors in a raceway or providing other suitable protection from physical damage.

430.75 Disconnect for Control Circuits.

(A) Control Circuit Disconnect. Motor control circuit conductors must have a disconnecting means that opens all sources of supply when the disconnecting means is in the open position. If the control circuit conductors are tapped from the controller disconnect, the controller disconnecting means can serve as the disconnecting means for the control circuit conductors [430.102(A)].

If the control circuit conductors aren't tapped from the controller disconnect, a separate disconnecting means is required for the control circuit conductors, and it must be located adjacent to the controller disconnect. Figure 430–18

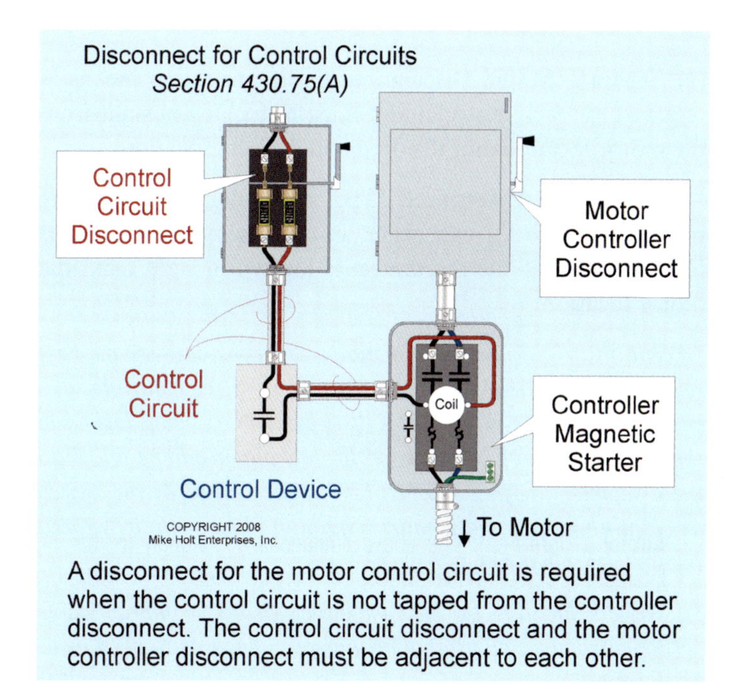

Figure 430–18

PART VII. MOTOR CONTROLLERS

430.83 Controller Rating.

(A) General. The controller must have one of the following ratings:

(1) Horsepower Rating. Controllers, other than circuit breakers and molded case switches, must have a horsepower rating not less than that of the motor.

(2) Circuit Breakers. A circuit breaker can serve as a motor controller [430.111].

> **Author's Comment:** Circuit breakers aren't required to be horsepower rated.

(3) Molded Case Switch. A molded case switch, rated in amperes, can serve as a motor controller.

> **Author's Comment:** A molded case switch isn't required to be horsepower rated.

(C) Stationary Motors of Two Horsepower or Less. For stationary motors rated at 2 hp or less, the controller can be:

(2) General-Use Snap Switch. A general-use ac snap switch, where the motor full-load current rating isn't more than 80 percent of the ampere rating of the switch.

> **Author's Comment:** A general-use snap switch is a general-use switch constructed for installation in device boxes or on box covers, or otherwise used in conjunction with wiring systems recognized by this *Code*.

430.84 Need Not Open All Conductors of the Circuit.

The motor controller can open only as many conductors of the circuit as necessary to start and stop the motor.

> **Author's Comment:** The controller is only required to start and stop the motor; it isn't a disconnecting means. See the disconnecting means requirement in 430.103 for more information.

430.87 Controller for Each Motor. Each motor must have its own individual controller.

PART IX. DISCONNECTING MEANS

430.102 Disconnect Requirement.

(A) Controller Disconnect. A disconnecting means is required for each motor controller, and it must be located within sight from the controller. Figures 430–19 and 430-20

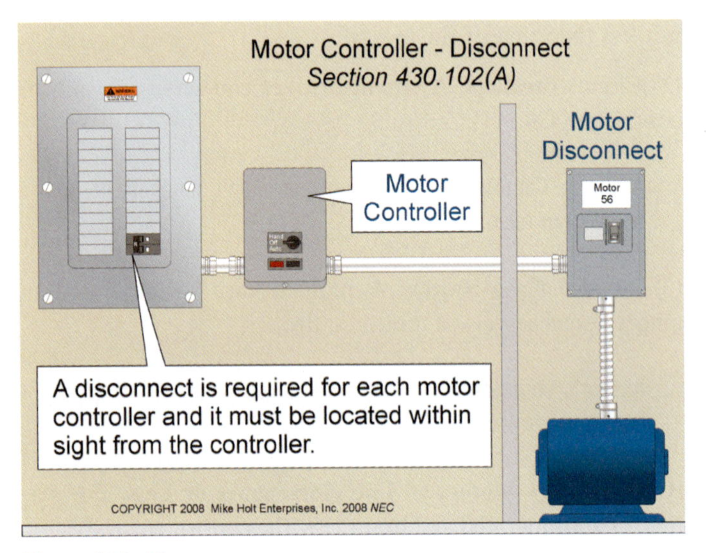

Figure 430–19

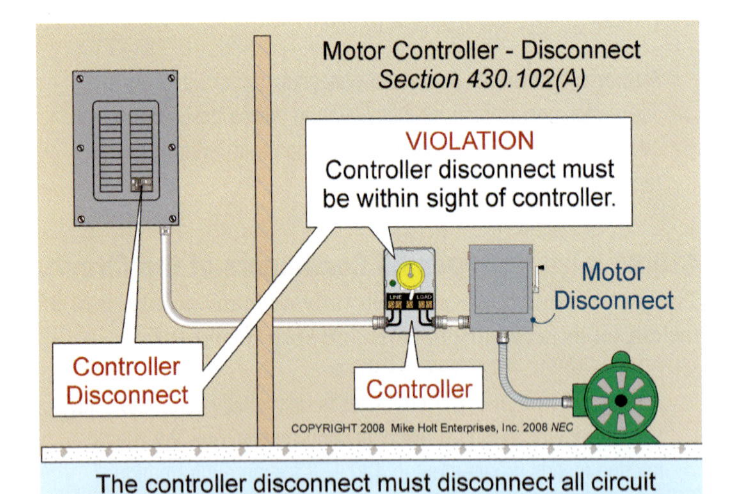

Figure 430–20

Author's Comment: "Within Sight" is visible and not more than 50 ft from each other [Article 100].

(B) Motor Disconnect. A motor disconnect must be provided in accordance with (B)(1) or (B)(2). Figure 430–21

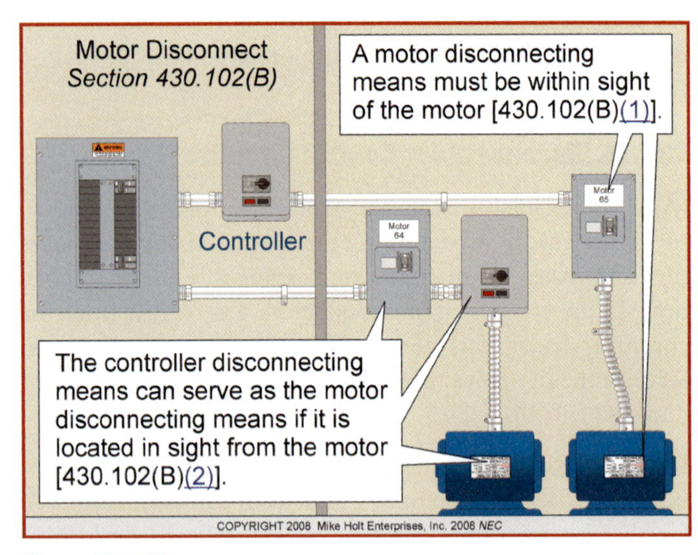

Figure 430–21

(1) Separate Motor Disconnect. A disconnecting means is required for each motor, and it must be located in sight from the motor location and the driven machinery location.

(2) Controller Disconnect. The controller disconnecting means [430.102(A)] can serve as the disconnecting means for the motor, if the disconnect is located in sight from the motor location.

Exception to (1) and (2): A motor disconnecting means isn't required under either condition (a) or (b), if the controller disconnecting means [430.102(A)] is capable of being locked in the open position. The provision for locking or adding a lock to the disconnecting means must be installed on or at the switch or circuit breaker, and it must remain in place with or without the lock installed. Figure 430–22

(a) Where locating the disconnecting means is impracticable or introduces additional or increased hazards to persons or property.

(b) In industrial installations, with written safety procedures, where conditions of maintenance and supervision ensure only qualified persons will service the equipment.

Author's Comment: See the definition of "Within Sight" in Article 100.

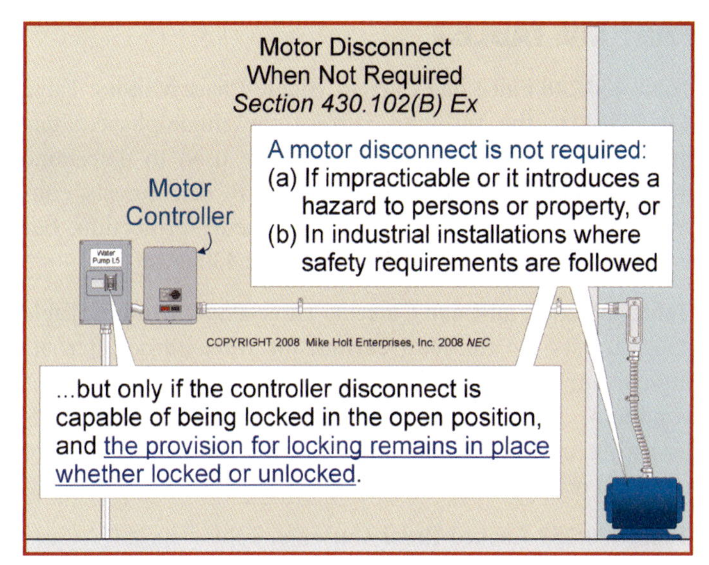

Figure 430–22

FPN No. 2: For information on lockout/tagout procedures, see NFPA 70E, *Standard for Electrical Safety in the Workplace*.

430.103 Operation of Disconnect. The disconnecting means for the motor controller and the motor must open all ungrounded supply conductors simultaneously, and it must be designed so that it will not close automatically. Figure 430–23

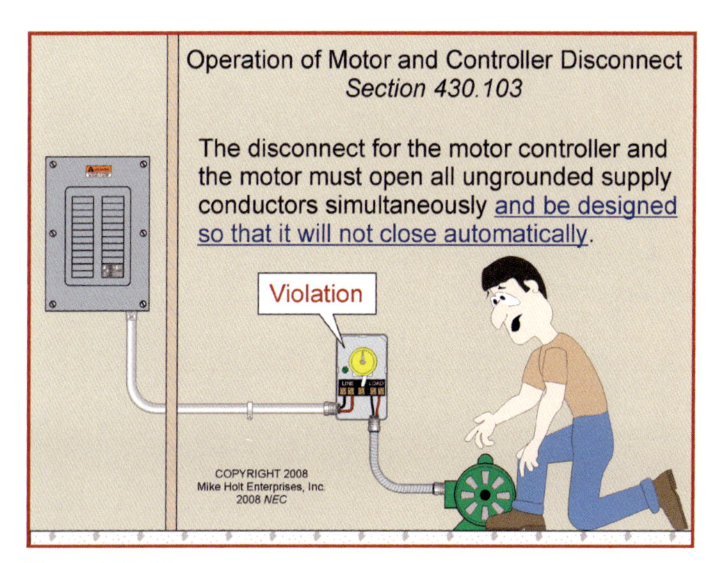

Figure 430–23

430.104 Marking and Mounting. The controller and motor disconnecting means must indicate whether they are in the "on" or "off" position.

Author's Comment: The disconnecting means must be legibly marked to identify its intended purpose [110.22 and 408.4], and when operated vertically, the "up" position must be the "on" position [240.81 and 404.6(C)].

430.107 Readily Accessible. Either the controller disconnecting means or the motor disconnecting means required by 430.102 must be readily accessible. Figure 430–24

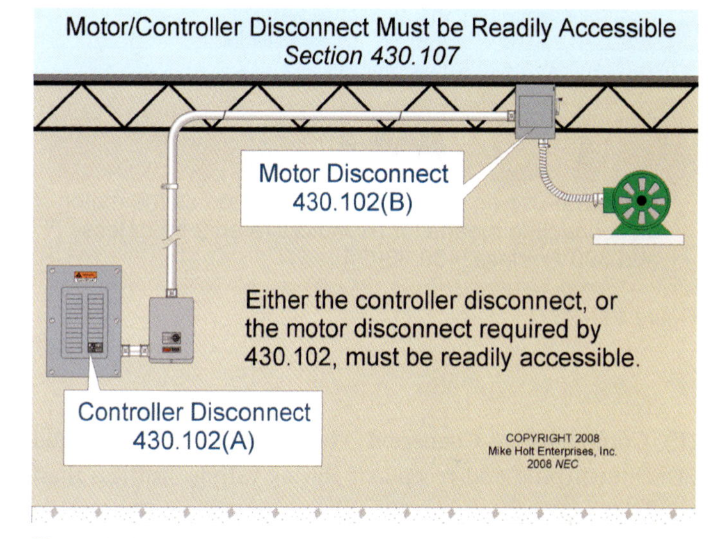

Figure 430–24

430.109 Disconnecting Means Rating.

(A) General. The disconnecting means for the motor controller and/or the motor must be a:

(1) Motor-Circuit Switch. A listed horsepower-rated motor-circuit switch.

(2) Molded Case Circuit Breaker. A listed molded case circuit breaker.

(3) Molded Case Switch. A listed molded case switch.

(6) Manual Motor Controller. A listed manual motor controller marked "Suitable as Motor Disconnect."

(C) Stationary Motors of Two Horsepower or Less.

(2) General-Use Snap Switch. A general-use ac snap switch, where the motor full-load current rating isn't more than 80 percent of the ampere rating of the switch. Figure 430–25

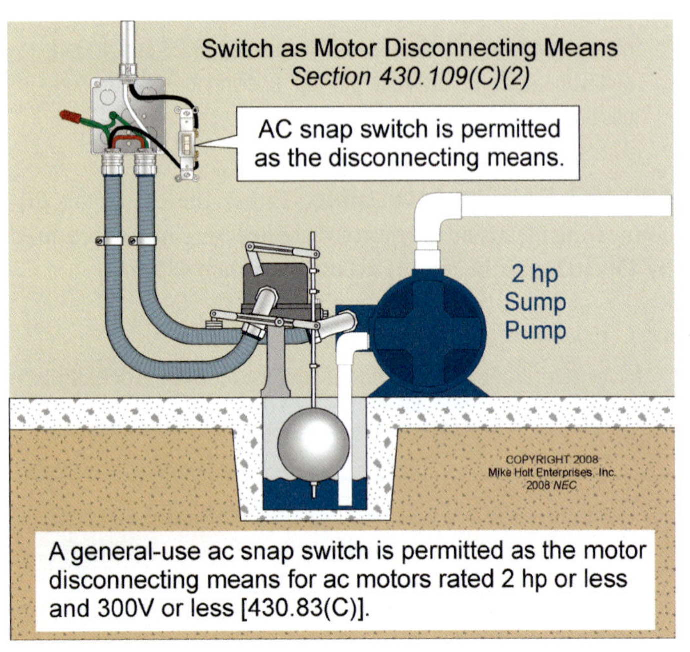

Figure 430–25

(F) Cord-and-Plug-Connected Motors. A horsepower-rated attachment plug and receptacle having ratings not less than the motor ratings.

430.111 Combination Controller and Disconnect.

A horsepower-rated switch or circuit breaker can serve as both a controller and disconnecting means if it opens all ungrounded conductors to the motor as required by 430.103.

PART XIV. TABLES

Table 430.248 Full-Load Current, Single-Phase Motors. Table 430.248 lists the full-load current for single-phase alternating-current motors. These values are used to determine motor conductor sizing, ampere ratings of disconnects, controller rating, and branch-circuit and feeder protection, but not overload protection [430.6(A)(1) and 430.6(A)(2)].

Table 430.250 Full-Load Current, Three-Phase Motors. Table 430.250 lists the full-load current for three-phase alternating-current motors. The values are used to determine motor conductor sizing, ampere ratings of disconnects, controller rating, and branch-circuit and feeder protection, but not overload protection [430.6(A)(1) and 430.6(A)(2)].

Table 430.251 Locked-Rotor Currents. Table 430.251(A) lists the locked-rotor current for single-phase motors, and Table 430.251(B) contains the locked-rotor current for three-phase motors. These values are used in the selection of controllers and disconnecting means when the horsepower rating isn't marked on the motor nameplate.

ARTICLE 430 Practice Questions

ARTICLE 430. MOTORS, MOTOR CIRCUITS, AND CONTROLLERS—PRACTICE QUESTIONS

1. Motor controllers and terminals of control circuit devices shall be connected with copper conductors unless identified for use with a different conductor.

   (a) True
   (b) False

2. Branch-circuit conductors supplying a single continuous-duty motor shall have an ampacity not less than _____ rating.

   (a) 125 percent of the motor's nameplate current
   (b) 125 percent of the motor's full-load current as determined by 430.6(A)(1)
   (c) 125 percent of the motor's full locked-rotor
   (d) 80 percent of the motor's full-load current

3. Overload devices are intended to protect motors, motor control apparatus, and motor branch-circuit conductors against _____.

   (a) excessive heating due to motor overloads
   (b) excessive heating due to failure to start
   (c) short circuits and ground faults
   (d) a and b

4. The ultimate trip current of a thermally protected motor with a full-load current not exceeding 9A shall not exceed _____ percent of the motor full-load current.

   (a) 140
   (b) 156
   (c) 170
   (d) 175

5. The motor branch-circuit short-circuit and ground-fault protective device shall be capable of carrying the _____ current of the motor.

   (a) varying
   (b) starting
   (c) running
   (d) continuous

6. A feeder supplying fixed motor load(s) shall have a protective device with a rating or setting _____ branch-circuit short-circuit and ground-fault protective device for any motor in the group, plus the sum of the full-load currents of the other motors of the group.

   (a) not greater than the largest rating or setting of the
   (b) 125 percent of the largest rating of any
   (c) equal to the largest rating of any
   (d) none of these

7. Motor control circuits shall be arranged so they will be disconnected from all sources of supply when the disconnecting means is in the open position.

   (a) True
   (b) False

8. For stationary motors of 2 hp or less and 300V or less on ac circuits, the controller can be an ac-rated only general-use snap switch where the motor full-load current rating is not more than _____ percent of the rating of the switch.

   (a) 50
   (b) 60
   (c) 70
   (d) 80

9. A _____ shall be located in sight from the motor location and the driven machinery location.

   (a) controller
   (b) protection device
   (c) disconnecting means
   (d) all of these

10. The disconnecting means for a motor controller shall be designed so that it does not _____ automatically.

    (a) open
    (b) close
    (c) restart
    (d) shut down

11. A motor disconnecting means can be a _____.

    (a) listed molded case circuit breaker
    (b) listed motor-circuit switch rated in horsepower
    (c) listed molded case switch
    (d) any of these

Air-Conditioning and Refrigeration Equipment

INTRODUCTION TO ARTICLE 440—AIR-CONDITIONING AND REFRIGERATION EQUIPMENT

This article applies to electrically driven air-conditioning and refrigeration equipment. The rules in this article add to, or amend, the rules in Article 430 and other articles.

Each equipment manufacturer has the motors for a given air-conditioning unit built to its own specifications. Cooling and other characteristics are different from those of nonhermetic motors. For each motor, the manufacturer has worked out all of the details and supplied the correct protection, conductor sizing, and other information on the nameplate.

The application itself—with the compressor motor often on the other side of an exterior building wall from the normal power sources so it can exchange heat with free air—poses additional problems, which the *NEC* addresses in Article 440.

PART I. GENERAL

440.1 Scope.

Article 440 applies to electrically driven air-conditioning and refrigeration equipment.

440.2 Definitions.

Hermetic Refrigerant Motor-Compressor. A compressor and motor enclosed in the same housing, operating in the refrigerant.

Rated-Load Current. The current resulting when the motor-compressor operates at rated load and rated voltage.

440.3 Other Articles.

(B) Equipment with No Hermetic Motor-Compressors. Air-conditioning and refrigeration equipment that do not have hermetic refrigerant motor-compressors, such as furnaces with evaporator coils, must comply with Article 422 for appliances, Article 424 for electric space-heating, and Article 430 for motors.

(C) Household Refrigerant Motor-Compressor Appliances. Household refrigerators and freezers, drinking water coolers, and beverage dispensing machines are listed as appliances, and their installation must also comply with Article 422 for appliances. Figure 440–1

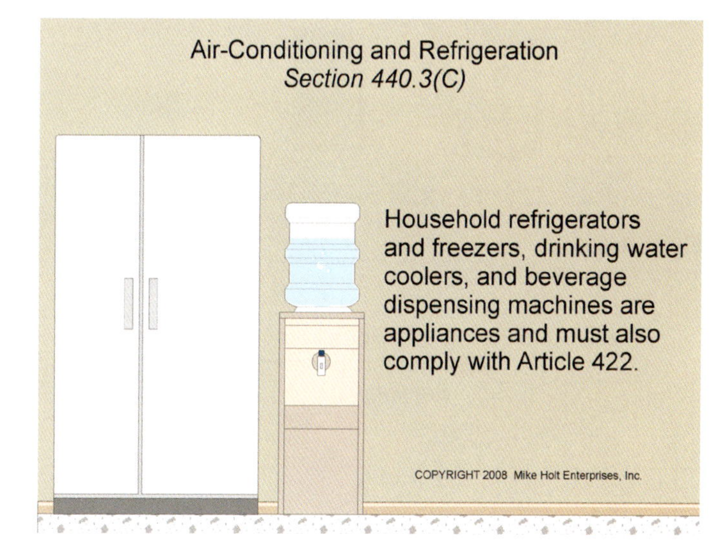

Figure 440–1

440.6 Ampacity and Rating.

(A) Hermetic Refrigerant Motor-Compressor. For a hermetic refrigerant motor-compressor, the rated-load current marked on the nameplate of the equipment is to be used in determining the rating of the disconnecting means, the branch-circuit conductors, the controller, and the branch-circuit short-circuit and ground-fault protection.

Exception No. 1: The branch-circuit selection current must be used instead of the rated-load current if provided on the equipment nameplate.

PART II. DISCONNECTING MEANS

440.14 Location. The disconnecting means for air-conditioning or refrigeration equipment must be located within sight from and readily accessible from the equipment. Figures 440–2 and 440–3

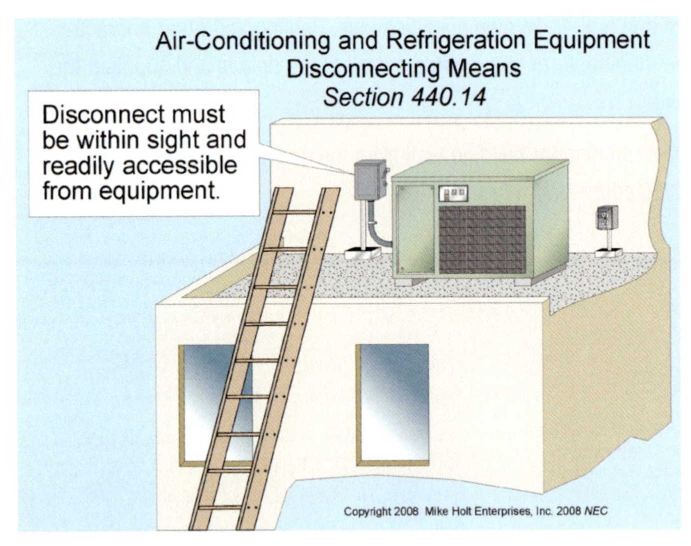

Figure 440–2

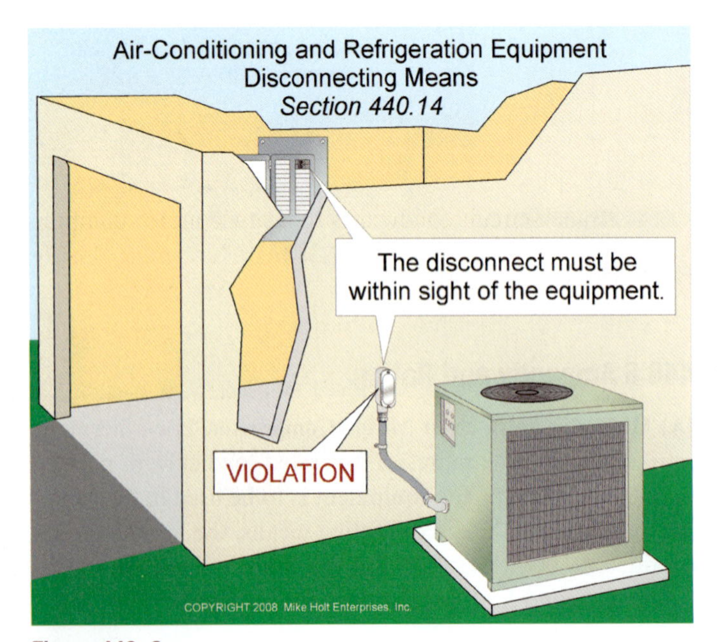

Figure 440–3

Author's Comment: "Within Sight" is visible and not more than 50 ft from each other [Article 100].

The disconnecting means can be mounted on or within the air-conditioning equipment, but it must not be located on panels designed to allow access to the equipment, or where it will obscure the equipment nameplate. Figure 440–4

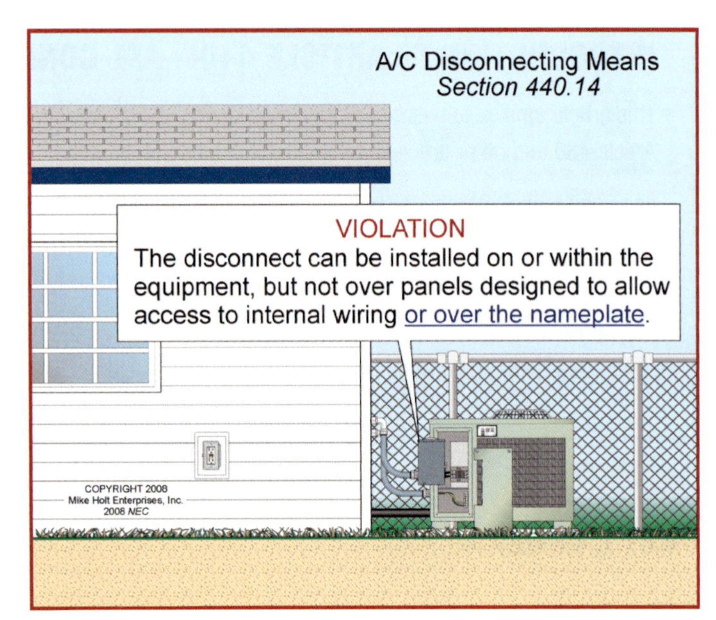

Figure 440–4

Exception No. 1: A disconnecting means isn't required to be within sight from the equipment, if the disconnecting means is capable of being individually locked in the open position, and if the equipment is essential to an industrial process in a facility that has written safety procedures, and where the conditions of maintenance and supervision ensure only qualified persons service the equipment. The provision for locking or adding a lock to the disconnecting means must be on the switch or circuit breaker, and it must remain in place with or without the lock installed.

Exception No. 2: An accessible attachment plug and receptacle can serve as the disconnecting means.

Author's Comment: The receptacle for the attachment plug isn't required to be readily accessible.

PART III. OVERCURRENT PROTECTION

440.21 General. The branch-circuit conductors, control apparatus, and circuits supplying hermetic refrigerant motor-compressors must be protected against short circuits and ground faults in accordance with 440.22.

> **Author's Comment:** If the equipment nameplate specifies "Maximum Fuse Size," then a one-time or dual-element fuse must be used.

440.22 Short-Circuit and Ground-Fault Overcurrent Device Size.

(A) Single Motor-Compressors. The short-circuit and ground-fault protective device must not be more than 175 percent of the motor-compressor current rating. If the protective device sized at 175 percent isn't capable of carrying the starting current of the motor-compressor, the next size larger protective device can be used, but in no case can it exceed 225 percent of the motor-compressor current rating.

> ***Question:*** *What size conductor and protection are required for a 24A motor-compressor connected to a 240V circuit?* **Figure 440–5**
>
> *(a) 10 AWG, 40A* *(b) 10 AWG, 60A*
> *(c) a or b* *(d) 10 AWG, 90A*
>
> ***Answer:*** *(a) 10 AWG, 40A*
>
> *Step 1: Determine the branch-circuit conductor [Table 310.16 and 440.32]:*
>
> *24A x 1.25 = 30A, 10 AWG, rated 30A at 75°C [Table 310.16]*
>
> *Step 2: Determine the branch-circuit protection [240.6(A) and 440.22(A)]:*
>
> *24A x 1.75 = 42A, next size down = 40A*
>
> *If the 40A short-circuit and ground-fault protective device isn't capable of carrying the starting current, then the protective device can be sized up to 225 percent of the equipment load current rating. 24A x 2.25 = 54A, next size down 50A*

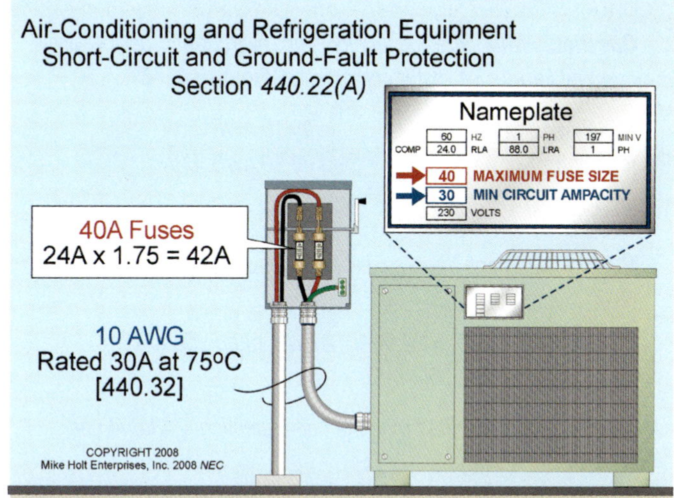

Figure 440–5

(B) Several Motor-Compressors. Where the equipment incorporates more than one hermetic refrigerant motor-compressor, or a hermetic refrigerant motor-compressor and other motors or other loads, the equipment short-circuit and ground-fault protection must be sized as follows:

(1) Motor-Compressor Largest Load. The rating of the branch-circuit short-circuit and ground-fault protective device must not be more than the largest motor-compressor short-circuit ground-fault overcurrent device, plus the sum of the rated-load currents of the other compressors.

> **Author's Comment:** The branch-circuit conductors are sized at 125 percent of the larger motor-compressor current, plus the sum of the rated-load currents of the other compressors [440.33].

PART IV. CONDUCTOR SIZING

440.32 Conductor Size for Single Motor-Compressors. Branch-circuit conductors to a single motor-compressor must have an ampacity not less than 125 percent of the motor-compressor rated-load current or the branch-circuit selection current, whichever is greater.

> **Author's Comment:** Branch-circuit conductors for a single motor-compressor must have short-circuit and ground-fault protection sized between 175 percent and 225 percent of the rated-load current [440.22(A)].

Question: *What size conductor and overcurrent device are required for an 18A motor compressor?* **Figure 440–6**

(a) 12 AWG, 30A
(b) 10 AWG, 50A
(c) a or b
(d) 10 AWG, 60A

Answer: *(a) 12 AWG, 30A*

Step 1: Determine the branch-circuit conductor [Table 310.16 and 440.32]:

18A x 1.25 = 22.50A, 12 AWG, rated 25A at 75°C [Table 310.16]

Step 2: Determine the branch-circuit protection [240.6(A) and 440.22(A)]:

18A x 1.75 = 31.50A, next size down = 30A

If the 30A short-circuit and ground-fault protection device isn't capable of carrying the starting current, then the protective device can be sized up to 225 percent of the equipment load current rating. 18A x 2.25 = 40.50A, next size down 40A

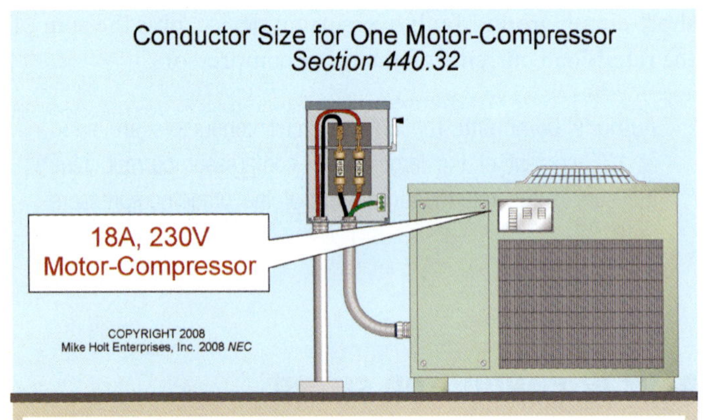

Figure 440–6

Author's Comment: A 30A or 40A overcurrent device is permitted to protect a 12 AWG conductor for an air-conditioning circuit. See 240.4(G) for details.

440.33 Conductor Size for Several Motor-Compressors.

Conductors that supply several motor-compressors must have an ampacity not less than 125 percent of the highest motor-compressor current of the group, plus the sum of the rated load or branch-circuit selection current ratings of the other compressors.

Author's Comment: These conductors must be protected against short circuits and ground faults in accordance with 440.22(B)(1).

PART VII. ROOM AIR CONDITIONERS

The requirements in this Part apply to a cord-and-plug-connected room air conditioner of the window or in-wall type that incorporates a hermetic refrigerant motor-compressor rated not over 40A, 250V, single-phase [440.60].

440.62 Branch-Circuit Requirements.

(A) Sizing Conductors and Protection. Branch-circuit conductors for a cord-and-plug-connected room air conditioner must have an ampacity not less than 125 percent of the rated-load currents [440.32].

(B) Separate Circuit. Where the room air conditioner is the only load on a circuit, the marked rating of the air conditioner must not exceed 80 percent of the rating of the circuit overcurrent device [210.3].

(C) Other Loads on Circuit. The total rating of a cord-and-plug-connected room air conditioner must not exceed 50 percent of the rating of a branch circuit where lighting outlets, other appliances, or general-use receptacles are also supplied. **Figure 440–7**

440.63 Disconnecting Means.

An attachment plug and receptacle can serve as the disconnecting means for a room air conditioner, provided: **Figure 440–8**

(1) The manual controls on the room air conditioner are readily accessible and within 6 ft of the floor, or

(2) A readily accessible disconnecting means is within sight from the room air conditioner.

Author's Comment: "Within Sight" is visible and not more than 50 ft from each other [Article 100].

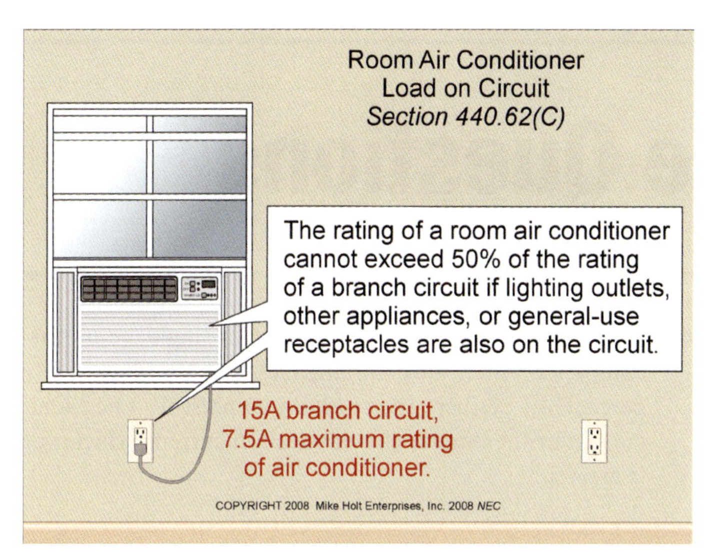

Figure 440–7

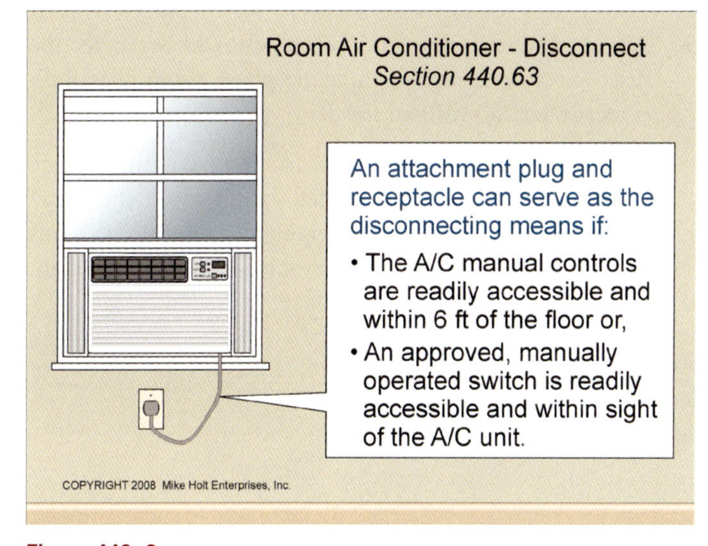

Figure 440–8

440.64 Supply Cords. Where a flexible cord is used to supply a room air conditioner, the cord must not exceed 10 ft for 120V units, or 6 ft for 208V or 240V units.

440.65 Leakage Current Detector-Interrupter and Arc-Fault Circuit Interrupter. Single-phase cord-and-plug-connected room air conditioners must be provided with a factory-installed leakage current detector, or with an arc-fault circuit-interrupter (AFCI).

ARTICLE 440 Practice Questions

ARTICLE 440. AIR-CONDITIONING AND REFRIGERATING EQUIPMENT—PRACTICE QUESTIONS

1. Article 440 applies to electric motor-driven air-conditioning and refrigerating equipment that has a hermetic refrigerant motor-compressor.

   (a) True
   (b) False

2. Equipment such as _____ shall be considered appliances, and the provisions of Article 422 apply in addition to Article 440.

   (a) room air conditioners
   (b) household refrigerators and freezers
   (c) drinking water coolers and beverage dispensers
   (d) all of these

3. Where the air conditioner disconnecting means is not within sight from the equipment, the provision for locking or adding a lock to the disconnecting means shall be on the switch or circuit breaker and remain in place _____ the lock installed.

   (a) with
   (b) without
   (c) with or without
   (d) none of these

4. Branch-circuit conductors supplying a single a/c motor-compressor shall have an ampacity not less than _____ percent of either the motor-compressor rated-load current or the branch-circuit selection current, whichever is greater.

   (a) 100
   (b) 125
   (c) 150
   (d) 200

5. An attachment plug and receptacle can serve as the disconnecting means for a single-phase room air conditioner rated 250 volts or less if _____.

   (a) the manual controls on the room air conditioner are readily accessible and located within 6 ft of the floor
   (b) an approved manually operable disconnecting means is installed in a readily accessible location within sight from the room air conditioner
   (c) a or b
   (d) a and b

ARTICLE

450 Transformers

INTRODUCTION TO ARTICLE 450—TRANSFORMERS

Article 450 opens by saying, "This article covers the installation of all transformers." Then it lists eight exceptions. So what does Article 450 really cover? Essentially, it covers power transformers and most kinds of lighting transformers.

A major concern with transformers is preventing overheating. The *Code* doesn't completely address this issue. Article 90 explains that the *NEC* isn't a design manual, and it assumes that the person using the *Code* has a certain level of expertise. Proper transformer selection is an important part of preventing transformer overheating.

The *NEC* assumes you have already selected a transformer suitable to the load characteristics. For the *Code* to tell you how to do that would push it into the realm of a design manual. Article 450 then takes you to the next logical step—providing overcurrent protection and the proper connections. But this article doesn't stop there; 450.9 provides ventilation requirements, and 450.13 contains accessibility requirements.

Part I of Article 450 contains the general requirements such as guarding, marking, and accessibility, and Part II contains the requirements for different types of transformers.

PART I. GENERAL

450.1 Scope. Article 450 covers the installation requirements of transformers and transformer vaults. Figure 450-1

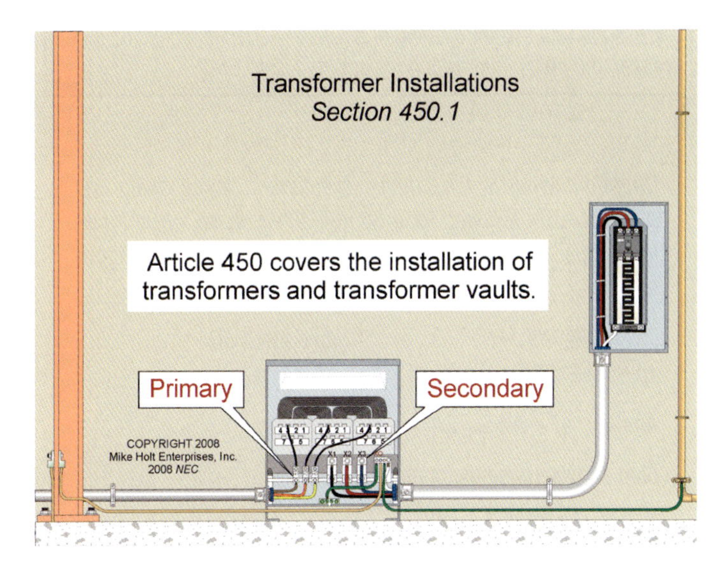

Figure 450–1

450.3 Overcurrent Protection.

FPN No. 2: Nonlinear loads on 4-wire, wye-connected secondary wiring can increase heat in a transformer without operating the primary overcurrent device. Figure 450-2

(B) Overcurrent Protection for Transformers Not Over 600V. The primary winding of a transformer must be protected against overcurrent in accordance with the percentages listed in Table 450.3(B) and all applicable notes.

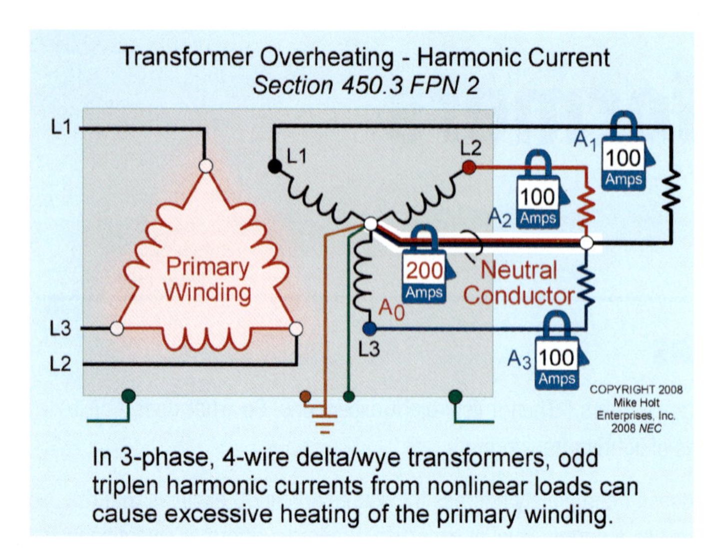

Figure 450–2

Table 450.3(B) Primary Protection Only	
Primary Current Rating	**Maximum Protection**
9A or More	125%, see Note 1
Less Than 9A	167%
Less Than 2A	300%

Note 1. Where 125 percent of the primary current doesn't correspond to a standard rating of a fuse or nonadjustable circuit breaker, the next higher rating is permitted [240.6(A)].

Question: *What is the primary overcurrent device rating and conductor size required for a 45 kVA, three-phase, 480V transformer that is fully loaded? The terminals are rated 75°C.* **Figure 450-3**

(a) 8 AWG, 40A *(b) 6 AWG, 50A*
(c) 6 AWG, 60A *(d) 4 AWG, 70A*

Answer: *(d) 4 AWG, 70A*

Step 1: Determine the primary current:

I = VA/(E x 1.732)
I = 45,000 VA/(480V x 1.732)
I = 54A

Step 2: Determine the primary overcurrent device rating [240.6(A)]:

54A x 1.25 = 68A, next size up 70A, Table 450.3(B), Note 1

Step 3: The primary conductor must be sized to carry 54A continuously (54A x 1.25 = 68A) [215.2(A)(1)] and be protected by a 70A overcurrent device [240.4(B)]. A 4 AWG conductor rated 85A at 75°C meets all of the requirements [110.14(C)(1) and 310.16].

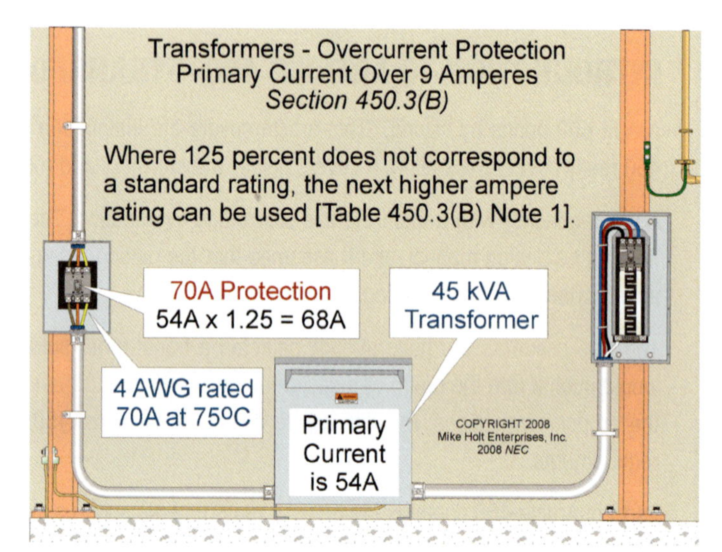

Figure 450–3

450.9 Ventilation. Transformers must be installed in accordance with the manufacturer's instructions, and their ventilating openings must not be blocked [110.3(B)].

FPN No. 2: Transformers can become excessively heated above their rating because nonlinear loads can increase heat in a transformer without operating its overcurrent protective device [450.3 FPN].

Author's Comment: The heating from harmonic currents is proportional to the square of the harmonic frequency. This means the 3rd order harmonic currents (180 Hz) will heat at nine times the rate of 60 Hz current. **Figure 450-4**

450.11 Marking. Transformers must be provided with a nameplate identifying the manufacturer of the transformer and indicating the transformer's rated kVA, primary and secondary voltage, impedance if 25 kVA or larger, and required clearances for transformers with ventilating openings.

450.13 Transformer Accessibility. Transformers must be readily accessible to qualified personnel for inspection and maintenance, except as permitted by (A) or (B).

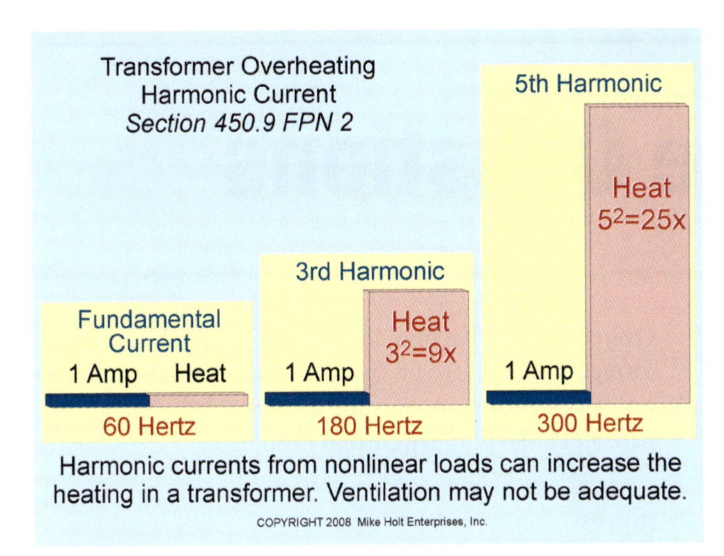

Figure 450–4

(A) Open Installations. Dry-type transformers can be located in the open on walls, columns, or structures. Figure 450-5

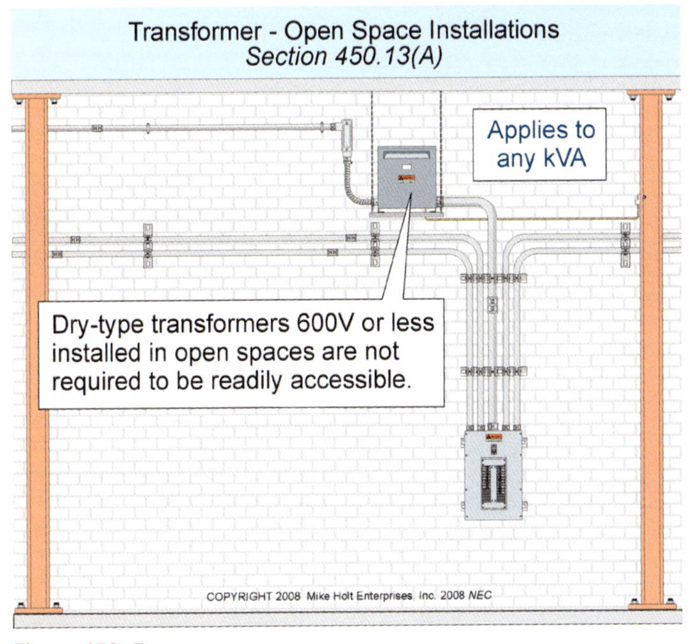

Figure 450–5

(B) Suspended Ceilings. Dry-type transformers, rated not more than 50 kVA, are permitted above suspended ceilings or in hollow spaces of buildings, if not permanently closed in by the structure. Figure 450-6

Author's Comment: Dry-type transformers not exceeding 50 kVA with a metal enclosure can be installed above a suspended ceiling space used for environmental air-handling purposes [300.22(C)(2)].

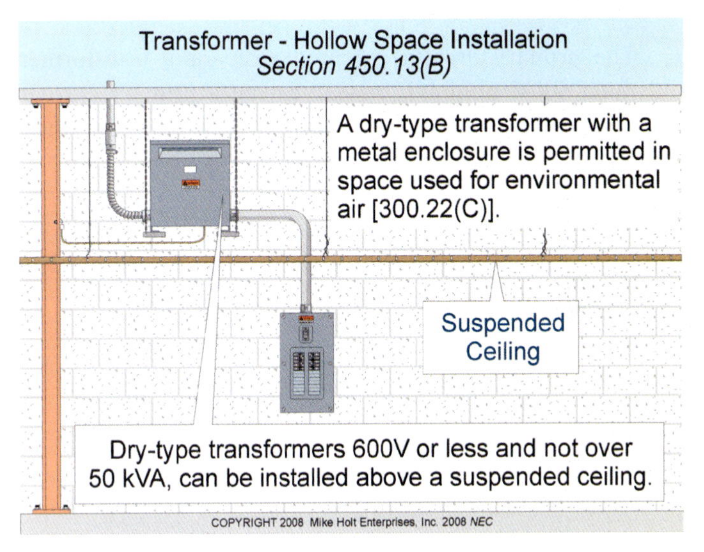

Figure 450–6

ARTICLE 450 Practice Questions

ARTICLE 450. TRANSFORMERS—PRACTICE QUESTIONS

1. The primary overcurrent protection for a transformer rated 600V, nominal, or less, having a primary current rating of over 9A or more must be set at not more than _____.

   (a) 125
   (b) 167
   (c) 200
   (d) 300

2. Overcurrent protection for the secondary of a transformer rated 600V or less having a primary current rating of 9A or more must be set at not more than _____ percent of the primary current rating.

   (a) 100
   (b) 125
   (c) 150
   (d) 170

Index

N

O

P

Notes

Notes